530

Z399

ANGLO-EUROPEAN COLLEGE OF CHIROPRACTIC

KU-016-954

101

ANGLO-EUROPEAN COLLEGE OF CHIROPRACTIC

ADVANCED LEVEL
PHYSICS

Seventh Edition

ANGLO-EUROPEAN COLLEGE OF CHIROPRACTIC

Books by Michael Nelkon
Published by Heinemann

ADVANCED LEVEL PRACTICAL PHYSICS (*with Prof. J. Ogborn*)
ADVANCED LEVEL PHYSICS EXERCISES AND EXAMPLES
SCHOLARSHIP PHYSICS

Published by Longman International Education
PRINCIPLES OF PHYSICS (*GCSE standard*)

ANGLO-EUROPEAN COLLEGE OF CHIROPRACTIC

ADVANCED LEVEL
PHYSICS

Seventh Edition

MICHAEL NELKON
M Sc CPhys F InstP AKC
formerly Head of the Science Department
William Ellis School, London

PHILIP PARKER
M Sc F InstP AMIEE
Late Senior Lecturer in Physics
The City University, London

AECC LIBRARY

J M L S	18/12/96
	£19.99

Heinemann Educational Publishers
A Division of Heinemann Publishers (Oxford) Ltd
Halley Court, Jordan Hill, Oxford OX2 8EJ

Heinemann: A Division of Reed Publishing (USA) Inc.
361 Hanover Street, Portsmouth, NH 03801-3912, USA

Heinemann Educational Books (Nigeria) Ltd
PMB 5205, Ibadan
Heinemann Educational Boleswa
PO Box 10103, Village Post Office, Gaborone, Botswana

Florence Prague Paris Madrid
Athens Melbourne Johannesburg
Auckland Singapore Tokyo
Chicago Sao Paulo

© M. Nelkon and Mrs. P. Parker 1958, 1964, 1970, 1977, 1982, 1987, 1995
First published as one volume 1958
Reprinted four times
Second edition 1964
Reprinted three times
Third edition (SI) 1970
Reprinted four times
Fourth edition 1977
Reprinted three times
Fifth edition 1982
Reprinted four times
Sixth edition 1987
Reprinted seven times
Seventh edition 1995
ISBN 0435 92303 X

British Library Cataloguing in Publication Data
A catalogue record for this book is available from the British Library.

Cover photograph by Alfred Pasieka/Science Photo Library.

Phototypeset by Datix International Limited, Bungay, Suffolk
Printed in England by Clays Ltd, St Ives plc

96 97 98 10 9 8 7 6 5 4 3

Preface to Seventh Edition

In this new edition the text has been revised and re-ordered to deal with the updated syllabus of the major examining boards. Summaries have been given at the end of main sections to assist students in their understanding of the principles of the subject and in revision. These are entitled *You Should Know*. More worked examples on basic points have also been added to the text.

Multiple choice questions are now included in most exercises, followed by longer questions graded in order of difficulty. To assist students limited in time for A-level study, parts of the text which may be omitted at a first reading have been indicated by a note or by a symbol ■, subject to the syllabus and to advice by the teacher.

To show students how physics is applied in the world outside, a brief account of a practical application is provided at the end of many chapters. I am very grateful to Dr. Andrew Gee, Research Fellow in Engineering, Queens' College, Cambridge, for the applications and for his valuable assistance in analogue and digital electronics. I am also grateful to A. C. Gibson of Bruel and Kjaer Limited for photographs showing a variety of engineering applications in the commercial world, to N. A. S. Nelkon, formerly of Imperial College of Science, London, for considerable assistance, and to T. Landy, formerly sixth-form science student, University College School, London, for work on the exercises. I also acknowledge with thanks the generous assistance from Kai Man Hung, King's College, London, and from M. H. Gaines and Mrs. M. D. Mole, National Radiological Protection Board, Chilton, UK. I am particularly grateful to Professor Michael Green FRS, Cambridge University, for his contribution on superstrings.

Some main changes in chapters are, briefly:

Simple Harmonic Motion—A fuller treatment has been given.

Gravitational and Electrostatic Fields—Field strength and potential have been compared and contrasted.

Analogue and Digital Electronics—Conversion between analogue and digital has been added.

I acknowledge with thanks the generous assistance in previous editions by: Dr. M. Crimes, Woodhouse College, London (for *Solid Materials*); Ian Lovat, Malvern College, Worcestershire (for *Analogue and Digital Electronics*); Prof. R. S. Ellis, Durham University; Dr. P. Gray, Anglo-Australian Observatory; Prof. R. S. Shaw, Royal Free Hospital, London; J. H. Avery, Stockport School; M. V. Detheridge, formerly Highgate School, London; Rev. M. D. Phillips, formerly Ampleforth School, York; Dr. D. Deutsch, Institute of Mathematical Studies; Dr. J. M. Graham, Imperial College, London; K. Danks, Brunel Technical College, Bristol; P. Betts, Barstable School, Essex; R. D. Harris, Ardingley College, Sussex; R. Croft, The City University, London; N. Phillips, Loughborough University; Dr. L. S. Julien, University of Surrey; C. F. Tolman, Whitgift School, Croydon; and Mrs. J. Pope, formerly Middlesex Polytechnic.

Publishers' Note

Since the first publication of *Advanced Level Physics*, the revisions for reprints and new editions have been undertaken by Michael Nelkon owing to the death of Philip Parker.

Acknowledgements

Thanks are due to the following Examining Boards for their kind permission to reprint past questions. Answers are the sole responsibility of the author, and not of the appropriate examining boards.

University of London Examinations and Assessment Council (*L.*)
Northern Examinations and Assessment Board (*N.*)
Oxford and Cambridge Schools Examination Board (*O. & C.*)
Cambridge Local Examinations Syndicate (*C.*)
Oxford Delegacy of Local Examinations (*O.*)
Associated Examining Board (*AEB.*)
Welsh Joint Education Committee (*W.*)

I am indebted to the following for kindly supplying photographs and permission to reprint them:

Ace Photo Agency, 15A; AEA Technology, 8A, 34C, 34F; Allsport/Clive Brunskill, 1A, 4A; AT&T (UK) Ltd, 15B, 31A; Barr & Stroud Ltd, 28A; Tim Beddow/Science Photo Library, 34A; Charles Belinky/Science Photo Library, 22A; J. Allan Cash Photo Library, 2A, 3A, 5A; The City University, London/ Physics Dept, 19C, 21A; CNRI/Science Photo Library, 34B; R. Croft, The City University, London, 20.2, 20D; Fred Espenak/Science Photo Library, 22C; Mary Evans Picture Library, 10A, 13C, 18A, 20A, 25A, 30D; Peter Gray/Epping Laboratory/Anglo-Australian Observatory, 16C; Hilger & Watts Ltd, 19A, 19D, 20B, 20C; Inmos Ltd, 31B; Instron, 6A; Kodak Ltd, 21C; Leybold-Heraeus Gmb & Co., 30C; Dr J. W. Martin, University of Oxford, 6C; Metropolitan Police Service, 22B; Hank Morgan/Science Photo Library, 29A; NASA/Science Photo Library, 7A, 16A; National Chemical Laboratory, 33C; National Physical Laboratory, 12A, 17A, 19B, 33B; Paragon Communications Ltd, 6B; David Parker/IMI/Science Photo Library, 12B; Philips Research Laboratories, 33A; Philippe Plailly/Science Photo Library, 20E; The Press Association, 13B; Science Photo Library, 13A, 34D; Prof. C. A. Taylor, Cardiff University, 23A; Teltron Ltd, 30A, 30B; University of London Observatory, Dept of Physics & Astronomy, University College, London, 16B; VG Micromass Ltd, 34E; The Worcester Royal Porcelain Company Ltd & Tom Biro, 29B.

Contents

Part One: Mechanics, Solid Materials

Linear motion. Velocity. Acceleration. Vectors, scalars. Equations of motion. Free-fall, g. Graphs. Vector addition and subtraction. Components. Projectiles.
  Laws of motion. Inertia, mass, weight. Force. Gravitational force. $F = ma$. Linear momentum. Impulse. Action, reaction in systems. Conservation of momentum. Inelastic and elastic collisions.
  Work, energy, power. Work calculations. Power of engine. Kinetic and gravitational potential, energy. Gravitational momentum and energy changes. Dimensions.

Circular motion. Angular speed. Acceleration in circle. Centripetal forces. Banked track. Conical pendulum. Bicycle motion.
  Simple harmonic motion. Formula for acceleration, velocity, displacement. Graphs of kinetic and potential energy. Oscillations in spring–mass system. Potential and kinetic energy exchanges. Springs in series and parallel. Simple pendulum. Oscillations of liquid in U-tube.

Forces in equilibrium. Adding forces—vector and parallelogram methods. Resolved components. Couple and moment. Triangle and polygon of forces. Equilibrium of forces. Centre of mass, centre of gravity.
  Forces in fluids. Pressure formula. Atmospheric pressure. Density. Archimedes' principle and proof. Flotation. Stokes' law. Terminal velocity.

Rotational dynamics. Torque and angular acceleration. Rotational kinetic energy. Work done in rotation. Angular momentum. Conservation of angular momentum and applications. Rolling motion down inclined plane.
  Fluid motion. Bernoulli principle. Filter pump, aerofoil lift, carburettor, Venturi meter. Pitot-static tube principle.

Part Three: Geometrical Optics, Waves, Wave Optics, Sound Waves

Part Five: Electrons, Electronics, Atomic Physics

Telecommunications. Optical fibre telecommunications. LED. Semiconductor laser. Photodiode. Radio telecommunications. Aerials and standing waves. Types of aerials. Amplitude modulation. Sidebands. AM radio receiver (trf).

ANGLO-EUROPEAN COLLEGE OF CHIROPRACTIC

PART ONE
Mechanics, Solid Materials

1 Dynamics

Dynamics *is the science of motion. It deals with the velocity, acceleration, force and energy of large objects such as cars and aeroplanes and tiny objects such as the electrons in your television set which produce the pictures. Dynamics also helps to investigate the motion of athletes or the motion of a ball bowled in cricket or hit in golf.*

Before you play a game like football or tennis, you need to learn the basic skills and how to apply them. In the same way, we start with the main points in dynamics and show how they are applied in velocity, acceleration, free-fall in gravity, and motion graphs.

Plate 1A *The study of dynamics helps to investigate and maximise the performance of athletes such as Sally Gunnell*

Linear Motion, Vectors, Free-fall, Graphs, Projectiles

We assume you have already met the meaning of velocity and acceleration and calculations of the distance travelled by a car moving with a uniform (constant) acceleration. We also assume you have met the motion of a ball moving downwards under the gravitational pull, called 'free-fall' if there is no resistance to its motion. The acceleration due to gravity is the same for heavy or light objects, about 9.8 m s^{-2} or 10 m s^{-2} in round figures for arithmetical calculations. (9.8 m s^{-2} is 9.8 m/s^2 and 10 m s^{-2} is 10 m/s^2.)

In mechanics, *direction* is important. In physics, quantities are divided into two classes: (1) *vectors*, which have both direction and size, and (2) *scalars*, which have no direction and only size. 'Velocity' is an example of a vector—it has direction. 'Speed' is an example of a scalar—it has no direction.

Formulae. In physics, we use symbols or letters to stand for different quantities, for example, v for velocity, a for acceleration and s for displacement or distance travelled. The relation between quantities is given in 'formulae' with symbols. The formula is a shorthand way of writing the relation and the quicker you can memorise and apply it, the better will be your handling of problems.

Motion in Straight Line, Velocity

If a runner, moving in a straight line, takes 10 s to run 100 m, the average *velocity* in this direction

$$= \frac{\text{distance}}{\text{time}} = \frac{100 \text{ m}}{10 \text{ s}} = 10 \text{ m/s or } 10 \text{ m s}^{-1}$$

The term 'displacement' (a vector) is given to the distance moved in a constant direction, for example, from L to C in Fig. 1.1 (i). So

velocity is the rate of change of displacement

or 'change in displacement/time taken'.

Velocity can be given in metres per second (m s^{-1}) or in kilometres per hour (km h^{-1}). By calculation, $36 \text{ km h}^{-1} = 10 \text{ m s}^{-1}$.

If a car moving in a straight line travels equal distances in equal times, no matter how small these distances may be, the car is said to be moving with a constant or *uniform* velocity. A falling stone increases in velocity, so this is a *non-uniform* velocity.

If, at any point of a journey, Δs is the small change in displacement made in a small time Δt, the velocity v is given by $v = \Delta s/\Delta t$. In the limit, using calculus notation,

$$v = \frac{ds}{dt}$$

We call ds/dt the *instantaneous velocity* at the time or place concerned. The term 'mean velocity' is the 'average velocity' over a particular time or distance.

Vectors and Scalars

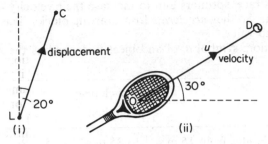

Figure 1.1 *Vectors*

Displacement and *velocity* are examples of a class of quantities called *vectors* which have both size and direction. They can therefore be represented to scale by a line drawn in a particular direction. For example, Cambridge is 80 km from London in a direction 20° E. of N. We can therefore represent the displacement between the cities in magnitude and direction by a straight line LC 4 cm long 20° E. of N., where 1 cm represents 20 km, Figure 1.1 (i). Similarly, we can represent the velocity u of a ball leaving a racket at an angle of 30° to the horizontal by a straight line OD drawn to scale in the direction of the velocity u, the arrow on the line showing the direction, Figure 1.1 (ii).

Unlike vectors, *scalars* are quantities which have size but no direction. A car moving along a winding road or a circular track at 80 km h^{-1} is said to have a *speed* of 80 km h^{-1}. 'Speed' is a quantity which has no direction but only size. 'Mass' or 'density' or 'temperature' are also examples of scalars.

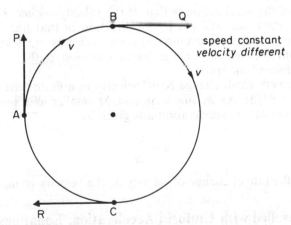

Figure 1.2 *Velocity (vector) and speed (scalar)*

The difference between speed and velocity can be made clear by considering a car moving round a circular track at say 80 km h^{-1}, Figure 1.2. At every point on the track the *speed* v is the same—it is 80 km h^{-1}. At every point, however, the *velocity* is different. At A, B or C, for example, the velocity is in the direction of the corresponding tangent AP, BQ or CR if we 'freeze' the motion of the car. So even though they have the same size, the three velocities are all *different* because their directions are different.

Acceleration in Linear Motion

In a 100 metres race, sprinters aim to increase their velocity to a maximum in the shortest time. So they *accelerate* from starting blocks. This acceleration can have a very high value.

The acceleration, symbol a, of an object is defined as the *rate of change of velocity* or

$$a = \frac{\text{velocity change}}{\text{time taken}}$$

So if a car accelerates from 15 m s^{-1} to 35 m s^{-1} in 5 s, then

$$\text{average acceleration, } a = \frac{(35 - 15) \text{ m s}^{-1}}{5 \text{ s}} = 4 \text{ m s}^{-2}$$

Note carefully that the time unit for acceleration is 's^{-2}' because the term 'second' occurs twice.

A car moving with constant (uniform) velocity has zero acceleration, from the definition. If a car brakes, its velocity may decrease from 30 m s^{-1} to 20 m s^{-1} in 5 s. In this case the car has a retardation or deceleration, or, mathematically, a *negative* acceleration. So

$$a = -\frac{30 - 20}{5} = -2 \text{ m s}^{-2}$$

Note carefully that

acceleration is a vector.

The direction of the acceleration is that of the *velocity change*. For motion in a straight line, the acceleration is in the direction of that line. We see later, however, that the velocity of a car moving round a circular track keeps changing in direction and the velocity change or acceleration is then in a direction towards the centre of the track.

With a finite very small change Δv of velocity in a finite time Δt, the mean acceleration $a = \Delta v/\Delta t$. As we make Δv and Δt smaller and smaller, then, in terms of the calculus, the acceleration a is given by

$$a = \frac{dv}{dt}$$

where dv/dt is the rate of change of velocity or the velocity change per second.

Distance Travelled with Uniform Acceleration, Equations of Motion

If the velocity changes by equal amounts in equal times, no matter how small the time-intervals may be, the acceleration is said to be *uniform*. Suppose that the velocity of a car moving in a straight line with uniform acceleration a increases from a value u to a value v in a time t. Then, from the definition of acceleration,

$$a = \frac{v - u}{t}$$

from which $\qquad\qquad v = u + at \qquad . \quad . \quad . \quad . \quad . \quad$ (1)

Suppose a train with a velocity u accelerates with a uniform acceleration a for a time t and reaches a velocity v. The distance s travelled by the object in the time t is given by

$$s = \text{average velocity} \times t$$
$$= \tfrac{1}{2}(u + v) \times t$$

But
$$v = u + at$$
$$\therefore s = \tfrac{1}{2}(u + u + at)t$$
$$\therefore s = ut + \tfrac{1}{2}at^2 \qquad . \qquad . \qquad . \qquad . \qquad . \qquad . \quad (2)$$

Also, $t = (v - u)/a$ from (1), so

$$s = \text{average velocity} \times t = \frac{(v + u)}{2} \times \frac{(v - u)}{a}$$

$$= (v^2 - u^2)/2a$$

Simplifying, $\qquad\qquad\qquad v^2 = u^2 + 2as \qquad . \qquad . \qquad . \qquad . \qquad . \quad (3)$

Equations (1), (2), (3) are the equations of motion of an object moving in a straight line with uniform acceleration. When an object undergoes a uniform *deceleration* or *retardation*, for example when brakes are applied to a car, a has a *negative* value.

Examples on Equations of Motion

1 An aeroplane lands on the runway with a velocity of 50 m s^{-1} and decelerates at 10 m s^{-2} to a velocity of 20 m s^{-1}. Calculate the distance travelled on the runway.

(*Analysis* No time is mentioned. So we use $v^2 = u^2 + 2as$.)
    Here $u = 50$ m s^{-1}, $v = 20$ m s^{-1}, $a = -10$ m s^{-2}
Using the formula $v^2 = u^2 + 2as$ to find s, then
$$20^2 = 50^2 + (2 \times -10 \times s) = 50^2 - 20s$$
So $\qquad\qquad 400 = 2500 - 20s$

$$s = \frac{2500 - 400}{20} = \frac{2100}{20} = 105 \text{ m}$$

2 A car moving with a velocity of 15 m s^{-1} accelerates uniformly at the rate of 2 m s^{-2} to reach a velocity of 20 m s^{-1}.
    Find (i) the time taken, (ii) the distance travelled in this time.

(*Analysis* (i) We need time t. So we use $v = u + at$. (ii) We need distance s. Knowing t, we can use $s = ut + \tfrac{1}{2}at^2$ or, without t, use $v^2 = u^2 + 2as$.)

(i) Using $\qquad\qquad\qquad v = u + at$
$$\therefore 20 = 15 + 2t$$
$$\therefore t = \frac{20 - 15}{2} = 2 \cdot 5 \text{ s}$$

(ii) Using $\qquad\qquad\qquad s = ut + \tfrac{1}{2}at^2$
$$s = (15 \times 2 \cdot 5) + \tfrac{1}{2} \times 2 \times 2 \cdot 5^2$$
$$= 37 \cdot 5 + 6 \cdot 25 = 43 \cdot 75 \text{ m}$$

Motion Under Gravity, Free-fall

When an object falls to the ground under gravitational pull, experiment shows that the object has a constant or uniform acceleration of about 9.8 m s^{-2} or 10 m s^{-2} approximately, while it is falling. The numerical value of this acceleration is usually denoted by the symbol g. Drawn as a vector quantity, g is always represented by a straight vertical line with an arrow on the line pointing downwards, Figure 1.3 (i).

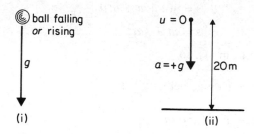

Figure 1.3 *Motion under gravity—free-fall*

Suppose that a ball is dropped from a height of 20 m above the ground, Figure 1.3 (ii). Then the initial velocity $u = 0$, and the acceleration $a = g = 10$ m s^{-2} (approx.). When the ball reaches the ground, $s = 10$ m. Substituting in $s = ut + \frac{1}{2}at^2$, then

$$s = 0 + \tfrac{1}{2}gt^2 = 5t^2$$

$$\therefore 20 = 5t^2 \quad \text{or} \quad t = 2 \text{ s}$$

So the ball takes 2 seconds to reach the ground.

If a cricket-ball is thrown vertically upwards, it slows down owing to the attraction of the earth. The magnitude of the deceleration is 9.8 m s^{-2}, or g. Mathematically, a deceleration can be regarded as a negative acceleration in the direction along which the object is moving; and so $a = -9.8$ m s^{-2} in this case.

Examples on Motion under Free-fall (Gravity)

1 A ball is thrown vertically upwards with an initial velocity of 30 m s^{-1}.
Find   (i) the time taken to reach its highest point,   (ii) the distance then travelled. (Assume $g = 10$ m s^{-2}.)

(*Analysis*   (i) We need time t. So we can use $v = u + at$.   (ii) We need distance s. So we can use $s = ut + \frac{1}{2}at^2$.)

(i) Here $u = 30$ m s^{-1}, $v = 0$ at highest point since ball is momentarily at rest, $a = -g = -10$ m s^{-2}. From $v = u + at$,

$$0 = 30 + (-10)t \quad \text{or} \quad 10t = 30 \text{ and so } t = 3 \text{ s}$$

(ii)   Distance $s = ut + \tfrac{1}{2}at^2$

$$= (30 \times 3) + \tfrac{1}{2} \times (-10) \times 3^2$$

$$= 90 - 45 = 45 \text{ m}$$

2   A lift is moving down with an acceleration of 3 m s^{-2}. A ball is released 1·7 m above the lift floor. Assuming $g = 9 \cdot 8$ m s^{-2}, how long will the ball take to hit the floor?

(*Analysis*   We need time t. So we have distance s, we can use $s = ut + \frac{1}{2}at^2$.)
  Acceleration of ball relative to lift, $a = 9 \cdot 8 - 3 = 6 \cdot 8$ m s^{-2}.
  Here $u = 0$, $a = 6 \cdot 8$ m s^{-2}, $s = 1 \cdot 7$ m. From $s = ut + \frac{1}{2}at^2$

$$1 \cdot 7 = 0 + \tfrac{1}{2} \times 6 \cdot 8 \times t^2 = 3 \cdot 4 t^2$$

So $t^2 = 1 \cdot 7 / 3 \cdot 4 = 0 \cdot 5$. Then $t = \sqrt{0 \cdot 5} = 0 \cdot 7$ s

Components of Vectors

In mechanics and other branches of physics, we often need to find the *component* of a vector in a certain direction.

  The component is the 'effective part' of the vector in that direction. We can illustrate it by considering a picture held up by two strings OP and OQ each at an angle of 60° to the vertical, Figure 1.4 (i). If the force or tension in OP is 6 N, its *vertical* component S acting upwards at P helps to support the weight W of the picture, which acts vertically downwards. The upward component T of the 6 N force in OQ acting in the direction QT, also helps to support W.

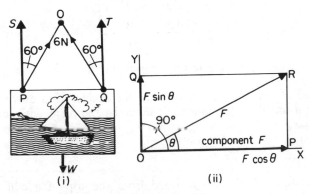

Figure 1.4 *Components of vectors*

  Figure 1.4 (ii) shows how the component value of a vector F can be found in a direction OX. If OR represents F to scale, we draw a rectangle OPRQ which has OR as its diagonal. Now F is the sum of the vectors OP and OQ. *The vector OQ has no effect in a direction OX at 90° to itself.* So the effective part or component of F in the direction OX is the vector OP.
  If θ is the angle between F and the direction OX, then

$$\text{OP} / \text{OR} = \cos \theta \quad \text{or} \quad \text{OP} = OR \cos \theta = F \cos \theta$$

So the component of any vector F in a direction making an angle θ to F is always given by

$$\textbf{component} = \textbf{\textit{F}} \cos \boldsymbol{\theta}$$

In a direction OY *perpendicular* to OX, F has a component $F \cos (90° - \theta)$

$$\textbf{which is } \textbf{\textit{F}} \sin \boldsymbol{\theta}$$

This component is represented by OQ in Figure 1.4 (ii).

Component Examples

(1) From Figure 1.4 (i), we can now see that

vertical component S of 6 N force in OP = $6 \cos 60° = 3$ N

This is also the value of the vertical component T of the 6 N force in OQ. Since the weight W of the picture is balanced by the two vertical components, then

$$\text{weight } W = 3 \text{ N} + 3 \text{ N} = 6 \text{ N}$$

(2) The acceleration due to gravity, g, acts vertically downwards. In free fall, an object has an acceleration g. An object sliding freely down an inclined plane ABC, however, has an acceleration due to gravity equal to the *component* of g down the plane, Figure 1.5. If the plane is inclined at 60° to the vertical, the acceleration down the plane is then $g \cos 60°$ or $9.8 \cos 60°$ m s^{-2}, which is 4.9 m s^{-2}.

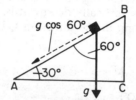

Figure 1.5 *Component of g*

Since $\cos 60° = \sin 30°$, we can say that the acceleration down the plane is also given by $g \sin 30°$, where the angle made by the plane with the horizontal is 30°.

Projectiles

Consider an object O thrown forward from the top of a cliff OA with a horizontal velocity u of 15 m s^{-1}, Figure 1.6. Since u is horizontal, it has no component in a *vertical* direction. Similarly, since g acts vertically, it has no component in a *horizontal* direction.

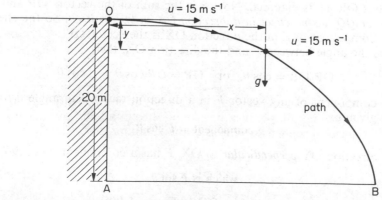

Figure 1.6 *Motion under gravity*

We can consider the vertical and horizontal motions of O independently. Consider the *vertical motion* from O. If OA is 20 m, the ball has an initial vertical velocity of zero and a vertical acceleration of g, which is 9·8 m s^{-2} (10 m s^{-2} approximately). So, from $s = ut + \frac{1}{2}at^2$, *the time t* to reach the bottom of the cliff is given, using $g = 10$ m s^{-2}, by

$$20 = \tfrac{1}{2} \times 10 \times t^2 = 5t^2 \quad \text{or} \quad t = 2\ \text{s}$$

So far as the horizontal motion is concerned, the ball continues to move forward with a constant horizontal velocity of 15 m s^{-1} since g has no component horizontally. In 2 seconds, therefore,

horizontal distance AB = distance from cliff = $15 \times 2 = 30$ m

Generally, in a time t the ball falls a *vertical* distance, y say, from O given by $y = \frac{1}{2}gt^2$. In the same time the ball travels a *horizontal* distance, x say, from O given by $x = ut$, where u is the velocity of 15 m s^{-1}. If t is eliminated by using $t = x/u$ in $y = \frac{1}{2}gt^2$, we obtain $y = gx^2/2u^2$. This is the equation of a *parabola*. It is the path OB in Figure 1.6. In our discussion air resistance has been ignored.

Motion of Projectiles and their Range

In Figure 1.7, a ball at O on the ground is thrown with a velocity u at an angle α to the horizontal. We consider the vertical and horizontal motion *separately* in motion of this kind and use components.

Vertical motion. The vertical component of u is $u \cos(90° - \alpha)$ or $u \sin \alpha$; the acceleration $a = -g$. When the projectile reaches the ground at B, the *vertical* distance s travelled is *zero*. So, from $s = ut + \frac{1}{2}at^2$, we have

$$0 = u \sin \alpha . t - \tfrac{1}{2}gt^2$$

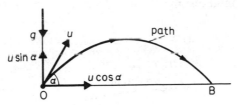

Figure 1.7 *Motion of projectiles*

Thus $$t = \frac{2u \sin \alpha}{g} \qquad . \qquad . \qquad . \qquad . \qquad . \qquad (1)$$

Horizontal motion. Since g acts vertically, it has no component in a horizontal direction. So the ball moves in a horizontal direction with an unchanged or constant velocity $u \cos \alpha$ because this is the component of u horizontally. So

Range $R = $ OB = velocity × time

$$= u \cos \alpha \times \frac{2u \sin \alpha}{g} = \frac{2u^2 \sin \alpha \cos \alpha}{g}$$

$$R = \frac{u^2 \sin 2\alpha}{g}$$

The *maximum range* is obtained when $\sin 2\alpha = 1$, or $2\alpha = 90°$. So $\alpha = 45°$ for maximum range with a given velocity of throw u. In this case, the range is u^2/g.

At the maximum height A of the path, the *vertical* velocity of the ball is zero. So, applying $v = u + at$ in a vertical direction, the time t to reach A is given by

$$0 = u \sin \alpha - gt \quad \text{or} \quad t = u \sin \alpha/g$$

From (1), we see that this is half the time to reach B.

Examples on Projectiles

A small ball A, suspended from a string OA, is set into oscillation, Figure 1.8. When the ball passes through the lowest point of the motion, the string is cut. If the ball is then moving with the velocity 0.8 m s^{-1} at a height of 5 m above the ground, find the horizontal distance travelled by the ball. (Assume $g = 10$ m s^{-2}.)

(*Analysis* (i) Horizontal distance = horizontal velocity (constant) × *time* (ii) Find time from vertical distance travelled, using g.)

When A is at the lowest point of the oscillation, it is moving horizontally with velocity 0.8 m s^{-1}. The ball lands at B on the ground.

To find the time of travel, consider the vertical motion. In this case the vertical distance s travelled is 5 m; the initial vertical velocity $u = 0$; and $a = g = 10$ m s^{-2}. From $s = ut + \frac{1}{2}at^2$, we have

$$5 = \frac{1}{2} \times 10 \times t^2$$

So
$$t^2 = 1 \quad \text{or} \quad t = 1 \text{ s}$$

To find the horizontal distance travelled, consider the horizontal motion. The velocity is 0.8 m s^{-1} and this is constant. So

$$\text{horizontal distance to B} = 0.8 \text{ m s}^{-1} \times 1 \text{ s} = 0.8 \text{ m}$$

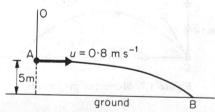

Figure 1.8 *Projectile example*

You should know:

1   A vector F has a component $F \cos \theta$ in a direction at angle θ to itself and a component $F \sin \theta$ in the perpendicular direction

2   In projectiles, (a) the horizontal velocity, $u \cos \theta$, is constant since g acts vertically, (b) for calculations use *either* a horizontal direction *or* a vertical direction, but not both. When the projectile reaches the ground again, the vertical distance travelled = 0

EXERCISES 1A Linear motion. Projectiles

Multiple Choice

1 A ball is thrown vertically upwards. The quantity which stays constant is

   A displacement **B** speed **C** kinetic energy **D** acceleration **E** velocity

2 A lunar landing module is descending to the moon's surface at a steady velocity of 10 m s^{-1}. At a height of 120 m, a small object falls from its landing gear. Taking the moon's gravitational acceleration as 1.6 m s^{-2}, at what speed, in m s^{-1}, does the object strike the moon?

   A 202 **B** 22 **C** 19.6 **D** 16.8 **E** 10 (*AEB.*)

3 A train travels on a straight track passing signal A at 20 m s^{-1}. It accelerates uniformly at 2 m s^{-2} and reaches signal B 100 m further than A. At B the velocity of the train in m s^{-1} is

   A 10 **B** 20 **C** 28 **D** 56 **E** 68

4 An aluminium ball X and an iron ball Y of the same volume are thrown horizontally with the same velocity from the top of a building. Neglecting air resistance, X reaches the ground
   A before Y and at the same distance from the building **B** at the same time as Y and at a nearer distance to the building **C** at the same time as Y and at the same distance from the building **D** at the same time as Y and further from the building **E** after Y and at the same distance from the building

Longer Questions

5 A car moving with a velocity of 10 m s^{-1} accelerates uniformly at 1 m s^{-2} until it reaches a velocity of 15 m s^{-1}. Calculate (i) the time taken, (ii) the distance travelled during the acceleration, (iii) the velocity reached 100 m from the place where the acceleration began.

6 A ball is thrown vertically upwards with an initial speed of 20 m s^{-1}. Calculate (i) the time taken to return to the thrower, (ii) the maximum height reached.

7 A ball is dropped from a height of 20 m and rebounds with a velocity which is 3/4 of the velocity with which it hit the ground. What is the time interval between the first and second bounces?

8 A small smooth object slides from rest down a smooth inclined plane inclined at $30°$ to the horizontal. What is (i) the acceleration down the plane, (ii) the time to reach the bottom if the plane is 5 m long?
   The object is now thrown up the plane with an initial velocity of 15 m s^{-1}. (iii) How long does the object take to come to rest? (iv) How far up the plane has the object then travelled?

9 A ball is thrown forward horizontally from the top of a cliff with a velocity of 10 m s^{-1}. The height of the cliff above the ground is 45 m. Calculate (i) the time to reach the ground, (ii) the distance from the cliff of the ball on hitting the ground, (iii) the direction of the ball to the horizontal just before it hits the ground.

10 Projectiles containing delicate electronic equipment may be damaged if they are subjected to high accelerations. For this reason, such projectiles may be fired from guns with long barrels but not from guns with short barrels.
   (a) State *two* factors which affect the range of a projectile fired from a gun.
   (b) Explain why a projectile fired from a long-barrelled gun is subject to less acceleration than a projectile fired from a short-barrelled gun if the range is the same in both cases.
   (c) Draw a free-body force diagram for a projectile just after it has left the barrel of a gun. (*L.*)

Graphs, Vector Addition and Subtraction

Distance–Time Graphs

When the distance s of a car moving in a constant direction from some fixed point is plotted against the time t, a *distance–time* (s–t) *graph* of the motion is obtained. The velocity of the car at any instant is given by the change in distance per second at that instant. So at E in Figure 1.9, if the change in distance s is Δs and this change is made in a time Δt,

$$\text{velocity at E} = \frac{\Delta s}{\Delta t}$$

In the limit, then, when Δt approaches zero, the velocity at E becomes equal to the *gradient of the tangent to the curve* at E. Using calculus notation, $\Delta s/\Delta t$ then becomes equal to ds/dt (p. 4). So the gradient of the tangent at E is the instantaneous velocity at E.

Velocity at E = gradient of s–t graph at E

If the distance–time graph is a straight line CD, the gradient is constant at all points; so the car is moving with a *uniform* velocity, Figure 1.9. If the distance–time graph is a curve CAB, the gradient varies at different points. The car then moves with non-uniform velocity. At the instant corresponding to A the velocity is zero, since the gradient at A of the curve CAB is zero.

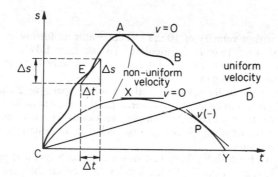

Figure 1.9 *Distance* (s)–*time* (t) *graphs* (constant direction)

When a ball is thrown upwards, the graph of the height s reached at any instant t is represented by the parabolic curve CXY in Figure 1.9. The gradient at X is zero, showing that the velocity of the ball at its maximum height is zero. From C to X the curve has a positive gradient (velocity upwards). From X to Y, as at P, the gradient is negative (velocity downwards).

Velocity–Time Graphs, Acceleration and Distance

When the velocity of a moving train is plotted against the time, a 'velocity–time (v–t) graph' is obtained. We can get useful information from this graph, as we shall see shortly.

If the velocity is uniform, the velocity–time graph is a straight line parallel to the time-axis, as shown by line (1) in Figure 1.10. If the train increases in velocity steadily from rest, the velocity–time graph is a straight line, line (2), inclined to the time-axis. If the velocity change is not steady, the velocity–time graph is curved. In Figure 1.10, the velocity–time graph OAB represents the velocity of a train starting from rest at O which reaches a maximum velocity at A, and then comes to rest at the time B.

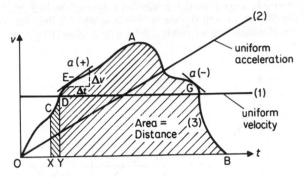

Figure 1.10 *Velocity* (v)–*time* (t) *curves*

Acceleration is the 'rate of change of velocity', that is, the change of velocity per second. The gradient to the curve at any point such as E is given by:

$$\frac{\text{velocity change}}{\text{time}} = \frac{\Delta v}{\Delta t}$$

where Δv represents a small change in v in a small time Δt. In the limit, the ratio $\Delta v/\Delta t$ becomes dv/dt, using calculus notation, or the gradient of the tangent at E. So

acceleration at E = gradient of velocity–time graph at E

At the peak point A of the curve OAB the gradient is zero. So the acceleration is then zero. From O to A, the gradient at any point such as E is upward or positive, so the train is accelerating.
At any point, such as G, between A, B the gradient to the curve is negative because the graph slopes downwards. Here the train has a *deceleration* or decrease in velocity with time.

Distance Travelled

We can also find the distance travelled from a velocity–time graph. In Figure 1.10, suppose the velocity increases in a very small time-interval XY from a value represented by XC to a value represented by YD. Since the small distance travelled = average velocity × time XY, the distance travelled is represented by the *area* between the curve CD and the time-axis, shown shaded in Figure 1.10. By considering every small time-interval between OB in the same way, it follows that

distance = AREA between v–t graph and time-axis

This result applies to any velocity–time graph, whatever its shape.

Example on Graphs

A rubber ball is thrown vertically upwards from the ground and falls on a horizontal smooth surface at the ground. The ball then bounces up and down with decreasing velocity.

(a) Draw the velocity–time graph of its motion.

(b) From your graph, show how the maximum height of the ball can be found when it is initially thrown up and the distance it falls when it first reaches the ground.

How can the acceleration due to gravity, g, be found from the graph?

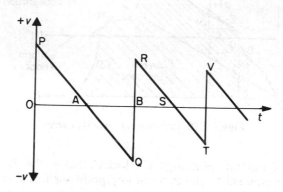

Figure 1.11 *Velocity–time graph*

(a) The graph is shown in Figure 1.11. The ball is thrown up with a velocity which we shall take as a positive velocity. As the ball rises, its velocity decreases to zero at A along the straight line PA.

Having reached its maximum height, the ball now falls. Its downward velocity is a negative one and it falls with the same numerical value of acceleration g. So the gradient of AQ is the same as PA and PAQ is a straight line.

At Q, the ball is about to hit the ground. A moment later the ball rebounds and its velocity is now high and positive. So the rebound velocity is represented by BR, where QBR is very nearly a vertical line. The velocity now varies along RSTV as explained. The lines PQ and RT are parallel because their gradients are equal to the acceleration g. The rebound velocity decreases as the ball continues to bounce, as shown.

(b) The maximum height above the ground is the area of triangle OAP. The height or distance s it falls is the area of triangle AQB, since AQ is the velocity–time graph as the ball falls.

The acceleration g is the gradient of PQ (or RT).

You should know:

In a velocity–time graph,

1   the acceleration = the *gradient* of the graph—a retardation has a negative gradient

2   the distance travelled = the *area* between the graph and the time-axis

EXERCISES 1B   Graphs

Multiple Choice

1 The acceleration of a moving object is equal to the

    A gradient of a displacement–time graph
    B gradient of a velocity–time graph
    C area below a speed–time graph
    D area below a displacement–time graph
    E area below a velocity–time graph

2

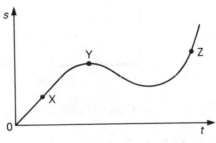

Figure 1A

The graph in Figure 1A shows how the distance, s, travelled by a body from its starting point varies with the time, t. The shape of the curve indicates that **1** the acceleration is increasing at X **2** the speed is increasing at Z **3** the body is at rest at Y.
Which of **A, B, C, D, E** show the correct answers?

    A 1, 2, 3   **B** 1, 2 only   **C** 2, 3 only   **D** 1 only   **E** 3 only (*L.*)

3 A ball is thrown vertically upwards and its velocity v varies with time t as shown in Figure 1B (i).

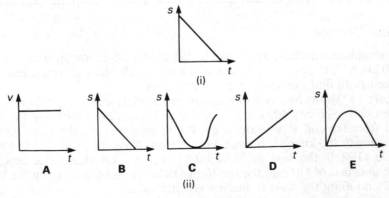

Figure 1B

Which of the graphs **A, B, C, D, E** in Figure 1B (ii) shows the displacement s–time t variation?

4 Figure 1C(i) shows the variation of displacement s with time t of a vehicle travelling along a straight road.

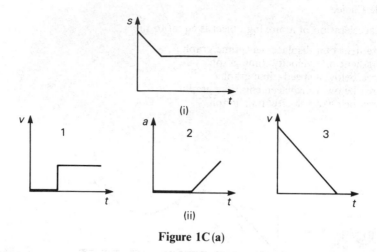

Figure 1C(a)

Which of the answers **A, B, C, D, E** are correct for the vehicle's velocity v or acceleration a in Figure 1C(ii)?

A 1 and 2   **B** 2 and 3   **C** 3 only   **D** 1 only   **E** 2 only

Longer Questions

5 A ball is thrown vertically upwards and caught by the thrower on its return. Sketch a graph of *velocity* (taking the upward direction as positive) against *time* for the whole of its motion, neglecting air resistance. How, from such a graph, would you obtain an estimate of the height reached by the ball? (*L.*)

6 A ball is thrown horizontally at a place above the ground. Show in sketches (a) how the vertical velocity v of the ball varies with time t,   (b) how its horizontal velocity u varies with time t,   (c) the angle at which the ball strikes the ground if the vertical and horizontal velocities are then 20 m s^{-1} and 5 m s^{-1} respectively.

Adding Vectors

If an aeroplane travelling at 400 km h^{-1} is blown off-course by a strong wind of 100 km h^{-1}, the pilot needs to know how to *add* the two velocities, which are vectors, to find the new aeroplane direction.

Figure 1.12 shows how two vectors can be added. Here OP represents to scale the aeroplane velocity 400 km h^{-1} in size and direction. Following the arrow round from the end P, we now draw PR to represent on the same scale the wind velocity 100 km h^{-1} in size and direction. Then the third side OR of the triangle OPR is the sum or *resultant* of the two velocities. The aeroplane direction is that of OR and the resultant OR is found by measuring the length of OR and using the scale to find its velocity value.

This is a triangle way of adding two vectors. A parallelogram way is to draw the two velocities along OP and OQ as shown and then complete the parallelogram OPRQ. The diagonal OR then represents the resultant. As you can see, the triangle or parallelogram way both give the same answer for the resultant or sum of the two velocities.

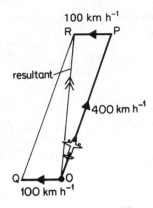

Figure 1.12 *Vector addition*

Subtracting Vectors

In a top-class sprint race, an athlete A may be running at 10 m s^{-1} and another athlete B may be running at 9 m s^{-1} in the same direction. The velocity of A relative to B is then $10 - 9 = 1$ m s^{-1}, which is the difference between the two velocities.

If the two velocities are not in the same direction, we can find the relative velocity again by subtraction, taking into account that we are dealing with vectors.

Suppose that a car X is travelling with a velocity v along a road 30° E. of N., and a car Y is travelling with a velocity u along a road due east, Figure 1.13 (i).

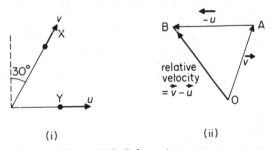

Figure 1.13 *Subtracting vectors*

Arrows above the velocity letters show they are vectors. So the velocity of X relative to Y = difference in velocities $= \vec{v} - \vec{u} = \vec{v} + (-\vec{u})$. Suppose OA represents the velocity, v, of X in magnitude and direction, Figure 1.13 (ii). Since Y is travelling due east, a velocity AB numerically equal to u but in the due *west* direction represents the vector $(-\vec{u})$. The vector sum of OA and AB is OB. So OB represents in magnitude and direction the velocity of X minus that of Y. By drawing an accurate scale diagram of the two velocities, OB can be found in size and direction.

EXERCISES 1C Vectors

Components

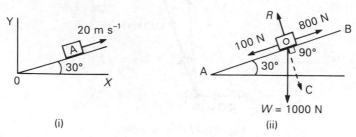

(i)   (ii)

Figure 1D

1 Figure 1D(i) shows a car A moving with a velocity of 20 m s^{-1} at 30° to a direction OX. What is the component of the velocity along OX and along the direction OY perpendicular to OX?

2 A model car O is moving up an inclined road AB at 30° to the horizontal, Figure 1D(ii). The weight W of the car is 1000 N, the frictional force down the plane is 100 N and the force on the car due to the engine is 800 N up the plane AB.

Calculate (a) the component of the weight down the plane AB, (b) the resultant force on the car up the plane, (c) the reaction R on the car at right angles to the road AB if this is balanced by the component of the weight in the opposite direction OC.

Vectors
Multiple Choice

3 Figure 1E(i) shows three force vectors. Which of the vectors **A, B, C, D, E** in Figure 1E(ii) would be most likely to represent their resultant?

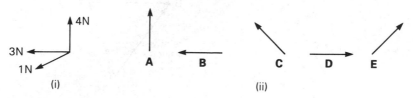

(i)   (ii)

Figure 1E

4 Figure 1F(i) shows the velocities 20 m s^{-1} and 25 m s^{-1} of two cars X and Y at one instant on a circular track. The velocity of Y relative to X in m s^{-1} is given by which vector in Figure 1F(ii)?

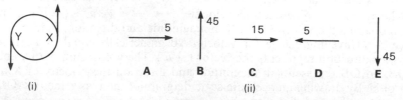

(i)   (ii)

Figure 1F

Longer Questions

5 A girl X runs with a velocity of 4 m s^{-1} across the deck of a ship Y moving in the sea at 3 m s^{-1} at $90°$ to the direction of X, Figure 1G (i).
   Find the resultant velocity of the girl relative to the sea.

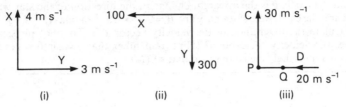

(i)   (ii)   (iii)

Figure 1G

6 An aeroplane Y moving due south at 300 km h^{-1} suddenly meets a wind X moving due west at 100 km h^{-1}, Figure 1G (ii). Find the resultant velocity of the aeroplane.

7 A car C is moving due north at 30 m s^{-1} at P, Figure 1G (iii). Another car D at Q is moving due west at 20 m s^{-1}. Find the velocity of C relative to D and show the relative direction in a diagram.

8 Figure 1H shows part of a steel bridge resting at one end on a support which has a vertical upward force of 400 N. The horizontal force at P is 300 N as shown. Calculate the force X if there is equilibrium at P.

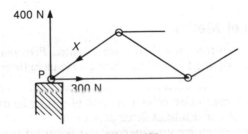

Figure 1H

9

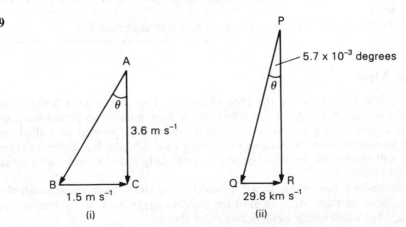

(i)   (ii)

Figure 1I

(a) The sides of the triangle ABC in Figure 1I (i) represent three velocities. AC (magnitude $3 \cdot 6 \text{ m s}^{-1}$) corresponds to the velocity of a boat as observed by a stationary tourist on a bridge. BC (magnitude $1 \cdot 5 \text{ m s}^{-1}$) corresponds to the velocity of a cyclist crossing the bridge, once again as seen by the tourist. Calculate the magnitude of the velocity vector AB, the velocity of the boat relative to the cyclist.

(b) QR, in the velocity triangle PQR in Figure 1I (ii), represents the velocity of the earth (29·8 m s⁻¹) moving at right angles to the direction of a star, QP, as seen from the earth. θ is the apparent change in the direction of the star which would occur if the earth were stationary and $\theta = 9.94 \times 10^{-5}$ radians (5.7×10^{-3} degrees). Calculate the magnitude of the velocity vector PQ. To what physical quantity does the velocity correspond? (Hint: remember that $\theta = \sin \theta = \tan \theta$ provided θ is a small angle measured in radians.) (*L.*)

Laws of Motion, Force and Momentum

In the last section, we discussed velocity and acceleration. If you kick a moving ball, or hit a ball with a tennis racket, you can see that the force produces a change in velocity or acceleration. *The first part of the next section will deal with the force on objects such as cars, for example, and the acceleration produced. After this section, we shall discuss the* momentum *of moving objects such as aeroplanes or trains, which is defined as* 'mass × velocity' *of a moving object. In this section we assume you know that a* force *is a push or a pull and that it produces a change in the velocity of the object on which it acts.*

Newton's Laws of Motion

In 1687 Sir ISAAC NEWTON published a work called *Principia Mathematica*, in which he set out clearly the Laws of Mechanics. He gave three 'laws of motion':

Law 1 Every body continues to be in a state of rest or to move with uniform velocity unless a resultant force acts on it.

Law 2 The change of momentum per second is proportional to the applied force and the momentum change takes place in the direction of the force.

Law 3 To every action there is an equal and opposite reaction.

Inertia, Mass

Newton's first law expresses the idea of *inertia*. The inertia of a body is its reluctance to start moving, and its reluctance to stop after it has begun moving. Thus an object at rest begins to move only when it is pushed or pulled, or when a *force* acts on it. An object O moving in a straight line with constant velocity will change its direction or move faster only if a new force acts on it, Figure 1.14 (i).

Passengers in a bus or car move forward when the vehicle stops suddenly. They *continue* in their state of motion until brought to rest by friction or collision. The use of safety belts reduces the shock.

Figure 1.14 (ii) shows a velocity change when an object O is whirled at constant speed by a string. This time the magnitude of the velocity v is constant but its *direction* changes. Since the velocity changes, O has an acceleration. So a *force* due to the string acts on O.

'Mass' is a measure of the inertia of a body. If an object changes its direction

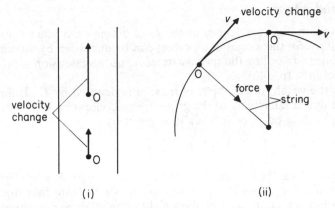

Figure 1.14 *Velocity changes:* (i) *magnitude,* (ii) *direction*

or its velocity slightly when a big force acts on it, its inertial mass is high. The mass of an object is constant all over the world; it is the same on the earth as on the moon. Mass is measured in kilograms (kg) by means of a chemical balance, where it is compared with standard masses based on the International Prototype Kilogram.

Force, the Newton

When an object X is moving it is said to have an amount of *momentum* given, by definition, by

$$momentum = mass \ of \ X \times velocity \quad . \quad . \quad . \quad (1)$$

Thus a runner of mass 50 kg moving with a velocity of 10 m s^{-1} has a momentum of 500 kg m s^{-1}. If another runner collides with X his velocity alters, and so the momentum of X alters.

From Newton's second law, a *force F* acts on X which is equal to its change in momentum per second. Using (1), it follows that if m is the mass of X,

$$F \propto m \times \text{change in velocity per second}$$

But the change in velocity per second is the *acceleration a* produced by the force.

$$\therefore F \propto ma$$

so
$$F = kma \quad . \quad . \quad . \quad . \quad . \quad (2)$$

where k is a constant.

With SI units, the **newton** (N) is the unit of force. It is defined as the force which gives a mass of 1 kg an acceleration of 1 m s^{-2}. Substituting $F = 1$ N, $m = 1$ kg and $a = 1 \text{ m s}^{-2}$ in the expression for F in (2), we obtain $k = 1$. So, with these units, $k = 1$.

$$\therefore F = ma \quad . \quad . \quad . \quad . \quad . \quad (3)$$

which is a standard equation in dynamics. If a mass of 0·2 kg is acted upon by a force F which produces an acceleration a of 4 m s^{-2}, then, since $m = 0.2$ kg,

$$F = ma = 0.2 \text{ (kg)} \times 4(\text{m s}^{-2}) = 0.8 \text{ N}$$

Weight and Mass

The *weight* of an object is defined as the *force* acting on it due to gravitational pull, or gravity. So the weight of an object can be measured by attaching it to a spring-balance and noting the spring extension, as the extension is proportional to the force on it (p. 142).

Suppose the weight of an object of mass m is denoted by W. If the object is released so that it falls freely to the ground, its acceleration is g. Now $F = ma$. So the force acting on it, or its weight, is given by

$$W = mg$$

If the mass is 1 kg, then, since $g = 9.8$ m s^{-2}, the weight $W = 1 \times 9.8 = 9.8$ N. The weight of a 5 kg mass is thus 5×9.8 N or 49 N. Note that the weight of a 100 g (0.1 kg) mass is about 1 N; the weight of an average-sized apple is about 1 N.

Gravitational Field Strength

The space round the earth where the mass of an object experiences a gravitational pull, or force due to gravity, is called the *gravitational field* of the earth. Molecules of air, or this book or the reader, are all in the earth's gravitational field.

We can see that on the surface of the earth, the value of g may be expressed as about 9.8 N kg^{-1}. The *force per unit mass* in a gravitational field is called the *gravitational field strength*. On the moon's surface this is only about 1.6 N kg^{-1}, so a mass of 1 kg has a gravitational pull on it of only 1.6 N.

You should note carefully that the *mass* of an object is constant all over the world, but its *weight* is a *force* whose magnitude depends on the value of g. The acceleration due to gravity, g, is slightly greater at the poles than at the equator, since the earth is not a perfectly spherical shape. So the weight of an object differs in different parts of the world. On the moon, which is smaller than the earth and has a smaller density, an object would have the same mass as on the earth. But it would weigh only about one-sixth of its weight on the earth, as the acceleration of free-fall on the moon is about $g/6$. For this reason astronauts tend to 'float' on the moon's surface.

Experimental Investigation of $F = ma$

An experimental investigation of $F = ma$ can be carried out by accelerating a trolley of mass m down a friction-compensated inclined plane using a constant force F due to a stretched piece of elastic, and measuring the acceleration a with a ticker tape. Details of the experiment can be obtained from GCSE texts, such as the author's *Principles of Physics* (Longman Educational, London). We assume the reader is familiar with the method. Figure 1.15 shows the results obtained. When the force is increased in the ratio 1:2:3, experiment shows that the acceleration increases in the ratio 2.4:4.8:7.3, which is approximately 1:2:3. So $a \propto F$ with constant mass.

The mass m can also be varied by placing similar trolleys on top of each other and the force F can be kept constant by stretching the elastic the same amount each time. One experiment shows that with 1, 2 and 3 trolleys the accelerations decrease in the ratio 7.5:4.9:2.5, which is 3:2:1 approximately. So $a \propto 1/m$ when F is constant.

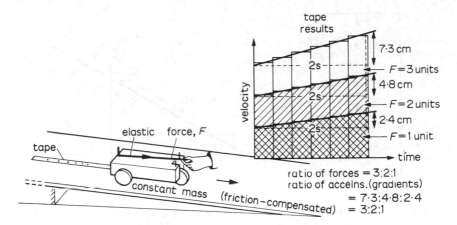

Figure 1.15 *Investigation of acceleration and force (mass constant)*

How to Apply $F = ma$

The following examples show you how to apply $F = ma$. You should carefully note that (i) if more than one force acts on a moving object, then F is the *resultant* force on the object, (ii) F must be in newtons (N) and m in kilograms (kg). Since F is a vector, the direction of the forces must be drawn in the diagram. Remember mass (m) is a scalar so this has no direction.

Examples on $F = ma$

1 *Two forces* A force of 200 N pulls a sledge of mass 50 kg and overcomes a constant frictional force of 40 N. What is the acceleration of the sledge?

$$\text{Resultant force } F = 200 - 40 = 160 \text{ N}$$

From $F = ma$,

$$160 = 50 \times a$$

$$\therefore a = 3 \cdot 2 \text{ m s}^{-2}$$

2 *Lift problem* An object of mass 2·00 kg is attached to the hook of a spring-balance, and the balance is suspended vertically from the roof of a lift. What is the reading on the spring-balance when the lift is (i) going up with an acceleration of $0 \cdot 2 \text{ m s}^{-2}$, (ii) going down with an acceleration of $0 \cdot 1 \text{ m s}^{-2}$, (iii) ascending with a uniform velocity of $0 \cdot 15 \text{ m s}^{-1}$ ($g = 10 \text{ m s}^{-2}$ or 10 N kg^{-1}).

Suppose T is the tension (force) in the spring-balance in N, Figure 1.16 (i).
(i) The object is acted on by two forces:
(a) the tension T in newtons in the spring balance which acts upwards,
(b) its weight mg or 20 N, which acts downwards.

Since the object accelerates *upwards*, T is greater than 20 N. So the net force, F, acting on the object $= T - 20$ N. Now

$$F = ma$$

where a is the acceleration in m s^{-2}, 0·2 m s^{-2}.

$$\therefore T - 20 = 2 \times a = 2 \times 0 \cdot 2$$

$$\therefore T = 20 \cdot 4 \text{ N} \qquad . \qquad . \qquad . \qquad . \qquad (1)$$

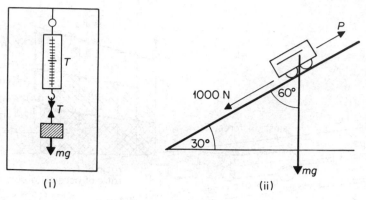

Figure 1.16 *Examples on force and acceleration*

(ii) When the lift *descends* with an acceleration of 0.1 m s^{-2}, its weight, 20 N, is now greater than T_1, the new tension in the spring-balance.

$$\therefore \text{ resultant force} = 20 - T_1$$

$$\therefore F = 20 - T_1 = ma = 2 \times 0.1$$

$$\therefore T_1 = 20 - 0.2 = 19.8 \text{ N}$$

(iii) When the lift moves with constant velocity, the acceleration is zero. Since the resultant force is zero, the reading on the spring-balance is exactly equal to the weight, 20 N.

3 *Inclined plane* A car of mass 1000 kg is moving up a hill inclined at 30° to the horizontal. The total frictional force on the car is 1000 N, Figure 1.16 (ii).

Calculate the force P due to the engine when the car is
(a) accelerating at 2 m s^{-2},
(b) moving with a steady velocity of 15 m s^{-1}.

(*Analysis* (a) Use $F = ma$. (b) There are three forces on the car—its weight mg, P and the frictional force. (c) Since the car is moving on an incline, we need to find the component of the weight mg, down the incline.)

(a) Weight of car $= mg = 10\,000$ N
   Component downhill $= mg \cos \theta = 10\,000 \cos 60° = 5000$ N

   So   resultant force uphill, $F = P - 5000 - 1000$

   From $F = ma$,

$$P - 5000 - 1000 = 1000 \times 2 = 2000$$

$$P = 8000 \text{ N}$$

(b) Since velocity is steady, acceleration $a = 0$
   So resultant force $F = 0$

   Then $P = 5000 + 1000 = 6000 \text{ N}$

Terminal Velocity

If we swim or drive a car fast, we can feel the frictional force or resistance due to motion through water or to air. The size of F increases with the velocity v

of the moving object and roughly $F = kv$, where k is a constant. Aerodynamic design of racing cars reduces the value of k.

Experiment Figure 1.17 (i) shows a simple experiment to investigate how F varies with increasing v. A tall measuring cylinder P is filled with a thick viscous liquid such as glycerine and narrow horizontal strips A, B, C, D, E are placed at equal distances down the cylinder. A small ball-bearing X is now gently allowed to fall into the liquid and the time taken to fall through distances AB, BC, CD, DE respectively is measured.

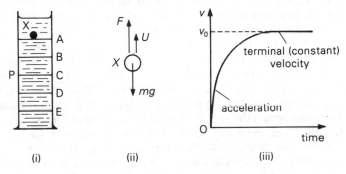

(i) (ii) (iii)

Figure 1.17 *Motion of falling sphere in glycerine*

The results show that X increases in velocity as it falls through AB and BC but reaches a constant velocity, called the *terminal velocity*, as it falls through CDE.

Explanation When X is in the liquid, forces on it are (1) its weight W or mg, (2) the frictional force F due to the liquid, (3) the liquid upthrust U (the upward force due to the liquid pressure), Figure 1.17 (ii). When X is first dropped into the liquid, its weight mg is greater than $(F + U)$. The resultant downward force then produces an acceleration and so the velocity v of X increases. The frictional force F on X, however, *increases* with the velocity v. So soon mg is *equal* to $(U + F)$. Since the resultant force on X is now zero, X then moves with a *constant velocity*, v_0, which is the terminal (end) velocity. A load dropped from a high aeroplane reaches a terminal (maximum) velocity because the frictional force of the air increases with the velocity of the object.

Figure 1.17 (iii) shows (a) how the velocity v of X increases with time to a maximum velocity v_0 and (b) how the acceleration a, the gradient to the velocity–time graph, decreases to zero in this time.

You should know:

1 **Newton's first law—an object continues to move in a straight line unless a force acts on it to change its direction (so seat belts necessary)**
2 **Force = mass × acceleration or $F = ma$**
 F is always the *resultant* force. So with a train moving up an incline, F moving the train along the incline = force due to engine − frictional force − component of weight down incline.
3 **Weight is the gravitational force mg, which always acts vertically downwards. Mass is a constant all over the world.**
 Weight is measured in newtons (N), mass in kilograms (kg). The weight of 1 kg mass = mg = 9·8 N or 10 N in round figures.

> **4** **Terminal velocity = maximum velocity of an object X falling from a height or moving from rest through a frictional medium such as liquid or air. Frictional force = kv, where v is velocity. At terminal velocity, resultant force = 0.**

Aerodynamics and Drag

The viscosity of a gas, a measure of its frictional effect on objects moving through the gas, is a very important factor in calculating the aerodynamic drag on a body moving through the gas. Let us take as an example a car moving through air: there are important economic and environmental reasons for reducing the drag, and so increasing the car's fuel efficiency. The air in contact with the car's surface moves with the same velocity as the car. This layer of moving air exerts a viscous force on air further away from the car, causing it to move also. In fact, you have to go some distance away from the car's surface until the air is stationary; the region around the car where the air has a non-zero velocity is called the *boundary layer*. For badly designed cars, the airflow passes over the windshield and then separates from the car's surface. This phenomenon, known as *boundary layer separation*, creates a low-pressure area in the car's wake, which exerts a retarding force on the car, adding to the aerodynamic drag. Sleek cars have better aerodynamic properties, in that the boundary layer does not separate until the rear of the car, and there is no corresponding low-pressure area to increase the drag. The reason why many modern cars look so similar is that they have all been designed with the same aerodynamic properties in mind.

There are times, though, when the shape of the moving object cannot be redesigned, and so drag must be reduced by other means. A good example is the golf-ball, which has to be spherical for obvious reasons. In this case, separation of the boundary layer can be delayed by causing the airflow around the ball to become *turbulent*. This is why golf-balls are pitted. The rough surface induces turbulence in the airflow, and the boundary layer stays attached much further around the ball, reducing the drag. If golf-balls were smooth, it would generally take more strokes to reach the green, and the par ratings of holes around the world would have to be changed!

EXERCISES 1D   Force and Acceleration

(Assume $g = 10 \ m \ s^{-2}$ or $10 \ N \ kg^{-1}$ unless otherwise stated)

Multiple Choice

1 A body is acted upon by a force F which varies with time t as in Figure 1J(i). Which graph in Figure 1J (ii) best represents the velocity (v)–time (t) variation of the body?

2 Figure 1K shows a force of 40 N acting at 30° to the horizontal on a body of mass 5 kg resting on a smooth horizontal surface. Assuming that the acceleration of free-fall is 10 m s^{-2}, which of the following statements **A, B, C, D, E** is (are) correct?
   1 The horizontal force acting on the body is 20 N
   2 The weight of the 5 kg mass acts vertically downwards
   3 The net vertical force acting on the body is 30 N

    A 1, 2, 3   **B** 1, 2   **C** 2, 3   **D** 1   **E** 3 (*L.*)

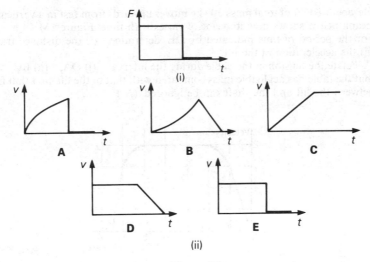

(i)

(ii)

Figure 1J

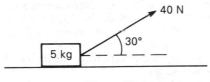

Figure 1K

Longer Questions

3 The diagrams in Figure 1L show a sphere, S, resting on a table and a free-body diagram on which the forces acting on the sphere have been marked.

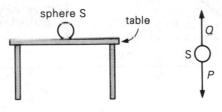

Figure 1L

We know from Newton's first law of motion that forces occur in equal and opposite pairs.
On which body does the force which pairs with force P act? Give its direction.
On which body does the force which pairs with force Q act? Give its direction.
A force F_1 acts on an object O and this same force F_1 forms a Newton's third law pair with a second force, F_2. State *two* ways in which F_1 and F_2 are similar and *two* ways in which they differ. (*L.*)

4 A car of mass 1000 kg is accelerating at 2 m s^{-2}. What resultant force acts on the car? If the resistance to the motion is 1000 N, what is the force due to the engine?

5 A box of mass 50 kg is pulled up from the hold of a ship with an acceleration of 1 m s^{-2} by a vertical rope attached to it. Find the tension in the rope. What is the tension in the rope when the box moves up with a uniform velocity of 1 m s^{-1}?

6 A lift moves (i) up and (ii) down with an acceleration of 2 m s^{-2}. In each case, calculate the reaction of the floor on a man of mass 50 kg standing in the lift.

7 A cable-operated lift of total mass 500 kg moves upwards from rest in a vertical shaft. The graph below shows how its velocity varies with time, Figure 1M.
(a) For the period of time indicated by DE, determine (i) the distance travelled, (ii) the acceleration of the lift.
(b) Calculate the tension in the cable during the interval (i) OA, (ii) BC. Assume that the table has negligible mass compared with that of the lift, and that friction between the lift and the shaft can be ignored. (N.)

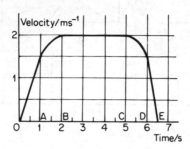

Figure 1M

8 A large cardboard box of mass 0·75 kg is pushed across a horizontal floor by a force of 4·5 N. The motion of the box is opposed by (i) a frictional force of 1·5 N between the box and the floor, and (ii) an air resistance force kv^2, where $k = 6·0 \times 10^{-2}$ kg m^{-1} and v is the speed of the box in m s^{-1}.
Sketch a diagram showing the directions of the forces which act on the moving box. Calculate maximum values for
(a) the acceleration of the box,
(b) its speed. (L).

9

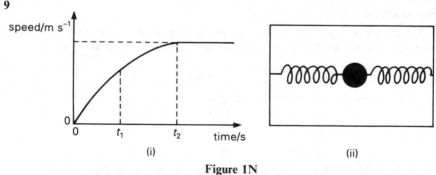

Figure 1N

The graph in Figure 1N(i) shows how the speed of a car varies with time as it accelerates from rest.
(a) State (i) the time at which the acceleration is a maximum, (ii) how you could use the graph to find the distance travelled between times t_1 and t_2.
(b) The driving force produced by the car engine can be assumed to be constant. Explain, in terms of the forces on the car, why (i) the acceleration is not constant, (ii) the car eventually reaches a constant speed.
Figure 1N(ii) shows a simple version of an instrument used to measure acceleration, in which a mass is supported between two springs in a box so that when one spring is extended, the other is compressed. At rest, the mass is in the position shown.
Redraw the diagram showing the position of the mass when the box accelerates to the right. Explain why the mass takes up this position. (N.)

10 An astronaut is outside her space capsule in a region where the effect of gravity can be neglected. She uses a gas gun to move herself relative to the capsule. The gas gun

fires gas from a muzzle of area 160 mm^2 at a speed of 150 m s^{-1}. The density of the gas is 0·800 kg m^{-3} and the mass of the astronaut, including her space suit, is 130 kg. Calculate
(a) the mass of gas leaving the gun per second,
(b) the acceleration of the astronaut due to the gun, assuming that the change in mass is negligible. (*N.*)

11 A light string carrying a small bob of mass 5·0 × 10^{-2} kg hangs from the roof of a **moving** vehicle.
(a) What can be said about the motion of the vehicle if the string hangs vertically?
(b) The vehicle moves in a horizontal straight line from left to right, with a constant acceleration of 2·0 m s^{-2}.
   (i) Show in a sketch the forces acting on the bob.
   (ii) By resolving horizontally and vertically or by scale drawing, determine the angle which the string makes with the vertical.
(c) The vehicle moves down an incline making an angle of 30° with the horizontal with a constant acceleration of 3·0 m s^{-2}. Determine the angle which the string makes with the vertical. (*N.*)

Linear Momentum, Impulse

Newton defined the force acting on an object as the rate of change of its momentum, the momentum being the *product mass × velocity* (p. 23). *Momentum is thus a vector quantity*; its direction is that of the velocity. Suppose an object of mass m changes in velocity from u to v when a force F acts on it. Then

$$\text{change of momentum} = mv - mu$$

and

$$F = \frac{mv - mu}{t}$$

$$\therefore F \times t = mv - mu = \textit{momentum change} \quad . \qquad . \qquad . \qquad (1)$$

The quantity $F \times t$ (force × time) is known as the *impulse* of the force on the object. From (1) we see that the units of momentum are the same as those of Ft, that is, *newton second* (N s). From 'mass × velocity', alternative units are 'kg m s^{-1}'. The impulse ($F \times t$) is the 'time effect' of a force on an object.

Force and Momentum Change

(1) A person of mass 50 kg who is jumping from a height of 5 metres will land on the ground with a velocity $= \sqrt{2gh} = \sqrt{2 \times 10 \times 5} = 10$ m s^{-1}, assuming $g = 10$ m s^{-2} approx. If he does not flex his knees on landing, he will be brought to rest very quickly, say in one tenth of a second. The force F acting is then given by

$$F = \frac{\text{momentum change}}{\text{time}}$$

$$= \frac{50 \times 10}{\frac{1}{10}} = 5000 \text{ N}$$

This is a force of about 10 times the person's weight and the large force has a severe effect on the body.

Suppose, however, that the person flexes his knees and is brought to rest much more slowly on landing, say in 1 second. Then, from above, the force F now acting is 10 times less than before, or 500 N. So much less damage is done to the body on landing.

Similarly, drawing the hands back when catching a fast-moving hard ball will increase the time to bring the ball to rest and so reduce the force on the hands.

(2) Suppose a ball of mass 0·1 kg hits a smooth wall normally with a velocity u of $10\ \text{m s}^{-1}$ four times per second and rebounds each time with a velocity $-u$ of $-10\ \text{m s}^{-1}$, Figure 1.18 (i). Each time, momentum change = mass × velocity change = $(0·1 \times 10) - (-0·1 \times 10) = 0·1 \times 20$. So

average force on wall = 4 × momentum change per second

$$= 4 \times 0·1 \times 20 = 8\ \text{N}$$

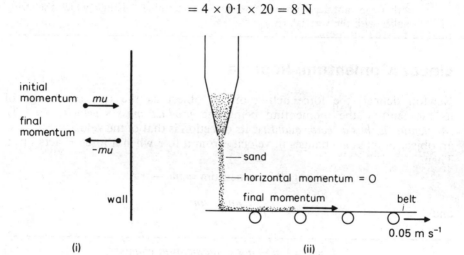

Figure 1.18 *Linear momentum changes*

Momentum due to Mass Change

Momentum is a *vector*. So we must always take account of its direction.

Suppose sand is allowed to fall vertically at a steady rate of $0·1\ \text{kg s}^{-1}$ on to a horizontal conveyor belt moving at a steady velocity of $0·05\ \text{m s}^{-1}$, Figure 1.18 (ii). The initial *horizontal* velocity of the sand is zero. The final horizontal velocity is $0·05\ \text{m s}^{-1}$. Now in one second in a horizontal direction,

mass = 0·1 kg, velocity gained = $0·05\ \text{m s}^{-1}$

∴ momentum change per second *horizontally* = $0·1 \times 0·05 = 0·005\ \text{N}$

The belt provides the extra force needed for this momentum increase per second of the sand. This is a case where the *mass* changes with time and the velocity gained is constant. We can find a useful formula for such cases. Consider a rocket moving upwards into the air and losing mass as its fuel is burnt. From

$$F = \text{mass} \times \frac{\text{velocity change}}{\text{time}}$$

we can write $$F = \frac{\text{mass}}{\text{time}} \times \text{velocity change.}$$

So

$$F = \text{mass per second} \times \text{velocity change}$$

In the case of the falling sand, Figure 1.18 (ii),

$$\text{mass per second} = 0{\cdot}1 \text{ kg s}^{-1}$$

and velocity change horizontally $= 0{\cdot}05 \text{ m s}^{-1}$

So $$F = 0{\cdot}1 \times 0{\cdot}05 = 0{\cdot}005 = 5 \times 10^{-3} \text{ N}$$

When passengers step on to an escalator moving upwards, they gain upward momentum. The belt driving the escalator rollers provides the extra force needed for the momentum change of the passengers.

Force due to Water Flow

When water from a horizontal hose-pipe strikes a wall at right angles, a force is exerted on the wall. Suppose the water comes to rest on hitting the wall. Then

force on wall = mass of water per second × its velocity change

Suppose the water flows out of the pipe at 2 kg s^{-1} and its velocity changes from 5 m s^{-1} to zero on hitting the wall. Then

$$\text{force } F = 2 \times (5 - 0) = 10 \text{ N}$$

A similar example is the case of raindrops falling on a flat roof. Suppose a steady mass of water, $0{\cdot}2 \text{ kg s}^{-1}$, falls vertically on the roof. If the raindrops have a velocity of 10 m s^{-1} just before hitting the roof and then come to rest, the velocity change $= 10 \text{ m s}^{-1}$. So

$$\text{force on roof} = \text{mass per second} \times \text{velocity change}$$

$$= 0{\cdot}2 \times 10 = 2 \text{ N}$$

Example on Force due to Water Flow

A hose ejects water at a speed of 20 cm s^{-1} through a hole of area 100 cm^2. If the water strikes a wall normally, calculate the force on the wall in newtons, assuming the velocity of the water normal to the wall is zero after collision.

The volume of water per second striking the wall $= 100 \times 20 = 2000 \text{ cm}^3$

$\therefore$ mass per second striking wall $= 2000 \text{ g s}^{-1} = 2 \text{ kg s}^{-1}$

Velocity change of water on striking wall $= 20 - 0 = 20 \text{ cm s}^{-1} = 0{\cdot}2 \text{ m s}^{-1}$

$\therefore$ momentum change per second $= 2 \text{ (kg s}^{-1}) \times 0{\cdot}2 \text{ (m s}^{-1}) = 0{\cdot}4 \text{ N}$

$\therefore$ Force on wall $= 0{\cdot}4 \text{ N}$

Newton's Third Law, Action and Reaction

Newton's third law—action and reaction are equal and opposite—means that if a body A exerts a force (action) on a body B, then B will exert an equal and opposite force (reaction) on A.

(1) These forces are produced between objects by direct contact when they touch, or by gravitational forces, for example, when they are apart. So if a ball is kicked upwards, the force on the ball by the kicker is equal and opposite to the force on the kicker by the ball. The initial upward acceleration of the ball is usually very much greater than the downward acceleration of the kicker because the mass of the ball is much less than that of the kicker.

(2) As the ball falls downwards towards the ground, the force of attraction on the ball by the earth is equal and opposite to the force of attraction on the earth by the ball. The upward acceleration of the earth is not noticeable since its mass is so large.

(3) In the case of a rocket, the downward force on the burning gases from the exhaust is equal to the upward force on the rocket. This is an application of the reaction force. Another is a water-sprinkler which spins backwards as the water is thrown forwards.

Action–reaction forces therefore always occur in pairs. It should be noted that the two forces act on *different* bodies. So only *one* of the forces is used in discussing the motion of one of the two bodies. In the case of a man standing in a lift moving upwards, for example, the upward reaction of the floor *on the man* is the force we need to take into account in applying $F = ma$ to the motion of the *man*. The equal downward force on the floor is *not* required.

This principle is also illustrated in the next example, where we take one object at a time.

Example on Truck and Trailer

Figure 1.19 shows a truck A of mass 1000 kg pulling a trailer of mass 3000 kg. The frictional force on A is 1000 N, on B it is 2000 N, and the truck engine exerts a force of 8000 N.

Calculate   (i) the acceleration of the truck and trailer,   (ii) the tension (force) T in the tow-bar connecting A and B.

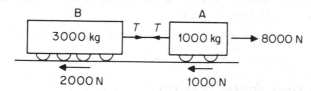

Figure 1.19 *Example on truck and trailer*

The forces T in the towbar are equal and opposite, from Newton's law.

For B only   From $F = ma$, where F is the resultant force

$$T - 2000 = 3000\,a \qquad\qquad (1)$$

For A only $8000 - 1000 - T = 1000\,a \qquad\qquad (2)$

Adding (1) and (2) to eliminate T, then

$$8000 - 1000 - 2000 = 4000\,a$$

So $5000 = 4000\,a$ and $a = 1\cdot25\ \mathrm{m\ s^{-2}}$

From (1), $T = (3000 \times 1\cdot25) + 2000 = 5750\ \mathrm{N}$

Force due to Rotating Helicopter Blades

When helicopter blades are rotating, they strike air molecules in a downward direction. The momentum change per second of the air molecules produces a *downward* force and by the Law of Action and Reaction, an equal *upward* force is exerted by the molecules on the helicopter blades. This upward force helps to keep the helicopter hovering in the air because it can balance the downward weight of the machine. This is illustrated in the example which follows.

Example on Rotating Helicopter Blades

A helicopter of mass 500 kg hovers when its rotating blades move through an area of $30 \ m^2$ and gives an average speed v to the air.

Estimate v assuming the density of air is $1 \cdot 3 \ kg \ m^{-3}$ and $g = 10 \ N \ kg^{-1}$.

(*Analysis* (i) The reaction of the downward force on the air = weight of helicopter, (ii) downward force = momentum change per second of air swept down, (iii) mass of air per second moving downwards = volume per second × density = area swept by blades × velocity of air × density.)

Volume of air per second moving downwards = area × velocity $v = 30v$

So   mass of air per second downwards = $30v \times 1 \cdot 3 = 39v$

∴ momentum change per second of air = mass per second × velocity change

$$= 39v \times v = 39v^2$$

So   reaction force upwards = $39v^2$ = helicopter weight 5000 N

$$v^2 = \frac{5000}{39}$$

$$v = \sqrt{\frac{5000}{39}} = 11 \ m \ s^{-1} \ (\text{approx.})$$

At lift-off, the fuselage of the helicopter would turn round the opposite way to the rotation of the blades, from Newton's law of Action and Reaction. A vertical rotor on the tail provides a counter-thrust.

You should know:

1 **Momentum = mass × velocity.**
   Impulse = force × time ($F.t$) = mass × velocity change
   Momentum or impulse unit: N s
2 **Newton's second law: force, F = rate of change of momentum**
   ($F = d(mv)/dt$)
3 *Mass constant:* **use $F = ma$ to find force**
   Mass varying: **to find force, use F = mass per second × velocity change**
4 **Newton's third law: Action and reaction forces are equal and opposite.**
   The law applies to forces between objects in contact *or* at a distance from each other, as in a gravitational field (or magnetic or electric field)

Conservation of Linear Momentum

We now consider what happens to the linear momentum of objects which *collide* with each other.

Experimentally, this can be investigated by several methods:

1   Trolleys in collision, with ticker-tapes attached to measure velocities.
2   Linear air-track, using Perspex models in collision and stroboscopic photography for measuring velocities.

As an illustration of the experimental results, the following measurements were taken in trolley collisions (Figure 1.20):

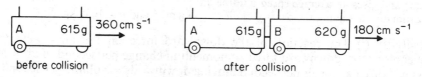

before collision   after collision

Figure 1.20 *Linear momentum experiment*

Before collision
Mass of trolley A = 615 g; initial velocity = 360 cm s^{-1}.

After collision
A and B coalesced and both moved with velocity of 180 cm s^{-1}.

So the total linear momentum of A and B before collision = 0·615 (kg) × 3·6 (m s^{-1}) + 0 = 2·20 kg m s^{-1} (approx.). The total momentum of A and B after collision = 1·235 × 1·8 = 2·20 kg m s^{-1} (approx.).
Within the limits of experimental accuracy, it follows that *the total momentum of A and B before collision = the total momentum after collision.*
Similar results are obtained if A and B are moving with different speeds after collision, or in opposite directions before collision.

Principle of Conservation of Linear Momentum

These experimental results can be shown to follow from Newton's second and third laws of motion (p. 22).

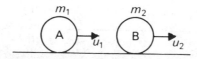

Figure 1.21 *Conservation of linear momentum*

Suppose that a moving object A, of mass m_1 and velocity u_1, collides with another object B, of mass m_2 and velocity u_2, moving in the same direction, Figure 1.21. By Newton's law of action and reaction, the force F exerted by A on B is equal and opposite to that exerted by B on A. Also, the time t during which the force acted on B is equal to the time during which the force of reaction acted on A. So the magnitude of the impulse, Ft, on B is equal and *opposite* to the magnitude of the impulse on A. From equation (1), p. 31 the impulse is equal to the change of momentum. It therefore follows that the *change* in the total momentum of the two objects is *zero*, i.e., the total momentum of the two objects is constant although a collision had occurred. So if A moves with a reduced velocity v_1 after collision, and B then moves with an increased velocity v_2,

$$m_1u_1 + m_2u_2 = m_1v_1 + m_2v_2$$

The principle of the conservation of linear momentum states that, if no external forces act on a system of colliding objects, the total momentum of the objects in a given direction before collision = total momentum in same direction after collision.

Examples on Conservation of Momentum

1 An object A of mass 2 kg is moving with a velocity of 3 m s^{-1} and collides head on with an object B of mass 1 kg moving in the opposite direction with a velocity of 4 m s^{-1}, Figure 1.22. After collision both objects stick, so that they move with a common velocity v. Calculate v.

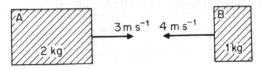

Figure 1.22 *Example*

Total momentum before collision of A and B in the direction of A

$$= (2 \times 3) - (1 \times 4) = 2 \text{ kg m s}^{-1}$$

Note that momentum is a vector and the momentum of B is of opposite sign to A.

After collision, momentum of A and B in the direction of A $= 2v + 1v = 3v$

$$\therefore 3v = 2$$

$$\therefore v = \tfrac{2}{3} \text{ m s}^{-1}$$

2 A bullet of mass 20 g, travelling horizontally at 100 m s^{-1}, embeds itself in the centre of a block of wood of mass 1 kg which is suspended by light vertical strings 1 m in length. Calculate the maximum inclination of the strings to the vertical. (Assume $g = 9 \cdot 8$ m s^{-2}.)

(*Analysis* (i) The angle of swing θ depends on the velocity of the bullet plus block after the bullet enters, (ii) the velocity v of the bullet on entering the block can be found by applying the conservation of momentum.)

Suppose A is the bullet, B is the block suspended from a point O, and θ is the maximum inclination to the vertical, Figure 1.23. If v m s^{-1} is the common velocity of block and bullet when the bullet is brought to rest relative to the block, then, from the principle of the conservation of momentum, since 20 g = 0·02 kg,

$$(1 + 0 \cdot 02)v = 0 \cdot 02 \times 100$$

$$\therefore v = \frac{2}{1 \cdot 02} = \frac{100}{51} \text{ m s}^{-1}$$

From the conservation of energy, loss in kinetic energy (see p. 52) of block and bullet = gain in their potential energy. So $\tfrac{1}{2}Mv^2 = Mgh$, where M is the total

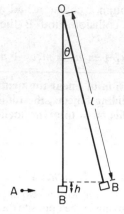

Figure 1.23 *Example*

and bullet and h is the vertical height risen. Cancelling M, $v^2 = 2gh$. Now

$$h = l - l \cos \theta = l(1 - \cos \theta)$$

$$\therefore v^2 = 2gl(1 - \cos \theta)$$

$$\therefore \left(\frac{100}{51}\right)^2 = 2 \times 9 \cdot 8 \times 1(1 - \cos \theta)$$

$$\therefore 1 - \cos \theta = \left(\frac{100}{51}\right)^2 \times \frac{1}{2 \times 9 \cdot 8} = 0 \cdot 1962$$

$$\therefore \cos \theta = 0 \cdot 8038, \text{ or } \theta = 37° \text{ (approx.)}$$

3 A snooker ball X of mass 0·3 kg, moving with velocity 5 m s^{-1}, hits a stationary ball Y of mass 0·4 kg. Y moves off with a velocity of 2 m s^{-1} at 30° to the initial direction of X, Figure 1.24.

Find the velocity v of X and its direction after hitting Y.

(*Analysis* We need two equations to find v and θ for X. So apply the momentum conservation (i) along the initial direction of X, (ii) perpendicular to the initial direction, direction Z.)

(i) In initial direction of X, from conservation of momentum,

$$0 \cdot 3v \cos \theta + 0 \cdot 4 \times 2 \cos 30° = 0 \cdot 3 \times 5$$

So
$$0 \cdot 3v \cos \theta = 1 \cdot 5 - 0 \cdot 8 \cos 30° = 0 \cdot 8 \quad . \quad . \quad (1)$$

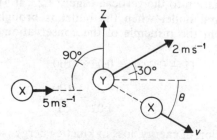

Figure 1.24 *Example on collision*

(ii) Along Z, 90° to initial X direction, initial momentum $= 0$.

So in this direction,

$$0\!\cdot\!4 \times 2 \sin 30° - 0\!\cdot\!3 \times v \sin \theta = 0$$

or $\qquad\qquad\qquad 0\!\cdot\!3v \sin \theta = 0\!\cdot\!4 \qquad\qquad . \quad . \quad . \quad . \quad . \qquad (2)$

Dividing (2) by (1),

$$\frac{\sin \theta}{\cos \theta} = \tan \theta = \frac{0\!\cdot\!4}{0\!\cdot\!8} = 0\!\cdot\!5$$

So $\qquad\qquad\qquad\qquad\qquad \theta = 27°$ (approx.)

Also, from (2) $\qquad\qquad\qquad v = \dfrac{0\!\cdot\!4}{0\!\cdot\!3 \sin 27°} = 3 \text{ m s}^{-1}$ (approx.)

EXERCISES 1E Momentum and Force

Multiple Choice

1 A force F acts on a ball initially at rest on a smooth surface for a time t. The variation of F with t is shown in Fig. 1O. The momentum of the ball in N s after 4·0 s is

A 10 B 20 C 30 D 40 E 80

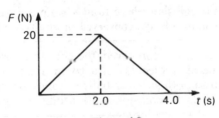

Figure 1O

Longer Questions

2 A ball of mass 0·2 kg falls from a height of 45 m. On striking the ground it rebounds in 0·1 s with two-thirds of the velocity with which it struck the ground. Calculate (i) the momentum change on hitting the ground, (ii) the force on the ball due to the impact.

3 A ball of mass 0·05 kg strikes a smooth wall normally four times in 2 seconds with a velocity of 10 m s^{-1}. Each time the ball rebounds with the same speed of 10 m s^{-1}. Calculate the average force on the wall.

Draw a sketch showing how the momentum varies with time over the 2 seconds.

4 The mass of gas emitted from the rear of a toy rocket is initially 0·1 kg s^{-1}. If the speed of the gas relative to the rocket is 50 m s^{-1}, and the mass of the rocket is 2 kg, what is the initial acceleration of the rocket?

5 A hose directs a horizontal jet of water, moving with a velocity of 20 m s^{-1}, on to a vertical wall. The cross-sectional area of the jet is 5×10^{-4} m^2. If the density of water is 1000 kg m^{-3}, calculate the force on the wall assuming the water is brought to rest there.

6 (a) State Newton's laws of motion. Explain how the *newton* is defined from these laws.

  (b) A rocket is propelled by the emission of hot gases. It may be stated that both the rocket and the emitted hot gases each gain kinetic energy and momentum during the firing of the rocket.

Discuss the significance of this statement in relation to the laws of conservation of energy and momentum, explaining the essential difference between these two quantities. (*Refer to pages 49–53 if you need to.*)

(c) *A bird of mass 0·05 kg hovers by beating its wings of effective area 0·03 m².*
(i) What is the upward force of the air on the bird? ($g = 9·8$ N kg⁻¹) (ii) What is the downward force of the bird on the air as it beats its wings? (iii) Estimate the velocity imparted to the air, which has a density of 1·3 kg m⁻³, by the beating of the wings.
Which of Newton's laws is applied in each of (i), (ii) and (iii) above? (*L.*)

7 12 passengers per minute on average move on to an elevator (lift) which is rising steadily at 2 m s⁻¹.
Assuming the initial velocity of the passengers is zero, calculate the extra tension (force) due to the passengers on the belt round the elevator rollers. Average passenger mass = 60 kg.

Inelastic and Elastic Collisions

In collisions, the total momentum of the colliding objects is always conserved. Usually, however, their total kinetic energy is not conserved. Some of it is changed to heat or sound energy, which is not recoverable. Such collisions are said to be *inelastic*. For example, when a lump of putty falls to the ground, the total momentum of the putty and earth is conserved, that is, the putty loses momentum and the earth gains an equal amount of momentum. But all the kinetic energy of the putty is changed to heat and sound on collision.

Inelastic collision = collision where total kinetic energy is *not* conserved (total momentum always conserved in *any* type of collision.)

If the total kinetic energy is conserved, the collision is said to be *elastic*. Gas molecules, such as the air molecules in a room, make elastic collisions. The collision between two smooth snooker balls is approximately elastic. Electrons may make elastic or inelastic collisions with atoms of a gas.

Elastic collision = collision when total kinetic energy is conserved.

As we shall prove soon, a mass moving with velocity v has kinetic energy equal to $\frac{1}{2}mv^2$. Suppose two masses m_1 and m_2, moving with different velocities u_1 and u_2 in the same direction, collide head-on with each other. After collision, their velocities change to v_1 and v_2 respectively. If the collision is an *elastic* one, their total kinetic energy is conserved. So

$$\tfrac{1}{2}m_1v_1{}^2 + \tfrac{1}{2}m_2v_2{}^2 = \tfrac{1}{2}m_1u_1{}^2 + \tfrac{1}{2}m_2u_2{}^2$$

Also, from the conservation of momentum, which applies to any type of collision,

$$m_1v_1 + m_2v_2 = m_1u_1 + m_2u_2$$

Note that the conservation of momentum equation applies even if the collision was not elastic; but the conservation of kinetic energy equation applies only if the collision was elastic.

Example on Elastic Collision

A ball A of mass 0·4 kg moving with velocity 5 m s⁻¹ collides head-on with a ball B of mass 0·2 kg moving with velocity 2 m s⁻¹, Figure 1.25 (i).

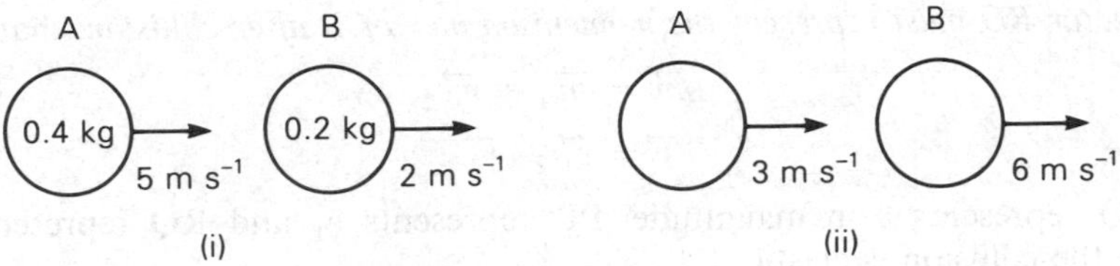

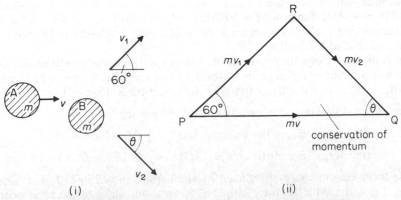

A   0.4 kg   5 m s^{-1} B   0.2 kg   2 m s^{-1}

(i)

A   3 m s^{-1} B   6 m s^{-1}

(ii)

Figure 1.25 *Elastic collision*

After the collision, A moves with velocity 3 m s^{-1} and B with velocity 6 m s^{-1}, Figure 1.25 (ii).

(i) Show that the collision obeys the law of conservation of momentum.

(ii) Is the collision elastic or inelastic?

(i) Before collision, total momentum $= (0.4 \times 5) + (0.2 \times 2) = 2.4 \text{ N s}$

After collision, total momentum $= (0.4 \times 3) + (0.2 \times 6) = 2.4 \text{ N s}$

So momentum is conserved.

(ii) Before collision, total kinetic energy $= (\frac{1}{2} \times 0.4 \times 5^2) + (\frac{1}{2} \times 0.2 \times 2^2)$

$$= 5 + 0.4 = 5.4 \text{ J}$$

After collision, total kinetic energy $= (\frac{1}{2} \times 0.4 \times 3^2) + (\frac{1}{2} \times 0.2 \times 6^2)$

$$= 1.8 + 3.6 = 5.4 \text{ J}$$

Since the total kinetic energy stays the same after collision, the collision is elastic.

Collision of Equal Masses

As an illustration of the mechanics associated with elastic collisions, consider a sphere A of mass m and velocity v incident on a stationary sphere B of equal mass m, Figure 1.26 (i). Suppose the collision is elastic, and after collision let A move with a velocity v_1 at an angle of $60°$ to its original direction and B move with the velocity v_2 at an angle θ to the direction of v.

Since momentum is a vector (p. 31), we may represent the momentum mv of A by the line PQ drawn in the direction of v, Figure 1.26 (ii). Likewise, PR represents the momentum mv_1 of A after collision. *Since momentum is conserved,*

v_1 $60°$

A m v B m

θ

v_2

(i)

R

mv_1 mv_2

P $60°$ mv θ Q

conservation of momentum

(ii)

Figure 1.26 *Conservation of momentum—equal masses*

the vector RQ must represent the momentum mv_2 of B after collision, that is

$$\vec{mv} = \vec{mv_1} + \vec{mv_2}$$

Hence

$$\vec{v} = \vec{v_1} + \vec{v_2}$$

or PQ represents v in magnitude, PR represents v_1 and RQ represents v_2. But if the collision is elastic,

$$\tfrac{1}{2}mv^2 = \tfrac{1}{2}mv_1{}^2 + \tfrac{1}{2}mv_2{}^2$$

$$\therefore v^2 = v_1{}^2 + v_2{}^2$$

Consequently, triangle PRQ is a right-angled triangle with angle R equal to 90°.

$$\therefore v_1 = v \cos 60° = \frac{v}{2}$$

Also, $\theta = 90° - 60° = 30°$, and $v_2 = v \cos 30° = \frac{\sqrt{3}}{2}v$

Bubble-chamber photographs show that the paths of an α-particle and a helium atom are 90° to each other after a collision. From our theory above, it follows that the two particles have the same mass. As we see later in Radioactivity (Chapter 34), an α-particle is a helium nucleus, a helium atom which has lost a few electrons.

Momentum and Explosive Forces

There are many cases where momentum changes are produced by *explosive* forces. An example is a bullet of mass $m = 50$ g, say, fired from a rifle of mass $M = 2$ kg with a velocity v of 100 m s^{-1}. Initially, the total momentum of the bullet and rifle is zero. From the principle of the conservation of linear momentum, when the bullet is fired the total momentum of bullet and rifle is still zero, since no external force has acted on them. So if V is the velocity of the rifle,

$$mv \text{ (bullet)} + MV \text{ (rifle)} = 0$$

$$\therefore MV = -mv \quad \text{or} \quad V = -\frac{m}{M}v$$

The momentum of the rifle is thus *equal and opposite* to that of the bullet. Also, $V/v = -m/M$. Since $m/M = 50/2000 = 1/40$, then $V = -v/40 = 2\cdot5$ m s^{-1}. This means that the rifle moves back or *recoils* with a velocity only about one fortieth of that of the bullet.

If it is preferred, one may also say that the explosive force produces the same numerical momentum change in the bullet as in the rifle. So $mv = MV$, where V is the velocity of the rifle in the *opposite* direction to that of the bullet.

The joule (J) is the unit of energy (p. 46). As we see later,

the kinetic energy, E_1, of the bullet $= \tfrac{1}{2}mv^2 = \tfrac{1}{2}.0\cdot05.100^2 = 250$ J

the kinetic energy, E_2, of the rifle $= \tfrac{1}{2}MV^2 = \tfrac{1}{2}.2.2\cdot5^2 = 6\cdot25$ J

So the total kinetic energy produced by the explosion $= 256\cdot25$ J. The kinetic energy E_1 of the bullet is then $250/256\cdot25$, or about 98%, of the total energy. This is explained by the fact that the kinetic energy depends on the *square* of the velocity. The high velocity of the bullet thus more than compensates for its small mass relative to that of the rifle when their kinetic energy is calculated.

We can get a general result for the way kinetic energy is shared in such explosions. Since momentum, $p = mv$, then kinetic energy

$$= \tfrac{1}{2}mv^2 = \tfrac{1}{2}\frac{m^2v^2}{m} = \frac{p^2}{2m}$$

So kinetic energy $\propto 1/m$ when the momentum is constant. This is the case when a stationary mass explodes into two masses m_1 and m_2. So in this case

$$\frac{\text{k.e. of mass } m_1}{\text{k.e. of mass } m_2} = \frac{1/m_1}{1/m_2} = \frac{m_2}{m_1}$$

We now see that after the explosion the kinetic energy is shared *inversely* as the ratio of the two masses, so the *lighter* mass has the greater share.

You should know:

1   **Conservation of momentum: the total momentum in a given direction before collision = the total momentum in the same direction after collision provided no external force acts in that direction**
2   **The conservation law is due to Newton's second and third laws**
3   **For an explosive force (such as bullet and rifle), the total momentum before collision = 0 = total momentum after collision**
4   **Conservation of momentum is true for *all* types of collisions, elastic and inelastic**
5   **Inelastic collision: total kinetic energy is not conserved—some is lost in heat or sound, for example**
    elastic collision: total kinetic energy conserved—air molecules have elastic collisions, an electron and an atom may have an elastic collision
6   **If a stationary object explodes into two parts, the total kinetic energy is shared *inversely* as the ratio of the masses of the parts—so the lighter part has the greater share of the energy.**

EXERCISES 1F   Force and Momentum. Conservation of Momentum

Multiple Choice

1   If a moving car X collides head-on with a moving car Y in the opposite direction, the conservation of momentum states

   A   the final momentum of X = the final momentum of Y
   B   the total momentum of X and Y is reversed by the collision
   C   the total momentum of X and Y stays constant
   D   the initial and final momentum of X is the same
   E   the initial and final momentum of Y is the same

2   A constant force F acts on a truck over a distance s and for a time t. The momentum gained by the truck is

   A $F \times s$ B $\tfrac{1}{2}F \times t$ C $\tfrac{1}{2}F \times s$ D $F \times t$ E $F \times t \times s$

3 A moving car X of mass 500 kg accelerates at 1 m s^{-2} when the force due to the engine is 600 N, Figure 1P(i). The average frictional force on the car is then

A 600 N **B** 500 N **C** 200 N **D** 150 N **E** 100 N

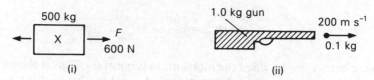

(i)   (ii)

Figure 1P

4 A gun of mass $1\cdot0$ kg has a bullet of mass $0\cdot1$ kg inside. The bullet leaves the pistol when fired at a velocity of 200 m s^{-1}, Figure 1P(ii). If U is the total momentum of the bullet and pistol after the bullet leaves the pistol and v the velocity of recoil of the pistol after firing, then

A $U = 200 \text{ kg m s}^{-1}$ and $v = 20 \text{ m s}^{-1}$ **B** $U = 20 \text{ kg m s}^{-1}$ and $v = 22\cdot2 \text{ m s}^{-1}$
C $U = 0$ and $v = 20 \text{ m s}^{-1}$ **D** $U = 0$ and $v = 22\cdot2 \text{ m s}^{-1}$
E $U = 2000 \text{ kg m s}^{-1}$ and $v = 200 \text{ m s}^{-1}$

5 If p is the momentum of an object of mass m, then the expression p^2/m has the same units as

A acceleration **B** energy **C** force **D** impulse **E** power

6 A body, initially at rest, explodes into two pieces of mass $2M$ and $3M$ respectively having a total kinetic energy E. The kinetic energy of the piece of mass $2M$ after the explosion is

A $E/3$ **B** $E/5$ **C** $2E/5$ **D** $3E/5$ **E** $4E/5$

Longer Questions

7 A ball of mass $0\cdot1$ kg, travelling at 4 m s^{-1}, hits a smooth billiard cushion normally and rebounds at 4 m s^{-1}.
(a) Is the collision elastic? (b) Calculate the force on impact if the time of impact is $0\cdot1$ s.
8 A block of mass 2 kg is on a smooth slope of $30°$. If it slides from rest at X, find its speed at Y after moving 10 m down the slope, Figure 1Q(i).
9 A bullet A of mass $0\cdot1$ kg travelling at 200 m s^{-1} embeds itself in a wooden block B of mass $0\cdot9$ kg moving in the opposite direction at 20 m s^{-1}, Figure 1Q(ii). Calculate the velocity of the block and bullet when the bullet comes to rest inside the block and write down the principle you used in the calculation.

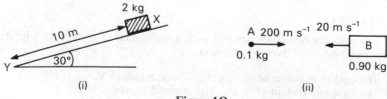

(i)   (ii)

Figure 1Q

10 A ball A of mass $0\cdot1$ kg, moving with a velocity of 6 m s^{-1}, collides directly with a ball B of mass $0\cdot2$ kg at rest. Calculate their common velocity if both balls move off together.
If A had rebounded with a velocity of 2 m s^{-1} in the opposite direction after collision, what would be the new velocity of B?

11 A bullet of mass 20 g is fired horizontally into a suspended stationary wooden block of mass 380 g with a velocity of 200 m s^{-1}. What is the common velocity of the bullet and block if the bullet is embedded in (stays inside) the block?

If the block and bullet experience a constant opposing force of 2 N, find the time taken by them to come to rest.

12 A small garden shed has a horizontal flat roof of area 2·0 m^2. During a heavy storm, 500 raindrops per second fall vertically onto the roof without bouncing. The rain runs off the roof and does not accumulate. The average mass of a drop is 3·0 × 10^{-5} kg; the average speed of a drop is 17·0 m s^{-1}.

(a) (i) What is the change in momentum of one drop as it hits the roof?

(ii) Hence calculate the mean pressure caused by rain hitting the roof.

(b) In another storm, wind causes raindrops to hit the roof at an angle of 30° to the vertical. The same number of drops land per second, still at the same speed relative to the roof and the rain runs off the roof. State and explain whether the pressure will be greater than, equal to, or less than the pressure caused by the vertical rainfall in part (a).

(c) A sudden drop in air temperature causes the raindrops to freeze and fall as hailstones which strike the roof at the same angle as in part (b) and at the same rate. However, they now bounce elastically off the roof and do not strike it a second time. Calculate by what factor this changes the pressure caused by the oblique rainfall in part (b). (O. & C.)

13 (a) A car of mass 1000 kg is initially at rest. It moves along a straight road for 20 s and then comes to rest again. The speed–time graph for the movement is shown in Figure 1R. (i) What is the total distance travelled? (ii) What resultant force acts on the car during the part of the motion represented by CD? (iii) What is the momentum of the car when it has reached its maximum speed? Use this momentum value to find the constant resultant accelerating force. (iv) During the part of the motion represented by OB on the graph, the constant resultant force found in (iii) is acting on the moving car although it is moving through air. Sketch a graph to show how the driving force would have to vary with time to produce this constant acceleration. Explain the shape of your graph.

(b) If, when travelling at this maximum speed, the 1000-kg car had struck and remained attached to a stationary vehicle of mass 1500 kg, with what speed would the interlocked vehicles have travelled immediately after collision? Calculate the kinetic energy of the car just prior to this collision and the kinetic energy of the interlocked vehicles just afterwards. Comment upon the values obtained.

Explain how certain design features in a modern car help to protect the driver of a car in such a collision. (L.)

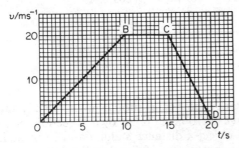

Figure 1R

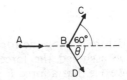

Figure 1S

14 In a nuclear collision, an alpha-particle A of mass 4 units is incident with a velocity v on a stationary helium nucleus B of 4 mass units, Figure 1S. After collision, A moves in the direction BC with a velocity $v/2$, where BC makes an angle of 60° with the initial direction AB, and the helium nucleus moves along BD.

Calculate the velocity of rebound of the helium nucleus along BD and the angle θ made with the direction AB. (A solution by drawing is acceptable.)

Work, Energy, Power

In this section we deal with the important topics of work, energy and power, which are applied to the performance of all kinds of engines or machines, such as in cars, aeroplanes or the human body. The efficiency, for example, of any machine can be calculated from the ratio

$$\frac{\text{work (or power) out}}{\text{work (or power) in}}$$

and being a ratio, it has no units.

Work

When an engine pulls a train with a constant force of 50 units through a distance of 20 units in the direction of the force, the engine is said by definition to do an amount of *work* equal to 50 × 20 or 1000 units, the product of the force and the distance.

So if W is the amount of work,

W × force × distance moved in direction of force

Work is a *scalar* quantity; it has no property of direction but only size. When the force is one newton and the distance moved is one metre, then the work done is one *joule* (J). So a force of 50 N, moving through a distance of 10 m in its own direction, does 50 × 10 or 500 J of work.

The force required to raise steadily a mass of 1 kg is equal to its weight, which is about 10 N (see p. 23–24). So if the mass of 1 kg is raised vertically through 1 m, then, approximately, work done = 10 (N) × 1 (m) = 10 J.

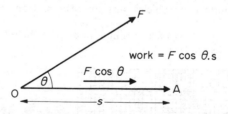

Figure 1.27 *Work and displacement*

Before leaving the topic of 'work', the reader should note carefully that we have assumed the force to move an object in its own direction. Suppose, however, that a force F pulls an object a distance s along a line OA acting at an angle θ to it, Figure 1.27. The component of F along OA is $F \cos \theta$ (p. 9), and this is the effective part of F pulling along the direction OA. The component of F along a direction perpendicular to OA has no effect along OA. So

work done = $F \cos \theta \times s$

In general, the work done by a force is equal to the product of the force and the *displacement in the direction of the force*.

Energy and Work

An engine does work when it pulls a train along a horizontal track. As a result of the work done, *energy* is transferred to the train. Assuming no energy losses, *the amount of energy of the moving train is equal to the work done*. The moving train has *kinetic energy*, or mechanical energy. Some of the chemical energy of the fuel used by the engine, or some of the electrical energy used by the engine if it is driven electrically, is thus transferred to mechanical energy.

When you wind a watch, you do some work. The work done is equal to the energy transferred to the moving parts of the watch. If a spring is wound up, the molecules in the metal spring are now closer together than before and they have gained molecular or 'elastic' energy. When you walk upstairs, you do work in moving your weight upwards. You have then gained gravitational *potential energy*, which is equal to the work done. The potential energy gained comes from the transfer of some chemical energy in your body.

Suppose an elastic band is pulled out steadily through a small distance of 1 cm or 0·01 m, and the force exerted increases steadily from zero to 10 N when you are pulling the band. Since the force is proportional to the extension of the band (p. 142), the work done = average force × distance moved. The average force during the stretching = $\frac{1}{2}(0 + 10)$ N. So

$$\text{work done} = \tfrac{1}{2}(0 + 10)\ \text{N} \times 0\cdot01\ \text{m}$$

$$= 5 \times 0\cdot01 = 0\cdot05\ \text{J}$$

This is the energy gained by the stretched elastic.

Power

When an engine does work quickly, it is said to be operating at a high *power*. If it does work slowly it is said to be operating at a lower power. 'Power' is defined as the *work done per second*, or energy spent per second. So

$$\text{power} = \frac{\text{work done}}{\text{time taken}}$$

The practical unit of power, the SI unit, is 'joule per second' or *watt* (W); the watt is defined as the rate of working at 1 joule per second. A 100 W lamp uses 100 J per second. A car engine of 1·2 kW uses 1200 J s^{-1}, since 1 kW (kilowatt) = 1000 W. 1 MW (megawatt) = 1 million (10^6) W, and is a large unit used in industry for power.

Suppose a car is moving steadily with a velocity of v metres per second and the force due to the engine is then F newtons. Then

$$\text{engine power} = \text{work done per second}$$

$$= F \times \text{distance per second} = F \times v$$

Power of engine $= F \times v$

Examples on Power

1 (a) A car of mass 1000 kg moving on a horizontal road with a steady velocity of 10 m s^{-1} has a total frictional force on it of 400 N. Find the power due to the engine, Figure 1.28 (i).

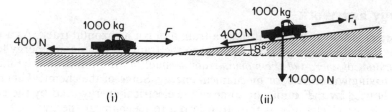

Figure 1.28 *Motion on inclined plane*

(b) The car now climbs a hill at an angle of 8° to the horizontal, Figure 1.28 (ii).
Assuming the frictional force stays constant at 400 N, what engine power is now needed to keep the car moving at 10 m s^{-1}?

(*Analysis* (i) Power $= F \times v$, so we need F due to engine.   (ii) In climbing the hill, the new force F_1 now has to overcome the component of the weight down the hill.)

(a) Since velocity is steady, acceleration $= 0$. So resultant force on car $= 0$. So if F is engine force,

$$F - 400 = 0 \quad \text{or} \quad F = 400 \text{ N}$$

$$\therefore \text{ power} = F \times v = 400 \times 10 = 4000 \text{ W} = 4 \text{ kW}$$

(b) Since velocity uphill is steady, acceleration $= 0 =$ resultant force uphill.
Now component of weight downhill $= 10000 \sin 8°$ (or $\cos 82°$) $= 1390$ N. So if new engine force uphill is F_1,

$$F_1 - 400 - 1390 = 0 \quad \text{or} \quad F_1 = 1790 \text{ N}$$

$$\therefore \text{ new power} = F_1 \times v = 1790 \times 10 = 17900 \text{ W} = 17\cdot9 \text{ W}$$

2 Sand drops vertically at the rate of 2 kg s^{-1} on to a conveyor belt moving horizontally with a velocity of 0·1 m s^{-1}. Calculate (i) the extra power needed to keep the belt moving, (ii) the rate of change of kinetic energy of the sand. Why is the power twice as great as the rate of change of kinetic energy?

(i) Force required to keep belt moving $=$ rate of increase of horizontal momentum of sand $=$ mass per second (dm/dt) $\times$ velocity change $= 2 \times 0\cdot1 = 0\cdot2$ newton.

$$\therefore \text{ power} = \text{work done per second} = \text{force} \times \text{rate of displacement}$$

$$= \text{force} \times \text{velocity} = 0\cdot2 \times 0\cdot1 = 0\cdot02 \text{ W}$$

(ii) Kinetic energy of sand $= \frac{1}{2}mv^2$

$$\therefore \text{ rate of change of energy} = \frac{1}{2}v^2 \times \frac{dm}{dt}, \text{ since } v \text{ is constant}$$

$$= \frac{1}{2} \times 0\cdot1^2 \times 2 = 0\cdot01 \text{ W}$$

Thus the power supplied is twice as great as the rate of change of kinetic energy. The extra power is due to the fact that the sand does not immediately assume the velocity of the belt, so that the belt at first moves relative to the sand. The extra power is needed to overcome the friction between the sand and belt.

Kinetic Energy

An object is said to possess *energy* if it can do work. When an object possesses energy because it is moving, the energy is said to be *kinetic*, for example, a flying stone has the energy to break a window. Suppose that an object of mass m is moving with a velocity v. To make it reach a velocity v from rest, the mass must have a force F on it which accelerates it from rest through a distance s. Then

$$\text{kinetic energy of mass} = \text{work done by } F = F \times s$$

Suppose a is the acceleration produced. Then $F = ma$. Also, from the equation of motion $v^2 = u^2 + 2as$, $v^2 = 0 + 2as$, so $as = v^2/2$. So

$$\text{kinetic energy} = F \times s = ma \times s = m \times as = m \times v^2/2 = \tfrac{1}{2}mv^2$$

Here v is the velocity or speed of the moving mass m. So

---

kinetic energy $= \tfrac{1}{2} \times$ mass $\times$ speed^2

Note that kinetic energy, like work or potential energy, is a *scalar*. Running along a road, it does *not* matter in which direction you run in calculating your kinetic energy. Also, 'mass' is a scalar and 'speed' is a scalar.

When m is in kg and u is in m s^{-1}, then $\tfrac{1}{2}mu^2$ is in *joules*, J. So a car of mass 1000 kg, moving with a velocity of 36 km h^{-1} or 10 m s^{-1}, has an amount W of kinetic energy given by

$$W = \tfrac{1}{2}mu^2 = \tfrac{1}{2} \times 1000 \times 10^2 = 50\,000\,\text{J}$$

Kinetic Energies due to Explosive Forces

Suppose that, due to an explosion or nuclear reaction, a particle of mass m breaks away from the total mass concerned and moves with velocity v, and that the mass M left moves with velocity V in the opposite direction.

The kinetic energy E_1 of the mass m is given by

$$E_1 = \tfrac{1}{2}mv^2 = \frac{(mv)^2}{2m} = \frac{p^2}{2m} \qquad . \qquad . \qquad . \qquad . \qquad (1)$$

where p is the momentum mv of the mass. Similarly, the kinetic energy E_2 of the mass M is given by

$$E_2 = \tfrac{1}{2}MV^2 = \frac{p^2}{2M} \qquad . \qquad . \qquad . \qquad . \qquad (2)$$

because numerically the momentum $MV = mv = p$, from the conservation of momentum. Dividing (1) by (2), we see that

$$\frac{E_1}{E_2} = \frac{1/m}{1/M} = \frac{M}{m}$$

Hence the energy is *inversely* proportional to the masses of the particles, that is, the smaller mass, m say, has the larger energy. Thus if E is the total energy of the two masses, the energy of the smaller mass $= ME/(M + m)$.

Suppose a bullet of mass $m = 50$ g is fired from a rifle of mass $M = 2$ kg = 2000 g and that the total kinetic energy produced by the explosion is 2050 J. Since the energy is shared inversely as the masses,

$$\text{kinetic energy of bullet} = \frac{2000}{2000 + 50} \times 2050 \text{ J}$$

$$= \frac{2000}{2050} \times 2050 \text{ J} = 2000 \text{ J}$$

So   kinetic energy of rifle $= 50$ J

An α-particle has a mass of 4 units and a radium nucleus a mass of 228 units. If disintegration of a stationary thorium nucleus, mass 232, produces an α-particle and radium nucleus, and a release of energy of 4·05 MeV, where 1 MeV $= 1·6 \times 10^{-13}$ J, then

$$\text{energy of α-particle} = \frac{228}{(4 + 228)} \times 4·05 = 3·98 \text{ MeV}$$

So the α-particle travels a relatively long distance before coming to rest compared with the comparatively heavy radium nucleus, which moves back or recoils a small distance.

Gravitational Potential Energy

A mass held stationary above the ground has energy, because, when released, it can raise another object attached to it by a rope passing over a pulley, for example. A coiled spring also has energy, which is released gradually as the spring uncoils. The energy of the weight or spring is called *potential energy*, because it arises from the position or arrangement of the body and not from its motion. In the case of the weight, the energy given to it is equal to the work done by the person or machine which raises it steadily to that position against the force of attraction of the earth. So this is *gravitational potential energy*. In the case of the spring, the energy is equal to the work done in displacing the molecules from their normal equilibrium positions against the forces of attraction of the surrounding molecules. So this is *molecular potential energy*, which is discussed later.

If the mass of an object is m, and the object is held stationary at a height h above the ground, the energy released when the object falls to the ground is equal to the work done

$$= \text{force} \times \text{distance} = \text{weight of object} \times h$$

Suppose the mass m is 5 kg, so that the weight is $5 \times 9·8$ N or 49 N, and h is 4 metres. Then

$$\text{gravitational potential energy, p.e.} = 49 \text{ (N)} \times 4 \text{ (m)} = 196 \text{ J}$$

Generally, when a mass m is moved through a height h,

---

change in gravitational potential energy $= mgh$

---

where m is in kg, h is in metres, $g = 9·8$ N kg^{-1} and the energy change is in joules, J.

This formula assumes that g is constant throughout the height h. Near the earth's surface g is fairly constant. But if a mass such as a space vehicle is sent up from the earth to an orbit high above the earth, then the gravitational field strength varies appreciably throughout the height h and '*mgh*' can not be used to find the gain in potential energy. We shall see later how the gain is calculated in this case.

You should know:

1 **Work = force × distance moved in the force direction. 1 N × 1 m = 1 J**
2 **Work is a scalar**
3 **An object has energy if it can do work. Energy is a scalar**
4 **Power = work done/time taken. Unit: watt, W ($=\text{J s}^{-1}$). Power is a scalar**
5 **Kinetic energy, $E = \frac{1}{2}mv^2$ (E in J when m in kg and v in m s^{-1})**
6 **If a stationary mass explodes into two parts X and Y, then kinetic energy of X/kinetic energy of Y = mass of Y/mass of X**
7 **Gravitational potential energy change = *mgh* where h is the vertical height change**
8 **A falling ball loses gravitational energy and gains an equal amount of kinetic energy, in the absence of air friction.**

Conservative Forces

If a ball of weight W is raised steadily from the ground to a point X at a height h above the ground, the work done is $W.h$. The potential energy, p.e., of the ball relative to the ground is thus $W.h$. Now whatever route is taken from ground level to X, the work done is the same—if a *longer* path is chosen, for example, the component of the weight in the particular direction must then be overcome and so the force required to move the ball is correspondingly smaller. The p.e. of the ball at X is thus independent of the route to X. This implies that if the ball is taken in a closed path round to X again, *the total work done against the force of gravity is zero*. Work or energy has been expended on one part of the closed path, and regained on the remaining part.

When the work done in moving round a closed path in a field to the original point is zero, the forces in the field are called *conservative forces*. The earth's gravitational field is an example of a field containing conservative forces, as we now show.

Suppose the ball falls from a place Y at a height h to another X at a height of x above the ground, Figure 1.29. Then, if W is the weight of the ball and m

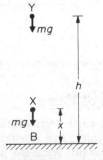

Figure 1.29 *Mechanical energy in a gravitational field*

its mass,

the potential energy, p.e. at $X = Wx = mgx$

and the kinetic energy, k.e. at $X = \frac{1}{2}mv^2 = \frac{1}{2}m.2g(h-x) = mg(h-x)$

using $v^2 = 2as = 2g(h-x)$. Hence

$$p.e. + k.e. = mgx + mg(h-x) = mgh$$

Thus at any point such as X, the total mechanical energy of the falling ball is equal to the original energy at Y. The mechanical energy is hence constant or conserved. This is the case for a conservative field.

Motion, Momentum and Energy Graphs in Gravitational Field

Figure 1.30 shows roughly the variation with *time t* of the speed, velocity, acceleration, distance, momentum and energy (kinetic and potential) of a ball thrown vertically upwards from the ground and then bouncing once on returning to the ground.

Note that (i) speed is a scalar but velocity is a vector, (ii) the deceleration of the rising ball and the acceleration of the falling ball are both numerically g; but on hitting the ground and rising, the acceleration changes at contact with the ground and becomes opposite in direction, as shown, (iii) momentum is a vector, (iv) energy is a scalar.

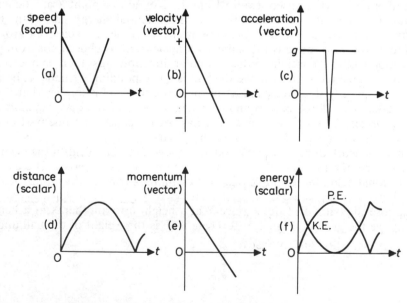

Figure 1.30 *Motion, momentum and energy graphs*

Principle of Conservation of Energy

When any object falls in the earth's gravitational field, a small part of the energy is used up in overcoming the resistance of the air. This energy is dissipated or lost as heat—it is not regained in moving the body back to its original position.

Although energy may be transferred from one form to another, such as from mechanical energy to heat energy as in this last example,

---

the total energy in a closed system is always constant.

If an electric motor is supplied with 1000 J of energy, for example, then 850 J of mechanical energy, 140 J of heat energy and 10 J of sound energy may be produced. This is called the *Principle of the Conservation of Energy* and is one of the key principles in science.

Momentum and Energy in Gravitational Field

Consider a ball B held stationary at a height above the earth E. Relative to each other, the momentum and kinetic energy of B and E are both zero.

Suppose the ball is now released. The gravitational force of E on B accelerates the ball. So its momentum increases as it falls. From the law of conservation of momentum, the equal and opposite gravitational force of B on E produces an equal but opposite momentum on the earth. The earth is so heavy, however, that its velocity V towards B is extremely small. For example, suppose the mass m is 0·2 kg and the mass M of E is 10^{25} kg, and the velocity v of the ball B is 10 m s^{-1} at an instant. Then, from the conservation of momentum,

$$MV = mv$$

or
$$V = \frac{m}{M} \times v = \frac{0·2}{10^{25}} \times 10 = 2 \times 10^{-25} \text{ m s}^{-1}$$

So the earth's velocity V is extremely small.

The total energy of B and E remains constant while the ball B is falling, since there are no external forces acting on the system. When B hits the ground the force (action) of B on E is equal and opposite to the force (reaction) of E on B, and the forces act for the same short time. So the total momentum of B and E is conserved. Thus when the ball rebounds, B moves upward with a momentum change equal and opposite to that of E. As B continues to rise its velocity and momentum decrease. So the momentum of E decreases. When B reaches its maximum height its momentum is zero. So the momentum of E is then zero.

Although momentum is conserved on collision with the earth, some mechanical energy is transformed to heat and sound. Hence the total mechanical energy of B and E is less after collision. As we showed on page 49, the ratio of the kinetic energy of the ball to the earth after collision is *inversely* proportional to their masses. So with the above figures,

$$\frac{\text{kinetic energy of earth}}{\text{kinetic energy of ball}} = \frac{0·2}{10^{25}} = 2 \times 10^{-26}$$

Hence the kinetic energy of the earth after collision is extremely small. Practically all the kinetic energy is transferred to the ball.

Dimensions

By the *dimensions* of a physical quantity we mean the way it is related to the fundamental quantities mass, length and time; these are usually denoted by M, L and T respectively. An area, length × breadth, has dimensions L × L or L^2;

a volume has dimensions L^3; density, which is mass/volume, has dimensions M/L^3 or ML^{-3}; an angle has no dimensions, since it is the ratio of two lengths.

As an area has dimensions L^2, the *unit* may be written in terms of the metre as m^2. Similarly, the dimensions of a volume are L^3 and hence the unit is m^3. Density has dimensions ML^{-3}. The density of mercury is thus written as $13\,600$ kg m^{-3}. If some physical quantity has dimensions $ML^{-1}T^{-1}$, its unit may be written as $kg\,m^{-1}\,s^{-1}$.

The following are the dimensions of some quantities in mechanics with their units in brackets:

Velocity. Since velocity $= \dfrac{\text{displacement}}{\text{time}}$, its dimensions are L/T or LT^{-1}
$(m\,s^{-1})$

Acceleration. The dimensions are those of velocity/time, or L/T^2 or LT^{-2}
$(m\,s^{-2})$

Force. Since force $=$ mass $\times$ acceleration, its dimensions are MLT^{-2}
$(kg\,m\,s^{-2}$ or $N)$

Work or Energy. Since work $=$ force $\times$ distance, its dimensions are ML^2T^{-2}
$(kg\,m^2\,s^{-2}$ or $J)$

Example on Dimensions

In the gas equation $\left(p + \dfrac{a}{V^2}\right)(V - b) = RT$, what are the dimensions of the constants a and b?

p represents pressure, V represents volume. The quantity a/V^2 must represent a pressure since it is added to p. The dimensions of $p = [\text{force}]/[\text{area}] = MLT^{-2}/L^2 = ML^{-1}T^{-2}$; the dimensions of $V = L^3$. So

$$\frac{[a]}{L^6} = ML^{-1}T^{-2} \quad \text{or} \quad [a] = ML^5T^{-2}$$

The constant b must represent a volume since it is subtracted from V. So

$$[b] = L^3$$

Application of Dimensions, Simple Pendulum

We can often use dimensions to solve problems. As an example, suppose a small mass is suspended from a long thread so as to form a simple pendulum. We may reasonably suppose that the period, T, of the oscillations depends only on the mass m, the length l of the thread, and the acceleration, g, of free-fall at the place concerned. Suppose then that

$$T = km^x l^y g^z \qquad . \qquad . \qquad . \qquad . \qquad (1)$$

where x, y, z, k are unknown numbers. The dimensions of g are LT^{-2} (see acceleration above). Now the dimensions of both sides of (i) must be the same.

$$\therefore T \equiv M^x L^y (LT^{-2})^z$$

Equating the indices of M, L, T on both sides, we have

$$x = 0$$

$$y + z = 0$$

and
$$-2x = 1$$

$$\therefore z = -\tfrac{1}{2}, \quad y = \tfrac{1}{2}, \quad x = 0$$

So, from (i), the period T is given by

$$T = kl^{\frac{1}{2}}g^{-\frac{1}{2}}$$

or
$$T = k\sqrt{\frac{l}{g}}$$

We cannot find the magnitude of k by the method of dimensions, since it is a number. A complete mathematical investigation shows that $k = 2\pi$ in this case, and hence $T = 2\pi\sqrt{l/g}$. (See p. 83.) Note that T is independent of the mass m.

■ Velocity of Transverse Wave in a String

As another illustration of the use of dimensions, consider a wave set up in a stretched string by plucking it. The velocity, c, of the wave depends on the tension, F, in the string, its length l, and its mass m, and we can therefore suppose that

$$c = kF^x l^y m^z \tag{1}$$

where x, y, z are numbers we hope to find by dimensions and k is a constant.

The dimensions of velocity, c, are LT^{-1}, the dimensions of tension, F, are MLT^{-2}, the dimensions of length, l, is L, and the dimensions of mass, m, is M. From (1), it follows that

$$LT^{-1} \equiv (MLT^{-2})^x + L^y \times M^z$$

Equating powers of M, L, and T on both sides,

$$0 = x + z \qquad . \qquad . \qquad . \qquad (2)$$

$$1 = x + y \qquad . \qquad . \qquad . \qquad (3)$$

and
$$-1 = -2x \qquad . \qquad . \qquad . \qquad (4)$$

$$\therefore x = \tfrac{1}{2}, \quad z = -\tfrac{1}{2}, \quad y = \tfrac{1}{2}$$

$$\therefore c = k . F^{\frac{1}{2}} l^{\frac{1}{2}} m^{-\frac{1}{2}}$$

or
$$c = k\sqrt{\frac{Fl}{m}} = k\sqrt{\frac{F}{m/l}} = k\sqrt{\frac{\text{Tension}}{\text{mass per unit length}}} = k\sqrt{\frac{T}{\mu}}$$

where μ is the mass per unit length. A complete mathematical investigation shows that $k = 1$.

The method of dimensions can be used to find the relation between quantities when the mathematics is too difficult. It has been extensively used in hydrodynamics, for example.

You should know:

1 **Principle of conservation of energy: In a closed system, the total energy is constant although energy may be transferred from one kind to another.**
2 **Mechanics dimensions are mass M, length L, time T**
3 **To show a formula is correct or homogeneous, dimensions or their units may be used. For mass M use 'kg', for length L use 'm', for time T use 's'.**
4 **A volume is L^3 (unit m^3), velocity is $L\,T^{-1}$ (unit $m\,s^{-1}$), force is $M\,L\,T^{-2}$ (kg m s^{-2}), work or energy is $M\,L^2\,T^{-2}$ (unit is kg m^2 s^{-2})**

Exploiting Wind Energy

Windmills have traditionally been used for grinding wheat to make flour. More recently, wind energy has been exploited as an alternative, environmentally friendly way of generating electricity. We can analyse the windmill in much the same way as we did the rotating helicopter blades on page 35, only this time we'll use an energy calculation to estimate how much electricity we can generate. Suppose the area swept by the blades is 300 m^2, and on a particular day the wind is blowing at 10 m s^{-1} (≈ 22 mph). This means that $10 \times 300 = 3000$ m^3 of air passes through the blades each second, of mass 3900 kg (density of air is 1·3 kg m^{-3}). Assuming 20% of the air is brought to rest by the blades, the loss of kinetic energy of the air per second is $\frac{1}{2} \times \frac{20}{100} \times 3900 \times 10^2 = 39\,000$ J s^{-1} ≈ 40 kW.

The windmill and its turbines will try to convert all this to electrical power, though the equipment will not be 100% efficient and the electrical output will be somewhat less than 40 kW. However, it is clearly possible to generate considerable amounts of electricity by wind power. *Wind farms* (large groups of windmills) have been placed in suitably windy locations around the world. Our calculation suggests that a farm of, say, 20 windmills could generate about 800 kW of electricity on a fairly windy day, enough to power a village. Along with tidal power, wind power offers the potential for cheap, clean electricity generation in the future. This is discussed more fully in Chapter 6.

EXERCISES 1G   Energy, Power

(Assume $g = 10$ m s^{-2} or 10 N kg^{-1} unless otherwise stated)

Multiple Choice

1 During the time between a parachutist leaving a plane and reaching the ground safely
   1 the sum of her gravitational potential energy and her kinetic energy is constant
   2 her kinetic energy depends upon her velocity
   3 her gravitational potential energy is proportional to her height above the ground

Which of A, B, C, D, E shows the correct answers?

A 1, 2, 3 B 1, 2 C 2, 3 D 1 only E 3 only (*L.*)

2 Starting from rest, a car of mass 1000 kg accelerates steadily to 20 m s^{-1} in 10 s. The average power developed in the time period is

A 0·2 kW B 4·0 kW C 10 kW D 15 kW E 20 kW

3 Which of the following pairs has one scalar and one vector quantity?
  A displacement, acceleration B potential energy, work
  C speed, power D kinetic energy, force E velocity, momentum
4 A ball is projected vertically upwards. Air resistance may be neglected. The ball rises to its maximum height *H* in a time *T*, the height being *h* after a time *t*.
  1 The graph of kinetic energy E_k of the ball against height *h* is shown in 1 (Figure 1T)
  2 The graph of height *h* against time *t* is shown in 2
  3 The graph of gravitational energy E_g of the ball against height *h* is shown in 3

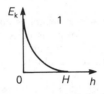
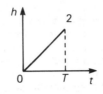
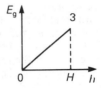

Figure 1T

Which of **A, B, C, D, E** shows the correct answers?

A 1, 2, 3 B 1, 2 C 2, 3 D 1 E 3 (*L.*)

5

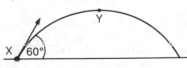

Figure 1U

A shell of mass 50 kg is fired at 60° to the horizontal with a speed of 200 m s^{-1} from X, Figure 1U. Neglecting air resistance, the kinetic energy of the shell in J at its highest point Y is

A 0 B 1·6 × 10^5 C 2·0 × 10^5 D 2·5 × 10^5 E 3·0 × 10^5

6 A force of 6 N acts horizontally on a stationary mass of 2 kg for 4 s. The kinetic energy gained by the mass in J is

A 12 B 24 C 48 D 72 E 144

Longer Questions

7 An object A of mass 10 kg is moving with a velocity of 6 m s^{-1}. Calculate its kinetic energy and its momentum.
   If a constant opposing force of 20 N suddenly acts on A, find the time it takes to come to rest and the distance through which it moves.
8 A ball of mass 0·1 kg is thrown vertically upwards with an initial speed of 20 m s^{-1}. Calculate (i) the time taken to return to the thrower, (ii) the maximum height reached, (iii) the kinetic and potential energies of the ball half-way up.

9 A 4 kg ball moving with a velocity of $10 \cdot 0$ m s^{-1} collides with a 16 kg ball moving with a velocity of $4 \cdot 0$ m s^{-1} (i) in the same direction, (ii) in the opposite direction. Calculate the velocity of the balls in each case if they coalesce on impact, and the loss of energy resulting from the impact. State the principle used to calculate the velocity.

10 A ball of mass $0 \cdot 1$ kg is thrown vertically upwards with a velocity of 20 m s^{-1}. What is the potential energy at the maximum height? What is the potential energy of the ball when it reaches three-quarters of the maximum height while moving upwards?

11 A stationary mass explodes into two parts of mass 4 units and 40 units respectively. If the larger mass has an initial kinetic energy of 100 J, what is the initial kinetic energy of the smaller mass? Explain your calculation.

12 A car of mass 1000 kg moves at a constant speed of 20 m s^{-1} along a horizontal road where the frictional force is 200 N. Calculate the power developed by the engine.
  If the car now moves up an incline at the same constant speed, calculate the new power developed by the engine. Assume that the frictional force is still 200 N and that $\sin \theta = 1/20$, where θ is the angle of the incline to the horizontal.

13 Which of the following are (i) scalars, (ii) vectors?
  A momentum B work C speed D force E energy G mass
  H acceleration

14 A horizontal force of 2000 N is applied to a vehicle of mass 400 kg which is initially at rest on a horizontal surface. If the total force opposing motion is constant at 800 N, calculate (i) the acceleration of the vehicle, (ii) the kinetic energy of the vehicle 5 s after the force is first applied, (iii) the total power developed 5 s after the force is first applied. (*AEB*.)

15 A railway truck of mass 4×10^4 kg moving at a velocity of 3 m s^{-1} collides with another truck of mass 2×10^4 kg which is at rest. The couplings join and the trucks move off together. What fraction of the first truck's initial kinetic energy remains as kinetic energy of the two trucks after the collision? Is energy conserved in a collision such as this? Explain your answer briefly. (*L*.)

16 An α-particle having a speed of $1 \cdot 00 \times 10^6$ m s^{-1} collides with a stationary proton which gains an initial speed of $1 \cdot 60 \times 10^6$ m s^{-1} in the direction in which the α-particle was travelling.
  What is the speed of the α-particle immediately after the collision? How much kinetic energy is gained by the proton in the collision?
  It is known that this collision is perfectly elastic. Explain what this means. (Mass of α-particle $= 6 \cdot 64 \times 10^{-27}$ kg. Mass of proton $= 1 \cdot 66 \times 10^{-27}$ kg.) (*L*.)

17 (a) A particle of mass m, initially at rest, is acted upon by a constant force until its velocity is v. Show that the kinetic energy of the particle is $\frac{1}{2}mv^2$.
  (b) A train of mass $2 \cdot 0 \times 10^5$ kg moves at a constant speed of 72 km h^{-1} up a straight incline against a frictional force of $1 \cdot 28 \times 10^4$ N. The incline is such that the train rises vertically $1 \cdot 0$ m for every 100 m travelled along the incline. Calculate (i) the rate of increase per second of the potential energy of the train, (ii) the necessary power developed by the train. (*N*.)

18 As shown in Figure 1V, two trolleys P and Q of masses $0 \cdot 50$ kg and $0 \cdot 30$ kg respectively are held together on a horizontal track against a spring which is in a state of compression. When the spring is released the trolleys separate freely and P moves to the left with an initial velocity of 6 m s^{-1}. Calculate

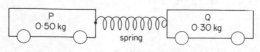

Figure 1V

(a) the initial velocity of Q,
(b) the initial total kinetic energy of the system.
  Calculate also the initial velocity of Q if trolley P is held still when the spring under the same compression as before is released. (*N*.)

19 Define linear momentum and state the principle of conservation of linear momentum. Explain briefly how you would attempt to verify this principle by experiment.

Sand is deposited at a uniform rate of 20 kilograms per second and with negligible kinetic energy on to an empty conveyor belt moving horizontally at a constant speed of 10 metres per minute. Find

(a) the force required to maintain constant velocity,

(b) the power required to maintain constant velocity, and

(c) the rate of change of kinetic energy of the moving sand.

Why are the latter two quantities unequal? (*O. & C.*)

2 Circular Motion, Simple Harmonic Motion

In the previous chapter we discussed the motion of an object moving in a straight line. There are many cases of objects moving in a curve or circular path about some point, such as bicycles or cars turning round corners or racing cars going round circular tracks. The earth and other planets move round the sun in roughly circular paths. In place of speed in linear motion, we then have to use 'angular speed'. This helps to find the 'period' or time to go once round the circle. We shall also find that objects moving at constant speed round a circle have an acceleration towards the centre of the circle.

Circular Motion

Angular Speed

Consider an object moving in a circle with a *uniform speed* round a fixed point O as centre, Figure 2.1.

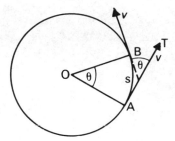

Figure 2.1 *Circular motion*

If the object moves from A to B so that the radius OA moves through an angle θ, its *angular speed*, ω, about O is defined as the *change of the angle per second*. So if t is the time taken by the object to move from A to B,

$$\omega = \frac{\theta}{t} \qquad . \qquad . \qquad . \qquad . \qquad . \qquad . \qquad (1)$$

The angle θ is measured in *radians*. (2π radians = 360°.) So angular speed is usually expressed in 'radians per second' (rad s^{-1}). From (1),

$$\theta = \omega t \qquad . \qquad . \qquad . \qquad . \qquad . \qquad . \qquad (2)$$

This is similar to the formula 'distance = uniform velocity × time' for motion in a straight line, It will be noted that the time T to describe the circle once, known as the *period* of the motion, is given by

$$T = \frac{2\pi}{\omega} \qquad . \qquad . \qquad . \qquad . \qquad . \qquad . \qquad (3)$$

since 2π radians is the angle in 1 revolution (360°).

If s is the length of the arc AB, then $s/r = \theta$, by definition of an angle in radians.

$$\therefore s = r\theta$$

Dividing by t, the time taken to move from A to B,

$$\therefore \frac{s}{t} = r\frac{\theta}{t}$$

But $s/t =$ the *speed*, v, of the rotating object, and θ/t is the angular velocity.

$$\therefore v = r\omega . \qquad . \qquad . \qquad . \qquad . \qquad (4)$$

Example on Circular Motion

A model car moves round a circular track of radius 0·3 m at 2 revolutions per second.
  What is
(a) the angular speed ω,
(b) the period T,
(c) the speed v of the car?
(d) Find also the angular speed of the car if it moves with a uniform speed of 2 m s^{-1} in a circle of radius 0·4 m.

(a) For 1 revolution, angle turned $\theta = 2\pi$ rad (360°). So

$$\omega = 2 \times 2\pi = 4\pi \text{ rad s}^{-1}$$

(b) Period $T =$ time for 1 rev $= \dfrac{2\pi}{\omega} = \dfrac{2\pi}{4\pi} = 0 \cdot 5$ s. (Or, $T = 1$ s/2 $= 0 \cdot 5$ s.)

(c) Speed $v = r\omega = 0 \cdot 3 \times 4\pi = 1 \cdot 2\pi = 3 \cdot 8$ m s^{-1}
(d) From $v = r\omega$

$$\omega = \frac{v}{r} = \frac{2 \text{ m s}^{-1}}{0 \cdot 4 \text{ m}} = 5 \text{rad s}^{-1}$$

Acceleration in a Circle

When a stone is attached to a string and whirled round at constant speed in a circle, one can feel the force (pull) in the string needed to keep the stone moving in its circular path. Although the stone is moving with a constant speed, the presence of the force implies that the stone has an *acceleration*.

The force on the stone acts *towards the centre* of the circle. We call it a *centripetal force*. The direction of the acceleration is the same as the direction of the force, that is, towards the centre. We now show that if v is the uniform speed in the circle and r is the radius of the circle,

$$\textbf{acceleration towards centre} = \frac{v^2}{r} \qquad . \qquad . \qquad (1)$$

or, since $v = r\omega$,

$$\textbf{acceleration towards centre} = \frac{r^2\omega^2}{r} = r\omega^2 . \qquad . \qquad (2)$$

The dimensions of v are LT^{-1} and the dimension of r is L. So v^2/r has the dimensions LT^{-2}, which is an acceleration. Also, the dimension of ω is T^{-1}, so $r\omega^2$ has the dimensions LT^{-2}, which is an acceleration.

Proof of Acceleration in a Circle

Consider an object moving with a constant speed v (a scalar which has no direction) round a circle of radius r, Figure 2.2 (i). At A, its velocity v_A (a vector) is in the tangent direction AC. A short time Δt later at B, its velocity v_B is in the direction of the tangent BD. Since their directions are different, the velocity v_B is different from the velocity v_A, although their magnitudes are both equal to v. So a *velocity change* or *acceleration* takes place from A to B.

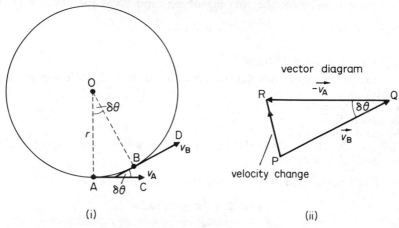

Figure 2.2 *Acceleration in a circle*

The velocity change from A to B $= \vec{v_B} - \vec{v_A} = \vec{v_B} + (-\vec{v_A})$. See p. 19. The arrows denote vector quantities. In Figure 2.2 (ii), PQ is drawn to represent $\vec{v_B}$ in magnitude (v) and direction (BD); QR is drawn to represent $(-\vec{v_A})$ in magnitude (v) and direction (CA). Then, as shown on p. 19,

$$\text{velocity change} = \vec{v_B} + (-\vec{v_A}) = \text{PR}$$

When Δt is small, the angle AOB or $\Delta\theta$ is small. Thus angle PQR, equal to $\Delta\theta$, is small. PR then points towards O, the centre of the circle. *The velocity change or acceleration is thus directed towards the centre.*
The magnitude of the acceleration, a, is given by

$$a = \frac{\text{velocity change}}{\text{time}} = \frac{\text{PR}}{\Delta t}$$

$$= \frac{v \cdot \Delta\theta}{\Delta t}$$

since PR $= v \cdot \Delta\theta$. In the limit, when Δt approaches zero, $\Delta\theta/\Delta t = d\theta/dt = \omega$, the angular speed. But $v = r\omega$ (p. 61). Hence, since $a = v\omega$,

$$a = \frac{v^2}{r} \quad \text{or} \quad r\omega^2 \quad . \qquad . \qquad . \qquad . \qquad . \qquad (3)$$

So an object moving in a circle of radius r with a constant speed v has an acceleration towards the centre equal to v^2/r or $r\omega^2$.

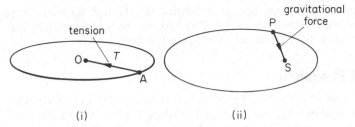

Figure 2.3 *Examples of centripetal forces*

Centripetal Forces

The centripetal force F required to keep an object of mass m moving in a circle of radius r is $ma = mv^2/r$. As already stated, F acts towards the centre of the circle. When a stone A is whirled in a horizontal circle of centre O by means of a string, the tension T provides the centripetal force, Figure 2.3 (i). For a racing car moving round a circular track, the friction at the wheels provides the centripetal force. Planets such as P, moving in a circular orbit round the sun S, have a centripetal force due to the gravitational attraction between S and P (p. 188), Figure 2.3.

If some water is placed in a bucket B attached to end of a string, the bucket can be whirled in a vertical plane without any water falling out. When the bucket is vertically above the point of support O, the weight mg of the water is less than the required force mv^2/r towards the centre and so the water stays in, Figure 2.4. The reaction R of the bucket base on the water provides the rest

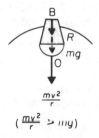

Figure 2.4 *Water in a rotated bucket*

Plate 2A *In this fairground ride, people are rotated in a large drum. They stick to the wall as the drum spins (centripetal force)*

of the force mv^2/r. If the bucket is whirled slowly and $mg > mv^2/r$, part of the weight provides the force mv^2/r. The rest of the weight causes the water to accelerate downward and hence to leave the bucket.

Conical Pendulum

Suppose a small object A of mass m is tied to a string OA of length l and then whirled round in a *horizontal* circle of radius r, with O fixed directly above the centre B of the circle, Figure 2.5. If the circular speed of A is constant, the string turns at a constant angle θ to the vertical. This is called a conical pendulum.

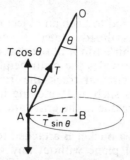

Figure 2.5 *Conical pendulum*

Since A moves with a constant speed v in a circle of radius r, there must be a centripetal force mv^2/r acting towards the centre B. The horizontal component, $T \sin \theta$, of the tension T in the string provides this force along AB. So

$$T \sin \theta = \frac{mv^2}{r} \qquad . \qquad . \qquad . \qquad . \qquad (1)$$

Also, since the mass does not move vertically, its weight mg must be counterbalanced by the vertical component $T \cos \theta$ of the tension. So

$$T \cos \theta = mg \qquad . \qquad . \qquad . \qquad . \qquad (2)$$

Dividing (1) by (2), then $\qquad \boldsymbol{\tan \theta = \dfrac{v^2}{rg}}$

A similar formula for θ was obtained for the angle of banking of a track which prevented side-slip.
If $v = 2 \text{ m s}^{-1}$, $r = 0.5 \text{ m}$ and $g = 10 \text{ m s}^{-2}$, then

$$\tan \theta = \frac{v^2}{rg} = \frac{2^2}{0.5 \times 10} = 0.8$$

So $\qquad\qquad\qquad \theta = 39°$

If $m = 2.0 \text{ kg}$, it follows from (2) that

$$T = \frac{mg}{\cos \theta} = \frac{2 \times 10}{\cos 39°} = 25.7 \text{ N}$$

A pendulum suspended from the ceiling of a train does not remain vertical while the train goes round a circular track. Its bob moves *outwards* away from the centre and the string becomes inclined at an angle θ to the vertical, as shown in Figure 2.5. In this case the centripetal force is provided by the horizontal component of the tension in the string, as we have already explained.

Motion of Car (or Train) round Banked Track

Suppose a car (or train) is moving round a banked track in a circular path of horizontal radius r, Figure 2.6 (i). If the only forces at the wheels A, B are the normal reaction forces R_1, R_2 respectively, that is, there is no side-slip or strain at the wheels, the force towards the centre of the track is $(R_1 + R_2) \sin \theta$, where θ is the angle of inclination of the plane to the horizontal.

$$\therefore (R_1 + R_2) \sin \theta = \frac{mv^2}{r} \quad . \quad . \quad . \quad . \quad (1)$$

The car does not move in a vertical direction. So, for vertical equilibrium,

$$(R_1 + R_2) \cos \theta = mg \quad . \quad . \quad . \quad . \quad (2)$$

Dividing (1) by (2), $$\tan \theta = \frac{v^2}{rg} \quad . \quad . \quad . \quad . \quad (3)$$

So for a given velocity v and radius r, the angle of inclination of the track for no side-slip must be given by $\tan \theta = (v^2/rg)$. As the speed v increases, the angle θ increases, from (3). A racing-track is made saucer-shaped because at high speeds the cars can move towards a part of the track which is steeper and sufficient to prevent side-slip, Figure 2.6 (ii). The outer rail of a curved railway track is raised above the inner rail so that the force towards the centre is largely provided by the component of the reaction at the wheels. It is desirable to bank a road at corners for the same reason as a racing track is banked.

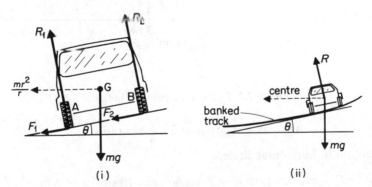

Figure 2.6 *Car on banked track*

An aeroplane with wings banked at an angle θ to the horizontal will make the aeroplane move with a speed v in a horizontal circular path of radius r, where $\tan \theta = v^2/rg$.

Examples on Circular Motion

1 A stone of mass 0·6 kg, attached to a string of length 0·5 m, is whirled in a horizontal circle at a constant speed.

If the maximum tension (force) in the string is 30 N before it breaks, calculate

(a) the maximum speed of the stone, (b) the maximum number of revolutions per second it can make.

The maximum tension or force in the string is given by

$$F = mv^2/r, \quad \text{from } F = ma$$

So

$$30 = 0.6v^2/0.5$$

Then

$$v^2 = 30 \times 0.5/0.6 = 25$$

(a)

$$v = 5 \text{ m s}^{-1}$$

(b)

$$\text{Period } T = 2\pi r/v = 2\pi \times 0.5/5 = \pi/5$$

So number of revs per second $= 1/T = 5/\pi = 1.6$.

2 A model aeroplane X has a mass of 0·5 kg and has a control wire OX of length 10 m attached to it when it flies in a horizontal circle with its wings horizontal, Figure 2.7. The wire OX is then inclined at 60° to the horizontal and fixed to a point O and X takes 2 s to fly once round its circular path.
 Calculate
(a) the tension T in the control wire,
(b) the upward force on X due to the air.

(*Analysis* (i) Force F towards centre of circle $= mr\omega^2$, (ii) $F =$ horizontal component of T, (iii) upward force due to air = weight of X + downward component of T.)

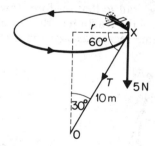

Figure 2.7 *Example on circular motion*

(a)

$$\text{Angular velocity } \omega = 2\pi/2 = \pi \text{ rad s}^{-1}$$

For motion in horizontal circle,

$$F = mr\omega^2, \text{ where } F = T\cos 60°, \; r = 10\sin 30° = 5 \text{ m}$$

So

$$T\cos 60° = 0.5 \times 5 \times \pi^2$$

$$T = \frac{2.5\pi^2}{0.5} = 50 \text{ N (approx.)}$$

(b) Upward force due to air $=$ weight of X $+ T\cos 30°$

$$= 5 \text{ N} + 50\cos 30° \text{ N}$$

$$= 48 \text{ N (approx.)}$$

You should know:

1 For steady circular motion, $v = r\omega$, $T = 2\pi/\omega$, $f = 1/T = \omega/2\pi$ (ω in rad s^{-1}, 2π rad $= 360°$)

2 Acceleration in circle, $a = v^2/r = r\omega^2$
V changes in direction but not in magnitude, so steady speed

3 Centripetal force $F = mv^2/r$ or $mr\omega^2$
F always acts *towards the centre* of the circle

4 Conical pendulum: resolve tension horizontally ($= mv^2/r$) and vertically ($= mg$)

5 $\tan\theta = v^2/rg$ for conical pendulum, banked track and banked wings of aeroplane moving in horizontal circle

EXERCISES 2A   Circular Motion

(Assume $g = 10\ m\ s^{-2}$ or $10\ N\ kg^{-1}$ unless otherwise given)

Multiple Choice

1 If a car moves round a circular road of radius r at a constant speed v,
   A its velocity changes and the acceleration is v/r^2
   B there is no force on the car since its speed is constant
   C there is no velocity change since the speed is constant
   D the force on the car is towards the centre and is mv^2/r
   E the force on the car is outwards from the centre and is v^2/r.

2 A small mass m is suspended from one end of a vertical string and then whirled in a horizontal circle at a constant speed v.
   A The string stays vertical
   B The string becomes inclined to the vertical
   C There is no force on m except its weight
   D The angle of inclination to the vertical of the string does not depend on v
   E The centripetal force is equal to the weight mg.

3 When a girl rides round the corner on her bicycle at a steady speed
   A there is no force when she turns
   B the force on her bicycle is only her weight when she turns
   C the centripetal force is due to the tyre friction at the ground
   D the force on the girl is always outwards
   E she stays upright (vertical).

4 When a stone of mass m at the end of a string is whirled in a vertical circle at a constant speed,
   A the tension (force) in the string stays constant
   B the tension is least when the stone reaches the bottom of the circle
   C the tension in the string is always mg
   D the weight mg is always the centripetal force
   E the tension is greatest when the stone is at the bottom of the circle.

5 A small mass of 0·2 kg is whirled round in a horizontal circle at the end of a string of length 0·5 m at a constant angular speed of 4 rad s^{-1}. The tension (force) in the string in N is

   A 0·4   B 0·6   C 0·8   D 1·0   E 1·6

Longer Questions

6 An object of mass 4 kg moves round a circle of radius 6 m with a constant speed of 12 m s^{-1}. Calculate   (i) the angular speed,   (ii) the force towards the centre.

7 An object of mass 10 kg is whirled round a horizontal circle of radius 4 m by a revolving string is inclined to the vertical. If the uniform speed of the object is 5 m s^{-1}, calculate (i) the tension in the string, (ii) the angle of inclination of the string to the vertical.

8 A racing-car of 1000 kg moves round a banked track at a constant speed of 108 km h^{-1}. Assuming the total reaction at the wheels is normal to the track, and the horizontal radius of the track is 100 m, calculate the angle of inclination of the track to the horizontal and the reaction at the wheels.

9 An object of mass 8·0 kg is whirled round rapidly in a vertical circle of radius 2 m with a constant speed of 6 m s^{-1}. Calculate the maximum and minimum tensions in the string.

10 Calculate the force necessary to keep a mass of 0·2 kg moving in a horizontal circle of radius 0·5 m with a period of 0·5 s. What is the direction of the force?

11 Calculate the mean angular speed of the earth assuming it takes 24·0 h to rotate about its axis.

 An object of mass 2·00 kg is (i) at the poles, (ii) at the equator. Assuming the earth is a perfect sphere of radius 6·4 × 10^6 m, calculate the change in weight of the mass when taken from the poles to the equator. Explain your calculation with the aid of a diagram.

12 A mass of 0·2 kg is whirled in a horizontal circle of radius 0·5 m by a string inclined at 30° to the vertical. Calculate (i) the tension in the string, (ii) the speed of the mass in the horizontal circle.

13 An object of mass 0·5 kg is rotated in a horizontal circle by a string 1 m long. The maximum tension in the string before it breaks is 50 N. What is the greatest number of revolutions per second of the object?

14 A mass of 0·4 kg is rotated by a string at a constant speed v in a vertical circle of radius 1 m. If the minimum tension of the string is 3 N, calculate (i) v, (ii) the maximum tension, (iii) the tension when the string is just horizontal.

15 What force is necessary to keep a mass of 0·8 kg revolving in a horizontal circle of radius 0·7 m with a period of 0·5 s? What is the direction of this force? (Assume that $\pi^2 = 10$.) (*L.*)

16 A spaceman in training is rotated in a seat at the end of a horizontal rotating arm of length 5 m. If he can withstand accelerations up to 9g, what is the maximum number of revolutions per second permissible? The acceleration of free-fall (g) may be taken as 10 m s^{-2}. (*L.*)

17 A stone of mass 150 g moves in a vertical circle on the end of a length of string. The radius of the circle is 70 cm and the speed of the stone at the topmost point P is 3·5 m s^{-1}, Figure 2A (i). For the instant at which the stone passes through the highest point P
(a) sketch and label a free-body force diagram for the stone,
(b) calculate the resultant force on the stone (i.e. the centripetal force), and
(c) calculate the tension in the string. (*L.*)

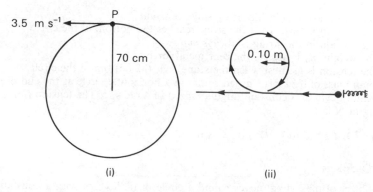

(i)   (ii)

Figure 2A

18 A compressed spring is used to propel a ball-bearing along a track which contains a circular loop of radius 0·10 m in a vertical plane, Figure 2A (ii). The spring obeys Hooke's law (see p. 142) and requires a force of 0·20 N to compress it by 1·0 mm.
   (a) The spring is compressed by 30 mm. Calculate the energy stored in the spring.
      (*Hint:* energy $= \frac{1}{2} \times$ force $\times$ distance compressed).
   (b) A ball-bearing of mass 0·025 kg is placed against the end of the spring which is then released. Calculate
      (i) the speed with which the ball-bearing leaves the spring,
      (ii) the speed of the ball at the top of the loop,
      (iii) the force exerted on the ball by the track at the top of the loop.
   Assume that the effects of friction can be ignored. (*N.*)

19 What is meant by *centripetal force*?
   Figure 2B(i) illustrates a child C of mass 20 kg attached to a rope OC of length 5·0 m and following a horizontal circular path of radius 3·0 m. The child is not in contact with the ground. Figure 2B(ii) is a free-body diagram showing the forces which act on the child. Calculate
   (a) the tension in the rope, and
   (b) the speed of the child.

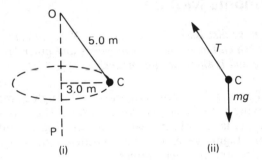

Figure 2B

20 (a) Write down an expression for the force required to maintain the motion of a body of mass *m* moving with constant speed *v* in a circle of radius *R*. In which direction does the force act?
   (b) Figure 2C shows a toy runway. After release from a point such as X, a small model car runs down the slope, 'loops the loop', and travels on towards Z. The radius of the loop is 0·25 m.
      (i) Ignoring the effect of friction outline the energy changes as the model moves from X to Z.
      (ii) What is the minimum speed with which the car must pass point P at the top of the loop if it is to remain in contact with the runway?
      (iii) What is the minimum value of *h* which allows the speed calculated in (ii) to be achieved?
   The effect of friction can again be ignored. (Assume that the acceleration of free-fall $g = 10$ m s^{-2}.) (*AEB.*)

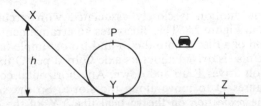

Figure 2C

21 Explain what is meant by *angular speed*. Derive an expression for the force required to make a particle of mass m move in a circle of radius r with uniform angular speed ω.

A stone of mass 500 g is attached to a string of length 50 cm which will break if they tension in it exceeds 20 N. The stone is whirled in a vertical circle, the axis of rotation being at a height of 100 cm above the ground. The angular speed is very slowly increased until the string breaks. In what position is this break must likely to occur, and at what angular speed? Where will the stone hit the ground? (*C.*)

22 A special prototype model aeroplane of mass 400 g has a control wire 8 cm long attached to its body. The other end of the control line is attached to a fixed point. When the aeroplane flies with its wings horizontal in a horizontal circle, making one revolution every 4 s, the control wire is elevated 30° above the horizontal. Draw a diagram showing the forces exerted on the plane and determine

(a) the tension in the control wire,

(b) the lift on the plane.

(Assume acceleration of free fall, $g = 10$ m s^{-2} and $\pi^2 = 10$.) (*AEB.*)

Simple Harmonic Motion

In the last section we discussed circular motion. Now we consider simple harmonic motion, which has application in waves and other branches of physics such as resonance and is therefore important.

When the bob of a pendulum moves to-and-fro through a small angle, the bob is said to be moving with *simple harmonic motion*. The prongs of a sounding tuning fork, and the layers of air near it, are moving to-and-fro with simple harmonic motion. Light waves can be considered due to simple harmonic variations of electric and magnetic forces.

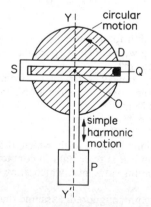

Figure 2.8 *Simple harmonic motion*

Simple harmonic motion is closely associated with circular motion. An example is shown in Figure 2.8. This illustrates an arrangement used to convert the circular motion of a disc D into the to-and-fro or simple harmonic motion of a piston P. The disc is driven about its axle O by a peg Q fixed near its rim. The vertical motion drives P up and down. Any horizontal component of the motion merely causes Q to move along the slot S. So the simple harmonic motion of P is the *projection* on the vertical line YY' of the circular motion of Q.

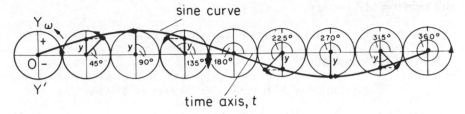

sine curve

time axis, t

Figure 2.9 *Simple harmonic and circular motion. The diagram shows eight positions of a particle moving round a circle through 360° at constant angular speed. The distances y from O of the foot of the projection on YOY' all lie on a sine curve as shown*

The projection of Q on YY' is the *foot* of the perpendicular from Q to the diameter passing through YY'. Figure 2.9 shows how the distance y from O of the projection varies as Q moves round the circular disc D with constant angular speed ω. In this rough sketch the horizontal axis represents angle of rotation or time, as the angle turned is proportional to the time. On one side of O, y has positive values; on the other side of O it has negative values. The graph of y against t is a *simple harmonic* curve or *sine (sinusoidal) curve* as we shall see shortly. The maximum value of y is called the *amplitude*. One complete set of values of y is called one *cycle* because the graph repeats itself after one cycle.

Formulae in Simple Harmonic Motion

Consider an object moving round a circle of radius r and centre Z with a uniform angular speed ω, Figure 2.10. As we have just seen, if CZF is a fixed diameter, the *foot* of the perpendicular from the moving object to this diameter moves from Z to C, back to Z and across to F, and then returns to Z, while the object moves once round the circle from O in an anti-clockwise direction. The to-and-fro motion along CZF of the foot of the perpendicular may be defined as *simple harmonic motion*.

Suppose the object moving round the circle is at A at some instant, where angle OZA $= \theta$, and suppose the foot of the perpendicular from A to CZ is M. The acceleration of the object at A is $\omega^2 r$, and this acceleration is directed along the radius AZ (see p. 61). Hence the acceleration of M towards Z

$$= \omega^2 r \cos AZC = \omega^2 r \sin \theta$$

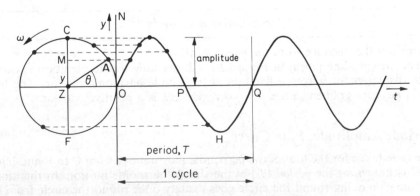

Figure 2.10 *Simple harmonic curve*

But $r \sin \theta = MZ = y$ say.

$$\therefore \text{ acceleration of M towards } Z = \omega^2 y$$

Now ω^2 is a constant.

$$\therefore \textit{acceleration of M towards Z} \propto \textit{distance of M from Z}$$

If we wish to express mathematically that the acceleration is always directed towards Z in simple harmonic motion, we must say

$$\text{acceleration towards } Z = -\omega^2 y \qquad . \qquad . \qquad . \qquad (1)$$

The minus indicates, of course, that the object begins to decelerate as it passes the centre, Z, of it motion. As we shall see later in discussing cases of simple harmonic motion, this is due to an opposing force. If the minus were omitted from equation (1) this would mean that the acceleration increases *in the direction* of y increasing, and the object would then never return to its original position.

We can now form a definition of **simple harmonic motion**.

It is the motion of a particle *whose acceleration is always* (i) *directed towards a fixed point,* (ii) *directly proportional to its distance from that point.*

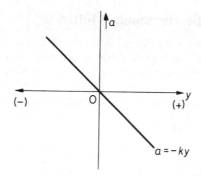

Figure 2.11 *Graph of acceleration* a *against displacement* y *for s.h.m.*

Mathematically if ω^2 is a constant

acceleration $a = -\omega^2 y$

where y is the distance from the fixed point.

The straight-line graph in Figure 2.11 shows how the acceleration a varies with displacement y from a fixed point for simple harmonic motion. The line has a *negative* gradient, since $a = -ky$, where k is a positive constant.

Period, Amplitude, Sine Curve

The time taken for the foot of the perpendicular to move from C to F and back to C is known as the *period* (T) of the simple harmonic motion. In this time, the object moving round the circle goes exactly once round the circle from C; and since ω is the angular speed and 2π radians (360°) is the angle described, the period T is given by

$$T = \frac{2\pi}{\omega} \qquad . \qquad . \qquad . \qquad . \qquad . \qquad (1\text{A})$$

The *frequency* f of a simple harmonic motion is the number of complete (to-and-fro) oscillations per second. Since one oscillation is made in a time T, then

$$f = \frac{1}{T} \qquad . \qquad . \qquad . \qquad . \qquad . \qquad (1\text{B})$$

f is measured in *hertz* (Hz). A frequency of 50 Hz = 50 oscillations per second. 1 kHz = 1 kilohertz = 1000 Hz. 1 MHz = 1 megahertz = 10^6 (1 million) Hz.

The distance ZC, or ZF, is the maximum distance from Z of the foot of the perpendicular, and is known as the *amplitude* of the motion. It is equal to r, the radius of the circle. So maximum acceleration, $a_{max} = -\omega^2 r$.

We have now to consider the variation with time t, of the distance, y, from Z of the foot of the perpendicular. The distance $y = \text{ZM} = r \sin \theta$. But $\theta = \omega t$, where ω is the angular speed.

$$\therefore y = r \sin \omega t \qquad . \qquad . \qquad . \qquad . \qquad . \qquad (2)$$

The graph of y against t is shown in Figure 2.10; ON represents the y-axis and OQ the t-axis. Since the angular speed of the object moving round the circle is constant, θ is proportional to the time t. So at X, the angle θ or ωt is equal to 90° or $\pi/2$ in radians; at P, the angle θ is 180° or π radians; and at Q, the angle θ is 360° or 2π radians. The simple harmonic graph is therefore a sine (sinusoidal) curve.

A cosine curve such as $y = r \cos \omega t$, has the same waveform as a sine curve. So this also represents simple harmonic motion. But as $y = r$ when θ or ωt is zero, the cosine curve starts at a *maximum value* instead of zero as in a sine curve.

The complete set of values of y from O to Q is known as a cycle. The number of cycles per second is called the *frequency*. The unit '1 cycle per second' is called '1 *hertz* (*Hz*)'. The mains frequency in Great Britain is 50 Hz or 50 cycles per second.

Velocity in Simple Harmonic Motion

If y is the displacement at an instant, then the velocity v at this instant is dy/dt, the rate of change of displacement (p. 4). Now $y = r \sin \omega t$, Figure 2.12. To find v or dy/dt from this graph we take the *gradient* of the curves at the time t considered. Figure 2.12 shows how v varies with time, t.

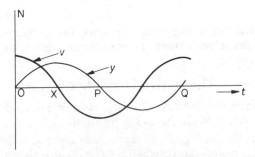

Figure 2.12 *Graph of velocity* v *and displacement* y *against time* t *in s.h.m.*

The velocity–time $(v–t)$ graph is a cosine curve. At $t = 0$, v has a maximum value. So $v = A \cos \omega t$ where A is the amplitude or maximum value of v. Now A is the gradient of the $y–t$ graph at $t = 0$. We can see by drawing different graphs of y against t that A depends on both r, the maximum value of y, and ω, the angular velocity (or number of cycles per second). Since $dy/dt = \omega r \cos \omega t = v$, we see that $A = \omega r$. So the velocity v is given by

$$v = \omega r \cos \omega t \qquad . \qquad . \qquad . \qquad . \qquad (1)$$

We can also express the velocity v in terms of y and r. From $y = r \sin \omega t$ and $v = r\omega \cos \omega t$, we have $\sin \omega t = y/r$ and $\cos \omega t = v/r\omega$. Now $\sin^2 \omega t + \cos^2 \omega t = 1$, from trigonometry. So

$$\frac{v^2}{r^2 \omega^2} + \frac{y^2}{r^2} = 1$$

Simplifying
$$v = \pm \omega \sqrt{r^2 - y^2} \qquad . \qquad . \qquad . \qquad . \qquad (2)$$

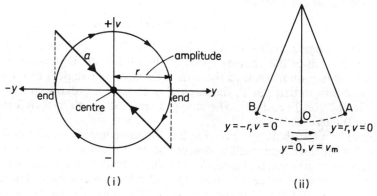

Figure 2.13 (i) shows the variation of v with *displacement* y. It is an ellipse. We can understand why this graph is obtained by considering the motion of a bob at the end of an oscillating simple pendulum, Figure 2.13 (ii). At the centre O ($y = 0$), the velocity v is a maximum. At the end A of the oscillation ($y = r$), $v = 0$. At the other end B ($y = -r$), $v = 0$. Note that v has an opposite direction on each half of the cycle. From (2), it follows that the maximum velocity v_m, when $y = 0$, is given numerically by

$$v_m = \omega r \qquad . \qquad . \qquad . \qquad . \qquad . \qquad (3)$$

When the velocity is a maximum ($y = 0$), the acceleration $a = 0$, since $a = -\omega^2 y$. When the velocity is zero ($y = r$), the acceleration a is a maximum.

S.H.M. Equations—Alternative Derivation

As we shall now show, all the equations used in s.h.m. can be derived by calculus without using the circle. With the usual notation,

$$\text{acceleration, } a = \frac{dv}{dt} = \frac{dy}{dt} \cdot \frac{dv}{dy} = v\frac{dv}{dy}$$

Now by definition of s.h.m., $a = -\omega^2 y$ (p. 72).

$$\therefore v \frac{dv}{dy} = -\omega^2 y$$

Integrating $\qquad\qquad \therefore \frac{v^2}{2} = -\omega^2 \frac{y^2}{2} + c \qquad . \qquad . \qquad . \qquad . \qquad (1)$

where c is a constant. Now $v = 0$ when $y = r$, the amplitude. So $c = \omega^2 r^2/2$, from (1). Substituting for c in (1) and simplifying,

$$v = \omega\sqrt{r^2 - y^2}$$

$$\therefore \frac{dy}{dt} = \omega\sqrt{r^2 - y^2}. \qquad . \qquad . \qquad . \qquad . \qquad (2)$$

$$\therefore \frac{1}{\omega} \int \frac{dy}{\sqrt{r^2 - y^2}} = \int dt$$

$$\therefore \frac{1}{\omega} \sin^{-1}\left(\frac{y}{r}\right) = t + C \qquad . \qquad . \qquad . \qquad . \qquad (3)$$

When $t = 0$, then $y = 0$; so $C = 0$, from (3).

$$\therefore \frac{1}{\omega} \sin^{-1}\left(\frac{y}{r}\right) = t$$

$$\therefore y = r \sin \omega t \qquad . \qquad . \qquad . \qquad . \qquad . \qquad (4)$$

When t increases to $t + 2\pi/\omega$, $y = r \sin(\omega t + 2\pi) = r \sin \omega t$, which is the same displacement value as at t. Hence the *period* T of the motion $= 2\pi/\omega$.

Learn these results

1   If the acceleration a of an object $= -\omega^2 y$, where y is the distance or displacement of the object from a fixed point, the motion is simple harmonic motion. The graph of a against y is a straight line through the origin with a negative gradient. Maximum acceleration, $a_{max} = -\omega^2 r$, where r is amplitude.

2   The *period*, T, of the motion $= 2\pi/\omega$, where T is the time to make a complete to-and-fro movement or cycle. The *frequency*, f, $= 1/T$ and its unit is 'Hz'. Note that $\omega = 2\pi/T = 2\pi f$.

3   The amplitude, r, of the motion is the maximum distance on either side of the centre of oscillation. The displacement $y = r \sin \omega t$ if $y = 0$ at $t = 0$.

4   The velocity at any instant, v, $= \pm\omega\sqrt{r^2 - y^2}$; the maximum velocity $= \omega r$. The graph of the variation of v with displacement y is an ellipse.

Examples on S.H.M.

1 A mass vibrates through an amplitude of 2·0 cm in simple harmonic motion with a period of 1·0 s. What distance is moved from the centre of oscillation in (a) 0·5 s, (b) 0·4 s?

(a) 0·5 s is half the period 1·0 s. So the mass moves a distance $2·0 + 2·0 = 4·0$ cm.
(b) We use $y = r \sin \omega t$. Here r = amplitude = 2·0 cm and $\omega = 2\pi/T = 2\pi/1 = 2\pi$.
   So when $t = 0·4$ s, $\omega t = 2\pi \times 0·4 = 0·8\pi$. Now π rad $= 180°$, so $0·8\pi = 0·8 \times 180° = 144°$.

   So
$$y = 2·0 \sin 144° = 1·2 \text{ cm.}$$

2 A steel strip, clamped at one end, vibrates with a frequency of 20 Hz and an amplitude of 5 mm at the free end, where a small mass of 2 g is positioned. Find
(a) the velocity of the end when passing through the zero position,
(b) the acceleration at maximum displacement,
(c) the maximum kinetic energy of the mass.

(a) When the end of the strip passes through the zero position $y = 0$, the speed is a maximum v_m given by
$$v_m = \omega r$$
   Now $\omega = 2\pi f = 2\pi \times 20$, and $r = 0·005$ m
$$\therefore v_m = 2\pi \times 20 \times 0·005 = 0·628 \text{ m s}^{-1}$$

(b) The acceleration $= -\omega^2 r$, where r is the amplitude
$$\therefore \text{acceleration} = (2\pi \times 20)^2 \times 0·005 \text{ numerically}$$
$$= 79 \text{ m s}^{-2}$$

(c) $m = 2 \text{ g} = 2 \times 10^{-3}$ kg, $v_m = 0·628$ m s^{-1}
$$\therefore \text{maximum k.e.} = \tfrac{1}{2}mv_m{}^2 = \tfrac{1}{2} \times (2 \times 10^{-3}) \times 0·628^2 = 3·9 \times 10^{-4} \text{ J (approx.)}$$

S.H.M. and g

If a small coin is placed on a horizontal platform connected to a vibrator, and the amplitude is kept constant as the frequency is increased from zero, the coin will be heard 'chattering' at a particular frequency f_0. At this stage the reaction of the platform with the coin becomes zero at some part of every cycle, so that it loses contact periodically with the surface, Figure 2.14.

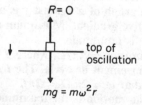

$R = 0$

top of
oscillation

$mg = m\omega^2 r$

Figure 2.14 *s.h.m. of coin on surface*

The maximum acceleration in s.h.m. occurs at the end of the oscillation because the acceleration is directly proportional to the displacement. This maximum acceleration $= \omega^2 r$, where r is the amplitude and ω is $2\pi f_0$.
The coin will lose contact with the surface when it is moving *down* with

acceleration g, Figure 2.14. Suppose the amplitude r is 0·08 m. Then

$$(2\pi f_0)^2 r = g$$

$$\therefore 4\pi^2 f_0^2 \times 0.08 = 9.8$$

$$\therefore f_0 = \sqrt{\frac{9.8}{4\pi^2 \times 0.08}} = 1.8 \text{ Hz}$$

Oscillating Systems—Spring and Mass

We now consider some oscillating systems in mechanics. A mass attached to a spring is a standard and useful case. For example, the body or chassis of a car is a mass attached to springs underneath and the oscillation needs study to provide a comfortable ride when travelling over ridges in the road surface. Also, when atoms or molecules in crystals vibrate, the molecular forces between the small masses can be represented by a 'spring'. In radio oscillators, one part of the basic circuit can be considered to behave like a 'mass' and another as a 'spring'.

Suppose that one end of a spring S of negligible mass is attached to a smooth object A, and that S and A are laid on a horizontal smooth table, Figure 2.15. If the free end of S is attached to the table and A is pulled slightly to extend the

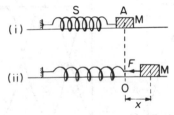

Figure 2.15 *Oscillating spring and mass*

spring and then released, the system vibrates with simple harmonic motion. The centre of oscillation O is the position of A at the end of the spring corresponding to its natural length, that is, when the spring is neither extended nor compressed.

Suppose the extension x of the spring is directly proportional to the force F in the spring (Hooke's law, p. 142). F acts in the opposite direction to x, so $F = -kx$, where k is known as the *force constant* of the spring or 'force per unit extension'. k is also called the spring *stiffness*. If m is the mass of A, the acceleration a is given by $F = ma$. So

$$ma = -kx$$

So

$$a = -\frac{k}{m} x = -\omega^2 x$$

where $\omega^2 = k/m$. So the motion of A is simple harmonic and the period T is given by

$$T = \frac{2\pi}{\omega} = 2\pi \sqrt{\frac{m}{k}}$$

Potential and Kinetic Energy Exchanges in Oscillating Systems

The energy of the stretched spring is *potential energy*, p.e.—its molecules are continually displaced or compressed relative to their normal distance apart. The p.e. for an extension $x = \int F . dx = \int kx . dx = \frac{1}{2}kx^2$.

The p.e. can also be found from

$$\text{work done} = average \text{ spring force} \times \text{distance } x$$

$$= \frac{1}{2}F \times x = \frac{1}{2}kx \times x = \frac{1}{2}kx^2$$

The energy of the mass is *kinetic energy*, k.e., or $\frac{1}{2}mv^2$, where v is the velocity. Now from $x = r \sin \omega t$, $v = dx/dt = \omega r \cos \omega t$

$$\therefore \text{total energy of spring plus mass} = \frac{1}{2}kx^2 + \frac{1}{2}mv^2$$

$$= \frac{1}{2}kr^2 \sin^2 \omega t + \frac{1}{2}m\omega^2 r^2 \cos^2 \omega t$$

But $\omega^2 = k/m$, or $k = m\omega^2$

$$\therefore \text{ total energy} = \frac{1}{2}m\omega^2 r^2 (\sin^2 \omega t + \cos^2 \omega t) = \frac{1}{2}m\omega^2 r^2 = constant$$

So the total energy of the vibrating mass and spring is constant. When the k.e. of the mass is a maximum (energy $= \frac{1}{2}m\omega^2 r^2$ and mass passing through the

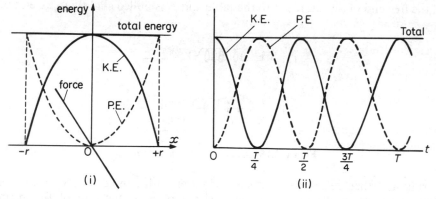

Figure 2.16 *Energy in s.h.m.*

centre of oscillation), the p.e. of the spring is then zero ($x = 0$). Conversely, when the p.e. of the spring is a maximum (energy $= \frac{1}{2}kr^2 = \frac{1}{2}m\omega^2 r^2$ and mass at the end of the oscillation), the k.e. of the mass is zero ($v = 0$). Figure 2.16 (i) shows the variation of p.e. and k.e. with displacement x; the force F extending the spring, also shown, is directly proportional to the displacement from the centre of oscillation. Figure 2.16 (ii) shows how the p.e. and k.e. vary with time t; the curves are simple harmonic or sine curves and T is the period.

The constant interchange of energy between potential and kinetic energies is essential for producing and maintaining oscillations, whatever their nature. In the case of the oscillating bob of a simple pendulum, for example, the bob loses kinetic energy after passing through the middle of the swing, and then stores the energy as potential energy as it rises to the top of the swing. The reverse occurs as it swings back. In the case of oscillating layers of air when a sound wave passes, kinetic energy of the moving air molecules is converted to potential energy when the air is compressed. In the case of electrical oscillations, a coil L and a capacitor C in the circuit constantly exchange energy; this is stored alternately in the magnetic field of L and the electric field of C.

Energy Method for Period

We can find the period T of the mass–spring oscillation in Figure 2.15 by an energy method.

When the oscillating mass reaches the *centre* of oscillation, it has maximum kinetic energy $\frac{1}{2}mr^2\omega^2$, where r is the amplitude (see p. 71). The spring then has a normal length and its potential energy $= 0$. When the mass reaches the end of its oscillation, the spring extension is then r and the spring energy $= \frac{1}{2}kr^2$. The kinetic energy of the mass is here zero. So from the conservation of energy,

$$\text{maximum k.e. of mass} = \text{maximum p.e. of spring}$$

So $$\frac{1}{2}mr^2\omega = \frac{1}{2}kr^2$$

Cancelling $\frac{1}{2}r^2$, $$m\omega^2 = k \quad \text{or} \quad \omega^2 = k/m$$

Then $$\omega = \sqrt{k/m} \quad \text{and} \quad T = 2\pi/\omega = 2\pi\sqrt{\frac{m}{k}}$$

Example on Spring Energy

A vertical spring fixed at one end has a mass of 2 kg attached at the other end. The spring constant $k = 5 \text{ N mm}^{-1}$.

Calculate   (a) the spring extension,   (b) the energy in the spring.

What is assumed in your calculations?

The mass loses gravitational energy mgx when it falls a distance x. Where is this energy transferred?

(a) From $F = kx$, $x = F/x = 20 \text{ N}/5000 \text{ N m}^{-1} = 4 \times 10^{-3} \text{ m}$
(b) Spring energy $= \frac{1}{2}kx^2 = \frac{1}{2} \times 5000 \times (4 \times 10^{-3})^2 = 0.04 \text{ J}$
Assumption: Hooke's law is obeyed.

Half of the energy, $mgx/2$ ($= \frac{1}{2}kx^2$) is transferred to the energy in the spring or to molecular energy. The other half is transferred to heat in the spring after the mass loses its kinetic energy on falling a distance x.

Oscillation of Mass Suspended from Helical Spring

Consider a helical spring or an elastic thread PA suspended from a fixed point P, Figure 2.17. When a mass m is placed on it, the spring stretches to O by a

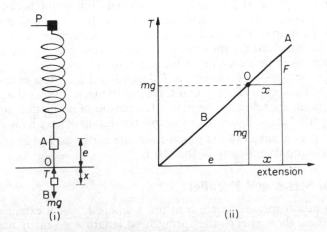

Figure 2.17 *Helical spring and s.h.m.*

length e given by

$$mg = ke \ . \qquad . \qquad . \qquad . \qquad . \qquad (1)$$

where k is the force constant (force per unit extension) of the spring, since the tension in the spring is then mg. If the mass is pulled down a little and then released, it vibrates up-and-down above and below O. Suppose at an instant that B is at a distance x below O. The tension T of the spring at B is then equal to $k(e + x)$. So the resultant force F downwards $= mg - k(e + x) = mg - ke - kx = -kx$, since $ke = mg$ from (1). From $F = ma$,

$$-kx = ma$$

$$\therefore a = -\frac{k}{m}x = -\omega^2 x$$

where $\omega^2 = k/m$. So the motion if simple harmonic about O, and the period T is given by

$$T = \frac{2\pi}{\omega} = 2\pi\sqrt{\frac{m}{k}} \ . \qquad . \qquad . \qquad . \qquad (2)$$

Also, since $mg = ke$, it follows that $m/k = e/g$.

$$\therefore T = 2\pi\sqrt{\frac{e}{g}} \ . \qquad . \qquad . \qquad . \qquad (3)$$

Figure 2.17 (ii) shows the straight-line variation of the tension T in the spring with the extension, assuming Hooke's law (p. 142). The point O on the line corresponds to the extension e when the weight mg is on the spring and $T = mg$. When the mass is pulled down and released as in Figure 2.17 (i), the tension values vary along the straight line AOB. So at a displacement x from O, the *resultant* force $F(T - mg)$ on m is proportional to x. From $F = ma$, the acceleration a of m is proportional to x. So the motion is simple harmonic about O. Also, from Figure 2.17 (ii), $F/x = mg/e$. Hence $F/m = a = gx/e$. So $\omega^2 = g/e$ and the period $= 2\pi/\omega = 2\pi\sqrt{e/g}$, as deduced in (3).

From (2), it follows that $T^2 = 4\pi^2 m/k$. So a graph of T^2 against m should be a straight line through the origin. In practice, when the load m is varied and the corresponding period T is measured, a straight line graph is obtained when T^2 is plotted against m, thus verifying indirectly that the motion of the load was simple harmonic. The graph does not pass through the origin, however, owing to the mass and the movement of the various parts of the spring. This has not been taken into account in the previous theory.

From (2), the period of oscillation T depends on the mass m and the force constant k of the spring. Since m and k are constants, it follows that if the same mass and spring are taken to the moon, the period of oscillation would be the same. The period of oscillation T of a simple pendulum of length l would change, however, if it were taken to the moon from the earth, as $T = 2\pi\sqrt{l/g}$ and the moon's gravitational intensity is about $g/6$.

Springs in Series and Parallel

Consider a helical spring of force constant k where $F = k \times$ extension. A mass m of weight mg then extends the spring by a length e given by $mg = ke$, and the period of oscillation of the mass is $T = 2\pi\sqrt{e/g}$ as previously obtained.

Suppose two identical helical springs are connected in *series*, each of force constant k, Figure 2.18 (i). The same weight mg will extend the springs twice as much as for a single spring since the total length is twice as much. So the extension is now $2e$. The period of oscillation of the mass m is therefore given by

$$T_1 = 2\pi\sqrt{\frac{2e}{g}}$$

So

$$T_1 = \sqrt{2}T$$

The mass therefore oscillates with a longer period at the end of the two springs.

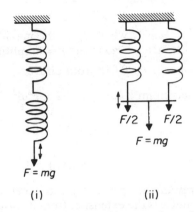

(i) (ii)

Figure 2.18 *Springs in series and parallel*

Now suppose the two springs are placed in *parallel* and the mass m is attached at the middle of a short horizontal connecting bar, Figure 2.18 (ii). This time the force on each spring is $mg/2$. So the extension is half as much as for a single spring, or $e/2$. So the period T_2 of the system is given by

$$T_2 = 2\pi\sqrt{\frac{e/2}{g}} = 2\pi\sqrt{\frac{e}{2g}} = \frac{1}{\sqrt{2}}T$$

The period of the parallel system is therefore less than for a single spring. Also, from above,

$$\frac{T_1}{T_2} = \frac{\sqrt{2}T}{T/\sqrt{2}} = \sqrt{4} = 2$$

Example on Spring–Mass Oscillations

A small mass of 0·2 kg is attached to one end of a helical spring and produces an extension of 15 mm or 0·015 m. The mass is now pulled down 10 mm and set into vertical oscillation of amplitude 10 mm. What is
(a) the period of oscillation,
(b) the maximum kinetic energy of the mass,
(c) the potential energy of the spring when the mass is 5 mm below the centre of oscillation? ($g = 9\cdot8$ m s^{-2}.)

(*Analysis* (a) Since $T = 2\pi\sqrt{m/k}$, we need to find k, (b) max k.e. $= \frac{1}{2}mv_m^2 = \frac{1}{2}mr^2\omega^2$, (c) p.e. $= \frac{1}{2}kx^2$.)

(a) The force constant k of the spring in N m^{-1} is given by

$$k = \frac{mg}{e} = \frac{0.2 \times 9.8}{0.015}$$

As we have previously shown,

$$T = 2\pi \sqrt{\frac{m}{k}} = 2\pi \sqrt{\frac{0.2 \times 0.015}{0.2 \times 9.8}}$$

$$= 2\pi \sqrt{\frac{0.015}{9.8}} = 0.25 \text{ s}$$

(b) The maximum k.e. $= \frac{1}{2}mv_m^2$, where v_m is the maximum velocity. Now for simple harmonic motion, $v_m = r\omega$ where $r =$ amplitude $= 10$ mm $= 0.01$ m. So, since $\omega = \sqrt{k/m} = \sqrt{9.8/0.015}$ from above

$$\text{maximum k.e.} = \frac{1}{2} \times 0.2 \times r^2\omega^2$$

$$= \frac{1}{2} \times 0.2 \times 0.01^2 \times \frac{9.8}{0.015}$$

$$= 6.5 \times 10^{-3} \text{ J}$$

(c) The potential energy of the spring is given generally by $\frac{1}{2}kx^2$, where k is the force constant and x is the extension from its *original* length. The centre of oscillation is 15 mm below the unstretched length, so 5 mm below the centre of oscillation corresponds to an extension x of 20 mm or 0.02 m. Since

$$k = (0.2 \times 9.8)/0.015$$

$$\text{Potential energy of spring} = \frac{1}{2}kx^2 = \frac{\frac{1}{2} \times 0.2 \times 9.8}{0.015} \times 0.02^2 \text{ J}$$

$$= 2.6 \times 10^{-2} \text{ J}$$

Simple Pendulum

We shall now study another case of simple harmonic motion. Consider a *simple pendulum*, which consists of a small mass m attached to the end of a length l of wire, Figure 2.19. If the other end of the wire is attached to a fixed point P and the mass is displaced slightly, it oscillates to-and-fro along the arc of a circle of centre P. We shall now show that the motion of the mass about its original position O is simple harmonic motion.

Suppose that the vibrating mass is at B at some instant, where OB $= y$ and angle OPB $= \theta$. At B, the force pulling the mass towards O is directed along the tangent at B, and is equal to $mg \sin \theta$. The tension, T, in the wire has no component in this direction, since PB is perpendicular to the tangent at B. Thus, since force $=$ mass $\times$ acceleration,

$$-mg \sin \theta = ma$$

where a is the acceleration along the arc OB; the minus indicates that the force is towards O, while the displacement, y, is measured along the arc from O in

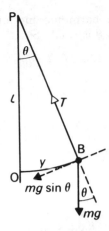

Figure 2.19 *Simple pendulum*

the opposite direction. *When θ is small*, $\sin \theta = \theta$ in radians; also $\theta = y/l$. Hence,

$$- mg\, \theta = - mg \frac{y}{l} = ma$$

$$\therefore a = -\frac{g}{l} y = -\omega^2 y$$

where $\omega^2 = g/l$. Since the acceleration is proportional to the distance y from a fixed point, the motion of the vibrating mass is simple harmonic motion (p. 72). Further, the period $T = 2\pi/\omega$.

$$\therefore T = \frac{2\pi}{\sqrt{g/l}} = 2\pi \sqrt{\frac{l}{g}} \qquad . \qquad . \qquad . \qquad . \qquad (1)$$

At a given place on the earth, where g is constant, the formula shows that the period T depends only on the length, l, of the pendulum. Moreover, the period remains constant even when the amplitude of the vibration diminishes owing to the resistance of the air. This result was first obtained by Galileo, who noticed a swinging lantern, and timed the oscillations by his pulse as clocks had not yet been invented. He found that the period remained constant although the swings gradually diminished in amplitude.

On the moon, g is about one-sixth that on the earth. From (1), we see that a pendulum of given length on the moon would have a period over twice as long as on the earth.

■ Energy method for period

We can find the period of oscillation T for a simple pendulum from an energy method, as we showed for the spring–mass period (p. 77).

The maximum gravitational potential energy of the oscillating mass is at B, the end of the swing, Figure 2.20 (i). Relative to the energy level at O, the lowest point, this energy is mgh, where h is the distance from O of the horizontal line BC. From the transfer of potential to kinetic energy,

maximum k.e. at O $= mgh$

But maximum k.e. in simple harmonic motion $= \frac{1}{2}mv_{max}^2 = \frac{1}{2}mr^2\omega^2$, where $r =$ amplitude $=$ line OB when θ is small. So

$$\frac{1}{2}mr^2\omega^2 = mgh$$

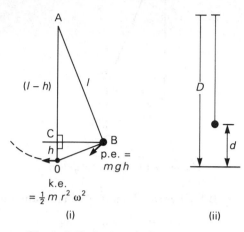

Figure 2.20 *Energy method for period*

In Figure 2.20 (i), $AC = l - h$, since $AO = l$. Also, $AB = l$. From the right-angled triangle ACB, using Pythagoras, $CB^2 = AB^2 - AC^2 = l^2 - (l - h)^2$. From the right-angled triangle OCB, $CB^2 = OB^2 - OC^2 = r^2 - h^2$. Using the two values for CB^2, then

$$r^2 - h^2 = l^2 - (l - h)^2 = l^2 - (l^2 - 2lh + h^2) = 2lh - h^2.$$

So $r^2 = 2lh$, and $h = r^2/2l$. From above,

$$\frac{1}{2}mr^2\omega^2 = mgh = mgr^2/2l$$

Cancelling, $$\omega^2 = g/l \quad \text{and} \quad \omega = \sqrt{g/l}$$

So $$T = 2\pi/\omega = 2\pi\sqrt{l/g}$$

Examples on Simple Pendulum

1   A simple pendulum has a bob of mass 0·3 kg and a length of 1 m. The bob is drawn aside through an angle of 6° and released from rest.
Calculate   (a) the maximum velocity of the bob,   (b) the maximum acceleration,   (c) the maximum tension in the string.

(a) From $T = 2\pi\sqrt{l/g} = 2\pi/\omega$, we see that $\omega = \sqrt{g/l} = \sqrt{9\cdot8/1} = \sqrt{9\cdot8}$, assuming $g = 9\cdot8$ m s^{-2}, since $l = 1$ m. The amplitude $r = l\sin\theta$ is a good approximation since θ is small. So $r = 1 \times \sin 6° = 0\cdot10$ m.

   Then   maximum velocity $= \omega r = \sqrt{9\cdot8} \times 0\cdot10 = 0\cdot31$ m s^{-1}

(b) Maximum acceleration $= \omega^2 r = 9\cdot8 \times 0\cdot10 = 0\cdot98$ m s^{-2}.

(c) The maximum tension in the string occurs at the centre of oscillation. Here the string is vertical. As the bob swings in a circle, the resultant force towards the centre $= mv^2/r$, where r is the *radius* of the circle which is 1 m. The resultant force $F = T - mg$, where T is the tension in the string. So

$$T - 0.3g = 0.3 \times 0.31^2/1$$
$$T - (0.3 \times 9.8) = 0.3 \times 0.31^2$$

Simplifying, $$T = 2.97 \text{ N}$$

2 In Figure 2.20 (ii), show that the period of oscillation of the simple pendulum bob is related to the distances D and d by

$$T^2 = 4\pi^2(D - d)/g$$

Show how, by measurements of d and T, a suitable graph can be drawn for finding the values of g and D.

Since $T = 2\pi\sqrt{l/g}$, and $l = D - d$, then

$$T = 2\pi\sqrt{(D - d)/g}$$

Squaring both sides, $T^2 = 4\pi^2(D - d)/g$.

We now have $$T^2 = \frac{4\pi^2}{g}D - \frac{4\pi^2}{g}d$$

A graph of T^2 against d is a straight line. The gradient of the line $= 4\pi^2/g$, so $g = 4\pi^2/\text{gradient}$ and can be calculated. Also, when $T^2 = 0$, $D = h = \text{OA}$ and so D can be found from the graph.

Oscillations of a Liquid in a U-Tube

If the liquid on one side of a U-tube T is depressed by blowing gently down that side, the levels of the liquid will oscillate for a short time about their respective initial positions O, C, before finally coming to rest, Figure 2.21.

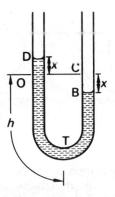

Figure 2.21 *s.h.m. of liquid*

At some instant, suppose that the level of the liquid on the left side of T is at D, at a height x above its original (undisturbed) position O. The level B of the liquid on the other side is then at a depth x below its original position C. So the excess pressure on the whole liquid, as shown on p. 105,

$$= \text{excess height} \times \text{liquid density} \times g = 2x\rho g$$

Since pressure = force per unit area,

force on liquid = pressure × area of cross-section of the tube = $2x\rho g \times A$

where A is the cross-sectional area of the tube. The mass of liquid in the U-tube = volume × density = $2hA\rho$, where $2h$ is the total length of the liquid in T. So, from $F = ma$ the acceleration, a, towards O or C is given by

$$- 2x\rho gA = 2hA\rho a$$

The minus indicates that the force towards O is opposite to the displacement measured from O at that instant.

$$\therefore a = -\frac{g}{h}x = -\omega^2 x$$

where $\omega^2 = g/h$. So the motion of the liquid about O (or C) is simple harmonic, and the period T is given by

$$T = \frac{2\pi}{\omega} = 2\pi \sqrt{\frac{h}{g}}$$

In practice the oscillations are heavily damped owing to friction, which we have ignored.

Combining Two Perpendicular S.H.M.s, Lissajous Figures

As we see later in the cathode-ray oscilloscope (p. 757), electrons can move under two simple harmonic forces of the same frequency *at right-angles* to each other. In this case the electron has an x-motion say given by $x = a \sin \omega t$ and a perpendicular or y-motion given by $y = b \sin (\omega t + \theta)$, where a and b are the respective amplitudes and θ is the phase angle between the oscillations in the x- and y-directions respectively.

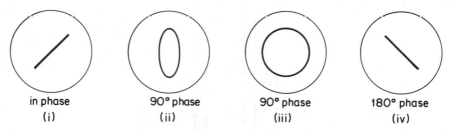

| in phase | 90° phase | 90° phase | 180° phase |
| (i) | (ii) | (iii) | (iv) |

Figure 2.22 *Lissajous figures*

If the oscillations are *in phase*, then $\theta = 0$. So $x = a \sin \omega t$ and $y = b \sin \omega t$. Then $\sin \omega t = x/a = y/b$. So $y = bx/a$. This is the equation of a *straight line* and this is the resultant motion of the electrons as shown on the oscilloscope screen, Figure 2.22 (i). This so-called 'Lissajous figure' is used to find out when two oscillations have the same phase (see p. 762).

If the two oscillations are 90° or $\pi/2$ out of phase, then $x = a \sin \omega t$ and $y = b \sin (\omega t + 90°) = b \cos \omega t$. So $\sin \omega t = x/a$ and $\cos \omega t = y/b$. Since, by trigonometry, $\sin^2 \omega t + \cos^2 \omega t = 1$, then $x^2/a^2 + y^2/b^2 = 1$. This is the equation of an *ellipse*, Figure 2.22 (ii). If the amplitudes a and b are equal, the equation is that of a circle, Figure 2.22 (iii). If the phase difference is 180° or π, then a straight line with a negative gradient is obtained, Figure 2.22 (iv). As explained in the cathode-ray oscilloscope section, two perpendicular oscillations which have a frequency ratio 2:1 produce a figure like the number eight on the screen.

You should know:

1 Definition of s.h.m. is $a = -\omega^2 x$ (the minus is necessary because a is always directed towards a fixed point when at distance x from centre of oscillation)

2 $T = 2\pi/\omega$, $\omega = 2\pi f$

3 velocity $v = \pm\sqrt{\omega^2(r^2 - x^2)}$. r = amplitude of oscillation

4 $v_{max} = \pm\omega r$ at $x = 0$ (centre), $v = 0$ at $x = r$ (end)
$a_{max} = -\omega^2 r$ at $x = r$ (end), $a = 0$ at $x = 0$ (centre)

5 kinetic energy: max ($\frac{1}{2}mr^2\omega^2$) at $x = 0$ (centre), zero at $x = \pm r$ (end)
potential energy: max at $x = r$, min at $x = 0$
total kinetic and potential energies = constant

6 *Spring and mass* On smooth horizontal table, $T = 2\pi\sqrt{m/k}$ ($F = -kx$ for spring)
Mass suspended from vertical spring, $T = 2\pi\sqrt{m/k} - 2\pi\sqrt{e/g}$, where $mg = ke$
Formula can be applied to vibrations of ions in crystals, to radio circuits and to mechanical vibrations.

7 Oscillations in mass–spring system are due to continuous exchange of energy. For spring, potential energy $= \frac{1}{2}kx^2$; for mass, kinetic energy $= \frac{1}{2}mv^2$; and total energy stays constant.

8 Simple pendulum: $T = 2\pi\sqrt{l/g}$ provided very small angle of oscillation and very small mass

9 At resonance, external frequency = natural frequency, amplitude large

---

EXERCISES 2B Simple Harmonic Motion

(*Assume g = 10 m s^{-2} or 10 N kg^{-1}*)

Multiple Choice

1 The graphs in Figure 2D show that a quantity y varies with displacement d in a system undergoing simple harmonic motion.

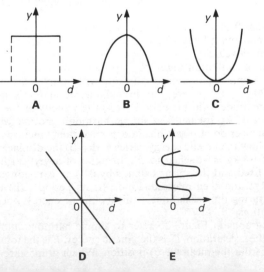

Figure 2D

Which graphs best represents the relationship obtained when y is
A the total energy of the system **B** the time **C** the unbalanced force acting on the system. (*L*.)

2 In simple harmonic motion with amplitude A and period T, the maximum velocity is

A TA **B** $A/2\pi T$ **C** $2\pi/AT$ **D** $2\pi A/T$ **E** $\pi A^2/T$

3 In a simple pendulum experiment, the length l is varied and the period T is measured. To find g due to gravity, it is best to plot

A l against T **B** l against T^2 **C** l/l against T **D** l^2 against T **E** $l^{1/2}$ against T^2

4 In simple harmonic motion, which graph in Figure 2E shows how the total energy E varies with time t during one cycle?

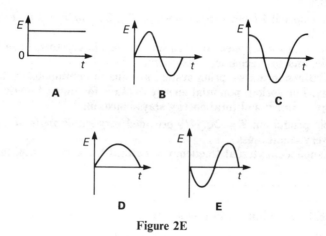

Figure 2E

Longer Questions

5 An object moving with simple harmonic motion has an amplitude of 0·02 m and a frequency of 20 Hz. Calculate (i) the period of oscillation, (ii) the acceleration at the middle and end of an oscillation, (iii) the velocities at the corresponding instants.

6 A body of mass 0·2 kg is executing simple harmonic motion with an amplitude of 20 mm. The maximum force which acts upon it is 0·064 N. Calculate
(a) its maximum velocity,
(b) its period of oscillation. (*L*.)

7 (a) The displacement y of a mass vibrating with simple harmonic motion is given by $y = 20 \sin 10\pi t$, where y is in millimetres and t is in seconds. What is
(i) the amplitude, (ii) the period, (iii) the velocity at $t = 0$?
(b) A mass of 0·1 kg oscillates in simple harmonic motion with an amplitude of 0·2 m and a period of 1·0 s. Calculate its maximum kinetic energy. Draw a sketch showing how the kinetic energy varies with (i) the displacement, (ii) the time.

8 A mass X of 0·1 kg is attached to the free end of a vertical helical spring whose upper end is fixed and the spring extends by 0·04 m. X is now pulled down a small distance 0·02 m and then released. Find (i) its period, (ii) the maximum force acting on it during the oscillations, (iii) its kinetic energy when X passes through its mean position.

9 Some of the graphs in Figure 2F refer to simple harmonic motion, where v is the velocity, a is the acceleration, E_k is the kinetic energy, E is the total energy and x is the displacement from the mean (zero) position. Which graphs are correct?

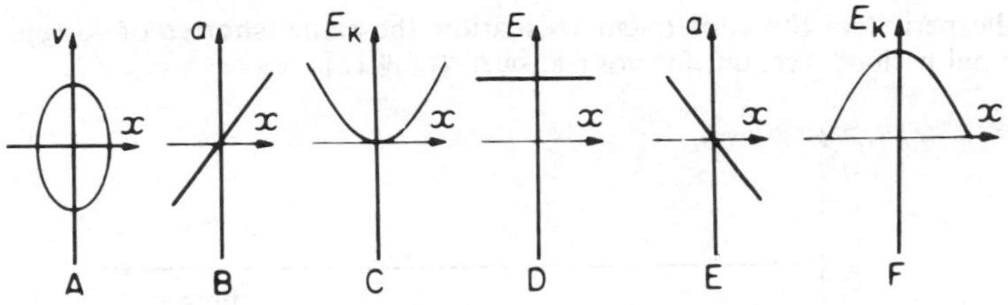

Figure 2F

10 A spring of force constant k of $5\,\mathrm{N\,m^{-1}}$ ($F = -kx$) is placed horizontally on a smooth table. One end of the spring is fixed and a mass X of 0·20 kg is attached to the free end. X is displaced a distance of 4 mm along the table and then released. Show that the motion of X is simple harmonic, and calculate (i) the period, (ii) the maximum acceleration, (iii) the maximum kinetic energy, (iv) the maximum potential energy of the spring.

11 A simple pendulum has a period of 4·2 s. When the pendulum is shortened by 1 m, the period is 3·7 s. From these measurements, calculate the acceleration of free-fall g and the original length of the pendulum.

   If the pendulum is taken from the earth to the moon where the acceleration of free-fall is $g/6$, what relative change, if any, occurs in the period T?

12 The bob of a simple pendulum moves simple harmonically with amplitude 8·0 cm and period 2·00 s. Its mass is 0·50 kg. The motion of the bob is undamped.

   Calculate maximum values for
   (a) the speed of the bob,
   (b) the kinetic energy of the bob. (*L.*)

13 (a) The displacement x, in m, from the equilibrium position of a particle moving with simple harmonic motion is given by

$$x = 0.05 \sin 6t$$

where t is the time, in s, measured from an instant when $x = 0$.
   (i) State the amplitude of the oscillations.
   (ii) Calculate the time period of the oscillations and the maximum acceleration of the particle.
   (b) A mass hanging from a spring suspended vertically is displaced a small amount and released. By considering the forces on the mass at the instant when the mass is released, show that the motion is simple harmonic and derive an expression for the time period. Assume that the spring obeys Hooke's law (see p. 142). (*N.*)

14 Explain what is meant by *simple harmonic motion*.

   Show that the vertical oscillations of a mass suspended by a light helical spring are simple harmonic and describe an experiment with the spring to determine the acceleration due to gravity.

   A small mass rests on a horizontal platform which vibrates vertically in simple harmonic motion with a period of 0·50 s. Find the maximum amplitude of the motion which will allow the mass to remain in contact with the platform throughout the motion. (*L.*)

15 A mass hangs from a light spring. The mass is pulled down 30 mm from its equilibrium position and then released from rest. The frequency of oscillation is 0·50 Hz.
   (a) Calculate
      (i) the angular frequency, ω, of the oscillation
      (ii) the magnitude of the acceleration at the instant it is released from rest.
   (b) Sketch a graph of the acceleration of the mass against time during the first 4·0 s of its motion. Put a scale on each axis of Figure 2G.
   (c) After a few oscillations half of the mass becomes detached when it is at the lowest point of its motion. The act of detachment still leaves the remaining half instantaneously at rest.

Is the period of the subsequent oscillation the same, shorter or longer than the original period? Account for your answer. (*O & C.*)

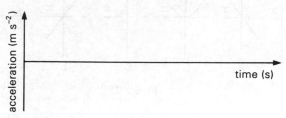

Figure 2G

16 Define *simple harmonic motion*. Explain what is meant by the *amplitude*, the *period* and the *phase* of such a motion.

A simple pendulum of length 1·5 m has a bob of mass 2·0 kg.
(a) State the formula for the period of small oscillations and evaluate it in this case.
(b) If, with the string taut, the bob is pulled aside a horizontal distance of 0·15 m from the mean position and then released from rest, find the kinetic energy and the speed with which it passes through the mean position.
(c) After 50 complete swings, the maximum horizontal displacement of the bob has become only 0·10 m. What fraction of the initial energy has been lost?
(d) Estimate the maximum horizontal displacement of the bob after a further 50 complete swings. (Take g to be 10 m s^{-2}.) (*O.*)

17

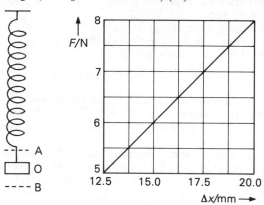

Figure 2H

Figure 2H shows a 0·714-kg mass suspended at the lower end of a helical spring. The graph shows how the tension, F, in the spring varies with the extension, Δx, of the spring. Use the graph to find a value for the spring constant, k, i.e. the force per unit extension of the spring.

The 0·714-kg mass oscillates vertically between the points A and B at which the extensions of the spring are 15·0 mm and 20·0 mm respectively. Calculate
(i) the period of oscillation of the mass,
(ii) the kinetic energy of the mass when it passes through the mid-point, O, and
(iii) the energy released from the spring when the mass rises from B to O.
What energy transformations occur when the mass rises from B to O? (*L.*)

3   Forces in Equilibrium, Forces in Fluids

Forces in Equilibrium

We now consider forces which are in equilibrium, *for example, forces which keep bridges stationary. As we shall see, forces are added together by a vector method. The effect of a force in a certain direction (its resolved component) depends on the size of the force and the angle between the force and the direction. This is often needed in problems of equilibrium . The turning-effects or* moments *of the forces must also be taken into account.*

Forces not only push or pull but also have a turning-effect or moment about an axis. In cases of equilibrium the moments have also to be considered. Objects are in equilibrium only if certain rules apply, as we discuss, and examples will show you how to use these rules.

Figure 3.1 shows two cases of equilibrium. In Figure 3.1 (i) is part of a bridge structure. Here the joint O resting on brickwork is in equilibrium under three

Plate 3A *The Middlesbrough Transporter Bridge, which is built using a cantilever system, with some parts in tension and some in compression*

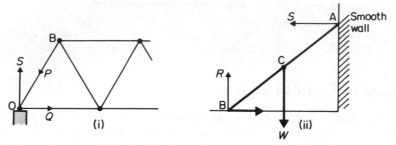

Figure 3.1 *Equilibrium of forces*

forces—*P* and *Q* are the forces or *tensions* in the metal beams, *S* is the *reaction force* on O due to the brickwork.

Figure 3.1 (ii) shows a uniform ladder AB with the top end A resting against a smooth wall and the bottom B resting on the rough ground. The forces on the ladder are its weight *W* acting at the midpoint C of the ladder, the frictional force *F* at the ground which stops the ladder sliding outwards, the *normal reaction R* of the ground at B acting at 90° to the ground and the normal reaction *S* at the wall acting at 90° to the wall. There is no frictional force here as the wall is smooth.

Adding Forces, Parallelogram of Forces

A force is a *vector* quantiy (p. 5). So it can be represented in size and direction by a straight line drawn to scale. The sum or *resultant R* of two forces *P* and *Q* can be added by one of two vector methods.

(1) Figure 3.2 (i) shows two forces *P* and *Q* acting at 60° to each other. To add them, draw a line *ab* to represent *P* and from *b* draw a line *bc* to represent *Q*—note that *ab* is parallel to *P* and *bc* is parallel to *Q*. Join *ac*. Then *ac* is the resultant *R* of *P* and *Q* in magnitude and direction. Note that the arrows on *ab* and *bc* follow each other.

We can find the size of *R*, and its direction *θ* to *P* either by accurate drawing or by calculation (using trigonometry in triangle *abc*). In branches of Physics,

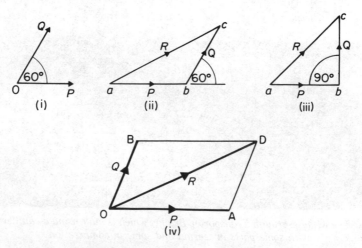

Figure 3.2 *Adding forces (vectors)*

P and Q are often at 90° to each other. In this case the vector triangle abc is a right-angled triangle, Figure 3.2(iii). Applying Pythagoras' theorem, then $R^2 = P^2 + Q^2$.

So
$$R = \sqrt{P^2 + Q^2}$$

Also, the angle θ R makes with P is given by $\tan \theta = bc/ab = Q/P$, so knowing P and Q, θ can be found.

(2) *Parallelogram of forces* The resultant R of P and Q can also be found by drawing a parallelogram of the forces. In Figure 3.2(iv), draw OA to represent P and OB to represent Q at the angle, say 60°, between P and Q as in Figure 3.2(i). Then complete the parallelogram OBDA. The resultant R is represented by *the diagonal OD through O*.

This gives the same result for R as in the previous method, since AD represents Q. If P and Q are at 90° to each other, the parallelogram becomes a rectangle.

Resolved Components

We often need to find the effect of a force F in a particular direction at an angle θ to F. This is called the 'resolved component' or simply *component* of F in this direction.

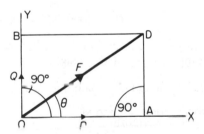

Figure 3.3 *Resolved components*

As we saw earlier on page 9, the resolved component P of a force F in a direction θ to itself (Figure 3.3) is given by

$$P = F \cos \theta$$

The cosine formula for the component should be memorised by the student. So the component Q in the direction OY is given by

$$Q = F \cos (90° - \theta) = F \sin \theta$$

So a force of 20 N has a component in a direction 60° to itself of $20 \cos 60° = 20 \times 0.5 = 10$ N. In a direction 30° to itself, its component is $20 \cos 30° = 20 \times 0.87 = 17.4$ N. Remember that a force has no component at 90° to itself ($\cos 90° = 0$).

Example on Components

Figure 3.4 shows a stationary car of weight W on a road sloping at 30° to the horizontal. The frictional force on the car is 4000 N acting up the road BC.

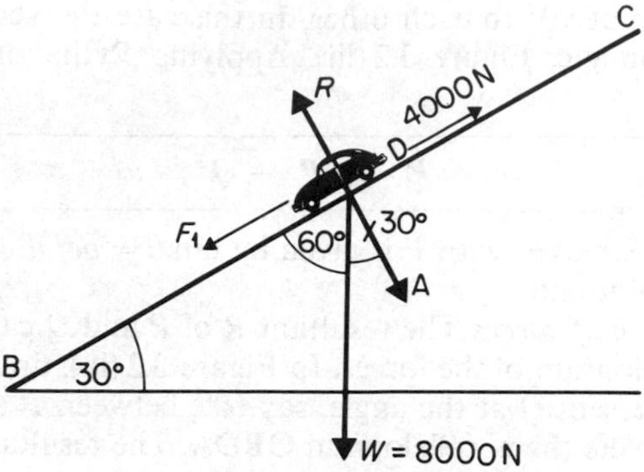

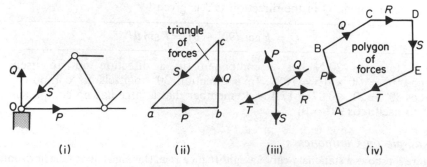

Figure 3.4 *Application of components*

What is   (i) the weight W,   (ii) the normal reaction force R of the road surface on the car?

(i) The weight W of the car acts vertically downwards. Since it makes an angle $60°$ to the road BC,

$$\text{component } F_1 \text{ down BC} = W \cos 60° = 0{\cdot}5\,W$$

So
$$0{\cdot}5W = \text{frictional force} = 4000 \text{ N}$$

$$\therefore\ W = 4000/0{\cdot}5 = 8000 \text{ N}$$

(ii) The car does *not* move in a direction AD at 90° to the road. So R must balance exactly the component F_2 of the weight in the direction DA. So

R = component of 8000 N along DA, which makes an angle of 30° with 8000 N

$= 8000 \cos 30° = 8000 \times 0{\cdot}87 = 6960$ N.

Forces in Equilibrium, Triangle and Polygon of Forces

We now consider forces *in equilibrium*. Figure 3.5 (i) shows forces P, Q and S acting on a joint O of a bridge structure which are in equilibrium or in balance.

In Figure 3.5 (ii), we draw *ab* to represent P and *bc* to represent Q. Then *ac* represents the resultant (sum) of P and Q. For equilibrium, S must balance

Figure 3.5 *Triangle and polygon of forces*

exactly the resultant. So *S* acts along *ca* opposite to *ac*, and has the same size as *ac*. In other words, the side *ca* of the triangle *abc* represents *S*.

This general result is stated in this way:

If three forces are in equilibrium, they can be represented in magnitude and direction by the three sides of a triangle taken in order.

This is called the *triangle of forces* theorem. Note that (i) the arrows showing the directions of the forces follow each other (or are in order) round the triangle, (ii) if the triangle does not close after drawing three forces acting on an object, then the forces are *not* in equilibrium.

We can extend this result to many forces in equilibrium. In Figure 3.5 (iii), five forces *P, Q, R, S, T* are in equilibrium at O. Starting with AB to represent *P*, all the forces taken in order form a *closed polygon* ABCDE, Figure 3.5 (iv). If the polygon did not close, the forces would not be in equilibrium.

Example on Triangle of Forces

A uniform ladder of weight 200 N and length 12 m is placed at an angle of 60° to the horizontal, with one end B leaning against a smooth wall and the other end A on the ground. Calculate the reaction force *R* of the wall at B and the force *F* of the ground at A.

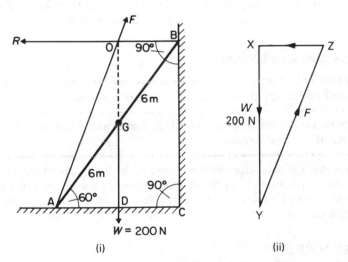

Figure 3.6 *Triangle of forces application*

The force *R* at B acts perpendicularly to the smooth wall. The weight *W* of the uniform ladder acts at its midpoint G. The forces *W* and *R* meet at O, as shown (Figure 3.6(i)). So the force *F* at A must pass through O to balance their resultant.

The *triangle of forces* can be used to find the unknown forces *R* and *F*.

First, we draw XY vertically to represent *W* = 200 N to scale, Figure 3.6 (ii). Then we draw a line from Y parallel to the force *F* and a line from X parallel to the force *R*, meeting at Z. Then XYZ is a triangle of the forces. From the scale used for *W*, 200 N, we find on measuring the lines XZ and YZ that *R* = 58 N and *E* = 208 N.

Moments

When the steering-wheel of a car is turned, the applied force is said to exert a *moment*, or turning-effect about the axle attached to the wheel. The magnitude of the moment of a force F about a point O is defined as *the product of the force F and the perpendicular distance OA from O to the line of action of F,* Figure 3.7.

Moment = force × *perpendicular* distance from axis

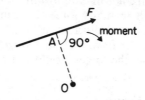

Figure 3.7 *Moment of force*

So

$$\text{moment} = F \times \text{AO}$$

The magnitude of the moment is expressed in *newton metres* (N m) when F is in newtons and AO is in metres. We can take an anticlockwise moment as positive in sign and a clockwise moment as negative in sign.

Moments and Equilibrium

The resultant of a number of forces *in equilibrium* is zero. So the moment of the resultant about any point is zero. It therefore follows that

---

The algebraic sum of the moments of all the forces about any point is zero when those forces are in equilibrium.

This means that the total clockwise moment of the forces about any point = the total anticlockwise moment of the remaining forces about the same point.

As the following Examples show, this rule or *principle of moments* is widely used to find unknown forces when an object is in equilibrium.

Examples on Moments

1 A horizontal rod AB is suspended at its ends by two strings, Figure 3.8 (i). The rod is 0·6 m long and its weight of 3 N acts at G where AG is 0·4 m and BG is 0·2 m.
Find the tensions X and Y in the string.

The forces X, Y and 3 N are parallel forces. So

$$X + Y = 3 \qquad . \qquad . \qquad . \qquad . \qquad . \qquad (1)$$

Clockwise moments about G = anticlockwise moments about G. Since 3 N has no moment about G,

$$X \times 0{\cdot}4 = Y \times 0{\cdot}2, \quad \text{so, cancelling,} \quad 2X = Y \qquad . \qquad . \qquad (2)$$

From (1), $X + 2X = 3$, so $X = 1$ N. Then $Y = 2X = 2$ N

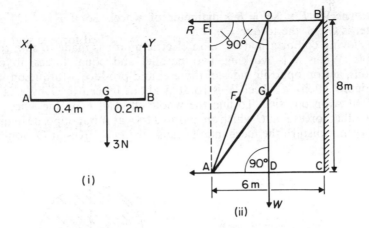

Figure 3.8 *Application of moments*

2 In Figure 3.8 (ii), a ladder rests against a smooth wall at B and rough ground at A. The weight W of 100 N acts at the midpoint G of the ladder, R is the normal reaction force of the wall at B and F is the total force at the ground at A. The height BC $= 8$ m and AC $= 6$ m.

Calculate R using the principle of moments and hence find F.

Take moments about A so as to eliminate F. Then, using perpendicular distances,

$$R \times AE = W \times AD$$

or $\qquad R \times 8 = 100 \times 3 \quad (AE = BC = 8 \text{ m}, AD = \tfrac{1}{2}AC = 3 \text{ m})$

So $\qquad R = \dfrac{100 \times 3}{8} = 37 \cdot 5 \text{ N}$

To find F, we see that F must be equal and opposite to the *resultant* of R and W. Now R and W meet at $90°$ at O. So, from page 93,

$$\text{resultant} = F = \sqrt{R^2 + W^2}$$

$$= \sqrt{37 \cdot 5^2 + 100^2} = 107 \text{ N}$$

Couple and its Moment or Torque

There are many examples in practice where two forces, acting together, exert a moment or turning-effect on some object. As a very simple case, suppose two strings are tied to a wheel at X, Y, and *two equal and opposite forces*, F, are exerted tangentially to the wheel, Figure 3.9 (i). If the wheel is pivoted at its centre, O, it begins to rotate about O in an anticlockwise direction. The total moment about O is then $(F \times OY) + (F \times OX) = F \times XY$.

Two equal and opposite forces whose lines of action do not coincide are said to form a *couple*. The two forces always have a turning-effect, or moment, called a *torque*, which is given by

---

$\qquad$ **torque** $=$ **one force** $\times$ **perpendicular distance between forces** . $\qquad$ (1)

Since XY is perpendicular to each of the forces F in Figure 3.9 (i), the torque on the wheel $= F \times XY = F \times$ diameter of wheel. So if $F = 10$ N and the diameter is 0·4 m, the torque $= 4$ N m.

The moving coil or armature in an electric motor is made to spin round by a couple. When it is working, two parallel and equal forces, in opposite directions, act on opposite sides of the coil and produce a torque on it.

In Figure 3.9 (ii), a wheel W is rotated about its centre O by a tangential force F at A on one side. To find the whole effect of F on the wheel, we can put two other forces F at O which are parallel to F at A but opposite in direction. This does not disturb the mechanics because the two forces at O would cancel

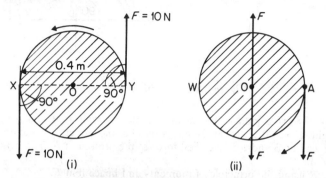

Figure 3.9 *Couples and moments*

and leave F at A. But the upward force F at O and the opposite parallel force F at A form a *couple* of moment $F \times OA$. This is actually the moment of F at A.

In addition, however, we are left with the downward force F at O. So when F is applied to A to turn the wheel, it produces a couple *plus* an equal force F at O, the centre of the wheel.

Examples on Couples

1 In Figure 3.10 (i), a beam AB is acted on by a force of 3 N at A and a parallel force in the opposite direction of 4 N at B. What is the effect on the beam if it is on a smooth table?

The force of 4 N at B can be considered as a force of 3 N and a force of 1 N. The 3 N force at B and the 3 N force at A together form a *couple*. So the beam rotates. The force of 1 N left over at B would also make the beam move forward. So the total effect on the beam is a rotation *and* forward movement.

2 In Figure 3.10 (ii), two parallel forces F act in opposite directions along the sides AD and CB of a rectangular horizontal plate ABCD. Two equal and opposite forces of 2 N act along the sides CD and AB.

Calculate F if the plate does not rotate and the sides of the plate are 0·4 m and 0·6 m as shown.

Since the plate does not rotate, the moments of the two couples must be equal and opposite. So

$$F \times 0{\cdot}4 = 2 \times 0{\cdot}6$$

and

$$F = 2 \times 0{\cdot}6/0{\cdot}4 = 3 \text{ N}$$

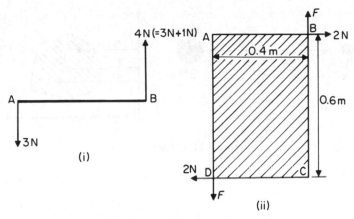

Figure 3.10 *Action of couples*

Conditions for Equilibrium

We conclude by listing the conditions which apply when any object is in equilibrium.

(1) **With three or more non-parallel forces acting on the object, a closed triangle or a closed polygon can be drawn to represent the forces in magnitude and direction.**
(2) **The algebraic sum of the moments of all the forces about any point is zero.**
(3) **The algebraic sum of the resolved components of all the forces in any direction is zero.**

Centre of Mass

Consider a smooth uniform rod on a horizontal surface with negligible friction such as ice. If a force is applied to the rod near one end, the rod will rotate as it accelerates. If it is applied at the centre, *it will accelerate without rotation.* The *centre of mass* of an object may be defined as the point at which an applied force produces acceleration but no rotation.

With two separated masses m_1 and m_2 connected by a rigid rod of negligible mass, the centre of mass C is at a distance x_1 from m_1 and a distance x_2 from m_2 given numerically by $m_1 x_1 = m_2 x_2$, Figure 3.11 (i). So if the mass m_1 is 1 kg and the mass m_2 is 2 kg, and the length of the rod is 3 m, the centre of mass is 2 m from the 1 kg mass and 1 m from the 2 kg mass.

In a molecule of sodium chloride, the sodium atom has a relative atomic mass of about 23·0 and the chlorine atom one of about 35·5. If the separation of the atoms is a, the centre of mass C has a distance x from the sodium atom given by

$$23 \times x = 35 \cdot 5 \times (a - x)$$

Solving for x,
$$x = \frac{35 \cdot 5a}{58 \cdot 5} = 0 \cdot 6a$$

Figure 3.11 (ii) shows the particles of masses $m_1, m_2, \ldots$ which together form the object of total mass M. If $x_1, x_2, \ldots$ are the respective x-coordinates of the

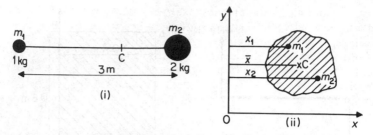

Figure 3.11 *Centre of mass*

particles relative to axes Ox, Oy, then generally the coordinate $\bar{x}$ of the centre of mass C is defined by

$$\bar{x} = \frac{m_1x_1 + m_2x_2 + \cdots}{m_1 + m_2 + \cdots} = \frac{\Sigma mx}{M}$$

Similarly, the distance $\bar{y}$ of the centre of mass C from Ox is given by

$$\bar{y} = \frac{\Sigma my}{M}$$

Centre of Gravity

Every particle is attracted towards the centre of the earth by the force of gravity.

The *centre of gravity* of a body is the point where the *resultant* force of attraction or *weight* of the body acts or appears to act.

In the simple case of a ruler, the centre of gravity is the point of support when the ruler is balanced.

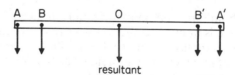

Figure 3.12 *Centre of gravity of rod*

An object can be considered to consist of many small particles. The forces on the particles due to the attraction of the earth are all parallel since they act vertically, and so their resultant is the sum of all the forces. The resultant is the *weight* of the whole object, of course. In the case of a rod of uniform cross-sectional area, the weight of a particle A at one end, and that of a corresponding particle A' at the other end, have a resultant which acts at the mid-point O of the rod, Figure 3.12. Similarly, the resultant of the weight of a particle B, and that of a corresponding particle B', have a resultant acting at O. In this way, i.e., by symmetry, it follows that the resultant of the weights of all the particles of the rod acts at O. Hence the centre of gravity of a uniform rod is at its mid-point.

The centre of gravity, c.g., of the curved surface of a *hollow cylinder* acts at the mid-point of the cylinder axis. This is also the position of the c.g. of a uniform *solid cylinder*. The c.g. of a *triangular plate or lamina* is two-thirds of the distance along a median from the corresponding vertex of the triangle. The c.g. of a uniform *right solid cone* is three-quarters along the axis from the apex.

Ideally, a *simple pendulum* has all its mass or weight concentrated at the centre of the swinging spherical bob. If the suspension has appreciable weight, the centre of gravity of the bob–suspension system would be higher than the centre of the spherical bob. The system would then no longer be a 'simple pendulum'.

You should know:

1 Forces (and other vectors) can be added by a triangle or parallelogram method.
2 Resolved component of $F = F \cos \theta$ in direction at angle θ to F or $F \sin \theta$ in perpendicular direction.
3 Forces in equilibrium have a *closed* triangle for three forces and a closed polygon for more than three forces.
   Three forces in equilibrium must meet at a point.
4 Forces in equilibrium: An unknown force may be calculated by taking moments about a suitable point—total moment = 0—or by resolving the forces in a suitable direction—total of resolved components = 0.
5 Couple = two equal and parallel forces acting in opposite directions. Moment of couple or *torque*—one force × perpendicular distance between forces.
6 Centre of mass = point where a force produces acceleration but no rotation.
   Centre of gravity = point where total weight appears to act.
   Take suitable moments to find centre of mass or gravity positions.

EXERCISES 3A Forces in Equilibrium

Multiple Choice

1

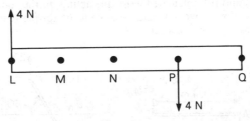

Figure 3A

A uniform bar LQ of weight 6 N is held horizontal by vertical forces. Two of the forces of 4 N each act at L and P as shown in Figure 3A. The points L, M, N, P and Q are at equal intervals along the bar. At which point must a vertical force of 6 N act to keep the bar in equilibrium?

A L B M C N D P E Q

2 A book of weight 4·0 N is gripped in a clamp at its top left-hand corner, Figure 3B (i). The centre of gravity of the book is at its centre, G, and T is the torque required to maintain its equilibrium in the position shown. The diagram below is a

free-body diagram of the book. In this situation

A $F = 4.0$ N and $T = 0.4$ N m **B** $F = 4.0$ N and $T = 0.6$ N m **C** $F = 4.0$ N and $T = 0.8$ N m **D** $F = 8.0$ N and $T = 0.6$ N m **E** $F = 8.0$ N and $T = 0.8$ N m. (*L.*)

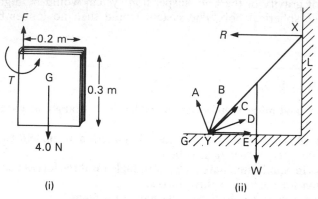

(i)   (ii)

Figure 3B

3 A uniform ladder XY rests with one end Y on the ground and the other end resting against a smooth wall L, Figure 3B (ii). W is its weight and R is the reaction of the wall. Which direction, A, B, C, D or E, is the reaction of the ground at Y?
4 Figure 3C shows three forces acting at a point. The lines drawn represent the forces roughly in magnitude and direction. Which diagram best represents equilibrium?

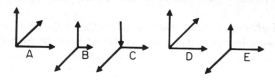

Figure 3C

Longer Questions

5 In Figure 3D (i) and (ii), calculate the torque acting on the rod AB, 0·4 m long.
   In Figure 3D (i), what work would be done if the torque remained constant and AB is rotated through two revolutions?

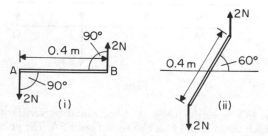

Figure 3D

6 The foot of a uniform ladder is on rough horizontal ground, and the top rests against a smooth vertical wall. The weight of the ladder is 400 N, and a man weighing 800 N stands on the ladder one-quarter of its length from the bottom. If the inclination of the ladder to the horizontal is 30°, find the reaction at the wall and the total force at the ground.

7 A rectangular plate ABCD has two forces of 100 N acting along AB and DC in opposite directions. If AB = 3 m, BC = 5 m, what is the moment of the couple or torque acting on the plate? What forces acting along BC and AD respectively are required to keep the plate in equilibrium?

8 A flat plate is cut in the shape of a square of side 20·0 cm, with an equilateral triangle of side 20·0 cm adjacent to the square. Calculate the distance of the centre of mass from the apex of the triangle.

9 The dimensions of torque are the same as those of energy. Explain why it would nevertheless be inappropriate to measure torque in joules. State an appropriate unit. (*C.*)

10 A trap-door 120 cm by 120 cm is kept horizontal by a string attached to the mid-point of the side opposite to that containing the hinge. The other end of the string is tied to a point 90 cm vertically above the hinge. If the trap-door weight is 50 N, calculate the tension in the string and the reaction at the hinge.

11 Three forces in one plane act on a rigid body. What are the conditions for equilibrium?

The plane of a kite of mass 6 kg is inclined to the horizonal at 60°. The thrust of the air acting normally on the kite acts at a point 25 cm above its centre of gravity, and the string is attached at a point 30 cm above the centre of gravity. Find the thrust of the air on the kite, and the tension in the string. (*C.*)

12 State and explain the conditions under which a rigid body remains in equilibrium under the action of a set of coplanar forces. Describe an experiment to determine the position of the centre of gravity of a flat piece of cardboard cut into the shape of a triangle. Use the conditions of equilibrium you have stated to justify your practical method.

A simple model of the ammonia molecule (NH_3) consists of a pyramid with the hydrogen atoms at the vertices of the equilateral base and the nitrogen atom at the apex. The N–H distance is 0·10 nm and the angle between two N–H bonds is 108°. Find the centre of gravity of the molecule. (Atomic weights: H = 1·0; N = 14; 1 nm = 1×10^{-9} m.) (*O.*)

13 Summarise the various conditions which are being satisfied when a body remains in equilibrium under the action of three non-parallel forces.

A wireless aerial attached to the top of a mast 20 m high exerts a horizontal force upon it of 600 N. The mast is supported by a stay-wire running to the ground from a point 6 m below the top of the mast, and inclined at 60° to the horizontal. Assuming that the action of the ground on the mast can be regarded as a single force, draw a diagram of the forces acting on the mast, and determine by measurement or by calculation the force in the stay-wire. (*C.*)

Forces in Fluids

Pressure

Liquids and gases are called *fluids*. Unlike solid objects, fluids can flow.

If a piece of cork is pushed below the surface of a pool of water and then released, the cork rises to the surface again. The liquid thus exerts an upward force on the cork and this is due to the *pressure* exerted on the cork by the surrounding liquid. Gases also exert pressures. For example, when a thin closed metal can is evacuated, it usually collapses with a loud bang. The surrounding air exerts a pressure on the outside which is no longer counter-balanced by the pressure inside, and so there is a resultant force.

Pressure is defined as the *average force per unit area* at the particular region of liquid or gas. In Figure 3.13, for example, X represents a small horizontal area, Y a small vertical area, and Z a small inclined area, all inside a vessel

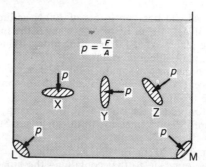

Figure 3.13 *Pressure in liquid*

containing a liquid. The pressure p acts normally to the planes of X, Y or Z. In each case

---

average pressure, $p = \dfrac{F}{A}$

---

where F is the normal force due to the liquid on one side of an area A of X, Y or Z. Similarly, the pressure p on the sides L or M of the curved vessel acts normally to L and M and has magnitude F/A. In the limit, when the area is very small, $p = \mathrm{d}F/\mathrm{d}A$.

At a given point in a liquid, the pressure can act in any direction. *Pressure is a scalar*, not a vector. The direction of the force on a particular surface, however, is normal to the surface.

Formula for Pressure

Observation shows that the pressure increases with the depth, h, below the liquid surface and with its density ρ (see p. 107).

To obtain a formula for the pressure, p, suppose that a horizontal plate X of area A is placed at a depth h below the liquid surface, Figure 3.14. By drawing vertical lines from points on the perimeter of X, we can see that the force on X due to the liquid is equal to the weight of liquid of height h and uniform cross-section A. Since the volume of this liquid is Ah, the mass of the liquid $= Ah \times \rho$.

$$\therefore \text{weight} = Ah\rho g \text{ N}$$

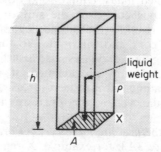

Figure 3.14 *Pressure and depth*

where g is $9\cdot8$ m s^{-2}, h is in m, A is in m^2, and ρ is in kg m^{-3}.

$$\therefore \text{ pressure, } p, \text{ on } X = \frac{\text{force}}{\text{area}} = \frac{Ah\rho g}{A}$$

$$\therefore p = h\rho g \qquad . \qquad . \qquad . \qquad . \qquad . \qquad (1)$$

When h, ρ, g have the units already mentioned, the pressure p is in *pascals* (Pa), where 1 Pa $= 1$ N m^{-2}.

Pressure is also expressed in terms of the pressure due to a height of mercury (Hg). One unit is the *torr* (after Torricelli):

$$1 \text{ torr} = 1 \text{ mmHg} = 133\cdot3 \text{ Pa (approx.)}$$

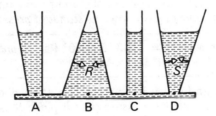

Figure 3.15 *Pressure and cross-section*

From $p = h\rho g$ it follows that *the pressure in a liquid is the same at all points on the same horizontal level in it.* Experiment also gives the same result. Thus a liquid filling the vessel shown in Figure 3.15 rises to the same height in each section if ABCD is horizontal. The cross-sectional area of B is greater than that of D. But the force on B is the sum of the weight of water above it together with the downward component of reaction R of the sides of the vessel, whereas the force on D is the weight of water above it *minus* the upward component of the reaction S of the sides of the vessel. This explains why the pressure in a vessel is independent of the cross-sectional area of the vessel such as those in Figure 3.15.

Atmospheric Pressure

A *barometer* is an instrument for measuring the pressure of the atmosphere, which is required in weather-forecasting, for example. An accurate form of barometer consists basically of a vertical barometer tube about a metre long containing mercury, with a vacuum at the closed top, Figure 3.16 (i). The other end of the tube is below the surface of mercury contained in a vessel B.

The pressure on the surface of the mercury in B is atmospheric pressure, A; and since the pressure is transmitted through the liquid, the atmospheric pressure supports the column of mercury in the tube. Suppose the column is a vertical height H above the level of the mercury in B. Now the pressure p at the bottom of a column of liquid of vertical height H and density ρ is given by $p = H\rho g$ (above). So if $H = 760$ mm $= 0\cdot76$ m and $\rho = 13\,600$ kg m^{-3},

$$p = H\rho g = 0\cdot76 \times 13\,600 \times 9\cdot8 = 1\cdot013 \times 10^5 \text{ Pa}$$

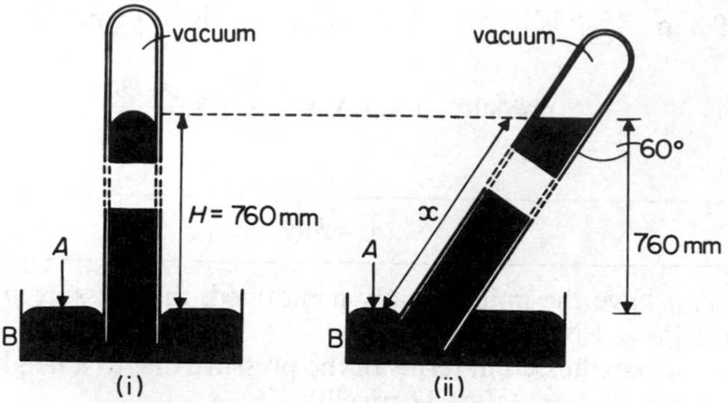

Figure 3.16 *Barometer* (*i*) *vertical and* (*ii*) *inclined*

The pressure at the bottom of a column of mercury 760 mm high for a particular mercury density and value of g is known as *standard pressure* or *one atmosphere*.

By definition, 1 atmosphere = 1.01325×10^5 Pa. *Standard temperature and pressure (s.t.p.)* is 0°C and 1.01325×10^5 Pa

---

The *bar* is 10^5 Pa and is nearly equal to one atmosphere.

It should be noted that the pressure p at a place X below the surface of a liquid is given by $p = H\rho g$, where H is the *vertical* distance of X below the surface. In Figure 3.16 (ii), a very long barometer tube is inclined at an angle of 60° to the vertical. The length of the mercury along the slanted side of a tube is x mm say. If the atmospheric pressure here is the same as in Figure 3.16 (ii), this means that the *vertical* height to the mercury surface is still 760 mm. So

$$x \cos 60° = 760$$

and

$$x = \frac{760}{\cos 60°} = \frac{760}{0.5} = 1520 \text{ mm}$$

Variation of Atmospheric Pressure with Height

The density of a liquid varies very slightly with pressure. The density of a gas, however, varies appreciably with pressure. Thus at sea-level the density of the atmosphere is about 1.2 kg m^{-3}; at 1000 m above sea-level the density is about 1.1 kg m^{-3}; and at 5000 m above sea-level it is about 0.7 kg m^{-3}. Standard atmospheric pressure is the pressure at the base of a column of mercury 760 mm high, a liquid which has a density about $13\,600 \text{ kg m}^{-3}$. Suppose air has a constant density of about 1.2 kg m^{-3}. Then the height of an air column of this density which has a pressure equal to standard atmospheric pressure

$$= \frac{760}{1000} \times \frac{13\,600}{1.2} = 8.6 \text{ km}$$

In fact, the air 'thins' the higher one goes, as explained above. The height of the air is thus much greater than 8·6 km.

Density

As we have seen, the pressure in a fluid depends on the density of the fluid. The *density* of a substance is defined as its *mass per unit volume*. So

$$\text{density, } \rho = \frac{\text{mass of substance}}{\text{volume of substance}} \qquad . \qquad . \qquad . \qquad (1)$$

The density of copper is about $9 \cdot 0$ g cm^{-3} or $9 \cdot 0 \times 10^3$ kg m^{-3}; the density of aluminium is $2 \cdot 7$ g cm^{-3} or $2 \cdot 7 \times 10^3$ kg m^{-3}; the density of water at 4°C is 1 g cm^{-3} or 1000 kg m^{-3}.

Substances which float on water have a density less than 1000 kg m^{-3} (p. 108). For example, ice has a density of about 900 kg m^{-3}; cork has a density of about 250 kg m^{-3}. Steel, of density 780 kg m^{-3}, will float on mercury, whose density is about 13 600 kg m^{-3} at 0°C.

Archimedes' Principle

An object immersed in a fluid experiences a resultant upward force owing to the pressure of fluid on it. This upward force is called the *upthrust* of the fluid on the object.

Archimedes stated that *the upthrust is equal to the weight of fluid displaced by the object*, **and this is known as** *Archimedes' principle*.

So if an iron cube of volume 400 cm^3 is totally immersed in water of density 1 g cm^{-3}, the upthrust on the cube = weight of 400×1 g = 4 N. If it is totally immersed in oil of density $0 \cdot 8$ g cm^{-3}, the upthrust on it = weight of $400 \times 0 \cdot 8$ g = $3 \cdot 2$ N.

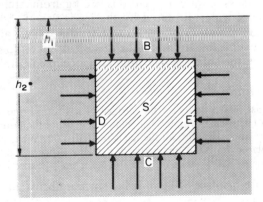

Figure 3.17 *Archimedes' principle*

Figure 3.17 shows why Archimedes' principle is true. If S is a solid immersed in a liquid, the pressure on the lower surface C is greater than on the upper surface B, since the pressure at the greater depth h_2 is more than that at h_1. The pressure on the remaining surfaces D and E act as shown. The *force* on each of the four surfaces is calculated by summing the values of *pressure × area* over every part, remembering that vector addition is needed to sum forces. With a simple *rectangular-shaped solid* and the sides, D, E vertical, it can be seen that (i) the resultant horizontal force is zero, (ii) the upward force on

C = pressure × area $A = h_2\rho gA$, where ρ is the liquid density, and the downward force on B = pressure × area $A = h_1\rho gA$. Thus

$$\text{resultant force on solid} = \text{upward force (upthrust)} = (h_2 - h_1)\rho gA$$

But
$$(h_2 - h_1)A = \text{volume of solid, } V$$

$$\therefore \textit{upthrust} = V\rho g = mg, \text{ where } m = V\rho$$

$$\therefore \textbf{\textit{upthrust = weight of liquid displaced}}$$

With a solid of irregular shape, taking into account horizontal and vertical components of forces, the same result is obtained. The upthrust is the weight of *liquid* displaced whatever the nature of the object immersed, or whether it is hollow or not. This is due primarily to the fact that the pressure on the object depends on the liquid in which it is placed.

Flotation

When an object *floats* in a liquid, the upthrust = the weight of the object, for equilibrium.

In *air*, for example, a balloon of constant volume 5000 m³ and mass 4750 kg rises to an altitude where the upthrust is 4750 g newtons, where g is the acceleration due to gravity at this height. From Archimedes' principle, the upthrust = the weight of air displaced = 5000 ρg newtons, where ρ is the density of air at this height. Thus

$$5000\,\rho g = 4750\,g$$

$$\therefore \rho = 0{\cdot}95 \text{ kg m}^{-3}$$

If a block of ice of volume 1 m³ and mass 900 kg floats in *water* of density 1000 kg m⁻³, the mass of water displaced is 900 kg, from Archimedes' principle. Thus the volume of water displaced by the ice is 0·9 m³. So the block floats with 0·1 m³ above the water surface. If the ice, mass 900 kg, all melts, the water formed has a volume of 0·9 m³. So all the melted ice takes up *exactly* the whole of the space which the solid ice had originally occupied below the water.

Example on Flotation

An ice cube of mass 50·0 g floats on the surface of a strong brine solution of volume 200·0 cm³ inside a measuring cylinder. Calculate the level of the liquid in the measuring cylinder (i) before and (ii) after all the ice is melted. (iii) What happens to the level if the brine is replaced by 200·0 cm³ water and 50·0 g of ice is again added? (Assume density of ice, brine = 900, 1100 kg m⁻³ or 0·9, 1·1 g cm⁻³ respectively.)

(i) Floating ice displaces 50 g of brine since upthrust equals weight of ice.

$$\therefore \text{volume displaced} = \frac{\text{mass}}{\text{density}} = \frac{50}{1{\cdot}1} = 45{\cdot}5 \text{ cm}^3$$

$$\therefore \text{level on measuring cylinder} = 245{\cdot}5 \text{ cm}^3$$

(ii) 50 g of ice forms 50 g of water when all of it is melted.

$$\therefore \text{level on measuring cylinder } \textit{rises} \text{ to } 250{\cdot}0 \text{ cm}^3$$

(iii) *Water*. Initially, volume of water displaced = 50 cm³, since upthrust = 50g.

$$\therefore \text{level on cylinder} = 250{\cdot}0 \text{ cm}^3$$

If 1 g of ice melts, volume displaced is 1 cm^3 less. But volume of water formed is 1 cm^3. Thus the net change in water level is zero. Hence the water level remains unchanged as the ice melts.

You should know:

1 Pressure in liquid, $p = h\rho g$, unit: Pa (1 Pa = 1 N m^{-2}).
2 Standard atmospheric pressure = $1\cdot01325 \times 10^5$ Pa = $101\cdot325$ kPa (1 kPa = 1000 Pa).
3 Archimedes' principle: Upthrust (upward force) on object immersed in fluid = weight of fluid displaced by object.
4 For a floating object in liquid, weight of object = weight of liquid displaced. So density of object is less than that of the liquid.

Submarines

We can use Archimedes' principle (page 107) to explain how submarines ascend and descend. Consider a submarine floating on the ocean surface as it leaves a port. Archimedes' principle tells us that the water displaced produces an upthrust equal to the weight of the submarine, producing an equilibrium of forces which keeps the submarine at a steady height. Within the submarine there are large tanks, initially full of air, which can be flooded with water from the ocean. If this is done, then the submarine becomes heavier, so its weight exceeds the upthrust, causing the submarine to dive. When the submarine is fully submerged it displaces a fixed volume of water, but since the water's density generally increases with depth, this volume will weigh more the deeper the submarine goes, producing an ever increasing upthrust. The submarine will therefore continue to dive until the upthrust at its current depth equals the weight of the submarine with flooded tanks.

The submarine can ascend again by replacing the water in the tanks with air. It does this using compressed air stored in cylinders, which is released into the tanks, forcing the water back out into the ocean. The submarine becomes lighter, so the upthrust exceeds the weight of the submarine, pushing it back towards the surface. The cruising depth of the submarine can be adjusted by flooding the tanks to varying degrees: the more water in the tanks, the greater the weight of the submarine, so the deeper the submarine will go to displace an equal weight of ocean water.

EXERCISES 3B Forces in Fluids

Multiple Choice

1 *Pressure* has dimensions

   A ML^{-1}T^{-1} B MLT^{-2} C ML^{2}T^{-2} D ML^{-1}T^{-2} E MT^{-2}

2 A hot-air balloon moving upwards has a total weight of 200 N and a volume of 20 m^3. Assuming the air density is $1\cdot2$ kg m^{-3}, the net upward force on the balloon in N is then about

   A 24 B 36 C 40 D 176 E 240

3 A ship, floating in clear water of density 1000 kg m^{-3}, moves to sea-water of density 1050 kg m^{-3} where it floats again. The upthrust on the ship then

A stays constant **B** decreases **C** increases **D** increases by 0·05 times
E decreases by 0·05 times

Longer Questions

4 An alloy of mass 588 g and volume 100 cm^3 is made of iron of density 8·0 g cm^{-3} and aluminium of density 2·7 g cm^{-3}. Calculate the proportion (i) by volume, (ii) by mass of the constituents of the alloy.

5 A string supports a solid iron object of mass 180 g totally immersed in a liquid of density 800 kg m^{-3}. Calculate the tension in the string if the density of iron is 8000 kg m^{-3}.

6 A hydrometer floats in water with 6·0 cm of its graduated stem above the water level, and in oil of density 0·8 g cm^{-3} with 4·0 cm of the stem above the oil level. What is the length of stem above the water level when the hydrometer is placed in a liquid of density 0·9 g cm^{-3}?

7 A uniform capillary tube contains air trapped by a mercury thread 40 mm long. When the tube is placed horizontally as in Figure 3E (i), the length of the air column is 36 mm. When placed vertically, with the open end of the tube downwards, the length of air column is now x mm, Figure 3E (ii). Calculate x if the atmospheric pressure is 760 mmHg, assuming that the air obeys Boyle's law, $pV =$ constant.

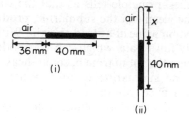

Figure 3E

8 A barometer tube, 960 mm long above the mercury in the reservoir, contains a little air above the mercury column inside it. When vertical, Figure 3F (i), the mercury column is 710 mm above the mercury in the reservoir. When inclined at 30° to the horizontal, Figure 3F (ii), the mercury column is now 910 mm along the barometer tube.

Assuming the air obeys Boyle's law, $pV =$ constant, calculate the atmospheric pressure.

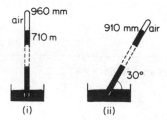

Figure 3F

9 State the principle of Archimedes and use it to derive an expression for the resultant force experienced by a body of weight W and density σ when it is totally immersed in a fluid of density ρ.

A solid weighs 237·5 g in air and 12·5 g when totally immersed in a liquid of density 0·9 g cm^{-3}. Calculate

(a) the density of the solid,

(b) the density of a liquid in which the solid would float with one-fifth of its volume exposed above the liquid surface. (*L.*)

10 State

(a) the laws of fluid pressure and

(b) the principle of Archimedes.

Show how (b) is a consequence of (a). Describe a simple experiment which verifies Archimedes' principle.

The volume of a hot-air balloon is 600 m^3 and the density of the surrounding air is $1 \cdot 25 \text{ kg m}^{-3}$. The balloon just hovers clear of the ground when the burner has heated the air inside to a temperature at which its density is $0 \cdot 80 \text{ kg m}^{-3}$.

(a) What is the total mass of the balloon, including the hot air inside it?

(b) What is the total mass of the envelope of the balloon and its load?

(c) Find the acceleration with which the balloon will start to rise when the density of the air inside is reduced to $0 \cdot 75 \text{ kg m}^{-3}$. (Take g to be 10 m s^{-2}.) (*O.*)

4   Rotational Dynamics, Fluids in Motion[1]

Rotational Dynamics

So far we have considered the equations of linear motion and other dynamical formulae connected with a particle or small mass m. *We now consider the dynamics of* large rotating *objects such as spinning wheels in machines, for example.*

We shall find that dynamical formulae in rotational dynamics are similar to those in linear or translational dynamics. For example, the kinetic energy of a mass m *moving with a velocity* v *is* $\frac{1}{2}mv^2$ *and the rotational kinetic energy of a spinning object is* $\frac{1}{2}I\omega^2$, *where* I *is called the 'moment of inertia' of the object about its axis and* ω *is the angular velocity.*

Plate 4A *By keeping her arms folded close to her body, the ice-skater can spin very fast*

Torque and Angular Acceleration

In linear motion, an object changing steadily from a velocity u to a velocity v in a time t has an acceleration a given by $a = $ velocity change/time $= (v - u)/t$.

In rotational motion, a wheel spinning about its centre may increase its angular velocity from ω_0 to ω in a time t. The *angular acceleration* α is given by

$$\alpha = \frac{\omega - \omega_0}{t}$$

So
$$\omega = \omega_0 + \alpha t \qquad . \qquad . \qquad . \qquad . \qquad . \qquad (1)$$

This is analogous to the linear relation $v = u + at$.

To make the wheel spin faster, a *couple* or *torque* T is applied to the wheel. We have already met forces which make a couple or a torque in a previous

[1] Students limited in time for A-level study may omit this chapter at a first reading, depending on the syllabus. Consult your teacher.

chapter. The turning-effect or torque T of a force F applied tangentially to a wheel of radius r spinning about its centre is given by $T = F \times r$ and the unit of T is N m (newton metre).

In linear dynamics, a force F produces an acceleration given by $F = ma$, where m is the mass of the object. In an analogous way, we soon see that a torque T, applied to a rotating wheel, gives it an angular acceleration given by

$$T = I\alpha \qquad . \qquad . \qquad . \qquad . \qquad . \qquad . \qquad (2)$$

where I is the *moment of inertia* of the wheel about its axis of rotation, which we now explain.

Moment of Inertia I

Consider a large rigid object X rotating about an axis O when a torque T acts on the object, Figure 4.1. At the instant shown, a small mass m_1 of X, distant r_1 from O, has a linear acceleration a perpendicular to r_1 given by $a = r_1\alpha$,

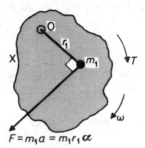

$$F = m_1 a = m_1 r_1 \alpha$$

Figure 4.1 *Moment of inertia*

where α is the angular acceleration about O. So the force F on m_1 is given by

$$F = m_1 a = m_1 r_1 \alpha$$

$$\therefore \text{ torque } T \text{ about O} = F \times r_1 = m_1 r_1^2 \alpha$$

Adding together all the torques on the masses which make up the object X, then

$$\text{total torque } T = m_1 r_1^2 \alpha + m_2 r_2^2 \alpha + \cdots$$

$$= \Sigma(m_1 r_1^2)\alpha = I\alpha$$

where $I = \Sigma(m_1 r_1^2)$, the sum of all the products 'mr^2' for the masses m of the object and the square of their distances, r^2, from the axis of rotation. So, as in equation (2), $T = I\alpha$.

We call I the 'moment of inertia about the axis'. Calculation shows that a uniform rod of mass M and length l has a moment of inertia $I = Ml^2/12$ when it rotates about an axis at one end perpendicular to the rod. A sphere of mass M and radius r has a moment of inertia $I = 2Mr^2/5$ when it spins about an axis through its centre. The unit of I is 'kg m^2' and you should note that the value of I depends not only on the mass and dimensions of the object but also on the position of the axis of rotation.

Examples on Torque and Angular Acceleration

1 A heavy flywheel of moment of inertia 0·3 kg m^2 is mounted on a horizontal axle of radius 0·01 m and negligible mass compared with the flywheel. Neglecting friction, find

(i) the angular acceleration if a force of 40 N is applied tangentially to the axle,   (ii) the angular velocity of the flywheel after 10 seconds from rest.

(i) Torque $T = 40 \,(\text{N}) \times 0{\cdot}01 \,(\text{m}) = 0{\cdot}4 \,\text{N m}$

From
$$T = I\alpha$$

$$\text{angular acceleration } \alpha = \frac{T}{I} = \frac{0{\cdot}4}{0{\cdot}3} = 1{\cdot}3 \text{ rad s}^{-2}$$

(ii) After 10 seconds, angular velocity $\omega = \alpha t$

$$= 1{\cdot}3 \times 10 = 13 \text{ rad s}^{-1}$$

2   The moment of inertia of a solid flywheel about its axis is $0{\cdot}1 \text{ kg m}^2$. It is set in rotation by applying a tangential force of 20 N with a rope wound round the circumference, the radius of the wheel being $0{\cdot}1$ m. Calculate the angular acceleration of the flywheel. What would be the angular acceleration if a mass of 2 kg were hung from the end of the rope? (*O. & C.*)

$$\text{Torque } T = I\alpha, \text{ and } T = 20 \times 0{\cdot}1 \,\text{N m}$$

So
$$\text{angular acceleration } \alpha = \frac{T}{I} = \frac{20 \times 0{\cdot}1}{0{\cdot}1} = 20 \text{ rad s}^{-2}$$

If a mass m of 2 kg, or weight 20 N assuming $g = \text{ m s}^{-2}$, is hung from the end of the rope, it moves down with an acceleration a, Figure 4.2. In this case, if F is the tension in the rope,

$$mg - F = ma \qquad . \qquad . \qquad . \qquad . \qquad . \qquad (1)$$

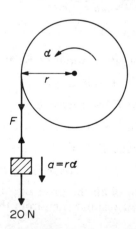

20 N

Figure 4.2 *Example on torque and angular acceleration*

For the flywheel, $\qquad F \cdot r = \text{torque} = I\alpha$. $\qquad . \qquad . \qquad . \qquad . \qquad$ (2)

where r is the radius of the wheel and α the angular acceleration about the centre. Now the mass descends a distance given by $r\theta$, where θ is the angle the flywheel has turned. Hence the acceleration $a = r\alpha$. Substituting in (1),

$$mg - F = mr\alpha$$

Multiplying by r,
$$mgr - F \cdot r = mr^2\alpha . \qquad . \qquad . \qquad . \qquad . \qquad (3)$$

Adding (2) and (3),

$$mgr = (I + mr^2)\alpha$$

$$\therefore \alpha = \frac{mgr}{I + mr^2} = \frac{2 \times 10 \times 0\cdot1}{0\cdot1 + 2 \times 0\cdot1^2}$$

$$= 16\cdot7 \text{ rad s}^{-2}$$

Angular Momentum and Relation to Torque

In linear or straight-line motion, an important property of a moving object is its linear momentum (p. 31). When an object spins or rotates about an axis, its *angular momentum* plays an important part in its motion.

Consider a particle A of a rigid object rotating about an axis O, Figure 4.3. The momentum of A = mass × velocity = $m_1 v = m_1 r_1 \omega$. The 'angular momentum'

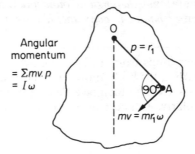

Angular
momentum

$= \Sigma mv.\,p$
$= I\omega$

$p = r_1$

$90°$ A

$mv = mr_1\omega$

Figure 4.3 *Angular momentum*

of A about O is defined as the *moment of the momentum* about O. Its magnitude is thus $m_1 v \times p$, where p is the perpendicular distance from O to the direction of v. So angular momentum of A $= m_1 vp = m_1 r_1 \omega \times r_1 = m_1 r_1^2 \omega$.

$\therefore$ **total angular momentum of whole body** $= \Sigma m_1 r_1^2 \omega = \omega \Sigma m_1 r_1^2$

$$= I\omega$$

---

where I is the moment of inertia of the body about O.

Angular momentum is analogous to 'linear momentum', mv, in the dynamics of a moving particle. In place of m we have I, the moment of inertia; in place of v we have ω, the angular velocity.

Torque × Time

In linear dynamics, a force F acting on an object for a time t produces a momentum change given by

$$F \times t \text{ (impulse)} = \textbf{momentum change}$$

In an analogous way, a torque T acting for a time t on a rotating object produces an angular momentum change given by

$$T \times t = \textbf{angular momentum change}$$

So if I is the moment of inertia about the axis concerned, and ω_1 and ω_2 are the initial and final angular velocities produced by a steady torque T, then

$$T \times t = I\omega_2 - I\omega_1$$

As an illustration, suppose a wheel of moment of inertia about its centre of 2 kg m^2 is spinning with an angular velocity of 15 rad s^{-1}. If it is brought to rest by a steady braking torque T in 5 s, the value of T is given by

$$T \times 5 = I\omega_2 - I\omega_1 = (2 \times 15) - 0 = 30$$

So $$T = 30/5 = 6 \text{ N m}$$

Conservation of Angular Momentum

The *conservation of angular momentum*, **which corresponds to the conservation of linear momentum, states that** *the angular momentum about an axis of a given rotating body or system of bodies is constant, if no external torque acts about that axis.*

Thus when a high diver jumps from a diving board, his moment of inertia, I, can be decreased by curling his body more, in which case his angular velocity ω is increased, Figure 4.4. He may then be able to turn more somersaults before striking the water. Similarly, a dancer on skates can spin faster by folding her arms.

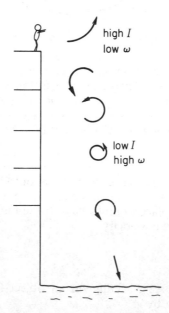

high I
low ω

low I
high ω

Figure 4.4 *Conservation of angular momentum*

The earth rotates about an axis passing through its geographic north and south poles with a period of one day. If it is struck by meteorites, then, since action and reaction are equal, no external couple acts on the earth and meteorites. Their total angular momentum is thus conserved. Neglecting the angular momentum of the meteorites about the earth's axis before collision compared with that of the earth, then

angular momentum of *earth plus meteorites* after collision = angular momentum of *earth* before collision

Since the effective mass of the earth has increased after collision, the moment of inertia has increased. So the earth will slow down slightly. Similarly, if a mass is dropped gently on to a turntable rotating freely at a steady speed, the conservation of angular momentum leads to a reduction in the speed of the table.

Angular momentum, and the principle of the conservation of angular momentum, have wide application in physics. They are used in connection with enormous rotating masses such as the earth, as well as minute spinning particles such as electrons, neutrons and protons found inside atoms.

Examples on Conservation of Angular Momentum

1 A ballet dancer spins about a vertical axis at 1 revolution per second with arms outstretched. With her arms folded, her moment of inertia about the vertical axis decreases by 60%. Calculate the new rate of revolution.

Suppose I is the initial moment of inertia about the vertical axis and ω is the initial angular velocity corresponding to 1 rev s^{-1}. The new moment of inertia $I_1 = 40\%$ of $I = 0.4I$.

Suppose the new angular velocity is ω_1. Then, from the conservation of angular momentum

$$I_1\omega_1 = I\omega$$

So
$$\omega_1 = \frac{I}{I_1}\omega = \frac{1}{0.4}\omega$$

Since angular velocity ∝ number of revs per second, the new number n of revs per second is given by

$$n = \frac{1}{0.4} \times 1 \text{ rev s}^{-1} = 2.5 \text{ rev s}^{-1}$$

2 A disc of moment of inertia 5×10^{-4} kg m^2 is rotating freely about axis O through its centre at 40 r.p.m., Figure 4.5. Calculate the new revolutions per minute (r.p.m.) if some wax W of mass 0.02 kg is dropped gently on to the disc 0.08 m from its axis.

Initial angular momentum of disc $= I\omega = 5 \times 10^{-4}\omega$

where ω is the angular velocity corresponding to 40 r.p.m.

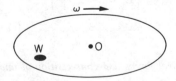

Figure 4.5 *Example on conservation of angular momentum*

When the wax of mass 0.02 kg is dropped gently on to the disc at a distance r of 0.08 m from the centre O, the disc slows down. Suppose the angular velocity is now ω_1. The total angular momentum about O of disc plus wax W

$$= I\omega_1 + mr^2\omega_1 = 5 \times 10^{-4}\omega_1 + 0.02 \times 0.08^2\omega_1$$
$$= 6.28 \times 10^{-4}\omega_1$$

From the conservation of angular momentum for the disc and wax about O

$$6.28 \times 10^{-4}\omega_1 = 5 \times 10^{-4}\omega$$

$$\therefore \frac{\omega_1}{\omega} = \frac{500}{628} = \frac{n}{40}$$

where n is the r.p.m. of the disc, because the angular velocity is proportional to the r.p.m.

$$\therefore n = \frac{500}{628} \times 40 = 32 \text{ (approx.)}$$

Central Forces and Conservation of Angular Momentum

Rotating objects sometimes have a force on them directed towards a particular point or axis. Figure 4.6 (i) shows a planet P moving in an *elliptical orbit* round the sun S under gravitational attraction. The force F on P is always directed towards S as it moves. So the moment of F about S is always *zero*.

Since the so-called central force or external force on P has no moment about S, the angular momentum about S is *constant*. At X, the nearest point to the sun, the angular momentum about $S = mv_1r_1$, where m is the mass of the planet, v_1 is the velocity at X and $SX = r_1$. At Y, the furthest distance from the sun, the angular momentum about $S = mv_2r_2$, where v_2 is the new velocity at Y and $SY = r_2$. So

$$mv_1r_1 = mv_2r_2$$

$$\therefore \frac{v_1}{v_2} = \frac{r_2}{r_1}$$

Since r_2 is greater than r_1, the velocity v_1 is greater than v_2. So the velocity of the planet P increases as it approaches its nearest distance to the sun.

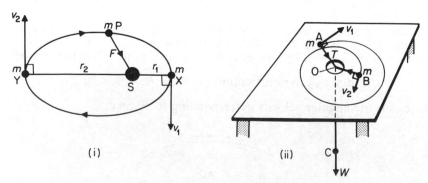

Figure 4.6 *Central forces and conservation of angular momentum*

Figure 4.6 (ii) shows a mass at A on a *smooth table*. It is connected by a string AOC through a hole O in the table to a weight W, hanging down below the table, so that OC is vertical.

With the string OA taut, the mass m is pushed at right angles to OA with an initial velocity v_1 of $6 \, \text{m s}^{-1}$. The length OA is then $0.4 \, \text{m}$. After the mass m rotates, the mass reaches a position at B on the table where OB is $0.3 \, \text{m}$ and the velocity of the mass has changed to a value v_2.

The tension T in the string may change as the mass rotates but T is always directed towards O on the table. So as the mass rotates, the external or central force T has no moment about O. So the angular momentum about O is *constant*. This helps us to find the velocity v_2 at B. We have, using angular momentum,

$$mv_1 \times 0.4 \, (\text{at A}) = mv_2 \times 0.3 \, (\text{at B})$$

So

$$v_1 \times 0.4 = v_2 \times 0.3$$

and

$$v_2 = \frac{v_1 \times 0.4}{0.3} = \frac{6 \times 0.4}{0.3}$$

$$= 8 \, \text{m s}^{-1}$$

So the velocity increases to $8 \, \text{m s}^{-1}$ when the string shortens to $0.3 \, \text{m}$, to keep the angular momentum constant.

Rotational Kinetic Energy, Work Done by Torque

We now consider the *kinetic energy* of a rotating object. In Figure 4.7 (i), the rotational kinetic energy of the object X about O

$$= \text{sum of kinetic energy of all its individual masses}$$

$$= \tfrac{1}{2}m_1v_1^2 + \tfrac{1}{2}m_2v_2^2 + \cdots$$

$$= \tfrac{1}{2}m_1r_1^2\omega^2 + \tfrac{1}{2}m_2r_2^2\omega^2 + \cdots$$

$$= \tfrac{1}{2}(m_1r_1^2 + \tfrac{1}{2}m_2r_2^2 + \cdots)\omega^2 = \tfrac{1}{2}I\omega^2$$

So
$$\textbf{rotational kinetic energy} = \tfrac{1}{2}I\omega^2 \qquad . \qquad . \qquad . \qquad (3)$$

If $I = 2 \, \text{kg m}^2$ and $\omega = 3 \, \text{rad s}^{-1}$, then

$$\text{rotational kinetic energy} = \tfrac{1}{2} \times 2 \times 3^2 = 9 \, \text{J}$$

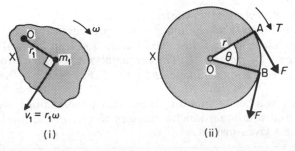

Figure 4.7 (i) *Rotational kinetic energy* (ii) *Work done by torque*

The *work done* W by a constant torque can be found from Figure 4.7 (ii). Here a force F is applied tangentially to a wheel X of radius r and X rotates through an angle θ as shown, while F stays tangential to the wheel. Then

$$\text{work done} = \text{force} \times \text{distance AB}$$

$$= F \times r\theta = F.r \times \theta = \text{torque} \times \theta$$

So **work done W = torque × angle of rotation**. . . (4)

When the torque is in N m and θ is in radians, the work done is in J. Suppose a torque of constant value 6 N m rotates a wheel through four revolutions. Since the angle $\theta = 4 \times 2\pi = 8\pi$ rad,

$$\text{work done } W = 6 \text{ (N m)} \times 8\pi \text{ (rad)} = 151 \text{ J}$$

Consider a wheel rotating about its centre with an angular velocity of 15 rad s^{-1} and with a moment of inertia I of 2 kg m^2 about its centre. If a steady braking torque T of 6 N m brings the wheel to rest in an angle of rotation θ, then

$$\text{work done by torque} = \text{change in kinetic energy}$$

So $$T \times \theta = \tfrac{1}{2}I\omega^2 - 0$$

$$\therefore 6 \times \theta = \tfrac{1}{2} \times 2 \times 15^2$$

$$\therefore \theta = 37{\cdot}5 \text{ rad}$$

Since 1 revolution $= 2\pi$ rad,

$$\text{number of revs} = \frac{37{\cdot}5}{2\pi} = 6 \text{ (approx.)}$$

So the wheel comes to rest after about six revolutions when the braking torque is applied.

Kinetic Energy of a Rolling Object

When an object such as a cylinder or ball rolls down a plane, the object is rotating as well as moving down the plane. So it has rotational energy in addition to translational energy.

Consider a uniform cylinder C rolling down an inclined plane without slipping, Figure 4.8. The forces on C are
(a) its weight Mg acting at its central axis O,
(b) the frictional force F at the plane which prevents slipping.

The force which produces linear acceleration and translational kinetic energy down the plane $= Mg \sin \alpha - F$. The *torque* about O which produces angular acceleration and rotational kinetic energy $= F.r$, where r is the radius of the cylinder.

Since energy is a scalar quantity (one with no direction), we can add the translational and rotational kinetic energies to obtain the total energy of the cylinder. So at a given instant,

total kinetic energy $= \tfrac{1}{2}Mv^2 + \tfrac{1}{2}I\omega^2$

where I is the moment of inertia about the axis O, ω is the angular velocity about O and v is the translational velocity down the plane. If the cylinder does

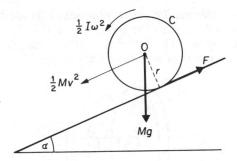

Figure 4.8 *Energy and acceleration of object rolling down-plane*

not slip, then $v = r\omega$. So

$$\text{total kinetic energy} = \tfrac{1}{2}Mv^2 + \tfrac{1}{2}I\left(\frac{v}{r}\right)^2$$

$$= \tfrac{1}{2}v^2\left(M + \frac{I}{r^2}\right)$$

Suppose the cylinder rolls from rest through a distance s along the plane. The loss of potential energy $= Mgs \sin \alpha =$ gain in kinetic energy $= \tfrac{1}{2}v^2(M + I/r^2)$ from above so.

$$v^2 = \frac{2Mgs \sin \alpha}{M + (I/r^2)}$$

But $v^2 = 2as$ where a is the *linear acceleration* down the plane. So

$$2as = \frac{2Mgs \sin \alpha}{M + (I/r^2)}$$

Thus
$$a = \frac{Mg \sin \alpha}{M + (I/r^2)} \qquad . \qquad . \qquad . \qquad . \qquad . \qquad (1)$$

A uniform *solid* cylinder of mass M and radius r has a moment of inertia $I = Mr^2/2$ about its axis. So $I/r^2 = M/2$. Substituting in (1), we find that the acceleration down the plane $a = 2g \sin \alpha/3$. A uniform *hollow* cylinder open at both ends has a moment of inertia about its axis given by $I = Mr^2$, where M is the mass and r is the radius. From (1), we find that the acceleration down the plane, $a, = g \sin \alpha/2$. So the solid cylinder would have a greater acceleration down the plane than a hollow cylinder of the same mass. If no other tests were available, we could distinguish between a solid cylinder and a hollow cylinder closed at both ends, both of the same mass, by allowing them to roll from rest down an inclined plane. Starting from the same place the cylinder which reached the bottom first would be the solid cylinder.

Measurement of Moment of Inertia of Flywheel

The moment of inertia of a flywheel W about a horizontal axle A can be determined by passing one end of some string through a hole in the axle, winding the string round the axle, and attaching a mass M to the other end of the string, Figure 4.9. The length of string is such that M reaches the floor, when released, at the same instant as the string is completely unwound from the axle.

M is released, and the number of revolutions, n, made by the wheel W up to the occasion when M strikes the ground is noted. The further number of revolutions n_1 made by W until it comes finally to rest, and the time t taken, are also observed by means of a chalk-mark W.

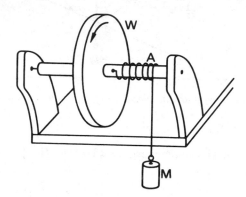

Figure 4.9 *Moment of inertia of flywheel*

Now the loss in potential energy of M = gain in kinetic energy of M + gain in kinetic energy of flywheel + work done against friction.

$$\therefore Mgh = \tfrac{1}{2}Mr^2\omega^2 + \tfrac{1}{2}I\omega^2 + nf. \qquad . \qquad . \qquad . \qquad (1)$$

where M is the mass of M, h is the distance M has fallen, r is the radius of the axle, ω is the angular velocity, I is the moment of inertia, and f is the energy per turn expended against friction. Since the energy of rotation of the flywheel when the mass M reaches the ground = work done against friction in n_1 revolutions, then

$$\tfrac{1}{2}I\omega^2 = n_1 f$$

$$\therefore f = \tfrac{1}{2}\frac{I\omega^2}{n_1}$$

Substituting for f in (1),

$$Mgh = \tfrac{1}{2}Mr^2\omega^2 + \tfrac{1}{2}I\omega^2\left(1 + \frac{n}{n_1}\right) \qquad . \qquad . \qquad . \qquad (2)$$

Since the angular velocity of the wheel when M reaches the ground is ω, and the final angular velocity of the wheel is zero after a time t, the average angular velocity $= \omega/2 = 2\pi n_1/t$. Thus $\omega = 4\pi n_1/t$. Knowing ω and the magnitude of the other quantities in (2), the moment of inertia I of the flywheel can be found.

Motor Vehicle Flywheel

In motor vehicles, the up-and-down motion of the engine piston is transferred to rotational energy of a crankshaft for turning the wheels. A flywheel, a heavy disc on the crankshaft, stores rotational energy $\tfrac{1}{2}I\omega^2$ and releases the energy when the engine piston is not delivering power. This helps to keep the engine running smoothly so the wheels turn steadily. The flywheel energy also helps to keep the wheels turning downhill if the engine is disconnected.

Summary

We conclude with a summary showing a comparison between formulae in rotational and linear motion.

Linear Motion	Rotational Motion
1. Velocity, v	Angular velocity, $\omega = v/r$
2. Momentum $= mv$	Angular momentum $= I\omega$
3. Energy $= \frac{1}{2}mv^2$	Rotational energy $= \frac{1}{2}I\omega^2$
4. Force, F, $= ma$	Torque, T, $= I\alpha$
5. $F \times t =$ momentum change	$T \times t =$ angular momentum change
6. $F \times s =$ work done $=$ k.e. change	$T \times \theta =$ work done $=$ k.e. change
7. Conservation of linear momentum on collision, if no external forces	Conservation of angular momentum on collision, if no external torques.

Example on Torque, Angular Momentum, Kinetic Energy

A uniform circular disc of moment of inertia $0.2 \, \text{kg m}^2$ and radius $0.15 \, \text{m}$ is mounted on a horizontal cylindrical axle of radius $0.015 \, \text{m}$ and negligible mass. Neglecting frictional losses in the bearings, calculate
(a) the angular velocity acquired from rest by the application, for 12 seconds of a force of 20 N tangential to the axle,
(b) the kinetic energy of the disc at the end of this period,
(c) the time required to bring the disc to rest if a constant braking force of 1 N were applied tangentially to its rim.

(a) Torque due to 20 N tangential to axle

$$= 20 \times 0.015 = 0.3 \, \text{N m}$$

Torque $\times t =$ angular momentum change

$$\therefore 0.3 \times 12 = 0.2 \times \omega$$

$$\omega = 0.3 \times 12/0.2 = 18 \, \text{rad s}^{-1}$$

(b) K.e. of disc after 12 seconds $= \frac{1}{2}I\omega^2$

$$= \frac{1}{2} \times 0.2 \times 18^2 = 32.4 \, \text{J}$$

(c) Decelerating torque $= 1 \times 0.15 = 0.15 \, \text{N m}$

Torque $\times t =$ angular momentum change

$$\therefore 0.15 \times t = 0.2 \times 18$$

$$\therefore t = 0.2 \times 18/0.15 = 24 \, \text{s}$$

EXERCISES 4A Rotational Dynamics

Multiple Choice

1 A wheel turns about its centre with a uniform angular speed of $10 \, \text{rad s}^{-1}$. Its moment of inertia about the centre is $2 \, \text{kg m}^2$. The kinetic energy of rotation of the wheel in J is

   A 10   B 12   C 20   D 40   E 100

2 A mass of 0.8 kg, at the end of a rope 0.5 m long is whirled round in a horizontal

circle at an angular speed of 10 rad s^{-1}. Its angular momentum about the centre in SI units is

A 0·1 **B** 0·2 **C** 0·4 **D** 2 **E** 4

3 A plastic horizontal disc is turning at a uniform angular speed ω_0 about its centre. When a small mass m is dropped gently on to the disc, the angular speed changes to ω. The moment of inertia of the disc about the centre is I. To find ω, we can apply

A $I\omega^0 = m\omega$ **B** the conservation of energy **C** $I = (\omega + \omega_0) m$
D $m\omega_0 = I\omega$ **E** the conservation of angular momentum

4 At the Olympic high-diving competition, a diver from the top board curves her body in order to

A dive cleanly into the water **B** spin more **C** increase her energy
D spin more slowly **E** increase her speed

Longer Questions

5 A disc of moment of inertia 10 kg m^2 about its centre rotates steadily about the centre with an angular velocity of 20 rad s^{-1}. Calculate   (i) its rotational energy, (ii) its angular momentum about the centre,   (iii) the number of revolutions per second of the disc.

6 A constant torque of 200 N m turns a wheel about its centre. The moment of inertia about this axis is 100 kg m^2. Find   (i) the angular velocity gained in 4 s,   (ii) the kinetic energy gained after 20 revs.

7 A flywheel has a kinetic energy of 200 J. Calculate the number of revolutions it makes before coming to rest if a constant opposing couple of 5 N m is applied to the flywheel.

   If the moment of inertia of the flywheel about its centre is 4 kg m^2, how long does it take to come to rest?

8 A constant torque of 500 N m turns a wheel which has a moment of inertia 20 kg m^2 about its centre. Find the angular velocity gained in 2 s and the kinetic energy gained.

9 A ballet dancer spins with 2·4 rev s^{-1} with her arms outstretched, when the moment of inertia about the axis of rotation is I. With her arms folded, the moment of inertia about the same axis becomes 0·6I. Calculate the new rate of spin.

   State the principle used in your calculation.

10 A disc rolling along a horizontal plane has a moment of inertia 2·5 kg m^2 about its centre and a mass of 5 kg. The velocity along the plane is 2 m s^{-1}.

   If the radius of the disc is 1 m, find   (i) the angular velocity,   (ii) the total energy (rotational and translational) of the disc.

11 A wheel of moment of inertia 20 kg m^2 about its axis is rotated from rest about its centre by a constant torque T and the energy gained in 10 s is 360 J. Calculate (i) the angular velocity at the end of 10 s,   (ii) T,   (iii) the number of revolutions made by the wheel before coming to rest if T is removed at 10 s and a constant opposing torque of 4 N m^{-1} is then applied to the wheel.

12 A uniform rod of length 3 m is suspended at one end so that it can move about an axis perpendicular to its length. The moment of inertia about the end is 6 kg m^2 and the mass of the rod is 2 kg. If the rod is initially horizontal and then released, find the angular velocity of the rod when   (i) it is inclined at 30° to the horizontal, (ii) it reaches the vertical.

13 A recording disc rotates steadily at 45 rev min^{-1} on a turntable. When a small mass of 0·02 kg is dropped gently onto the disc at a distance of 0·04 m from its axis and sticks to the disc, the rate of revolution falls to 36 rev min^{-1}. Calculate the moment of inertia of the disc about its centre.

   Write down the principle used in your calculation.

14 A disc of moment of inertia 0·1 kg m^2 about its centre and radius 0·2 m is released from rest on a plane inclined at 30° to the horizontal. Calculate the angular velocity after it has rolled 2 m down the plane if its mass is 5 kg.

15 A flywheel with an axle 1·0 cm in diameter is mounted in frictionless bearings and set in motion by applying a steady tension of 2 N to a thin thread wound tightly round the axle. The moment of inertia of the system about its axis of rotation is $5·0 \times 10^{-4}$ kg m^{-2}. Calculate
   (a) the angular acceleration of the flywheel when 1 m of thread has been pulled off the axle,
   (b) the constant retarding couple which must then be applied to bring the flywheel to rest in one complete turn, the tension in the thread having been completely removed. (*JMB.*)

16 Define the moment of inertia of a body about a given axis. Describe how the moment of inertia of a flywheel can be determined experimentally.
   A horizontal disc rotating freely about a vertical axis makes 100 r.p.m. A small piece of wax of mass 10 g falls vertically on to the disc and adheres to it at a distance of 9 cm from the axis. If the number of revolutions per minute is thereby reduced to 90, calculate the moment of inertia of the disc. (*N.*)

17 (a) For a rigid body rotating about a fixed axis, explain with the aid of a suitable diagram what is meant by *angular velocity*, *kinetic energy* and *moment of inertia*.
   (b) In the design of a passenger bus, it is proposed to derive the motive power from the energy stored in a flywheel. The flywheel, which has a moment of inertia of $4·0 \times 10^2$ kg m^2, is accelerated to its maximum rate of rotation $3·0 \times 10^3$ revolutions per minute by electric motors at stations along the bus route. (i) Calculate the maximum kinetic energy which can be stored in the flywheel. (ii) If, at an average speed of 36 kilometres per hour, the power required by the bus is 20 kW, what will be the maximum possible distance between stations on the level? (*N.*)

18 (a) Explain what is meant by (i) a *couple*, (ii) the *moment of a couple*. Show that a force acting along a given line can always be replaced by a force of the same magnitude acting along a parallel line, together with a couple.
   (b) A flywheel of moment of inertia 0·32 kg m^2 is rotated steadily at 120 rad s^{-1} by a 50 W electric motor. (i) Find the kinetic energy and angular momentum of the flywheel. (ii) Calculate the value of the frictional couple opposing the rotation. (iii) Find the time taken for the wheel to come to rest after the motor has been switched off. (*O.*)

Fluids in Motion

Streamlines and Velocity

A stream or river flows slowly when it runs through open country and faster through narrow openings or constrictions. As shown shortly, this is due to the fact that water is practically an incompressible fluid, that is, changes of pressure cause practically no change in fluid density at various parts.

Figure 4.10 shows a tube of water flowing steadily between X and Y, where X has a bigger cross-sectional area A_1 than the part Y, of cross-sectional area A_2. The *streamlines* of the flow represent the directions of the velocities of the

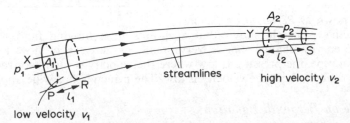

Figure 4.10 *Bernoulli's equation*

particles of fluid and the flow is uniform or laminar. Assuming the liquid is incompressible, then, if it moves from PQ to RS, the volume of liquid between P and R is equal to the volume between Q and S. So $A_1l_1 = A_2l_2$, where l_1 is PR and l_2 is QS, or $l_2/l_1 = A_1/A_2$. Hence l_2 is greater than l_1. Consequently the *velocity* of the liquid at the narrow part of the tube, where the streamlines are closer together, is greater than at the wider part Y, where the streamlines are further apart. For the same reason, slow-running water from a tap can be made into a fast jet by placing a finger over the tap to narrow the exit.

Pressure and Velocity, Bernoulli's Principle

In about 1740, Bernoulli obtained a relation between the pressure and velocity at different parts of a moving incompressible fluid. If the viscosity of the fluid is negligibly small, there are no frictional forces to overcome. In this case the work done by the pressure difference per unit volume of a fluid flowing along a pipe steadily is equal to the gain in kinetic energy per unit volume plus the gain in potential energy per unit volume.

Now the work done by pressure in moving a fluid through a distance = force × distance moved = (pressure × area) × distance moved = pressure × volume moved, assuming the area is constant at a particular place for a short time of flow. At the beginning of the pipe where the pressure is p_1, the work done per unit volume on the fluid is thus p_1; at the other end, the work done per unit volume by the fluid is likewise p_2. So the net work done *on* the fluid per unit volume $= p_1 - p_2$.

The kinetic energy per unit volume $= \frac{1}{2}$ mass per unit volume × velocity^2 = $\frac{1}{2}\rho$ × velocity^2, where ρ is the density of the fluid. Thus if v_2 and v_1 are the final and initial velocities respectively at the end and the beginning of the pipe, the kinetic energy gained per unit volume $= \frac{1}{2}\rho(v_2{}^2 - v_1{}^2)$. Further, if h_2 and h_1 are the respective heights measured from a fixed level at the end and beginning of the pipe, the potential energy gained per unit volume = mass per unit volume × g × $(h_2 - h_1) = \rho g(h_2 - h_1)$.

So from the conservation of energy,

$$p_1 - p_2 = \tfrac{1}{2}\rho(v_2{}^2 - v_1{}^2) + \rho g(h_2 - h_1)$$

$$\therefore p_1 + \tfrac{1}{2}\rho v_1{}^2 + \rho g h_1 = p_2 + \tfrac{1}{2}\rho v_2{}^2 + \rho g h_2$$

$$\therefore p + \tfrac{1}{2}\rho v^2 + \rho g h = \textbf{constant}$$

where p is the pressure at any part and v is the velocity there. So for streamline motion of an incompressible non-viscous fluid,

the sum of the pressure at any part plus the kinetic energy per unit volume plus the potential energy per unit volume there is always constant.

This is known as *Bernoulli's principle*.

Bernoulli's principle shows that at points in a moving fluid where the potential energy change $\rho g h$ is very small, or zero as in flow through a horizontal pipe, the pressure is low where the velocity is high. Conversely, the pressure is high where the velocity is low. The principle has wide applications.

Example on Bernoulli Equation

As a numerical illustration, suppose the area of cross-section A_1 of X in Figure 4.10 is 4 cm^2, the area A_2 of Y is 1 cm^2, and water flows past each section in laminar flow at the rate of 400 cm^3 s^{-1}. Then

$$\text{at X, speed } v_1 \text{ of water} = \frac{\text{vol. per second}}{\text{area}} = 100 \text{ cm s}^{-1} = 1 \text{ m s}^{-1}$$

$$\text{at Y, speed } v_2 \text{ of water} = 400 \text{ cm s}^{-1} = 4 \text{ m s}^{-1}$$

The density of water, $\rho = 1000$ kg m^{-3}. So, if p is the pressure difference,

$$p = \tfrac{1}{2}\rho(v_2{}^2 - v_1{}^2) = \tfrac{1}{2} \times 1000 \times (4^2 - 1^2) = 7 \cdot 5 \times 10^3 \text{ N m}^{-2}$$

If h is in metres, $\rho = 1000$ kg m^{-3} for water, $g = 9 \cdot 8$ m s^{-2}, then, from $p = h\rho g$

$$h = \frac{7 \cdot 5 \times 10^3}{1000 \times 9 \cdot 8} = 0 \cdot 77 \text{ m (approx.)}$$

The pressure head h is thus equivalent to $0 \cdot 77$ m of water.

Applications of Bernoulli's Principle

1. A suction effect is experienced by a person standing close to the platform at a station when a fast train passes. The fast-moving air between the person and train produces a decrease in pressure and the excess air pressure on the other side pushes the person towards the train.

2. *Filter pump.* A filter pump has a narrow section in the middle, so that a jet of water from the tap flows faster here, Figure 4.11 (i). This causes a drop in pressure near it and air therefore flows in from the side tube to which a vessel is connected. The air and water together are forced through the bottom of the filter pump.

A similar principle is used in the engine *carburettor* for vehicles. At one stage of its cycle, the engine draws in air. This rushes past the fine nozzle of a pipe connected to the petrol tank and lowers the air pressure at the nozzle, Figure 4.11 (ii). Some petrol is then forced out of the tank by atmospheric pressure

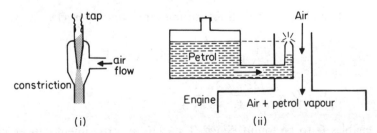

Figure 4.11 *Principle of* (i) *filter pump* (ii) *carburettor*

through the nozzle in a fine spray. The petrol vapour mixes with the air and so provides the air–petrol mixture required for the engine.

3. *Aerofoil lift.* The curved shape of an aerofoil creates a faster flow of air over its top surface than the lower one, Figure 4.12. This is shown by the closeness of the streamlines above the aerofoil compared with those below. From Bernoulli's principle, the pressure of the air below is greater than that above, and this produces the lift on the aerofoil.

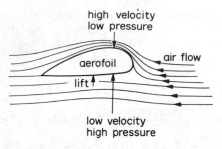

Figure 4.12 *Fluid velocity and pressure*

4. *Venturi meter*. This meter measures the volume of gas or liquid per second flowing through gas pipes or oil pipes.

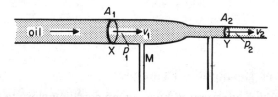

Figure 4.13 *Principle of Venturi meter*

Figure 4.13 shows the principle. A manometer M is connected between a wide section X, area A_1, and a narrow section Y, area A_2, of the horizontal pipe carrying a steady flow of oil, for example. Since the velocity v_2 at Y is greater than the velocity v_1 at X, the pressure p_2 at Y is less than the pressure p_1 at X. The manometer then has a difference in levels H of a liquid of density ρ' say.

Suppose Q is the volume per second of oil flowing at X or at Y. Then $Q = A_1v_1 = A_2v_2$. Also, from the Bernoulli principle, if ρ is the density of the oil,

$$p_1 + \tfrac{1}{2}\rho v_1{}^2 = p_2 + \tfrac{1}{2}\rho v_2{}^2$$

So
$$p_1 - p_2 = H\rho'g = \tfrac{1}{2}\rho(v_2{}^2 - v_1{}^2). \qquad . \qquad . \qquad . \quad (1)$$

But $v_2 = Q/A_2$ and $v_1 = Q/A_1$. Substituting for v_2 and v_1 in (1),

$$H\rho'g = \tfrac{1}{2}\rho\left(\frac{Q^2}{A_2{}^2} - \frac{Q^2}{A_1{}^2}\right) = \tfrac{1}{2}\rho Q\left(\frac{A_1{}^2 - A_2{}^2}{A_1{}^2 A_2{}^2}\right)$$

So
$$Q = \sqrt{\frac{2H\rho'g A_1{}^2 A_2{}^2}{\rho(A_1{}^2 - A_2{}^2)}} \qquad . \qquad . \qquad . \qquad . \qquad . \quad (2)$$

This enables Q to be found. Since $Q \propto \sqrt{H}$, an experiment can be carried out to calibrate the difference in levels H of the manometer, using (2), in terms of volume per second rate of flow.

Measurement of Fluid Velocity, Pitot-static Tube

The velocity at a point in a fluid flowing through a horizontal tube can be measured by the application of the Bernoulli equation on p. 126. In this case h is zero and so

$$p + \tfrac{1}{2}\rho v^2 = \text{constant}$$

Here p is the static pressure at a point in the fluid, that is, the pressure unaffected by its velocity. The pressure $p + \frac{1}{2}\rho v^2$ is the total or dynamic pressure, that is, the pressure which the fluid would exert if it were brought to rest by striking a surface placed normally to the velocity at the point concerned.

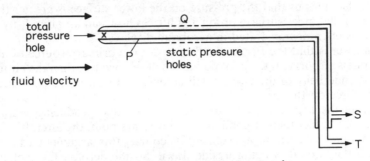

Figure 4.14 *Principle of Pitot-static tube*

Figure 4.14 illustrates the principle of a *Pitot-static tube*. The inner or Pitot tube P, named after its inventor, has an opening X at one end *normal* to the fluid velocity. A manometer connected to T would measure the total pressure $p + \frac{1}{2}\rho v^2$. The outer or static tube has holes Q in its side which are *parallel* to the fluid velocity. A manometer connected to S would measure the static pressure p. The difference in pressure, $h\rho'g$, in the two sides of a single manometer joined respectively to T and S would hence be equal to $\frac{1}{2}\rho v^2$. Thus v can be calculated from $v = \sqrt{2h\rho'g/\rho}$. In practice, corrections are applied to the manometer readings to take account of differences from the simple theory outlined.

Example on Fluid Motion in Pipe

(i) Water flows steadily along a horizontal pipe at a volume rate of $8 \times 10^{-3}\ \text{m}^3\ \text{s}^{-1}$. If the area of cross-section of the pipe is $40\ \text{cm}^2$ ($40 \times 10^{-4}\ \text{m}^2$), calculate the flow velocity of the water. (ii) Find the total pressure in the pipe if the static pressure in the horizontal pipe is 3.0×10^4 Pa, assuming that water is incompressible, non-viscous and its density is $1000\ \text{kg m}^{-3}$. (iii) What is the new flow velocity if the total pressure is 3.6×10^4 Pa?

(i) Velocity of water $= \dfrac{\text{volume per second}}{\text{area}}$

$$= \frac{8 \times 10^{-3}}{40 \times 10^{-4}} = 2\ \text{m s}^{-1}$$

(ii) Total pressure $=$ static pressure $+ \frac{1}{2}\rho v^2$

$$= 3.0 \times 10^4 + \frac{1000 \times 2^2}{2}$$

$$= 3.0 \times 10^4 + 0.2 \times 10^4 = 3.2 \times 10^4\ \text{Pa}$$

(iii) $\frac{1}{2}\rho v^2 =$ total pressure $-$ static pressure

So $\frac{1}{2} \times 1000 \times v_2 = 3.6 \times 10^4 - 3.0 \times 10^4 = 0.6 \times 10^4$

$$v = \sqrt{\frac{0.6 \times 10^4}{500}} = 3.5\ \text{m s}^{-1}$$

Aerobatics

On page 129 we show how Bernoulli's principle can be used to explain how an aerofoil section produces lift. The shape of the aerofoil section is such that the air moves faster over its upper surface than its lower one. Bernoulli's principle then tells us that the pressure on the lower surface is greater than the pressure on the upper surface, producing lift. So how can an aircraft fly upside down, as is often demonstrated in aerobatic displays?

The answer is that the Bernoulli effect is not the primary source of lift once the aircraft is airborne. If the pilot points the nose upwards, then the air meeting the tilted underbelly of the aircraft will be deflected downwards, producing an upwards force on the aircraft. In addition, with the aircraft at this angle, the thrust from the engines will be directed downwards, producing yet another source of lift. In fact there is enough total lift to maintain the aircraft's altitude, even without any aerofoil assistance. Of course, this argument can also be applied if the aircraft is flying upside down. So the aerofoil lift explained by Bernoulli's principle only really applies if the aircraft is moving horizontally. For other angles of flight, lift from the engines and deflection of the oncoming airstream can maintain flight even with the aerofoil upside-down.

EXERCISES 4B Fluid Motion

Multiple Choice

1 A steady flow of water passes along a horizontal tube from a wide section X to the narrower section Y, Figure 4A (i). Manometers are placed at P and Q at the sections. Which of the statements **A, B, C, D, E** is most correct?
   A water velocity at X is greater than at Y
   B the manometer at P shows lower pressure than at Q
   C kinetic energy per m³ of water at X = kinetic energy per m³ at Y
   D the manometer at P shows greater pressure than at Y
   E pressure at X is less than at Y

2 In Figure 4A (ii), the height of the liquid, density ρ, at X is kept constant while the liquid flows out of the narrow tube at a depth h below X. The velocity v of the liquid from the narrow tube is

   A $h\rho g$ **B** $2gh$ **C** $\sqrt{2gh}$ **D** gh **E** $\sqrt{2gh\rho}$

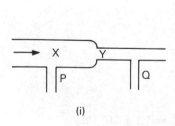

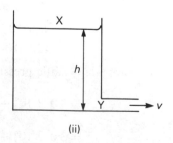

(i)   (ii)

Figure 4A

Longer Questions

3 An open tank holds water 1·25 m deep. If a small hole of cross-section area 3 cm² is made at the bottom of the tank, calculate the mass of water per second initially flowing out of the hole. ($g = 10$ m s^{-2}, density of water = 1000 kg m^{-3}.)

4 A lawn sprinkler has 20 holes of cross-section area $2 \cdot 0 \times 10^{-2}$ cm^2 and is connected to a hose-pipe of cross-section area $2 \cdot 4$ cm^2. If the speed of the water in the hose-pipe is $1 \cdot 5$ m s^{-1}, estimate the speed of the water as it emerges from the holes.

5 Show that the term $\frac{1}{2}\rho v^2$ which enters into the Bernoulli equation has the same dimensions as pressure p.

    A fluid flows through a horizontal pipe of varying cross-section. Assuming the flow is streamline and applying the Bernoulli equation $p + \frac{1}{2}\rho v^2 = $ constant, show that the pressure in the pipe is greatest where the cross-sectional area is greatest.

6 Water flows along a horizontal pipe of cross-sectional area 48 cm^2 which has a constriction of cross-sectional area 12 cm^2 at one place. If the speed of the water at the constriction is 4 m s^{-1}, calculate the speed in the wider section.

    The pressure in the wider section is $1 \cdot 0 \times 10^5$ Pa. Calculate the pressure at the constriction. (Density of water $= 1000$ kg m^{-3}.)

7 Water flows steadily along a uniform flow tube of cross-section 30 cm^2. The static pressure is $1 \cdot 20 \times 10^5$ Pa and the total pressure is $1 \cdot 28 \times 10^5$ Pa.

    Calculate the flow velocity and the mass of water per second flowing past a section of the tube. (Density of water $= 1000$ kg m^{-3}.)

8 (a) Distinguish between *static pressure*, *dynamic pressure* and *total pressure* when applied to streamline (laminar) fluid flow and write down expressions for these three pressures at a point in the fluid in terms of the fluid velocity v, the fluid density ρ, pressure p, and the height h, of the point with respect to a datum.

  (b) Describe, with the aid of a labelled diagram, the Pitot-static tube and explain how it may be used to determine the flow velocity of an incompressible, non-viscous fluid.

  (c) The static pressure in a horizontal pipeline is $4 \cdot 3 \times 10^4$ Pa, the total pressure is $4 \cdot 7 \times 10^4$ Pa, and the area of cross-section is 20 cm^2. The fluid may be considered to be incompressible and non-viscous and has a density of 10^3 kg m^{-3}. Calculate (i) the flow velocity in the pipeline, (ii) the volume flow rate in the pipeline. (*N*.)

5 Sources of Energy[1]

In this section we discuss basic basic topics in Energy. Only the core principles are given. For further details reference must be made to specialist books.

Solar Energy

There are many sources of energy in the world. The chemical energy from burning coal, oil and gas, called *fossil fuels*, is widely used. These fuels are non-renewable after they are burnt. Nuclear energy, wind power and wave power from the sea are other sources of energy. Wind and wave power are renewable sources.

All our energy comes primarily from solar energy. As stated on page 630, the sun's ultraviolet rays are absorbed in the green matter of plants and make them grow. The plants and trees centuries ago were turned into coal and oil. Water power comes from the sun. Water is evaporated by the sun and this produces the rains which fill the lakes and reservoirs. Wind power also comes from the sun. Unequal heating of air masses world-wide results in wind movement or kinetic energy.

In hot areas of the world, *solar energy* is collected by large concave mirrors and concentrated on water to produce steam to drive turbines, for example. Solar energy is also collected by solar panels on the roofs of houses for domestic heating purposes. As explained later, the sun's energy comes from nuclear reactions of the sun's elements.

Nuclear Energy, Fossil Fuels, Geothermal Energy

In the final chapter we explain how *nuclear energy* is used in nuclear reactors for generating electrical power. Heat exchangers pass the heat produced in fission to boilers, which then produce steam to drive the turbines in electrical power stations. In coal-fired or oil-fired power stations, these fossil fuels are burnt to produce the heat needed for the boilers.

Geothermal energy appears to come from nuclear energy changes deep in the earth, which produces hot dry rock. In the United States of America in California, and in Russia in the Arctic Circle, deep holes are tunnelled into the earth through hot rocks below. A depth of over 10 km has been reached. Brine or water is pumped under pressure through the holes and hot brine or steam at about 350°C can be obtained at the surface. About 24 MW (24×10^6 W) has been produced in this way for use in surrounding areas in California.

Figure 5.1 shows in block form the system needed for generating electrical power, which is widely used in industry and the home. The final power output depends on the *efficiencies* of all the machines used. For example:

(1) *Boiler.* Owing to unburnt fuel and the heat absorbed by the vessels used,

---

[1] Students limited in time for A-level study may omit this chapter at a first reading depending on the syllabus. Consult your teacher.

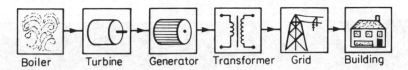

Figure 5.1 *Electrical power generation*

only a small percentage of chemical energy of the fuel–air mixture provides useful heat for changing water to steam in the boiler.

(2) *Turbine.*

(a) Wasted energy is due to frictional forces at the moving parts.

(b) The Second Law of Thermodynamics limits the maximum energy obtainable. An ideal engine has a maximum efficiency of $(T_2 - T_1)/T_2$, where T_2 is the kelvin steam temperature and T_1 is that of the condenser for the steam condensed during the cycles. See p. 662. High pressure steam at 500°C or 773 K, and a condenser at 25°C or 228 K, produces a maximum efficiency = $(773 - 288)/773 = 0.61$ or 61%. With special design, 45% efficiency may be reached in turbines, taking losses into account.

(3) *Transformer.* Although losses occur, over 90% efficiency can be reached.

(4) *Grid system.* Heat losses due to current are produced in the cables (p. 295). These may be reduced in the future, following recent promising research for superconductors at normal temperatures using ceramics.

Wind Power

Blades on a horizontal or vertical axis can be rotated like windmills by wind power. By connecting the axle to turbines, generators can be driven to produce electrical power. High wind speeds near the coast round the British Isles produce sufficient power for local areas. In Scotland and the Isle of Man, wind turbines provide a back-up for electrical power supplies and save fossil energy.

To estimate roughly the *available wind power*, suppose a fast-rotating vertical blade of 20 metres diameter or span is rotating about a horizontal axis O at its centre, and a horizontal wind of 13 m s^{-1} (30 mph) is blowing horizontally towards the blade, Figure 5.2.

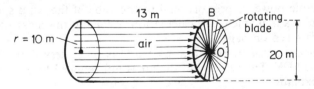

Figure 5.2 *Available wind power*

The cylindrical column of air moving to the circular area of the rotating blade in 1 second has a volume

$$= \pi \times 10^2 \times 13 = 4084 \text{ m}^3$$

Assuming the density of air is 1·3 kg m^{-3}, the mass per second reaching the blade = $4084 \times 1.3 = 5300$ kg (approx).

The kinetic energy per second of the air = $\frac{1}{2}mv^2$ per second

$$= \frac{1}{2} \times 5300 \times 13^2 = 450\,000 \text{ J s}^{-1} = 0.45 \text{ MW}$$

Plate 5A *These wind turbines are sited in the San Gorgonio Pass in California, USA*

Assuming the velocity of the air is reduced to zero at the blade, the available power would be 0·45 MW.

If r is the radius of the rotating blade, v is the wind speed and ρ is the air density, then generally

$$\text{power available} = \pi r^2 \rho v^3$$

So the available power is proportional to the cube of the wind speed and the square of the blade diameter. A calculation involving the mass per second of moving air is given on page 35.

The *power extracted* by the rotating blade is much less than the available power of 0·45 MW. The velocity of the air is not reduced to zero at the blade, as we now explain.

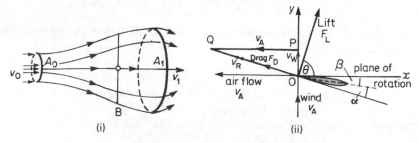

Figure 5.3 (i) *Air streamlines and rotor* (ii) *Forces on rotor*

Power Extracted, Forces on Rotor

Power extracted. Figure 5.3 (i) shows an ideal situation. The air streamlines move with velocity v_0 through a cross-sectional area A_0 some distance from the rotor or blade and leave the blade with a final velocity v_1 through a greater cross-section A_1. As we now show, the broadening of the air stream passing through the blade is due to a decrease in energy.

Assuming a steady state and an incompressible fluid, then $v_1 A_1 = v_0 A_0$, since the volume per second through each area is the same. See *Bernoulli's principle*, p. 126. So v_1 is *less* than v_0. The power extracted by the blade is therefore $(\frac{1}{2}mv_0^2 - \frac{1}{2}mv_1^2)$, where m is the mass of air per second through the areas.

Betz showed that the power extracted has a maximum value of about 60% of the available power, discussed earlier. The electrical power output is also reduced by frictional forces at the turbine and alternator. In design, too, the alternator power must be matched to the mechanical power extracted.

Forces on rotor. Figure 5.3 (ii) shows a section of the rotating blade looking along it towards the axis O. The blade is at small angle β to the plane of rotation Ox and the wind velocity v_W at the blade is in the direction Oy of the axis. The rotating blade produces an air flow of velocity v_A in the plane of rotation which is many times greater than v_W. The tip-speed at the end of the rotor is taken as numerically equal to v_A.

From the triangle of velocities OPQ, the resultant air velocity or 'relative wind' velocity v_R is in a direction OQ at O. As in an aerofoil, the *lift* force F_L on the blade is 90° to v_R. See p. 127. There is also a *drag* force F_D on the blade in the direction v_R of the relative wind velocity. The net force or thrust on the blade in the plane of rotation is therefore $(F_L \cos \theta - F_D \sin \theta)$, where θ is the angle shown. In design, by varying the shape of the aerofoil and the angle α between the blade and the relative wind velocity, the ratio F_L/F_D is made as high as possible for maximum thrust and hence maximum power output, without the blade stalling.

A modern wind turbine has a high tip-speed to v_W ratio, typically 5:1, and a small pitch angle β. In practice, v_W is the wind velocity taking into account its slowing at the blade and the blade velocity v_A is slightly increased by an induced swirl behind the rotating blade.

Design of Wind Power Generators

Wind power turbines have long blades whose cross-sections may be shaped similarly to an aerofoil. The wind is deflected by the blade and the sideways component of the reaction force causes the blade to rotate about its axis. See p. 134 and above.

The two main types of wind turbines are horizontal-axis and vertical-axis types. The *horizontal-axis* turbine has two or more long vertical blades rotating about a horizontal axis, Figure 5.4 (i). This machine needs to be turned into the wind to extract the wind energy effectively, which is a disadvantage in view of the cost of the device needed. The alternator is placed at the top of the supporting tower.

In the *vertical-axis* turbine, the blades are long and vertical and can accept wind from any direction, Figure 5.4 (ii). This is an advantage over the horizontal-axis type. The alternator can be placed on the ground at the base of the tower supporting the blades, which is another advantage over the horizontal-axis turbine. Due to centripetal forces, there are varying stresses on the blades as they rotate. Musgrove of Reading University has overcome the problem of

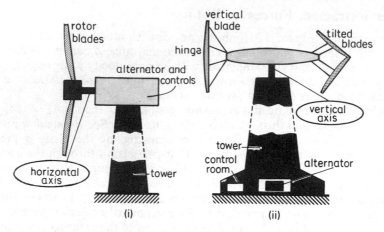

Figure 5.4 *Wind power generators*

limiting the power in very high winds by making the blade in two halves hinged at the middle. At high winds an operator tilts the two halves into an arrow-head shape, which reduces the stress.

The design of wind power systems is complex. For example, the generator power must be matched to the wind power extracted, and sensors and computers are needed in the control room for wind direction and speed. The world's most powerful wind turbine generator is installed on Orkney at one of the windiest places in the British Isles. Built by British Aerospace Wind Energy Group, the rotor has a span of 60 metres. It will turn at 34 rev min^{-1} at wind speeds between 7 m s^{-1} (15 mph) and 27 m s^{-1} (60 mph) and will produce 3 MW of electrical power. Smaller models, with a 20 m span, are operating in North Devon and in California, United States. An alternative to a large machine is a cluster of up to 100 medium machines (about 30 m blade diameter and 300 kW power) on so-called 'wind farms', which operate in Wales and Scotland.

Tidal Power

Tides are due to the gravitational pull of the moon on the waters surrounding the earth. The pull varies during the monthly cycle of rotation of the moon round the earth and the tides vary from high to low tide twice per day.

To harness tidal power, it is first necessary to build a dam across the tidal region of water. Sluice gates allow water to flow in at high tide. As the tide falls the gates are shut and water is allowed to run back through turbines to generate electricity.

Figure 5.5 shows roughly the rise and fall of the trapped water at high (H) and low (L) tide during a 24 h period and the time of operation T of the turbines.

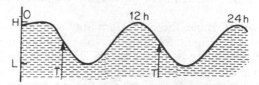

Figure 5.5 *Tidal power*

Suppose the water is trapped in a basin of area 40 km² or 40×10^6 m², and the maximum height of water is 10 m. Then, using water density = 1000 kg m⁻³ and $g = 10$ m s⁻²,

$$\text{weight of water, } mg = \text{volume} \times \text{density} \times g$$

$$= 40 \times 10^6 \times 1000 \times 10 = 4 \times 10^{12} \text{ N}$$

If the maximum height of water above low tide is h, the centre of gravity is then at a height $h/2$ above low tide. So

$$\text{gravitational potential energy change from high to low tide}$$

$$= mg \times h/2 = 4 \times 10^{12} \times 10/2 = 2 \times 10^{13} \text{ J}$$

From high to low tide, about six hours, ideally the average power obtained would be

$$\text{average power} = \frac{2 \times 10^{13} \text{ J}}{6 \times 3600 \text{ s}} = 9 \times 10^8 \text{ W (approx)}$$

$$= 900 \text{ MW}$$

With system effiencies taken into account, the available power is much less than that calculated.

In the tidal power system used in the Severn Estuary in the UK, the maximum height of the tide is about 10 m above low tide and the area of water is about 70 km² or 70×10^6 m². Ideally, this produces an energy change of 35×10^{12} J over a period of six hours, so about 1500 MW of power is obtained. Tidal power in the Bristol Channel between Cardiff and Weston-Super-Mare can produce about 6000 MW.

Wave Power

Waves in the sea have kinetic and gravitational potential energy as they rise and fall. Various systems have been designed to use wave energy and power.

To see what order of magnitude of energy and power may be obtained for a water wave, consider an ideal wave with straight wavefronts of width 1 m, Figure 5.6 (i). Suppose it is a sine wave of amplitude 1 m and wavelength 100 m, which we approximate to the *rectangular wave* in Figure 5.6 (ii) for a basic treatment.

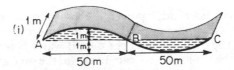

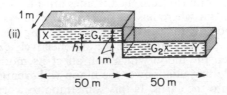

Figure 5.6 *Wave power*

When the wave crest in X falls into a trough at Y, the gravitational potential energy of the water changes by mgh, where m is the mass of water and h is the change in height of the centre of gravity G_1 to G_2. In this case $h = 1$ m. Since the density of water is 1000 kg m^{-3}, and assuming $g = 10$ m s^{-2},

$$mgh = \text{volume} \times \text{density} \times g \times h = (50 \times 1 \times 1) \times 1000 \times 10 \times 1$$

$$= 5 \times 10^5 \text{ J}$$

The wave has kinetic energy as it moves forward in addition to potential energy. It can be shown that the kinetic energy of a wave in deep water is equal to the potential energy change. So

$$\text{total energy} = 2 \times 5 \times 10^5 \text{ J} = 10^6 \text{ J}$$

Suppose the waves have a period of 5 s. In this time the water crest returns to its original height about the normal water level. So

$$\text{power} = \frac{\text{energy}}{\text{time}} = \frac{10^6 \text{ J}}{5 \text{ s}} = 200\,000 \text{ W} = 200 \text{ kW}$$

This is the power per metre wavefront. For a 1 km or 1000 m wavefront,

$$\text{power} = 1000 \times 200 \text{ kW} = 200 \text{ MW}$$

since 1 MW = 1000 kW.

The power from a sine wave is much less than this simplified rectangular wave. Further, although the wave travels with a constant speed, the water particles appear to have a circular motion which decreases rapidly with depth. More than 95% of the energy in a deep-water wave is contained in a depth $\lambda/4$, where λ is the wavelength. In winter, when waves are higher than in summer, more power power is obtained.

Britain's first wave power scheme will be tested on Islay, an island in the Hebrides. The oscillating wave will flood into a special chamber and pump air to drive a turbine generator. The plant will produce about 200 kW of electrical power for a small village at very low cost.

Tidal Power

On page 136 we consider tidal power as environment friendly, alternative energy source. The gravitational pull of the moon on the waters of the earth causes the tides to rise and fall twice per day. If water is trapped in an estuary at high tide, and then allowed to run back to the sea at low tide through turbines, electricity can be generated at apparently no cost!

In Chapter 1 we discussed the conservation of energy, which tells us that the electrical energy obtained from the turbines must have originally come from some other system. But where has it come from? The system which generates the tides is the moon orbiting the earth. By trapping the waters at high tide for a longer period of time than usual, we are effectively exerting an unnatural gravitational pull on the moon which will affect its motion. Does this mean that if we keep exploiting tidal power the moon will eventually lose its orbiting energy and crash into the earth? Is this something to worry about?

Let us do some sums. On page 199 we shall see that the total energy of a satellite of mass m in orbit around the earth of mass M at a radius of r_0 is given by $E = -GMm/2r_0$. Substituting values for the moon and the earth gives $E \approx -10^{37}/r_0$. The moon currently orbits at a radius of 400×10^3 km. If through the action of tidal power exploitation, the radius was to change by as little as 1 km, then the amount of energy lost by the moon is $(-10^{37}/400\,000 + 10^{37}/399\,999) = 6\cdot25 \times 10^{25}$ joules. A hypothetical tidal power system in the Bristol Channel could produce about 6000 MW of electricity. At this rate it would take more than 300 million years to consume $6\cdot26 \times 10^{25}$ joules and cause the moon to move 1 km nearer the earth. So, even with many such schemes operating on the earth, it would appear that tidal power is nothing to worry about!

EXERCISES 5   Energy

1 (a) In the early days of electricity generation and distribution each town in the United Kingdom had its own electricity station. Nowadays there is a National Grid.
   (i) What is meant by a National Grid? At what voltages is electricity generally transmitted over the National Grid in the United Kingdom?
   Explain briefly the advantages and disadvantages of the modern large-scale system.
   (ii) A power station delivers 50 MW of electrical power to the grid. If this is transmitted over a distance of 100 km by cables with a total resistance of $0\cdot08\,\Omega\,\text{km}^{-1}$, calculate the minimum input transmission voltage needed to ensure that no more than 2% of the power is lost in heating the cables.
   (b) The UK Department of Energy in 1988 described as 'Promising but uncertain' the following renewable energy technologies:

   —tidal energy,
   —shoreline wave energy,
   —land-based wind energy.

   Choose one of these. Describe briefly, with the aid of a sketch, the way in which the energy is harnessed and give an approximate calculation to illustrate the average power which might be available using the system you have described. (*L.*)
2 (a) 'With few exceptions, mankind derives all its energy ultimately from the sun.'
   (i) Discuss briefly three methods by which energy is derived from the sun, explaining the role of the sun in each process. Your choices should be as diverse as possible.
   Classify each energy resource you mention as renewable or non-renewable.
   (ii) State two energy resources which are independent of solar energy.
   (iii) Indicate the approximate percentage of Britain's energy need which is supplied by each resource to which you have referred in parts (i) and (ii).
   (b) The solar constant at the top of the earth's atmosphere is $1\cdot37\,\text{kW m}^{-2}$. Explain the meaning of this statement.
   Discuss quantitatively the feasibility of siting a 600 MW solar power station on land, at latitude $51°$ N, in close proximity to areas of high population density. (The maximum conversion efficiency from solar energy to electrical energy is about 10%.) (*L.*)
3 (a) The U values of four construction components are given below.

Component	U value/$\text{W m}^{-2}\,\text{K}^{-1}$
Single-glazed window	5·6
Double-glazed window	3·2
Uninsulated roof	1·9
Well-insulated roof	0·4

What do you understand by the *U value* of a component?

A house has

windows of area	24 m²
a roof of area	60 m²

The occupier heats the house for 3000 hours per year to a temperature which on average is 14 K above that of the air outside. Calculate the energy lost per year through (i) the single-glazed windows, and (ii) the uninsulated roof, expressing your answers in kWh.

If electricity costs 5·5p per unit, calculate the annual savings the occupier could make by (iii) installing double-glazing, and (iv) insulating the roof.

If double-glazing costs £3000 and roof insulation costs £100, which if either, of the two energy-saving steps would you advise the occupier to take?

(b) To cope with sudden surges in power demands, gas turbine generators are sometimes used, running permanently, but only switched into the grid when required—perhaps only a few hours each year. A far cheaper proposed solution is to switch off the gas generators and to keep a heavy flywheel spinning at high speed. When extra power is required, the wheel drives a generator and its angular speed drops to about one-third of its initial value in about three minutes. By this time the gas generators can be started up to meet the demand.

(i) Give an example of an event which might cause a major power surge.

(ii) The mass, m, of the proposed wheel is $1·6 \times 10^5$ kg and its radius, r, is 1·7 m. It will be given an angular speed, ω, of 240 rad s^{-1}. Calculate its kinetic energy, E_k, where $E_k = \frac{1}{4}mr^2\omega^2$. Comment on the use of this flywheel to cope with a sudden surge in the electrical power demand.

(iii) Name one other type of power station which is capable of being brought on-line quickly to meet a power surge. Name one type which is not capable of such a fast response. Explain the reasons for the difference in response times. (L.)

6 Elasticity, Molecular Forces, Solid Materials

Elasticity

Metals and other solids can be classified as crystalline, glassy, amorphous or polymeric. All these solid materials are widely used in engineering, industry and everyday life. In this chapter we start with metals and show how they react to forces which stretch them. We then deal generally with molecular forces. Finally, we consider the microscopic or molecular behaviour of metals and the ideas of dislocations and slip planes in a more detailed account of solid materials and their uses.

Elasticity of Metals

A bridge used by traffic is subjected to loads or forces of varying amounts. Before a steel bridge is constructed, therefore, samples of the steel are sent to a research laboratory. Here they undergo tests to find out whether the steel can withstand the loads likely to be put on it.

Figure 6.1 illustrates a simple laboratory method of investigating the property of steel we are discussing. Two long thin steel wires, P, Q, are suspended beside each other from a rigid support B, such as a girder at the top of the ceiling. The wire P is kept taut by a weight A attached to its end and carries a scale M graduated in millimetres. The wire Q carries a vernier scale V alongside the scale M. V measures the small extension e, or change in length of Q, when the load W is increased, which increases the force F in the wire.

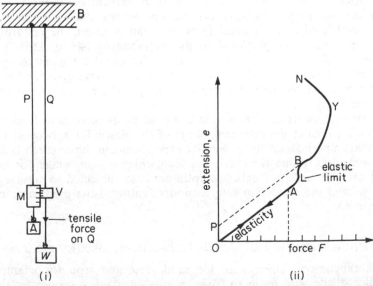

Figure 6.1 (i) *Elasticity experiment and* (ii) *result—extension against load*

Proportion and Elasticity Limits

When the extension, e, is plotted against the force F in the wire, a graph is obtained which is a *straight line* OA, followed by a curve ABY rising slowly at first and then very sharply, Figure 6.1 (ii). Up to A, about 50 N, the results show that the extension increased by 0·014 mm per newton added to the wire. A is the *proportional limit* of the wire as it is the end of the straight line OA.

Along OA, and up to L, just beyond A, the wire returned to its original length when the load was removed. The force at L is called the *elastic limit*. Along OL the metal is said to undergo changes called *elastic deformation*. Later we show that any energy stored in the metal during elastic deformation is recovered when the load is removed.

Beyond the elastic limit L, however, the wire has a permanent extension such as OP when the force is removed at B, for example, Figure 6.1 (ii). Beyond L, therefore, the wire is no longer elastic. The extension increases rapidly along the curve ABY as the force on the wire is further increased and at N the wire thins and breaks. Molecular theory, discussed on p. 154, explains why this occurs.

As we see later, when the elastic limit is passed, the energy stored in the metal is transferred to heat and is not recovered when the load is removed.

Hooke's Law

From the straight line graph OA, we deduce that

---

the extension is proportional to the force or tension in a wire if the proportional limit is not exceeded.

This is known as *Hooke's law*, after ROBERT HOOKE, founder of the Royal Society, who discovered the relation in 1676.

The extension of a wire is due to the displacement of its molecules from their mean (average) positions. So the law shows that when a molecule of the metal is slightly displaced from its mean position, the restoring force on the molecule is proportional to its displacement (see p. 156). We may therefore conclude that the molecules of a solid metal are undergoing simple harmonic motion (p. 70). Up to the elastic limit the energy gained or stored by a stretched wire is molecular potential energy, which is recovered when the load is removed.

The measurements also show that it would be dangerous to load the wire with weights greater than the magnitude of the elastic limit, because the wire then suffers a permanent strain. Similar experiments in the research laboratory enable scientists to find the maximum load which a steel bridge, for example, should carry for safety. Rubber samples are also subjected to similar experiments, to find the maximum safe tension in rubber belts used in machinery. See Plate 6A.

Yield Point, Ductile and Brittle Substances, Breaking Stress

Careful experiments show that, for mild steel and iron for example, the molecules of the wire begin to 'slide' across each other soon after the load exceeds the elastic limit. We say that the material now becomes *plastic*. This is indicated by the slight 'kink' at B beyond L in Figure 6.1 (ii), and it is called

Plate 6A *A metal specimen of a gas transmission pipe at the point of failure following a dynamic test on a fully-digital Instron 8506 Servohydraulic Testing System (rated at 1.5 MN static/1 MN dynamic)*

the *yield point* of the wire. The change from an elastic to a plastic stage is often shown by a sudden increase in the extension. In the plastic stage, the energy gained by the stretched wire is dissipated as heat and unlike the elastic stage, the energy is not recovered when the load is removed.

As the load is increased further the extension increases rapidly along the curve YN and the wire then becomes narrower and finally breaks. The *breaking stress* of the wire is the corresponding force per unit area of the narrowest cross-section of the wire, Figure 6.2.

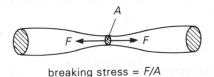

breaking stress = *F/A*

Figure 6.2 *Yield point, breaking stress*

Substances such as those just described, which lengthen considerably and undergo plastic deformation until they break, are known as *ductile* substances. Lead, copper and wrought iron are ductile. Other substances, however, break just after the elastic limit is reached; they are known as *brittle* substances. Glass and high carbon steels are brittle.

Brass, bronze, and many alloys appear to have no yield point. These materials increase in length beyond the elastic limit as the load is increased without the sudden appearance of a plastic stage.

The strength and ductility of a metal, its ability to flow, depend on defects in the metal crystal lattice. This is discussed later (p. 165).

Tensile Stress and Tensile Strain, Young Modulus

We have now to consider the technical terms used in the subject of elasticity of wires. When a force or tension F is applied to the end of a wire of cross-sectional area A along its length, Figure 6.3 (i),

$$\text{the } \textit{tensile stress} = \textit{force per unit area} = \frac{F}{A}. \qquad (1)$$

If the extension of the wire is e, and its original length is l,

$$\text{the } \textit{tensile strain} = \textit{extension per unit length} = \frac{e}{l} \qquad (2)$$

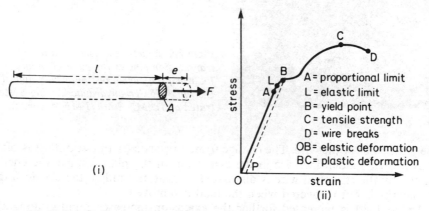

(i)

A = proportional limit
L = elastic limit
B = yield point
C = tensile strength
D = wire breaks
OB = elastic deformation
BC = plastic deformation

(ii)

Figure 6.3 (i) *Tensile stress and tensile strain* (ii) *Stress against strain, ductile material*

Suppose a 2 kg mass is attached to the end of a vertical wire of length 2 m and diameter 0·64 mm, and the extension is 0·60 mm. Then

$$F = 2 \times 9 \cdot 8 \text{ N}, \quad A = \pi \times 0 \cdot 032^2 \text{ cm}^2 = \pi \times 0 \cdot 032^2 \times 10^{-4} \text{ m}^2$$

$$\therefore \text{ tensile stress} = \frac{2 \times 9 \cdot 8}{\pi \times 0 \cdot 032^2 \times 10^{-4}} = 6 \times 10^7 \text{ Pa (N m}^{-2}) . \qquad (3)$$

and

$$\text{tensile strain} = \frac{0 \cdot 6 \times 10^{-3} \, m}{2 \, m} = 0 \cdot 3 \times 10^{-3} \qquad (4)$$

Note that 'stress' has units Pa (1 pascal, $\text{Pa} = 1 \text{ N m}^{-2}$); 'strain' has no units because it is the ratio of two lengths. Figure 6.3 (ii) shows the general stress–strain graph for a ductile material.

Under elastic conditions, a *modulus of elasticity* of the wire, called the **Young modulus** (E), is defined as the ratio

$$E = \frac{\text{tensile stress}}{\text{tensile strain}} \qquad (5)$$

So

$$E = \frac{F/A}{e/l}$$

Using (3) and (4), when the elastic limit is not exceeded,

$$E = \frac{6 \times 10^7}{0.3 \times 10^{-3}}$$

$$= 2.0 \times 10^{11} \text{ Pa} = 2.0 \times 10^5 \text{ MPa}$$

where $1 \text{ MPa} = 10^6$ (million) Pa. $1 \text{ kPa} = 10^3$ Pa.

Dimensions of Young Modulus

As stated before, the 'strain' of a wire has no dimensions of mass, length, or time, since, by definition, it is the ratio of two lengths. Now

$$\text{dimensions of stress} = \frac{\text{dimensions of force}}{\text{dimensions of area}}$$

$$= \frac{MLT^{-2}}{L^2} = ML^{-1}T^{-2}$$

$\therefore$ dimensions of the Young modulus, E,

$$= \frac{\text{dimensions of stress}}{\text{dimensions of strain}} = ML^{-1}T^{-2}$$

Measuring Young Modulus

The magnitude of the Young modulus for a material in the form of a wire can be found with the apparatus illustrated in Figure 6.1 (i), p. 141, to which the reader should now refer. The following practical points should be specially noted, remembering that the elastic limit must not be exceeded:

(1) The use of *two wires*, P, Q, *of the same material and length*, eliminates the correction for (i) the yielding of the support when loads are added to Q, (ii) changes of temperature.

(2) The wire is made *thin* so that a moderate load of several kilograms produces a large tensile stress. The wire is also made *long* so that a measurable extension is produced and the error in the measurement is then small.

(3) Both wires should be free of kinks, otherwise the increase in length cannot be accurately measured. The wires are straightened by attaching suitable weights to their ends, as shown in Figure 6.1 (i).

(4) A vernier scale is necessary to measure the extension of the wire since this is always small. The 'original length' of the wire is measured from the top B *to the vernier V* by a ruler, since an error of 1 millimetre is negligible compared with an original length of several metres. For very accurate work, the extension can be measured by using a spirit level between the two wires, and adjusting a vernier screw to restore the spirit level to its original reading after a load is added.

(5) The diameter of the wire must be found by a micrometer screw gauge at several places, and the average value then calculated. The area of cross-section, A, $= \pi r^2$, where r is the radius.

(6) The readings on the vernier are also taken when the load is gradually removed in steps of 1 kilogram; they should be very nearly the same as the readings on the vernier when the weights were added, showing that the elastic limit was not exceeded.

Calculation and Magnitude of Young Modulus

From the measurements, a graph can be plotted of the force F in newtons against the average extension s in metres. A straight line graph AB passing through the origin is drawn through all the points, Figure 6.4.

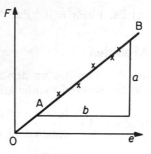

Figure 6.4 *Calculation of E*

Now
$$E = \frac{F/A}{e/l} = \frac{F}{e} \times \frac{l}{A}$$

with the usual notation. The value of F/e is the gradient, a/b, of the straight line AB and this can be found. So knowing F/e, the original length l of the wire and the cross-sectional area A ($\pi d^2/4$, where d is the diameter of the wire), E can be calculated.

Mild steel (0·2% carbon) has a Young modulus value of about $2\cdot0 \times 10^{11}$ Pa, copper has a value of about $1\cdot2 \times 10^{11}$ Pa; and brass a value about $1\cdot0 \times 10^{11}$ Pa.

The breaking stress of cast-iron is about $1\cdot5 \times 10^8$ Pa or 150 MPa; the breaking stress of mild steel is about $4\cdot5 \times 10^8$ Pa or 450 MPa.

At Royal Ordnance and other Ministry of Supply factories, tensile testing is carried out by placing a sample of the material in a machine known as an *extensometer*, which applies stresses of increasing value along the length of the sample and automatically measures the slight increase in length. See Plate 6A. When the elastic limit is reached, the pointer on the dial of the machine flickers, and soon after the yield point is reached the sample becomes thin at some point and then breaks. A graph showing the load against extension is recorded automatically by a moving pen while the sample is undergoing test.

Example on Loaded Wire

Find the maximum load which may be placed on a steel wire of diameter 1·0 mm if the permitted strain must not exceed $\frac{1}{1000}$ and the Young modulus for steel is $2\cdot0 \times 10^{11}$ Pa.

We have
$$\frac{\text{max. stress}}{\text{max. strain}} = 2 \times 10^{11}$$

$$\therefore \text{max. stress} = \tfrac{1}{1000} \times 2 \times 10^{11} = 2 \times 10^8 \text{ Pa}$$

Now area of cross-section in m^2 = $\dfrac{\pi d^2}{4} = \dfrac{\pi \times 1\cdot0^2 \times 10^{-6}}{4}$

and
$$\text{stress} = \frac{\text{load } F}{\text{area}}$$

$$\therefore F = \text{stress} \times \text{area} = 2 \times 10^8 \times \frac{\pi \times 1 \cdot 0^2 \times 10^{-6}}{4} \text{ N}$$

$$= 157 \text{ N}$$

Force in Bar due to Contraction or Expansion

When a metal rod is heated, and then prevented from contracting as it cools, a considerable force is exerted at the ends of the rod. We can derive a formula for the force if we consider a rod of Young modulus E, a cross-sectional area A, a linear expansivity of magnitude α, and a decrease in temperature of $\theta°C$ when the rod cools. Then, if the original length of the rod is l, the decrease in length e if the rod were free to contract $= \alpha l \theta$ since, by definition, α is the change in length per unit length per degree temperature change.

Now
$$E = \frac{F/A}{e/l}$$

$$\therefore F = \frac{EAe}{l} = \frac{EA\alpha l\theta}{l}$$

$$\therefore F = EA\alpha\theta$$

As an illustration, suppose a steel rod of cross-sectional area $2 \cdot 0 \text{ cm}^2$ is heated to $100°C$, and then prevented from contracting when it is cooled to $10°C$. The linear expansivity of steel $= 12 \times 10^{-6} \text{ K}^{-1}$ and Young modulus $= 2 \cdot 0 \times 10^{11}$ Pa. Then
$$A = 2 \text{ cm}^2 = 2 \times 10^{-4} \text{ m}^2, \quad \theta = 90°C$$

$$\therefore F = EA\alpha\theta = 2 \times 10^{11} \times 2 \times 10^{-4} \times 12 \times 10^{-6} \times 90 \text{ N}$$

$$= 43\,200 \text{ N}$$

Energy Stored in a Stretched Wire

Suppose that a wire has an original length l and is stretched by a length e when a force F is applied at one end. If the elastic limit is not exceeded, the extension is directly proportional to the applied load (p. 142). So the force *in the wire* has increased uniformly in magnitude from zero to F. The average force in the wire while stretching was therefore $F/2$. Now

$$\text{work done} = \text{force} \times \text{distance in direction of force}$$

$$\therefore \textbf{work} = \textbf{average force} \times \textbf{extension}$$

$$= \tfrac{1}{2}Fe \qquad . \qquad . \qquad . \qquad . \qquad . \qquad . \qquad (1)$$

This is the amount of energy stored in the wire. It is the gain in molecular potential energy of the molecules due to their displacement from their mean positions. The formula $\frac{1}{2}Fe$ gives the energy in *joules* when F is in newtons and e is in metres.

Further, since $F = EAe/l$, with the usual symbols

$$\text{energy } W = \tfrac{1}{2}EA\,\frac{e^2}{l}$$

Suppose that a vertical wire, suspended from one end, is stretched by attaching a weight of 20 N to the lower end. If the weight extends the wire by 1 mm or 1×10^{-3} m, then

$$\text{energy gained by wire} = \tfrac{1}{2}Fe = \tfrac{1}{2} \times 20 \times 1 \times 10^{-3}$$

$$= 10^{-2}\text{ J} = 0\cdot01\text{ J}$$

The gravitational potential energy (mgh) lost by the weight in dropping a distance of 1 mm $= 20 \times 1 \times 10^{-3}$ J $= 0\cdot02$ J. Half of this energy, $0\cdot01$ J, is the molecular energy gained by the wire; the remainder is the energy dissipated as heat in the wire when the weight loses its kinetic energy after falling and comes to rest.

Graph of *F* Against *e* and Energy Measurement

The energy in the wire when it is stretched can also be found from the graph of F against e, Figure 6.5. Suppose the wire extension is e_1 when a force F_1 is applied.

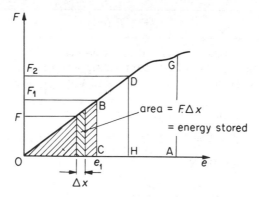

Figure 6.5 *Energy in stretched wire*

At some stage before the extension e_1 is reached suppose that the force in the wire is F and that the wire now extends by a very small amount Δx, as shown. Then over this small extension,

$$\text{energy in wire} = \text{work done} = F \cdot \Delta x$$

Now $F \cdot \Delta x$ is represented by the small *area* between the axis of e and the graph, shown shaded in Figure 6.5. So the **total work done between zero extension and e_1 is the area OBC between the graph and the axis of e.** The area of the triangle OBC $= \tfrac{1}{2}$ base $\times$ height $= \tfrac{1}{2}F_1e_1$, which is in agreement with our formula on p. 147 for the energy stored in the wire.

The area result is a general one. It can be used for both the linear (elastic) and the non-linear (non-elastic) parts of the force F against e graph. So in Figure 6.5, the work done when the force F_1 (extension e_1) is increased to F_2 (extension e_2) is the area of the trapezium BDHC. If the extension occurs from

O to A, which is beyond the elastic limit, the work done is still equal to the area of OGA.

It should be noted that the energy in the wire is equal to the area between the graph and the e-axis because F is plotted vertically and e is plotted horizontally. If e is plotted vertically and F is plotted horizontally, the energy in the wire would then be the area between the graph and the *vertical* or e-axis.

Energy per Unit Volume of Wire

When the elastic limit is not exceeded, the energy per unit volume of a stretched wire is given by a useful simple formula, as we now see.

The energy stored $= \frac{1}{2}Fe$ and the volume of the wire $= Al$, where A is the cross-sectional area and l is the length of the wire. So

$$\text{energy per unit volume} = \frac{1}{2}\frac{F \cdot e}{A \cdot l} = \frac{1}{2} \times \left(\frac{F}{A}\right) \times \left(\frac{e}{l}\right)$$

So **energy per unit volume $= \frac{1}{2}$ stress $\times$ strain**

So if the stress in a wire is 2×10^7 Pa and the strain is 10^{-2}, then

$$\text{energy per unit volume} = \frac{1}{2} \times 2 \times 10^7 \times 10^{-2}$$

$$= 10^5 \text{ J m}^{-3}$$

Examples on Young Modulus

Energy
1 A uniform steel wire of length 4 m and area of cross-section 3×10^{-6} m^2 is extended 1 mm. Calculate the energy stored in the wire if the elastic limit is not exceeded. (Young modulus $= 2 \cdot 0 \times 10^{11}$ Pa.)

(*Analysis* Energy stored $= \frac{1}{2}F \times e$)

$$\text{Stretching force } F = EA\frac{e}{l}$$

So $\qquad$ energy stored $= \frac{1}{2}Fe = \frac{1}{2}\dfrac{EAe^2}{l}$

$$= \frac{1}{2} \times \frac{2 \times 10^{11} \times 3 \times 10^{-6} \times (1 \times 10^{-3})^2 \text{ J}}{4}$$

$$= 0 \cdot 075 \text{ J}$$

Young modulus
2 Two vertical wires X and Y, suspended at the same horizontal level, are connected by a light rod XY at their lower ends, Figure 6.6. The wires have the same length l and cross-sectional area A. A weight of 30 N is placed at O on the rod, where XO:OY $= 1:2$. Both wires are stretched and the rod XY then remains horizontal.

If the wire X has a Young modulus E_1 of $1 \cdot 0 \times 10^{11}$ N m^{-2}, calculate the Young modulus E_2 of the wire Y assuming the elastic limit is not exceeded for both wires.

(*Analysis* (i) Since the rod remains horizontal, extension e_1 of X = extension e_2 of Y. (ii) Forces F_1 and F_2 on wires can be found by moments. (iii) Use $e = Fl/EA$.)

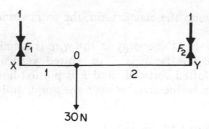

Figure 6.6 *Example on Young modulus*

By moments about X, $F_2 \times 3 = 30 \times 1$, so $F_2 \times 10$ N
So force at X, $F_1 = 30 - 10 = 20$ N
Since rod XY remains horizontal when wires are stretched, extension e_1 of
X = extension e_2 of Y = e say.

Now, from $\qquad F = EAe/l, \quad = Fl/EA$

So
$$\frac{F_1 l}{E_1 A} = \frac{F_2 l}{E_2 A}$$

$$\therefore E_2 = \frac{F_2}{F_1} \times E_1 = \frac{10}{20} \times 1.0 \times 10^{11}$$

$$= 5.0 \times 10^{10} \text{ Pa}$$

3 A rubber cord of a catapult has a cross-sectional area of 2 mm² and and initial length of 0.20 m, and is stretched to 0.24 m to fire a small object of mass 10 g (0.01 kg). Calculate the initial velocity of the object when it just leaves the catapult.

Assume the Young modulus for rubber is 6×10^8 Pa and that the elastic limit is not exceeded.

(*Analysis* (i) Kinetic energy of object $= \frac{1}{2}mv^2 =$ energy stored in stretched rubber. (ii) Energy stored $= \frac{1}{2}Fe$.)

$$\text{Force stretching rubber, } F = EA \frac{e}{l} = \frac{6 \times 10^8 \times 2 \times 10^{-6} \times 0.04}{0.20} = 240 \text{ N}$$

since $A = 2 \text{ mm}^2 = 2 \times 10^{-6} \text{ m}^2$ and $e = 0.24 - 0.20 = 0.04$ m

$$\therefore \text{ energy stored in rubber} = \frac{1}{2}Fe \times 240 \times 0.04 = 4.8 \text{ J}$$

$$\text{Kinetic energy of object} = \frac{1}{2}mv^2 = \frac{1}{2} \times 0.01 \times v^2$$

$$\therefore \frac{1}{2} \times 0.01 \times v^2 = 4.8$$

$$\therefore v = \sqrt{\frac{4.8 \times 2}{0\partial01}} = 31 \text{ m s}^{-1}$$

You should know:

1 Hooke's law: extension $\propto$ force in wire, provided proportional limit not exceeded.
2 Tensile stress $= F/A$, unit Pa (N m^{-2}). Tensile strain $= e/l$ (no unit) Young modulus, $E =$ stress/strain $= F/e \times l/A$. Unit of $E =$ Pa or kPa or MPa.
3 If proportional limit not exceeded, force in wire $F = EAe/l$.
Energy in stretched wire $= area$ between F against e graph and e-axis
Energy per unit volume $= \frac{1}{2}$ stress $\times$ strain if Hooke's law obeyed.
4 Measuring Young modulus for steel: Two similar wires needed to eliminate temperature effects and yielding of support, vernier between wires, micrometer gauge for diameter of wire. Hooke's law must be obeyed.
5 *Elastic* behaviour when metal returns to original length after load removed; energy then recovered. *Plastic* behaviour when metal permanently strained after load removed; energy transferred to heat after elastic limit exceeded.
6 Breaking stress = stress when wire thins and breaks. Brittle materials break at low strains, ductile materials have large plastic behaviour.

EXERCISES 6A   Young Modulus and Energy in Stretched Metals

Multiple Choice

1 Figure 6A shows how the extension e of a wire varies with the force F applied. If the original length of the wire is l, its cross-sectional area A and its Young modulus is E, the gradient of the graph is

A El/A B EA^2/l C El^2A D EA/l E A/lE

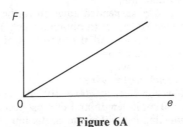

Figure 6A

2 A steel wire of length 2 m and cross-sectional area 0.80×10^{-6} m^2 has a Young modulus of 2.0×10^{11} Pa. The upper end of the wire is fixed and a load of 4 N is tied at the other end. The extension in mm of the wire is

A 0·002   B 0·050   C 0·50   D 1·00   E 5·00

3 A vertical wire fixed at one end is stretched by a load of 40 N at the lower end. The extension is 3×10^{-3} m, the cross-sectional area of the wire is 2.0×10^{-6} m^2, and the original length of the wire is 2·0 m. Assuming Hooke's law holds, the molecular potential energy in J of the stretched wire is

A 0·024   B 0·048   C 0·060   D 6·00   E 24

Longer Questions

4 A wire 2 m long and cross-sectional area 10^{-6} m² is stretched 1 mm by a force of 50 N in the elastic region. Calculate (i) the strain, (ii) the Young modulus, (iii) the energy stored in the wire.

5 Figure 6B shows the variation of F, the load applied to two wires X and Y, and their extension e. The wires are both iron and have the same length. (i) Which

Figure 6B

wire has the smaller cross-section? (ii) Explain how you would use the graph for X to obtain a value for the Young modulus of iron, listing the additional measurements needed.

6 Define *tensile stress, tensile strain, Young modulus*. What are the units and dimensions of each?

A force of 20 N is applied to the ends of a wire 4 m long, and produces an extension of 0·24 mm. If the diameter of the wire is 2 mm, calculate the stress on the wire, its strain, and the value of the Young modulus.

7 What force must be applied to a steel wire 6 m long and diameter 1·6 mm to produce an extension of 1 mm? (Young modulus for steel = $2·0 \times 10^{11}$ Pa.)

8 Find the extension produced in a copper wire of length 2 m and diameter 3 mm when a force of 30 N is applied. (Young modulus for copper = $1·1 \times 10^{11}$ Pa.)

9 A spring is extended by 30 mm when a force of 1·5 N is applied to it. Calculate the energy stored in the spring when hanging vertically supporting a mass of 0·20 kg if the spring was unstretched before applying the mass. Calculate the loss in potential energy of the mass. Explain why these values differ. (*L.*)

10 In an experiment to measure the Young modulus for steel a wire is suspended vertically and loaded at the free end. In such an experiment,
(a) why is the wire long and thin,
(b) why is a second steel wire suspended adjacent to the first?
Sketch the graph you would expect to obtain in such an experiment showing the relation between the applied load and the extension of the wire. Show how it is possible to use the graph to determine
(a) the Young modulus for the wire,
(b) the work done in stretching the wire.
If the Young modulus for steel is $2·00 \times 10^{11}$ Pa, calculate the work done in stretching a steel wire 100 cm in length and of cross-sectional area 0·030 cm² when a load of 100 N is slowly applied within the elastic limit being reached. (*N.*)

11 A muscle exerciser consists of two steel ropes attached to the ends of a strong spring contained in a telescopic tube, Figure 6C(i). When the ropes are pulled sideways in opposite directions, as shown in the simplified diagram, the spring is compressed.

The spring has an uncompressed length of 0·80 m. The force F (in N) required to compress the spring to a length x (in m) is calculated from the equation $F = 500(0·80 - x)$.

The ropes are pulled with equal and opposite forces, P, so that the spring is compressed to a length of 0·60 m and the ropes make an angle of 30° with the length of the spring.
(a) Calculate (i) the force, F, (ii) the work done in compressing the spring.
(b) By considering the forces at A or B, calculate the tension in each rope.
(c) By considering the forces at C or D, calculate the force, P. (*N.*)

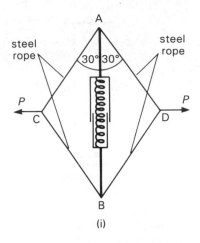

(i)

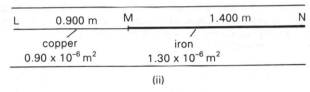

(ii)

Figure 6C

12 A copper wire LM is fused at one end, M, to an iron wire MN, Figure 6C (ii). The copper wire has length 0·900 m and cross-section 0.90×10^{-6} m². The iron wire has length 1·400 and cross-section 1.30×10^{-6} m². The compound wire is stretched; its total length increases by 0·0100 m. Calculate
(a) the ratio of the extensions of the two wires, (b) the extension of each wire, (c) the tension applied to the compound wire.
(The Young modulus of copper = 1.30×10^{11} Pa. The Young modulus of iron = 2.10×10^{11} Pa.) (L.)

13 What is meant by saying that a substance is 'elastic'?
A vertical brass rod of circular section is loaded by placing a 5 kg weight on top of it. If its length is 50 cm, its radius of cross-section 1 cm, and the Young modulus of the material 3.5×10^{10} Pa, find
(a) the contraction of the rod, (b) the energy stored in it. (C.)

14 Explain the terms *stress, strain, modulus of elasticity* and *elastic limit*. Derive an expression in terms of the tensile force and extension for the energy stored in a stretched rubber cord which obeys Hooke's law.
The rubber cord of a catapult has a cross-sectional area 1·0 mm² and a total unstretched length 10·0 cm. It is stretched to 12·0 cm and then released to project a missile of mass 5·0 g. From energy considerations, or otherwise, calculate the velocity of projection, taking the Young modulus for the rubber as 5.0×10^{8} Pa. State the assumptions made in your calculation. (L.)

15 (a) Describe an experiment using two long, parallel, identical wires to determine the Young modulus for steel. Explain why it is necessary to use two such wires. Indicate what quantities you would measure and what measuring instrument you would use in each case. State what graph you would plot, and show how it is used to calculate the Young modulus.
(b) A light rigid bar is suspended horizontally from two vertical wires, one of steel and one of brass, as shown in Figure 6D. Each wire is 2·00 m long. The diameter of the steel wire is 0·60 mm and the length of the bar AB is 0·20 m. When a mass of 10·0 kg is suspended from the centre of AB the bar remains horizontal.

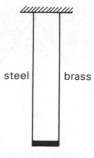

Figure 6D

   (i) What is the tension in each wire?
  (ii) Calculate the extension of the steel wire and the energy stored in it.
 (iii) Calculate the diameter of the brass wire.
 (iv) If the brass wire were replaced by another brass wire of diameter 1·00 mm, where should the mass be suspended so that AB would remain horizontal? the Young modulus for steel = $2·0 \times 10^{11}$ Pa, the Young modulus for brass = $1·0 \times 10^{11}$ Pa. (*N*.)

Molecular Forces

Particle Nature of Matter, Molecules

Matter is made up of many millions of molecules, which are tiny particles of linear dimensions about 3×10^{-10} m. Evidence for the existence of molecules is given by experiments demonstrating *Brownian motion*, with which we assume the reader is familiar. One example of Brownian motion is the random motion of smoke particles in air, which, using a microscope, can be seen moving to-and-fro. This is due to continuous bombardment of a tiny smoke particle by numerous air molecules all round it. The air molecules move with different velocities in different directions. The resultant force on the small smoke particle is therefore unbalanced, and irregular in magnitude and direction. Larger particles do not show Brownian motion when struck on all sides by air molecules. The resultant force is then relatively negligible.

    More evidence of the existence of molecules is supplied by the successful predictions made by the *kinetic theory of gases*. This theory assumes that a gas consists of millions of separate particles or molecules moving about in all directions (p. 669). *X-ray diffraction patterns* of crystals also provide evidence for the particle nature of matter (p. 853). The symmetrical patterns of spots obtained are those which one would expect from a three-dimensional grating or lattice formed from particles. A smooth continuous medium would not give a diffraction pattern of spots.

Size and Separation of Molecules

The size of atoms and molecules can be estimated in several different ways. By allowing an oil drop of known volume to spread on water, for example, and measuring the area covered after the oil has spread, an upper limit of about 5×10^{-9} m is obtained for the size of an oil molecule. X-ray diffraction experiments enable the interatomic spacing between atoms in a crystal to be accurately found. The results are of the order of 3×10^{-10} m or 0.3 nm (nanometre).

A simple calculation shows the order of magnitude of the enormous number of molecules present in a small volume. One gram of water occupies 1 cm^3. One mole has a mass of 18 g, and so occupies a volume of 18 cm^3 or 18×10^{-6} m^3. Assuming the diameter of a molecule is 3×10^{-10} m, its volume is roughly $(3 \times 10^{-10})^3$ or 27×10^{-30} m^3. So the number of molecules in one mole $= 18 \times 10^{-6}/(27 \times 10^{-30}) = 7 \times 10^{23}$ approximately.

The *Avogadro constant*, N_A, is the number of molecules in one mole of a substance. Accurate values show that $N_A = 6.02 \times 10^{23}$ mol^{-1}, or 6.02×10^{26} kmol^{-1}, where 'kmol' represents a kilomole, 1000 moles.

The order of separation of molecules in liquids is about the same as in solids. We can calculate the separation of gas molecules at standard pressure from the fact that a mole of any gas occupies about 22.4 litres or 22.4×10^{-3} m^3 at $0°$C and 10^5 Pa pressure. Since one mole contains about 6×10^{23} molecules, then, roughly, taking the cube root of the volume per molecule,

$$\text{average separation} = \sqrt[3]{\frac{22.4 \times 10^{-3}}{6 \times 10^{23}}} \text{ m}$$

$$= 33 \times 10^{-10} \text{ m (approx.)}$$

This is about 10 times the separation of molecules in solids or liquids.

The lightest atom is hydrogen. Since about 6×10^{23} hydrogen molecules have a mass of 2 g or 2×10^{-3} kg, and each hydrogen molecule consists of two atoms, then

$$\text{mass of hydrogen atom} = \frac{2 \times 10^{-3}}{2 \times 6 \times 10^{23}} = 1.7 \times 10^{-27} \text{ kg (approx.)}$$

Heavier atoms have masses in proportion to their relative atomic masses.

Example on Molecular Separation

Estimate the order of separation of atoms in aluminium metal, given the density is 2700 kg m^{-3}, the molar mass of aluminium is 27 and the Avogadro constant is 6×10^{23} mol^{-1}.

1 mole of aluminium has a mass of 27 g. From the density value,

$$1 \text{ m}^3 \text{ of aluminium has } \frac{2700 \times 10^3}{27} \text{ or } 10^5 \text{ moles}$$

Now 1 mole contains 6×10^{23} molecules or atoms of aluminium

So $\quad\quad 6 \times 10^{23} \times 10^5$ atoms occupy a volume of 1 m^3

Then $\quad\quad$ volume per atom $= \dfrac{1}{6 \times 10^{28}} = 1.7 \times 10^{-29}$ m^3

The volume occupied per atom is of the order of d^3, where d is the separation of the atoms. So, approximately,

$$d = \sqrt[3]{1.7 \times 10^{-29}} \text{ m}$$

$$= 2.6 \times 10^{-10} \text{ m}$$

Intermolecular Forces

The forces which exist between molecules can explain many of the bulk properties of solids, liquids and gases. These intermolecular forces arise from two main causes:

(1) The *potential energy* of the molecules, which is due to interactions with surrounding molecules (this is principally electrical in origin).

(2) The *thermal energy* of the molecules—this is the kinetic energy of the molecules and it depends on the temperature of the substance concerned.

We shall see later that the particular state or phase in which matter appears—that is, solid, liquid or gas—and the properties it then has, are determined by the relative magnitudes of these two energies.

Potential Energy and Force

In bulk, matter consists of numerous molecules. To simplify the situation, Figure 6.7 shows the variation of the mutual potential energy V between two molecules at a distance r apart.

Along the part BCD of the curve, the potential energy V is negative. Along the part AB, the potential energy V is positive. Generally, V can be written approximately as

$$V = \frac{a}{r^p} - \frac{b}{r^q} \qquad . \qquad . \qquad . \qquad . \qquad (1)$$

where p and q are powers of r, and a and b are constants. The positive term with the constant a indicates a repulsive force and the negative term with the constant b an attractive force, as discussed shortly.

There are different kinds of bonds or forces between atoms and molecules in solids, depending on the nature of the solid. In an *ionic solid*, for example sodium chloride, V can be approximated by

$$V = \frac{a}{r^9} - \frac{b}{r} \qquad . \qquad . \qquad . \qquad . \qquad (2)$$

The force F between molecules is generally given by $F = -dV/dr$, the negative *potential gradient* in the molecular field (see p. 228). By differentiating (2), it follows that, for the two ions,

$$F = \frac{9a}{r^{10}} - \frac{b}{r^2} \qquad . \qquad . \qquad . \qquad . \qquad (3)$$

The $+$ve term in (3) indicates a *repulsive* force between two molecules since the force is in the direction of increasing r. This is the force along LM in Figure 6.7. The $-$ve term in (3) indicates an *attractive* force since the force is opposite to the direction of increasing r. This force acts along MPQ in Figure 6.7. As shown, F decreases along PQ with increasing separation, r, of the two molecules.

Properties of Solids from Molecular Theory

Several properties of a model solid can be deduced or calculated from the potential–separation ($V - r$) graph or the force–separation ($F - r$) graph.

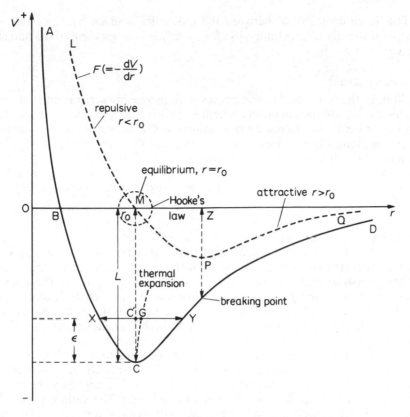

Figure 6.7 *Intermolecular potential energy and force*

Equilibrium spacing of molecules

The value of r when the potential energy V is a minimum corresponds to the stable or equilibrium spacing between the molecules. At the absolute zero, where the thermal energy is zero, this corresponds to C in Figure 6.7, or a separation $r = r_0$. At this separation or spacing OM, the repulsive and attractive forces balance, that is, $F = 0$ at M. Hence, from (3), r_0 is given, for the ions concerned, by

$$r_0 = \left(\frac{9a}{b}\right)^{1/8}$$

The value of r_0 for solids is about 2 to 5×10^{-10} m.

If the separation r of the molecules is slightly increased beyond M from r_0, the attractive force along MPQ between them will restore the molecules to their equilibrium position after the external force is removed. If the separation is decreased from r_0, the repulsive force along LM will restore the molecules to their equilibrium position after the external force is removed. So the molecules of a solid *oscillate* about their equilibrium or mean position.

Elasticity and Hooke's law

Near the equilibrium position r_0, the graph of F against r is approximately a straight line, Figure 6.7. This means that the extension is proportional to the applied force (Hooke's law, p. 142).

The 'force constant', k, between the molecules is given by $F = -k(r - r_0)$, where r is slightly greater than r_0. So $k = -dF/dr = -$ gradient of the tangent to the curve at $r = r_0$.

Breaking strain

So long as the restoring force increases with increasing separation from $r = r_0$, the molecules will remain bound together. This is the case from $r = r_0$ or OM to $r = $ OZ in Figure 6.7. Beyond a separation $r = $ OZ, however, the restoring force *decreases* along PQ with increasing separation. OZ is, therefore, the separation between the molecules at the *breaking point* of the solid (see p. 144). It corresponds to the value of r for which $dF/dr = 0$, that is, to the point below Z on the $V - r$ curve. The *breaking strain* = extension$/r_0 = (x - r_0)/r_0$, where $x = $ OZ.

Thermal expansion

Molecules remain stationary at absolute zero, since their thermal energy is then zero. This corresponds to the point C of the energy curve in Figure 6.7. At a higher temperature, the molecules have some energy, ε, above the minimum value, as shown. Hence they oscillate between points such as X and Y. Since the V–r curve is not symmetrical, the mean position G of the oscillation is on the right of C', as shown. This corresponds to a greater separation than r_0. So the solid *expands* when its thermal energy is increased.

At a slightly higher temperature, the mean position moves further to the right of G and so the solid expands further. When the energy equals CM, the latent heat value, the energy enables the molecules to break completely the bonds of attraction which keep them together in a bound state. The molecules then have little or no interaction and so form a *gas* with separation very much greater than OQ.

Latent Heat of Vaporisation

Inside a liquid, molecules continually break and reform bonds with neighbours. The 'latent heat of vaporisation L' of a liquid is the energy needed to break all the bonds between its molecules.

Suppose ε is the energy required to separate a particular molecule X from its nearest neighbour, that is, the energy per pair of molecules. If there are n nearest neighbours per molecule, and we neglect the effect of the other molecules, then the energy to break the bonds between X and its neighbours is $n\varepsilon$.

With a mole of liquid, there are N_A molecules inside it, where N_A is the Avogadro constant. The number of pairs of molecules is $\frac{1}{2}N_A$. So the energy required to break the bonds of all the molecules at the boiling point is roughly $\frac{1}{2}N_A n\varepsilon$. So the latent heat of vaporisation per mol, $L = \frac{1}{2}N_A n\varepsilon$. In molecular terms, it corresponds roughly to the energy difference between C and D in the V–r curve in Figure 6.7, assuming C is about the equilibrium separation for two liquid molecules.

Bonds Between Atoms and Molecules

The atoms and molecules in solids, liquids and gases are held together by so-called *bonds* between them. There are different types of bonds. All are due to electrostatic forces which arise from the +ve charge on the nucleus of an atom and its surrounding electrons which carry −ve charges.

Briefly, the different types of bonds are:

(a) *Ionic bonds*. Sodium chloride in the solid state consists of positive sodium ions and negative chloride ions held together by electrostatic attraction between the opposite charges.

(b) *Covalent bonds*. The electron in one atom of a hydrogen molecule, H_2, for example, wanders to the other atom, and the two atoms then attract each other as a result of their unlike charges. These covalent bonds, which are due to shared electrons between atoms, are very strong.

(c) *Metallic bonds*. In solid metals such as sodium or copper, one or more electrons in the outermost part of the atom may leave and occupy the orbit of another atom. These so-called 'free' electrons wander through the metal crystal structure, which consists of fixed +ve ions. The metallic bond is similar to a covalent bond except that electrons are not attached to any particular atoms; it keeps the metal in its solid state. The metallic bond is not as strong as the ionic and covalent bonds.

(d) *Van der Waals' bonds*. Over a long time-interval, the 'centre' of an electron cloud round the nucleus is at the nucleus itself. At any instant, however, more electrons may appear on one side of the nucleus than the other. In this case the 'centre' of the electron cloud, or −ve charge, is slightly displaced from the +ve charge on the nucleus. The two charges now form an 'electric dipole'. A dipole attracts the electrons in neighbouring atoms, forming other dipoles.

The electric dipoles have weak forces between them, called van der Waals' forces because similar attractive forces were predicted by van der Waals in connection with the molecules of gases. Solid neon, an inert element, is kept in this state by these bonds; the low melting point of solid neon shows that the bonds are weak.

You should know:

1 **Intermolecular forces: at equilibrium separation r_0 of molecules, attractive forces balance repulsive forces. At $r > r_0$, force is attractive; at $r < r_0$, force is repulsive.**

2 **Graphs of F–r and V–r have + and − values due to repulsive and attractive forces. Graphs are related since**
$$F = -dV/dr = -\text{gradient of } V - r \text{ graph.}$$

3 **F–r graph: $F = 0$ (equilibrium) at $r = r_0$. Less than r_0, F is +ve (repulsive force = $+kr$). Greater than r_0, F is −ve (attractive force = $-kr$)**
Round $r = r_0$, straight line graph shows Hooke's law. At graph maximum, F = breaking force.

4 **V–r graph: Minimum V at $r = r_0$. At higher temperatures V increases; and non-symmetry of graph shows increase in r of average molecular separation and so thermal expansion. As temperature increases, latent heat of vaporisation produces large separation r and a gas with negligible F and V for its molecules.**

EXERCISES 6B Molecular Forces and Energy

Multiple Choice

1 Figure 6E (i) shows how the potential energy V for two neighbouring atoms varies with their separation r. Which statement is correct?

A the atoms are in equilibrium at separation OX
B the slope of the curve at Z is related to Hooke's law
C the force between atoms is repulsive for separation less than OX
D the force between atoms is repulsive for separations greater than OY
E the atoms have maximum energy E_0

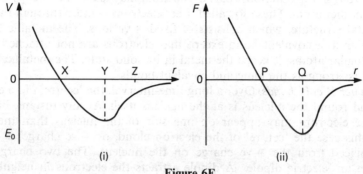

Figure 6E

2 Figure 6E (ii) shows roughly how the force F between two adjacent atoms in a solid varies with separation r. Which statements are correct?

1 OQ is the equilibrium separation
2 Hooke's law is obeyed round P
3 The potential energy of the atoms is the gradient of the graph at all points
4 The energy to separate the atoms completely is obtained from the area enclosed below the axis of r

Answer: **A** if 1 and 2 only   **B** if 2 and 4 only   **C** if 1 and 3 only
D if 1, 2, 3 only   **E** if 1, 2 and 4 only

Longer Questions

3

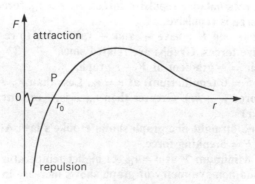

Figure 6F

The graph, Figure 6F, shows how the force, F, between a pair of molecules might vary with their separation, r.

Describe in words how the force between the pair of molecules varies:

(a) as r *decreases* from the value r_0, (b) as r *increases* from the value r_0.

What is the significance of the separation value r_0? (*L.*)

4 Explain how you would use the V–r curve in Figure 6G.

(a) to obtain the equilibrium separation of the molecules,

(b) to find the energy needed to completely separate two molecules initially at the equilibrium separation,

(c) to show that a solid usually expands when its thermal energy is increased.

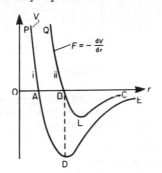

Figure 6G

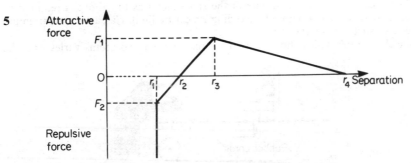

Figure 6H

The graph in Figure 6H shows a much simplified model of the force between two atoms plotted against their distance of separation. Express (i) the maximum restoring force between the atoms when they are pulled apart, (ii) the equilibrium separation of the atoms, and, (iii) the energy required to separate the two atoms in terms of the forces and distances given on the graph.

With the aid of the graph explain why solids show resistance to both stretching and compressing forces. Explain over what region you would expect Hooke's law to apply. What would this model predict about the elastic limit and the yield point for a material whose atoms followed the model? (*L.*)

6 In the model of a crystalline solid the particles are assumed to exert both attractive and repulsive forces on each other. Sketch a graph of the potential energy between two particles as a function of the separation of the particles. Explain how the shape of the graph is related to the assumed properties of the particles.

The force F, in N, of attraction between two particles in a given solid varies with their separation d, in m, according to the relation

$$F = \frac{7 \cdot 8 \times 10^{-20}}{d^2} - \frac{3 \cdot 0 \times 10^{-96}}{d^{10}}$$

State, giving a reason, the resultant force between the two particles at their equilibrium separation. Calculate a value for this equilibrium separation.

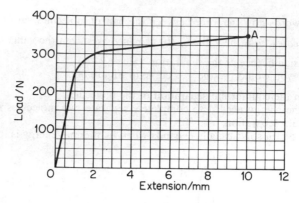

Figure 6I

The graph displays a load against extension plot for a metal wire of diameter 1·5 mm and original length 1·0 m, Figure 6I. When the load reached the value at A the wire broke. From the graph deduce values of

(a) the stress in the wire when it broke,
(b) the work done in breaking the wire,
(c) the Young modulus for the metal of the wire.

Define *elastic* deformation. A wire of the same metal as the above is required to support a load of 1·0 kN without exceeding its elastic limit. Calculate the minimum diameter of such a wire. (*O. & C.*)

7 (a) Sketch a graph which shows how the force between two atoms varies with the

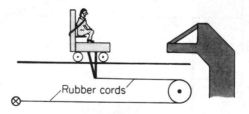

Rubber cords

Figure 6J

distance between their centres. With reference to this graph explain why (i) any reversible changes in volume of a solid is always a small fraction of the unstressed volume, (ii) a metal wire obeys Hooke's law for small extensions.

(b) (i) Sketch a graph which shows how the length of a rubber cord varies with the tension in the cord as the tension increases from zero until the cord breaks. Account for the shape of the curve you have drawn in terms of the molecular structure of rubber. (ii) In what important way does the structure of a metal at the molecular level differ from that of a rubber? How do you account for the large extension of a rubber cord compared with the extension of a mild steel wire of the same dimensions and acted on by the same force?

(c) Figure 6J shows a trolley of total mass 560 kg which is used for testing seat belts. The trolley runs on rails and is attached to six identical, parallel rubber cords whose unstretched lengths are 40 m each. When the trolley is pulled back far enough to extend the cords by 21 m each and then released, it reaches a speed of 15 m s^{-1} just as the cords begin to slacken. If the cords are assumed to obey Hooke's law over the full range of their extension in this application, if the system is assumed to be free of friction and the Young modulus for the rubber is $2·2 \times 10^7$ Pa, calculate (i) the maximum force applied to each cord, (ii) the area of cross-section of each cord when stretched. (*L.*)

Plate 6B *Sir Cyril Smith being lifted off the ground by a high tenacity polypropylene rope*

■ Solid Materials

Industry uses many different kinds of solid materials. For example, metals such as iron, steel and aluminium; glassy solids such as Perspex and window glass; and organic materials called polymers for making rubbers, plastics and resins.

Classification of Solids

Since the atoms of a solid occupy fixed positions, it is convenient to classify them according to the way their three-dimensional structure is built.

For many solids, and in particular metals, the crystalline state is the preferred one. Here the atoms are arranged in a regular, repetitive manner forming a three-dimensional lattice. With such an ordered packing system, the greatest number of atoms may be arranged in the smallest volume and the potential energy of the system tends to a minimum for stability.

Many solids, however, such as organic materials like rubber, are unable to adopt a structure such as a crystal state which has long-range order. *Polymers*, for example, are organic solids with very large and irregular molecules which are not capable of forming large-scale regular structures.

Other solids such as glass have no ordered structure on account of the way they are made. In making glass, molten material is cooled and its viscosity increases. The disordered liquid structure is then 'frozen in'. Solids which have their atoms arranged in a completely irregular structure are called *amorphous solids*. However, most non-crystalline solids do show some short-range order.

The way in which atoms are arranged in solids will obviously have a strong effect on the physical properties of the material. For example, diamond and graphite are both different structural forms of solid carbon. Diamond is transparent, hard and non-conducting but graphite is black, soft and conducting. Further, the nature of any solid structure will be largely determined by the nature of the interatomic forces. These are summarised below and were discussed on page 159.

Type	Strength	Nature	Example
Ionic	Strong	Electron transfer	Sodium salt
Covalent	Strong	Electron sharing	diamond
Metallic	Fairly strong	Electron sharing	copper
Van der Waals	Weak	Dipole interaction	solid neon

Solids can be classified into (i) crystalline—ordered structure, (ii) amorphous —irregular structure, (iii) glassy—disordered structure, (iv) polymer—organic irregular structure

Crystalline Solids

The simplest crystalline systems are those in pure metals. Here we are dealing with identical atoms linked through the fairly strong metallic bond (p. 159).

In any crystal, a particular group of atoms is repeated many times, like the pattern in some wallpapers or textiles. The unit of pattern is known as a *unit cell* and the whole structure is called a 'space lattice', with atoms or ions at the lattice corners. As shown in Figure 6.8 many atomic planes such as P, Q, R, S, T can be drawn through the crystal which are rich in atoms.

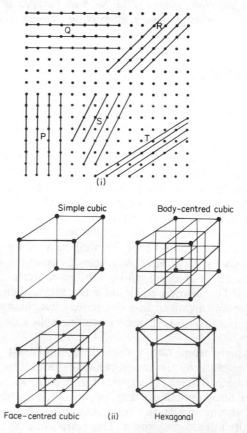

Figure 6.8 (i) *Atomic planes in crystal* (ii) *Types of crystal structure*

There are different types of unit cells. Figure 6.8 shows four types of *structure*. The *simple cubic* has an atom or ion at the eight corners of a cube. More common in nature is the *body-centred cube* (BCC), which has one atom at the centre of each cube in addition to the eight atoms at the corners, and the *face-centred cube* (FCC), which has one atom at the centre of the six faces of the cube in addition to eight atoms at the corners. Alternatively, crystals may have layers of atoms arranged with *hexagonal* rather than cubic symmetry. This appears to be a very efficient way of packing layers of atoms. The table below compares the packing efficiencies of these crystal structures. Here the atoms are considered to be touching spheres and calculations of the volumes occupied by the atoms as a fraction of the available volume give a good guide to the efficiency of packing.

Structure	Packing fraction	Occurrence
Simple cubic	0·52	very rare
Body-centred cubic (BCC)	0·68	fairly common
Face-centred cubic (FCC)	0·74	very common
Hexagonal close packed (HCP)	0·74	very common

In crystalline solids such as metals, the atoms are grouped in a lattice structure with many planes rich in atoms.

Imperfections in Crystals

Real crystals are rarely perfect. Although less than one crystal site (place) in ten thousand may be imperfect, the existence of lattice defects can have considerable influence on the mechanical and electrical properties of a material. The industrial development of electronics was due to the control of the electrical properties of silicon and other semiconductors by adding impurity atoms in very low concentrations. Small impurities are also responsible for the characteristic colours of many gemstones. As we see later, the mechanical properties of solids are determined to a great extent by imperfections.

Broadly, crystal defects can be classified into two groups; either imperfections in the *occupation* of sites or in the *arrangement* of sites.

Imperfections in Occupation of Sites

Here the main imperfections are commonly *point defects*, especially
(a) the presence of foreign atoms and
(b) the existence of vacancies (unoccupied lattice sites).

Foreign atoms may exist among the 'host' atoms in several ways. They may be grouped in clusters in the host crystal or they may be dispersed through the crystal as single atoms. If sufficiently small, they may exist on *interstitial* sites, that is, they may occupy non-lattice sites between the host atoms. For example, pure iron can be made into steel, an engineering material widely used, by adding carbon, whose atoms reside between the iron atoms.

If foreign atoms are comparable in size to host atoms and valency requirements are met, they can exist as substitutes for host atoms on a lattice site. For example, one type of brass consists of 70% copper and 30% zinc whose atoms exist as substitutes for copper atoms in a cubic lattice.

Vacancies exist in all crystals and occur naturally when the crystal solidifies. They may also be present by diffusion into the crystal from the surface. Irradiation by α-particles may create a vacancy by knocking a host atom off its usual site and this is important to nuclear engineers.

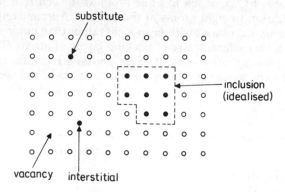

Figure 6.9 *Imperfections in crystals*

Figure 6.9 shows diagrammatically how these imperfections occur.

Imperfections in Arrangements of Sites, Dislocations

Imperfections in crystals can occur in the arrangements of sites, where atoms in the lattice structure should exist. These imperfections can be large and affect thousands of millions of atoms in the crystal.

From the viewpoint of the mechanical behaviour of a solid, the most important defects are the *dislocations* in the crystal. These are defects along a line of atoms. An *edge* dislocation can be considered as an extra part of a layer of atoms either removed from, or inserted into, a perfect crystal structure, Figure 6.10 (i).

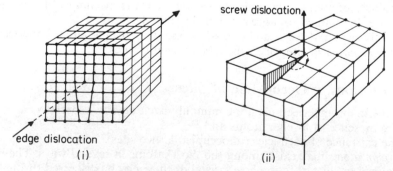

Figure 6.10 (*i*) *Dislocation is edge of plane of atoms*   (*ii*) *A screw dislocation*

Another type of dislocation is called a *screw* dislocation. Here the atoms are displaced so that they can be imagined to be on the spiral of a screw. The crystal then has the appearance of having been partially sliced and the two exposed faces displaced vertically, Figure 6.10 (ii). Details of the dislocations are given on pages 169–170.

The atomic bonds along the line of a dislocation are strained. Where the

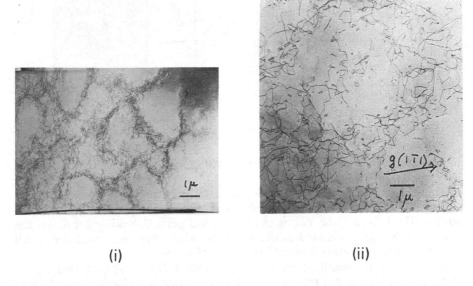

(i)   (ii)

Plate 6C *Transmission electron micrographs of thin foils of aluminium showing disloca-tions* (i). *Further strain causes dislocation multiplication and entanglement, leading to work hardening* (ii)

dislocations meet a surface, these strained bonds can be made visible by etching the surface of the crystal and studying the surface in a microscope.

Crystals are imperfect. The most important defects are *dislocations*.

Polycrystalline Materials—Grains and Boundaries

Many crystal materials are *polycrystalline*, that is, they exist as a large collection of tiny crystals all pointing in different (random) directions. Each tiny crystal is known as a *grain* and the groups of crystals are connected together at the *grain boundaries*.

The existence of grains appears to be due to the solidification process in manufacture which begins simultaneously at several places. The surface is the first to solidify or freeze. When the solid is completely solidified, the structure consists of grains pointing in different directions, as shown diagrammatically in Figure 6.11. The grain boundaries assist the strength of a material.

Mechanical Behaviour of Solids

An engineer must select the right material to use in a project. He or she must then be confident that the final structure will perform safely and within the design limits which have been set. To achieve this, the engineer must know and

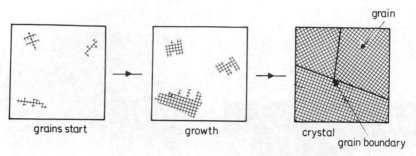

Figure 6.11 *Growth of grain boundaries (diagrammatic)*

understand how the materials available will respond or react to the stresses they may meet.

The section on Elasticity, page 144, discusses the strain in solid metals when stresses are applied. If required, the reader should refer to this section for topics such as Hooke's law, the Young modulus, yield point, breaking stress and the energy stored in a strained material, all of which are important for a full understanding of the mechanical behaviour of solids.

Here it may be useful to recall that if Hooke's law is obeyed, that is, the proportional limit is not reached, the material is said to deform *elastically* and the energy stored is recovered on removing the load. In this case, the strain is typically less than $\frac{1}{2}\%$ for metals.

In elastic deformation, when Hooke's law is obeyed and the atoms undergo small displacements, the energy stored is fully recovered when the load is removed.

Plastic Deformation, Breaking Stress

Beyond the elastic limit, a material undergoes *plastic deformation*. This occurs by movement of dislocations in the solid, discussed shortly. Figure 6.12 (i) shows a typical stress–strain curve for a specimen. If B is the elastic limit, the region BE corresponds to plastic deformation. In this case the energy stored in the

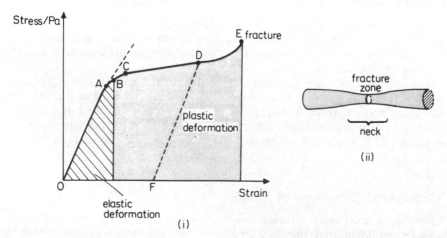

Figure 6.12 *(i) Stress-strain curve   (ii) Breaking at fracture*

solid is *not* recovered but is transferred to heat. Unloading from D, only the elastic energy is recovered and so DF is parallel to OA.

E corresponds to the point of *fracture*. The strain at fracture may be as high as 50% for a metal such as steel. Fracture usually occurs at E after a 'neck' has been produced because this has minimum cross-sectional area and so maximum stress, Figure 6.12 (ii). During necking, the material appears to 'flow' like a viscous liquid. The *breaking stress* at E (or *ultimate tensile stress*) is taken as a measure of the *tensile strength* of a material. The Young modulus value is a measure of the *stiffness* of the material and is the value of the gradient of the line OA (stress/strain). The table shows the tensile strength of some materials.

Material	Tensile strength/10^8 Pa
Steel	4–10
Cast iron	1
Aluminium	0·8
Glass	1
Rubber	0·2

Plastic deformation is due to movement of dislocations. The energy in plastic deformation is converted to heat.
Stiffness = Young modulus. Tensile strength = breaking point stress.

Ductile and Brittle Materials

Materials which show a large amount of plastic deformation under stress are called *ductile*. Ductility is an important property and allows metals to be drawn into wires. Metals may be *malleable* or beaten into sheets and then rolled and shaped. Metals are very useful engineering materials because many have high tensile strength, malleability and ductility.

Materials such as glass and ceramics fracture (break) close to the elastic limit without any appreciable plastic flow. They are called *brittle* materials. In these cases fracture occurs at low strains.

The absence of any significant plastic flow means that the fractured pieces may be fitted together to recreate the original shape. This is not the case with ductile materials. The difference between the two classes can be illustrated by considering what happens when a china teapot (brittle material) and a metal teapot (ductile material) are both dropped on a hard floor. The china pot will probably break but the pieces can be glued together to form the original shape. The metal pot will not break but is likely to be permanently dented.

We now discuss the behaviour inside ductile and brittle materials.

Ductile materials show a large amount of plastic deformation. Brittle materials fracture at low strains close to their elastic limit.

Plastic (Ductile) Behaviour of Solids

In an earlier section we discussed a model showing how the forces between atoms or molecules varied with their separation (p. 155). The theoretical

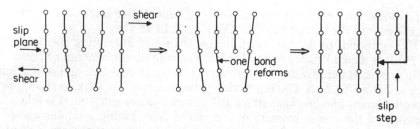

Figure 6.13 *Bond breaking and re-forming at edge dislocation*

breaking stress of ductile solids can be calculated using this model but the result is many times greater than the value obtained in practice.

The reason for the relatively small value of breaking stress is due to a process called *slip*. Slip is due to a movement of dislocations throughout the crystal. Here *one bond at a time* is broken, so the process occurs at a much lower stress than that calculated theoretically, Figure 6.13. With a large number of dislocations in operation we can account for the plastic behaviour of ductile materials at the stresses observed. Large scale slip, requiring a whole plane of atoms to move bodily relative to an adjacent plane, would involve the simultaneous breaking of a very large number of bonds and require very much larger stresses than that obtained in practice.

A slip plane is generally in a direction in which the atoms are most closely packed. Several such planes, pointing in different directions, may exist, depending on the crystal structure. Slipping preserves the crystal structure and results in a permanent extension of length, which is not recovered when the stress is removed, see Figure 6.14. Slip 'steps' at the surface, often visible to the naked eye, provide an indication of the mechanism involved. Slip steps may be several thousand atoms in height and occur close together to form slip bands.

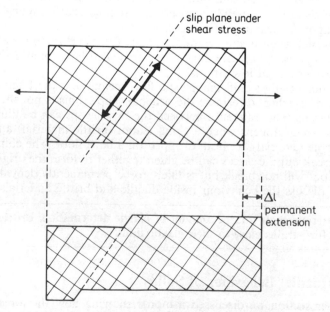

Figure 6.14 *Slip plane and extension of length*

In plastic deformation, *slip* occurs due to the movement of dislocations. Bonds between atoms are broken one at a time. Due to slip, the tensile strength of a metal is very much lower than the value calculated from the interatomic forces.

Work Hardening, Annealing, Iron and Steel

As we have seen, the ductility of metals is due to the existence of dislocations and their ability to move under a local shear stress. Materials with a large number of very mobile dislocations are therefore expected to be ductile.

Under repeated stress, however, dislocations move and intersect, and become entangled. They now *pin* each other and so become immobile (fixed). Dislocations between the pinning points can produce more dislocations, which also become immobilised due to intersections. This explains why copper wire, for example, can be broken by flexing it to-and-fro in the hand. The many dislocations produced by the repeated stress on the wire all become pinned and immobilised. Ductile behaviour is then not possible and the wire fractures in a brittle manner.

This is an example of *work hardening* a metal, a process which increases the strength of the metal but at the expense of ductility because plastic flow is then considerably reduced. *Cold rolling*, when the thickness of a metal is reduced by pressure, usually strengthens a metal by work hardening.

During cold working processes, the crystal structure is plastically deformed and there is lower ductility and toughness (see p. 170). The crystal grain boundaries (p. 167) are particularly affected. *Annealing* helps to restore the metal to its ductile state. In this process the metal is heated to a high temperature (below its melting point) and maintained at this temperature for a length of time. This increases the thermal vibrations of the crystal lattice and relaxes the internal strains. In this way the solid is *recrystallised* and returns to a ductile state.

Pure iron is usually too ductile for use in load-bearing applications such as bridges. Steel is made by adding a small percentage of carbon to pure iron. The carbon atoms reside between the atoms of the iron lattice and are very effective in pinning dislocations and reducing their mobility. So steel has less ductility and greater strength than pure iron.

Other elements, added with carbon, produce specialist steels. For example, stainless steel has 20% chromium and possesses very good corrosion resistance and hardness.

In *work hardening*, moving dislocations are pinned or entangled and the metal may fracture as a brittle material. *Annealing*, heating to a suitable high temperature, restores the crystalline state and ductility.

Brittle Materials, Cracks

Materials which exhibit little or no plastic flow before failure are *brittle*. The absence of plastic flow implies that dislocations are absent or their ability to move under stress is much less than in metals.

In glass, for example, there is no concept of a 'dislocation'. A dislocation is a region of disorder within an otherwise ordered structure. Glass is amorphous and has no ordered structure. So we cannot define a dislocation for such a material. Dislocations are rare in ionic materials such as sodium chloride, where electrostatic forces are concerned in the lattice structure.

The theoretical strength of brittle materials is much higher than that obtained in practice. The explanation for the low breaking stress was due to Griffith. He suggested that microscopic flaws or *cracks* in the surface (or just below) act as *stress concentrators*. Figure 6.15 illustrates the uniformly distributed stress in unflawed materials and the concentration of stress at the tip of the crack in flawed materials.

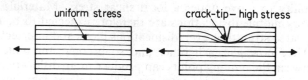

Figure 6.15 *Unflawed solid (uniform stress). Flawed solid with crack (high stress at tip)*

It can be shown that the stress at the crack tip is greater than the nominal uniform applied stress by a factor $k = 2\sqrt{l/r}$ approximately, where l is the crack length and r is the crack tip radius, which is of atomic dimensions. A scratch about 10^{-3} mm deep and a tip radius of 2×10^{-10} m would cause the stress at the tip to be about 140 times the nominal (applied) value. As the crack gets longer, the stress lines become more concentrated at the tip. Crack growth may then suddenly accelerate, leading to the characteristic sudden failure of brittle materials.

Griffith showed by experiment that freshly drawn glass fibres, with no surface flaws or cracks, came very close to their theoretical high strength. The fibres quickly lose their strength, however, as the cooling process introduces microscopic flaws which reduces the strength to normal values.

The sensitivity of brittle materials to cracks is shown by the cutting of glass. A fine line is scratched on the surface with a glass cutter. Slight pressure on either side of the crack causes the glass to fracture cleanly along the line. Most brittle materials are much stronger in compression because the surface cracks will then tend to be closed and unable to spread. Pillars made of cement, for example, are strong in compression. Cracks develop if the cement is in tension and it then breaks.

Cracks in the surface, or just below, produce high stress concentration at the tip. Rapid crack growth produces brittle behaviour (sudden fracture). Brittle materials are stronger in compression than in tension as this prevents cracks spreading.

Toughness and Hardness

The *toughness* of a material is a measure of its ability to resist crack growth. This is not to be confused with 'strength'. Plasticine, for example, is a tough material but not strong. Glass is much stronger than Plasticine but not as tough.

Metals will contain surface cracks of microscopic dimensions. In general, however, metals are tough. This is due to the ability of dislocations to move and blunt the crack tip. The stress concentration is then relieved and the crack does not spread.

The *hardness* of a material is a measure of its resistance to plastic deformation. An 'indenter' such as a hardened metal sphere is pressed into the surface of the test material for a certain time and the 'hardness' is calculated by dividing the applied force by the contact area left in the surface by the indentation. A ductile material will produce a large area indentation and have low hardness.

The stress below the indenter is compressive, so it is possible to measure the hardness of brittle materials such as glass without causing brittle fracture.

Most cracks which occur in metals are due to *fatigue*, when the metal undergoes failure after many cycles of normal stress. Aircraft accidents can occur with metal fatigue starting at a rivet hole, for example. Metal *creep* is the gradual growth of plastic strain under a static, not cyclic, load.

Toughness = ability to resist crack growth
Hardness = resistance to plastic deformation

Composite Materials

In engineering design, it is often difficult to meet the mechanical properties needed by using a single material. *Composite materials*, where two materials are used, have wide application. Reinforced concrete, for example, is made by setting the concrete round steel wires or mesh. This improves the tensile (tension) properties of the concrete for use in structures or buildings. Concrete itself is a mixture of cement, sand and small stones. Figure 6.16 shows pre-stressed concrete.

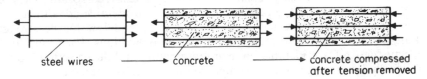

steel wires   concrete   concrete compressed
after tension removed

Figure 6.16 *Pre-stressed concrete*

In fibre-reinforced materials, which are used in the plastics industry, long straight fibres are embedded in a tough *matrix*. Polymers and metals have been used for matrix materials. Glass-fibre reinforced plastic (GRP), used in construction materials for many years, have glass fibres embedded in a matrix such as polyester resin. Carbon-fibre reinforced plastic (CFRP) is also used.

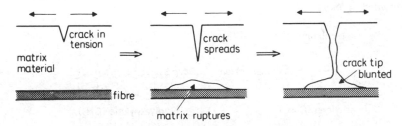

crack in tension   crack spreads   crack tip blunted

matrix material   fibre   matrix ruptures

Figure 6.17 *Preventing cracks by matrix*

As we saw previously, *cracks* which spread in a material will weaken it. The principle of the toughening process in fibre-reinforced materials is illustrated in Figure 6.17. The matrix separates from the fibres and the tip of the crack is blunted by it and stopped from spreading. The matrix also helps to transfer the load to the fibres as these are bonded to the matrix surface. The GRP is used for making small boats and canoes, storage tanks and some car bodies. The CFRP is used in the aircraft industry as it has excellent strength/weight ratio property, like the GRP, and is much stiffer than steel.

Car windscreens are also made from composites. Thin sheets of laminated glass, bonded by resin, can prevent pieces of glass flying about dangerously after an accident. On impact, the cracks in the glass spread parallel to the surface but not through the windscreen, owing to the crack-blunting at the glass–resin boundary. Laminated glass is an example of a *layer composite*.

Composite materials improve mechanical properties. Glass-fibre and carbon-fibre reinforced materials prevent cracks spreading, are strong in relation to their weight and may be stiffer than steel.

Polymers, Structure and Mechanical Properties

We now discuss a class of organic materials generally described as *polymers* which have wide applications. Polymers consist of very long chains of carbon atoms bonded to hydrogen and other atoms.

Polymers occur naturally in materials such as rubber, resin, cotton-wool and wood. They are used widely in the plastics industry. In the home, for example, there may be plastic dustbins, washing-up bowls, light fittings and wrapping paper, in addition to nylon socks. Plastics also make good thermal and electrical insulators, have low density, great toughness and resist corrosion. They are easy to mould and cheap to produce, which is a considerable advantage.

Their disadvantages are a low Young modulus and low tensile strength, making them unsuitable for many load-bearing applications. Their mechanical properties depend considerably on their temperature and they also tend to melt at relatively low temperatures accompanied by dangerous fumes.

Structure of Polymers

The basic structure of a polymer can be illustrated by considering polyethylene, better known as *polythene*. By a chemical process called *polymerisation*, the double bonds of a large number of ethylene molecules (C_2H_4) are broken to allow them to form a giant molecule, the polymer. Figure 6.18 illustrates the formation of a polyethylene $(CH_2)_n$ molecule. The basic unit ethylene is called a *monomer* or *mer*.

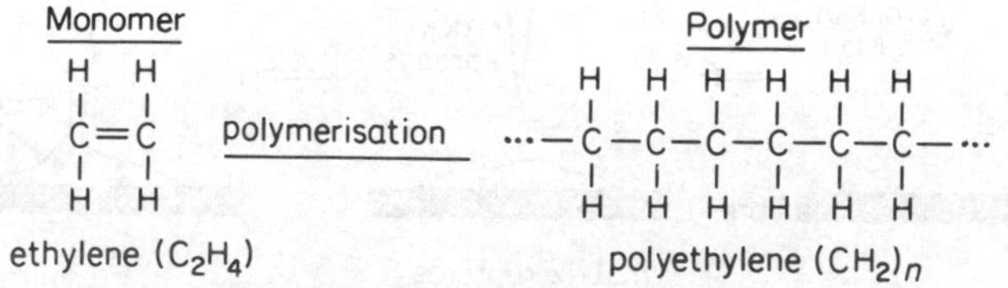

Figure 6.18 *Polymerisation*

The polymer molecule may contain thousands of carbon atoms so that n is very large. A wide range of materials can be made starting with different monomers. The table illustrates how polyvinyl chloride (PVC), which uses chloride atoms, and polystyrene, which uses benzene rings, are made from their monomers, see Figure 6.19.

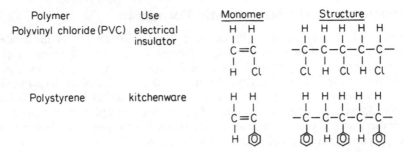

Figure 6.19 *Polymer molecule from monomer*

Polymers, **widely used in the plastics industry, are made chemically with monomers. They consist of very long chains of carbon atoms bonded to hydrogen and other atoms.**

Branching and Cross-linking of Polymer Molecules

During chemical manufacture, polymer molecules may form *branches* or become *cross-linked* like the rungs of a rope ladder, Figure 6.20. Polyethylene molecules, for example, usually contain more than a thousand carbon atoms and may have about seventy branches. As branching or cross-linking develops, freedom of movement of the molecules becomes less because they entangle.

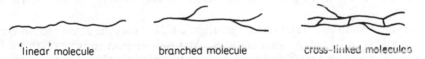

'linear' molecule   branched molecule   cross-linked molecules

Figure 6.20 *Linear, branched and cross-linked molecules*

Cross-linking is used in the manufacture of some materials. Natural rubber molecules, for example, can be cross-linked with sulphur atoms. 1% sulphur produces a soft but solid rubber called *vulcanite* and 4% sulphur produces the hard material called *ebonite*, both used as insulators in the electrical industry. Cross-linking in rubber also occurs with oxygen atoms from the atmosphere. The rubber then tends to harden and become brittle with age, as old rubber bands show.

In polymers, the ease with which molecules slide over each other depends on the shape and size of the groups of atoms attached to the carbon 'backbone'. With polystyrene, the mechanical interference between molecules is large, so this material is inflexible and glasslike. In PVC, the chlorine atoms produce interference, so PVC is much less flexible than polyethylene. PVC, however, can be made flexible for use as electrical insulators round copper wires by adding a 'plasticiser', which acts as an internal lubricant separating the PVC molecules.

Polymer molecules can be linear, branched or cross-linked. Cross-linking with other atoms is made in the manufacture of some materials.

Thermosetting and Thermoplastic Polymers

Polymers which form cross-links between their long chains of molecules in manufacture are called *thermosetting polymers*. At ordinary temperatures the cross-links keep these polymers solid and rigid. When heated, however, the agitation of the molecules destroys the cross-links and chemical decomposition occurs. So thermosetting polymers cannot be re-moulded to a new shape by heating. Bakelite, melamine and epoxide resin are examples of thermosets.

Polymers with few cross-links are called *thermoplastic polymers*. When heated, these polymers become softer and can be re-moulded, unlike thermosetting polymers. Polythene, polystyrene, polyvinyl chloride (PVC) and nylon are examples of thermoplastics. The mechanical properties of these materials are

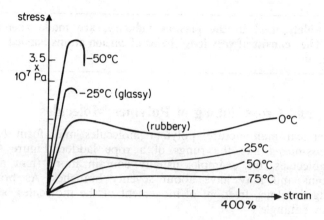

Figure 6.21 *Stress–strain graphs (diagrammatic) for polythene at different temperatures*

more sensitive to temperature change. Figure 6.21 shows the stress–strain graph of polythene at different temperatures. At lower temperatures polythene has a 'glassy' behaviour (stiff and fracturing at low strain); at higher temperatures it has a 'rubbery' behaviour (less stiff and able to stretch much more).

Thermosetting **polymers have cross-links between chains of molecules. They are rigid and cannot be re-moulded by heating.**
Thermoplastic **polymers have no cross-links. They become soft on reheating and can be re-moulded.**

Comparison of Mechanical Properties

The table shows some of the mechanical properties of a wide range of materials, including plastics. Compared with metals and brittle materials such as glass, the plastics have much lower values of Young modulus (less stiff) and very large stretching or elongation. The elongation is a measure of the permanent extension of a length of the specimen after failure. In the majority of non-polymeric solids, the elastic strains are very small. Larger stresses result in either rapid brittle fracture (glass, for example) or plastic deformations (metals) which rarely produce elongations higher than 50%.

In plastics, the low Young modulus and enormous elongations occur because the long irregular molecules can uncoil and straighten out under tensile stresses. This process can be done without straining the individual bonds within the

		Young modulus 10^9 Pa	Tensile strength 10^6 Pa	Elongation %
Metals	Steel	200	250	35
	Copper	120	150	45
	Aluminium	70	60–120	45
Woods	Oak (parallel to grain)	5–9	21	
Brittle materials	Glass	71	100 (about)	0 (about)
	Concrete	20–40	4	
Thermosets	Bakelite	6–8	50	0·6
	Melamine	9	70	
	Epoxide resin	1–5	30–80	
Thermoplastics	Perspex	3·4	55–70	2–10
	PVC (rigid)	2·5	60	2
	Polystyrene	3·5	40	2·5
	Nylon		70	60–300
Rubber	Natural	1 (25% elongn)	32	850

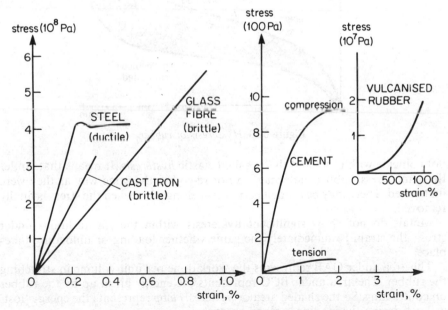

Figure 6.22 *Stress–strain graphs of some industrial materials*

molecules and so it can take place at much lower stresses than metals, for example. In general, then, due to uncoiling of molecules, polymeric materials do not obey Hooke's law. The molecules in a cross-linked thermoset are much more difficult to uncoil than those in a thermoplastic. For example, bakelite (a thermoset) is much stiffer and shows less elongation than polythene (a thermoplastic). Figure 6.22 shows roughly stress–strain curves for various materials used in engineering.

Metals and glass have high Young modulus and low elastic strain. Plastics have lower Young modulus (less stiff), large elongation and do not obey Hooke's law like metals do.

Rubber, Hysteresis

Unlike other materials, rubber shows an increase in Young modulus when its temperature rises. The increase in temperature produces more agitation of the molecules and increases their tendency to coil up. It is then more difficult to uncoil, or produce more order in the molecules, by tensile forces. So the Young modulus increases.

This also explains why rubber *contracts* when heated. The molecules become more coiled and the rubber shortens.

Figure 6.23 shows the stress–strain curve for a specimen of rubber when it is first loaded within the elastic limit, OAB, and then unloaded, BCA. BCA does

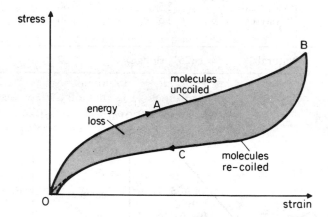

Figure 6.23 *Hysteresis of rubber*

not coincide with OAB and this is called elastic *hysteresis*. It means that, under the same stress, rubber molecules do not re-coil in the same way as they were first coiled. There may be a small permanent elongation when the stress is finally removed.

Metals do not show significant hysteresis within the elastic limit—under stress, the strain is immediately the same whether loading or unloading takes place.

The area under OAB represents the work done per unit volume in stretching the rubber. The area under BCO represents the energy given up by the rubber on contracting. So the shaded area or *hysteresis loop* represents the energy 'lost' as heat during the loading–unloading cycle.

Effects of Hysteresis

The hysteresis of rubber enables it to convert mechanical energy to heat. It is therefore used as shock absorber material, for example.

A rolling car tyre is taken through many cycles of loading and unloading during a journey. A large-area hysteresis loop for such a rubber is not desirable because the heat produced may lead to dangerously high tyre temperatures. It would also increase petrol consumption due to conversion of mechanical energy to heat. In practice, therefore, tyres are made with synthetic rubbers, which have small-area hysteresis loops. Materials with small hysteresis effects are called *resilient*.

Conventional plastics, notably thermoplastics, show hysteresis effects and have about one-tenth the resilience of steel, for example. Plastic gear wheels are less noisy than metal ones as they are able to dissipate vibrational energy more effectively as heat.

Rubber molecules are coiled. Under increased tension the molecules uncoil and high strain (800%) can be produced. On removing the load a hysteresis loop is obtained. The 'host' energy is converted to heat and is proportional to the area of the hysteresis loop. *Resilient* materials have low hysteresis.

Wood

Wood contains a natural polymer based on the cellulose molecule. The grain of the wood is the line of the cellulose fibres. Wood is a composite material—the fibres are bonded together in a *lignin* matrix which consists of carbohydrates and is non-polymer.

Wood has strength and stiffness parallel to the grain but is weak across the grain. It is also a much weaker material in compression than in tension. In *plywood*, the wood is made stronger by glueing together alternately sheets whose grains go in perpendicular directions. The plywood is then equally strong in the two directions and so is less likely to warp than the single wood with grains in one direction.

Examples on Young Modulus and Solid Materials

1 Figure 6.24 shows the stress–strain curve for a metal alloy. Fracture occurred at an extension of 15%.

With reference to the diagram, estimate

(a) the Young modulus of the alloy,
(b) the ultimate tensile stress,
(c) the elongation (permanent extension remaining in specimen after fracture),
(d) the extension and breaking stress at fracture had the material been brittle rather than ductile.

(a) The Young modulus is the gradient of the linear (elastic) part of the stress–strain curve. So

$$\text{Young modulus, } E = \frac{\text{stress}}{\text{strain}} = \frac{300 \times 10^6 \, \text{N m}^{-2}}{0 \cdot 005}$$

$$= 6 \times 10^{10} \, \text{Pa}$$

(At stress 300×10^6 Pa, extension $= 0 \cdot 5\%$, so strain $= 0 \cdot 005$)

(b) Ultimate tensile stress = maximum stress material can withstand without fracture. So

$$\text{ultimate tensile stress} = 380 \times 10^6 \, \text{Pa}$$

(c) At fracture, extension is 15% but on fracturing, the *elastic* deformation, $0 \cdot 5\%$, is recovered. So

$$\text{elongation} = 14 \cdot 5\%$$

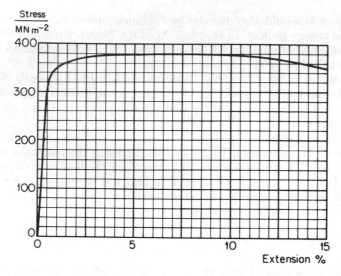

Figure 6.24 *Stress–strain curve*

(d) If the metal were brittle it would have fractured at, or just after, the end of the elastic deformation region, so no plastic deformation would have occurred. So

extension at fracture = 0·5% (brittle)

and
breaking stress = 300–320 MPa

2 Figure 6.25 (i) shows a cross-section through a reinforced concrete beam. It is supported at its ends and vertically loaded in the middle.

From the properties of concrete and of steel, explain the purpose of the steel reinforcement rods and why the steel rods are placed as shown.

In this loading situation, the top and bottom faces of the beam are in compression and tension respectively, Figure 6.25 (ii). Brittle materials such as

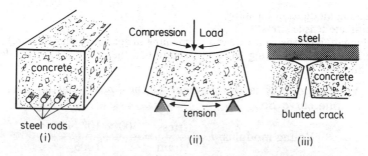

Figure 6.25 *Example on steel rods in concrete*

concrete tend to fail at, or near, the elastic limit. A surface crack, under tension, then suddenly elongates and runs through the material.

To stop this happening, steel rods are laid in the concrete close to the *bottom* face and running *parallel* with the direction of tension. The Young modulus of steel is many times greater than that of concrete, that is, stiffer than concrete. So the tension in the bottom face is supported largely by the steel rather than by the concrete. This lessens the possibility of brittle fracture of the concrete, increases the beam strength and reduces the 'sag' under the load. Also, any cracks which do elongate in the concrete will be 'blunted' when they meet the steel rods, Figure 6.25 (iii). This improves the toughness (resistance to crack growth) of the beam. The stress at a crack tip increases with crack length and also increases with a decrease in crack-tip radius. So blunting the crack tip, when the steel–concrete interface ruptures, reduces the stress at the crack tip and stops it spreading.

3 Explain why sheets of postage stamps are *perforated*. What is the most suitable shape of a perforation?

The sheets are perforated to help tearing between the stamps and to prevent accidental tearing of the stamps themselves. Paper, like all material, will tear where the stress is greatest and the stress is concentrated round high curvatures as in the case of a crack. The high curvature of the perforations helps tearing.

An ellipse is a better perforation shape than a circle. The narrower end of the ellipse has a higher curvature than a circle of the same area.

4 Explain briefly why carbon in the form of a diamond is used for diamond bits to drill hard rocks but carbon in the form of graphite is used in lead (graphite) pencils.

Since carbon atoms in the form of diamond have very strong covalent bonds, diamond has a very large value of stiffness or Young modulus. Diamond bits are therefore used to drill hard rocks.

In the form of graphite, however, carbon atoms have a layer lattice structure in which the atoms are bonded mainly by relatively weak van der Waals' forces. This is why, when a lead (graphite) pencil is pressed on paper when writing, some of the graphite is transferred to the paper.

You should know:

1 Solid materials: (a) metals—crystalline structure, (b) polythene—amorphous or disordered structure, (c) glass—disordered structure, (d) rubber—a polymer with organic molecules.

2 Metals have elastic deformation to elastic limit—energy recovered on load removal, then plastic deformation with greater load—energy not recovered on load removal but transferred to heat.

3 Ductile metals have large plastic deformation due to movement of dislocations. Brittle metals fracture at low strains due to pinned dislocations.

4 Window glass has no definite melting point; at high temperatures it flows like a liquid.

5 Polymers have long chains of carbon molecules.

6 Rubber has coiled molecules. With increased tension coils unwind and so there is high strain at low stress. After molecules unwind, low strain as tension increased. Rubber shows hysteresis in stress–strain graph; energy lost as heat is proportional to area enclosed. Synthetic rubber has low hysteresis and is used for car tyres.

7 Reinforced concrete has steel rods at bottom to reduce cracks spreading when the concrete is loaded. Composite materials such as car windscreens prevent cracks spreading.

Adhesives

Modern adhesives are finding more and more applications as new materials are introduced, since many of the high-performance composite materials (like CFRP—page 173) cannot be bolted or riveted together. It may surprise you to know that modern aircraft are increasingly held together with glue. But don't worry—the bonds formed using adhesives are often far stronger than those obtained with rivets or bolts. But how do adhesives work? What properties are required of a good adhesive? Adhesives form strong intermolecular bonds with the materials they come into contact with, but then so do most liquids. For example, water also forms strong bonds with many materials, but makes a poor adhesive, since its *strength* (page 170) is negligible. So an effective adhesive must also set to a strong solid, preferably with a final strength comparable with that of the material being adhered to.

Why are adhesives usually liquids in the first place? This is so that they can flow closely over the rough surface of the material, getting close enough so that intermolecular bonds form between the adhesive and the material over a large area. From the considerations on page 159, we know that the adhesive must approach the material within $2-5 \times 10^{-10}$ m for these bonds to form. It *is* possible for two solids to spontaneously adhere, but only if they can come into close contact with each other over a large surface area. To do this, they must be flat and free of surface dust. Highly polished, flat pieces of metal, used in engineering workshops to measure lengths accurately, often stick together if placed on top of each other and need quite a lot of force to separate them again. Most solids, however, are not this flat or clean, so spontaneous adhesion is a rare phenomenon!

EXERCISES 6C   Solid Materials

Multiple Choice

1 When stretched beyond its elastic limit, a metal rod such as steel

   A becomes plastic   B has no energy   C obeys Hooke's law
   D becomes plastic   E becomes colder

2 A metal teapot X and a china teapot Y break when they fall on a concrete floor. Then

   A X and Y can each be put together again
   B X is broken because it is brittle
   C Y is broken because it is brittle
   D the metal of C has no dislocations
   E the china of Y has many dislocations

3 Rubber and glass are widely used in industry. Which of the statements A to E is correct?

   A the stress–strain curves of the two materials are similar
   B glass has a definite melting-point
   C rubber has a low strain when initially stretched
   D glass has no dislocations
   E rubber has no energy when stretched

4 Figure 6K shows the stress–strain variation of a sample of rubber when loaded and

then unloaded. Which of the statements **A** to **E** is correct?

A the molecules are uncoiled along ORQ
B the energy gained by the rubber is proportional to the area OPQRO
C no energy is gained by the rubber
D the molecules are coiled along OPQ
E the area OPQRO is proportional to the strain

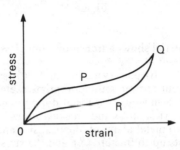

Figure 6K

Longer Questions

5 The measured strengths of brittle and ductile solids are very much smaller than calculations based on the forces between individual atoms in the solid.
   Explain why this is the case.
6 The breaking strain of rubber may be as high as 800%, yet most metals break at strains which rarely are greater than 50%. With reference to the molecular or atomic structure of these materials, account for the difference in the breaking strain.
7 What is meant by the following terms:
   (a) stiffness,
   (b) strength,
   (c) toughness, of a solid material?
   Explain briefly how these properties might be affected if the material were fibre reinforced.
8 In terms of their structure and likely physical properties, distinguish between a *thermoplastic* and a *thermoset*. Name one example of each type of plastic.
9 Rubber shows large hysteresis. What is meant by 'hysteresis'?
   Give one application of the use of rubber where a large hysteresis would be
   (a) an advantage and
   (b) a disadvantage.
10 Aluminium and glass have almost the same values of the Young modulus, tensile strength and density.
   Why is glass, which is much cheaper to produce, not used in place of aluminium in load-bearing applications?
11 Explain the difference between *elastic deformation* and *plastic deformation*.
   Plastic deformation is the result of a process called *slipping*. What do you understand by this term and how does it take place in a typical metal?
12 Using the same axes, sketch approximate stress–strain curves for (i) a metal, (ii) glass, and (iii) rubber.
   With reference to the stress–strain curves, describe and explain what would happen if three identical hollow spheres made respectively from a metal, glass and rubber were dropped from a great height onto a hard surface.
13 Figure 6L shows the stress–strain curves for a thermoplastic at two different temperatures A and B. Which is the higher temperature? At which temperature would the plastic behaviour be best described as (a) glassy, (b) rubbery?

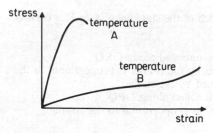

Figure 6L

Explain why this material shows a transition from glassy to rubbery behaviour as the temperature changes.

14 (a) Distinguish between (i) polycrystalline, and (ii) single crystal specimens of metals. Illustrate your answer with appropriate sketches.

(b) By considering the behaviour of a metal during elongation, explain the terms *strain, tensile stress, yield stress* and *breaking stress*.

Test pieces of two metal alloys of identical size and shape were subjected to tensile strength tests up to fracture (X), and the stress–strain diagrams (drawn to the same scale) which were obtained are shown in Figure 6M.

Explain which of the two materials (i) has the greater tensile strength, (ii) is the more ductile, (iii) exhibits greater toughness.

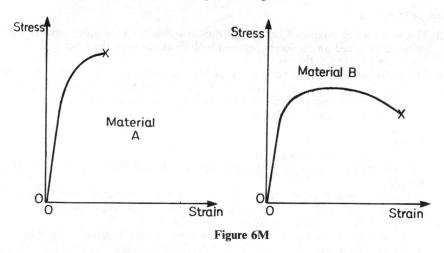

Figure 6M

(c) What is a composite material? Distinguish between layer and particle composite materials. Give an example of each type and state its engineering application. (*L.*)

15 (a) Figure 6N is a graph showing how the extension of a steel wire of length 1·2 m and area of cross-section 0·012 mm² alters as a stretching force is applied.

(i) Use the graph to calculate the Young modulus for steel. (ii) Draw a labelled diagram of an experimental arrangement suitable for obtaining such a set of results.

(b) Figure 6O shows the results of a similar experiment done with a copper wire. In this case the wire has been stretched until it broke. (i) The graph drawn in this instance is a stress–strain curve. Explain *one* advantage of representing the results in this way. (ii) Account in molecular terms for the behaviour of the wire as it is stretched from A to B. (iii) The copper wire used was 2·0 m long and 0·25 mm² in cross-section. Calculate the tension in the wire at A and an approximate value for the work done in producing a strain of 0·1.

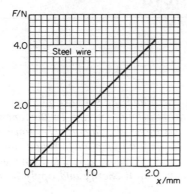

Figure 6N

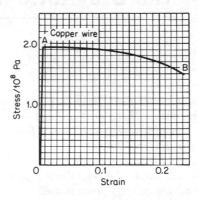

Figure 6O

(c) A length of rubber cord is suspended from a rigid support and stretched by means of weights attached to its lower end. (i) Sketch a stress–strain curve to represent the behaviour of such a cord as it is first loaded then unloaded. (ii) Suppose the cord were continuously stretched and relaxed at a rapid rate. What might you notice? How would this be explained by the stress–strain graph? (*L.*)

7 The Gravitational Field

Gravitational fields, such as those round the earth and the sun, are places where masses have gravitational forces and gravitational energy. In this chapter we deal first with gravitational forces. We shall apply the famous Newton law of gravitational force between masses to (a) the motion of planets round the sun, (b) TV satellites placed round the earth, and (c) moon satellites launched from the earth. We then discuss gravitational potential energy and apply it to the escape velocity from planets.

Kepler's Laws

Kepler (1571–1630) had studied for many years the records of observations on the planets made by Tycho Brahe, and discovered three laws now known by his name. *Kepler's laws* state:

Law 1   The planets describe ellipses about the sun as one focus.
Law 2   The line joining the sun and the planet sweeps out equal areas in equal times.
Law 3   The squares of the periods of revolution of the planets are proportional to the cubes of their mean distances from the sun.

Newton's Investigation on Planetary Motion

About 1666, at the early age of 24, Newton investigated the motion of a planet moving in a circle round the sun S as centre, Figure 7.1 (i). The force acting on the planet of mass m is $mr\omega^2$, where r is the radius of the circle and ω is the angular speed of the motion (p. 61). Since $\omega = 2\pi/T$, where T is the period of the motion,

$$\text{force on planet} = mr\left(\frac{2\pi}{T}\right)^2 = \frac{4\pi^2 mr}{T^2}$$

This is equal to the force of attraction of the sun on the planet. *Assuming an inverse-square law* for the distance r, then, if k is a constant,

$$\text{force on planet} = \frac{km}{r^2}$$

$$\therefore \frac{km}{r^2} = \frac{4\pi^2 mr}{T^2}$$

$$\therefore T^2 = \frac{4\pi^2}{k} r^3$$

$$\therefore T^2 \propto r^3$$

since k, π are constants.

Now Kepler had announced that the squares of the periods of revolution of the planets are proportional to the cubes of their mean distances from the sun (see above). Newton thus suspected that *the force between the sun and the planet was inversely proportional to the square of the distance between them.*

Motion of Moon round Earth

Newton now tested the inverse-square law by applying it to the case of the moon's motion round the earth, Figure 7.1 (ii). The moon has a period of

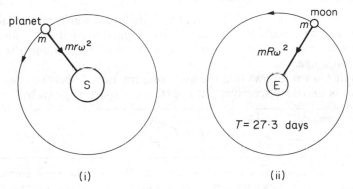

Figure 7.1 *Satellites*

revolution, T, about the earth of approximately 27·3 days, and the force on it $= mR\omega^2$, where R is the radius of the moon's orbit and m is its mass.

$$\therefore \text{force} = mR\left(\frac{2\pi}{T}\right)^2 = \frac{4\pi^2 mR}{T^2}$$

If the moon were at the earth's surface, the force of attraction on it due to the earth would be mg, where g is the acceleration due to gravity, Figure 7.1 (ii). Assuming that the force of attraction varies as the inverse square of the distance between the earth and the moon, then, by ratio of the two forces,

$$\frac{4\pi^2 mR}{T^2} : mg = \frac{1}{R^2} : \frac{1}{r_E^2}$$

where r_E is the radius of the earth. Cancelling m and simplifying,

$$\frac{4\pi^2 R}{T^2 g} = \frac{r_E^2}{R^2}$$

$$\therefore g = \frac{4\pi^2 R^3}{r_E^2 T^2} \qquad . \qquad . \qquad . \qquad . \qquad (1)$$

Newton substituted the then known values of R, r_E and T, but was disappointed to find that the answer for g was not near to the observed value, 9·8 m s^{-2}. Some years later, he heard of a new estimate of the radius of the earth, and we now know that r_E is about $6·4 \times 10^6$ m. The radius R of the moon's orbit is about $60·1 r_E$ and the period T of the moon is about 27·3 days or

27·3 × 24 × 3600 s. So

$$g = \frac{4\pi^2 R^3}{r_E^2 T^2} = \frac{4\pi^2 \times (60 \cdot 1 r_E)^3}{r_E^2 T^2} = \frac{4\pi^2 \times 60 \cdot 1^3 r_E}{T^2}$$

$$= \frac{4\pi^2 \times 60 \cdot 1^3 \times 6 \cdot 4 \times 10^6}{(27 \cdot 3 \times 24 \times 3600)^2} = 9 \cdot 9 \text{ m s}^{-2}$$

The result is very close to the measured value of g.

Newton's Law of Gravitation, G

Newton saw that a universal law could be stated for the gravitational attraction between any two particles of matter. He suggested that: *The force of the attraction between two given particles is inversely proportional to the square of their distance apart.*

From this law it follows that the force of attraction, F, between two particles of masses m and M respectively, at a distance r apart, is given by

$$F = G\frac{mM}{r^2} \qquad \qquad (2)$$

where G is a universal constant known as the **gravitational constant**. This expression for F is **Newton's law of gravitation**. It is a universal law because it appears to be true for masses all over the world.

From (2), $G = Fr^2/mM$. So G can be expressed in $\text{N m}^2 \text{ kg}^{-2}$. Careful measurement shows that $G = 6 \cdot 67 \times 10^{-11} \text{ N m}^2 \text{ kg}^{-2}$. The dimensions of G are

$$[G] = \frac{\text{MLT}^{-2} \times \text{L}^2}{\text{M}^2} = \text{L}^3\text{M}^{-1}\text{T}^{-2}$$

So the unit of G may also be expressed as $\text{m}^3 \text{ kg}^{-1} \text{ s}^{-2}$.

A celebrated experiment to measure G was carried out by C. V. Boys in 1895, using a method similar to one of the earliest determinations of G by Cavendish in 1798. Two identical balls, a, b, of gold, 5 mm in diameter, were suspended by a long and a short fine quartz fibre, respectively, from the ends, C, D, of a highly-polished bar CD, Figure 7.2. Two large identical lead spheres, A, B,

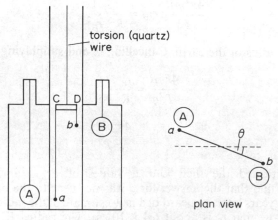

Figure 7.2 *Experiment on G*

115 mm in diameter, were brought into position near a, b respectively. As a result of the attraction between the masses, two equal but opposite forces acted on CD. The two forces form a couple, which has a turning effect or torque. The bar was thus deflected, and the angle of deflection, θ, was measured by a lamp and scale method by light reflected from CD. The high sensitivity of the quartz fibres enabled the small deflection to be big enough to be measured accurately. The small size of the apparatus allowed it to be screened considerably from air convection currents.

Calculation for G

Suppose d is the distance between a, A, or b, B, when the deflection is θ. Then if m, M are the respective masses of a, A,

$$\text{torque of couple on CD} = G\frac{mM}{d^2} \times \text{CD}$$

But $$\text{torque} = c\theta$$

where c is the torque in the torsion wire per unit radian of twist.

$$\therefore G\frac{mM}{d^2} \times \text{CD} = c\theta$$

$$\therefore G = \frac{c\theta d^2}{mM \times \text{CD}} \qquad . \qquad . \qquad . \qquad . \qquad (1)$$

The constant c was determined by allowing CD to oscillate through a small angle and then observing its period of oscillation, T, which was of the order of 3 minutes. If I is the known moment of inertia of the system about the torsion wire, then

$$T = 2\pi\sqrt{\frac{I}{c}}$$

$c = 4\pi^2 I/T^2$ and can now be calculated.

Gravitational Force on Masses, Relation between g and G

On the earth's surface, an object of mass m has a gravitational force or weight mg on it, where g is the acceleration of free-fall. So a mass of 1 kg has a weight of $1g$ or about 10 N, assuming g is 10 m s^{-2} at the earth's surface.

To find the gravitational force on masses on the earth or outside it, we can consider that the whole mass M_E of the earth is concentrated at its centre. Assuming the earth is a sphere of radius r_E, a mass m on the surface is then at a distance r_E from the mass M_E. If the same mass is taken above the earth to a distance $2r_E$ from the centre, the force between M_E and m is reduced to $1/2^2$ or 1/4, since the force between given masses is inversely proportional to the square of their distance apart. So now

$$\text{gravitational force} = \frac{1}{4} \times 10\ \text{N} = 2\cdot5\ \text{N}$$

For a mass m on the earth's surface of radius r_E, gravitational force $=$ $GMm/r_E^2 = mg$. Cancelling m on both sides, then

$$g = \frac{GM}{r_E^2}$$

As it is widely used, this relation between g and G should be memorised. From it, GM can be replaced in any formula by gr_E^2, where $g = 9.8$ m s^{-2} and $r_E = 6.4 \times 10^6$ m approximately.

Gravitational Field Strength E_G

'Field strength' is a technical term widely used in connection with fields, such as electric fields. The *gravitational field strength*, symbol E_G, at a point X in the field is defined as the *force per unit mass* or the *force per kilogram* on a mass at X. When a force F acts on a mass m at X, then, by definition,

$$E_G = F/m, \quad \text{so} \quad F = E_G m$$

The unit of E_G is N kg^{-1}. So if the field strength above the earth is 4 N kg^{-1}, the force F on a mass of 2 kg placed there is $F = 4 \times 2 = 8$ N.

The earth's gravitational field strength is connected with the acceleration of free fall, g. On the earth's surface, for example, the gravitational force F on a mass $m =$ weight of mass $= mg$. So E_G on the earth's surface $= F/m = g$. So if $g = 9.8$ m s^{-2}, then $E_G = 9.8$ N kg^{-1}. Similarly, if the acceleration of free fall g' high above the earth $= 2.5$ m s^{-2}, then E_G there $= g' = 2.5$ N kg^{-1}. At this place, the weight of a 2 kg mass $= 2 \times 2.5 = 5$ N. Generally, then, if the acceleration of free fall is g' at a point,

earth's field strength $= g'$ in N kg^{-1}

As we showed before, using Newton's law of gravitation,

$$g' = GM/r^2,$$

where r is the distance from the earth's centre of the point concerned.

Variation of Acceleration of Free-Fall or Field Strength

For points *outside* the earth, the gravitational force obeys an inverse-square law. So the acceleration of free-fall, g', $\propto 1/r^2$, where r is the distance to the centre of the earth, Figure 7.3. The maximum value of g' is obtained at the earth's surface, where $r = r_E$.

Inside the earth, the value of g' is *not* inversely-proportional to the square of the distance from the centre. Assuming a uniform earth density, which is not true in practice, theory shows that g' varies linearly with the distance from the centre, as shown in Figure 7.3.

Since the gravitational force F on a mass m is given generally by $F = mg'$, then $g' = F/m$. We see that g' can be expressed in 'newtons per kilogram' (N kg^{-1}). The *force per unit mass* in the gravitational field of the earth is called its *gravitational field strength*. We see that, on the earth, $g = 9.8$ m s$^{-2} = 9.8$ N kg^{-1}.

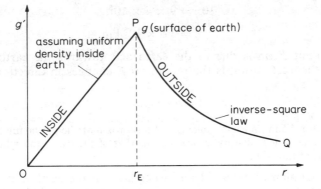

Figure 7.3 *Variation of g′, acceleration of free fall*

Example on Gravitation

1 *Earth–moon system*
The mass of the earth is 81 times that of the moon and the distance from the centre of the earth to that of the moon is about 4.0×10^5 km.

Calculate the distance from the centre of the earth where the resultant gravitational force becomes zero when a spacecraft is launched from the earth to the moon. Draw a sketch showing roughly how the gravitational force on the spacecraft varies in its journey.

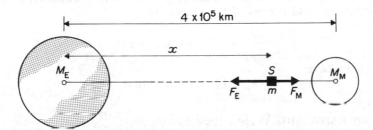

Figure 7.4 *Earth–moon gravitational force*

(*Analysis* (i) The gravitational force on the spacecraft S due to the earth is opposite in direction to that of the moon, (ii) $F = Gm/r^2$.)

Suppose the spacecraft S is a distance x in km from the centre of the earth and a distance $(4 \times 10^5 - x)$ from the moon when the resultant force is zero, Figure 7.4. If m is the spacecraft mass, then

$$\frac{GM_E m}{x^2} = \frac{GM_M m}{(4 \times 10^5 - x)^2}$$

Cancelling G and m and re-arranging,

$$\frac{M_E}{M_M} = \frac{81}{1} = \frac{x^2}{(4 \times 10^5 - x)^2}$$

Taking the square root of both sides,

$$9 = \frac{x}{4 \times 10^5 - x}$$

So
$$10x = 9 \times 4 \times 10^5$$
$$x = 3\cdot6 \times 10^5 \text{ km}$$

The resultant F on m due to the earth acts towards the earth until S is reached. It then acts towards the moon. So F changes in direction after S is passed.

2 *Variation of g*

A man can jump 1·5 m on earth. Calculate the approximate height he might be able to jump on a planet whose density is one-quarter that of the earth and whose radius is one-third that of the earth.

Suppose the man of mass m leaps a height h on the earth and a height h_1 on the planet. Assuming he can give himself the same initial kinetic energy on the two planets, the potential energy gained is the same at the maximum height. So

$$mg_1h_1 = mgh$$

where g_1 and g are the respective gravitational intensities on the planet and earth. So

$$h_1 = \frac{g}{g_1} \times h \qquad . \qquad . \qquad . \qquad . \qquad . \qquad (1)$$

But for the earth, $g = GM/r_E{}^2$ (p. 190) $= G.\frac{4}{3}\pi r_E{}^3 \rho_E / r_E{}^2 = G.\frac{4}{3}\pi r_E \rho_E$, where ρ_E is the density of the earth. Similarly, $g_1 = G.\frac{4}{3}\pi r_1 \rho_1$, where r_1, ρ_1 are the respective radius and density of the planet. So

$$\frac{g}{g_1} = \frac{r_E \rho_E}{r_1 \rho_1} = 4 \times 3 = 12$$

From (1), we have

$$h_1 = 12 \times 1\cdot5 \text{ m} = 18 \text{ m}$$

Force on Astronaut, Weightlessness

When a rocket is fired to launch a spacecraft and astronaut into orbit round the earth, the initial thrust must be very high owing to the large initial acceleration required. This acceleration, a, is of the order of $15g$, where g is the gravitational acceleration at the earth's surface.

Suppose S is the reaction of the couch to which the astronaut is initially strapped, Figure 7.5 (i). Then, from $F = ma$, $S - mg = ma = m.15g$, where m is the mass of the astronaut. Thus $S = 16mg$. This force is 16 times the weight of the astronaut and so, initially, he experiences a large force.

In orbit, however, the state of affairs is different. This time the acceleration of the spacecraft and astronaut are both g' in magnitude, where g' is the acceleration due to gravity at the particular height of the orbit, Figure 7.5 (ii). If S' is the reaction of the surface of the spacecraft in contact with the astronaut, then, for circular motion,

$$F = mg' - S' = ma = mg'$$

Thus $S' = 0$. The astronaut now experiences no reaction at the floor when he walks about, for example, and so he experiences the sensation of being 'weightless' although he has a gravitational force mg' acting on him.

At the earth's surface we feel the reaction at the ground and are thus conscious of our weight. Inside a lift which is falling fast, the reaction at our feet

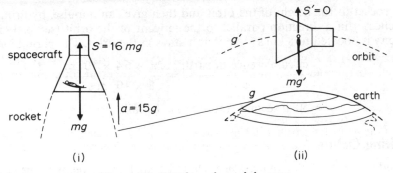

Figure 7.5 *Weight and weightlessness*

diminishes. If the lift falls freely, the acceleration of objects inside is the same as that outside and hence the reaction on them is zero. This produces the sensation of 'weightlessness'. In orbit, as in Figure 7.5 (ii), objects inside a spacecraft are also in 'free-fall' because they have the same acceleration g' as outside the spacecraft.

Earth Satellites

Satellites can be launched from the earth's surface to circle the earth. They are kept in their orbit by the gravitational attraction of the earth. Consider a satellite of mass m which just circles the earth of mass M close to its surface in an orbit 1, Figure 7.6. Then, if r_E is the radius of the earth,

$$\frac{mv^2}{r_E} = G\frac{Mm}{r_E{}^2} = mg$$

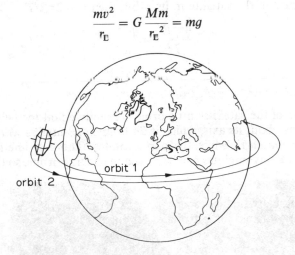

Figure 7.6 *Orbits round earth*

where g is the acceleration due to gravity at the earth's surface and v is the speed of m in its orbit. Thus $v^2 = r_E g$, and hence, using $r_E = 6.4 \times 10^6$ m and $g = 8.8$ m s^{-2},

$$v = \sqrt{r_E g} = \sqrt{6.4 \times 10^6 \times 9.8} = 8 \times 10^3 \text{ m s}^{-1} \text{ (approx.)}$$
$$= 8 \text{ km s}^{-1}$$

The speed v in the orbit is thus about 8 m s^{-1}. In practice, the satellite is carried by a rocket to the height of the orbit and then gives an impulse, by firing jets, to deflect it in a direction parallel to the tangent of the orbit (see p. 195). Its velocity is boosted to 8 km s^{-1} so that it stays in the orbit. The period in orbit

$$= \frac{\text{circumference of earth}}{v} = \frac{2\pi \times 6\cdot4 \times 10^6 \text{ m}}{8 \times 10^3 \text{ m s}^{-1}}$$

$$= 5000 \text{ seconds (approx.)} = 83 \text{ min}$$

Parking Orbits

Consider now a satellite of mass m circling the earth in the plane of the equator in orbit 2 concentric with the earth, Figure 7.6. Suppose the direction of rotation is the same as the earth and the orbit is at a distance R from the centre of the earth. Then if v is the speed in orbit,

$$\frac{mv^2}{R} = \frac{GMm}{R^2}$$

But $GM = gr_E^2$, where r_E is the radius of the earth.

$$\therefore \frac{mv^2}{R} = \frac{mgr_E^2}{R^2}$$

$$\therefore v^2 = \frac{gr_E^2}{R}$$

If T is the period of the satellite in its orbit, then $v = 2\pi R/T$

$$\therefore \frac{4\pi^2 R^2}{T^2} = \frac{gr_E^2}{R}$$

$$\therefore T^2 = \frac{4\pi^2 R^3}{gr_E^2} \quad . \quad . \quad . \quad . \quad . \quad (1)$$

If the period of the satellite in its orbit is exactly equal to the period of the earth as it turns about its axis, which is 24 hours, *the satellite will stay over the same place on the earth* while the earth rotates. This is sometimes called a 'parking orbit'. Relay satellites can be placed in parking orbits, so that television

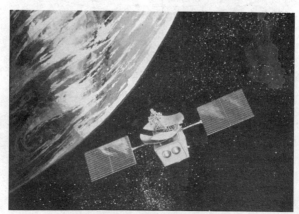

Plate 7A *Communications satellite above the earth*

programmes can be transmitted continuously from one part of the world to another.

Since $T = 24$ hours, the radius R can be found from (1). Its value is

$$R = \sqrt[3]{\frac{T^2 g r_E^2}{4\pi^2}} \quad \text{and} \quad g = 9\cdot8 \text{ m s}^{-2}, \ r_E = 6\cdot4 = 10^6 \text{ m}$$

$$\therefore R = \sqrt[3]{\frac{(24 \times 3600)^2 \times 9\cdot8 \times (6\cdot4 \times 10^6)^2}{4\pi^2}} = 42\,400 \text{ km}$$

The height above the earth's surface of the parking orbit

$$= R - r_E = 42\,400 - 6400 = 36\,000 \text{ km}$$

In the orbit, assuming it is circular, the speed of the satellite

$$= \frac{2\pi R}{T} = \frac{2\pi \times 42\,400 \text{ km}}{24 \times 3600 \text{ s}} = 3\cdot1 \text{ km s}^{-1}$$

The satellite, with the necessary electronic equipment inside, rises vertically from the equator when it is fired. At a particular height the satellite is given a horizontal momentum by firing rockets on its surface and the satellite then turns into the required orbit. This is illustrated in the next example.

Example on Satellite in Orbit

A satellite is to be put into orbit 500 km above the earth's surface. If its vertical velocity after launching is 2000 m s^{-1} at this height, calculate the magnitude and direction of the impulse required to put the satellite directly into orbit, if its mass is 50 kg. Assume $g = 10$ m s^{-2}, radius of earth, $r_E = 6400$ km.

Suppose u is the velocity required for orbit, radius R. Then, with usual notation,

$$\text{Force on satellite} = \frac{mu^2}{R} = \frac{GmM}{R^2} = \frac{g r_E^2 m}{R^2}, \text{ as } \frac{GM}{r_E^2} = g$$

$$\therefore u^2 = \frac{g r_E^2}{R}$$

Now $r_E = 6400$ km, $R = 6900$ km, $g = 10$ m s^{-2}

$$\therefore u^2 = \frac{10 \times (6400 \times 10^3)^2}{6900 \times 10^3}$$

$$\therefore u = 7700 \text{ m s}^{-1} \text{ (approx.)}$$

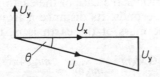

Figure 7.7 *Example on satellite*

At this height, vertical momentum

$$U_y = mv = 50 \times 2000 = 100\,000 \text{ kg m s}^{-1}$$

Horizontal momentum required $U_x = mu = 50 \times 7700 = 385\,000$ kg m s^{-1}

$\therefore$ *impulse needed*, U, $= \sqrt{U_y^2 + U_x^2} = \sqrt{100\,000^2 + 385\,000^2}$ (Figure 7.7)

$$= 4{\cdot}0 \times 10^5 \text{ kg m s}^{-1}$$

Direction. The angle θ made by the total impulse with the horizontal or orbit tangent is given by $\tan \theta = U_y/U_x = 100\,000/385\,000 = 0{\cdot}260$. Thus $\theta = 14{\cdot}6°$.

Mass and Density of Earth

At the earth's surface the force of attraction on a mass m is mg, where g is the acceleration due to gravity. Now it can be shown in this case that we can assume that the mass, M, of the earth is concentrated at its centre, if it is a sphere (p. 189). Assuming that the earth is spherical of radius r_E, it then follows that the force of attraction of the earth on the mass m is GmM/r_E^2. So

$$G\frac{mM}{r_E^2} = mg$$

$$\therefore g = \frac{GM}{r_E^2}$$

$$\therefore M = \frac{gr_E^2}{G}$$

Now, $g = 9{\cdot}8$ m s^{-1}, $r_E = 6{\cdot}4 \times 10^6$ m, $G = 6{\cdot}7 \times 10^{-11}$ N m^2 kg^{-2}

$$\therefore M = \frac{9{\cdot}8 \times (6{\cdot}4 \times 10^6)^2}{6{\cdot}7 \times 10^{-11}} = 6{\cdot}0 \times 10^{24} \text{ kg}$$

The volume of a sphere is $4\pi r^3/3$, where r is the radius. So the mean density, ρ, of the earth is approximately given by

$$\rho = \frac{M}{V} = \frac{gr_E^2}{4\pi r_E^3 G/3} = \frac{3g}{4\pi r_E G}$$

By substituting known values of g, G and r_E, the mean density of the earth is found to be about 5500 kg m^{-3}. The density of the earth is actually non-uniform and may approach a value of 10000 kg m^{-3} towards the interior.

Mass of Sun

The mass M_S of the sun can be found from the period of a satellite and its distance from the sun. Consider the case of the earth. Its period T is about 365 days or $365 \times 24 \times 3600$ seconds. Its distance r_S from the centre of the sun is about $1{\cdot}5 \times 10^{11}$ m. If the mass of the earth is m, then, for circular motion round the sun,

$$\frac{GM_S m}{r_S^2} = mr_S\omega^2 = \frac{mr_{ES}4\pi^2}{T^2}$$

$$\therefore M_S = \frac{4\pi^2 r_S^3}{GT^2} = \frac{4\pi^2 \times (1{\cdot}5 \times 10^{11})^3}{6{\cdot}7 \times 10^{-11} \times (365 \times 24 \times 3600)^2} = 2 \times 10^{30} \text{ kg}$$

In the equation $GM_S m/r_S^2 = mr_S\omega^2$ above, we see that the mass m of the satellite cancels on both sides and does not appear in the final equation for ω. So ω, the angular speed in the orbit, is *independent* of the mass of the satellite. The angular speed ω (and the period) depends only on the value of r_S, the orbit distance from the sun. This is true for all planets, that is

the angular speed of a planet depends only on the radius of the orbit and is independent of the mass of the planet.

Gravitational Potential

The *potential, V,* at a point due to the gravitational field of the earth is defined as numerically equal to the work done in taking a unit mass from infinity to that point. The potential at *infinity* is conventionally taken as *zero*. Points in electric fields have 'electric potential', as we see later.

For a point outside the earth, assumed spherical, we can imagine the whole mass M of the earth concentrated at its centre. The force of attraction on a unit mass outside the earth is thus GM/r^2, where r is the distance from the centre. The work done by the gravitational force in moving a distance Δr towards the earth = force × distance = $GM \cdot \Delta r/r^2$. So the potential at a point distant R from the centre outside the earth is given by

$$V_a = \int_{\infty}^{R} \frac{GM}{r^2}\, dr = -\frac{GM}{R} \qquad . \qquad . \qquad . \qquad (1)$$

if the potential at infinity is taken as zero by convention. The negative sign indicates that the potential at infinity (zero) is *higher* than the potential close to the earth.

On the earth's surface, of radius r_E, we therefore obtain

$$V = -\frac{GM}{r_E} \qquad . \qquad . \qquad . \qquad (2)$$

Substituting $G = 6\cdot7 \times 10^{11}\,\mathrm{N\,m^2\,kg^{-2}}$, $M = 6\cdot0 \times 10^{24}\,\mathrm{kg}$, $r_E = 6\cdot4 \times 10^6\,\mathrm{m}$, the potential V at the earth's surface is about $-6\cdot3 \times 10^7\,\mathrm{J\,kg^{-1}}$. Above the earth, the value of V will be *smaller* numerically than at the earth's surface since the distance to the earth's centre will then be greater than r_E.

Potential Changes

For large distances from the earth, for example, when a rocket travels from the earth to the moon, the change in potential energy of a mass can only be calculated by using *mass* × $(GM/R - GM/R_1)$, where R_1 and R are the distances from the centre of the earth. For small distances above the earth, however, the gravitational force on a mass is fairly constant. So the change in potential energy in this case can be calculated using *force* × *distance* or *mgh*.

Figure 7.8 shows roughly how the resultant force F on a satellite varies after it is launched from the earth E towards the moon M. The direction of F changes from F_1 at S_1, where the earth's gravitational pull is greater than that of the moon, to F_2 at S_2 near the moon, where the pull of this body is now stronger than that of the earth. At O, the gravitational pull of the earth is balanced by that of the moon.

The potential energy V of the satellite is the sum of its negative potential

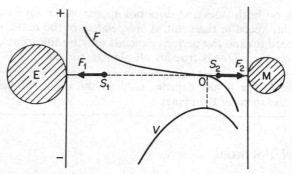

Figure 7.8 *Force (F) and potential (V) for moon satellite*

values due to the earth and the moon and is shown roughly in Figure 7.8. The maximum value of V occurs just below O. Here the resultant force F is zero. Since $F = -dV/dr$, the gradient dV/dr of the potential curve is then zero.

Velocity of Escape

Suppose a rocket of mass m is fired from the earth's surface Q so that it just escapes from the gravitational influence of the earth. Then work done = $m \times$ potential difference between infinity and Q

$$= m \times \frac{GM}{r_E}$$

$\therefore$ kinetic energy of rocket = $\frac{1}{2}mv^2 = m \times \dfrac{GM}{r_E}$

$$v = \sqrt{\frac{2GM}{r_E}} = \text{velocity of escape}$$

Now $$GM/r_E{}^2 = g$$

$$\therefore v = \sqrt{2gr_E}$$

$$\therefore v = \sqrt{2 \times 9 \cdot 8 \times 6 \cdot 4 \times 10^6} = 11 \times 10^3 \text{ m s}^{-1} = 11 \text{km s}^{-1} \text{ (approx.)}$$

With an initial velocity, then, of about 11 km s^{-1}, a rocket will completely escape from the gravitational attraction of the earth. It can be made to travel towards the moon, for example, so that eventually it comes under its gravitational attraction. At present, 'soft' landings on the moon have been made by firing retarding or retro rockets.

Summarising, with a velocity of about 8 km s^{-1}, a satellite can describe a circular orbit close to the earth's surface (p. 194). With a velocity greater than 8 km s^{-1} but less than 11 km s^{-1}, a satellite describes an elliptical orbit round the earth. Its maximum and minimum height in the orbit depends on its particular velocity. Figure 7.9 illustrates the possible orbits of a satellite launched from the earth.

The molecules of air at normal temperatures and pressures have an average velocity of the order of 480 m s^{-1} or 0·48 km s^{-1} which is much less than the velocity of escape. Many molecules move with higher velocity than 0·48 km s^{-1} but gravitational attraction keeps the atmosphere round the earth. The

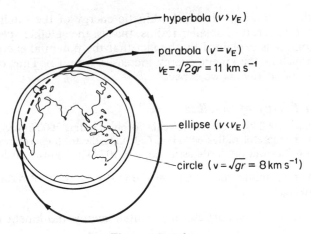

Figure 7.9 *Orbits*

gravitational attraction of the moon is much less than that of the earth and this accounts for the lack of atmosphere round the moon.

P.E. and K.E. of Satellite

A satellite of mass m in orbit round the earth has both kinetic energy, k.e., and potential energy, p.e. The k.e. $= \frac{1}{2}mv^2$, where v is the speed in the orbit. Now for circular motion in an orbit of radius r_0, if M is the mass of the earth,

$$\text{force towards centre} = \frac{mv^2}{r_0} = G\frac{Mm}{r_0{}^2}$$

$$\therefore \text{k.e.} = \frac{1}{2}mv^2 = G\frac{Mm}{2r_0} \qquad . \qquad . \qquad . \qquad . \qquad (1)$$

Assuming the zero of potential energy in the earth's field is at infinity (p. 197),

$$\textbf{p.e. of mass in orbit} = -G\frac{Mm}{r_0}. \qquad . \qquad . \qquad . \qquad (2)$$

So, from (1), the potential energy of the mass in orbit is numerically *twice* its kinetic energy and of opposite sign.

From (1) and (2),

$$\text{total energy in orbit} = -\frac{GMm}{r_0} + \frac{GMm}{2r_0}$$

$$= -\frac{GMm}{2r_0} \qquad . \qquad . \qquad . \qquad . \qquad (3)$$

Owing to friction in the earth's atmosphere, the satellite energy diminishes and the radius of the orbit decreases to r_1 say. The total energy in this orbit, from above, is $-GMm/2r_1$. Since this is *less* than the initial energy in (3), it follows that

$$\frac{GMm}{2r_1} > \frac{GMm}{2r_0}$$

From (1), these two quantities are the kinetic energy values in the respective orbits of radius r_1 and r_0. Hence the kinetic energy of the satellite *increases* when it falls to an orbit of smaller radius, that is, the satellite speeds up. This apparent anomaly is explained by the fact that the potential energy decreases by twice as much as the kinetic energy increases, from (2). Thus on the whole there *is* a loss of energy, as we expect.

Example on Energy of Satellite

A satellite of mass 1000 kg moves in a circular orbit of radius 7000 km round the earth, assumed to be a sphere of radius 6400 km. Calculate the total energy needed to place the satellite in orbit from the earth, assuming $g = 10$ N kg^{-1} at the earth's surface.

To launch the satellite, mass m, from the earth's surface of radius r_E into an orbit of radius r_0,

energy needed W = increase in potential energy and kinetic energy

$$= \frac{GMm}{r_E} - \frac{GMm}{r_0} + \tfrac{1}{2}mv^2$$

$$= \frac{GMm}{r_E} - \frac{GMm}{2r_0}$$

from equation (3) of the previous section. But $GM/r_E^2 = g$, or $GM/r_E = gr_E$.

So
$$W = mgr_E - \frac{mgr_E^2}{2r_0} = mg\left(r_E - \frac{r_E^2}{2r_0}\right)$$

$$= 1000 \times 10\left(6{\cdot}4 \times 10^6 - \frac{6{\cdot}4^2 \times 10^{12}}{2 \times 7 \times 10^6}\right)$$

$$= 3{\cdot}5 \times 10^{10} \text{ J}$$

Relation between Field Strength and Potential Gradient

Suppose a mass in a gravitational field is taken from A to B through a small distance Δr, Figure 7.10 (i). Then the work done per kg = force per kg $\times \Delta r$ = change in potential energy per kg from A to B.

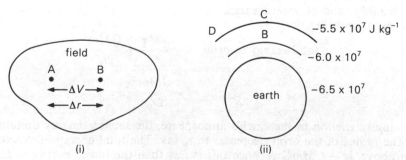

Figure 7.10 *Field strength and potential gradient*

Now force per kg = average field strength E_G in the region AB

So
$$E_G \times \Delta r = \Delta V$$

where ΔV is the potential energy change from A to B

Then
$$E_G = \frac{\Delta V}{\Delta r}$$

$\Delta V / \Delta r$ is the called the *potential gradient* in the region AB So

---
field strength = potential gradient numerically

The potential *decreases* when r increases. So, we write

$$E_G = \text{ — potential gradient} = -dV/dr$$

where dV/dr is the gradient at the point on the graph of V against r in the particular field.

Equipotentials

The gravitational potential outside a spherical mass M at a point distance r from its centre is given by $V = -GM/r$ (p. 197). So all points at a distance r have the same potential. These points lie on a sphere of radius r. So we call this sphere an *equipotential surface*.

Figure 7.10 (ii) shows part of some equipotentials round the earth and the values of V, their potentials. The distance from A to B is *less* than from B to C, although the work done in taking a mass from A to B = the work done from B to C from the equipotential values. This is because the field strength (force per kg) decreases further from the centre and so a greater distance is needed to do the same work.

If a satellite of mass 1000 kg moves from A to B,

$$\text{work done} = (V_B - V_A) \times 1000 = [(-5 \cdot 5 \times 10^7) - (-6 \cdot 0 \times 10^7)] \times 1000$$

$$= 0 \cdot 5 \times 10^7 \times 1000 = 5 \times 10^9 \text{ J}$$

The work done in taking a mass from D to C = 0, since D and C lie on an equipotential.

Black Holes

Gravitational attraction occurs throughout the universe. It is used in theories of galaxies and stars in addition to explaining the motion of planets and earth satellites as we have done.

Stars appear to have formed after a 'big bang' or explosion of matter when the universe first began. Initially, the hydrogen gas atoms produced moved towards each other under gravitational attraction, the speed and energy of the atoms increased and their temperature rose. Many atoms then had sufficient energy to fuse together and form helium atoms. The fusion of hydrogen atoms to form helium atoms produces nuclear energy (p. 899) and the high temperature rise produces light. A star reaches a stable size or radius when the outward pressure of the gas is balanced by the inward gravitational attraction between the atoms.

When all the hydrogen in the star is used up in the fusion process, which may occur after many millions of years, the unbalanced gravitational attraction between the atoms will result in the star becoming smaller and smaller. Eventually the star will collapse to a point of enormous density of matter. The light from the star is also affected as the star collapses. From the wave–particle duality of light (p. 861), light can be considered as particles (photons). So as the star collapses in size and becomes denser, the gravitational attraction on light from its surface increases. This makes the light bend more and more towards the inside of the star and soon no light escapes from the star.

So a collapsed star has such a strong gravitational field that not even light can escape from it, which is why it appears black. The region round the star from which no light can escape is called a *black hole*. Fortunately, our star, the sun, is estimated to continue producing nuclear energy for about 5000 million years. In 1994 the Hubble Space Telescope discovered one of the biggest black holes in the universe, about 500 light years across and in a galaxy 52 million light years away from the earth. (1 light year = 9.46×10^{12} km approx.)

You should know:

Gravitational Force

1 Newton's Law: $F = Gm_1 m_2/r^2$, where m_1, m_2 are small masses r apart. $G = 6.7 \times 10^{-12} Mm^2$ kg^{-2}.

2 Gravitational field strength E = force per kg at point. Unit: N kg^{-1}
   (a) On earth's surface, $E = g$ ($= 9.8$ N kg^{-1}) $= GM_E/r_E^2$. So $gr_E^2 = GM_E$.
   (b) Outside earth, $g'/g = r_E^2/r^2$, since $E \propto 1/r^2$.

3 Earth orbit: (a) $GM_E m/r^2 = mv^2/r = mr\omega^2 = mg'$.
   (b) $T = 2\pi r/v = 2\pi/\omega = 24$ hours for parking (TV) orbit.
   (c) Orbit of radius r depends only on speed of satellite and not on its mass.

4 'Weightlessness' means no reaction force on astronaut on contact with spacecraft.

Gravitational Potential

5 For mass m distance r from M, potential energy $= -GMm/r$ ($V_\infty = 0$, hence minus for V).

6 In orbit, kinetic energy of satellite $= GMm/2r$,

$$\text{total k.e.} + \text{p.e.} = -GMm/2r$$

a spacecraft returning to earth loses more p.e. than gain in k.e. and needs a heatshield for frictional heat from atmosphere.

7 Velocity of escape from earth $= \sqrt{2gr_E} = 11$ km s^{-1} (approx.).
   This is greater than average speed of gas molecules round earth, and hence earth has 'atmosphere'.
   On moon, low escape velocity produces no atmosphere.

8 Equipotential surfaces round the earth are concentric with the earth assumed spherical. Since $E = -$ potential gradient, and E increases near earth, equipotential surfaces are closer together near earth.

Satellites for TV and Surveillance

Satellite TV is available in the UK and many other countries. But what is the advantage of satellite TV over conventional, ground-based transmission systems? A single satellite sends pictures to a very large area of the earth's surface. For instance, the Astra satellite transmits to the whole of Europe. In contrast, a ground-based system needs many transmitters to cover even a single country, since hills and other obstacles easily obstruct the path of the signal.

As we saw on page 194, the satellite must be in a *parking orbit* if it is to sit over the same part of the earth all the time, which is clearly necessary for TV transmission. If you are ever lost in the Northern hemisphere, try to find a satellite TV receiving dish and see which way it is pointing—it will be pointing towards the south, where the satellite is parked over the equator. Surveillance satellites, for spying or looking at weather systems, ideally need to be in a parking orbit for continuous monitoring of the same area. Unfortunately, as we saw on page 195, the height of a parking orbit is 36 000 km above the earth's surface, which is too far away for high-resolution monitoring. Such satellites are therefore placed in lower level orbits, and pass over particular areas of the earth for only small amounts of time each orbit. If continuous monitoring of a single location is necessary, then there is no choice but to use a series of low-level satellites whose times over that location overlap, as the USA did over Moscow during the Cold War.

EXERCISES 7 Gravitation

(*Assume $g = 10$ N kg^{-1} unless otherwise stated*)

Multiple Choice

1 Figure 7A shows three uniform masses in a row. The force on the 1 kg mass is zero if the distance x in metres is

A 2 **B** 3 **C** 4 **D** 5 **E** 6

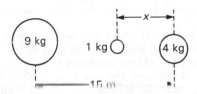

Figure 7A

2 A communications satellite has a circular orbit round the earth of three times the radius of the earth. The weight of the satellite on earth is 90 N. Its weight in N in orbit is

A 40 **B** 30 **C** 20 **D** 10 **E** 9

3 A satellite weighs 80 N at the earth's surface. If R is the earth's radius, at what distance from the earth would the weight of the satellite be 20 N?

A $R/4$ **B** $R/2$ **C** R **D** $2R$ **E** $3R$

4 The gravitational potential at the earth's surface is $-6\cdot3 \times 10^7$ J kg^{-1}. The earth's radius is $6\cdot4 \times 10^6$ m. Assuming the density of the earth is uniform, the gravitational potential at a distance $8\cdot0 \times 10^6$ m from the centre of the earth in J kg^{-1} is

A $+5\cdot0 \times 10^7$ **B** $+7\cdot9 \times 10^7$ **C** $-5\cdot0 \times 10^7$ **D** $-8\cdot0 \times 10^7$ **E** $-5\cdot0 \times 10^8$

5 Satellites P and Q are in the same circular orbit round the earth. The mass of P is greater than the mass of Q. Which is greater for P than Q in 1, 2, 3 and 4?
1 speed   2 gravitational force   3 kinetic energy   4 potential energy

A 1 only **B** 1 and 2 only **C** 1 and 4 only **D** 1, 2, 3 only **E** 2, 3, 4 only

Longer Questions

6 A satellite in a stable orbit contains two closed vessels: one of these is filled with water while the other is filled with hot steam. Explain why the water exerts very little pressure on its container but the steam exerts almost the same pressure as it would on earth at the same temperature.

7 The gravitational force on a mass of 1 kg at the earth's surface is 10 N. Assuming the earth is a sphere of radius R, calculate the gravitational force on a satellite of mass 100 kg in a circular orbit of radius $2R$ from the centre of the earth.

8 Assuming the earth is a uniform sphere of mass M and radius R, show that the acceleration of free-fall at the earth's surface is given by $g = GM/R^2$.

What is the acceleration of a satellite moving in a circular orbit round the earth of radius $2R$?

9 A planet of mass m moves round the sun of mass M in a circular orbit of radius r with an angular speed ω. Show (i) that ω is independent of the mass m of the planet, (ii) that in a circular orbit of radius $4r$ round the sun, the angular speed decreases to $\omega/8$.

10 Obtain the dimensions of G.

The period of vibration T of a star under its own gravitational attraction is given by $T = 2\pi/\sqrt{G\rho}$, where ρ is the mean density of the star. Show that this relation is dimensionally correct.

11 A satellite X moves round the earth in a circular orbit of radius R. Another satellite Y of the same mass moves round the earth in a circular orbit of radius $4R$. Show that (i) the speed of X is twice that of Y, (ii) the kinetic energy of X is greater than that of Y, (iii) the potential energy of X is less than that of Y.

Has X or Y the greater total energy (kinetic plus potential energy)?

12 Find the period of revolution of a satellite moving in a circular orbit round the earth at a height of 3.6×10^6 m above the earth's surface. Assume the earth is a uniform sphere of radius 6.4×10^6 m, the earth's mass is 6×10^{24} kg and G is 6.7×10^{-11} N m^2 kg^{-1}.

13 If the acceleration of free fall at the earth's surface is 9.8 m s^{-2}, and the radius of the earth is 6400 km, calculate a value for the mass of the earth. ($G = 6.7 \times 10^{-11}$ N m^2 kg^{-2}.) Give the theory.

14 Assuming the mean density of the earth is 5500 kg m^{-3}, that G is 6.7×10^{-11} N m^2 kg^{-2}, and that the earth's radius is 6400 km, find a value for the acceleration of free-fall at the earth's surface. Derive the formula used.

15 Two stars, masses 10^{20} kg and 2×10^{20} kg respectively, rotate about their common centre of mass with an angular speed ω. Assuming that the only force on a star is the mutual gravitational force between them, calculate ω. Assume that the distance between the stars is 10^6 km and that G is 6.7×10^{-11} N m^2 kg^{-2}.

16 A preliminary stage of spacecraft *Apollo 11*'s journey to the moon was to place it in an earth parking orbit. This orbit was circular, maintaining an almost constant distance of 189 km from the earth's surface. Assuming the gravitational field strength in this orbit is 9.4 N kg^{-1}, calculate

(a) the speed of the spacecraft in this orbit and

(b) the time to complete one orbit. (Radius of the earth = 6370 km.) (*L.*)

17 Explorer 38, a radio-astronomy research satellite of mass 200 kg, circles the earth in an orbit of average radius $3R/2$ where R is the radius of the earth. Assuming the gravitational pull on a mass of 1 kg at the earth's surface to be 10 N, calculate the pull on the satellite. (*L.*)

18 A satellite of mass 66 kg is in orbit round the earth at a distance of $5.7R$ above its surface, where R is the value of the mean radius of the earth. If the gravitational field strength at the earth's surface is 9.8 N kg^{-1}, calculate the centripetal force acting on the satellite.

Assuming the earth's mean radius to be 6400 km, calculate the period of the satellite in orbit in hours. (*L.*)

19 (a) Explain what is meant by *gravitational field strength*. In what units is it measured?

Starting with Newton's law of gravitation, derive an expression for g, the acceleration of free-fall on the surface of the earth, stating clearly the meaning of each symbol used. (Assume that the earth may be considered as a point mass located at its centre.)

(b) g may be found by measuring the acceleration of a freely falling body. Outline how you would measure g in this way, indicating the measurements needed and how you would calculate a value for g from them.

(c) At one point on the line between the earth and the moon, the gravitational field caused by the two bodies is zero. Briefly explain why this is so.
If this point is 4×10^4 km from the moon, calculate the ratio of the mass of moon to the mass of earth. (Distance from earth to moon $= 4 \cdot 0 \times 10^5$ km.) (L.)

20 (a) (i) State Newton's law of gravitation. Give the meaning of any symbol you use.
   (ii) Define *gravitational field strength*.
   (iii) Use your answers to (i) and (ii) to show that the magnitude of the gravitational field strength at the earth's surface is GM/R^2, where M is the mass of the earth, R is the radius of the earth and G is the gravitational constant.

(b) Define *gravitational potential*. Use the data below to show that its value at the earth's surface is approximately -63 M J kg^{-1}.

(c) A communications satellite occupies an orbit such that its period of revolution about the earth is 24 h. Explain the significance of this period and show that the radius, R_0, of the orbit is given by

$$R_0 = \sqrt[3]{\frac{GMT^2}{4\pi^2}}$$

where T is the period of revolution and G and M have the same meanings as in (a)(iii).

(d) Calculate the least kinetic energy which must be given to a mass of 2000 kg at the earth's surface for the mass to reach a point a distance R_0 from the centre of the earth. Ignore the effect of the earth's rotation.

$G = 6 \cdot 7 \times 10^{-11} \text{ N m}^2 \text{ kg}^{-2}$ $M = 6 \cdot 0 \times 10^{24} \text{ kg}$ $R = 6 \cdot 4 \times 10^6 \text{ m}$.   (N.)

21 State the law of universal gravitation.
When a planet moves in a circular orbit of radius R about the sun, the centripetal force is provided by the gravitational attraction. Use this to show that the period T of such a planet is given by

$$T^2 = \frac{4\pi^2}{GM} R^3 \qquad . \qquad . \qquad . \qquad . \qquad . \qquad (1)$$

where G is the gravitational constant and M is the mass of the sun.
The graph in Figure 7B shows how log T varies with log R for the various planets.

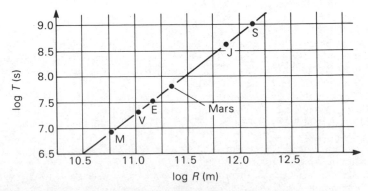

Figure 7B

Use equation (1) to find the equation for this graph. Explain how the graph verifies the relation $T^2 = \text{constant} \times R^3$.

Use the graph to find the radius and the time period for the orbit of the planet Mars. Use these values to calculate the mass of the sun.

(Gravitational constant, $G = 6 \cdot 67 \times 10^{-11}\,\text{N}\,\text{m}^{-2}\,\text{kg}^{-2}$.) (L.)

22 (a) The circle shown in Figure 7C represents a planet of radius of $4 \cdot 8 \times 10^6$ m with a gravitational field strength at the surface of $16\,\text{N}\,\text{kg}^{-1}$.

(i) Draw on the diagram a set of solid lines to represent the gravitational field of the planet. Add a set of dotted lines to represent gravitational equipotentials.

(ii) Calculate the mass of the planet.

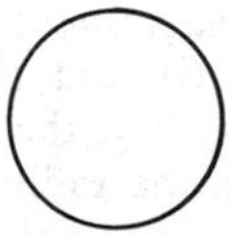

Figure 7C

(b) Two points, P and Q, lie above the surface of the planet. The gravitational potential at P is $-5 \cdot 2 \times 10^7\,\text{J}\,\text{kg}^{-1}$ and that at Q is $-6 \cdot 9 \times 10^7\,\text{J}\,\text{kg}^{-1}$.

(i) By comparison of these gravitational potentials, deduce which point, P or Q, is nearer to the surface of the planet. Explain your reasoning.

(ii) Calculate the difference in height of points P and Q above the surface of the planet. (O & C.)

8 Electrostatics, The Electric Field

In this chapter we begin with an account of the basic phenomena about static (stationary) electric charges. We then show that charges in an electrostatic field are analogous to masses in a gravitational field—they have forces acting on them and have electric potential energy. The ideas here are widely used in many branches of electricity, for example, in solid state physics which deals with diodes and amplifiers, in the cathode-ray tube and in the theory of the atom.

General Phenomena

If a rod of ebonite is rubbed with fur, or a fountain-pen with a coat-sleeve, it gains the power to attract light bodies, such as pieces of paper or aluminium foil or a piece of cork. The discovery that rubbed amber could attract silk was mentioned by THALES (640–548 B.C.). The Greek work for amber is *elektron*, and a body made attractive by rubbing is said to be 'electrified' or *charged*. This branch of electricity, the earliest discovered, is called *electrostatics*.

Briefly, we assume that the reader knows the following:

1 There are two kinds of electric charge (a 'charge' is an amount of electricity)—positive (+) charge and negative (−) charge.
2 Charging a rod by rubbing does not create or make electricity. The rubbing simply transfers, for example, negative charge from the rubber to the rod. This leaves the rubber with an equal amount of positive charge because the rubber and rod were originally neutral.
3 Like (similar) charges repel. Unlike charges attract.
4 Electrons carry *negative* charge. Metals are conductors because the free electrons in them can move along the metal. Insulators such as plastics have no free electrons in them.
5 An 'induced' charge on a metal X is one obtained without touching X. The induced charge is due to electrons moving along X when a charge is brought near to one end of X, as discussed shortly.

Conductors and Insulators, Positive and Negative Charges

A metal rod can be charged by rubbing with fur or silk, but only if it is held in a handle of glass or amber. The rod could not be charged if it were held directly in the hand. This is because electric charges can move along the metal and pass through the human body to the earth. The human body, metals and water are examples of *conductors*. Glass and amber are examples of *insulators*. Rubber is an insulator. Due to friction with the ground when a car is moving, charges build up on the car's rubber tyres. This produces a charge on the car's metal body which may affect passengers such as young children. Metal chains hanging from the car bumper to the ground can provide a discharge.

A suspended piece of cork is attracted to an electrified ebonite rod E, Figure 8.1 (i). But when we bring an electrified glass rod G towards the ebonite rod, the cork falls away, Figure 8.1 (ii). So the charge on the glass rod *opposes* the effect of the charge on the ebonite rod.

Benjamin Franklin, a pioneer of electrostatics, gave the name of 'positive electricity' to the charge on a glass rod rubbed with silk, and 'negative

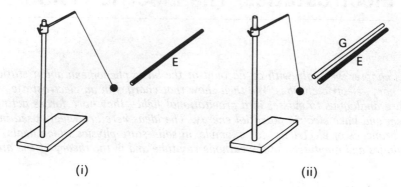

Figure 8.1 *Demonstrating that an electrified glass or acetate rod* G *tends to oppose the effect of electrified ebonite or polythene rod* E

electricity' to that on an ebonite rod rubbed with fur. Rubbed by a duster, a cellulose acetate rod obtains a positive charge and a polythene rod obtains a negative charge.

Experiment shows that two positive, or two negative, charges repel each other but a positive and a negative charge attract each other. So a fundamental law of electrostatics is:

Like (similar) charges repel. Unlike charges attract.

Electrons and Electrostatics

Towards the end of the nineteenth century Sir. J. J. Thomson discovered the existence of the *electron* (p. 739). This is a particle of very low mass—about 1/1840 of the mass of the hydrogen atom—and experiments show that it carries a tiny quantity of *negative* charge. Later experiments showed that electrons are present in all atoms.

The detailed structure of atoms is complicated, but generally, electrons move round a very tiny core or nucleus carrying *positive* charge. Normally, atoms are electrically neutral, that is, there is no surplus of charge on them. So the total negative charge on the electrons is equal to the positive charge on the nucleus. In insulators, all the electrons appear to be firmly 'bound' to the nucleus under the attraction of the unlike charges. In metals, however, some of the electrons appear to be relatively 'free'. These free electrons play an important part in electrical phenomena concerning metals, as we shall see.

Electrostatics Today

The discovery of the electron led to a considerable increase in the practical importance of electrostatics. In cathode-ray tubes and in electron microscopes, for example, electrons are moved by electrostatic forces. The problems of preventing sparks and the breakdown of insulators are essentially electrostatic. These problems occur in high voltage electrical engineering. Later, we shall also describe an electrostatic generator used to provide a million volts or more for X-ray work and nuclear bombardment. Such generators work on principles of electrostatics discovered over a hundred years ago.

Charging by Induction

Figure 8.2 shows how a conductor can be given a permanent charge by induction without touching it. We first bring a charged polythene rod C, say, near to the conductor XY, (i); next we connect the conductor to earth by touching it momentarily, (ii). Finally we remove the polythene. We then find that the conductor is left with a positive charge, (iii). If we use a charged

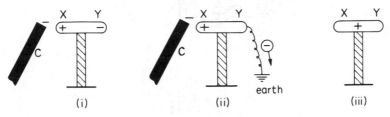

Figure 8.2 *Charging permanently by induction*

acetate rod, we find that the conductor is left with a negative charge. The charge left, called the induced charge, has always the *opposite* sign to the inducing charge.

This phenomenon of induction can be explained by the movement of electrons. In Figure 8.2 (i), the inducing charge C repels electrons to Y, leaving an equal positive charge at X as shown. When we touch the conductor XY, electrons are repelled from it to earth, as shown in Figure 8.2 (ii), and a positive charge is left on the conductor. If the inducing charge is positive, then the electrons are attracted up *from the earth to the conductor*, which then becomes negatively charged.

The Action of Points, Van de Graaff Generator

Electrostatic generators are machines for continuously separating charges by induction and so building up very great charges and high potential difference or voltage. Figure 8.3 is a simplified diagram of one such machine due to Van de Graaff and shown in Figure 8.4. A hollow metal sphere is supported on an insulating tube T, many metres high. A silk belt B runs over the pulleys shown, of which the lower is driven by an electric motor. Near the bottom and top of its run, the belt passes close to the electrodes E, which are sharply pointed combs, pointing towards the belt. The electrode E_1 is made about 10 000 volts positive with respect to the earth by a battery A.

As shown later, the high electric field at the points ionises the air there; and so positive charges are repelled on to the belt, which carries it up into the sphere. There it induces a negative charge on the points of electrode E_2 and a positive charge on the sphere to which the blunt end of E_2 is connected. The high electric field at the points ionises the air there, and negative charges, repelled to the belt, discharge the belt before it passes over the pulley. In this way the sphere gradually charges up positively, until its potential is about a million volts relative to the earth.

Large machines of this type are used with high-voltage X-ray tubes, and for atom-splitting experiments. They have more elaborate electrode systems, stand about 15 m high, and have 4 m spheres. They can produce potential differences or voltages up to 5 000 000 volts and currents of about 50 microamperes. The *electrical energy* which they deliver comes from the work done by the motor in

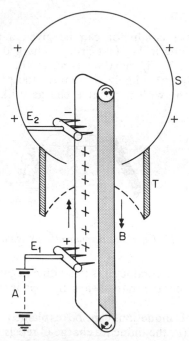

Figure 8.3 *Principle of Van de Graaff generator*

Plate 8A *Van de Graaff electrostatic generator at Aldermaston, England. The dome is the high-voltage terminal. The insulated rings are equipotentials, and provide a uniform potential gradient down the column. Beams of protons or deuterons, produced in the dome, are accelerated down the column to bombard different materials at the bottom, so producing nuclear reactions which can be studied*

drawing the positively charged belt towards the positively charged sphere, which repels it.

In all types of high-voltage equipment sharp corners and edges must be avoided, except where points are deliberately used as electrodes. Otherwise, electric discharges called 'corona' discharges may occur at the sharp places.

The Electrostatic Field

Electric Fields and Lines of Electric Force

As we saw in Chapter 7 on Gravitation, masses in the earth's gravitational field have forces on them due to the earth's gravitational attraction. In a similar way, the *electric field* in the space round a charge Q produces a force on a charge placed near Q.

To show the magnetic field pattern round a magnet, we draw magnetic lines of force round the magnet. To show the electric field pattern round charges, we draw electric lines of force. An arrow on the line shows the direction of movement of a *positive* charge in the field. And the closeness or density of the lines at a place shows how strong the field is there. Figure 8.4 (i) and (ii) shows some lines of force between positive and negative charges when an ebonite rod is charged by rubbing with fur and the effect of the wall of a room. Note that the lines of force radiate outwards from a positive (+) charge and inwards to a negative (−) charge.

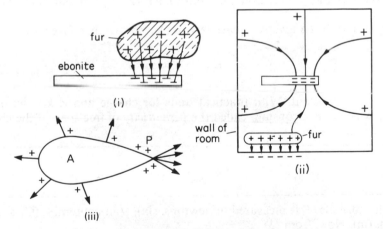

Figure 8.4 *Field lines of force*

The charge per unit area over a region of the body is called the *surface-density* of the charge in that region. We find that

---

the surface-density increases with the curvature of the body,

---

as shown roughly in Figure 8.4 (iii) on a pear-shaped charge conductor A. The point P has a high density of charge because it has high curvature. Other parts with less curve have lower charge density.

Generally, a charged conductor with a sharp point (such as a lightning conductor) has a high surface density of charge at that point. For this reason pointed conductors are used in the Van de Graaff generator described earlier.

We now consider in more detail the electrostatic fields due to charges. You will meet formulae for forces and potentials in these fields. These are needed, for example, in designing a television tube. The tube has static charges on metals inside. The charges produce electric fields which are needed to make electrons move in a special way in order to get a clear picture on the screen.

Law of Force Between Two Charges

The magnitude of the force between two electrically charged bodies was studied by COULOMB in 1875. He showed that, if the bodies were small compared with the distance between them, then the force F was inversely proportional to the square of the distance r,

$$F \propto \frac{1}{r^2} \qquad . \qquad . \qquad . \qquad . \qquad . \qquad (1)$$

This result is known as the *inverse square law*, or Coulomb's law.

Fundamental Law of Force

The SI unit of charge is the *coulomb* (C.). The coulomb is defined as that quantity of charge which passes a section of a conductor in one second when the current flowing is one ampère. The ampère is defined later.

By measuring the force F between two charges when their respective magnitudes Q and Q' are varied, we find that F is proportional to the product QQ'.

Together with the inverse-square law in (1), we can therefore write

$$F = k \frac{QQ'}{r^2} \qquad . \qquad . \qquad . \qquad . \qquad (2)$$

where k is a constant. With practical units for charge and r, k is written as $1/4\pi\varepsilon_0$, where ε_0 is a constant called the *permittivity* of free space if the charges are in a vacuum. So

$$F = \frac{1}{4\pi\varepsilon_0} \frac{QQ'}{r^2} \qquad . \qquad . \qquad . \qquad . \qquad (3)$$

In this formula, F is measured in newtons, (N), Q in coulombs, (C) and r in metres, (m). Now, from (3),

$$\varepsilon_0 = \frac{QQ'}{4\pi F r^2}$$

So the units of ε_0 are coulomb[2] newton[-1] metre[-2] ($C^2\,N^{-1}\,m^{-2}$). Another unit of ε_0, more widely used, is *farad metre*[-1] ($F\,m^{-1}$). See p. 244.

We shall see later that ε_0 has the numerical value of $8 \cdot 854 \times 10^{-12}$, and $1/4\pi\varepsilon_0$ then has the value 9×10^9 approximately.

So in free space we can write (3) as approximately

$$F = 9 \times 10^9 \frac{QQ'}{r^2} \qquad . \qquad . \qquad . \qquad . \qquad (4)$$

This is useful to simplify calculations, as we shall see.

Permittivity, Relative Permittivity

So far we have considered charges in a vacuum. If charges are in other media such as water, then the force between the charges is *reduced*. Equation (3) on page 212 is true only in a vacuum. In general, we write

$$F = \frac{1}{4\pi\varepsilon} \frac{QQ'}{r^2} \qquad . \qquad . \qquad . \qquad . \qquad (1)$$

where ε is the *permittivity* of the medium. The permittivity of air at normal pressure is only about 1·005 times that of a vacuum. For most purposes, therefore, we may assume that value of ε_0 for the permittivity of air. The permittivity of water is about eight times that of a vacuum. So the force between charges situated in water is eighty times less than if they were situated the same distance apart in a vacuum.

For this reason common salt (sodium chloride) dissolves in water. The electrostatic forces of attraction between the positive sodium ions and the negative chlorine ions, which keep the solid crystal structure in equilibrium, are reduced considerably by the water and the solid structure collapses.

The *relative permittivity*, ε_r, of a medium is the ratio of its permittivity ε to that of a vacuum, ε_0. So

$$\varepsilon_r = \varepsilon/\varepsilon_0$$

Although ε and ε_0 have dimensions, ε_r is a number and has no dimensions.

Examples on Force Between Charges

1 Figure 8.5 shows three small charges A, B and P in a line. The charge at A is positive, that at B is negative and that at P is positive. The values are those shown.
(a) Calculate the force on the charge at P due to A and B.
(b) At what point X on the line AB could there be *no* force on the charge P due to A and B if P were placed there?

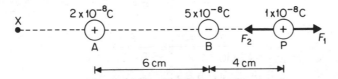

Figure 8.5 *Force on charges*

(a) The distance from A to P is 10 cm or 0·1 m. So charge at A *repels* charge at P with a force F_1 given by

$$F_1 = 9 \times 10^9 \frac{Q_1 Q_2}{r^2}$$

$$= \frac{9 \times 10^9 \times 2 \times 10^{-8} \times 1 \times 10^{-8}}{0\cdot 1^2}$$

$$= 1\cdot 8 \times 10^{-4}\,\text{N}$$

The distance from B to P is 4 cm or 4×10^{-2} m. So charge at B *attracts* charge at P with a force F_2 given by

$$F_2 = 9 \times 10^9 \frac{Q_1 Q_2}{r^2}$$

$$= \frac{9 \times 10^9 \times 5 \times 10^{-8} \times 1 \times 10^{-8}}{(4 \times 10^{-2})^2}$$

$$= 2 \cdot 8 \times 10^{-3} \, \text{N}$$

So resultant force towards B

$$= F_2 - F_1 = 2 \cdot 8 \times 10^{-3} - 1 \cdot 8 \times 10^{-4}$$

$$= 2 \cdot 8 \times 10^{-3} - 0 \cdot 18 \times 10^{-3} = 2 \cdot 62 \times 10^{-3} \, \text{N}$$

(b) If the charge at P were taken to a point X to the *left* of A on the line AB, there would be no force on the charge.

In this case the smaller charge at P would repel the positive charge at X and the negative charge at B would attract the positive charge at X. Although the charge at A is smaller than the charge at B, it is *nearer* the charge at X. So at some point such as X the two forces would be equal and opposite and cancel each other.

2   In Figure 8.6, two small equal charges 2×10^{-8}C are placed at A and B, one positive and the other negative. AB is 6 cm.

Find the force on a charge $+1 \times 10^{-8}$C placed at P, where P is 4 cm from the line AB along the perpendicular bisector XP.

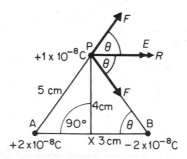

Figure 8.6 *Resultant force on charge*

From triangle APX, using Pythagoras, AP $= \sqrt{4^2 + 3^2} = \sqrt{25} = 5$ cm $= 5 \times 10^{-2}$ m. The charge at A repels the charge at P with a force F given by

$$F = \frac{9 \times 10^9 \times 2 \times 10^{-8} \times 1 \times 10^{-8}}{(5 \times 10^{-2})^2}$$

$$= 7 \cdot 2 \times 10^{-4} \, \text{N, in direction AP.}$$

The charge at B attracts the charge at P with a force also equal to F, because BP = 5 cm = AP. But this force acts in the direction PB. So the *resultant* force R on the charge at P is along the bisector PE of the angle between the two equal forces, as shown.

To find R, we can use the components of the two forces F along PE.

Then $\qquad\qquad\qquad R = F \cos\theta + F \cos\theta = 2F \cos\theta$

Now $\qquad\qquad\qquad \cos\theta = \text{BX/BP} = 3/5$

So $\qquad\qquad\qquad R = 2F \cos\theta = 2 \times 7\cdot2 \times 10^{-4} \times 3/5$

$$= 8\cdot64 \times 10^{-4}\,\text{N}$$

Electric Field-strength or Intensity, Field Patterns of Lines of Force

An 'electric field' can be defined as a region where an electric force is obtained. As in magnetism, electric fields can be mapped out by lines of electric force. A line of force may be defined as a line such that the tangent to it at a point X is in the direction of the force on a small *positive* charge placed at X. Arrows on the lines of force show the direction of the force on a positive charge; the force on a negative charge is in the opposite direction. Figure 8.7 shows the lines of force, also called *electric flux*, in some electrostatic fields of charges.

The force exerted on a charged body in an electric field depends on the charge on the body and on the *strength* or *intensity* of the field. If we wish to explore the variation in strength of an electric field, then we must place a test charge Q' at the points concerned which is small enough not to upset the field by its introduction. The strength E of the electric field at any point is defined as the *force per unit charge* which it exerts at X. The direction of E is that of the force on a *positive* charge. Note carefully that field strength E is a *vector* and so its direction is important when adding two or more values of E.

From this definition, if F is the force on a charge Q' at a point, then at this point

$$E = \frac{F}{Q'} \quad \text{so } F = EQ'. \qquad . \qquad . \qquad . \qquad (1)$$

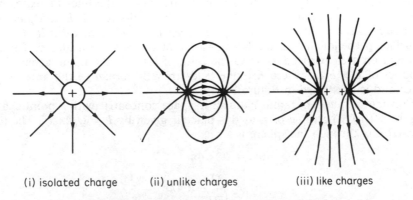

(i) isolated charge   (ii) unlike charges   (iii) like charges

Figure 8.7 *Field pattern of electric lines of force*

Since F is measured in newtons and Q' in coulombs, it follows that field-strength E *has units of newton per coulomb* (N C^{-1}). We shall see later that a more practical unit of E is *volt metre*$^{-1}$ (V m^{-1}) (p. 229).

Suppose a force F of 8×10^{-14} N is exerted on an electron of charge $1 \cdot 6 \times 10^{-19}$ C at a point in a television tube. The field strength E at the point $= F/Q' = 8 \times 10^{-14}$ N$/1 \cdot 6 \times 10^{-19}$ C $= 5 \times 10^5$ N C^{-1}.

Field-strength E due to Point Charge

We can easily find the strength E of the electric field due to a very small or point charge Q situated in a vacuum (Figure 8.8). We start from the equation for the force between two such charges:

$$F = \frac{1}{4\pi\varepsilon_0} \frac{QQ'}{r^2}$$

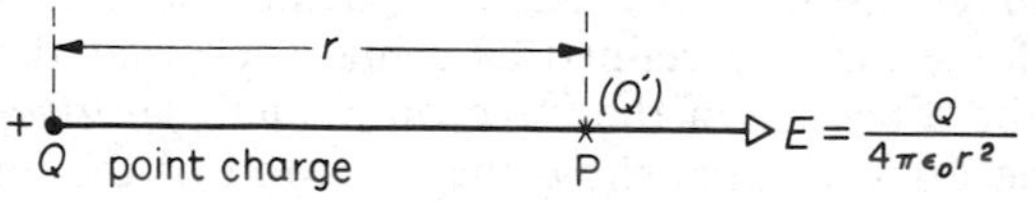

Figure 8.8 *Electric field-strength due to point charge*

If the test charge Q' is situated at the point P in Figure 8.8, the electric field-strength at P is

$$E = \frac{F}{Q'} = \frac{Q}{4\pi\varepsilon_0 r^2} \qquad \qquad (2)$$

The direction of the field is radially outward if the charge Q is positive (Figure 8.7 (i)); it is radially inward if the charge Q is negative. If the charge were surrounded by a material of permittivity ε instead of being in a vacuum, then

$$E = \frac{Q}{4\pi\varepsilon r^2} \qquad \qquad (3)$$

Flux from a Point Charge

We have already shown how electric fields can be described by lines of force. From Figure 8.7 (i) it can be seen that the density of the lines increases near the charge where the field-strength is high. The field-strength E at a point can thus be represented by *the number of lines per unit area* or *flux density* through a surface perpendicular to the lines of force at the point considered. The *flux* through an area perpendicular to the lines of force is the name given to the product of $E \times area$, where E is the field-strength *normal* to the area at that place and is illustrated in Figure 8.9 (i).

Consider a sphere of radius r drawn in space concentric with a point charge, Figure 8.9 (ii). The value of E at this place is given by $E = Q/4\pi\varepsilon r^2$. The total normal flux through the sphere is

$$E \times area = E \times 4\pi r^2$$

$$= \frac{Q}{4\pi\varepsilon r^2} \times 4\pi r^2 = \frac{Q}{\varepsilon}$$

So $\qquad \qquad E \times area = \dfrac{\text{charge inside sphere}}{\text{permittivity}} \qquad \qquad (1)$

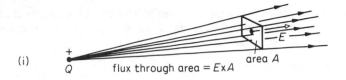

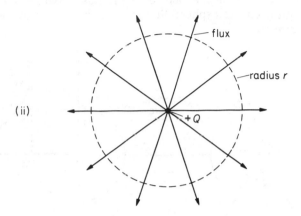

Figure 8.9 *Flux from a point charge*

This demonstrates the important fact that the total flux crossing normally any sphere drawn outside and concentrically around a point charge is a constant. It does not depend on the distance from the charged sphere.

It should be noted that this result is only true if the inverse square law is true. To see this, suppose some other force law were valid, so that $E = Q/4\pi\varepsilon r^n$ where n is a number. Then the total flux through the area

$$= \frac{Q}{4\pi\varepsilon r^n} \times 4\pi r^2 = \frac{Q}{\varepsilon} r^{(2-n)}$$

This is only independent of r if $n = 2$.

Field due to Charged Sphere and Plane Conductor

Equation (1) can be shown to be generally true. So the total flux passing normally through any *closed* surface whatever its shape, is always equal to Q/ε, where Q is the total charge enclosed by the surface. This relation, called *Gauss's theorem*, can be used to find the value of E in other common cases.

(1) *Outside a charged sphere*
The flux across a spherical surface of radius r, concentric with a small sphere carrying a charge Q (Figure 8.10 (ii)), is given by

$$\text{flux} = \frac{Q}{\varepsilon}$$

$$\therefore E \times 4\pi r^2 = \frac{Q}{\varepsilon}$$

$$\therefore E = \frac{Q}{4\pi\varepsilon r^2}$$

This is the same answer as that for a point charge. This means that *outside* a charged sphere, the field behaves as if all the charge on the sphere were concentrated at the centre.

(2) Inside a charged empty sphere

Suppose a spherical surface A is drawn *inside* a charged sphere, as shown in Figure 8.10 (i). Inside this sphere there are no charges. So Q in equation (1) above is zero. This result is independent of the radius drawn, provided that it is less than that of the charged sphere. So from (1), **E must be zero everywhere inside a charged sphere.**

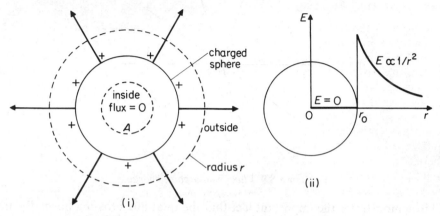

Figure 8.10 *Electric field of a charged sphere*

Figure 8.10 (ii) shows how E varies with the distance r from the *centre* of the sphere of radius r_0. E is zero from $r = 0$ to $r = r_0$. Beyond $r = r_0$, $E \propto 1/r^2$.

(3) Outside a charged plane conductor

Now consider a charged *plane* conductor S, with a surface charge density of σ coulomb metre^{-2}. Figure 8.11 shows a plane surface P, drawn outside S, which is parallel to S and has an area A square metre. Applying equation (1),

$$\therefore E \times \text{area} = \frac{\text{charge inside surface}}{\varepsilon}$$

Now by symmetry, the intensity in the field must be perpendicular to the surface. Further, the charges which produce this field are those in the projection of the

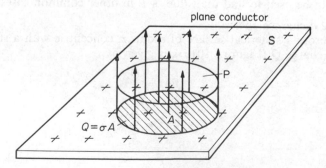

Figure 8.11 *Field of a charged plane conductor*

area P on the surface S, or those within the shaded area A in Figure 8.11. The total charge is here σA coulomb.

$$\therefore E \cdot A = \frac{\sigma A}{\varepsilon}$$

$$\therefore E = \frac{\sigma}{\varepsilon}$$

So high charge density shows a strong field.

Electrostatic Shielding

The fact that there is no electric field inside a closed conductor, when it contains no charged bodies, was demonstrated by Faraday in a spectacular manner. He made for himself a large wire cage, supported it on insulators, and sat inside it with electroscopes, which are electric field detectors. He then had the cage charged by an induction machine—a forerunner of the type we described on p. 209—until painful sparks could be drawn from its outside. Inside the cage Faraday sat in safety and comfort, however, and there was no deflection to be seen on even his most sensitive electroscope.

If we wish to protect any persons or instruments from intense electric fields, therefore, we enclose them in hollow conductors. These are called 'Faraday cages', and are widely used in high-voltage measurements in industry.

We may also wish to prevent charges in one place from setting up an electric field beyond their immediate neighbourhood. To do this we surround the charges with a Faraday cage, and connect the cage to earth (Figure 8.12). The

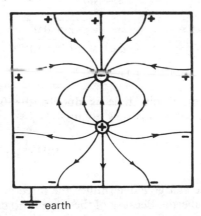

Figure 8.12 *Lines of force round charges*

charge induced on the outside of the cage then runs to earth, and there is no external field. (When a cage is used to shield something *inside* it, it does not have to be earthed.)

Field Round Points

On p. 211 we saw that the surface-density of charge (charge per unit area) round a point of a conductor is very great. Consequently, the strength of the electric field near the point is very great. The intense electric field breaks down the insulation of the air, and sends a stream of charged molecules away from the point. The mechanism of the breakdown, which is called a 'corona discharge', is complicated , and we shall not discuss it here.

Corona breakdown starts when the electric field-strength E in air is about 3 million volts per metre. The corresponding surface-density of charge σ is about $2 \cdot 7 \times 10^{-5}$ coulombs per square metre calculated from $E = \sigma/\varepsilon_0$.

Example on Electron Motion in Strong Field

An electron of charge $e = 1 \cdot 6 \times 10^{-19}$ C is situated in a uniform electric field of intensity or field-strength 120 000 V m^{-1}. Find the force on it, its acceleration, and the time it takes to travel 20 mm from rest (electron mass, $m = 9 \cdot 1 \times 10^{-31}$ kg).

$$\text{Force on electron } F = EQ = Ee$$

Now $E = 120\,000$ V m^{-1}:

$$\therefore F = 1 \cdot 6 \times 10^{-19} \times 1 \cdot 2 \times 10^5$$

$$= 1 \cdot 92 \times 10^{-14} \text{ N}$$

Acceleration, $\qquad a = \dfrac{F}{m} = \dfrac{1 \cdot 92 \times 10^{-14}}{9 \cdot 1 \times 10^{-31}}$

$$= 2 \cdot 12 \times 10^{16} \text{ m s}^{-2}$$

Time for 20 mm or 0·02 m travel is given by

$$s = \tfrac{1}{2}at^2$$

$$\therefore t = \sqrt{\frac{2s}{a}} = \sqrt{\frac{2 \times 0 \cdot 02}{2 \cdot 12 \times 10^{16}}}$$

$$= 1 \cdot 37 \times 10^{-9} \text{ s}$$

The extreme shortness of this time is due to the fact that the ratio of charge-to-mass for an electron is very great:

$$\frac{e}{m} = \frac{1 \cdot 6 \times 10^{-19}}{9 \cdot 1 \times 10^{-31}} = 1 \cdot 8 \times 10^{11} \text{ C kg}^{-1}$$

In an electric field, the charge e determines the force on an electron, while its mass m determines its inertia. Because of the large ratio e/m, the electron moves almost instantaneously, and requires very little energy to displace it. Also it can respond to changes in an electric field which take place even millions of times per second. So it is the large value of e/m for electrons which makes electronic tubes, for example, useful in electrical communication and remote control, explained later.

Electric Potential

Potential in Fields

When an object is held at a height above the earth it is said to have gravitational *potential energy*. A heavy body tends to move under the force of attraction of the earth from a point of great height to one of less, and we say that points in the earth's gravitational field have potential values depending on their height.

Electric potential is analogous to gravitational potential, but this time we think of points in an electric field. So in the field round a positive charge, for example, a positive charge moves from points near the charge to points further away. Points round the charge are said to have an 'electric potential'.

Potential Difference, Work, Energy of Charges

In mechanics we are always concerned with differences of height; if a point A on a hill is h metres higher than a point B, and our weight is w newtons, then we do wh joules of work in climbing from B to A, Figure 8.13 (i). Similarly in electricity we are often concerned with differences of potential; and we define these also in terms of work.

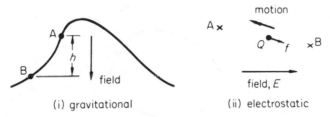

(i) gravitational   (ii) electrostatic

Figure 8.13 *Work done, in gravitational and electrostatic fields*

Let us consider two points A and B in an electrostatic field of strength E, and let us suppose that the force on a positive charge Q has a component f in the direction AB, Figure 8.13 (ii). Then if we move a positively charged body from B to A, we do work against this component of the field E. *We define the potential difference between A and B as the work done in moving a unit positive charge from B to A.* We use the symbol V_{AB} for this potential difference. So

potential difference V_{AB} = work per coulomb in moving charge from B to A

The work done will be measured in joules (J). The unit of potential difference is called the volt and may be defined as follows: *The potential difference between two points A and B is one volt if the work done in taking one coulomb of positive charge from B to A is one joule.* So

1 volt = 1 joule per coulomb (1 V = 1 J/C or 1 J C^{-1})

From this definition, if a charge of Q coulomb is moved through a p.d. of V volt, then the work done W, in joules, is given by

$$W = QV . \qquad . \qquad . \qquad . \qquad . \qquad . \qquad (1)$$

This is one of the key formulae in electricity. Work done produces *energy*. For example, as we see later, electric cookers provide heat energy, and filament lamps provide light energy, by charges on electrons moving from one end to the other in metals joined to the potential difference at the mains. So remember that QV = work done = energy produced.

Potential and Energy

Let us consider two points A and B in an electrostatic field, A being at a higher potential than B. The potential difference between A and B we denote as usual by V_{AB}. If we take a positive charge Q from B to A, we do work on it of amount QV_{AB}: the charge gains this amount of potential energy. If we now let the charge go back from A to B, it loses that potential energy: work is done on it by the electrostatic force, in the same way as work is done on a falling stone by gravity. This work may become kinetic energy, if the charge moves freely, or external work if the charge is attached to some machine, or a mixture of the two.

The work which we must do in first taking the charge from B to A does *not* depend on the path along which we carry it, just as the work done in climbing a hill does not depend on the route we take. If this were not true, we could devise a perpetual motion machine, in which we did less work in carrying a charge from B to A via X than it did for us in returning from A to B via Y, Figure 8.14.

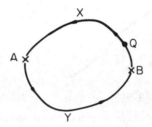

Figure 8.14 *A closed path in an electrostatic field*

The fact that the potential difference between two points is a constant, independent of the path chosen between the points, is a most important property of potential in general. This property can be conveniently expressed by saying that the work done in carrying a charge round any closed path in an electrostatic field, such as BXAYB in Figure 8.14, is zero.

As we also stress later, it shows that potential, symbol V, has magnitude but no direction. So

---

electric potential is a scalar.

This is in contrast to the electric field-strength E, which has both size and direction and is therefore a vector.

The Electron-Volt

The kinetic energy gained by an electron which has been accelerated through a potential difference of 1 volt is called an *electron-volt* (eV). Since the energy gained in moving a charge Q though a p.d. $V = QV$, and the electron charge $= 1.6 \times 10^{-19}$ C,

$$1 \text{ eV} = \text{electronic charge} \times 1 = (1.6 \times 10^{-19} \times 1) \text{ joule} = 1.6 \times 10^{-19} \text{ J}$$

The electron-volt is a useful unit of energy in atomic physics. For example, the work necessary to extract a conduction electron from tungsten is 4.52 eV. This quantity determines the size of the thermionic emission from the metal at a given temperature (p. 742); it is analogous to the latent heat of evaporation of a liquid. An electron in an X-ray tube moving through a p.d. of 50 000 V will gain energy equal to 50 000 eV.

Potential Difference due to Point Charge

We can now calculate the potential difference between two points in the field of a *single positive charge*, Q in Figure 8.15. For simplicity we will assume that the points, A and B, lie on a line of force at distances a and b respectively from the charge. When a unit positive charge is at a distance r from the charge Q in

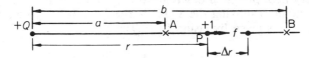

Figure 8.15 *Calculation of potential*

free space the force f on it is

$$f = \frac{Q \times 1}{4\pi\varepsilon_0 r^2}$$

The work done in taking the charge from B to A, against the force f over a short distance Δr is

$$\Delta W = f\Delta r$$

Over the whole distance AB, therefore, the work done by the force on the unit charge is

$$\int_A^B \Delta W = \int_{r=a}^{r=b} f \, dr = \int_a^b \frac{Q}{4\pi\varepsilon_0 r^2} \, dr$$

$$= -\left[\frac{Q}{4\pi\varepsilon_0 \phi}\right]_a^b = \frac{Q}{4\pi\varepsilon_0 a} - \frac{Q}{4\pi\varepsilon_0 b}$$

This, then, is the value of the work which an external agent must do to carry a unit positive charge from B to A. The work per coulomb is the potential difference V_{AB} between A and B.

$$\therefore V_{AB} = \frac{Q}{4\pi\varepsilon_0}\left(\frac{1}{a} - \frac{1}{b}\right) \qquad \cdot \qquad \cdot \qquad \cdot \qquad (1)$$

V_{AB} will be in volts if Q is in coulombs, a and b are in metres and ε_0 is taken as $8\cdot85 \times 10^{-12}$ F m^{-1} or $1/4\pi\varepsilon_0$ as 9×10^9 m F^{-1} approximately (see p. 212).

Example on Potential Difference and Work Done

Two positive point charges, of 12 and 8 µC (microcoulomb) respectively, are 10 cm apart. Find the work done in bringing them 4 cm closer so they are 6 cm apart. (Assume $1/4\pi\varepsilon_0 = 9 \times 10^9$ m F^{-1}.)

Suppose the 12 µC charge is fixed in position. Since 6 cm $= 0\cdot06$ m and 10 cm $= 0\cdot1$ m, then the potential difference between points 6 and 10 cm from it is given by (1).

$$\therefore V = \frac{12 \times 10^{-6}}{4\pi\varepsilon_0}\left(\frac{1}{0\cdot06} - \frac{1}{0\cdot1}\right)$$

$$= 12 \times 10^{-6} \times 9 \times 10^9(16\tfrac{2}{3} - 10)$$

$$= 720\,000 \text{ V}$$

(Note the very high potential difference due to quite small charges.)

The work done in moving the 8 μC charge from 10 cm to 6 cm away from the 12 μC charge is given by, using $W = QV$,

$$W = 8 \times 10^{-6} \times V$$

$$= 8 \times 10^{-6} \times 720\,000 = 5\cdot8 \text{ J}$$

Zero Potential, Potential at a Point

Instead of speaking continually of potential differences between pairs of points, we may speak of the potential at a single point—provided we always refer it to some other, agreed, reference point as a 'zero' of potential. This procedure is analogous to referring the heights of mountains to sea-level.

For practical purposes we generally choose as our zero reference point the electric potential of the surface of the *earth*. Although the earth is large it is all at the same potential, because it is a good conductor of electricity; if one point on it were at a higher potential than another, electrons would flow from the lower to the higher potential. As a result, the higher potential would fall, and the lower would rise; the flow of electricity would stop only when the potentials became equal.

In general, it is difficult to calculate the potential of a point relative to the earth. This is because the electric field due to a charged body near a conducting surface is complicated, as shown by the lines of force diagram in Figure 8.16. In theoretical calculations, therefore, we often find it convenient to consider charges so far from the earth that the effect of the earth on their field is negligible; we call these 'isolated' charges.

Figure 8.16 *Electric field of positive charge near earth*

In gravitation, the gravitational potential of the earth is taken as zero at an infinite distance from the earth. Theoretically, we take the electric potential due to a charge as *zero* at points which are an infinite distance from the charge. So we define the potential at a point A as

the work done per coulomb in bringing a positive charge from infinity to A.

Potential due to Point Charge and to Charged Sphere

Point charge

Equation (1), p. 223, gives the potential difference between two points A and B in the field of an isolated point charge Q:

$$V_{AB} = \frac{Q}{4\pi\varepsilon_0}\left(\frac{1}{a} - \frac{1}{b}\right)$$

If B is at infinity, then since b is very much greater than a, $1/b$ is negligible compared with $1/a$. So the potential at A is:

$$V_A = \frac{Q}{4\pi\varepsilon_0 a}$$

So at a distance r from a point charge Q, the potential V is:

$$V = \frac{Q}{4\pi\varepsilon_0 r} \qquad . \quad . \quad . \quad . \quad . \quad (1)$$

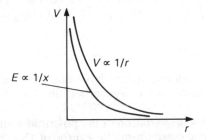

Figure 8.17 *Variation of V with r*

Figure 8.17 shows how V varies with the distance r from the point charge Q. As shown, the curve does not fall as rapidly as the curve of E, the field-strength, with r, since $E \propto 1/r^2$.

When Q is a positive charge, V is positive. This means that work is done by an *external* force in moving a positive charge to the point concerned since Q repels the charge. When Q is a negative charge, V is *negative*. This means that the *field itself* does work or loses energy when a positive charge is moved to the point concerned, since it is now attracted by Q.

Charged sphere

Suppose the charge on a spherical conductor is $+Q$ and the radius of the sphere is r_0, Figure 8.18 (i). The lines of force spread out radially from the surface and so we can imagine them starting from a charge Q concentrated at the centre point O. We have just seen that the potential at a distance r from a point charge Q is $V = Q/4\pi\varepsilon_0 r$. *Outside the sphere*, then, the potential at a point such as A, distance r from the centre, is

$$V = \frac{Q}{4\pi\varepsilon_0 r} \qquad . \quad . \quad . \quad . \quad . \quad (1)$$

So $V \propto 1/r$ along BC in Figure 8.18 (ii). At the *surface* S, where $r = r_0$, the

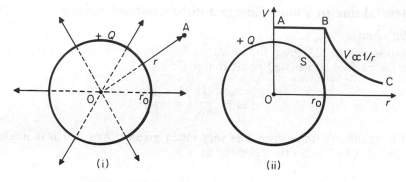

Figure 8.18 *Potential due to charged sphere*

radius, the potential is

$$V = \frac{Q}{4\pi\varepsilon_0 r_0} \qquad \cdot \qquad \cdot \qquad \cdot \qquad \cdot \qquad (2)$$

Inside the sphere, the electric field-strength $E = 0$ (p. 218). So no work is done when a charge is taken from *any* point inside to a point on the surface S. Therefore there is no potential difference between any point inside and S. But the potential of S $= Q/4\pi\varepsilon_0 r_0$. So the potential at any point inside the sphere is

$$V = \frac{Q}{4\pi\varepsilon_0 r_0} \qquad \cdot \qquad \cdot \qquad \cdot \qquad \cdot \qquad (3)$$

Note that *all* points inside the sphere have this same potential value, because $E = 0$ for all points inside, as we saw earlier. So the inside of the sphere is an *equipotential volume*.

Figure 8.18 (ii) shows the variation of the potential V due to a charged sphere with the distance r measured from the *centre* of the sphere. Note that V is constant ($= Q/4\pi\varepsilon_0 r_0$) from $r = 0$ to $r = r_0$ along the horizontal line AB.

Example on Potential Variation due to Charges

In Figure 8.19, a positively-charged sphere C is near a long insulated conductor AB.

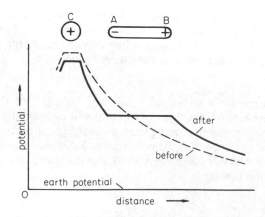

Figure 8.19 *Potential near positive charge before and after bringing up uncharged conductor*

Draw sketches to show how the potential V all round C varies with the distance from C measured along AB and beyond   (i) before and   (ii) after AB is placed in position.
    Draw a sketch showing the new variation of V with distance if AB is earthed.

(i) *Before* AB is placed in position, the variation of V with distance r is similar to the isolated spherical charged conductor in Figure 8.18(ii).
(ii) *After* AB is placed in position, there are now induced equal and opposite charges at A and B as shown. Since the charge on A is nearer C than B and opposite to that on C, the potential between C and A is now *less* than before.
    Further, the potential of a conductor such as AB is constant at all points on it. So this part of the graph is a horizontal straight line. Beyond the +ve charge at B, the potential decreases as shown.
Earthed conductor to AB.   There is now only a −ve charge at the end A and the potential falls rapidly to zero as shown in Figure 8.20. Beyond B, C has a small potential.

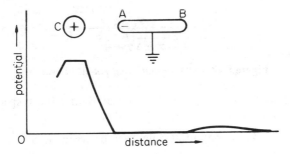

Figure 8.20 *Potential near positive charge in the presence of an earthed conductor*

Example on Potential Energy

In Figure 8.21, an alpha-particle A of charge $+3.2 \times 10^{-19}$ C and mass 6.8×10^{-27} kg is travelling with a velocity v of 1.0×10^7 m s^{-1} directly towards a nitrogen nucleus N which has a charge of $+11.2 \times 10^{-19}$ C.

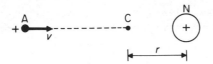

Figure 8.21 *Example on potential energy*

Calculate the closest distance of approach of A to N, assuming that A is initially a very long way from N compared with the closest distance of approach.

(*Analysis*   The alpha-particle loses kinetic energy as it approaches the nitrogen nucleus and this is transferred to electrical potential energy in the field of N.)

Initial kinetic energy of A $= \frac{1}{2}mv^2 = \frac{1}{2} \times 6.8 \times 10^{-27} \times (1.0 \times 10^7)^2$

$$= 3.4 \times 10^{-13} \text{ J}$$

If r is the closest distance of approach at C, potential energy at r due to N

$$= \frac{Q_1 Q_2}{4\pi\varepsilon_0 r} = \frac{9 \times 10^9 \times 3.2 \times 10^{-19} \times 11.2 \times 10^{-19}}{r}$$

$$= \frac{3.2 \times 10^{-27}}{r}$$

assuming $1/4\pi\varepsilon_0 = 9 \times 10^9$ and the initial potential energy of A is zero.

So $$3.4 \times 10^{-13} = \frac{3.2 \times 10^{-27}}{r}$$

$$\therefore r = \frac{3.2 \times 10^{-27}}{3.4 \times 10^{-13}} = 9.4 \times 10^{-15} \text{ m}$$

Potential Gradient and Field-strength (Intensity)

We shall now see how potential difference in an electric field is related to its field-strength or intensity. Suppose A, B are two neighbouring points on a line of force, so close together that the electric field-strength between them is constant and equal to E (Figure 8.22).

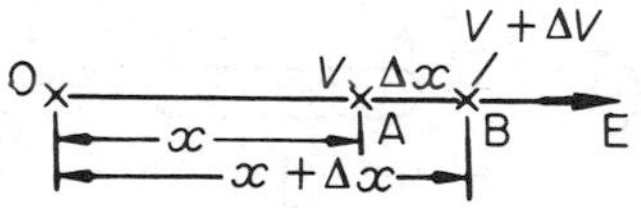

Figure 8.22 *Field-strength and potential gradient*

If V is the potential at A, $V + \Delta V$ is that at B, and the respective distances of A, B from the origin are x and $x + \Delta x$, then

$$V_{AB} = \text{potential difference between A, B}$$

$$= V_A - V_B = V - (V + \Delta V) = -\Delta V$$

The work done in taking a unit charge from B to A

$$= \text{force} \times \text{distance} = E \times \Delta x = V_{AB} = -\Delta V$$

So $$E = -\frac{\Delta V}{\Delta x}$$

or, in the limit,

$$E = -\frac{dV}{dx} \qquad . \qquad . \qquad . \qquad . \qquad . \qquad (1)$$

The quantity dV/dx is the rate at which the potential rises with distance, and is called the *potential gradient*. Equation (1) shows that the strength of the electric field is equal to the negative of the potential gradient.

Potential Variation in Fields, Unit of E

Strong fields have a high potential gradient and weak fields a low potential gradient. Strong and weak fields are illustrated in Figure 8.23.

Uniform fields, those with a constant field strength *and* a constant direction, are often needed in experiments with charges such as electrons. Parallel plates connected to a battery provide a uniform linear field if they are close to each other, as shown in Figure 8.24. Here the electric field-strength $= V/h$, the potential gradient, and this is uniform in magnitude in the middle of the plates. At the edge of the plates the field becomes non-uniform, as shown by the curved lines of force.

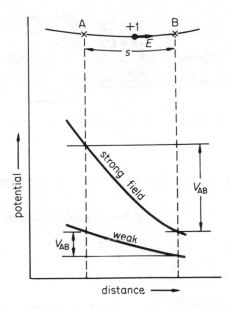

Figure 8.23 *Relationship between potential and field-strength*

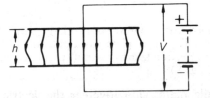

Figure 8.24 *Electric linear field between parallel plates; in middle, E = V/h*

We can now see why E can be given in units of 'volts per metre' (V m^{-1}). From (1), $E = -(\mathrm{d}V/\mathrm{d}x)$. Since V is measured in volts and x in metres, then E will be in volts per metre (V m^{-1}). From the original definition of $E(=F/Q)$, the units of E were newton coulomb^{-1} (N C^{-1}). To show these two units are equivalent, we have 1 joule = 1 newton × 1 metre from mechanics and so

$$1 \text{ volt} = 1 \text{ joule coulomb}^{-1}$$

$$= 1 \text{ newton metre coulomb}^{-1}$$

$$\therefore 1 \text{ volt metre}^{-1} = 1 \text{ newton coulomb}^{-1}$$

The examples which follow will show how forces on charges are calculated when the charges are in the electric linear field between parallel plates.

Examples on Potential Gradient and Field-strength

1 An oil drop of mass 2×10^{-14} kg carries a charge Q. The drop is stationary between two parallel plates 20 mm apart with a p.d. of 500 V between them, Figure 8.25. Calculate Q.

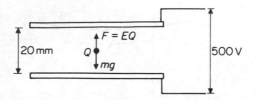

Figure 8.25 *Oil drop in electric linear field*

Since the drop is stationary,

upward force on charge, F = weight of drop, mg

Now $\qquad F = EQ = \dfrac{V}{d}Q$, since E = potential gradient $= \dfrac{V}{d}$

So $\qquad\qquad \dfrac{V}{d}Q = mg$

$$\therefore Q = \frac{mgd}{V} = 2 \times 10^{-14} \times 10 \times 20 \times 10^{-3}$$

$$= 8 \times 10^{-18} \text{ C}$$

2 An electron is liberated from the lower of two large parallel metal plates separated by a distance $h = 20$ mm. The upper plate has a potential of $+2400$ V relative to the lower. How long does the electron take to reach it? (Assume charge–mass ratio, e/m, for electron $= 1\cdot8 \times 10^{11}$ C kg^{-1}.)

Between large parallel plates, close together, the electric linear field is uniform except near the edges of the plates, as shown in Figure 8.24. Except near the edges, therefore, the potential gradient between the plates is uniform; its magnitude is V/h, where $h = 0\cdot02$ m, so

electric field-strength E = potential gradient

$$= 2400/0\cdot02 \text{ V m}^{-1}$$

$$= 1\cdot2 \times 10^5 \text{ V m}^{-1}$$

Force on electron of charge e is given by $F = Ee$.

$$\text{Acceleration, } a = \frac{F}{m} = \frac{Ee}{m}$$

$$= 1\cdot2 \times 10^5 \times 1\cdot8 \times 10^{11}$$

$$= 2\cdot16 \times 10^{16} \text{ m s}^{-2}$$

Then, from $s = \frac{1}{2}at^2$,

$$t = \sqrt{\frac{2s}{a}} = \sqrt{\frac{2 \times 20 \times 10^{-3}}{2\cdot16 \times 10^{16}}}$$

$$= 1\cdot4 \times 10^{-9} \text{ s}$$

Equipotentials

We have already said that the earth must have the same potential all over, because it is a conductor. In any conductor there can be no differences of potential. Otherwise these would set up a potential gradient or electric field and electrons would then redistribute themselves throughout the conductor, under the influence of the field, until they have destroyed the field. This is true whether the conductor has a net charge, positive or negative, or whether it is uncharged.

Any surface or volume over which the potential is constant is called an *equipotential*. The space inside a hollow charged conductor has the same potential as the surface at all points and so is an equipotential volume. The surface of a conductor of *any* shape, even pear-shaped, is an equipotential surface.

Equipotential surfaces can be drawn throughout any space in which there is an electric field. Figure 8.26 (i) shows the field of an isolated point charge Q.

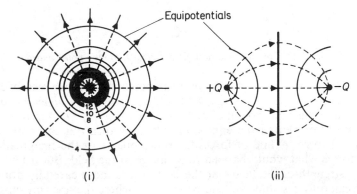

Figure 8.26 *Equipotentials and lines of force round* (i) *a point charge* (ii) *two opposite charges*

At a distance r from the charge, the potential is $Q/4\pi\varepsilon_0 r$; a sphere of radius r and centre at Q is therefore an equipotential surface, of potential $Q/4\pi\varepsilon_0 r$. In fact, all spheres centred on the charge are equipotential surfaces, whose potentials are inversely proportional to their radii, Figure 8.26 (i). Values proportional to the potentials are shown.

Figure 8.26 (ii) shows two equal and opposite point charges $+Q$ and $-Q$, and some typical lines of force and equipotentials in their field. The equipotential lines meet the lines of force at 90°.

An equipotential surface has the property that, along any direction lying in the surface, there is no electric field; for there is no potential gradient. *Equipotential surfaces are therefore always at right angles to lines of force*, as shown in Figure 8.26. Since conductors are always equipotentials, if any conductors appear in an electric-field diagram the lines of force must always be drawn to meet them at right angles.

The electric equipotentials in Figure 8.26 (i) are analogous to the gravitational equipotentials round the earth (p. 201). In both cases, $V \propto 1/r$ where r is greater than the radius of the earth. Figure 8.26 (ii) would be analogous to the case of two attracting masses in place of $+Q$ and $-Q$ respectively. But since gravitational masses do *not* repel, there is no gravitational analogy to the case of electric equipotentials due to two positive or to two negative charges.

Potential due to a System of Charges

When we consider the electric field due to more charges than one, we see the advantages of the idea of potential over the idea of field-strength. If we wish to find the field-strength E at the point P in Figure 8.27, due to the two positive charges Q_1 and Q_2, we have first to find the force exerted by each on a unit charge at P, and then to add these forces by a *vector* method such as the parallelogram method, shown in Figure 8.27.

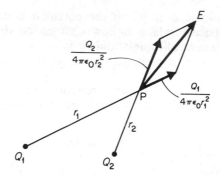

Figure 8.27 *Finding resultant field-strength of two point charges*

On the other hand, if we wish to find the potential at P, we merely calculate the potential due to each charge, and *add the potentials algebraically*, since potentials are scalar. So potential at P, $V = Q_1/4\pi\varepsilon_0 r_1 + Q_2/4\pi\varepsilon_0 r_2$.

When we have plotted the equipotentials, they turn out to be more useful than lines of force. A line of force diagram appeals to the imagination, and helps us to see what would happen to a charge in the field. But it tells us little about the strength of the field—at the best, if it is more carefully drawn than most, we can only say that the field is strongest where the lines are closest. But equipotentials can be labelled with the values of potential they represent; and from their spacing we can find the actual value of the potential gradient, and hence the field-strength. The direction of the field-strength is always at right angles to the equipotential curves.

Comparison of Static and Current Phenomena

Broadly speaking, we may say that in electrostatic phenomena we meet small quantities of charge, but great differences of potential. On the other hand in the phenomena of current electricity discussed later, the potential differences are small but the amounts of charge transported by the current are great. Sparks and shocks are common in electrostatics, because they require great potential differences; but they are rarely dangerous, because the total amount of energy available is usually small. On the other hand, shocks and sparks in current electricity are rare, but, when the potential difference is great enough to cause them, they are likely to be dangerous.

These quantitative differences make problems of insulation much more difficult in electrostatic apparatus than in apparatus for use with currents. The high potentials met in electrostatics make leakage currents relatively great, and the small charges therefore tend to disappear rapidly. Any wood, for example, ranks as an insulator for current electricity, but a conductor in electrostatics. In electrostatic experiments we sometimes wish to connect a charged body to earth; all we have then to do is to touch it.

Comparison between Electrostatic and Gravitational Fields

We conclude with a comparison between electrostatic (electric) fields and gravitational fields. Scientists consider that gravitational forces are the weakest in the universe and electric forces are much stronger.

Like charges repel and unlike charges attract in electric fields, so electric forces may be repulsive or attractive. In the gravitational field, masses always attract each other and no repulsive force has yet been detected. So electric potential may be positive or negative but gravitational potential is only negative (zero potential is at infinity in both cases). Further, the earth is such a large sphere that the gravitational field strength, g, near the earth's surface is fairly uniform for a height above the surface which is small compared to its radius.

The following table summarises some other points:

		Electric	Gravitational
1	Field-strength Unit	F/Q $N\,C^{-1}$	F/m $N\,kg^{-1}$
2	Force formula Force direction	$F = Q_1Q_2/4\pi\varepsilon_0 r^2$ attractive/repulsive	$F = Gm_1m_2/r^2$ attractive only
3	Strength outside isolated sphere—radial field	$E = \pm Q/4\pi\varepsilon_0 r^2$	$E = GM/r^2$
4	Potential outside isolated sphere	$V = \pm Q/4\pi\varepsilon_0 r$	$V = -GM/r$

You should know:

1 The Coulomb law for the force between two point charges in vacuum: $F = Q_1Q_2/4\pi\varepsilon_0 r^2$.
2 Electric field strength E at point X = force per coulomb at X. E is a vector. Unit: $N\,C^{-1}$. E is direction of a +ve charge at X.
3 *Outside* a spherical charged conductor, $E = Q/4\pi\varepsilon_0 r^2$, where r is the distance from the centre of sphere. Field is radial. *Inside* the sphere, $E = 0$.
4 Between parallel plates with charges $+Q$ and $-Q$, field is uniform ($E = \sigma/\varepsilon_0$). At plate edges, field is non-uniform (edge-effect).
5 The electric potential V at a point X in a field = work done per coulomb in moving a positive charge from infinity to X. Earth potential is a practical 'zero' of potential. V is a scalar, unit volt, V.
6 At a distance r from a point charge Q, $V = Q/4\pi\varepsilon_0 r$. *Outside* a charged sphere at distance r from centre, $V = Q/4\pi\varepsilon_0 r$. *Inside* a charged sphere, V has the same value everywhere and $V = Q/4\pi\varepsilon_0 r_0$, where r_0 is the radius of the sphere.

7 **Potential gradient $= V/d = E$, numerically. Use for E between parallel plates.**
   Unit of potential gradient $= \text{V m}^{-1} = \text{N C}^{-1}$ ($=$ unit of E).
8 **Equipotentials in fields are lines or surfaces or volumes which have the same potential along them. Surface of a charged conductor of any shape is an equipotential. Since no work is done in moving along an equipotential, the lines of force cut the equipotentials at 90° and so meet the surface of a conductor at 90°.**

Electrical Storms

Perhaps the most spectacular displays of electrostatic phenomena are the lightning strikes accompanying electrical storms. Thunderclouds usually have a positively charged upper portion, a negatively charged lower portion (the magnitude of each charge being about 30 coulombs) and a neutral central area. The negatively charged base of the cloud induces a corresponding positive charge on the earth below the cloud. This in turn sets up a strong electric field between the cloud and the earth. Forked lightning flashes occur when the electrical field is so great that the air *ionises*, or breaks up into positively and negatively charged ions. The positive ions flow up to the base of the cloud, neutralising the negative charge there, while the negative ones flow down to earth, neutralising the induced positive charge there: this ionisation is visible as the lightning flash.

Such a large transfer of charge in a short amount of time produces temperatures in the air as high as 30 000 K. The sudden heating causes the air around the strike to expand rapidly, setting into motion an intense longitudinal vibration of the air which can be heard as thunder. Since the sound of the thunder travels relatively slowly compared with the image of the lightning, to a distant observer the thunder arrives much later than the lightning strike, though if you're unfortunate enough to be very close to the strike the lightning and thunder will occur almost simultaneously. When lightning strikes a building it will flow to earth through the masonry and produce rapid heating. Any moisture in the lightning's path will boil, producing pockets of high pressure steam which can easily crack the masonry, often resulting in severe structural damage. This is why tall buildings are usually protected by *lightning conductors*, copper strips reaching from the highest point of the building to the earth below. A lightning charge will flow to earth down the lightning conductor, and so the surrounding masonry will be unaffected.

EXERCISES 8   Electric Field

Multiple Choice

1 In Coulomb's law, the force between two point charges is proportional to

   1 the square of the distance between them
   2 the product of the charges
   3 the permittivity of the medium

   A 1 and 2 only   **B** 1 and 3 only   **C** 2 and 3 only   **D** 1 only   **E** 2 only

2 Figure 8A(i) shows a charge $+Q$ a distance $2d$ from a charge $-Q$ and a point X distance d from $-Q$. The field strength at X is numerically, using SI units,

   A $Q/4\pi\varepsilon_0 d^2$ **B** $Q/36\pi\varepsilon_0 d^2$ **C** $3Q/4\pi\varepsilon_0 d^2$ **D** $2Q/9\pi\varepsilon_0 d^2$ **E** $Q/12\pi\varepsilon_0 d^2$

3 Figure 8A (ii) shows a small charge $+Q$ coulombs between the two parallel plates R, S separated 20 mm. A p.d. of 100 V is applied to R and S as shown. The force on the charge in N is

A $Q/5$ upwards   B $50Q$ downwards   C $200Q$ upwards
D $2000Q$ downwards   E $5000Q$ downwards

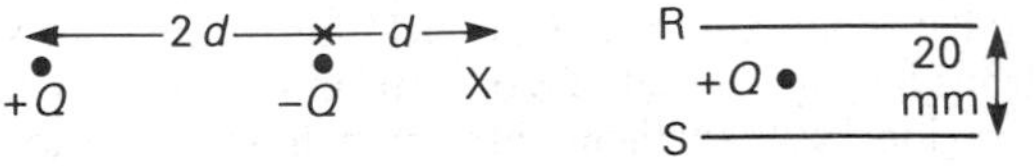

Figure 8A

4 In Figure 8A (i), the potential at X due to the charges $+Q$ and $-Q$ is, using SI units,

A $-2Q/9\pi\varepsilon_0 d$ B $-Q/6\pi\varepsilon_0 d$ C $+3Q/4\pi\varepsilon_0 d$ D $+Q/6\pi\varepsilon_0 d$ E $-2Q/9\pi\varepsilon_0 d$

Longer Questions

5 What is the *potential gradient* between two parallel plane conductors when their separation is 20 mm and a p.d. of 400 V is applied to them? Calculate the force on an oil drop between the plates if the drop carries a charge of 8×10^{-19} C.
6 Using the same graphical axes in each case, draw sketches showing the variation of potential   (i) inside and outside an isolated hollow spherical conductor A which has a positive charge,   (ii) between A and an insulated sphere B brought near to A, (iii) between A and B if B is now earthed.
7 A charged oil drop remains stationary when situated between two parallel horizontal metal plates 25 mm apart and a p.d. of 1000 V is applied to the plates. Find the charge on the drop if it has a mass of 5×10^{-15} kg. (Assume $g = 10$ N kg^{-1}.)
    Draw a sketch of the electric field between the plates and state if the field is everywhere uniform.
8 How do
   (a) the magnitude of the gravitational field, and
   (b) the magnitude of the electrostatic field, vary with distance from a point mass and a point charge respectively?
       Sketch a graph illustrating the variation of electrostatic field-strength E with distance r from the *centre* of a uniformly solid metal sphere of radius r_0 which is positively charged. Explain the shape of your graph   (i) for $r > r_0$, and (ii) for $r < r_0$. (L.)
9 (a) For each of the following, state whether it is a scalar or a vector and give an appropriate unit.
       (i) electric potential   (ii) electric field strength.
   (b) Points A and B are 0·10 m apart. A point charge of $+3·0 \times 10^{-9}$ C is placed at A and a point charge of $-1·0 \times 10^{-9}$ C is placed at B.
       (i) X is the point on the straight line through A and B, between A and B, where the electric potential is zero. Calculate the distance AX.
       (ii) Show on a diagram the approximate position of a point, Y, on the straight line through A and B where the electric field strength is zero. Explain your reasoning, but no calculation is expected. (N.)
10 The word 'field' is used in physics in connection with electrostatics and gravitation. For each of these explain what is meant by *field* and *field strength*.
    An electrostatic field may be produced by a small charge and a gravitational field by a small mass. Explain clearly
    (i) *two* ways in which these fields behave similarly, and
    (ii) *two* ways in which these fields behave differently. (L.)
11 Define
   (a) electric field strength
   (b) difference of potential.

How are these quantities related?

A charged oil-drop of radius 1.3×10^{-6} m is prevented from falling under gravity by the vertical field between two horizontal plates charged to a difference of potential of 8340 V. The distance between the plates is 16 mm, and the density of oil is 920 kg m^{-3}. Calculate the magnitude of the charge on the drop ($g = 9.81$ m s^{-1}). (*O. & C.*)

12 What is an *electric field*? With reference to such a field define *electric potential*.

Two plane parallel conducting plates are held horizontal, one above the other, in a vacuum. Electrons having a speed of 6.0×10^6 m s^{-1} and moving normally to the plates enter the region between them through a hole in the lower plate which is earthed. What potential must be applied to the other plate so that the electrons just fail to reach it? What is the subsequent motion of these electrons? Assume that the electrons do not interact with one another.

(Ratio of charge to mass of electron is 1.8×10^{11} C kg^{-1}. (*N.*)

13 (a) Define the terms *potential* and *field-strength* at a point in an electric field. Figure 8B shows two horizontal parallel conducting plates in a vacuum.

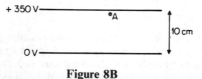

Figure 8B

A small particle of mass 4×10^{-12} kg, carrying a positive charge of 3.0×10^{-14} C is released at A close to the upper plate. What *total* force acts on this particle?

Calculate the kinetic energy of the particle when it reaches the lower plate.

(b) Figure 8C shows a positively charged metal sphere and a nearby uncharged metal rod.

Figure 8C

Explain why a redistribution of charge occurs on the rod when the charged metal sphere is brought close to the rod.

Copy this diagram and show on it the charge distribution on the rod. Sketch a few electric field lines in the region between the sphere and the rod.

Sketch graphs which show how (i) the potential relative to earth, and (ii) the field-strength vary along the axis of the rod from the centre of the charged sphere to a point beyond the end of the rod furthest from the sphere. How is graph (i) related to graph (ii)?

How will the potential distribution along this axis be changed if the rod is now earthed? (*L.*)

14 Define the *electric potential V* and the *electric field-strength E* at a point in an electrostatic field. How are they related? Write down an expression for the electric field-strength at a point close to a charged conducting surface, in terms of the surface density of charge.

Corona discharge into the air from a charged conductor takes place when the potential gradient at its surface exceeds 3×10^6 V m^{-1}; a potential gradient of this magnitude also breaks down the insulation afforded by a solid dielectric. Calculate the greatest charge that can be placed on a conducting sphere of radius 20 cm supported in the atmosphere on a long insulating pillar; also calculate the corresponding potential of the sphere. Discuss whether this potential could be achieved if the pillar of insulating dielectric was only 50 cm long. (Take ε_0 to be 8.85×10^{-12} F m^{-1}.) (*O.*)

9 Capacitors

Capacitors are important components in the electronics and telecommunications industries. They are essential, for example, in radio and television receivers and in transmitter circuits. We shall describe how charges and energy are stored in capacitors, series and parallel capacitors, and the charge and discharge of a capacitor through a resistor which occurs in many practical circuits.

A capacitor is a device for storing charge. The earliest capacitor was invented—almost accidentally—by van Musschenbroek of Leyden, in about 1746, and became known as a Leyden jar. One form of it is shown in Figure 9.1 (i); J is a glass jar, FF are tin-foil coatings over the lower parts of its walls,

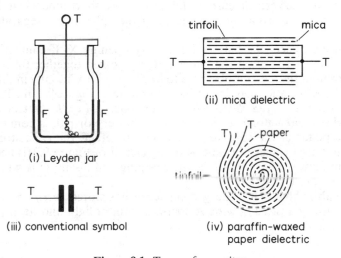

(i) Leyden jar

(ii) mica dielectric

(iii) conventional symbol

(iv) paraffin-waxed paper dielectric

Figure 9.1 *Types of capacitor*

and T is a knob connected to the inner coating. Modern forms of capacitor are shown at (ii) and (iv) in the figure. Essentially

all capacitors consist of two metal plates separated by an insulator.

The insulator is called the *dielectric*; in some capacitors it is polystyrene, oil or air. Figure 9.1 (iii) shows the circuit symbol for such a capacitor; T, T are terminals joined to the plates.

Charging and Discharging Capacitor

Figure 9.2 (i) shows a circuit which may be used to study the action of a capacitor. C is a large capacitor such as 500 microfarad (see later), R is a large

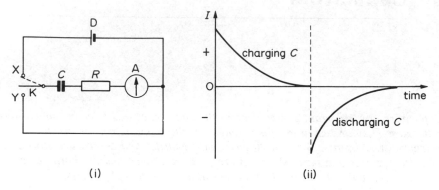

Figure 9.2 *Charging and discharging capacitor*

resistor such as 100 kilohms ($10^5 \, \Omega$), A is a current meter reading up to 100 microamperes (100 µA), K is a two-way key, and D is a 6 V d.c. supply.

When the battery is connected to C by contact at X, the current I in the meter A is seen to be initially about 60 µA. Then as shown in Figure 9.2 (ii), it slowly decreases to zero. So current flows to C for a short time when the battery is connected to it, even though the capacitor plates are separated by an insulator.

We can disconnect the battery from C by opening X. If contact with Y is now made, so that in effect the plates of C are joined together through R and A, the current in the meter is observed to be about 60 µA initially in the opposite direction to before and then slowly decreases to zero, Figure 9.2 (ii). This current flow shows that C *stored charge when it was connected to the battery* originally.

Generally, a capacitor is *charged* when a battery or p.d. is connected to it. When the plates of the capacitor are joined together, the capacitor becomes *discharged*. Large values of C and R in the circuit of Figure 9.2 (i) help to slow the current flow, so that we can see the charging and discharging which occurs, as explained more fully later.

We can also show that a charged capacitor has stored *energy* by connecting the terminals by a piece of wire. A spark, a form of light and heat, passes just as the wire makes contact.

Charging and Discharging Processes

When we connect a capacitor to a battery, electrons flow from the negative terminal of the battery on to the plate A of the capacitor connected to it (Figure 9.3). At the same rate, electrons flow from the other plate B of the capacitor towards the positive terminal of the battery. Equal positive and negative charges therefore appear on the plates, and opposite the flow of electrons which causes them. As the charges increase, the potential difference between the plates increases, and the charging current falls to zero when the potential difference becomes equal to the battery voltage V_0. The charges on the plates B and A are now $+Q$ and $-Q$, and the capacitor is said to have stored a charge Q in amount.

Alternatively, if we think of the battery as an 'electric pump', it pushes electrons or negative charge from B round to A. So when the electrons stop moving, A has a negative charge $-Q$ and B is left with an *equal* positive charge $+Q$.

When the battery is disconnected and the plates are joined by a wire,

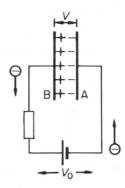

Figure 9.3 *A capacitor charging (resistance is shown because some is always present, even if only that of the connecting wires)*

wire, electrons flow back from plate A to plate B until the positive charge on B is completely neutralised. So a current flows for a time in the wire, and at the end of the time the charges on the plates become zero. So the capacitor is *discharged*. Note that a charge Q flows from one plate to the other during the discharge.

Capacitors and A.C. Circuits

Capacitors are widely used in alternating current and radio circuits, because they can transmit alternating currents. To see how they do so, let us consider

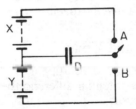

Figure 9.4 *Reversals of voltage applied to capacitor*

the circuit of Figure 9.4, in which the capacitor D may be connected across either of the batteries X, Y. When the key is closed at A, current flows from the battery X, and charges the plate D of the capacitor positively. If the key is now closed at B instead, current flows from the battery Y; the plate D loses its positive charge and becomes negatively charged. So if the key is rocked rapidly between A and B, current surges backwards and forwards along the wires connected to the capacitor. An alternating voltage, as we shall see later, is one which reverses many times a second. When such a voltage is applied to a capacitor, therefore, an alternating current flows in the connecting wires, even though the capacitor has an insulator between its plates.

Variation of Charge with P.D., Vibrating Reed Switch

Figure 9.5 shows a circuit which may be used to investigate how the charge Q stored on a capacitor C varies with the p.d. V applied. The d.c. supply D can be altered in steps from a value such as 10 V to 25 V, C is a capacitor consisting of two large square metal plates separated by small pieces of polythene at the

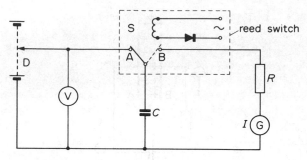

Figure 9.5 *Variation of Q with V—vibrating reed switch*

corners, G is a sensitive current meter, and R is a protective high resistor in series with G.

The capacitor can be charged and discharged rapidly by means of a *vibrating reed switch* S. The vibrator charges C by contact with A and discharges C through G by switching to B. When the vibrator frequency f is high, such as 50 or 100 Hz, the rapid flow of charge from C produces a steady current in G. Its magnitude is given by

$$I = \text{charge per second} = fQ$$

where Q is the charge on the capacitor each time it is charged, since f is the number of times per second it is charged. So, for a given value of f, *the charge Q is proportional to the current I* in G.

When V is varied and values of I are measured, results show that $I \propto V$. So experiment shows that, for a given capacitor, $Q \propto V$.

Capacitance Definition and Units

Since $Q \propto V$, the ratio Q/V is a constant for the capacitor. The ratio of the charge on either plate to the potential difference between the plate is called the *capacitance*, C, of the capacitor:

$$C = \frac{Q}{V} \qquad . \qquad . \qquad . \qquad . \qquad . \qquad (1)$$

So
$$Q = CV \qquad . \qquad . \qquad . \qquad . \qquad . \qquad (2)$$

and
$$V = \frac{Q}{C} \qquad . \qquad . \qquad . \qquad . \qquad . \qquad (3)$$

When Q is in coulombs (C) and V in volts (V), the capacitance C is in farads (F). One farad (1 F) is the capacitance of a very large capacitor. In practical circuits, such as in radio receivers, the capacitance of capacitors used are therefore expressed in *microfarads* (μF). One microfarad is one millionth part of a farad, that is $1\,\mu F = 10^{-6}$ F. It is also quite usual to express small capacitors, such as those used on hi-fi systems, in picofarads (pF). A picofarad is one millionth part of a microfarad, that is $1\,pF = 10^{-6}\,\mu F = 10^{-12}$ F. A nanofarad (nF) = 10^{-9} F.

$$1\,\mu F = 10^{-6}\,F \qquad 1\,nF = 10^{-9}\,F \qquad 1\,pF = 10^{-12}\,F$$

Comparison of Capacitances, Measurement of C

The vibrating reed circuit shown in Figure 9.5 can be used to compare large capacitances (of the order of microfarads) or to compare small capacitances. With large capacitances, a meter with a suitable range of current of the order of milliamperes may be required. With smaller capacitances, a sensitive galvanometer may be more suitable, as the current flowing is then much smaller. In both cases suitable values for the applied p.d. V and the frequency f must be chosen.

Suppose two large, or two small, capacitances, C_1 and C_2, are to be compared. Using C_1 first in the vibrating reed circuit, the current flowing is I_1 say. When C_1 is replaced by C_2, suppose the new current is I_2.

Now $Q_1 = C_1 V$ and $Q_2 = C_2 V$; so $Q_1/Q_2 = C_1/C_2$. But from p. 240, $Q \propto I$. So $Q_1/Q_2 = I_1/I_2$.

$$\therefore \frac{C_1}{C_2} = \frac{I_1}{I_2}$$

So the ratio C_1/C_2 can be found from the current readings I_1 and I_2.

An *unknown capacitor* C can also be found using the vibrating reed circuit. Suppose I is the current measured in G when the applied p.d. is V. Using a low voltage from the a.c. mains for the switch, C is charged and discharged 50 times per second, the mains frequency. Since the current I is the charge flowing per second, then

$$I = 50\, CV$$

So
$$C = \frac{I}{50\, V}$$

With I in amperes and V in volts, then C is in *farads*.

Factors Determining Capacitance

As we have seen, a capacitor consists of two metal plates separated by an insulator called a 'dielectric'. We now find out by experiment what factors affect capacitance.

Distance between plates. Figure 9.6 (i) shows two parallel metal plates X and Y

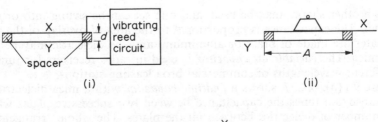

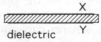

Figure 9.6 *Factors affecting capacitance*

separated by a distance d equal to the thickness of the polythene spacers shown. The capacitance C can be varied by separating the plates a distance $2d$ and then $3d$ and $4d$, using more spacers.

The capacitance can be found each time using the vibrating reed circuit described before. As we have shown, the current I in G is proportional to C for a given applied p.d. V. Experiment shows that, plotting I against $1/d$ and allowing for error, $C \propto 1/d$, where d is the *separation* between the plates. So halving the separation will double the capacitance.

Area between plates. By placing a weight on the top plate X and moving X sideways, Figure 9.6 (ii), the area A of overlap, or *common area* between the plates, can be varied while d is kept constant. Alternatively, pairs of plates of different area can be used which have the same separation d. By using the vibrating reed circuit, experiment shows that $C \propto A$.

Dielectric. Let us now replace the air between the plates by completely filling the space with a 'dielectric' such as polystyrene or polythene sheets or glass, Figure 9.6 (iii). In this case the area A and distance d remain constant. The vibrating reed experiment then shows that the capacitance has *increased* appreciably when the dielectrics is used in place of air.

Some Practical Capacitors

As we have just seen, the simplest capacitor consists of two flat parallel plates with an insulating medium between them. Practical capacitors have a variety of forms but basically they are all forms of parallel-plate capacitors.

A capacitor in which the effective area of the plates can be adjusted is called a *variable capacitor*. In the type shown in Figure 9.7, the plates are semicircular

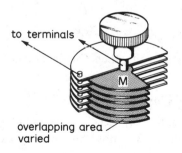

to terminals

overlapping area
varied

Figure 9.7 *Variable air capacitor*

although other shapes may be used, and one set can be swung into or out of the other. The capacitance is proportional to the area of overlap of the plates. The plates are made of brass or aluminium, and the dielectric may be air or oil or mica. The *variable air capacitor* is used in radio receivers for tuning to the different wavelengths of commercial broadcasting stations.

Figure 9.1 (ii), p. 237, shows a *multiple capacitor* with a mica dielectric. The capacitance is n times the capacitance between two successive plates where n is the number of dielectrics between all the plates. The whole arrangement is sealed into a plastic case.

Figure 9.1 (iv), p. 237, shows a *paper* capacitor—it has a dielectric of paper impregnated with paraffin wax or oil. Unlike the mica capacitor, the papers can be rolled and sealed into a cylinder of relatively small volume. To increase the stability and reduce the power losses, the paper is now replaced by a thin layer of *polystyrene*.

Electrolytic capacitors are widely used. Basically, they are made by passing a direct current between two sheets of aluminium foil, with a suitable electrolyte or liquid conductor between them, Figure 9.8. A very thin film of aluminium oxide is then formed on the anode plate, which is on the *positive* side of the d.c. supply as shown. This film is an insulator. It forms the dielectric between the two plates, the electrolyte being a good conductor, Figure 9.8 (i). Since the

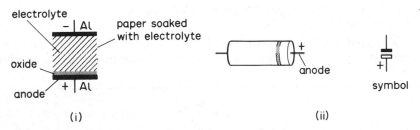

(i)   (ii)

Figure 9.8 *Electrolytic capacitor*

dielectric thickness d is so very small, and $C \propto 1/d$, the capacitance value can be very high. Several thousand microfarads may easily be obtained in a capacitor of small volume. To maintain the oxide film, the anode terminal is marked in red or by a + sign, Figure 9.8 (ii). This terminal must be connected to the positive side of the circuit in which the capacitor is used, otherwise the oxide film will break down. It is represented by the unblacked rectangle in the symbol for the electrolytic capacitor shown in Figure 9.8 (ii).

Parallel Plate Capacitor

We now obtain a formula for the capacitance of a parallel-plate capacitor which is widely used.

Suppose two parallel plates of a capacitor each have a charge numerically equal to Q, Figure 9.9. The surface density σ is then Q/A where A is the area of either plate, and the field-strength between the plates, E, is given, from p. 219, by

$$E = \frac{\sigma}{\varepsilon}$$

$$= \frac{Q}{\varepsilon A}$$

Now E is numerically equal to the potential gradient V/d, p. 228.

$$\therefore \frac{V}{d} = \frac{Q}{\varepsilon A}$$

and

$$\frac{Q}{V} = \frac{\varepsilon A}{d}$$

$$\therefore C = \frac{\varepsilon A}{d} \quad . \quad . \quad . \quad . \quad . \quad (1)$$

It should be noted that this formula for C is approximate, as the field becomes non-uniform at the edges. See Figure 8.25, p. 229.

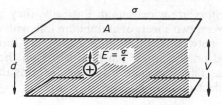

Figure 9.9 *Parallel-plate capacitor*

So a capacitor with parallel plates, having a vacuum (or air, if we assume the permittivity of air is the same as a vacuum) between them, has a capacitance given by

$$C = \frac{\varepsilon_0 A}{d}$$

where C = capacitance in farads (F), A = area of overlap of plates in metre^2, d = distance between plates in metres and $\varepsilon_0 = 8.854 \times 10^{-12}$ farads metre^{-1}.

Capacitance of Isolated Sphere

Suppose a sphere of radius r metres situated in air is given a charge of Q coulombs. We assume, as on p. 225, that the charge on a sphere gives rise to potentials *on and outside* the sphere as if all the charge were concentrated at the *centre*. From p. 225, the surface of the sphere thus has a potential relative to that 'at infinity' (or, in practice, to that of the earth) given by:

$$V = \frac{Q}{4\pi\varepsilon_0 r}$$

$$\therefore \frac{Q}{V} = 4\pi\varepsilon_0 r$$

$$\therefore \text{ capacitance, } C = 4\pi\varepsilon_0 r \qquad . \qquad . \qquad . \qquad . \qquad (2)$$

The other 'plate' of the capacitor is the earth.
    Suppose $r = 10$ cm $= 0.1$ m. Then

$$C = 4\pi\varepsilon_0 r$$

$$= 4\pi \times 8.85 \times 10^{-12} \times 0.1 \text{ F}$$

$$= 11 \times 10^{-12} \text{ F (approx.)}$$

$$= 11 \text{ pF}$$

Concentric Spheres

Faraday used two concentric spheres to investigate the relative permittivity (p. 246) of liquids. Suppose a, b are the respective radii of the inner and outer spheres, Figure 9.10. Let $+Q$ be the charge given to the inner sphere and let the outer sphere be earthed, with air between them.
    The induced charge on the outer sphere is $-Q$ (see p. 209). The potential V_a

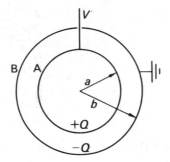

Figure 9.10 *Concentric spherical capacitor*

of the inner sphere = potential due to $+Q$ plus potential due to $-Q$ = $+\dfrac{Q}{4\pi\varepsilon_0 a} - \dfrac{Q}{4\pi\varepsilon_0 b}$, since the potential due to the charge $-Q$ is $-Q/4\pi\varepsilon_0 b$ everywhere inside the larger sphere (see p. 226).

But $V_b = 0$, as the outer sphere is earthed.

$\therefore$ potential difference, $V = V_a - V_b = \dfrac{1}{4\pi\varepsilon_0}\left(\dfrac{Q}{a} - \dfrac{Q}{b}\right)$

$$\therefore V = \dfrac{Q}{4\pi\varepsilon_0}\left(\dfrac{b-a}{ab}\right)$$

$$\therefore \dfrac{Q}{V} = \dfrac{4\pi\varepsilon_0 ab}{b-a}$$

$$C = \dfrac{4\pi\varepsilon_0 ab}{b-a} \qquad . \qquad . \qquad . \qquad . \qquad . \qquad (3)$$

As an example, suppose $b = 10$ cm $= 0.1$ m and $a = 9$ cm $= 0.09$ m.

$$\therefore C = \dfrac{4\pi\varepsilon_0 ab}{b-a}$$

$$= \dfrac{4\pi \times 8.85 \times 10^{-12} \times 0.1 \times 0.09}{(0.1 - 0.09)} \text{ F}$$

$$= 100 \text{ pF (approx.)}$$

Note that the inclusion of a nearby second sphere in the capacitor increases the capacitance. For an *isolated* sphere of radius 10 cm, the capacitance was 11 pF (p. 244).

The same effect is obtained for a metal plate A which has a charge $+Q$, Figure 9.11 (i). If the plate is isolated, A will then have some potential V relative to earth and its capacitance $C = Q/V$.

Now suppose that another metal plate B is brought near to A, as shown, Figure 9.11 (ii). Induced charges $-q$ and $+q$ are then obtained on B. Now the charge $-q$ is nearer A than the charge $+q$. This *lowers* the potential V of A to a value V_1. So the value of C changes from $C = Q/V$ to $C_1 = Q/V_1$ and since V_1 is less than V, the new capacitance is *greater* than C.

If B is *earthed*, only the negative charge $-q$ is left on B. This lowers the potential of A more than before. So the capacitance C is again *increased*.

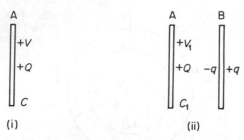

Figure 9.11 *Increasing capacitance of plate*

Relative Permittivity and Dielectric Strength

The ratio of the capacitance with and without the dielectric between the plates is called the *relative permittivity* of the material used. The expression 'without a dielectric' strictly means 'with the plates in a vacuum'; but the effect of air on the capacitance of a capacitor is so small that for most purposes it may be neglected. The relative permittivity of a substance is denoted by the letter ε_r. So

$$\varepsilon_r = \frac{C_d}{C_v}$$

where C_d is the capacitance with a dielectric completely filling the space between the plates and C_v is the capacitance with a vacuum between the plates. An experiment to measure relative permittivity is given on page 249.

The following table gives the value of relative permittivity, and also of *dielectric strength*, for various substances. The strength of a dielectric is the potential gradient at which its insulation breaks down, and a spark passes through it. A solid dielectric is ruined by such a breakdown, but a liquid or gaseous one heals up as soon as the applied potential difference is reduced.

Water is not suitable as a dielectric in practice, because it is a good insulator only when it is very pure, and to remove all matter dissolved in it is almost impossible.

PROPERTIES OF DIELECTRICS

Substance	Relative permittivity	Dielectric strength, kilovolts per mm
Glass.	5–10	30–150
Mica	6	80–200
Ebonite	2·8	30–110
Ice*	94	—
Paraffin wax	2	15–50
Paraffined paper	2	40–60
Methyl alcohol*	32	—
Water*	81	—
Air (*normal pressure*). . .	1·0005	—

* Polar molecules (see p. 247).

Action of Dielectric

We regard a molecule as a collection of atomic nuclei, positively charged, and surrounded by a cloud of negative electrons. When a dielectric is in a charged

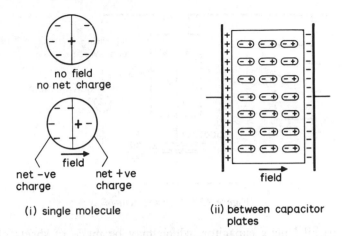

no field
no net charge

net −ve
charge

field

net +ve
charge

field

(i) single molecule

(ii) between capacitor plates

Figure 9.12 *Polarisation of dielectric*

capacitor, its molecules are in an electric field; the nuclei are urged in the direction of the field, and the electrons in the opposite direction, Figure 9.12 (i). So each molecule is distorted, or *polarised*: one end has an excess of positive charge, the other an excess of negative. At the surfaces of the dielectric, therefore, charges appear, as shown in Figure 9.12 (ii). These charges are of opposite sign to the charges on the plates. So they *reduce* the electric field strength E between the plates. Since E = potential difference/plate separation (V/d), the potential difference between the plates is reduced. From $C = Q/V$, where Q is the charge on the plates and V is the p.d. between the plates, it follows that C is *increased*.

If the capacitor is connected to a battery, then its potential difference is constant; but the surface charges on the dielectric still increase its capacitance. They do so because they offset the charges on the plates, and so enable greater charges to accumulate there before the potential difference rises to the battery voltage.

Some molecules, we believe, are permanently polarised: they are called *polar molecules*. Water has polar molecules. The effect of this, in a capacitor, is to increase the capacitance in the way already described. The increase is, in fact, much greater than that obtained with a dielectric which is polarised merely by the action of the field.

ε_0 and its Measurement

We can now see how the units of ε_0 may be stated in a more convenient manner and how its magnitude may be measured.

Unit. From $C = \dfrac{\varepsilon_0 A}{d}$, we have $\varepsilon_0 = \dfrac{Cd}{A}$.

Thus the unit of $\varepsilon_0 = \dfrac{\text{farad} \times \text{metre}}{\text{metre}^2}$

$$= \text{farad metre}^{-1}, \text{F m}^{-1}$$

Measurement. In order to find the magnitude of ε_0, the circuit in Figure 9.13 is used.

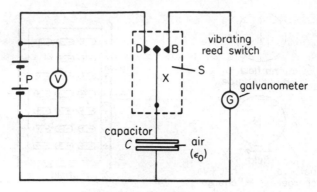

Figure 9.13 *Measurement of ε_0*

C is a parallel plate capacitor, which may be made of sheets of glass or Perspex coated with aluminium foil. The two conducting surfaces are placed facing inwards, so that only air is present between these plates. The area A of the plates in metre^2, and the separation d in metres, are measured. P is a high voltage supply capable of delivering about 200 V, and G is a calibrated sensitive galvanometer. S is a *vibrating reed switch* unit, energised by a low a.c. voltage from the mains. When working, the vibrating bar X touches D and then B, and the motion is repeated at the mains frequency, fifty times a second.

As we explained previously, when the circuit is on, the vibrating reed switch charges and discharges the capacitor 50 times per second. The average steady current I in G is then read.

Charged once, the charge Q on C is

$$Q = CV = \frac{\varepsilon_0 VA}{d}$$

The capacitor is discharged fifty times per second. Since the current is the charge flowing per second,

$$\therefore I = \frac{\varepsilon_0 VA \cdot 50}{d} \text{ A}$$

$$\therefore \varepsilon_0 = \frac{Id}{50VA} \text{ farad metre}^{-1}$$

The following results were obtained in one experiment.

$$A = 0{\cdot}0317 \text{ m}^2, \ d = 1{\cdot}0 \text{ cm} = 0{\cdot}010 \text{ m}, \ V = 159 \text{ V}, \ I = 0{\cdot}21 \times 10^{-6} \text{ A}$$

$$\therefore \varepsilon_0 = \frac{Id}{50VA}$$

$$= \frac{0{\cdot}21 \times 10^{-6} \times 0{\cdot}01}{50 \times 150 \times 0{\cdot}0317}$$

$$= 8{\cdot}8 \times 10^{-12} \text{ F m}^{-1}$$

As very small currents are concerned, care must be taken to make the apparatus of high quality insulating material, otherwise leakage currents will lead to serious error.

Relative Permittivity of Glass and Oil

The same method can be used to find the relative permittivity of various solid materials such as *glass*. If the glass completely fills the space between the two plates, and the current in G is I with the glass and I_0 with air between the plates, then, for the glass,

$$\varepsilon_r = \frac{G_{glass}}{C_{air}} = \frac{I}{I_0}$$

So ε_r for glass can be found from the ratio of the two currents.

If ε_r of an insulating liquid such as an *oil* is required, a similar method can be used. This time, however, two parallel metal plates can be used in a large tall vessel as the capacitor. If I_0 is the current with air between the plates and I is the current when oil completely fills the space between the plates, then ε_r for oil is the ratio I/I_0.

Capacitors in Parallel and Series

In radio circuits, capacitors are used in arrangements whose total capacitance C must be known. To find C, we need the equations

$$C = \frac{Q}{V}, \qquad V = \frac{Q}{C}, \qquad Q = CV$$

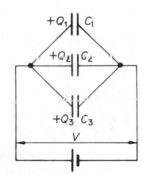

Figure 9.14 *Capacitors in parallel*

In Parallel. Figure 9.14 shows three capacitors, having all their left-hand plates connected, and also all their right-hand plates. They are said to be connected *in parallel* across the same potential difference V. The charges on the individual capacitors are respectively

$$Q_1 = C_1 V \qquad Q_2 = C_2 V \qquad Q_3 = C_3 V . \qquad . \qquad . \qquad (1)$$

The total charge on the system of capacitors is

$$Q = Q_1 + Q_2 + Q_3 = (C_1 + C_2 + C_3)V$$

So the system is equivalent to a single capacitor, of capacitance

$$C = \frac{Q}{V} = C_1 + C_2 + C_3$$

So when capacitors are connected in parallel, their resultant capacitance C is the *sum* of their individual capacitances. C is greater than the greatest individual one.

In Series. Figure 9.15 shows three capacitors having the right-hand plate of one connected to the left-hand plate of the next, and so on—connected *in series*.

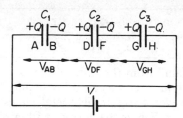

Figure 9.15 *Capacitors in series*

When a cell is connected across the ends of the system, a charge Q is transferred from the plate H to the plate A, a charge $-Q$ being left on H. This charge induces a charge $+Q$ on plate G; similarly, charges appear on all the other capacitor plates, as shown in the figure. (The induced and inducing charges are equal because the capacitor plates are very large and very close together; in effect, either may be said to enclose the other.) The potential differences across the individual capacitors are, therefore, given by

$$V_{AB} = \frac{Q}{C_1}, \qquad V_{DF} = \frac{Q}{C_2}, \qquad V_{GH} = \frac{Q}{C_3} \ . \qquad . \qquad . \qquad (2)$$

The sum of these is equal to the applied potential difference V because the work done in taking a unit charge from H to A is the sum of the work done in taking it from H to G, from F to D, and from B to A. So

$$V = V_{AB} + V_{DF} + V_{GH}$$

$$= Q\left(\frac{1}{C_1} + \frac{1}{C_2} + \frac{1}{C_3}\right) \qquad . \qquad . \qquad . \qquad . \qquad (3)$$

The resultant capacitance of the system is the ratio of the charge stored to the applied potential difference, V. The charge stored is equal to Q, because, if the battery is removed, and the plates HA joined by a wire, *a charge Q will pass through that wire*, and the whole system will be discharged. The resultant capacitance is therefore given by

$$C = \frac{Q}{V} \quad \text{or} \quad \frac{1}{C} = \frac{V}{Q}$$

so, by equation (3),

$$\frac{1}{C} = \frac{1}{C_1} + \frac{1}{C_2} + \frac{1}{C_3} \ . \qquad . \qquad . \qquad . \qquad (4)$$

So, to find the resultant capacitance C of capacitors in series, we must add the reciprocals of their individual capacitances. C is less than the smallest individual.

Comparison of Series and Parallel Arrangements. Let us compare Figures 9.14 and 9.15. In Figure 9.15, where the capacitors are in series, all the capacitors carry the same charge, which is equal to the charge carried by the system as a *whole*, Q. So to find the charge Q on each capacitor, use

$$Q = CV$$

where C is the *resultant* or *total* capacitance given by the $1/C$ formula in (4). The potential difference applied to the system, however, is divided amongst the capacitors, in inverse proportion to their capacitances (equations (2)).

In Figure 9.14, where the capacitors are in *parallel*, they all have the same potential difference. The charge stored is divided amongst them, in direct proportion to the capacitances (equations (1)).

Examples on Capacitors in Series and Parallel

1 In Figure 9.16 (i), C_1 (3 µF) and C_2 (6 µF) are in series across a 90 V d.c. supply. Calculate the charges on C_1 and C_2 and the p.d. across each.

Total capacitance C is given by $1/C = 1/C_1 + 1/C_2$

$$\frac{1}{C} = \frac{1}{3} + \frac{1}{6} = \frac{3}{6}$$

$$\therefore C = 6/3 = 2\,µF$$

The charges on C_1 and C_2 are the same and equal to Q on C.

So $\qquad Q = CV = 2 \times 10^{-6} \times 90 = 180 \times 10^{-6}\,C$

Then $\qquad V_1 = Q/C_1 = 180 \times 10^{-6}/3 \times 10^{-6} = 60\,V$

and $\qquad V_2 = Q/C_2 = 180 \times 10^{-6}/6 \times 10^{-6} = 30\,V$

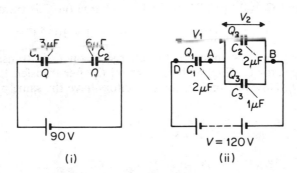

Figure 9.16 *Examples on capacitors*

2 Find the charges on the capacitors in Figure 9.16 (ii) and the potential differences across them.

Capacitance between A and B,

$$C' = C_2 + C_3 = 3\,µF$$

Overall capacitance B to D, since C_1 and C' are in series, is, from $1/C = 1/C_1 + 1/C'$,

$$C = \frac{C_1 C'}{C_1 + C'} = \frac{2 \times 3}{2 + 3} = 1{\cdot}2\,µF$$

Charge stored in this capacitance C

$$= Q_1 = Q_2 + Q_3 = CV = 1\cdot2 \times 10^{-6} \times 120$$

$$= 144 \times 10^{-6}\,\text{C}$$

$$\therefore V_1 = \frac{Q_1}{C_1} = \frac{144 \times 10^{-6}}{2 \times 10^{-6}} = 72\,\text{V}$$

So

$$V_2 = V - V_1 = 120 - 72 = 48\,\text{V}$$

$$Q_2 = C_2 V_2 = 2 \times 10^{-6} \times 48 = 96 \times 10^{-6}\,\text{C}$$

$$Q_3 = C_3 V_2 = 10^{-6} \times 48 \times 10^{-6}\,\text{C}$$

Measuring Charge and Capacitance

To measure a charge, a capacitor C_i such as $0\cdot01\,\mu\text{F}$ or $0\cdot1\,\mu\text{F}$ is first connected to a digital voltmeter V with an electronic amplifier, which has a very high input impedance or resistance, Figure 9.17.

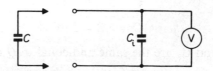

Figure 9.17 *Measuring Q and C*

The charge Q on a small capacitor C such as an insulated metal sphere is then transferred to C_i as shown and the voltmeter reading V is taken. Suppose this is $0\cdot3\,\text{V}$ and C_i is $0\cdot1\,\mu\text{F}$. Then if all the charge on C is transferred to C_i,

$$Q = C_i V = 0\cdot1 \times 10^{-6} \times 0\cdot3 = 3 \times 10^{-8}\,\text{C}$$

We can now see how much of the charge Q on C is transferred to the uncharged capacitor C_i. If Q_i is the charge on C_i after transfer, the charge left on $C = Q - Q_i$. Now on contact, the capacitors have the same p.d. V. So

$$V = \frac{Q_i}{C_i} = \frac{Q - Q_i}{C}$$

Simplifying,

$$Q_i = \frac{C_i}{C + C_i} Q$$

So if $C_i = 20 \times C$, then $Q_i = (20/21) \times Q = 95\%$ of Q. Therefore C_i must be very large compared with C in order to transfer practically all the charge to C_i.

A small capacitor C can be measured by charging it to a suitable known voltage V, and then transferring its charge Q to C_i as we have just described. Then $C = Q/V$.

Energy of a Charged Capacitor

A charged capacitor is a store of electrical energy, as we may see from the vigorous spark it can give on discharge. This can also be shown by charging a large electrolytic capacitor C, such as $10\,000\,\mu\text{F}$, to a p.d. of 6 V, and then

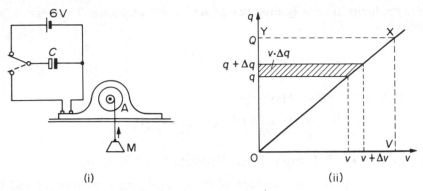

Figure 9.18 *Energy in charged capacitor*

discharging it through a small or toy electric motor A, Figure 9.18 (i). A small mass M such as 10 g, suspended from a thread tied round the motor wheel, now rises as the motor functions. Some of the stored energy in the capacitor is then transferred to gravitational potential energy of the mass; the remainder is transferred to kinetic energy and heat in the motor.

To find the energy stored in the capacitor, we note that since q (charge) is proportional to v (p.d. across the capacitor) at any instant, the graph OX showing how q varies with v is a straight line, Figure 9.18 (ii). We may therefore consider that the final charge Q on the capacitor moved from one plate to the other through an *average* p.d. equal to $\frac{1}{2}(0 + V)$, since there is zero p.d. across the plates at the start and a p.d. V at the end. So

$$\text{work done, } W = \text{energy stored} = \text{charge} \times \text{p.d.} = Q \times \tfrac{1}{2}V$$

So $$W = \tfrac{1}{2}QV$$

From $Q = CV$, other expressions for the energy stored are

$$W = \tfrac{1}{2}CV^2 = \frac{Q^2}{2C}$$

$$\textbf{Energy } W = \tfrac{1}{2}CV^2 = \tfrac{1}{2}\frac{Q^2}{C} = \tfrac{1}{2}QV$$

If C is measured in farads, Q in coulombs and V in volts, then the formulae will give the energy W in joules.

Alternative Proof of Energy Formulae

We can also calculate the energy stored in a charge capacitor by a calculus method.

At any instant of the charging process, suppose the charge on the plates is q and the p.d. across the plates is then v. If an additional tiny charge Δq now flows from the negative to the positive plate, we may say that the charge Δq has moved through a p.d. equal to v. So

$$\text{work done in displacing the charge } \Delta q = v \cdot \Delta q$$

and $$\text{total work done} = \text{energy stored} = \int_0^Q v \cdot dq$$

where the limits are $q = Q$, final charge, and $q = 0$, as shown. To integrate, we substitute $v = q/C$. Then

$$\text{energy stored } W = \int_0^Q \frac{q \cdot dq}{C} = \frac{1}{C}\left[\frac{q^2}{2}\right]_0^Q = \frac{Q^2}{2C}$$

Using $Q = CV$, other expressions for W are

$$W = \tfrac{1}{2}CV^2 \quad \text{or} \quad W = \tfrac{1}{2}QV$$

Energy and Q–V Graph, Heat Produced in Charging

Figure 9.18 (ii) shows the variation of the charge q on the capacitor and its corresponding p.d. v while the capacitor is charged to a final value Q. The small shared area shown $= v \cdot \Delta q$. So the area represents the small amount of work done or energy stored during a change from q to $q + \Delta q$. It therefore follows that the total energy stored by the capacitor is represented by the area of the triangle OXY. This area $= \tfrac{1}{2}QV$, as previously obtained.

If a high resistor R is included in the charging circuit, the rate of charging is slowed. When the charging current ceases to flow, however, the final charge Q on the capacitor is the same as if negligible resistance was present in the circuit, since the whole of the applied p.d. V is the p.d. across the capacitor when the current in the resistor is zero. So the energy stored in the capacitor is $\tfrac{1}{2}QV$ *whether the resistor is large or small.*

It is important to note that the energy in the capacitor comes from the battery. This supplies an amount of energy equal to QV during the charging process. Half of the energy, $\tfrac{1}{2}QV$, goes to the capacitor. The other half is transferred to *heat* in the circuit resistance. If this is a high resistance, the charging current is low and the capacitor gains its final charge after a long time. If it is a low resistance, the charging current is higher and the capacitor gains its final charge in a quicker time. In *both* cases, however, the total amount of heat produced is the same, $\tfrac{1}{2}QV$.

Connected Capacitors, Loss of Energy

Consider a capacitor C_1 of 2 μF charged to a p.d. of 50 V, and a capacitor C_2 of 3 μF charged to a p.d. of 100 V, Figure 9.19 (i). Then

$$\text{charge } Q_1 \text{ on } C_1 = C_1 V_1 = 2 \times 10^{-6} \times 50 = 10^{-4}\,\text{C}$$

and $\quad$ charge Q_2 on $C_2 = C_2 V_2 = 3 \times 10^{-6} \times 100 = 3 \times 10^{-4}\,\text{C}$

$$\therefore \text{total charge} = 4 \times 10^{-4}\,\text{C} \quad . \quad . \quad . \quad (1)$$

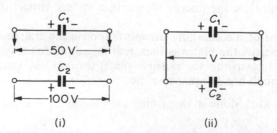

(i)   (ii)

Figure 9.19 *Loss of energy in connected capacitors*

Suppose the capacitors are now joined with plates of like charges connected together, Figure 9.19 (ii). Then some charges will flow from C_1 to C_2 until the p.d. across each capacitor becomes *equal* to some value V. Further, since charge is conserved, the total charge on C_1 and C_2 after connection = the total charge before connection. Now after connection,

$$\text{total charge} = C_1 V + C_2 V = (C_1 + C_2)V = 5 \times 10^{-6} \text{ V} \qquad . \qquad (2)$$

Hence, from (1),

$$5 \times 10^{-6}V = 4 \times 10^{-4}$$

$$\therefore V = 80 \text{ V}$$

$\therefore$ total energy of C_1 and C_2 after connection

$$= \tfrac{1}{2}(C_1 + C_2)V^2$$

$$= \tfrac{1}{2} \times 5 \times 10^{-6} \times 80^2 = 0 \cdot 016 \text{ J} \qquad . \qquad . \qquad . \qquad (3)$$

The total energy of C_1 and C_2 *before* connection

$$= \tfrac{1}{2}C_1 V_1^2 + \tfrac{1}{2}C_2 V_2^2$$

$$= \tfrac{1}{2} \times 2 \times 10^{-6} \times 50^2 + \tfrac{1}{2} \times 3 \times 10^{-6} \times 100^2$$

$$= 0 \cdot 0025 + 0 \cdot 015 = 0 \cdot 0175 \text{ J} \qquad . \qquad . \qquad . \qquad (4)$$

Comparing (4) with (3), we can see that a *loss of energy* occurs when the capacitors are connected. This loss of energy is transferred to *heat* in the connecting wires.

The heat is produced by *flow of current* in the wires connecting the two capacitors when they are joined.

When two capacitors are connected, in calculations always use:
1 After connection, the p.d. V across both capacitors is the *same*.
2 The total charge before connection = the total charge after connection.

Discharge in C–R Circuit

We now consider in more detail the *discharge* of a capacitor C through a resistor R, which is widely used in electronic circuits. Suppose the capacitor is initially charged to a p.d. V_0 so that its charge is then $Q = CV_0$. At a time t after the discharge through R has begun, the current I flowing $= V/R_0$ where V is then the p.d. across C, Figure 9.20 (i). Now

$$V = \frac{Q}{C} \text{ and } I = -\frac{dQ}{dt} \text{ (the minus shows } Q \text{ decreases with increasing } t\text{)}.$$

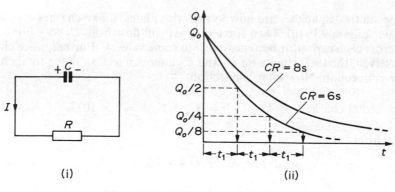

Figure 9.20 *Discharge in C–R circuit*

So, from $I = V/R$, we have

$$-\frac{dQ}{dt} = \frac{1}{CR} Q$$

Integrating,

$$\int_0^Q \frac{dQ}{Q} = -\frac{1}{CR} \int_0^t dt$$

$$\therefore \ln\left(\frac{Q}{Q_0}\right) = -\frac{t}{CR}$$

$$\therefore Q = Q_0 \, e^{-t/CR} \qquad . \qquad . \qquad . \qquad . \qquad (1)$$

So Q decreases exponentially with time t, Figure 9.20 (ii). Since the p.d. V across C is proportional to Q, it follows that $V = V_0 \, e^{-t/CR}$. Further, since the current I in the circuit is proportional to V, then $I = I_0 \, e^{-t/CR}$, where I_0 is the initial current value, V_0/R.

From (1), Q decreases from Q_0 to half its value, $Q_0/2$, in a time t given by

$$e^{-t/CR} = \tfrac{1}{2} = 2^{-1}$$

Taking logs to the base e $-t/CR = -\ln 2$

So $t = CR \ln 2$

Similarly, Q decreases from $Q_0/2$ to half this value, $Q_0/4$, in a time $t = CR \ln 2$. This is the same time from Q_0 to $Q_0/2$. So the time for a charge to diminish to half its initial value, no matter what the initial value may be, is always the same. See Figure 9.20 (ii). This is true for fractions other than one-half. It is typical of an exponential variation or 'decay' which also occurs in radioactivity (p. 873).

Time Constant

The *time constant* T of the discharge circuit is defined as CR seconds, where C is in farads and R is in ohms. So if $C = 4 \ \mu F$ and $R = 2 \ M\Omega$, then $T = (4 \times 10^{-6}) \times (2 \times 10^6) = 8$ seconds. Now, from (1), if $t = CR$, then, using e = 2·72 (approx.)

$$Q = Q_0 \, e^{-1} = \frac{1}{e} Q_0 = 0{\cdot}37 Q_0 \text{ (approx.)}$$

So the time constant may be defined as the time for the charge to decay to about 0·37 times its initial value Q_0. If the time constant CR is high, then the charge will diminish slowly; if the time constant is small, the charge will diminish rapidly. See Figure 9.20 (ii).

Since $Q \propto V$, the p.d. across the capacitor, note that in CR seconds, the p.d. decreases from its initial value V_0 to about 0·37V_0.

Charging C through R

Consider now the *charging* of a capacitor C through a resistance R in series, and suppose the applied battery has an e.m.f. E and a negligible internal resistance, Figure 9.21 (i).

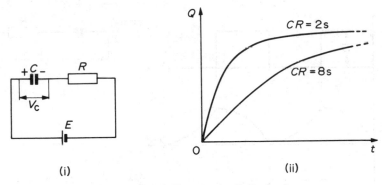

Figure 9.21 *Charging in C–R circuit*

At the instant of making the circuit, there is no charge on C and so no p.d. across it. So the p.d. across $R = E$, the applied circuit p.d. Then the initial current flowing, $I_0 = E/R$. Suppose I is the current flowing after a time t. Then, if V_C is the p.d. now across C,

$$I = \frac{E - V_C}{R}$$

Now $I = dQ/dt$ and $V_c = Q/C$. Substituting in the above equation and simplifying,

$$CR\frac{dQ}{dt} = CE - Q = Q_0 - Q$$

where $Q_0 = CE =$ final charge on C, when no further current flows.
Integrating,

$$\frac{1}{CR}\int_0^t dt = \int_0^Q \frac{dQ}{Q_0 - Q}$$

$$\therefore \frac{t}{CR} = -\ln\left(\frac{Q_0 - Q}{Q_0}\right)$$

$$\therefore Q = Q_0(1 - e^{-t/CR}) \qquad . \qquad . \qquad . \qquad . \qquad (2)$$

As in the case of the discharge circuit, the *time constant* T is defined as CR seconds with C in farads and R in ohms. If T is high, it takes a long time for C

to reach its final charge, that is, C charges slowly. If T is small, C charges rapidly. See Figure 9.21 (ii). The voltage V_C follows the same variation as Q, since $V_C \propto Q$.

Rectangular Pulse Voltage and C–R Circuit

We can apply our results to find how the voltages across a capacitor C and resistor R vary when a *rectangular pulse voltage*, shown in Figure 9.22 (i), is applied to a C–R series circuit. This type of circuit is used in analogue computers.

On one half of a cycle, the p.d. is constant along AB at a value E say. We can therefore consider that this is similar to the case of *charging* a C–R circuit

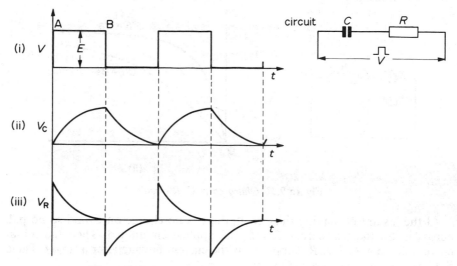

Figure 9.22 *Rectangular pulse voltage and C–R circuit*

by a battery of e.m.f. E. The p.d. V_C across the capacitor hence *rises* along an exponential curve, Figure 9.22 (ii). During the same time, the p.d. across R, V_R, falls exponentially as shown in Figure 9.22 (iii), since $V_R = E - V_C$; that is, the curves for V_R and V_C together *add up to* the straight line graph AB in Figure 9.22 (i).

Similarly, during the time when $V = 0$, the curves for V_C and V_R add up to zero. This helps to check the drawings of the two curves.

You should know:

1 Capacitors store charge on both plates ($+Q$ and $-Q$) due to electron flow.

2 $C = Q/V, \qquad Q = CV, \qquad V = Q/C$
  C in farads (F) when Q in coulombs and V in volts
  $1 \, \mu F = 10^{-6} \, F, \qquad 1 \, nF = 10^{-9} \, F, \qquad 1 \, pF = 10^{-12} \, F.$

3 Parallel-plate capacitor $C = \varepsilon_r \varepsilon_0 A/d$, where ε_r is relative permittivity of insulator between plates, ε_0 is permittivity of free space (vacuum), A is common area in m^2 and d in m.

4 **Parallel capacitors:** $C_T = C_1 + C_2$
 Series capacitors: (a) $1/C_T = 1/C_1 + 1/C_2$ (b) Q is the *same* on each
 capacitor = charge on total capacitor C_T with total p.d. V.
5 **Energy in capacitor** $= \frac{1}{2}QV = \frac{1}{2}CV^2 = \frac{1}{2}Q^2/C$.
 (The '$\frac{1}{2}$' is due to the average p.d., $\frac{1}{2}V$, across the capacitor while it is
 charged.)
 Heat in wires = energy from battery − energy in capacitors =
 $QV - \frac{1}{2}QV = \frac{1}{2}QV$.
6 A resistance R in series with C will slow the charging rate. But the
 final charge $Q_0 = CE$, where E is the battery e.m.f. and is independent
 of R.
 The heat in the wires is then also independent of R.
7 *Time-constant* of charging or discharging $= CR$ seconds, where C in
 farads and R in ohms.
 (a) **Discharging C:** In CR seconds, Q (and V) on capacitor *decreases*
  by about 63%.
 (b) **Charging C:** In CR seconds, Q (and V) on capacitor *increases* by
  about 63% of final charge Q_0 ($Q_0 = CE$, where E is the battery
  e.m.f.).
 In discharge, the area below the current (I) − time (t) graph gives the
 initial charge Q_0 on C since generally Q_0 = integral of $I . dt$.

Examples on Capacitors

1 Energy
A capacitor of capacitance C is fully charged by a 200 V battery. It is then discharged
through a small coil of resistance wire embedded in a thermally insulated block of
specific heat capacity 2.5×10^2 J kg^{-1} K^{-1} and of mass 0·1 kg. If the temperature of
the block rises by 0·4 K, what is the value of C? (*L*.).

(*Analysis* Energy (heat) through coil = energy in capacitor.)

Energy in capacitor $= \frac{1}{2}CV^2 = \frac{1}{2} \times C \times 200^2 = 20\,000C$

Energy through coil $= mc\theta = 0.1 \times 2.5 \times 10^2 \times 0.4 = 10$ J

So $\qquad 20\,000C = 10$

$$C = \frac{10}{20\,000} = \frac{1}{2000} \text{ F}$$

$$= 500 \text{ μF}$$

2 Vibrating reed switch, Parallel-plate capacitor
In a vibrating reed experiment, two parallel plates have an area 0·12 m^2 and are
separated 2 mm by a dielectric. The battery of 150 V charges and discharges the capacitor
at a frequency of 50 Hz, and a current of 20 μA is produced. Calculate the relative
permittivity of the dielectric if the permittivity of free space is 8.9×10^{-12} F m^{-1}.
 What is the new capacitance if the dielectric is half withdrawn from the plates?

(*Analysis* (i) Use $I = 50CV$, (ii) $C \propto A$, common area between plates.)

Suppose C is the capacitance between the plates. Then, with the usual notation,

$$\text{current } I = 50CV$$

So

$$C = \frac{I}{50V} = \frac{20 \times 10^{-6}}{50 \times 150} = \frac{4 \times 10^{-8}}{15} \qquad . \qquad . \qquad (1)$$

But

$$C = \frac{\varepsilon_r \varepsilon_0 A}{d} = \frac{\varepsilon_r \times 8 \cdot 9 \times 10^{-12} \times 0 \cdot 12}{2 \times 10^{-3}} \qquad . \qquad . \qquad (2)$$

So, from (1) and (2),

$$\varepsilon_r = \frac{4 \times 10^{-8} \times 2 \times 10^{-3}}{15 \times 8 \cdot 9 \times 10^{-12} \times 0 \cdot 12} = 5$$

If the dielectric is half withdrawn, the common area of each of the two capacitors formed is now $0 \cdot 5A$. One capacitor, with air dielectric, has a capacitance given by $0 \cdot 5\varepsilon_0 A/d$. The other, with dielectric of $\varepsilon_r = 5$, has a capacitance given by $2 \cdot 5\varepsilon_0 A/d$. These capacitors are in parallel, so adding,

$$\text{total capacitance, } C = \frac{3\varepsilon_0 A}{d}$$

$$= \frac{3 \times 8 \cdot 9 \times 10^{-12} \times 0 \cdot 12}{2 \times 10^{-3}} = 1 \cdot 6 \times 10^{-9} \text{ F}$$

3 Connected capacitors

The plates of a parallel plate air capacitor consisting of two circular plates, each of 10 cm radius, placed 2 mm apart, are connected to the terminals of an electrostatic voltmeter. The system is charged to give a reading of 100 on the voltmeter scale. The space between the plates is then filled with oil of dielectric constant 4·7 and the voltmeter reading falls to 25. Calculate the capacitance of the voltmeter. You may assume that the voltage recorded by the voltmeter is proportional to the scale reading.

(*Analysis*   (i) Total charge is constant,   (ii) p.d. is same for both capacitors after connection.)

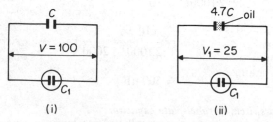

Figure 9.23 *Example on capacitors*

Suppose V is the initial p.d. across the air capacitor and voltmeter, and let C_1 be the voltmeter capacitance and C the plates capacitance, Figure 9.23 (i).

Then

$$\text{total charge} = CV + C_1 V = (C + C_1)V \quad . \qquad . \qquad . \qquad (1)$$

When the plates are filled with oil the capacitance increases to $4 \cdot 7C$, and the p.d. falls to V_1, Figure 9.23 (ii). But the total charge remains constant.

$$\therefore 4 \cdot 7CV_1 + C_1 V_1 = (C + C_1)V, \quad \text{from (1)}$$

$$\therefore (4 \cdot 7C + C_1)V_1 = (C + C_1)V$$

$$\therefore \frac{4 \cdot 7C + C_1}{C + C_1} = \frac{V}{V_1} = \frac{100}{25} = 4$$

$$\therefore 0 \cdot 7C = 3C_1$$

$$\therefore C_1 = \frac{0 \cdot 7C}{3} = \frac{7}{30}C$$

Now $C = \varepsilon_0 A/d$, where A is in metre^2 and d is in metres.

$$\therefore C = \frac{8 \cdot 85 \times 10^{-12} \times \pi \times (10 \times 10^{-2})^2}{2 \times 10^{-3}} \text{ F}$$

$$= 1 \cdot 4 \times 10^{-10} \text{ F (approx.)}$$

$$\therefore C_1 = \frac{7}{30} \times 1 \cdot 4 \times 10^{-10} \text{ F} = 3 \cdot 3 \times 10^{-11} \text{ F}$$

Stray Capacitances

On page 242 we saw how practical capacitors are designed for electrical circuits. The aim is to pack as much surface area of overlapping conductors into as small a space as possible, so producing a compact device of high capacitance. However, it is important to realise that while the parallel plate capacitor is a highly efficient design, it is by no means the only way of producing a capacitance. On page 237 we saw that 'all capacitors consist of two metal plates separated by an insulator', so *any* two pieces of conducting metal in air will have a capacitance between them, although a small one.

So, in any electrical circuit, there are bound to be many *stray capacitances* between the circuit elements, which were not intentionally included in the design. For example, there would be a small stray capacitance between the metal pins of any integrated circuit package (the plastic package containing a silicon chip), which could cause the circuitry inside to behave in an unexpected way. For this reason it is common practice to identify which inputs to the circuit would be most sensitive to stray capacitances, and place these inputs as far away as possible on the package (at diagonally opposite corners), so minimising the stray capacitance.

EXERCISES 9 Capacitors

Multiple Choice

1 The three capacitors in Figure 9A store a total energy in μJ of

   A 12 **B** 36 **C** 48 **D** 80 **E** 120

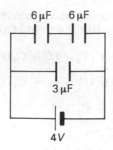

Figure 9A

2 A battery is permanently connected to a parallel-plate capacitor P and the energy stored in P is X joules. When one plate is moved so that the separation of the plates is doubled, the energy now stored in joules is

A $4X$ **B** $2X$ **C** X **D** $X/2$ **E** $X/4$

3 In Figure 9B(i) the time-constant is 4 s. The time-constant in s in Figure 9B(ii) is

A 8 **B** 4 **C** 2 **D** 1 **E** 1/2

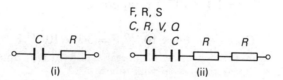

Figure 9B

4 A 4 μF capacitor is charged by a 50 V battery and then discharged through a 2 MΩ resistor. In 8 s the p.d. across the capacitor decreases to a value in V of about

A 63 **B** 50 **C** 32 **D** 18 **E** 12

5 Figure 9C shows two parallel plates, R, S joined to a battery of voltage V and with charges $+Q$ and $-Q$.
1 The energy stored is QV.
2 The electric field strength between the plates increases uniformly from S to R.
3 The electric potential between the plates decreases uniformly from R to S.
Which of the statements **A** to **E** is/are correct?

Answer: **A** if 1, 2, 3 **B** if 1 and 2 only **C** if 2 and 3 only
       D if 1 only **E** if 3 only

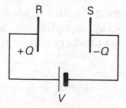

Figure 9C

Longer Questions

6 A capacitor from a 50 V d.c. supply is discharged across a charge-measuring instrument and found to have carried a charge of 10 μC. What was the capacitance of the capacitor and how much energy was stored in it? (*L.*)

7 A 300 V battery is connected across capacitors of 3 µF and 6 µF
(a) in parallel,
(b) in series.
Calculate the charge and energy stored in each capacitor in (a) and (b).

8 A parallel-plate capacitor with air as the dielectric has a capacitance of 6×10^{-4} µF
and is charged by a 100 V battery. Calculate
(a) the charge,
(b) the energy stored in the capacitor,
(c) the energy supplied by the battery.
What accounts for the difference in the answers for (b) and (c)?
   The battery connections are now removed, leaving the capacitor charged, and a
dielectric of relative permittivity 3 is then carefully placed between the plates. What
is the new energy stored in the capacitor?

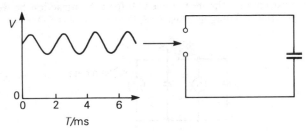

Figure 9D

9 The graph in Figure 9D shows how the potential difference, V, applied between the
terminals in the above circuit varies with time t.
   Using the same time axis, and assuming that the circuit resistance is negligible,
sketch
(a) a graph of how the charge on the capacitor changes over 6 ms, and
(b) a graph of the current in the circuit, marking clearly on this graph the $I = 0$
value. (*L.*)

10 Explain the meaning of *capacitance*.
   Two capacitors C_1 and C_2 are connected in series and then charged with a
battery. The battery is disconnected and C_1 and C_2, still in series, are discharged
through an 80 kΩ resistor. The time constant for the discharge is found to be 4·8
seconds. Calculate
(a) the capacitance of C_1 and C_2 in series, and
(b) the capacitance of C_1 if C_2 has a capacitance of 100 µF. (*L.*)

11 Explain what is meant by *dielectric constant (relative permittivity)*. State two
physical properties desirable in a material to be used as the dielectric in a capacitor.
   A sheet of paper 40 mm wide and $1·5 \times 10^{-2}$ mm thick between metal foil of the
same width is used to make a 2·0 µF capacitor. If the dielectric constant (relative
permittivity) of the paper is 2·5, what length of paper is required? ($\varepsilon_0 = 8·85 \times 10^{-12}$ F m^{-1}. (*N.*)

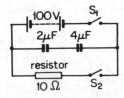

Figure 9E

12 If S_2 is left open and S_1 is closed, calculate the quantity of charge on each capacitor,
Figure 9E.

If S_1 is now opened and S_2 is closed, how much charge will flow through the 10 Ω resistor?

If the entire process were repeated with the 10 Ω resistor replaced by one of much larger resistance what effect would this have on the flow of charge? (*L.*)

13 A capacitor consists of two parallel metal plates in air. The distance between the plates is 5·00 mm and the capacitance is 72 pF. The potential difference between the plates is raised to 12·0 V with a battery.

(a) Calculate the energy stored in the capacitor.

(b) The battery is then disconnected from the capacitor; the capacitor retains its charge.

Calculate the energy stored in the capacitor if the distance between the plates is now increased to 10·00 mm.

The answers to (a) and (b) are different. Why is this so?

Use your answers to estimate the average force of attraction between the plates while the separation is being increased. (*L.*)

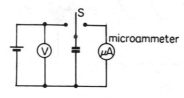

Figure 9F

14 In the circuit shown in Figure 9F, S is a vibrating reed switch and the capacitor consists of two flat metal plates parallel to each other and separated by a small air-gap. When the number of vibrations per second of S is n and the potential difference between the battery terminals is V, a steady current I is registered on the microammeter.

(a) Explain this and show that $I = nCV$, where C is the capacitance of the parallel plate arrangement.

(b) Describe how you would use the apparatus to determine how the capacitance C depends on   (i) the area of overlap of the plates,   (i) their separation, and show how you would use your results to demonstrate the relationships graphically.

(c) Explain how you could use the measurements made in (b) to obtain a value for the permittivity of air.

(d) In the above arrangement, the microammeter records a current I when S is vibrating. A slab of dielectric having the same thickness as the air-gap is slid between the plates so that one-third of the volume is filled with dielectric. The current is now observed to be $2I$. Ignoring the edge effects, calculate the relative permittivity of the dielectric. (*JMB.*)

15 (a) Two capacitors of capacitance C_1 and C_2 respectively are connected in series. Derive an expression for the capacitance of a single equivalent capacitor.

(b)

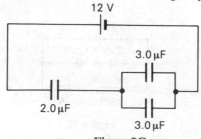

Figure 9G

For the circuit shown in Figure 9G, calculate   (i) the capacitance of this combination of capacitors, (ii) the total energy stored in the capacitors, (iii) the p.d. across the 2·0 µF capacitor. (*JMB.*)

16 In an experiment to investigate the discharge of a capacitor through a resistor, the circuit shown in Figure 9H was set up. The battery had an e.m.f. of 10 V and negligible internal resistance. The switch was first closed and the capacitor allowed to charge fully. The switch was then opened (at time $t = 0$), and Figure 9I shows how the milliammeter reading subsequently changed with time.

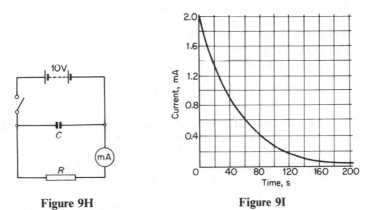

Figure 9H   **Figure 9I**

(a) Use the graph to estimate the initial charge on the capacitor. Explain how you arrived at your answer.
(b) Use your answer to (a) to estimate the capacitance of C.
(c) Calculate the resistance of R. (*AEB.*)

17 (a) Describe a method for measuring the relative permittivity of a material. Your account should include a labelled circuit diagram, brief details of the procedure and the method used to calculate the result.

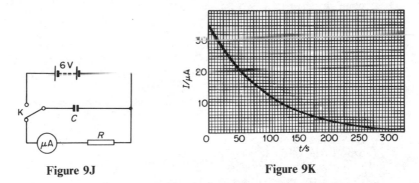

Figure 9J   **Figure 9K**

(b) In the capacitor above, Figure 9J, the capacitor C is first fully charged by using the two-way switch K. The capacitor C is then discharged through the resistor R. The graph (Figure 9K) shows how the current in the resistor R changes with time. Use the graph to help you answer the following questions.

Calculate the resistance R. (The resistance of the microammeter can be neglected.)

Find an approximate value for the charge on the capacitor plates at the beginning of the discharging process and hence calculate (i) the energy stored by the capacitor at the beginning of the discharging process, and (ii) the capacitance C. (*L.*)

18 (i) Figure 9L shows the apparatus used by a student to measure the capacitance, C, of a capacitor. The resistor R is a wire-wound resistor whose value is 100·0 kΩ. The time constant for the circuit is about 2 minutes.

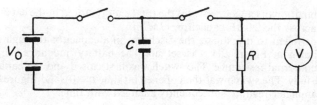

Figure 9L

What do you understand by the term *time constant*?

The potential difference, V, across the capacitor varies with the discharge time, t, according to the equation

$$\ln V = \ln V_0 - \frac{t}{CR}.$$

State what measurements you would make in order to obtain a value for C. How would you use a graph to find the result?

(ii) A student has a 360 kΩ resistor and two capacitors. One capacitor is known to have a capacitance of 300 μF. The two capacitors in parallel discharge through the resistor with a time constant of 180 s. Calculate a value for the capacitance of the second capacitor.

What will be the time constant if the two capacitors are connected in series with the resistor? (*L*.)

10 Current Electricity

We begin current electricity with a study of conduction in metals and the formula for current in terms of the drift velocity of charges. Series and parallel circuits, and ammeters and voltmeters, are then fully discussed, followed by ohmic and non-ohmic conductors. We then discuss the complete circuit with e.m.f. and internal resistance of batteries and their terminal p.d., and the general Kirchhoff laws. Finally, we deal with the important topic of electrical energy and power.

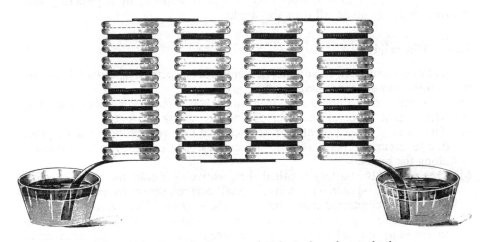

Plate 10A. *An early diagram of Volta's first electrical pile*

Ohm's and Joule's Laws: Resistance and Power

Discovery of Electric Current

By the middle of the eighteenth century, electrostatics was a well-established branch of physics. Machines had been invented which could produce by friction great amounts of charge, giving sparks and electric shocks. The momentary current (as we would now call it) carried by the spark or the body was called a 'discharge'.

In 1799 Volta discovered how to obtain from two metals a *continuous supply* of electricity: he placed a piece of cloth soaked in brine between copper and zinc plates, Figure 10.1 (i). The arrangement is called a *voltaic cell*, and the metal plates its 'poles'; the copper is known as the positive pole, the zinc as the negative. Volta increased the power by building a pile of cells or *battery*, with the zinc of one cell resting on the copper of the other, Figure 10.1 (ii). From this pile he obtained sparks and shocks similar to those given by electrostatic machines.

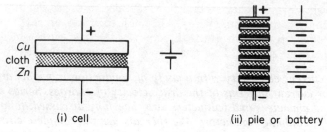

(i) cell

(ii) pile or battery

Figure 10.1 *Voltaic cell and pile, with conventional symbols*

Shortly after, it was found that water was decomposed into hydrogen and oxygen when connected to a voltaic pile. This was the earliest discovery of the chemical effect of an electric current. The heating effect was also soon found, but the magnetic effect, the most important effect, was discovered some twenty years later. Today, current electricity is widely applied, in engineering and electronics, for example, all over the world.

Basic Electricity

It would be useful here to summarise some basic electricity points which you may know already. More detail will be given later.
(a) A current flows along a metal or wire when a battery is connected to it.
(b) The current is due to free electrons moving along the metal.
(c) The battery has a potential difference (p.d.) or voltage between its poles due to chemical changes inside the battery. The p.d. pushes the electrons along the metal.
(d) One pole of the battery is called the positive (+) pole, the other is called the negative (−) pole. The 'conventional' current, shown by an arrow, flows in a circuit connected from the + to the − pole. The electrons carrying the current along the circuit wires actually move in the opposite direction to the conventional current but this need not be taken into account in calculations or circuit formulae.
(e) Current, I, is measured in amperes (A), p.d., V, in volts (V), electrical resistance, R, in ohms (Ω).

$$V = I \times R, \qquad I = V/R, \qquad R = V/I$$

We shall revise and extend some of these points as electric circuits are discussed.

Conduction in Metals, Heating Effect of Current

The conduction of electricity in metals is due to *free electrons*. Free electrons have thermal energy which depends on the metal temperature, and they wander randomly through the metal from atom to atom.

When a battery is connected across the ends of the metal, an electric field is set up between the ends. As we saw in the electrostatics section, charges in fields have forces on them. So the electrons are accelerated in the metal and gain velocity and energy. When they 'collide' with an atom vibrating about its fixed mean position (called a 'lattice site'), they give up some of their energy to it.

The amplitude of the atom vibrations is then increased and so the temperature of the metal rises.

After slowing or stopping on collision, the electrons are again accelerated by the field and again give up energy. Although their movement is erratic, on the average the electrons drift in the direction of the field with a mean or average speed calculated shortly. This one-way drift of the charges on the electrons is an 'electric current'.

The heat generated by the electron collisions is obtained whichever way the electrons flow. So the heating effect of a current—called *Joule heating* (p. 293)—is irreversible, that is, it still occurs when the current in a wire is reversed.

Drift Velocity of Electrons

A simple calculation enables the average drift speed to be estimated. Figure 10.2 shows a part of a copper wire of cross-sectional area A through which a current I is flowing. We suppose that there are n electrons per unit volume, and that each electron carries a charge e. Now, in one second all those electrons

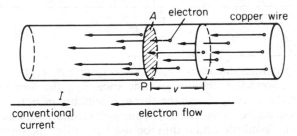

Figure 10.2 *Theory of metallic conduction*

within a distance v to the right of the plane at P, that is, in a volume Av, will flow through this plane, as shown. This volume contains nAv electrons and, therefore, a charge $nAve$. So a charge of $nAve$ *per second* passes P. The current I is the quantity of charge per second, dQ/dt, passing a section of the metal. So

$$I = nAve \qquad . \qquad . \qquad . \qquad . \qquad . \qquad (1)$$

To find the order of magnitude of v, suppose $I = 10$ A, $A = 1$ mm$^2 = 10^{-6}$ m^2, $e = 1·6 \times 10^{-19}$ C, and $n = 10^{28}$ electrons m^{-3}. Then, from (1),

$$v = \frac{I}{nAe} = \frac{10}{10^{28} \times 10^{-6} \times 1·6 \times 10^{-19}}$$

$$= \frac{1}{160} \text{ m s}^{-1} \text{ (approx.)}$$

This is a surprisingly slow drift compared with the average thermal speeds of electrons, which are of the order of several hundred metres per second (p. 673).

Current Density

When a wire of uniform cross-sectional area A carries a current I, the *current density*, J, the current per unit area, is given by $J = I/A$. As shown, $I = nevA$ when the charge carriers are considered. So

$$J = nev$$

Current densities may be used in charge carrier calculations.

Comparison with Gas Molecules

The molecules in a gas have random motion and have an average high speed called *thermal speed* because it depends on the gas temperature. The free electrons behave in a similar way to molecules in a gas and have thermal speeds due to their temperatures. But in a metal carrying a current, the electrons are slowed considerably by the frequent collisions with the metal atoms. So their average drift velocity, which produces the current, is very small as we have shown.

Resistance

The *resistance R* of a conductor is *defined* as the ratio V/I, where V is the p.d. across the conductor and I is the current flowing in it. So if the same p.d. V is applied to two conductors A and B, and a smaller current I flows in A, then the resistance of A is greater than that of B. We write, then,

$$\frac{V}{I} = R \qquad . \qquad . \qquad . \qquad . \qquad . \qquad (2)$$

The unit of potential difference, V, is the *volt*, symbol V; the unit of current, I, is the *ampere*, symbol A; the unit of resistance, R, is the *ohm*, symbol Ω. The ohm is the resistance of a conductor through which a current of one ampere flows when a potential difference (p.d.) of one volt is maintained across it. Figure 10.3 shows some symbols which may be used for different types of resistors, and for ammeters, voltmeters and galvanometers (sensitive current-measuring meters).

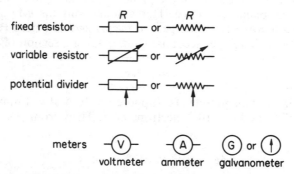

Figure 10.3 *Symbols for resistors and meters*

From the above equation, it also follows that

$$V = IR \quad \text{and} \quad I = \frac{V}{R} . \qquad . \qquad . \qquad . \qquad . \qquad (3)$$

Smaller units of current are the milliampere (one-thousandth of an ampere or 10^{-3} A), symbol mA, and the microampere (one-millionth of an ampere or 10^{-6} A), symbol μA. Smaller units of p.d. are the millivolt (10^{-3} V) and the microvolt (10^{-6} V). A small unit of resistance is the microhm ($1/10^6$ or 10^{-6} Ω); larger units are the kilohm (1000 Ω), symbol kΩ, and the megohm (10^6 ohms), symbol MΩ.

Conductance G is defined as the ratio I/V, and is, therefore, the inverse of resistance, or $1/R$ in numerical value. The unit of conductance is the *siemens*, symbol S. Engineers may use resistance R or conductance G in circuit calculations.

Examples on Basic Formulae

1 A current of 2 mA flows in a radio resistor R when a p.d. of 4 V is connected. What are the values of the resistance and the conductance?

$$R = V/I = 4/(2 \times 10^{-3}) = 2 \times 10^3 = 2000 \ \Omega = 2 \ k\Omega$$

$$G = \text{conductance} = I/V = (2 \times 10^{-3})/4 = 10^{-3}/2 = 5 \times 10^{-4} \ \text{S (siemens)}$$

Note In formulae, I must be in amps, V in volts, R in ohms.

2 What current flows in a resistor of 2 kΩ when (a) a p.d. of 6 V and (b) a p.d. of 8 mV is connected in turn?

(a) $I = V/R = 6/2000 = 3 \times 10^{-3} \ \text{A} = 3 \ \text{mA}$
(b) $I = V/R = 8 \times 10^{-3}/(2 \times 10^3) = 4 \times 10^{-6} \ \text{A} = 4 \ \mu\text{A}$

Series Resistors

The resistors of an electric circuit may be arranged in series, so that the charges carrying the current flow through each in turn (Figure 10.4). Or they may be arranged in parallel, so that the flow of charge divides between them as in Figure 10.5, p. 272.

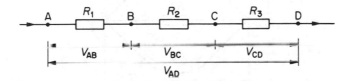

Figure 10.4 *Resistances in series*

Figure 10.4 shows three passive resistors in series, carrying a current I. If V_{AD} is the potential difference across the whole system, the electrical energy supplied to the system per second is IV_{AD} (p. 295). This is equal to the electrical energy per second in all the resistors.

So
$$IV_{AD} = IV_{AB} + IV_{BC} + IV_{CD}$$

from which
$$V_{AD} = V_{AB} + V_{BC} + V_{CD} \qquad . \qquad . \qquad . \qquad . \qquad (1)$$

The individual potential differences are given, from previous, by

$$V_{AB} = IR_1, \qquad V_{BC} = IR_2, \qquad V_{CD} = IR_3 . \qquad . \qquad . \qquad (2)$$

So, by equation (1),
$$V_{AD} = IR_1 + IR_2 + IR_3$$

$$= I(R_1 + R_2 + R_3) \qquad . \qquad . \qquad . \qquad . \qquad (3)$$

And the effective resistance of the system is

$$R = \frac{V_{AD}}{I} = R_1 + R_2 + R_3 \quad . \qquad . \qquad . \qquad . \qquad (4)$$

Summarising:

> (i) *Current same through all resistors in series.*
> (ii) *Total potential difference = sum of individual potential differences (equation (1)).*
> (iii) *Individual potential differences directly proportional to individual resistances (equation (2)).*
> (iv) *Total resistance = sum of individual resistances.*

Resistors in Parallel

Figure 10.5 shows three passive resistors connected in parallel, between the points A, B. This means that the potential difference between A and B is the *same* for each resistor. For this reason lamps in houses are in parallel with the mains potential difference or voltage. A passive device is one which produces

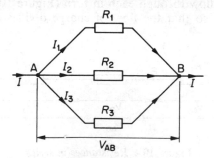

Figure 10.5 *Resistances in parallel*

no energy by itself. A current I enters the system at A and leaves at B, setting up a potential difference V_{AB} between those points. The current branches into I_1, I_2, I_3, through the three elements, and

$$I = I_1 + I_2 + I_3 \quad . \qquad . \qquad . \qquad . \qquad (5)$$

Now
$$I_1 = \frac{V_{AB}}{R_1}, \qquad I_2 = \frac{V_{AB}}{R_2}, \qquad I_3 = \frac{V_{AB}}{R_3}$$

$$\therefore I = V_{AB}\left(\frac{1}{R_1} + \frac{1}{R_2} + \frac{1}{R_3}\right)$$

$$\therefore \frac{I}{V_{AB}} = \frac{1}{R} = \frac{1}{R_1} + \frac{1}{R_2} + \frac{1}{R_3} \quad . \qquad . \qquad . \qquad (6)$$

where R is the effective resistance (V_{AB}/I) of the system.

Summarising:

> (i) *Potential difference same across each resistor in parallel.*
> (ii) *Total current = sum of individual currents* **(equation (5)).**
> (iii) *Individual currents inversely proportional to individual resistances.*
> (iv) *Effective resistance less than least individual resistance* **(equation (6)).**

The Potential Divider

Two resistances in series are often used to provide a known fraction of a given p.d. The arrangement is known as a *potential divider*.

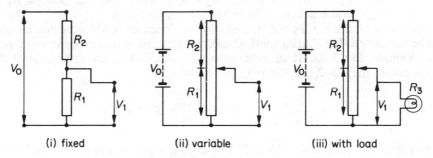

|(i) fixed|(ii) variable|(iii) with load|

Figure 10.6 *Potential divider*

Figure 10.6 (i) shows a potential divider with resistances R_1 and R_2 across a p.d. V_0. The current flowing, I, is given by

$$I = \frac{V_0}{R_1 + R_2}$$

$$\therefore V_1 = IR_1 = \frac{R_1}{R_1 + R_2} V_0 \qquad . \qquad . \qquad . \qquad . \qquad (7)$$

So the fraction of V_0 obtained across R_1 is $R_1/(R_1 + R_2)$. If R_1 is 10 Ω and R_2 is 1000 Ω, then

$$V_1 = \frac{10}{10 + 1000} V_0 = \frac{10}{1010} V_0 = \frac{1}{101} V_0$$

A resistor with a sliding contact can similarly be used, as shown in Figure 10.6 (ii), to provide a continuously variable potential difference V_1, from zero to the full supply value V_0. This is a convenient way of controlling the voltage applied to a load such as a lamp, Figure 10.6 (iii). The resistance of the load, R_3, however, acts in parallel with the resistance R_1. So equation (7) is no longer true, and the voltage V_1 must be measured with a voltmeter. It can be calculated, as in the following example, if R_3 is known. If the load is a lamp its resistance R_3 varies with the current through it, because its temperature then varies.

Sharing P.D. in Potential Divider

In Figure 10.6 (i), suppose $R_1 = 2\,\Omega$, $R_2 = 3\,\Omega$. Since the same current I flows through the series resistors, then, from $V = IR$, $V \propto R$. This means that the potential difference across each resistance is in proportion to its resistance value. So *the ratio of the potential differences* across R_1 and $R_2 = 2:3$. The p.d. across both resistors is V_0. So

$$\text{p.d. across } R_1 = \frac{2}{2+3}\,V_0 = \frac{2}{5}\,V_0$$

$$\text{p.d. across } R_2 = \frac{3}{2+3}\,V_0 = \frac{3}{5}\,V_0$$

Similarly, suppose $R_1 = 4\,\Omega$ and $R_2 = 8\,\Omega$. Then the p.d. across $R_1 = 4V_0/12 = V_0/3$ and p.d. across $R_2 = 8V_0/12 = 2V_0/3$.

This way of sharing the p.d. across both resistors does not need the current to be calculated. The sharing method can be extended to three resistors in series, for example, $2\,\Omega$, $3\,\Omega$, $5\,\Omega$, in series across a p.d. of 20 V. Since the ratio of the p.d.s across them is $2:3:5$, the p.d. across the $2\,\Omega$

$$= \frac{2}{2+3+5} \times 20\text{ V} = \frac{2}{10} \times 20\text{ V} = 4\text{ V}$$

The p.d. across the $3\,\Omega = 3/10 \times 20\text{ V} = 6\text{ V}$. The p.d. across the $5\,\Omega = 5/10 \times 20 = 10\text{ V}$. As a check, total p.d. $= 4 + 6 + 10 = 20\text{ V}$, which is correct.

Rotary Rheostat as Potential Divider

Figure 10.7 shows a *rotary rheostat* used as a potential divider. It has rotary contact with a circular resistance wire connected between X, Y when a spindle S is turned. So a variable resistance is obtained between A and X.

A potential difference V_0 is connected to X and Y. A variable p.d. V is obtained across X and A when S is turned, so a lamp L between X and A glows brighter when S is turned clockwise and the resistance R_1 increases. The p.d. V across the lamp L in Figure 10.7 $= R_1 V_0/(R_1 + R_2)$, where $(R_1 + R_2)$ is the total resistance across X and Y. So if a $100\,\Omega$ rheostat is used and $R_1 = 30\,\Omega$, the lamp p.d. $= 30\,V_0/100 = 0.3\,V_0$.

This simple relation, however, does not take into account the resistance of the lamp L, which is in parallel with the varied resistance. The following example shows how the p.d. is calculated accurately.

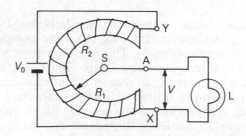

Figure 10.7 *Rotary rheostat as potential divider*

Example on Potential Divider

A load of 2000 Ω is connected, via a potential divider of resistance 4000 Ω, to a 10 V supply, Figure 10.8. What is the potential difference across the load when the slider is
(a) one-quarter,
(b) half-way up the divider?

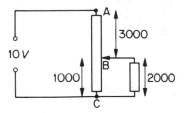

Figure 10.8 *A loaded potential divider*

(a) Since

$$\frac{1}{R_{BC}} = \frac{1}{2000} + \frac{1}{1000}$$

$$R_{BC} = \frac{2000 \times 1000}{2000 + 1000} = \frac{2000}{3}\ \Omega$$

$$\therefore R_{AC} = R_{AB} + R_{BC} = 3000 + \frac{2000}{3} = \frac{11\,000}{3}\ \Omega$$

$$\therefore V_{BC} = \frac{R_{BC}}{R_{AC}} V_{AC}$$

$$= \frac{2000/3}{11\,000/3} \times 10 = \frac{2}{11} \times 10$$

$$= 1{\cdot}8\ V$$

If the load were removed, V_{BC} would be (1000/4000) of 10 V or 2·5 V.
(b) It is left for the reader to show similarly that $V_{BC} = 3{\cdot}3$ V if the slider is half-way up the divider. Without the load V_{BC} would be 5 V.

Ohm's Law

Ohm investigated how the current I in a given metal varied with the p.d. V across it and came to a conclusion about their relationship, stated later, called *Ohm's law*.

Figure 10.9 (i) shows a circuit with a suitable voltmeter V and a suitable ammeter or milliammeter A for measuring the potential difference (p.d.) and current in a resistor Q. S is the battery and P is a rheostat for varying the current I and p.d. V. Provided there is no temperature change in Q and Q is not under strain, the results obtained are shown in Figure 10.9 (ii). The graph of V against I is a *straight line OA passing through the origin*. When the battery is turned round, so that the p.d. V is reversed and the current is reversed, the graph between $-V$, the reversed p.d., and $-I$, the reversed current, is the straight line OB which is the continuation of OA. So we say

$$V \propto I$$

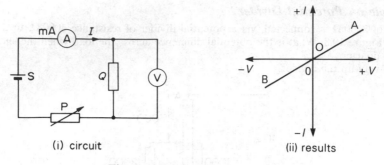

(i) circuit
(ii) results

Figure 10.9 *Measuring resistance R*

This relation was found by Ohm to hold for many conductors even when the potential difference V was reversed. So their resistance R, which is the ratio V/I, is a constant independent of V or I. This is known as *Ohm's law*. Taking into account that resistance depends on temperature and other physical conditions such as mechanical strain, Ohm's law for these type of conductors can be stated as follows:

Under constant physical conditions, the resistance V/I is a constant independent of V or I and their directions.

Ohmic and Non-ohmic Conductors

Ohm's law is obeyed by the most important class of conductors, metals. For example, copper and tungsten, used respectively in cables and lamp filaments, obey Ohm's law. These are called *ohmic conductors*. In this type of conductor the current I is reversed in direction when the p.d. V is reversed but the magnitude of I is unchanged. So the characteristic or I–V graph is a straight line *passing through the origin*, as shown in Figure 10.10 (i). An electrolyte such as copper sulphate solution with copper electrodes obeys Ohm's law, Figure 10.10 (ii).

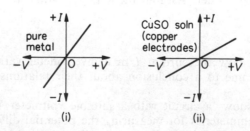

(i)
(ii)

Figure 10.10 *Characteristics of ohmic conductors*

Non-ohmic conductors are those which do not obey Ohm's law, $V \propto I$. Many useful components in the electrical industry must be non-ohmic; for example, a non-ohmic component is essential in a radio receiver circuit. A non-ohmic characteristic or I–V graph may have a curve instead of a straight line; or it may not pass through the origin as in the ohmic characteristic; or it may conduct poorly or not at all when the p.d. is reversed ($-V$). Figure 10.11

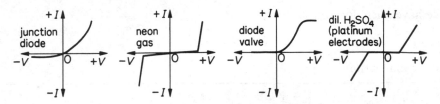

Figure 10.11 *Characteristics of some non-ohmic conductors*

illustrates the non-ohmic characteristics of a junction (semiconductor) diode, neon gas, a diode valve, and the electrolyte dilute sulphuric acid with tungsten electrodes where, unlike Figure 10.11 (ii), an e.m.f. or voltage is produced at the electrodes by the chemicals liberated there.

Some Non-ohmic Conductors

At this stage it would be helpful to list some useful non-ohmic components which play an important part in electronic and other circuits.

Diode. This component, symbol shown in Figure 10.12 (i), is used in radio and TV receivers for changing the mains a.c. (alternating) voltage to d.c. voltage (d.c. = direct current, so d.c. voltage is a voltage in one direction, such as the voltage or p.d. from a battery).

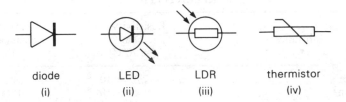

diode   LED   LDR   thermistor
(i)   (ii)   (iii)   (iv)

Figure 10.12 *Symbols for some non-ohmic conductors*

The diode conducts well in one direction of p.d. but conducts poorly when the p.d. or V is reversed. This helps to change a.c. to d.c. voltage.

LED or light-emitting diode. This glows when a suitable current is passed through it due to energy changes and this can produce luminous figures on wrist-watches, for example. Figure 10.12 (ii) shows the LED symbol.

LDR or light-dependent resistance. The metal used has a high resistance in the dark and a lower resistance in the light. It can be used in an electronic circuit for a burglar alarm. The symbol is shown in Figure 10.12 (iii).

Thermistor. The metal used has a low resistance at high temperature and a much higher resistance at low temperatures. With a suitable electronic circuit, it can be used to control or monitor the temperature of hot liquids. The thermistor symbol is shown in Figure 10.12 (iv). See also p. 308.

Resistivity

Ohm showed, by using wires of different length and diameter, that the resistance of a wire, R, is proportional to its length, l, and inversely proportional to its cross-sectional area A. The truth of this can easily be demonstrated today by measuring R using a voltmeter–ammeter method and suitable lengths of wire.

We have, then, for a given wire,

$$R \propto \frac{l}{A}$$

So we may write

$$R = \rho \frac{l}{A} \qquad . \qquad . \qquad . \qquad . \qquad . \qquad (1)$$

where ρ is a constant for the material of the wire called the *resistivity* of that material. So

$$\rho = R \frac{A}{l} \qquad . \qquad . \qquad . \qquad . \qquad . \qquad (2)$$

and resistivity has units

$$\text{ohm} \times \frac{\text{metre}^2}{\text{metre}} = \text{ohm} \times \text{metre or } \Omega\,\text{m}$$

In equation (1), R is in ohm when l is in metres, A is in metre^2 and ρ is in ohm metres.

RESISTIVITIES

Substance	Resistivity ρ, Ω m (at 20°C)	Temperature coefficient α, K^{-1}
Aluminium	$2 \cdot 82 \times 10^{-8}$	0·0039
Constantan[1]	$c.\ 49 \times 10^{-8}$	0·00001
Copper	$1 \cdot 72 \times 10^{-8}$	0·0043
Iron	$c.\ 9 \cdot 8 \times 10^{-8}$	0·0056
Manganin[2]	$c.\ 44 \times 10^{-8}$	$c.$ 0·00001
Mercury	$95 \cdot 77 \times 10^{-8}$	0·00091
Nichrome[3]	$c.\ 100 \times 10^{-8}$	0·0004
Silver	$1 \cdot 62 \times 10^{-8}$	$c.$ 0·0039
Tungsten[4]	$5 \cdot 5 \times 10^{-8}$	0·0058
Carbon (graphite)	$.33$ to 185×10^{-8}	$-0 \cdot 0006$ to $-0 \cdot 0012$

[1] Also called eureka; 60% Cu, 40% Ni.
[2] 84% Cu, 12% Mn, 4% Ni; used for resistance boxes and shunts.
[3] Ni–Cu–Cr; used for electric fires—does not oxidise at 1000°C.
[4] Used for lamp filaments—melts at 3380°C.

The resistivity of a metal is increased by even small amounts of impurity. Alloys, such as constantan, may have resistivities far greater than any of their constituents as the table of resistivities above shows.

The temperature coefficient values listed in the table are a measure of the fractional rise of the materials' resistances when their temperature rises (temperature coefficient is defined later on page 307). The low values for constantan and manganin show that their resistance remains practically constant when they are warmed by moderate currents. These materials are, therefore, used to make high-quality resistances of known values called *standard resistances.*

Tungsten, used for filaments in electric lamps, increases in resistance when heated. So when a tungsten lamp is switched on in a d.c. voltage supply, the initial current flowing is greater than the current when the filament becomes hot because the tungsten resistance is initially lower.

Conductivity

Conductivity, symbol σ, is the opposite property to resistivity, ρ. So $\sigma = 1/\rho$.

Earlier we met conductance, G, which is defined as I/V (resistance $R = V/I$). The conductance of a conductor increases when its uniform cross-section area A increases and its length l decreases. So

$$G = \sigma \frac{A}{l}$$

whereas $R = \rho l/A$ as we showed before. Since the unit of ρ, resistivity, is Ω m, the unit of σ, conductivity, is $\Omega^{-1}\,m^{-1}$.

Example on Resistivity and Conductivity

A p.d. of 2 V is connected to a uniform resistance wire of length 2·0 m and cross-sectional area $8 \times 10^{-9}\,m^2$. A current of 0·10 A then flows in the wire.

Find   (a) the current flowing if the length and diameter of the wire are doubled and the same p.d. is connected to it,   (b) the resistivity of the material,   (c) the conductivity of the material.

(a) The resistance $R \propto l/A$, where l is the length and A is the cross-sectional area. For a wire of circular cross-section, $A = \pi d^2/4$, where d is the diameter, so A increases by 4 times when the diameter is doubled. So R changes by 2(length)/4(area) = 1/2. If R is halved, the current is *doubled*. So new current $= 2 \times 0{\cdot}10$ A $= 0{\cdot}20$ A.

(b) Resistance R of wire $= V/I = 2$ V$/0{\cdot}10$ A $= 20\,\Omega$. From $R = \rho l/A$,

$$\rho = R \times A/l = (20 \times 8 \times 10^{-9})/2{\cdot}0 = 8 \times 10^{-8}\,\Omega\,m$$

(c) Conductivity $\sigma = 1/\rho = 1/(8 \times 10^{-8}) = 1{\cdot}25 \times 10^7\,\Omega^{-1}\,m^{-1}$.

Examples on Circuit Calculations

1   A battery C of 2·2 V and negligible internal resistance, is connected to the combination of resistors shown in Figure 10.13. What is the effective value of the resistance connected across the terminals of the cell? What are the values of the currents I_1, I_2 and I_3?

Resistance along DEF $= 10 + 5 = 15\,\Omega$

Since DEF is in parallel with the $5\,\Omega$ resistor between D, G (G and F are

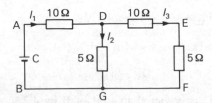

Figure 10.13 *Example on circuits*

connected together), the combined resistance R between DG is given by

$$\frac{1}{R} = \frac{1}{15} + \frac{1}{5} = \frac{4}{15}, \quad \text{so } R = 15/4 = 3{\cdot}75 \ \Omega$$

So total resistance between terminals of C $= 10 + 3{\cdot}75 = 13{\cdot}75 \ \Omega$.   (1)

So
$$I_1 = \frac{V}{R} = \frac{2{\cdot}2}{13{\cdot}75} = 0{\cdot}16 \text{ A} \qquad . \qquad . \qquad . \qquad . \qquad (2)$$

Also, since DEF (15 Ω) is in parallel with DG (5 Ω),

$$I_2 = \frac{15}{5 + 15} \times 0{\cdot}16 \text{ A} = 0{\cdot}12 \text{ A}$$

and
$$I_3 = I_1 - I_2 = 0{\cdot}16 - 0{\cdot}12 = 0{\cdot}04 \text{ A}$$

2 In the circuit shown in Figure 10.14, calculate the p.d. between B and D, assuming the battery of 12 V has negligible internal resistance.
   Has B or D the higher potential?

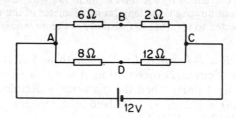

Figure 10.14 *Example on potential difference*

In Figure 10.14, the series 6 Ω and 2 Ω form a potential divider across the 12 V. Similarly, the series 8 Ω and 12 Ω form a potential divider across the 12 V.

So p.d. across A and B, $V_{AB} = \dfrac{6}{(6 + 2)} \times 12 \text{ V} = 9 \text{ V}$

and p.d. across A and D, $V_{AD} = \dfrac{8}{(8 + 12)} \times 12 \text{ V} = 4{\cdot}8 \text{ V}$

By subtraction, p.d. across B and D, $V_{BD} = 9 - 4{\cdot}8 = 4{\cdot}2 \text{ V}$

Higher potential $V_{AB} = V_A - V_B = 9 \text{ V}$.   .   .   .   .   (1)

$$V_{AD} = V_A - V_D = 4{\cdot}8 \text{ V} \qquad . \qquad . \qquad . \qquad . \qquad . \qquad (2)$$

Subtracting (2) from (1),

$$V_D - V_B = 4{\cdot}2 \text{ V}$$

So the potential of D is higher than that of B by 4·2 V.

Electromotive Force

E.M.F. and Internal Resistance

An electrical generator provides energy and power. This is considered later. Here we consider the current and potential difference, p.d., in circuits connected to a generator such as a battery.

If a high resistance voltmeter is connected across the terminals of a dry battery B, the meter may read about 1·5 V, Figure 10.15 (i). Since practically no current flows from the battery in this case we say it is on 'open circuit'. The p.d. across the terminals of a battery (or any other generator) on open circuit is called its *electromotive force* or *e.m.f.*, symbol E. We define e.m.f. in terms of energy later (p. 297).

When a resistor is connected to the battery, the current flows through the resistor and through the *internal resistance, r,* of the battery to complete the circuit flow.

The e.m.f. of a battery depends on the nature of the chemicals used and not on its size. A tiny battery has the same e.m.f. as a large battery made of the same chemicals. The internal resistance of the tiny battery, however, is much less than that of the large battery. Provided only a small current is taken from a battery, its e.m.f. and internal resistance are fairly constant.

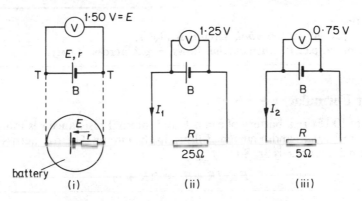

Figure 10.15 *E.m.f. and internal resistance*

Any electrical generator, then, has two important properties, an e.m.f. E and an internal resistance r. As shown in Figure 10.15 (i), E and r may be represented separately in a diagram, though in practice they are together between the terminals T, T. So we can think of the battery as an 'electric pump', with its e.m.f. E pushing the current round the circuit through both the external (outside) resistor R and the internal resistance r. As we emphasise again later, the e.m.f. E is the p.d. across the external resistor R plus the p.d. across the internal resistance r.

Circuit Principles, Terminal P.D.

In Figure 10.15 (ii), a resistor of 25 Ω is connected to the battery B so that a current I_1 flows in the circuit. The voltmeter reading across the battery terminals, or terminal p.d., may then be 1·25 V, although the e.m.f. is 1·5 V. When the resistor is replaced by one of 5 Ω, Figure 10.15 (iii), a larger current I_2 flows and the voltmeter reading or terminal p.d. is now 0·75 V.

To understand why the terminal p.d. varies when a current flows from a battery, it is important to realise that the voltmeter is connected across the *external* or outside resistance in Figure 10.15 (ii). So 1·25 V is the p.d. across the 25 Ω resistor. Now the *e.m.f.*, 1·5 V, maintains the current in the *whole* circuit, that is, through the external *and* internal resistance r. So we deduce that the p.d. across the internal resistance $r = 1·5 - 1·25 = 0·25$ V.

Similarly, in Figure 10.15 (iii) 0·75 V is the p.d. across the external resistance 5 Ω. So in this case the p.d. across the internal resistance $r = 1·5 - 0·75 = 0·75$ V. A common error is to think that the voltmeter across the terminals reads the e.m.f. This is not the case here as there is a p.d. across the internal resistance when a current flows, and *the voltmeter can only read the p.d. across the external resistance R, which is the terminal p.d.*

In Figure 10.15 (ii), the p.d. across the 25 Ω external resistor R is 1·25 V and the p.d. across the internal resistance r is 0·25 V. Since the same current flows in R and r it follows that $R = 5r$, or $r = R/5 = 5$ Ω.

Similarly, the p.d. across the external resistor R of 5 Ω in Figure 10.15 (iii) is 0·75 V and that across the internal resistance r is 0·75 V. So $r = R = 5$ Ω, as previously calculated.

You should now see that as the external resistance R increases, the terminal p.d. increases. When R is an infinitely high value, so that $I = 0$, the terminal p.d. is equal to the e.m.f. E.

Summarising:
1. **E = p.d. across the *whole* circuit, R plus r.**
2. **Terminal p.d. V, when current flows, = p.d. across R only.**

Circuit Formulae

In Figure 10.15 (ii), a battery of e.m.f. E and internal resistance r is joined to an external resistor R, and a current I flows in the circuit. The p.d. across R is IR and the p.d. across r is Ir. So

$$E = IR + Ir = I(R + r) \quad . \quad . \quad . \quad . \quad (1)$$

or
$$I = \frac{E}{R + r} \quad . \quad . \quad . \quad . \quad . \quad . \quad (2)$$

Note carefully that when the e.m.f. E is used to find the current I, the resistance $(R + r)$ of the *whole* circuit is required.

On the other hand, the terminal p.d., V = p.d. across external resistor R. So

$$\text{terminal p.d. } V = IR = \frac{ER}{R + r}$$

Further, from (1),
$$V = E - Ir \quad . \quad . \quad . \quad . \quad (3)$$

This is a useful formula for the terminal p.d. when the e.m.f. E and internal resistance r are known. For example, suppose a current of 0·5 A flows from a battery of e.m.f. E of 3 V and internal resistance 4 Ω. Then

$$\text{terminal p.d. } V = E - Ir = 3 - (0·5 \times 4) = 1 \text{ V}$$

The internal resistance r can be found from the e.m.f. E, the terminal p.d. V and the current I. From (1),

$$Ir = E - IR = E - V$$

So
$$r = \frac{E - V}{I}$$
. . . . . (4)

If I is needed, we may use $I = V/R$.

Example on Circuit Calculation

A battery of e.m.f. 1·50 V has a terminal p.d. of 1·25 V when a resistor of 25 Ω is joined to it. Calculate the current flowing, the internal resistance r and the terminal p.d. when a resistor of 10 Ω replaces the 25 Ω resistor.

The terminal p.d. = the p.d. across the external resistor, 25 Ω.

So
$$I = \frac{V}{R} = \frac{1·25}{25} = 0·05 \text{ A}$$

Also,
$$r = \frac{\text{p.d.}}{\text{current}} = \frac{E - V}{I}$$

$$= \frac{1·50 - 1·25}{0·05} = \frac{0·25}{0·05} = 5 \text{ Ω}$$

When the external resistor is 10 Ω, the current I flowing is

$$I = \frac{E}{R + r} = \frac{1·50}{10 + 5} = 0·1 \text{ A}$$

So terminal p.d. $V = IR = 0·1 \times 10 = 1 \text{ V}$

Terminal P.D. with Current in Opposition to E.M.F.

So far we have considered the terminal p.d. when the battery e.m.f. maintains the current. Suppose, however, that a current is passed through a battery in *opposition* to its e.m.f., a case which occurs in re-charging an accumulator, for example.

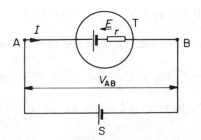

Figure 10.16 *Terminal p.d.*

Figure 10.16 shows a supply S sending a current I through a battery T in opposition to its e.m.f. E. The terminal p.d. V_{AB} must be greater than E in this case. Since the net p.d., $V_{AB} - E$, across the terminals must maintain the current

I in r, then the net p.d. $= Ir$. So

$$V_{AB} - E = Ir$$

Hence
$$V_{AB} = E + Ir$$

In contrast, when the battery e.m.f. itself maintains a current in the same direction as the e.m.f., then the terminal p.d. $V = E - Ir$, as we have already seen.

Example on Terminal P.D.

Figure 10.17 shows a circuit with two batteries in opposition to each other. One has an e.m.f. E_1 of 6 V and internal resistance r_1 of 2 Ω and the other an e.m.f. E_2 of 4 V and internal resistance r_2 of 8 Ω. Calculate the p.d. V_{XY} across XY.

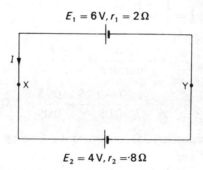

$$E_1 = 6\,V, r_1 = 2\,\Omega$$

$$E_2 = 4\,V, r_2 = 8\,\Omega$$

Figure 10.17 *Example*

Net e.m.f. in circuit $= E_1 - E_2 = 6 - 4 = 2$ V

So $\qquad$ current, $I = \dfrac{E_1 - E_2}{r_1 + r_2} = \dfrac{6 - 4}{2 + 8} = 0.2$ A

The e.m.f. E_1 is in the *same* direction as the current I. So

terminal p.d., $V_{XY} = E_1 - Ir_1 = 6 - (0.2 \times 2) = 5.6$ V

If we consider the battery of e.m.f. E_2, we see that I flows in *opposition* to E_2. In this case,

terminal p.d., $V_{XY} = E_2 + Ir_2 = 4 + (0.2 \times 8) = 5.6$ V

This result agrees with the terminal p.d. value obtained by using E_1.

Cells in Series and Parallel

When cells or batteries are in series and assist each other, then the total e.m.f.

$$E = E_1 + E_2 + E_3 + \cdots \qquad . \qquad . \qquad . \qquad . \qquad (1)$$

and the total internal resistance

$$r = r_1 + r_2 + r_3 + \cdots \qquad . \qquad . \qquad . \qquad . \qquad (2)$$

where E_1, E_2, E_3 are the individual e.m.f.s and r_1, r_2, r_3 are the corresponding internal resistances. If one cell, e.m.f. E_2 say, is turned round 'in opposition' to the others, then $E = E_1 - E_2 + E_3 + \cdots$; but the total internal resistance remains unaltered.

When *similar cells are in parallel*, the total e.m.f. = E, the e.m.f. of any one of them. The internal resistance r is here given by

$$\frac{1}{r} = \frac{1}{r_1} + \frac{1}{r_1} + \cdots \qquad . \qquad . \qquad . \qquad . \qquad . \qquad (3)$$

where r_1 is the internal resistance of each cell. If different cells are in parallel, there is no simple formula for the total e.m.f. and the total internal resistance, and any calculations involving circuits with such cells are dealt with by applying Kirchhoff's laws, discussed later.

Examples on Circuits

1 Two similar cells A and B are connected in series with a coil of resistance 9·8 Ω. A voltmeter of very high resistance connected to the terminals of A reads 0·96 V and when connected to the terminals of B it reads 1·00 V, Figure 10.18. Find the internal resistance of each cell. (Take the e.m.f. of a cell as 1·08 V.) (*L.*)

The p.d. across both cells = 0·96 + 1·00 = 1·96 V

$$= \text{p.d. across } 9\cdot8 \ \Omega$$

$$\therefore \text{ current flowing, } I, = \frac{V}{R} = \frac{1\cdot96}{9\cdot8} = 0\cdot2 \text{ A}$$

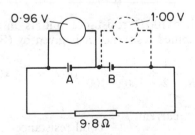

Figure 10.18 *Example*

Now terminal p.d. across each cell = $E - Ir$

$$\therefore \text{ for cell A, } 0\cdot96 = 1\cdot08 - 0\cdot2r, \text{ or } r = 0\cdot6 \ \Omega$$

$$\text{for cell B, } 1\cdot00 = 1\cdot08 - 0\cdot2r, \text{ or } r = 0\cdot4 \ \Omega$$

2 What is meant by the *electromotive force* of a cell?

A voltmeter is connected in parallel with a variable resistance, R, which is in series with an ammeter and a cell, Figure 10.19 (i). For one value of R the meters read 0·3 A and 0·9 V. For another value of R the readings are 0·25 A and 1·0 V. Find the values of R, the e.m.f. of the cell, and the internal resistance of the cell. What assumptions are made about the resistance of the meters in the calculation?

If in this experiment the ammeter had a resistance of $10\,\Omega$ and the voltmeter a resistance of $100\,\Omega$ and R was $2\,\Omega$, what would the meters read? (*L.*)

(i) The voltmeter reads the terminal p.d. across the cell if the resistances of the meters are neglected. So, with the usual notation (Figure 10.19 (ii)),

$$E - Ir = 0{\cdot}9, \text{ or } E - 0{\cdot}3r = 0{\cdot}9 \qquad . \qquad . \qquad . \qquad (1)$$

and
$$E - 0{\cdot}25r = 1{\cdot}0 \qquad . \qquad . \qquad . \qquad . \qquad (2)$$

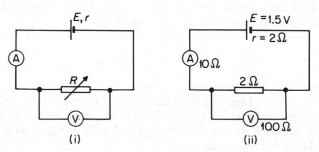

Figure 10.19 *Example*

Subtracting (1) from (2),

$$0{\cdot}05r = 0{\cdot}1, \text{ i.e. } r = 2\,\Omega$$

Also, from (1),

$$E = 0{\cdot}3r + 0{\cdot}9 = 0{\cdot}6 + 0{\cdot}9 = 1{\cdot}5 \text{ V}$$

Further,
$$R_1 = \frac{V}{I} = \frac{0{\cdot}9}{0{\cdot}3} = 3\,\Omega$$

and
$$R_2 = \frac{1{\cdot}0}{0{\cdot}25} = 4\,\Omega$$

(ii) If the voltmeter has $100\,\Omega$ resistance and is in parallel with the $2\,\Omega$ resistance, the combined resistance R is given by (Figure 10.19 (ii)),

$$\frac{1}{R} = \frac{1}{2} + \frac{1}{100} = \frac{51}{100}, \text{ or } R = \frac{100}{51}\,\Omega$$

$$\therefore \text{ current, } I = \frac{E}{\text{Total resistance}}$$

$$= \frac{1{\cdot}5}{\dfrac{100}{51} + 10 + 2} = 0{\cdot}11 \text{ A}$$

Also, voltmeter reading $V = IR = 0{\cdot}11 \times \dfrac{100}{51} = 0{\cdot}21 \text{ V}$

Kirchhoff's Laws

A 'network' is usually a complicated system of electrical conductors. KIRCHHOFF (1824–87) extended Ohm's law to networks, and gave two laws, which together enabled the current in any part of the network to be calculated.

The *first law* refers to any junction in the network, such as A in Figure 10.20 (i); it states that the total current flowing into the junction is equal to the total current flowing out of it:

$$I_1 = I_2 + I_3$$

The law follows from the fact that electric charges do not accumulate at the points of a network. It is often put in the form that

the algebraic sum of the currents at a junction of a circuit is zero,

where a current, I, is reckoned positive if it flows towards the point, and negative if it flows away from it. So at A in Figure 10.20 (i),

$$I_1 - I_2 - I_3 = 0$$

Kirchhoff's first law gives a set of equations which help towards solving the network. In practice we can shorten the work by putting the first law straight into the diagram, as shown in Figure 10.20 (ii) for example, since

current along AC $= I_1 - I_g$

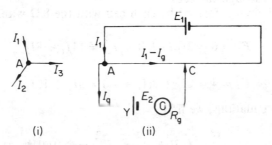

(i)   (ii)

Figure 10.20 *Kirchhoff's laws*

Kirchhoff's second law connects the e.m.f. and p.d. in a complete circuit. It refers to any *closed loop*, such as AYCA in Figure 10.20 (ii). It states that

round such a loop, *the algebraic sum of the e.m.f.s is equal to the algebraic sum of all the p.d.s in that circuit.*

So, going clockwise round the loop AYCA in Figure 10.20 (ii),

$$E_2 = R_{AC}(I_1 - I_g) - R_g I_g$$

Note carefully that a p.d. is *positive* if it is in the *same* direction as the net e.m.f. Since I_g is opposite to E_2, the p.d. $R_g I_g$ is *negative*.
Briefly, we now say that

Kirchhoff's first law is a statement of the conservation of charge.
Kirchhoff's second law is a statement of the conservation of energy.

Example on Kirchhoff's laws

Figure 10.21 shows a network in which the currents I_1, I_2, can be found from Kirchhoff's laws. Here two batteries, one of e.m.f. 6 V and internal resistance 3 Ω and the other of e.m.f. 4 V and internal resistance 2 Ω, are in parallel across an external resistance R of 8 Ω.

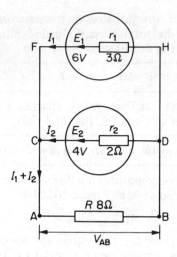

Figure 10.21 *Load R across cells in parallel*

From the first law, the current in the $8\,\Omega$ wire is $(I_1 + I_2)$, assuming I_1, I_2 are the currents through the cells.

Taking closed circuits formed by each cell with the $8\,\Omega$ wire, we have, from the second law,

$$E_1 = 6 = 3I_1 + 8(I_1 + I_2) = 11I_1 + 8I_2$$

and

$$E_2 = 4 = 2I_2 + 8(I_1 + I_2) = 8I_1 + 10I_2$$

Solving the two equations, we find

$$I_1 = \frac{14}{23} = 0\cdot61 \text{ A}, \; I_2 = -\frac{2}{23} = -0\cdot09 \text{ A}$$

The minus sign indicates that the current I_2 flows in the opposite direction to that shown in the diagram; so it flows against the e.m.f. of the generator E_2.

You should know:

1 $I = nAve$, where v is the electron drift velocity in a current-carrying metal and n is the number of electrons per metre^3. If the metal becomes narrower at a point, A decreases and v increases (n stays constant).

2 Series resistors R_1, R_2: $R_T = R_1 + R_2$; total p.d. = sum of separate p.ds. The p.d. is proportional to resistance, so $V_1/V_2 = R_1/R_2$.

3 Potential divider: With R_1, R_2 in series across p.d. V, p.d. across $R_1 = (R_1/(R_1 + R_2)) \times V$. When R_1 varied, p.d. obtained across R_1 can be varied from 0 to V.

4 Parallel resistors R_1, R_2: $1/R_T = 1/R_1 + 1/R_2$; I_T (main current) $= I_1 + I_2$.
   Main current divides *inversely* as resistance values.

5 $I = V/R, \qquad R = V/I, \qquad V = IR.$

6 *Ohmic* conductor if $I-V$ graph is straight line through origin for $+V$ and $-V$.
Non-ohmic conductor if $I-V$ graph is not a straight line through origin.
7 $I = \rho l / A$, where ρ is resistivity of material. Unit of ρ: Ω m.
8 *Complete circuit*: Battery e.m.f. E, internal resistance r, external resistance R,

$$I = E/(R + r) = V/R, \text{ where } V \text{ is terminal p.d. across } R.$$

9 Terminal p.d. when battery supplies current, $V = E - Ir$.
Terminal p.d. when current in opposition to e.m.f., $V = E + Ir$.
10 Kirchhoff's laws: (a) algebraic sum of currents at a *junction* = 0 (conservation of charge), (b) net e.m.f. E in *closed* circuit = algebraic sum of p.d.s of resistances in that circuit (conservation of energy).

Electrocution

Electricity can be dangerous and needs to be treated with a great deal of respect. Why can electrocution so often prove fatal? When a person touches a high voltage wire, current will start to flow through the person to earth. The amount of current depends on the resistance offered by the person between the wire and earth. Wearing rubber boots will decrease the current but having wet skin will increase the current (this is why extra care has to be taken with bathroom electrical fittings, which should be operated by pull-cords so that the actual electrical switch is nowhere near your wet hand).

It does not take much current to have seriously harmful effects: as little as 0·02 A is very painful, whereas 0·1 A causes *fibrillation*, uncontrolled contractions of the heart which will cause death if not rapidly stopped (it is interesting to note that the only way to stop fibrillation is to apply a controlled electric shock through *defibrillation* equipment). Another common phenomenon when people are electrocuted is that they find themselves stuck to the wire, unable to let go. This is because the current passing through the hand causes the muscles to contract around the wire. (*WARNING: Always take care with mains electricity.*)

In many schools a common experiment is to charge a person up to many thousands of volts by having them hold on to a Van de Graff generator (page 209). Why does this have no more serious effect than making the person's hair stand on end? This is because the person usually stands on an insulated stool, so while he or she becomes charged, no current flows through his or her body to earth. It is *current*, not *voltage*, which is dangerous.

EXERCISES 10A Electric Circuits

Multiple Choice

1 Figure 10A (i) shows a battery of 12 V and negligible internal resistance connected to three resistors. The p.d. V across AB and the main current I are

A $V = 7.5$ V $\quad I = 6$ A $\qquad$ B $V = 4.5$ V $\quad I = 2.5$ A $\quad$ C $V = 2.4$ V $\quad I = 2.4$ A
D $V = 2.4$ V $\quad I = 1.2$ A $\qquad$ E $V = 1.2$ V $\quad I = 1.2$ A

2 In Figure 10A (ii), the voltmeter V reads 5 V when the switch S is closed. The e.m.f. of the battery is 8 V. The internal resistance r in Ω is

A 10 B 8 C 6 D 4 E 2

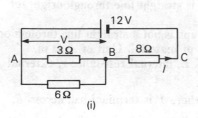 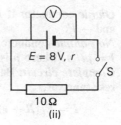

Figure 10A

3 In the circuit shown in Figure 10B (i), a rheostat R is placed across the 12 Ω resistor to produce a variable p.d. V across BC. R is varied from 0 to its maximum value 6 Ω. Then V varies from

   A 12 V to 24 V **B** 6 V to 12 V **C** 0 to 12 V **D** 0 to 9·6 V **E** 0 to 4·7 V

4 Tungsten obeys Ohm's law because
   A the wire becomes hot when current flows
   B the current is carried by electrons
   C the resistance is proportional to temperature
   D the current is proportional to p.d. at constant temperature
   E the current is proportional to the number of charge carriers.

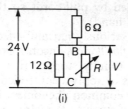 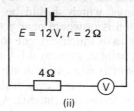

Figure 10B

5 By error, a student places a moving-coil voltmeter V in series in a circuit in order to read the current, as shown in Figure 10B (ii). The voltmeter reading will be practically

   A 0 **B** 4 V **C** 6 V **D** 9 V **E** 12 V

6 A uniform copper wire carries a current i amperes and has p carriers per metre³. The length of the wire is l metres and its cross-section area is s metre². If the charge on a carrier is q coulombs, the drift velocity in m s⁻¹ is given by

   A ilp **B** i/lsq **C** i/psq **D** psq/i **E** $i/pslq$

Longer Questions

7 A battery of e.m.f. 4 V and internal resistance 2 Ω is joined to a resistor of 8 Ω. Calculate the terminal p.d.
   What additional resistance in series with the 8 Ω resistor would produce a terminal p.d. of 3·6 V?

8 A battery of e.m.f. 24 V and internal resistance r is connected to a circuit having two parallel resistors of 3 Ω and 6 Ω in series with an 8 Ω, Figure 10C (i). The current flowing in the 3 Ω resistor is then 0·8 A. Calculate (i) the current in the 6 Ω resistor, (ii) r, (iii) the terminal p.d. of the battery.

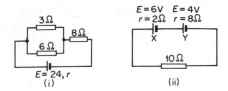

Figure 10C

9 A battery X of e.m.f. 6 V and internal resistance 2 Ω is in series with a battery Y of e.m.f. 4 V and internal resistance 8 Ω so that the two e.m.f.s act in the same direction, Figure 10C (ii). A 10 Ω resistor is connected to the batteries. Calculate the terminal p.d. of each battery.

If Y is reversed so that the e.m.f.s now oppose each other, what is the new terminal p.d. of X and Y?

10 Two resistors of 1200 Ω and 800 Ω are connected in series with a battery of e.m.f. 24 V and negligible internal resistance. What is the p.d. across each resistor?

A voltmeter V of resistance 600 Ω is now connected firstly across the 1200 Ω resistor, and then across the 800 Ω resistor. Find the p.d. recorded by the voltmeter in each case.

11 Copper wire of cross-section area 2·0 mm² carries a current of 1·5 A. Find the drift velocity of the electrons, assuming $9·0 \times 10^{28}$ conduction electrons per metre³ and the electron charge $1·6 \times 10^{-19}$ C.

Why is the drift velocity slow?

12 (a) State Ohm's law.
(b) The following are four electrical components:
   A a component which obeys Ohm's law
   B another component which obeys Ohm's law but which has higher resistance than A
   C a filament lamp
   D a component, other than a filament lamp, which does not obey Ohm's law.
(i) For each of these components, sketch current–voltage characteristics, plotting current on the *vertical* axis, and showing both positive and negative values. Use one set of axes for A and B, and separate sets of axes for C and for D. Label your graphs clearly.
(ii) Explain the shape of the characteristic for C.
(iii) Name the component you have chosen for D. (*N.*)

13

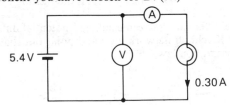

Figure 10D

The battery in the circuit in Figure 10D has e.m.f. 5·4 V and drives a current 0·30 A through a lamp. The voltmeter reading is 4·8 V.

Explain why the voltmeter reading is less than the e.m.f. of the cell.

Calculate values for (a) the internal resistance of the battery, and (b) the energy transformed per second in the lamp. State *two* assumptions you made in order to complete these calculations. (*L.*)

14 The light-dependent resistor (LDR) in Figure 10E is found to have resistance 800 Ω in moonlight and resistance 160 Ω in daylight. Calculate the voltmeter reading, V_m, in moonlight with the switch S open.

If the reading of the voltmeter in daylight with the switch S closed is also equal to V_m, what is the value of the resistance R? (*L.*)

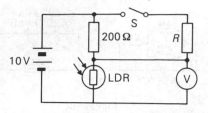

Figure 10E

15 In this question you are asked to analyse the simple circuit shown in Figure 10F (i). The box represents the circuit of a d.c. motor running at a steady speed. The motor behaves as a source of e.m.f. E (opposing that of the 120 V battery), and an internal resistance 4 Ω. These are in parallel with a fixed 240 Ω resistor. A current of 5·5 A is drawn from the 120 V battery (which has negligible internal resistance).

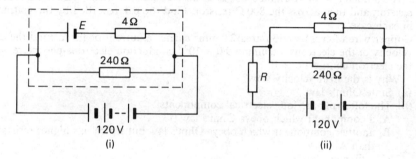

Figure 10F

(a) Calculate (i) the current through the 4 Ω internal resistance, (ii) the e.m.f. E.

(b) When the motor is not running, the e.m.f. E is zero. To limit the current on starting, an extra resistor R is included as shown in Figure 10F (ii). Calculate a value for its resistance which will limit the starting current drawn from the battery to 20 A. (*O. & C.*)

16 Explain clearly the difference between e.m.f. and potential difference.

Write down the Kirchhoff network laws and point out that each is essentially a statement of a conservation law.

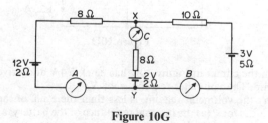

Figure 10G

Find, for the above circuit, Figure 10G, (i) the readings on the ammeters A, B and C (assumed to have effectively zero resistances), (ii) the potential difference between X and Y, (iii) the power dissipated as heat in the circuit, (iv) the power delivered by the 12 V cell.

Account carefully for the difference between (iii) and (iv). (*W.*)

Heat and Power

Electrical Heating, Joule's Laws

In 1841 Joule studied the heating effect of an electric current by passing it through a coil of wire in a jar of water, Figure 10.22. He used various currents, measured by an early form of galvanometer G, and various lengths of wire, but always the same mass of water. The rise in temperature of the water, in a given time, was then proportional to the heat developed by the current in that time. Joule found that the heat produced in a given time, with a given wire, was proportional to I^2, where I is the current flowing. If H is the heat produced per second, then

$$H \propto I^2 \qquad . \qquad . \qquad . \qquad . \qquad . \qquad (1)$$

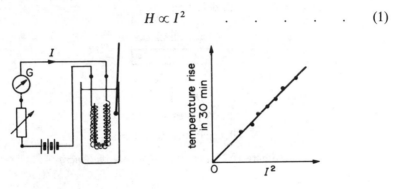

Figure 10.22 *Joule's experiment on heating effect of current*

Joule also made experiments on the heat produced by a given current in different wires. He found that the rate of heat production was proportional to the *resistance* of the wire:

$$H \propto R \qquad . \qquad . \qquad . \qquad . \qquad . \qquad (2)$$

Relationships (1) and (2) together give

$$H \propto I^2 R \qquad . \qquad . \qquad . \qquad . \qquad . \qquad (3)$$

How Current Produces Heat in Metals

Heat is a form of energy. The heat produced per second by a current in a wire is, therefore, a measure of the energy which it liberates in one second, as it flows through the wire.

The heat is produced, we suppose, by the free electrons as they move through the metal. On their way they collide frequently with atoms. *At each collision they lose some of their kinetic energy*, and give it to the atoms which they strike. Thus, as the current flows through the wire, it increases the kinetic energy of vibration of the metal atoms: it generates heat in the wire. The electrical resistance of the metal is due, we say, to its atoms obstructing the drift of the electrons past them: it is analogous to mechanical friction. As the current flows through the wire, the energy lost per second by the electrons is the electrical power supplied by the battery which maintains the current. That power comes, as we shall see later, from chemical energy in the battery.

Potential Difference and Energy

On p. 221 we defined the potential difference V_{AB} between two points, A and B, as the work done by an external agent in taking a unit positive charge from B to A, Figure 10.23 (i). This definition applies equally well to points on a conductor carrying a current.

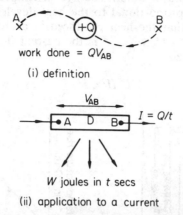

work done = QV_{AB}

(i) definition

V_{AB}

$I = Q/t$

W joules in t secs

(ii) application to a current

Figure 10.23 *Potential difference and energy*

In Figure 10.23 (ii), D represents any electrical device or circuit element: a lamp, motor, or battery on charge, for example. A current of I amperes flows through it from the terminal A to the terminal B. If it flows for t seconds, the charge Q which it carries from A to B is, since a current is the quantity of electricity per second flowing,

$$Q = It \text{ coulombs} \qquad . \qquad . \qquad . \qquad . \qquad (1)$$

Let us suppose that the device D liberates a total amount of energy W joules in the time t; this total may be made up of heat, light, sound, mechanical work, chemical transformation, and any other forms of energy. Then W is the amount of *electrical energy* given up by the charge Q in passing through the device D from A to B. From the definition of p.d.,

$$W = QV_{AB} \qquad . \qquad . \qquad . \qquad . \qquad (2)$$

where V_{AB} is the potential difference between A and B in volts.

The work, in all its forms, which the current I does in t seconds as it flows through the device, is therefore

$$W = IV_{AB}t \qquad . \qquad . \qquad . \qquad . \qquad (3)$$

by equations (1) and (2). W, the energy produced, is in joules if I is in amperes, V_{AB} in volts and t in seconds.

Electrical Power

The energy liberated per second in the device is defined as its electrical *power*.

The electrical power, P, supplied is given, from above, by

$$P = \frac{W}{t} = \frac{IV_{AB}t}{t}$$

(Figure 10.24 (i)), or

$$P = IV_{AB} \qquad . \qquad . \qquad . \qquad . \qquad (1)$$

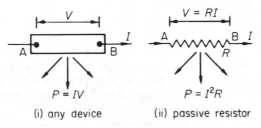

$P = IV$

(i) any device

$P = I^2R$

(ii) passive resistor

Figure 10.24 *Power equations*

When an electric current flows through a wire or 'passive' resistor, all the power which it conveys to the wire appears as *heat*. If I is the current, R is the resistance, then $V_{AB} = IR$, Figure 10.24 (ii).

$$\therefore P = I^2R \qquad . \qquad . \qquad . \qquad . \qquad (2)$$

Also,

$$P = \frac{V_{AB}^2}{R} \qquad . \qquad . \qquad . \qquad . \qquad (3)$$

The power, P, is in *watts* (W) when I is in amperes, R is in ohms, and V_{AB} is in volts. 1 kilowatt (kW) = 1000 watts.

The formulae for power, $P = I^2R$ or V^2/R, is true only when all the electrical power supplied is dissipated as heat. As we shall see, the formulae do not hold when part of the electrical energy supplied is converted into mechanical work, as in a motor, or into chemical energy, as in an accumulator being charged. A device which converts all the electrical energy supplied to it into heat is called a 'passive' resistor; it may be a wire, or a strip of carbon, or a liquid which conducts electricity but is not decomposed by it. Since the joule (J) is the unit of heat, it follows that, for a resistor, the heat H in it in joules is given by

$$H = IVt$$

or by

$$H = I^2Rt \qquad . \qquad . \qquad . \qquad . \qquad (4)$$

or by

$$H = \frac{V^2t}{R}$$

The units of I, V, R are ampere (A), volt (V), ohm (Ω) respectively.

High-tension (High-voltage) Transmission

When electricity has to be transmitted from a source, such as a power station, to a distant load, such as a factory, the two must be connected by cables. These cables have resistance, which is in effect added to the internal resistance of the generator. Power is wasted in them as heat. If r is the total resistance of the cables, and I the supply current, the power wasted is $I^2 r$.

The power delivered to the factory is IV, where V is the potential difference at the factory. Economy requires the waste power, $I^2 r$, to be small; but it also requires the cables to be thin, and therefore cheap to buy and connect. The thinner the cables, however, the higher their resistance r. So the most economical way to transmit the power is to make the current, I, as small as possible; this means making the potential difference V as *high* as possible. When large amounts of power are to be transmitted, therefore, very high voltages are used: 400 000 volts on the main lines of the National Grid in the UK, 23 000 volts on subsidiary lines.

These voltages are much too high to be brought into a house, or even a factory. They are stepped down by transformers, in a way which we shall describe later; stepping-down in that way is possible only with alternating current, which is one of the main reasons why alternating current is used all over the world.

Summary of Formulae Related to Power

In any device
 Electrical power consumed = power developed in other forms,

$$P = IV$$

watts = amperes × volts

In a passive resistor

(i)
$$V = IR \qquad I = \frac{V}{R} \qquad R = \frac{V}{I}$$

(ii) Power consumed = heat developed per second, in watts.

$$P = I^2 R = IV = \frac{V^2}{R}$$

(iii) Heat developed in time t:
 Electrical energy consumed = heat developed in joules

$$I^2 Rt = IVt = \frac{V^2}{R} t$$

Commercial energy unit = kilowatt hour (kWh) = kilowatt × hour

$$= 3{\cdot}6 \times 10^6 \text{ joule}$$

Example on Heating Effect of Current

An electric heating element to dissipate 480 watts on 240 V mains is to be made from Nichrome ribbon 1 mm wide and thickness 0·05 mm. Calculate the length of ribbon required if the resistivity of Nichrome is $1{\cdot}1 \times 10^{-6}$ ohm metre.

Power,
$$P = \frac{V^2}{R}$$

$$\therefore R = \frac{V^2}{P} = \frac{240^2}{480} = 120\,\Omega$$

The area A of cross-section of the ribbon $= 1 \times 0.05 \text{ mm}^2 = 0.05 \times 10^{-6} \text{ m}^2$.

From
$$R = \frac{\rho l}{A}$$

$$l = \frac{R.A}{\rho} = \frac{120 \times 0.05 \times 10^{-6}}{1.1 \times 10^{-6}} = 5.45 \text{ m}$$

Electromotive Force and Energy

We can now get a definition of electromotive force E from energy principles.

We can define the e.m.f. E of a battery or any other generator as the *total energy per coulomb it delivers round a circuit joined to it*.

So if a device has an electromotive force E, then, in passing a charge Q round a circuit joined to it, it liberates an amount of electrical energy equal to QE. If a charge Q is passed through the source against its e.m.f., then the work done against the e.m.f. is QE. The above definition of e.m.f. does not depend on any assumptions about the nature of the generator.

If a device of e.m.f. E passes a steady current I for a time t, then the charge it circulates is

$$Q = It$$

So, from the definition of E,

total electrical energy liberated, $W, = QE = IEt$ . . (1)

and **total electrical power generated, $P = \dfrac{W}{t} = EI$** . . (2)

We can also define e.m.f. in terms of power and current, and, therefore, in a way suitable for dealing with circuit problems. From equation (2)

$$P = EI$$

or
$$E = \frac{P}{I}$$

So *the e.m.f. of a device is the ratio of the electrical power which it generates to the current which it delivers*. If current is forced through a device in opposition to its e.m.f., then equation (2) gives the power used in overcoming the e.m.f.

Electromotive force resembles potential difference in that both can be defined as the ratio of power to current. The unit of e.m.f. is, therefore, 1 watt per ampere, or 1 volt; and *the e.m.f. of a source, in volts, is numerically equal to the power which it generates when it delivers a current of 1 ampere*.

Current Formula

We can apply the definition of E in terms of power to the circuit in Figure 10.25. The total power supplied by the source is EI. The power delivered to the

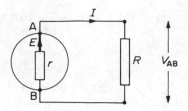

Figure 10.25 *A complete circuit*

external resistor is called the *output power*. Its value $= IV_{AB} = I \times IR = I^2R$. The power delivered to the internal resistance $r = I^2r$. So

$$EI = I^2R + I^2r \qquad . \qquad . \qquad . \qquad . \qquad (1)$$

Dividing by I, $\qquad\qquad\qquad E = IR + Ir \qquad . \qquad . \qquad . \qquad . \qquad (2)$

and $\qquad\qquad\qquad\qquad I = \dfrac{E}{R + r} \qquad . \qquad . \qquad . \qquad . \qquad (3)$

The same results in (2) and (3) were obtained earlier in the chapter.

Output Power and Efficiency

As we have just seen, the power delivered to the external resistor R, often called the *load* of a battery or other generator, is given by $P_{out} = IV_{AB}$ in Figure 10.25. The power supplied by the source or generator $P_{gen} = IE$. The difference between the power generated and the output is the power wasted as heat in the source: I^2r. The ratio of the power output to the power generated is the efficiency, η, of the circuit as a whole:

$$\eta = \frac{P_{out}}{P_{gen}} \qquad . \qquad . \qquad . \qquad . \qquad (1)$$

So $\qquad\qquad\qquad\qquad \eta = \dfrac{P_{out}}{P_{gen}} = \dfrac{IV_{AB}}{IE} = \dfrac{V_{AB}}{E}$

Now $V_{AB} = IR = ER/(R + r)$. So

$$\eta = \frac{R}{R + r} \qquad . \qquad . \qquad . \qquad . \qquad (2)$$

This shows that the efficiency tends to unity (or 100 per cent) as the load resistance R tends to *infinity*. For high efficiency the load resistance must be several times the internal resistance of the source. When the load resistance is equal to the internal resistance, the efficiency is 50 per cent. (See Figure 10.26 (i).)

Power Variation, Maximum Power

Now let us consider how the *power output* varies with the load resistance. Since the power output $= IV_{AB} = I \times IR$, then

$$P_{out} = I^2R \qquad .$$

Also
$$I = \frac{E}{R + r}$$

so
$$P_{out} = \frac{E^2 R}{(R + r)^2}$$

If we take fixed values of E and r, and plot P_{out} as a function of R, we find that it passes through a maximum when $R = r$, Figure 10.26 (i). So

---

power output to R is a maximum when $R = r$, internal resistance.

We shall explain this result shortly. Physically, this result means that the power output is very small when R is either very large or very small compared with r. When R is very large, the terminal potential difference, V_{AB}, approaches a constant value equal to the e.m.f. (Figure 10.26 (ii)); as R is increased the current falls, and the power IV_{AB} falls with it. When R is very small, the current approaches the constant value E/r, but the potential difference (which is equal to IR) falls steadily with R; the power output therefore falls likewise. Consequently the power output is greatest for a moderate value of R; the mathematics shows that this value is actually $R = r$, the internal resistance.

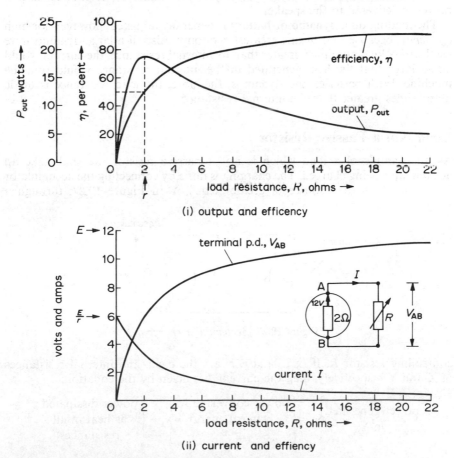

(i) output and effciency

(ii) current and effiency

Figure 10.26 *Effects of varying load resistance in circuit*

To prove $R = r$ for maximum power, we have, from before,

$$P_{out} = \frac{E^2 R}{(R + r)^2}$$

Now

$$(R + r)^2 = R^2 + 2Rr + r^2 = (R - r)^2 + 4Rr$$

So

$$P_{out} = \frac{E^2 R}{(R - r)^2 + 4Rr} = \frac{E^2}{(R - r)^2/R + 4r}$$

When $R = r$, the denominator of the fraction is *least* and so P_{out} is then a maximum. We see that the maximum output power $= E^2/4r$, which agrees with the power value when $R = r$ calculated from $P_{out} = E^2 R/(R + r)^2$.

Examples of Loads in Electrical Circuits

Loading for greatest *power output* is common in communications engineering. For example, the last transistor in a receiver delivers electrical power to the loudspeaker, which the speaker converts into mechanical power as sound waves.

To get the loudest sound, the speaker resistance (or impedance) is 'matched' to the internal resistance (or impedance) of the transistor, so that maximum power is delivered to the speaker.

The loading on a dynamo or battery is generally adjusted, however, for high *efficiency*, because that means greatest economy. Also, if a large dynamo were used with a load not much greater than its internal resistance, the current would be so large that the heat generated in the internal resistance would ruin the machine. With batteries and dynamos, therefore, the load resistance is made many times greater than the internal resistance.

Load Not a Passive Resistor

As an example of a load which is not a passive resistor, we shall take an accumulator being charged. The charging is done by connecting the accumulator X in *opposition* to a source of greater e.m.f., Y in Figure 10.27, through a

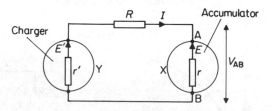

Figure 10.27 *Accumulator charging*

controlling resistor R. If E, E' and r, r' are the e.m.f. and internal resistances of X and Y respectively, then the current I is given by the equation:

$$\left\{ \begin{array}{c} \text{power generated} \\ \text{in Y} \end{array} \right\} = \left\{ \begin{array}{c} \text{power converted to} \\ \text{chemical energy} \\ \text{in X} \end{array} \right\} + \left\{ \begin{array}{c} \text{power dissipated} \\ \text{as heat in all} \\ \text{resistances} \end{array} \right\}$$

$$E'I \quad = \quad EI \quad + \quad I^2 R + I^2 r' + I^2 r \quad (1)$$

So $$(E' - E)I = I^2(R + r' + r),$$

from which $$I = \frac{E' - E}{R + r' + r} = \frac{\text{net e.m.f.}}{\text{total resistance}} \qquad (2)$$

The potential difference across the accumulator itself, V_{AB}, is given by

$$\left. \begin{matrix} \text{power delivered} \\ \text{to X} \end{matrix} \right\} = \left\{ \begin{matrix} \text{power converted to} \\ \text{chemical energy} \end{matrix} \right\} + \left\{ \begin{matrix} \text{power dissipated} \\ \text{as heat} \end{matrix} \right.$$

So $$I V_{AB} \qquad = \qquad IE \qquad + \qquad I^2 r$$

and $$V_{AB} = E + Ir \qquad . \qquad . \qquad . \qquad . \qquad . \qquad (3)$$

Equation (3) shows that, when current is driven through a generator in opposition to its e.m.f., then the potential difference across the generator is equal to the *sum* of its e.m.f. and the voltage drop across its internal resistance.

The Thermoelectric Effect

Seebeck Effect

The heating effect of the current transfers electrical energy into heat, but we have not so far described any mechanism which transfers heat into electrical energy. This was discovered by SEEBECK in 1822.

In his experiments he connected a plate of bismuth between copper wires leading to a galvanometer, as shown in Figure 10.28 (i). He found that if one of the bismuth–copper junctions was heated, while the other was kept cool, then a current flowed through the galvanometer. The direction of the current was from the copper to the bismuth at the cold junction. We can easily repeat Seebeck's experiment, using copper and iron wires and a meter capable of indicating a few microamperes, Figure 10.28 (ii).

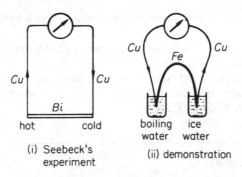

(i) Seebeck's experiment

(ii) demonstration

Figure 10.28 *The thermoelectric effect*

Thermocouples

Seebeck went on to show that a current flowed, without a battery, in any circuit containing two different metals, with their two junctions at different temperatures. Currents obtained in this way are called *thermoelectric currents*, and a pair of metals, with their junctions at different temperatures, are said to form a *thermocouple*. The following is a list of metals, such that if any two of them form a thermocouple, then the current will flow from the higher to the lower in the list, across the cold junction:

Antimony, Iron, Zinc, Lead, Copper, Platinum, Bismuth.

Thermoelectric currents often appear when they are not wanted. They may occur from small differences in purity of two samples of the same metal, and from small differences of temperature—due, perhaps, to the warmth of the hand. They can cause a great deal of trouble in circuits used for precise measurements, or for detecting other small currents, not of thermal origin. As sources of electrical energy, thermoelectric currents are neither convenient nor economical, but they have been used in solar batteries with semiconductor materials. Thermocouples are used in the measurement of temperature, and of other quantities, such as radiant energy, which can be measured by a temperature rise at one junction.

Variation of Thermoelectric E.M.F. with Temperature

Later we shall see how thermoelectric e.m.f.s are measured. When the cold junction of a given thermocouple is kept constant at 0°C, and the hot junction temperature θ°C is varied, the e.m.f. E is found to vary as $E = a\theta + b\theta^2$, where a, b are constants. This is a parabola-shaped curve (Figure 10.29). The temperature A corresponding to the maximum e.m.f. is known as the *neutral*

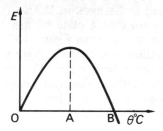

Figure 10.29 *Thermoelectric e.m.f. variation with temperature*

temperature; it is about 250°C for a copper–iron thermocouple. Beyond the temperature B, known as the *inversion temperature*, the e.m.f. reverses. Thermo-electric thermometers, which use thermocouples, are used only as far as the neutral temperature, because the same e.m.f. is obtained at two different temperatures, from Figure 10.29.

You should know:

1   **Energy $= IVt$ (joules). Power $= IV$ (watts).**
2   **For a resistor R, energy = heat $= IVt = I^2Rt = V^2t/R$**
    or heat per second $= IV = I^2R = V^2/R$.
3   **The e.m.f. E of a battery = total energy per coulomb = total power per ampere.**
    So $EI = I^2R + I^2r$, where R and r are external and internal resistances.
4   **In a circuit with a battery, maximum power is obtained for a load R when $R = r$.**

5   **A thermoelectric e.m.f. is obtained in two wires (thermocouple) for a temperature difference between the two junctions.
When the cold junction is at 0°C and the hot junction is θ°C,
$E = a\theta + b\theta^2$, so E varies as a parabola with θ.**

EXERCISES 10B   Power and Heat

Multiple Choice

1   In the simple circuit shown in Figure 10H (i), the power produced in the resistor R is one-quarter that produced in the 3 Ω. Then R in Ω is

   A 6 **B** 9 **C** 12 **D** 15 **E** 18

2   A filament lamp L is connected to a battery of e.m.f. E and internal resistance r. Figure 10H (ii). The lamp has a resistance R when working. The fraction of its power supplied by the battery to the lamp is

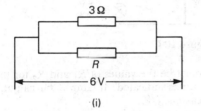
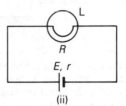

Figure 10H

   A R/r **B** r/R **C** $(R+r)/r$ **D** $R/(R+r)$ **E** $R^2/(R+r)^2$

3   An electrical heater supplies 320 J in 1 minute to a 4·00 Ω resistor. The steady current flowing in **A** is then about

   A 1·2 **B** 0·67 **C** 0·34 **D** 0·16 **E** 0·10

4   A current I in a lamp varies with voltage V as shown in Figure 10I (i). The variation of power P with current I is best shown in Figure 10I (ii) by which graph from **A** to **E**?

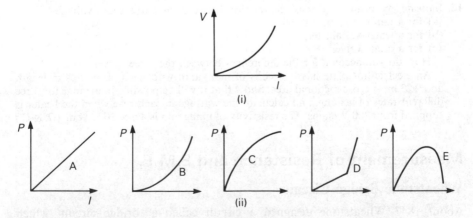

Figure 10I

Longer Questions

5 The maximum power dissipated in a $10\,000\,\Omega$ resistor is 1 W. What is the maximum current?

6 Two heating coils A and B, connected in parallel in a circuit, produce powers of 12 W and 24 W respectively. What is the ratio of their resistances, R_A/R_B, when used?

7 A heating coil of power rating 10 W is required when the p.d. across it is 20 V. Calculate the length of nichrome wire needed to make the coil if the cross-sectional area of the wire used is $1 \times 10^{-7}\,\text{m}^2$ and the resistivity of nichrome is $1 \times 10^{-6}\,\Omega\,\text{m}$.
What length of wire would be needed if its diameter was half that previously used?

8 The running temperature of the filament of a 12 V, 48 W tungsten filament lamp is 2700°C and the average temperature coefficient of resistance for tungsten from 0°C to 2700°C is $6\cdot4 \times 10^{-3}\,\text{K}^{-1}$. Calculate the resistance of the filament at 0°C. (*L.*)

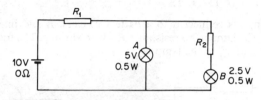

Figure 10J

9 In the above circuit, Figure 10J, what must be the values of R_1 and R_2 for the two lamps *A* and *B* to be operated at the ratings indicated? If lamp *A* burns out, what would be the effect on lamp *B*? Give reasons. (*W.*)

10 A thin film resistor in a solid-state circuit has a thickness of 1 μm and is made of nichrome of resistivity $10^{-6}\,\Omega\,\text{m}$. Calculate the resistance available between opposite edges of a 1 mm² area of film
(a) if it is square shaped,
(b) if it is rectangular, 20 times as long as it is wide. (*C.*)

11 State the laws of the development of heat when an electric current flows through a wire of uniform material.
An electrical heating coil is connected in series with a resistance of $X\,\Omega$ across the 240 V mains, the coil being immersed in a kilogram of water at 20°C. The temperature of the water rises to boiling-point in 10 minutes. When a second heating experiment is made with the resistance X short-circuited, the time required to develop the same quantity of heat is reduced to 6 minutes. Calculate the value of X. (Heat losses may be neglected.) (*L.*)

12 Indicate, by means of graphs, the relation between the current and voltage
(a) for a uniform manganin wire;
(b) for a water voltameter;
(c) for a diode valve.
How do you account for the differences between the three curves?
An electric hot plate has two coils of manganin wire, each 20 metres in length and 0·23 mm² cross-sectional area. Show that it will be possible to arrange for three different rates of heating, and calculate the wattage in each case when the heater is supplied from 200 V mains. The resistivity of manganin is $4\cdot6 \times 10^{-7}\,\Omega\,\text{m}$. (*O. & C.*)

Measurement of Resistance and E.M.F.

Wheatstone Bridge Circuit

About 1843 Wheatstone designed a circuit called a 'bridge circuit' which gave an accurate method for measuring resistance. We shall deal later with

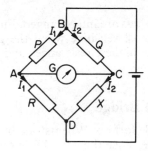

Figure 10.30 *Wheatstone bridge*

the practical aspects. In Figure 10.30, X is the unknown resistance, and P, Q, R are resistance boxes. One of these—usually R—is adjusted until the galvanometer G between A and C shows no deflection, a so-called 'balance' condition. In this case the current I_g in G is zero. Then, as we shall show,

$$\frac{P}{Q} = \frac{R}{X}$$

so

$$X = \frac{Q}{P} R$$

Wheatstone Bridge Proof

At balance, since no current flows through the galvanometer, the points A and C must be at the same potential, Figure 10.30. Therefore

$$V_{AB} = V_{CB} \text{ and } V_{AD} = V_{CD}$$

Using ratios,

$$\frac{V_{AB}}{V_{AD}} = \frac{V_{CB}}{V_{CD}} \quad . \quad . \quad . \quad . \quad (1)$$

Also, since $I_g = 0$, P and R carry the same current, I_1, and X and Q carry the same current, I_2. So from $V = IR$,

$$\frac{V_{AB}}{V_{AD}} = \frac{I_1 P}{I_1 R} = \frac{P}{R}$$

and

$$\frac{V_{CB}}{V_{CD}} = \frac{I_2 Q}{I_2 X} = \frac{Q}{X} \quad . \quad . \quad . \quad . \quad (2)$$

From equations (1) and (2),

$$\frac{P}{R} = \frac{Q}{X}$$

So

$$X = \frac{Q}{P} R$$

Exactly the same relationship between the four resistances is obtained if the galvanometer and cell positions are interchanged. Further analysis of the circuit shows that the bridge is most sensitive when the galvanometer is connected between the junction of the highest resistances and the junction of the lowest resistances.

The Slide-wire (Metre) Bridge

Figure 10.31 shows a simple form of Wheatstone bridge; it is sometimes called a slide-wire or metre bridge, since the wire AB is often a metre long. The wire is uniform. So a slider S can be used to measure a length on it.

The unknown resistance X and a known resistance R are connected as shown in the figure; heavy brass or copper strip is used for the connections AD, FH, KB, whose resistances are generally negligible. When the slider is at a point C in the wire it divides the wire into two parts, of resistances R_{AC} and R_{CB}; these,

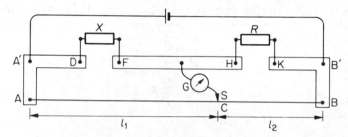

Figure 10.31 *Slide wire (metre) bridge*

with X and R, form a Wheatstone bridge. (The galvanometer and battery are interchanged relative to the circuits we have given earlier; that enables the slider S to be used as the galvanometer key. We have already seen that the interchange does not affect the condition for balance in G.) The connections are checked by placing S first on A, then on B. The balance-point is found by trial and error—not by scraping S along AB. At balance,

$$\frac{X}{R} = \frac{R_{AC}}{R_{CB}}$$

Since the wire is uniform, the resistances R_{AC} and R_{CB} are proportional to the lengths of wire, l_1 and l_2. Therefore

$$\frac{X}{R} = \frac{l_1}{l_2} \qquad . \qquad . \qquad . \qquad . \qquad . \qquad (1)$$

The resistance R should be chosen so that the balance-point C comes fairly near to the centre of the wire—within, say, its middle third. If either l_1 or l_2 is small, the resistance of its end connection AA' or BB' in Figure 10.31 is not negligible in comparison with its own resistance; equation (1) then does not hold. Some idea of the accuracy of a particular measurement can be got by interchanging R and X, and balancing again. If the new ratio agrees with the old within about 1%, then their average may be taken as the value of X.

Since the galvanometer G is a sensitive current-reading meter, a high protective resistor (not shown) is required in series with it until a *near* balance is found on the wire. At this stage the high resistor is shunted or removed and the final balance-point found.

The lowest resistance which a bridge of this type can measure with reasonable accuracy is about 1 ohm. Resistances lower than about 1 ohm cannot be measured accurately on a Wheatstone bridge, because of the resistances of the wires connecting them to the X terminals, and of the contacts between those wires and the terminals to which they are, at each end, attached. This is the reason why the potentiometer method is more satisfactory for comparing and measuring low resistances.

Temperature Coefficient of Resistance

We have already seen that the resistance of a wire varies with its temperature. If we put a coil of fine copper wire into a water bath, and use a Wheatstone

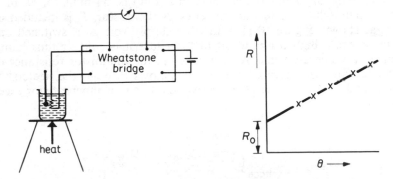

Figure 10.32 *Measurement of temperature coefficient*

bridge to measure its resistance at various moderate temperatures θ, we find that the resistance, R_θ, increases with the temperature, Figure 10.32. We may, therefore, define a *temperature coefficient of resistance*, α, such that

$$R_\theta = R_0(1 + \alpha\theta) \quad . \quad . \quad . \quad . \quad (1)$$

where R_0 is the resistance at 0°C. In words, starting with the resistance at 0°C,

$$\alpha = \frac{\text{increase of resistance per K rise of temperature}}{\text{resistance at } 0°C}$$

If R_1 and R_2 are the resistances at θ_1°C and θ_2°C, then, from (1),

$$\frac{R_1}{R_2} = \frac{1 + \alpha\theta_1}{1 + \alpha\theta_2} \quad . \quad . \quad . \quad . \quad (2)$$

Values of α for pure metals are of the order of 0·004 K^{-1}. They are much less for alloys than for pure metals, a fact which makes alloys useful materials for resistance boxes and shunts on electric meters.

Equation (1) represents the change of resistance with temperature fairly well, but not as accurately as it can be measured. More accurate equations are given later in the Heat section of this book, where resistance thermometers are discussed.

Thermistors

A *thermistor* is a heat-sensitive resistor usually made from semiconductors. One type of thermistor has a high positive temperature coefficient of resistance. So when it is placed in series with a battery and a current meter and warmed, the current is observed to decrease owing to the rise in resistance. Another type of thermistor has a *negative* temperature coefficient of resistance, that is, its resistance rises when its temperature is decreased, and falls when its temperature is increased. Thus when it is placed in series with a battery and a current meter and warmed, the current is observed to increase owing to the decrease in resistance.

Thermistors with a high negative temperature coefficient are used for resistance thermometers in very low temperature measurement of the order of 10 K, for example. The higher resistance at low temperature enables more accurate measurements to be made.

Thermistors with negative temperature coefficient may be used to safeguard against current surges in circuits where this could be harmful, for example, in a radio circuit where heaters are in series. A thermistor, T, is included in the circuit, as shown, Figure 10.33. When the supply voltage is switched on, the thermistor has a high resistance at first because it is cold. It thus limits the current to a moderate value. As it warms up, the thermistor resistance drops appreciably and an increased current then flows through the heaters. Thermistors are also used in transistor receiver circuits to compensate for excessive rise in current.

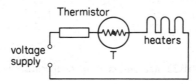

Figure 10.33 *Use of thermistor*

Example on Temperature Coefficient

In an experiment carried out at 0°C, A was 1·20 m of Nichrome wire of resistivity 100×10^{-8} Ω m and diameter 1·20 mm, and B a German silver wire 0·80 mm diameter and resistivity 28×10^{-8} Ω m. The ratio of the resistances A/B was 1·20. What was the length of the wire B?

If the temperature coefficient of Nichrome is 0·000 40 K^{-1} and of German silver is 0·000 30 K^{-1}, what would the ratio of resistances become if the temperature were raised by 100 K?

With usual notation,

for A,
$$R_1 = \frac{\rho_1 l_1}{A_1}$$

and for B,
$$R_2 = \frac{\rho_2 l_2}{A_2}$$

$$\therefore \frac{R_1}{R_2} = \frac{\rho_1}{\rho_2} \cdot \frac{l_1}{l_2} \cdot \frac{A_2}{A_1} = \frac{\rho_1}{\rho_2} \cdot \frac{l_1}{l_2} \cdot \frac{d_2^{\,2}}{d_1^{\,2}}$$

where d_2, d_1 are the respective diameters of B and A.

$$\therefore 1\cdot20 = \frac{100}{28} \times \frac{1\cdot20}{l_2} \times \frac{0\cdot8^2}{1\cdot20^2}$$

$$\therefore l_2 = \frac{100 \times 1\cdot20 \times 0\cdot64}{1\cdot20 \times 28 \times 1\cdot44} = 1\cdot59 \text{ m} \quad . \qquad . \qquad . \qquad (1)$$

When the temperature is raised by 100 K, the resistance increases according to the relation $R_\theta = R_0(1 + \alpha\theta)$. So

new Nichrome resistance, $R_A = R_1(1 + \alpha\cdot100) = R_1 \times 1\cdot04$

and new German silver resistance, $R_B = R_2(1 + \alpha'\cdot100) = R_2 \times 1\cdot03$

$$\therefore \frac{R_A}{R_B} = \frac{R_1}{R_2} \times \frac{1\cdot04}{1\cdot03} = 1\cdot20 \times \frac{1\cdot04}{1\cdot03} = 1\cdot21 . \qquad . \qquad . \qquad (2)$$

Strain gauges

On page 144 we introduced the concept of *strain* as the extension per unit length of a material under tensile stress. For hard materials this strain can be very small, and yet engineers often need to measure such small strains accurately to investigate the deformation of structures or components under load. This is usually done using a *strain gauge*, a small, flat coil of wire which is attached firmly to the material under strain. When the material deforms, the strain gauge will also deform by the same amount, causing the wire in the coil to become thinner or fatter and so changing its electrical resistance. So a strain in the material shows itself as a change in the resistance of the strain gauge.

This change is usually detected by placing the strain gauge in one arm of a Wheatstone bridge: for example, we can imagine the resistor P in Figure 10.30 as being the strain gauge. If the bridge is initially balanced when the gauge is not strained, then any subsequent change in the gauge's resistance will unbalance the bridge, and a current will be detected in the galvanometer G. In addition, the magnitude of the current will show the magnitude of the strain. If the galvanometer is initially calibrated using some known strains, then such a system is capable of making accurate, simple measurements of very small strains.

EXERCISES 10C Wheatstone bridge, Resistance

Multiple Choice

1 The wire on a metre bridge is 100 cm long. For the most accurate measurement of an unknown resistance, the balance-point on the wire in cm should best be in the range

   A 0–20 B 20–40 C 40–60 D 60–80 E 80–100

2 For making standard resistances such as $1\cdot00\,\Omega$, the wire used should best have a temperature coefficient of resistance which is

   A high B moderate C low D zero E negative

3 In a metre bridge experiment, an unknown resistance X is compared with a known resistance R. Then X should best be

A much higher in value than R
B much lower in value than R
C the same order as R
D on the right of R in the bridge circuit
E on the left of R in the bridge circuit

4 A $2\cdot00\,\Omega$ wire X at $0°C$ has a temperature coefficient $0\cdot004\ K^{-1}$. A $3\cdot00\,\Omega$ wire Y at $0°C$ has a temperature coefficient $0\cdot002\ K^{-1}$. At a temperature of $100°C$, X and Y are compared in resistance on a $100\ cm$ long metre bridge. The ratio of the balance-lengths on the wire is

A $0\cdot67$ B $0\cdot78$ C $C0\cdot84$ D $1\cdot26$ E $1\cdot45$

Longer Questions

5 A copper coil has a resistance of $20\cdot0\,\Omega$ at $0°C$ and a resistance of $28\cdot0\,\Omega$ at $100°C$. What is the temperature coefficient of resistance of copper?
   Used in a circuit, the p.d. across the coil is 12 V and the power produced in it is 6 W. What is the temperature of the coil?
6 A tungsten coil has a resistance of $12\cdot0\,\Omega$ at $15°C$. If the temperature coefficient of resistance of tungsten is $0\cdot004\ K^{-1}$, calculate the coil resistance at $80°C$.
7 A heating coil is to be made, from Nichrome wire, which will operate on a 12 V supply and will have a power of 36 W when immersed in water at 373 K. The wire available has an area of cross-section of $0\cdot10\ mm^2$. What length of wire will be required? (Resistivity of Nichrome at 273 K $= 1\cdot08 \times 10^{-6}\ \Omega$ m. Temperature coefficient of resistivity of Nichrome $= 8\cdot0 \times 10^{-5}\ K^{-1}$.) (*L.*)
8 Describe how you would measure the temperature coefficient of resistance of a metal.
   A steady potential difference of 12 V is maintained across a wire which has a resistance of $3\cdot0\,\Omega$ at $0°C$; the temperature coefficient of resistance of the material is $4 \times 10^{-3}\ K^{-1}$. Compare the rates of production of heat in the wire at $0°C$ and at $100°C$.
   The wire is embedded in a body of constant heat capacity $600\ J\ K^{-1}$. Neglecting heat losses, and taking the thermal conductivity of the body to be large, find the time taken to increase the temperature of the body from $0°C$ to $100°C$. (*O.*)
9 Define temperature coefficient of resistance. Describe how you would measure the average temperature coefficient of resistance for an iron wire across the temperature range $0°C$ to $100°C$, using an iron wire of resistance about $4\,\Omega$ at $0°C$ and $6\cdot5\,\Omega$ at $100°C$ wound on an insulating former and provided with copper leads. State approximate values for the circuit components which you would use. (*L.*)
10 (a) In the Wheatstone bridge arrangement in Figure 10K, X is the resistance of a length of constantan wire and Y is the resistance of a standard resistor. Derive from first principles the equation relating X, Y, L_1 and L_2 when no current flows through the galvanometer, stating any assumptions you make.

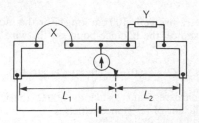

Figure 10K

(b) A student plans to use the apparatus described in (a) to determine the resistance of several different lengths of constantan wire of diameter 0·50 mm. If a 2 Ω standard resistor is available, suggest the range of lengths he might use given that the resistivity of constantan wire is approximately 5×10^{-7} Ω m. Give reasons for your answers. (*N*.)

11 Derive the balance condition for a Wheatstone bridge. Describe a practical form of Wheatstone bridge and explain how you would use it to determine the resistance of a resistor of nominal value 20 Ω.

An electric fire element consists of 4·64 m of Nichrome wire of diameter 0·500 mm, the resistivity of Nichrome at 15°C being 112×10^{-8} Ω m. When connected to a 240 V supply the fire dissipates 2·00 kW and the temperature of the element is 1015°C. Determine a value for the mean temperature coefficient of resistance of Nichrome between 15°C and 1015°C. (*L*.)

E.M.F. Measurement and Applications

Variation of P.D. with Wire Length

As we explained earlier, the e.m.f. of a cell Y is the p.d. at its terminals when *no current* flows from the cell, that is, Y is on 'open circuit'. If a current flows, the p.d. at its terminals is less than the e.m.f. (p. 281). When a moving-coil voltmeter is joined to the terminals of Y, a small current flows in the meter. So the voltmeter reading is slightly less than the e.m.f.

Figure 10.34 (i) shows a uniform wire AB, about one metre long, with a connected accumulator joined to it so that a steady current I flows in AB. Since the wire is uniform, its resistance per centimetre, R, is constant. The potential difference across 1 cm of the wire, RI, is, therefore, also constant. The potential difference between the end A of the wire V_{AC} and any point C upon it, is then proportional to the length of wire l between A and C since $R \propto l$:

$$V_{AC} \propto l \qquad \cdots \qquad \cdots \qquad (1)$$

Comparing E.M.F.s of Two Cells

Suppose we now take a cell Y whose e.m.f. we have to find. As shown in Figure 10.34 (ii), we join its positive terminal to the point A (to which the positive

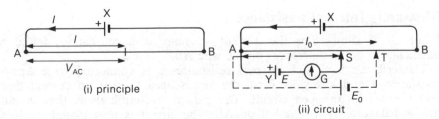

(i) principle

(ii) circuit

Figure 10.34 *Measurement of e.m.f.*

terminal of X is also joined). We connect the negative terminal of Y through a sensitive galvanometer G to a slider S, which we can press on to any point in the wire.

Let us suppose that the cell Y has an e.m.f. E which is less than the potential difference V_{AB} across the whole of the wire AB. Then if we press the slider on

B, a current will flow through Y in opposition to its e.m.f. This current will deflect the galvanometer G—let us say to the *right*. If we now press the slider on A, the cell Y will be connected straight across the galvanometer G and will deliver a current in the same direction as its e.m.f. The galvanometer will, therefore, show a deflection to the *left*.

When the slider is moved from A along AB, we can find a point S on the wire such that, when the slider is pressed on it, the galvanometer shows *no* deflection. *The potential difference V_{AS} is then equal to the e.m.f. E of Y.* No current flows through the galvanometer because E and V_{AS} act in opposite directions in the galvanometer circuit.

Because no current flows from cell Y, its full e.m.f. E, therefore, appears between the points A and S, and is balanced by V_{AS}, the p.d. along the wire, that is,

$$E = V_{AS}$$

If we now take another cell of e.m.f. E_0 and balance it in the same way at a point T, Figure 10.34 (ii), then

$$E_0 = V_{AT}$$

So, by ratios,

$$\frac{E}{E_0} = \frac{V_{AS}}{V_{AT}}$$

But we showed before that the potential differences V_{AS}, V_{AT} are proportional to the lengths l and l_0 from A to S and from A to T respectively. So

$$\frac{E}{E_0} = \frac{l}{l_0} \qquad . \qquad . \qquad . \qquad . \qquad . \qquad (2)$$

So the ratio of the e.m.f.s is proportional to the ratio of the balancing lengths and can, therefore, be calculated by measuring l and l_0. For example, if $l = 62 \cdot 5$ cm and $l_0 = 80 \cdot 0$ cm, then $E/E_0 = 62 \cdot 5/80 = 0 \cdot 78$. If the e.m.f. of one cell is known, say $E_0 = 1 \cdot 40$ V, the e.m.f. $E = 0 \cdot 78 \times 1 \cdot 40 = 1 \cdot 09$ V.

Standard cells, cells whose e.m.f.s are known accurately, are available.

Measuring Internal Resistance

Figure 10.35 shows how the *internal resistance* of a cell can be found by balancing potential differences on the wire AB.

The cell, e.m.f. E and internal resistance r, is connected to a known resistance R and a key K. First, the key is *open* so that no current flows and the cell is on open circuit. The p.d. at its terminals is then E, and this is balanced by a length l on AB. The circuit is now closed by K so that a current flows through R. The terminal p.d., V, is now balanced by a *smaller* length l' on AB.

Now
$$E = I(R + r) \quad \text{and} \quad V = IR.$$

Dividing to eliminate I,
$$\frac{E}{V} = \frac{R + r}{R} = \frac{l}{l'}$$

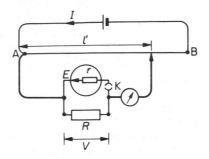

Figure 10.35 *Measurement of internal resistance*

So
$$Rl' + rl' = Rl$$

and
$$r = \frac{l - l'}{l'} R$$

Knowing l, l' and R, then r can be found. For example, suppose $l = 75\cdot2$ cm, $l' = 55\cdot8$ cm, $R = 10\cdot0\ \Omega$. Then

$$r = \frac{75\cdot2 - 55\cdot8}{55\cdot8} \times 10\cdot0 = 2\cdot7\ \Omega$$

Measurement of Thermoelectric E.M.F.

The e.m.f. of a thermocouple (p. 301) is small—of the order of millivolts. If we tried to measure such an e.m.f. on a wire AB as before, we should find the balance-point very near one end of the wire and the error in the small length would be considerable if we tried to measure it.

Figure 10.36 shows a circuit for measuring thermoelectric e.m.f. A suitable high resistance R is needed in series with the wire. Suppose the wire has a resistance of $3\cdot0\ \Omega$ and we assume that the accumulator D has an e.m.f. 2 V

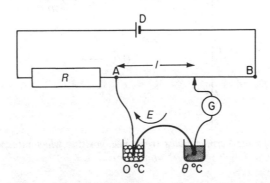

Figure 10.36 *Measurement of thermoelectric e.m.f.*

and negligible internal resistance. If the p.d. across the whole wire needs to be, say, 4 mV or 0·004 V, then the p.d. across $R = 2 - 0\cdot004 = 1\cdot996$ V. Since the p.d. across resistors in a series circuit is proportional to the resistance, then R

is given by

$$\frac{R}{3} = \frac{1 \cdot 996}{0 \cdot 004}$$

So $R = 1497 \,\Omega$ on calculation.

Assuming the p.d. across the wire AB is $4 \cdot 0$ mV, we can now measure the thermoelectric e.m.f. E of a thermocouple at various temperatures $\theta °C$ of the hot junction, the other junction being kept constant at $0°C$, Figure 10.36. Suppose the balance length on the wire at a particular temperature is $62 \cdot 4$ cm. Then, from above,

$$E = \frac{62 \cdot 4}{100} \times 4 \cdot 0 \,\text{mV} = 2 \cdot 5 \,\text{mV}$$

Thermoelectric E.M.F. and Temperature

Figure 10.37 shows results of measuring the e.m.f. E when the cold junction is at $0°C$ and the hot junction is at various temperatures θ in $°C$. The curves approximate to parabolas:

$$E = a\theta + b\theta^2 \quad . \quad . \quad . \quad . \quad . \quad (1)$$

Since the same value of E is obtained at *two* different temperatures θ, the thermocouple is never used for measuring temperature greater than the value corresponding to its maximum e.m.f.

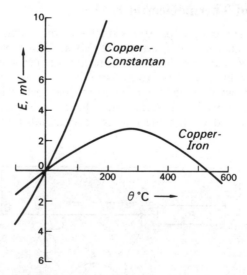

Figure 10.37 *E.m.f.s of thermocouples (reckoned positive when into copper at the cold junction)*

To find the values of a and b, we see from (1) that

$$E/\theta = a + b\theta \quad . \quad . \quad . \quad . \quad . \quad (2)$$

A graph of E/θ against θ is then a straight line whose *gradient* is the value of b. The value of a is the *intercept* on the E/θ-axis.

Examples on E.M.F. Measurements

1 Figure 10.38 shows a circuit *wrongly* set up to balance the e.m.f. *E* on the wire AB.
(a) What is the error in the circuit?
(b) When the error is corrected, the e.m.f. *E* is balanced by a length 50·0 cm on AB. Another cell of e.m.f. $E_1 = 1·5$ V is balanced by a length 75·0 cm. Calculate *E*.

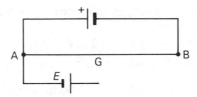

Figure 10.38 *Example*

(a) The cell of e.m.f. *E* should be turned round, so that its *positive* (+) pole is joined to A. Otherwise no balance can be obtained on AB.

(b) $\dfrac{E}{E_1} = \dfrac{l}{l_1}$ so $\dfrac{E}{1·5 \text{ V}} = \dfrac{50·0}{75·0}$

$$E = \frac{50}{75} \times 1·5 = 1·0 \text{ V}$$

2 In measuring the internal resistance *r* of a cell, the e.m.f. *E* is first balanced by a length 90·0 cm on the wire carrying a constant current. When a resistance of 5·0 Ω is connected to the cell, the terminal p.d. of the cell is now balanced by a length 45·0 cm.
(a) Why is the balancing length smaller when the resistance is connected to the cell?
(b) Calculate *r*.

From $E = I(R + r)$ and $V = IR$,
(a) the balance length for *V* is less than for *E* because some p.d. is needed to push current through the internal resistance *r*.
(b) $E = I(R + r)$ and $V = IR$ where $R = 5·0$ Ω.

So $\dfrac{E}{V} = \dfrac{R + r}{R} = \dfrac{90·0}{45·0}$

Cross-multiplying to clear fractions,

$$45(R + r) = 90R$$

Simplifying, $r = R = 5·0$ Ω

3 In the circuit shown in Figure 10.39, the e.m.f. E_s of a standard cell is 1·02 V and this is balanced by the p.d. across a resistance of 2040 Ω in series with a wire AB. If AB is 1·00 m long and has a resistance of 4 Ω, calculate the length AC on it which balances the e.m.f. 1·2 mV of the thermocouple XY.

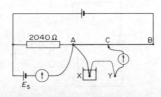

Figure 10.39 *Example*

Since 1·02 V is the p.d. across 2040 Ω, and the 4 Ω wire AB is in series with 2040 Ω, then

$$\text{p.d. across AB} = \frac{4}{2040} \times 1{\cdot}02 \text{ V} = \frac{4}{2000} \text{ V} = 2 \text{ mV}$$

So thermocouple e.m.f., 1·2 mV, is balanced by a length AC on AB (1 m) given by

$$\frac{AC}{AB} = \frac{1{\cdot}2 \text{ mV}}{2 \text{ mV}} = \frac{3}{5}$$

$$\therefore AC = \tfrac{3}{5} \times AB = \tfrac{3}{5} \times 1{\cdot}00 \text{ m} = 0{\cdot}60 \text{ m}$$

You should know:

1 *Resistance.* At a balance, Wheatstone bridge gives ratio relation $P/Q = R/X$, where X is the fourth resistance in the bridge.
2 A metre (slide-wire) bridge can measure (a) an unknown resistance X using a known resistance, (b) resistivity, (c) temperature coefficient of resistance.
3 Temperature coefficient of resistance, α: $R_\theta = R_0(1 + \alpha\theta)$, where R_0 is the resistance at melting ice and θ is in °C.
4 *Potentiometer.* Advantage: At a balance, no current taken from p.d. source. So, unlike using a moving-coil voltmeter, p.d. is undisturbed.
5 P.d. $\propto$ balance-length on potentiometer wire. Using a cell of known e.m.f. E_0 and an unknown e.m.f. E, $E/E_0 = l_1/l_2$. So E can be found.
6 To measure internal resistance r, balance the e.m.f. E, length l_0; then balance the terminal p.d. V with known external resistance R, length l. Then

$$r = (E/V - 1)R = (l_0/l - 1)R$$

7 To measure a thermoelectric (very small) e.m.f., a high resistance is placed in series with the potentiometer wire to reduce its p.d. considerably. The small e.m.f. can then be balanced on the wire.

EXERCISES 10D   Potentiometer

Multiple Choice

1 In Figure 10L, the e.m.f. E of the battery X is balanced across the 2 Ω resistor using the centre-zero meter G. If the 12 V supply has negligible internal resistance, the

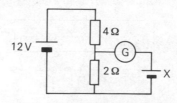

Figure 10L

value of E in V is

A 10 **B** 8 **C** 6 **D** 4 **E** 2

2 A thermocouple has an e.m.f. of 3 mV. It can *not* be balanced directly on a potentiometer wire of length 100 cm connected to a driver (supply) of 2·0 V because

  A the current in the wire is too low
  B the balance-length would be too high
  C the wire p.d. is too high
  D the balance-length would be near the middle of the wire
  E the thermocouple e.m.f. needs to be lower.

Longer Questions

3 The e.m.f. of a battery A is balanced by a length of 75·0 cm on a potentiometer wire. The e.m.f. of a standard cell, 1·02 V, is balanced by a length of 50·0 cm. What is the e.m.f. of A?
   Calculate the new balance length if A has an internal resistance of 2 Ω and a resistor of 8 Ω is joined to its terminals.

4 A 1·0 Ω resistor is in series with an ammeter M in a circuit. The p.d. across the resistor is balanced by a length of 60·0 cm on a potentiometer wire. A standard cell of e.m.f. 1·02 V is balanced by a length of 50·0 cm. If M reads 1·10 A, what is the error in the reading?

5 The driver cell of a potentiometer has an e.m.f. of 2 V and negligible internal resistance. The potentiometer wire has a resistance of 3 Ω. Calculate the resistance needed in series with the wire if a p.d. of 5 mV is required across the whole wire.
   The wire is 100 cm long and a balance length of 60 cm is obtained for a thermocouple e.m.f. E. What is the value of E?

6 (a) Figure 10M (i), in which AB is a uniform resistance wire, is a simple potentiometer circuit. Explain why a point X may be found on the wire which gives zero galvanometer deflection.
   When the circuit was first set up it was impossible to find a balance point. State and and explain two possible causes of this.
   How would you use the circuit to compare the e.m.f.s of two cells with minimum error? Why is this circuit not suitable for the comparison of an e.m.f. of a few millivolts with an e.m.f. of about a volt?

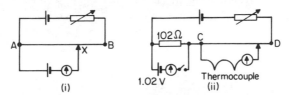

Figure 10M

(b) The second circuit, Figure 10M (ii), may be used to measure the e.m.f. of a thermocouple provided that the resistance of CD is known. Describe how you would use it. If the resistance of CD were 2·00 Ω, its length were 1·00 m and the balance length were 79 cm, what would be the e.m.f. of the thermocouple? (*L.*)

7 A simple potentiometer circuit is set up as in Figure 10N, using a uniform wire AB,

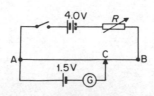

Figure 10N

1·0 m long, which has a resistance of 2·0 Ω. The resistance of the 4-V battery is negligible. If the variable resistor R were given a value of 2·4 Ω, what would be the length AC for zero galvanometer deflection?

If R were made 1·0 Ω and the 1·5 V cell and galvanometer were replaced by a voltmeter of resistance 20 Ω, what would be the reading of the voltmeter if the contact C were placed at the mid-point of AB? (*L.*)

11 Magnetic Field and Force on Conductor

In this chapter we introduce magnetic fields. We shall see the difference in pattern of the lines of force or magnetic flux round magnets, and round current-carrying straight and coiled conductors. The patterns and the flux density both play an important part in magnetic applications widely used in industry.

The force on a current-carrying conductor in a magnetic field is due to the 'interaction' between two magnetic fields. We discuss in detail how the force on a conductor is used in the moving-coil meter and the electric motor. Finally, we consider the force on moving charges in a magnetic field and show how it is applied in the Hall effect.

Magnetism

We shall assume you know:

(a) bar magnets have their magnetism concentrated mainly at the ends or poles;
(b) like poles repel (two S- or two N-poles repel) but unlike poles attract (a N- and a S-pole attract);
(c) permanent magnets are made of steel because steel keeps its magnetism, but temporary magnets, such as electromagnets, are made of soft (pure) iron because this loses its magnetism easily.

Natural magnets were known some thousands of years ago, and in the eleventh century A.D. the Chinese invented the magnetic compass. This consisted of a magnet, floating on a buoyant support in a dish of water. The respective ends of the magnet, where iron filings are attracted most, are called the north and south poles.

In the thirteenth century the properties of magnets were studied by Peter Peregrinus. He showed that

like poles repel and *unlike poles attract.*

His work was forgotten, however, and his results were rediscovered in the sixteenth century by Dr Gilbert, who is famous for his researches in magnetism and electrostatics.

Ferromagnetism

About 1823 STURGEON placed an iron core into a coil carrying a current, and found that the magnetic effect of the current was increased enormously. When the current was switched off the iron lost nearly all its magnetism. Iron, which can be magnetised strongly, is called a *ferromagnetic* material. Steel, made by adding a small percentage of carbon to iron, is also ferromagnetic. It retains its magnetism, however, after removal from a current-carrying coil, and is more difficult to magnetise than iron.

Nickel and cobalt are the only other ferromagnetic elements in addition to iron, and are widely used for modern magnetic apparatus. A modern alloy for permanent magnets, called *Alnico*, has the composition 54 per cent iron, 18 per cent nickel, 12 per cent cobalt, 6 per cent copper, 10 per cent aluminium. It retains its magnetism extremely well, and, by analogy with steel, is, therefore, said to be magnetically very hard. Alloys which are very easily magnetised, but do not retain their magnetism, are said to be magnetically soft. An example is *Mumetal*, which contains 76 per cent nickel, 17 per cent iron, 5 per cent copper, 2 per cent chromium.

Magnetic Fields

The region round a magnet, where a magnetic force occurs, is called a *magnetic field*. The appearance of a magnetic field is quickly obtained by iron filings, and accurately plotted with a small compass, as the reader knows. The *direction* of a magnetic field is taken as the direction of the force on a *north* pole if placed in the field.

Figure 11.1 shows a few typical fields. The field round a bar magnet is 'non-uniform', that is, its strength and direction vary from place to place, Figure 11.1 (i). The earth's field locally, however, is uniform, Figure 11.1 (ii). A bar of soft iron placed north–south becomes magnetised by induction by the earth's field, and the lines of force, also called lines of *flux*, become concentrated in the soft iron, Figure 11.1 (iii). The *tangent* to a line of force at a point gives the direction of the magnetic field at that point.

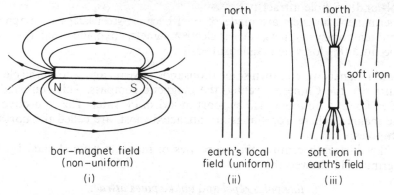

Figure 11.1 *Magnetic fields*

Oersted's Discovery

The magnetic effect of the electric current was discovered by OERSTED in 1820. Like many others, Oersted suspected a relationship between electricity and magnetism, and was deliberately looking for it. In the course of his experiments, he happened to lead a wire carrying a current over, but parallel to, a compass-needle, as shown in Figure 11.2 (i); the needle was deflected. Oersted then found that if the wire was led under the needle, it was deflected in the opposite sense, Figure 11.2 (ii).

From these observations he concluded that the magnetic field was *circular* round the wire. We can see this by plotting the lines of force of a long vertical wire, as shown in Figure 11.3. To get a clear result a strong current is needed, and we must work close to the wire, so that the effect of the earth's field is negligible. It is then seen that the lines of force are *circles*, concentric with the wire.

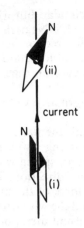

Figure 11.2 *Deflection of compass needle by electric current*

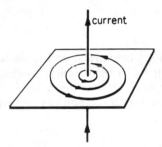

Figure 11.3 *Magnetic field of long straight conductor*

Directions of Current and Field: Rules

The relationship between the direction of the lines of force and of the current is expressed in Maxwell's *corkscrew rule*: if we imagine ourselves driving a corkscrew in the direction of the current, then the direction of rotation of the corkscrew is the direction of the lines of force. Figure 11.4 illustrates this rule, the small, heavy circle representing the wire, and the large light one a line of force. At (i) the current is flowing into the paper; its direction is indicated by a cross, which stands for the tail of an arrow moving away from the reader. At (ii) the current is flowing out of the paper; the dot in the centre of the wire stands for the point of an approaching arrow.

If we plot the magnetic field of a circular coil carrying a current, we get the result shown in Figure 11.4. Near the circumference of the coil, the lines of force

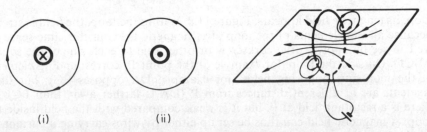

Figure 11.4 *Illustrating corkscrew rule*   **Figure 11.5** *Magnetic field of narrow coil*

are closed loops, which are not circular, but whose directions are still given by the corkscrew rule, as in Figure 11.5. Near the centre of the coil, the lines are almost straight and parallel. Their direction here is again given by the corkscrew rule, but the current and the lines of force are interchanged, that is, if we turn the screw in the direction of the current, then its point travels in the direction of the lines.

The *clenched first rule* is an alternative to the corkscrew rule: Hold the right hand so that

(a) the fist is tightly clenched with the fingers curled, and

(b) the thumb is straight and pointing away from the fingers.

(1) With a straight conductor, grasp the wire with the clenched right hand, pointing the thumb in the current direction. Then the curled fingers give the direction of the circular lines of force of the magnetic field.

(2) For a coiled conductor, hold the wire with the clenched right hand so that the fingers curl round it in the current direction. Then the straight thumb gives the direction of the magnetic field. The reader should verify this rule with Figures 11.4 and 11.6.

The Solenoid

The magnetic field of a long cylindrical coil is shown in Figure 11.6. Such a coil is called a *solenoid*; it has a field similar to that of a bar magnet, whose poles are indicated in the figure. If an iron or steel core were put into the coil, it would become magnetised with the polarity shown.

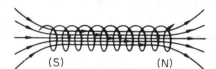

(S)   (N)

Figure 11.6 *Magnetic field of solenoid*

If the terminals of a battery are joined by a wire which is simply doubled back on itself, as in Figure 11.7, there is no magnetic field at all. Each element of the outward run, such as AB, in effect cancels the field of the corresponding element of the inward run, CD. But as soon as the wire is opened out into a

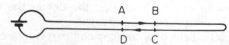

A   B

D   C

Figure 11.7 *A doubled-back current has no magnetic field*

loop, its magnetic field appears, Figure 11.8. Within the loop, the field is strong, because all the elements of the loop give magnetic fields in the same sense, as we can see by applying the corkscrew or other rule to each side of the square ABCD. Outside the loop, for example at the point P, corresponding elements of the loop give opposing fields (for example, DA opposes BC); but these elements are at different distances from P (DA is farther away than BC). So there is a resultant field at P, but it is weak compared with the field inside the loop. A magnetic field can thus be set up either by wires carrying a current, or by the use of permanent magnets.

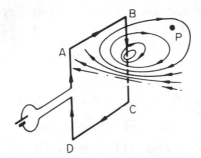

Figure 11.8 *An open loop of current has magnetic field*

You should know:

1 *Straight* conductor—field lines round it are *circular*.
2 *Solenoid*—in air, field lines are straight in middle and curved at ends.
3 *Field direction*—it is the direction a *north* pole would move in the field. The corkscrew rule, or the clenched fist of the right hand, can be used for the straight conductor and the solenoid.

Force on Conductor, Fleming's Left-hand Rule

When a conductor carrying a current is placed in a magnetic field due to some source other than itself, the conductor experiences a *mechanical force*. As we see later, this led to the invention of electric motors and to the design of electrical meters. To demonstrate the force, a short brass rod R is connected across a pair of brass rails, as shown in Figure 11.9. A horseshoe magnet is placed so that

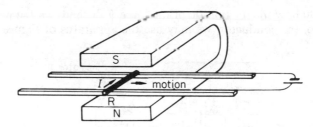

Figure 11.9 *Force on current in a magnetic field*

the rod lies in the vertically upward field between its N, S poles. When we pass a current *I* through the rod, from an accumulator, the rod rolls along the rails.

The relative directions of the current, the applied field, and the motion are shown in Figure 11.10 (i). They are the same as those of the middle finger, the forefinger and the thumb of the *left* hand when held all at right angles to one another. If we place the magnet so that its field *B* lies in the *same* direction as the current *I*, then the rod R experiences no force, Figure 11.10 (ii).

Experiments like this were first made by Ampère in 1820. As a result of them, he concluded that

---

the force on a conductor is always at right angles to the plane which contains both the conductor and the direction of the field in which it is placed.

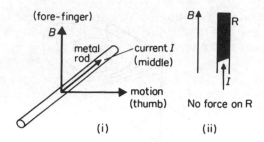

Figure 11.10 *Left-hand rule*

He also showed that, if the conductor makes an angle α with the field, the force on it is proportional to sin α. So the maximum force is exerted when the conductor is *perpendicular* to the field, when sin $\alpha = 1$.

Dependence of Force on Physical Factors

Since the magnitude of the force on a current-carrying conductor is given by

$$F \propto \sin \alpha \qquad . \qquad . \qquad . \qquad . \qquad . \qquad (1)$$

where α is the angle between the conductor and the field, it follows that F is zero when the conductor is parallel to the field direction. This defines the direction of the magnetic field. To find which way it points, we can apply Fleming's rule to the case when the conductor is placed at right angles to the field. The direction of the field then corresponds to the direction of the forefinger.

Variation of *F* with *I*

To investigate how the magnitude of the force F depends on the current I and the length l of the conductor, we may use the apparatus of Figure 11.11.

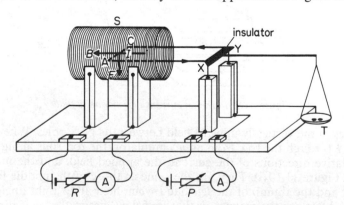

Figure 11.11 *Experiment to show F varies with I*

Here the conductor AC is situated in the field B of a solenoid S. The current flows into, and out of, the wire via the pivot points Y and X. The scale pan T is placed at the same distance from the pivot as the straight wire AC, which is perpendicular to the axis of the coil. The frame is first balanced with no current flowing in AC. A current is then passed, and the extra weight needed to restore the frame to a horizontal position is equal to the force on the wire AC. By varying the current in AC with the rheostat P, for example, by doubling or halving the circuit resistance, it may be shown that:

$$F \propto I \qquad (2)$$

If different frames are used so that the length, l, of AC is changed, it can be shown that, with constant current and field,

$$F \propto I \qquad (3)$$

Effect of B

The magnetic field due to the solenoid will depend on the current flowing in it. If this current is varied by adjusting the rheostat R, it can be shown that the larger the current *in the solenoid*, S, the larger is the force F. It is reasonable to suppose that a larger current in S produces a stronger magnetic field. Thus the force F increases if the magnetic field strength is increased. The magnetic field is represented by a *vector* quantity which is given the symbol B and is defined shortly. This is called the *flux density* in the field. We assume that:

$$F \propto B \qquad (4)$$

Magnitude of F

From the results expressed in equations (1) to (4), we obtain

$$F \propto BIl \sin \alpha$$

or $\qquad F = kBIl \sin \alpha \qquad (5)$

where k is a constant.

In the SI system of units, the unit of B is the tesla (T). One tesla may be defined as the flux density of a uniform field when the force on a conductor 1 metre long, placed perpendicular to the field and carrying a current of 1 ampere, is 1 newton. Substituting $F = 1$, $B = 1$, $l = 1$ and $\sin \alpha = \sin 90° = 1$ in (5), then $k = 1$. So in Figure 11.12 (i), with the above units,

$$F = BIl \sin \alpha \qquad (6)$$

When the whole length of the conductor is *perpendicular* to the field B,

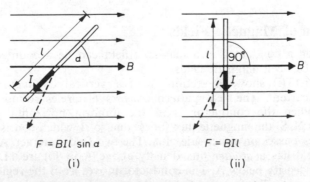

$$F = BIl \sin \alpha \qquad\qquad F = BIl$$

(i) $\qquad\qquad$ (ii)

Figure 11.12 *Magnitude of F which acts towards reader*

Figure 11.12 (ii), then, since $\alpha = 90°$ in this case,

$$F = BIl \qquad . \quad . \quad . \quad . \quad . \quad (7)$$

It may be noted that the apparatus of Figure 11.11 can be used to determine the flux density B of the field in the solenoid. In this case, $\alpha = 90°$ and $\sin \alpha = 1$. So measurement of F, I and l enables B to be found from (7).

You should know:

1 When a current-carrying conductor XY is turned in a uniform magnetic field of flux density B until no force acts on it, then XY points in the direction of B.
2 When a straight conductor of length l carrying a current I is placed perpendicular to a uniform field and a force F acts on the conductor, then the magnitude B of the flux density is *defined* by

$$B = \frac{F}{Il}$$

and since F, I and l can all be measured, B can be calculated.
3 B is a vector. So its component in a direction at an angle θ to B is $B \cos \theta$.

Example on Force on Conductor

A wire carrying a current of 10 A and 2 metres in length is placed in a field of flux density 0·15 T. What is the force on the wire if it is placed
(a) at right angles to the field,
(b) at 45° to the field,
(c) along the field?

From (6) $F = BIl \sin \alpha$

(a) $F = 0·15 \times 10 \times 2 \times \sin 90°$

                    $= 3 \text{ N}$

(b) $F = 0·15 \times 10 \times 2 \times \sin 45°$

                    $= 2·12 \text{ N}$

(c) $F = 0$, since $\sin 0° = 0$

Interaction of Magnetic Fields

The force on a conductor in a magnetic field can be accounted for by the interaction between magnetic fields.

Figure 11.13 (i) shows a section A of a vertical conductor carrying a downward current. The field pattern consists of circles round A as centre (p. 321). When the conductor is in the uniform horizontal field B due to the poles N, S, the magnetic flux (lines) due to B, which consists of straight parallel lines, passes on either side of A. The two fields interact. As shown, the resultant field has a *greater* flux density above A in Figure 11.13 (ii) and a *smaller* flux density below A. The conductor moves from the region of greater flux density to smaller flux density. So A moves downwards as shown. As the

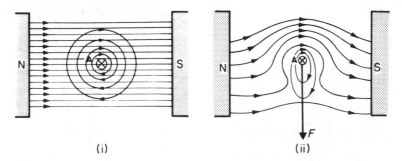

(i)   (ii)

Figure 11.13 *Interaction of magnetic fields. A force F is produced*

reader should verify, the direction of the force F on the conductor is given by Fleming's left-hand rule.

If a current-carrying conductor is placed in the *same* direction as a uniform magnetic field, the flux-density on both sides of the conductor is the same, as the reader should verify. The conductor is now not affected by the field, that is, no force acts on it in this case.

Moving-coil Loudspeaker

The moving-coil loudspeaker, widely used for sound reproduction in radio receivers or television or public address systems in sports grounds, is an application of the force on a current-carrying conductor in a magnetic field.

Basically, the loudspeaker consists of a light coil of wire A, known as the *speech coil*, wound tightly round a cylindrical former, a cylinder to which a large thin cardboard cone C is rigidly attached, Figure 11.14 (i). The coil is situated centrally in the uniform radial field of a permanent magnet M with circular poles P, as shown in section in Figure 11.14 (ii).

When the loudspeaker in a radio receiver, for example, is working normally, *alternating current* (a.c.) of the same frequency as the sound, called audio-frequency, flows continuously in the speech coil A. So the current in A *reverses* every half-cycle (p. 377). A mechanical force then acts on every part of the coil such as a, b, c, d, since the magnetic field is 90° to the coil at every part. By Fleming's left-hand rule, the force on the coil in Figure 11.14 (ii) is downward into the page. When the current reverses on the other half of the same cycle, the force on the coil reverses in direction.

So the coil *vibrates* along its axis at a frequency equal to the audio-frequency alternating current flowing in A. The former and cone vibrate at the same frequency and produce oscillations in the large mass of air in contact with the cone. A loud sound is then heard. The greater the electrical energy supplied to the coil A, the louder will be the sound produced.

We can easily calculate the axial force F on the coil. Suppose the coil has a radius of 3 cm, 10 turns and the radial field is 0·4 T. The coil circumference of one turn is $2\pi r$ or $2\pi \times 0·03$ m. If a current of 5 mA or 0·005 A flows in the coil, then

$$F = BIl = 0·4 \times 0·005 \times 10 \times 2\pi \times 0·03 = 4 \times 10^{-3} \text{ N (approx.)}$$

Baffle board. As shown in Figure 11.14 (i), the speaker has a circular board around it, called the baffle board, which is essential to make it work efficiently. When the cone moves forward during its vibration, a compression or increase of pressure is set up in the air in front. Simultaneously, a rarefaction or decrease

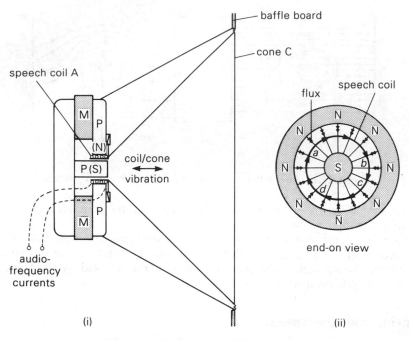

Figure 11.14 *Moving-coil loudspeaker*

of pressure is set up behind the cone. So the sound wave produced in front of the cone is 180° or $\lambda/2$ out of phase with that produced behind. If the wave behind travels round to the front, the two waves interfere destructively. With 180° phase difference the resultant sound wave has a low intensity (see *Interference of waves*, p. 493.) The baffle board is used to overcome this effect. It reduces the interference by effectively stopping the wave behind it travelling round to the front.

Torque on Rectangular Coil in Uniform Field

A rectangular coil of insulated copper wire is used in the moving-coil meter, which we discuss shortly. Measurements of current and p.d. are made mainly with moving-coil meters.

Consider a rectangular coil situated with its plane *parallel* to a uniform magnetic field of flux density B. Suppose a current I is passed into the coil, Figure 11.15(i). Viewed from above, the coil appears as shown in Figure 11.15(ii).

The side PS of length l is perpendicular to B. So the force on it is given by $F = BIl$. If the coil has N turns, the length of the conductor is increased N times and so the force on the side PS, $F_1 = BIlN$.

The force on the opposite side QR is also given by $F = BIlN$, but its direction is *opposite* to that on PS. There are no forces on the sides PQ and SR although they carry currents because PQ and SR are parallel to the field B.

The two forces F on the sides PS and QR tend to turn the coil about an axis XY passing through the middle of the coil. The two forces together are called a *couple* and their moment (turning-effect) or *torque* T is given, by definition, by

$$T = F \times p$$

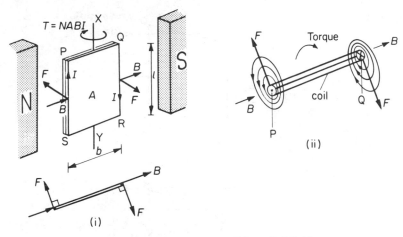

Figure 11.15 *Torque on coil in radial field*

where p is the *perpendicular* distance between the two forces. See p. 97. Now from Figure 11.15 (i), $p = b$, the width PQ or SR of the coil. So,

$$T = F \times p = BIlN \times b$$

But $l \times b = $ area A of the coil. So

$$\textit{torque } T = BANI \quad . \qquad . \qquad . \qquad . \qquad . \qquad (1)$$

The unit of torque (force $\times$ distance) is the newton metre, symbol N m. In using $T = BANI$, B must be in units of T (tesla), A in m^2 and I in A (amperes).

If there were no opposition to the torque, the coil PQRS would turn around and settle with its plane normal to B, that is, facing the poles N, S in Figure 11.15 (i). As we see later, springs can control the amount of rotation of the coil.

Figure 11.15 (ii) is a plan view PQ of the rectangular coil with its plane in the same direction as the uniform magnetic field of the magnet N, S. As we explained previously, the magnetic field of the current in the straight sides PS, QR of the coil interacts with the field of the magnet. Figure 11.15 (ii) shows roughly the appearance of the resultant field around the vertical sides of the conductors whose tops are P and Q respectively. The current is downward in Q and upward towards the reader in P. The forces F act from the dense to the less dense flux and together they produce a torque on the coil.

Torque on Coil at Angle to Uniform Field

Suppose now that the plane of the coil is at an angle θ to the field B when it carries a current I. Figure 11.16 (i) shows the forces F_1 on its vertical sides PS and QR; these two forces set up a torque which rotate the coil. The forces F_2 on its horizontal sides merely compress the coil and are resisted by its rigidity.

The forces F_1 on the sides PS and QR are still given by $F_1 = BIlN$ because PS and QR are perpendicular to B. But now the forces F_1 are not separated by a perpendicular distance b, the coil breadth. The perpendicular distance p is less than b and is given by (Figure 11.16 (ii))

$$p = b \cos \theta$$

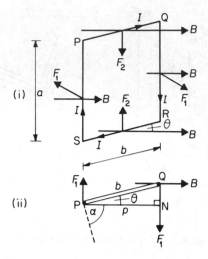

Figure 11.16 *Torque on coil at angle to uniform field*

So this time

$$\text{torque } T = F_1 \times p = BIlN \times b \cos \theta$$

So $$T = BANI \cos \theta \qquad . \qquad . \qquad . \qquad . \qquad . \qquad (2)$$

When the plane of the coil is *parallel* to B, then $\theta = 0°$ and $\cos \theta = 1$. So the torque $T = BANI$ as we have already shown. If the plane of the coil is *perpendicular* to B, then $\theta = 90°$ and $\cos \theta = 0$. So the torque $T = 0$ in this case.

If α is the angle between B and the *normal* to the plane of the coil, then $\theta = 90° - \alpha$. From (2), the torque T is then given by

$$T = BANI \sin \alpha \qquad . \qquad . \qquad . \qquad . \qquad . \qquad (3)$$

Magnetism is due to circulating and spinning electrons inside atoms. The moving charges are equivalent to electric currents. Consequently, like a current-carrying coil, permanent magnets also have a torque acting on them when they are placed with their axis at an angle to a magnetic field. Like the coil, they turn and settle in equilibrium with their axis along the field direction.

Example on Torque

A vertical rectangular coil of sides 5 cm by 2 cm has 10 turns and carries a current of 2 A. Calculate the torque on the coil when it is placed in a uniform horizontal magnetic field of 0·1 T with its plane
(a) parallel to the field,
(b) perpendicular to the field,
(c) 60° to the field.

The area A of the coil $= 5 \times 10^{-2}\,\text{m} \times 2 \times 10^{-2}\,\text{m} = 10^{-3}\,\text{m}^2$

So (a) $$T = BANI = 10 \times 10^{-3} \times 0·1 \times 2$$
$$= 2 \times 10^{-3}\,\text{N m}$$

(b) Here $$T = 0$$

(c) $$T = BANI \cos 60° \text{ or } BANI \sin 30°$$
$$= 2 \times 10^{-3} \times 0·5 = 10^{-3}\,\text{N m}$$

> ## You should know:
>
> 1 *Straight conductor.* When 90° to field, $F = BIl$ (F in N, B in T, I in A, l in m). Field strength B can be defined from $B = F/Il$. When parallel to field, $F = 0$.
> 2 *Direction* of force F—use Fleming LEFT hand rule, Forefinger = Field, mIddle or seCond finger = current, thuMb = Motion or force.
> 3 *Rectangular coil* (a) with B *parallel* to plane of coil, torque $T = BANI$ (N m)
> > (b) with B 90° to plane of coil, $T = 0$
> > (c) with B at angle θ to plane of coil, $T = BANI \cos \theta$.
>
> The force on a current-carrying wire is produced by the *interaction* between the two magnetic fields.

The Moving-coil Meter

All current measurements except the most accurate are made today with a moving-coil meter. In this instrument a rectangular coil of fine insulated copper wire is suspended in a strong magnetic field, Figure 11.17 (i). The field is set up between soft iron pole-pieces, NS, attached to a powerful permanent magnet.

The pole-pieces are curved to form parts of a cylinder coaxial with the suspension of the coil. And between them lies a cylindrical core of soft iron, C. It is supported on a brass pin, T in Figure 11.17 (ii), which is placed so that it does not foul the coil. As the diagram shows, the magnetic field B is *radial* to the core and pole-pieces, over the region in which the coil can swing. In this case the deflected coil *always* comes to rest with its plane *parallel* to the field in which it is then situated, as shown in Figure 11.17 (ii).

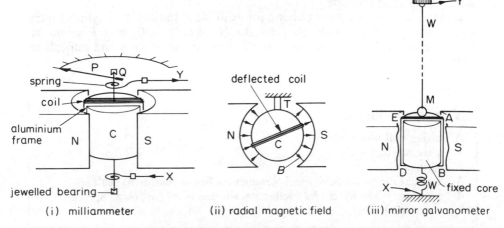

Figure 11.17 *Moving-coil meters*

The moving-coil milliammeter or ammeter has hair-springs and jewelled bearings. The coil is wound on a rigid but light aluminium frame, which also carries the pivots. The pivots are insulated from the former if it is aluminium, and the current is led in and out through the springs. The framework, which carries the springs and jewels, is made from brass or aluminium—if it were steel it would affect the magnetic field. An aluminium pointer, P, shows the deflection of the coil; it is balanced by a counterweight, Q, Figure 11.17 (i).

In the more sensitive mirror galvanometer, the coil is suspended on a phosphor-bronze wire, WM, which is kept taut, Figure 11.17 (iii). The current is led into and out of the coil EABD through the suspension, at X and Y, and the deflection of the coil is shown by a beam of light, reflected by a mirror M to a scale in front of the instrument.

Theory of Moving-coil Instrument

The rectangular coil is situated in the radial field B. When a current is passed into it, the coil rotates through an angle θ which depends on the strength of the springs. *No matter where the coil comes to rest*, the field **B** in which it is situated always lies along the *plane* of the coil because the field is radial. As we have previously seen, the torque T on the coil is then always given by $BANI$. So the torque $T \propto I$, since B, A, N are constant.

In equilibrium, the deflecting torque T on the coil is equal to the opposing torque due to the elastic forces in the springs. The opposing torque $= c\theta$, where c is a constant of the springs which depends on their elasticity under twisting forces and on their dimensions. So

$$BANI = c\theta$$

and
$$I = \frac{c}{BAN}\theta \qquad . \qquad . \qquad . \qquad . \qquad . \qquad (1)$$

Equation (1) shows that the deflection θ is proportional to the current I. So the scale showing current values is a *uniform* one, that is, equal divisions along the calibrated scale represent equal steps in current. This is an important advantage of the moving-coil meter. It can be accurately calibrated and its subdivisions read accurately.

If the radial field were not present, for example, if the soft iron cylinder were removed, the torque would then be $BANI \cos \theta$ (p. 330) and I would be proportional to $\theta/\cos \theta$. The scale would then be *non-uniform* and difficult to calibrate or to read accurately.

You should know:

A moving-coil meter has
1   a rectangular coil
2   springs
3   a radial magnetic field which produces a linear (uniform) scale
4   a current given by $BANI$ (deflection torque) $= c\theta$ (opposing spring torque).

Sensitivity of Current Meter

The *sensitivity* of a current meter is the *deflection per unit current*, or θ/I. Small currents must be measured by a meter which gives an appreciable deflection.

From $BANI = c\theta$, we have $\theta/I = BAN/c$. So greater sensitivity is obtained with a stronger field B, a low value of c, that is, *weak* springs, and a greater value of N and A. The size and number of turns of a coil would increase the resistance of the meter, which is not desirable. The elastic constant c of the springs can be varied, however.

When a galvanometer is of the suspended-coil type (Figure 11.17 (iii)), its sensitivity is generally expressed in terms of the displacement of the spot of light reflected from the mirror on to the scale. A Scalamp or Edspot, a form of light beam galvanometer, may give a deflection of 25 mm per microampere.

All forms of moving-coil galvanometer have one disadvantage: they are easily damaged by overload. A current much greater than that which the instrument is intended to measure will burn out its hair-springs or suspension.

Sensitivity of Voltmeter

The sensitivity of a voltmeter is the deflection per unit p.d., or θ/V, where θ is the deflection produced by a p.d. V.

If the resistance of a moving-coil meter is R, the p.d. V across its terminals when a current I flows through it is given by $V = IR$. From our expression for I given previously,

$$V = \frac{cR}{BAN}\theta$$

So   **voltage sensitivity** $= \dfrac{\theta}{V} = \dfrac{BAN}{cR}$

So unlike the current sensitivity, the voltage sensitivity depends on the resistance R of the meter coil.

Example on Sensitivity of Meter

A moving-coil meter X has a coil of 20 turns and a resistance $10\,\Omega$. Another moving-coil meter Y has a coil of 10 turns and a resistance of $4\,\Omega$. If the area of each coil, the strength of the springs and the field B are the same in each meter, which has
(a) the greater current sensitivity,
(b) the greater voltage sensitivity?

(a) The current sensitivity is given by

$$\frac{\theta}{I} = \frac{BAN}{c}$$

Since the sensitivity $\propto N$, with A, c and B constant, then X (20 turns) has a greater sensitivity than Y (10 turns).
(b) The voltage sensitivity $= BAN/cR$. So with A, B, c constant,

$$\text{sensitivity} \propto \frac{N}{R}$$

Now $N/R = 20/10 = 2$ numerically for X, and $N/R = 10/4 = 2.5$ for Y. So Y has the greater voltage sensitivity.

Converting a Milliammeter to a Voltmeter

We will now see how to use a milliammeter, which measures current, as a voltmeter, which measures potential difference. Suppose we have a moving-coil meter which requires 5 milliamperes, which is 5 mA or 5×10^{-3} A or 0·005 A, for full-scale deflection. And let us suppose that the resistance of its coil, r, is 20 Ω, Figure 11.18.

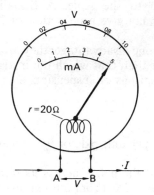

Figure 11.18 *P.d. across moving-coil ammeter*

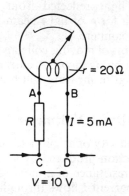

Figure 11.19 *Milliammeter converted to 0–10 V voltmeter*

Then, when it is fully deflected, the potential difference across it is

$$V = rI$$
$$= 20 \times 5 \times 10^{-3} = 100 \times 10^{-3} \text{ V}$$
$$= 0·1 \ V$$

So if the coil resistance is constant, the instrument can be used as a *voltmeter*, giving full-scale deflection for a potential difference of 0·1 V, or 100 mV. Its scale could be engraved as shown at the top of Figure 11.18.

The potential differences to be measured in the laboratory are usually greater than 100 mV, however. To measure such a potential difference we put a resistor R *in series* with the coil, as shown in Figure 11.19. If we wish to measure up to 10 V we must choose the resistance R so that, when 10 V is applied between the terminals CD, then a *full-scale* current of 5 mA (5×10^{-3} A) flows through the moving coil.

Now
$$V = (R + r)I$$
$$\therefore 10 = (R + 20) \times 5 \times 10^{-3}$$

or
$$R + 20 = \frac{10}{5 \times 10^{-3}} = 2 \times 10^3 = 2000 \ \Omega$$

$$\therefore R = 2000 - 20$$
$$= 1980 \ \Omega \qquad . \qquad . \qquad . \qquad . \qquad . \qquad . \qquad (8)$$

The resistance R is called a *multiplier*. Many voltmeters contain a series of multipliers of different resistances. The range of potential difference measured by the meter can then be varied, for example, from 0–10 mV to 0–150 V.

Converting a Milliammeter to an Ammeter

Moving-coil meters give full-scale deflection for currents smaller than those generally met in the laboratory. If we wish to measure a current of the order of an ampere or more we connect a *low resistance S*, called a *shunt*, across the terminals of a moving-coil meter, Figure 11.20. The shunt diverts most of the

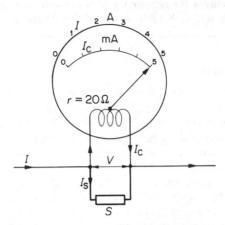

Figure 11.20 *Conversion of milliammeter to ammeter*

current to be measured, I, away from the coil—hence its name. Let us suppose that, as before, the coil of the meter has a resistance r of 20 Ω and is fully deflected by a current, I_C, of 5 mA (0·005 A).

Suppose we wish to shunt it so that a full-scale deflection is obtained for a current I of 5 A, and the meter is then converted to a range 0–5 A. Then the shunt resistance S must shunt a current I_S of (5–0·005) A or 4·995 A through itself.

The potential difference across the shunt is the *same* as that across the coil, which is

$$V = rI_C = 20 \times 0.005 = 0.1 \text{ V}$$

The resistance of the shunt must, therefore, be

$$S = \frac{V}{I_S} = \frac{0.1}{4.995} = 0.020\,02 \ \Omega \quad . \qquad . \qquad . \qquad . \qquad (9)$$

The ratio of the current measured to the current through the coil is

$$\frac{I}{I_C} = \frac{5}{5 \times 10^{-3}} = 1000$$

This ratio is the same whatever the current I, because it depends only on the resistances S and r; the reader may easily show that its value is $(S + r)/S$. The deflection of the coil is, therefore, proportional to the measured current, as shown in Figure 11.20.

Multimeters, widely used in the radio and electrical industries, are moving-coil meters which can read potential differences or currents on the same scale, by switching to series or shunt resistors at the back of the meter.

> ## You should know:
> 1 A milliammeter can be made into
>   (a) a *voltmeter* by adding a *high* resistance in *series*
>   (b) an *ammeter* by adding a *low* resistance in *parallel*.
> 2 The springs control the sensitivity of a current moving-coil meter.
> $NABI = c\theta$, so $\theta/I = NAB/c$ = sensitivity. Small c = high sensitivity.

Magnetic Materials

Pages 319–20 contained a brief introduction to some of the magnetic materials currently available. When choosing a magnetic material it is important to bear in mind the end application, which generally falls into one of two categories. We either want a material to be a permanent magnet, perhaps for use in a motor or dynamo, or we want a core material to carry an externally induced field, as in a transformer.

Permanent magnets are made from the so called *hard* magnetic materials. The aim would be to maximise the magnetic flux density B available from a unit volume of material, at lowest cost. Alloys of nickel and aluminium, sometimes including cobalt, copper and titanium, are very good hard magnetic materials: two of the best are Columax and Alcomax III.

For the other type of application, carrying an externally induced field, we have a very different aim. Here we wish to maximise the amount of magnetic flux density B available per ampere-turn of the coil wrapped around the material. Not surprisingly, a completely different range of materials, the *soft* magnetic materials, are best. The cheapest such materials include grain orientated silicon–iron, 4% silicon–iron and cast steel. More expensive and superior materials include mumetal, radiometal and rhometal, all based on alloys of nickel and iron. Finally, specialist soft magnetic materials have been developed for applications where eddy current losses (page 384) have to be minimised. These materials, which include manganese–zinc, nickel–zinc and barium ferrites, are non-metallic and, therefore, do not conduct electricity very well, so reducing eddy current losses considerably. However, their purely magnetic properties are inferior to those of the other soft materials, and so they are usually only used for high-frequency applications where eddy currents would be a serious problem.

EXERCISES 11A Magnetic Fields, Force on Conductors

Multiple Choice

1 A long air-core solenoid PQ has a flux density B of 60 μT in the middle, X, Figure 11A (i). An exactly similar solenoid is connected to Q between Q and R coaxial with the solenoid PQ. The flux density in μT at Q is now

A 120 B 100 C 60 D 40 E 30

2 In Figure 11A (ii), a horizontal straight conductor PQ is 2 m long and is placed in a uniform perpendicular field of flux density B of 0·4 T. When a current flows in PQ the force on PQ is 1·6 N vertically upwards. The current in the conductor is

A 2·8 A from P to Q B 2·4 A from Q to P C 2·0 A from P to Q
D 2·0 A from Q to P E 1·6 A from P to Q

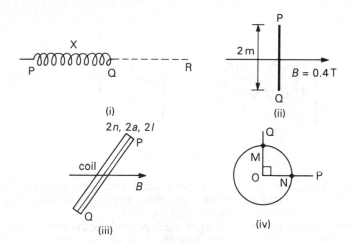

Figure 11A

3 A rectangular coil PQ has $2n$ turns, an area $2a$ and carries a current $2I$, Figure 11A (iii). The plane of the coil is 60° to a horizontal uniform field of flux density B. The torque (moment of couple) on the coil is

A $8naI \sin 60°$ B $8naI\theta \cos 60°$ C $4naIB \sin 60°$
D $2naIB \cos 60°$ E $naI \sin 60°$

4 A 0–10 mA moving-coil meter of 5 Ω resistance can be converted into a 0–2 A meter by a resistance R with the meter when

A $R = 0.025\,\Omega$ in parallel   B $R = 0.025\,\Omega$ in series   C $R = 190\,\Omega$ in series
D $R = 0.050\,\Omega$ in parallel   E $R = 0.10\,\Omega$ in parallel

5 In Figure 11A (iv), OP and OQ are two perpendicular directions in a horizontal plane. To produce a vertically upward flux density at O, we need

A a vertically downward current at M   B a vertically downward current at N
C a vertically upward current at M   D a vertically upward current at N
E a current at M parallel to PO

Longer Questions

6 A vertical straight conductor X of length 0·5 m is situated in a uniform horizontal magnetic field of 0·1 T.   (i) Calculate the force on X when a current of 4 A is passed into it. Draw a sketch showing the directions of the current, field and force. (ii) Through what angle must X be turned in a vertical plane so that the force on X is halved?

7 A straight horizontal rod X, of mass 50 g and length 0·5 m, is placed in a uniform horizontal magnetic field of 0·2 T perpendicular to X. Calculate the current in X if the force acting on it just balances its weight. Draw a sketch showing the directions of the current, field and force. ($g = 10\,\text{N kg}^{-1}$.)

8 A narrow vertical rectangular coil is suspended from the middle of its upper side with its plane parallel to a uniform horizontal magnetic field of 0·02 T. The coil has 10 turns, and the lengths of its vertical and horizontal sides are 0·1 m and 0·05 m respectively. Calculate the torque on the coil when a current of 5 A is passed into it. Draw a sketch showing the directions of the current, field and torque.
   What would be the new value of the torque if the plane of the vertical coil was initially at 60° to the magnetic field and a current of 5 A was passed into the coil?

9 Show how a 0–10 mA moving-coil meter of 5 Ω resistance can be converted to (a) a 0–5 V voltmeter, (b) a 0–2 A ammeter.

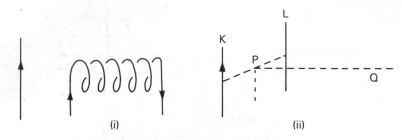

Figure 11B

10 (i) Figure 11B(i) shows a long straight vertical wire and a solenoid. The current directions are indicated by arrows. Copy these diagrams and sketch the magnetic fields generated by each of these conductors alone.

(ii) Figure 11B(ii) shows two long straight parallel conductors, K and L. P is a point midway between the wires and PQ is a line perpendicular to the plane containing the wires. The direction of the current through wire K is shown.

The resultant field at P is zero. Explain, with the aid of a diagram, what you can deduce about the current through the wire L.

What is the direction of the resultant field along the line from P towards Q? Sketch a graph to show how you would expect the magnitude of this field to vary with distance along PQ. (*L*)

11 Figure 11C represents a cylindrical aluminium bar A resting on two horizontal aluminium rails which can be connected to a battery to drive a current through A. A magnetic field, of flux density 0·10 T, acts perpendicularly to the paper and into it. In which direction will A move if the current flows?

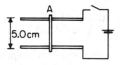

Figure 11C

Calculate the angle to the horizontal to which the rails must be tilted to keep A stationary if its mass is 5·0 g, the current in it is 4·0 A and the direction of the field remains unchanged. (Acceleration of free-fall, $g = 10 \text{ m s}^{-2}$). (*L.*)

12 Describe a moving-coil type of galvanometer and deduce a relation between its deflection and the steady current passing through it.

A galvanometer, with a scale divided into 150 equal divisions, has a current sensitivity of 10 divisions per milliampere and a voltage sensitivity of 2 divisions per millivolt. How can the instrument be adapted to serve
(a) as an ammeter reading to 6 A,
(b) as a voltmeter in which each division represents 1 V? (*L.*)

Forces on Charges Moving in Magnetic Fields

We now consider the forces acting on *charges* moving through a magnetic field. The forces are used to focus the moving electrons on to the screen of a television receiver using a magnetic field. The forces due to the earth's magnetic field make electrical particles bunch together near the north pole of the earth and produce a glow in the sky called the Northern Lights.

As we explained earlier, an electric current in a wire can be regarded as a drift of electrons in the wire, superimposed on their random thermal motions. If the electrons in the wire drift with average velocity *v*, and the wire lies at right angles to the field, then the force on *each* electron, as we soon show, is given by

$$F = Bev \qquad . \qquad . \qquad . \qquad . \qquad . \qquad (1)$$

Generally, the force F on a charge Q moving at right angles to a field of flux density B is given by

$$F = BQv \qquad . \qquad . \qquad . \qquad . \qquad . \qquad (2)$$

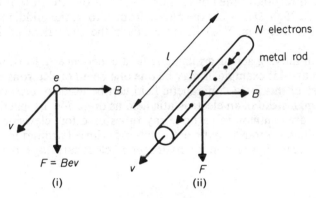

Figure 11.21 *Force on moving electron in magnetic field (v at right angles to page)*

If B is in tesla (T), e or Q is in coulombs (C) and v in metres second^{-1} (m s^{-1}), then F will be in newtons (N) (Figure 11.21 (i)).

The proof of equation (1) can be obtained as follows. Suppose a current I flows in a straight conductor of length l when it is perpendicular to a uniform field of flux density B. From p. 269, $I = nvAe$, where n is the number of electrons per unit volume, v is the drift velocity of the electrons, A is the area of cross-section of the conductor and e is the electron charge. Then the force F' on the conductor is given by

$$F' = BIl = BnevAl = Bev \times nAl$$

Now Al is the volume of the wire. So nAl is the number N of electrons in the conductor, Figure 11.21 (ii).

So $\qquad\qquad$ force on one electron, $F = \dfrac{F'}{N} = Bev$

Generally, a charge Q moving *perpendicular* to a magnetic field B with a velocity v has a force on it given by

$$F = BQv$$

If the velocity v and the field B are inclined to each other at an angle θ,

$$F = BQv \sin \theta$$

Force Direction, Energy in Magnetic Field

It should be carefully noted that the force F acts *perpendicular* to v and to B. This means that F is a *deflecting force*, that is, it changes the direction of motion of the moving charge when the charge enters the field B but does not alter the magnitude of v.

Further, since F is perpendicular to the direction of motion or displacement of the charge, *no work* is done by F as the charge moves in the field. So *no energy* is gained by a charge when it enters a magnetic field and forces act on it.

The *direction* of F is given by Fleming's left-hand rule. The middle finger points in the direction of the conventional current or direction of motion of a *positive* charge. If a *negative* charge moves from X to Y, the middle finger points in the opposite direction, Y to X, since this is the equivalent positive charge movement.

An electron moving across a magnetic field experiences a force whether it is in a wire or not—for example, it may be one of a beam of electrons in a vacuum tube. Because of this force, a magnetic field can be used to *focus* or deflect an electron beam, instead of an electrostatic field as on p. 746. Magnetic deflection and focusing are common in cathode ray tubes used for television. In nuclear energy machines, protons may be deflected and whirled around in a circle by a strong magnetic field. A proton is a hydrogen nucleus carrying a positive charge (p. 883).

Hall Effect

In 1879, Hall found that an e.m.f. or voltage is set up *transversely* or *across* a current-carrying conductor when a perpendicular magnetic field is applied. This is called the *Hall effect*.

To explain the Hall effect, consider a slab of metal carrying a current, Figure 11.22. The flow of electrons is in the opposite direction to the conventional current. If the metal is placed in a magnetic field B at right angles to the face AGDC of the slab and directed out of the plane of the paper, a force Bev then acts on each electron in the direction from CD to AG. *Thus electrons collect along the side AG of the metal*, which will make AG negatively charged and

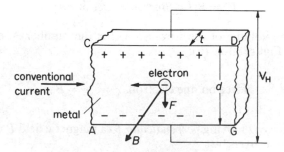

Figure 11.22 *Hall voltage*

lower its potential with respect to CD. So a potential difference or e.m.f. opposes the electron flow. The flow ceases when the e.m.f. reaches a particular value V_H called the *Hall voltage* as shown in Figure 11.22, which may be measured by using a high impedance voltmeter.

Magnitude of Hall Voltage

Suppose V_H is the magnitude of the Hall voltage and d is the width of the slab. Then the electric field strength E set up across the slab is numerically equal to the potential gradient and hence $E = V_H/d$. So the force on each electron = $Ee = V_H e/d$.

The force, which is directed upwards from AG to CD, is equal to the force produced by the magnetic field when the electrons are in equilibrium.

$$\therefore Ee = Bev$$

$$\therefore \frac{V_H e}{d} = Bev$$

$$\therefore V_H = Bvd \quad . \quad . \quad . \quad . \quad . \quad (1)$$

From p. 269, the drift velocity of the electrons is given by

$$I = nevA \quad . \quad . \quad . \quad . \quad . \quad (2)$$

where n is the number of electrons per unit volume and A is the area of cross-section of the conductor. In this case $A = td$ where t is the thickness. Hence, from (2),

$$v = \frac{I}{netd}$$

Substituting in (1),

$$\therefore V_H = \frac{BI}{net} \quad . \quad . \quad . \quad . \quad . \quad (3)$$

We now take some typical values for copper to see the order of magnitude of V_H. Suppose $B = 1$ T, a field obtained by using a large laboratory electro magnet. For copper, $n \simeq 10^{29}$ *electrons* per metre^3, and the charge on the electron is 1.6×10^{-19} coulomb. Suppose the specimen carries a current of 10 A and that its thickness is about 1 mm or 10^{-3} m. Then

$$V_H = \frac{1 \times 10}{10^{29} \times 1.6 \times 10^{-19} \times 10^{-3}} = 0.6 \, \mu V \text{ (approx.)}$$

This e.m.f. is very small and would be difficult to measure. The importance of the Hall effect becomes apparent when semiconductors are used, as we now see.

Hall Effect in Semiconductors

In semiconductors, the charge carriers which produce a current when they move may be positively or negatively charged (see p. 766). The Hall effect helps us to find the sign of the charge carried. In Figure 11.22, p. 340, suppose that electrons were not responsible for carrying the current, and that the current was due to the movement of positive charges in the *same* direction as the conventional current. The magnetic force on these charges would also be *downwards*, in the same direction as if the current were carried by electrons. This is because the sign *and* the direction of movement of the charge carriers have both been reversed. Thus AG would now become *positively* charged, and the polarity of the Hall voltage would be reversed.

Experimental investigation of the polarity of the Hall voltage hence tells us whether the current is predominantly due to the drift of positive charges or to the drift of negative charges. In this way it was shown that the current in a metal such as copper is due to movement of negative charges, but that in impure semiconductors such as germanium or silicon, the current may be predominantly due to movement of either negative or positive charges (p. 766).

The magnitude of the Hall voltage V_H in metals was shown as above to be very small. In semiconductors it is much larger because the number n of charge carriers per metre^3 is much *less* than in a metal and $V_H = BI/net$. Suppose that n is about 10^{25} per metre^3 in a semiconductor, and $B = 1$ T, $t = 10^{-3}$ m, $e = 1·6 \times 10^{-19}$ C, as above. Then

$$V_H = \frac{1 \times 10}{10^{25} \times 1·6 \times 10^{-19} \times 10^{-3}} = 6 \times 10^{-3} \text{ V (approx.)} = 6 \text{ mV}$$

The Hall voltage is thus much more measurable in semiconductors than in metals.

Use of Hall Effect

Apart from its use in semiconductor investigations, a *Hall probe* may be used to measure the flux density B of a magnetic field. A simple Hall probe is shown in Figure 11.23. Here a wafer of semiconductor has two contacts on opposite sides which are connected to a high impedance voltmeter, V. A current,

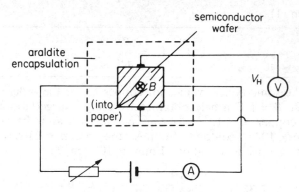

Figure 11.23 *Measurement of B by Hall voltage*

generally less than one ampere, is passed through the semiconductor and is measured on the ammeter, A. The Araldite glue prevents the wires from being detached from the wafer. Now, from (3) on p. 341,

$$V_H = \frac{BI}{net}$$

$$\therefore B = \frac{V_H net}{I}$$

Now *net* is a constant for the given semiconductor, which can be determined previously. So from the measurement of V_H and I, B can be found.

In practice, the voltmeter scale is calibrated in tesla (T) by the manufacturer and so the flux density B of the magnetic field is read directly from the scale.

Note that the direction of B must be *perpendicular* to the semiconductor probe when measuring B. Later we shall use the Hall probe to measure the flux density B round a straight current-carrying conductor and inside a current-carrying solenoid (p. 346).

You should know:

1 With B perpendicular to a conductor S, a Hall voltage is obtained on the sides of S normal to the current flowing through S.
2 Hall voltage $V_H = BI/net$.
3 The Hall voltage is used
  (a) in semiconductors to find whether the current flow is due mainly to positive or negative charges,
  (b) to measure n, the charge density,
  (c) as a basis of a Hall probe, for measuring the flux density B of a magnetic field.

EXERCISES 11B *B*-field force on charges, Hall voltage

1 An electron beam moving with a velocity of $10^6\,\mathrm{m\,s^{-1}}$, moves through a uniform magnetic field of $0.1\,\mathrm{T}$ which is perpendicular to the direction of the beam. Calculate the force on an electron if the electron charge is $-1.6 \times 10^{-19}\,\mathrm{C}$. Draw a sketch showing the directions of the beam, field and force.
2 A current of $0.5\,\mathrm{A}$ is passed through a rectangular section of a semiconductor 4 mm thick which has majority carriers of negative charges or free electrons. When a magnetic field of $0.2\,\mathrm{T}$ is applied perpendicular to the section, a Hall voltage of $6.0\,\mathrm{mV}$ is produced between the opposite edges.

   Draw a diagram showing the directions of the field, charge carriers and Hall voltage, and calculate the number of charge carriers per unit volume.
3 Define the coulomb. Deduce an expression for the current I in a wire in terms of the number of free electrons per unit volume, n, the area of cross-section of the wire, A, the charge on the electron, e, and its drift velocity, v.

   A copper wire has 1.0×10^{29} free electrons per cubic metre, a cross-sectional area of $2.0\,\mathrm{mm^2}$ and carries a current of $5.0\,\mathrm{A}$. Calculate the force acting on each electron if the wire is now placed in a magnetic field of flux density $0.15\,\mathrm{T}$ which is perpendicular to the wire. Draw a diagram showing the directions of the electron velocity, the magnetic field and this force on an electron.

   Explain, without experimental detail, how this effect could be used to determine whether a slab of semiconducting material was n-type or p-type. (Charge on electron $= -1.6 \times 10^{-19}\,\mathrm{C}$.) (*L.*) (*You may need to refer to Chapter 31.*)
4 Explain the origin of the Hall effect. Include a diagram showing clearly the directions of the Hall voltage and other relevant vector quantities for a specimen in which electron conduction predominates.

   A slice of indium antimonide is $2.5\,\mathrm{mm}$ thick and carries a current of $150\,\mathrm{mA}$. A magnetic field of flux density $0.5\,\mathrm{T}$, correctly applied, produces a maximum Hall voltage of $8.75\,\mathrm{mV}$ between the edges of the slice. Calculate the number of free charge carriers per unit volume, assuming they each have a charge of $-1.6 \times 10^{-19}\,\mathrm{C}$. Explain your calculation clearly.

   What can you conclude from the observation that the Hall voltage in different conductors can be positive, negative or zero? (*C.*)

5 (a) Figure 11D shows a rectangular piece of semiconductor material with leads attached to metal end faces.

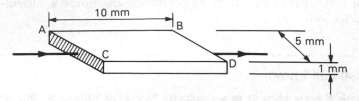

Figure 11D

   (i) If the resistance of the specimen is approximately 100 Ω, show that the resistivity of the material is about 0·050 Ωm.
   (ii) Describe how you would determine experimentally the resistivity of the material of a similar specimen using apparatus normally available in a school or college laboratory. You can assume that the specimen is mounted in such a way that its dimensions can be measured.
(b) For a specimen similar to that shown in the diagram, a magnetic field is applied perpendicular to the face ABCD. When a current flows, a potential difference, called the Hall p.d., develops across the specimen and is perpendicular to both field and current.
   (i) Explain how this effect occurs.
   (ii) Explain why the Hall p.d. is much larger in a semiconductor than in a metal specimen of the same dimensions under the same conditions. You may assume that the current, I, in a specimen of cross-sectional area, A, is given by

$$I = nAvq$$

   where n is the number of charge carriers per unit volume, q is the charge of each carrier and v is the drift velocity of the charge carriers.
(c) You are given two pieces of semiconductor material which are identical in appearance. One is p-type material and the other is n-type. Outline an experiment using the Hall effect to distinguish between them, explaining carefully how you would use the observations to identify the n-type material. (N.) (*You may need to refer to Chapter 31.*)

12 Magnetic Fields and Interaction of Current-Carrying Conductors

In this chapter we deal more fully with the magnetic fields due to currents in three types of conductor: the solenoid, the straight conductor (wire) and a narrow circular coil.

Solenoids are widely used, particularly with soft iron inside, in the electrical and radio industries. The force between two straight conductors 1 metre apart is used to define the ampere and this is the basis of measuring current accurately in an ampere-meter.

We shall first give the values of the flux density B for the solenoid and straight wire when they carry currents. If required, proof of these formulae will be found at the end of the chapter.

Solenoids

Solenoids, or relatively long coils of wire, are widely used in industry. For example, solenoids are used in telephone earpieces to carry the speech current and in magnetic relays used in telecommunications.

The magnetic field inside an infinitely-long solenoid is constant in magnitude. A form of coil which gives a very nearly uniform field is shown in Figure 12.1 (i).

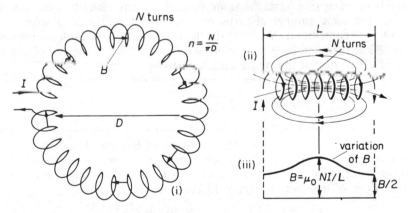

Figure 12.1 *A toroid and solenoid*

It is a solenoid of N turns and length L metres wound on a circular support instead of a straight one, and is called a *toroid*. If its average diameter D is several times its core diameter, then the turns of wire are almost equally spaced around its inside and outside circumferences; their number per metre is, therefore,

$$n = \frac{N}{L} = \frac{N}{\pi D} \qquad . \qquad . \qquad . \qquad . \qquad . \qquad (1)$$

The magnetic field within a toroid is very nearly uniform, because the coil

345

has no ends. The coil is equivalent to an infinitely long solenoid. If I is the current, the flux density B at all points within it is given by

$$B = \mu_0 nI \qquad . \qquad . \qquad . \qquad . \qquad (2)$$

and μ_0 is a constant known as the *permeability of free space* which has the value $4\pi \times 10^{-7}$ H m^{-1}. (H is a unit called a 'henry' and is discussed later.) The constant μ_0 is necessary to make the units correct, that is, B is then in tesla (T) when I is in amperes (A) and l is in metres (m).

Note that the flux density B in the middle of a current-carrying solenoid does *not* depend on the cross-sectional area (or radius) of the cylindrical coil. B depends only on the *number of turns per metre length* of the solenoid.

Solenoids of Finite Length

In practice, solenoids cannot be made infinitely long. But if the length L of a solenoid is about ten times its diameter, the field near its middle is fairly uniform, and has the value given by equation (2). Figure 12.1 (ii) shows a solenoid of length L and N turns, so that $n = N/L$. The flux density in the *middle* of the coil is given approximately by

$$B = \mu_0 nI = \mu_0 \frac{NI}{L} \qquad . \qquad . \qquad . \qquad . \qquad (3)$$

If a long solenoid is imagined cut at any point R near the middle, the two solenoids on each side have the same field B at their respective centres since each has the same number of turns per unit length as the long solenoid. So each solenoid contributes equally to the field at R. Hence each solenoid provides a field $B/2$ at their end R. We therefore see that the field at the *end* of any long solenoid is *half* that at the centre and is given by

$$B = \frac{1}{2} \mu_0 \frac{NI}{L} \qquad . \qquad . \qquad . \qquad . \qquad (4)$$

Figure 12.1 (iii) shows roughly the variation of B along the solenoid.

As was explained on p. 321, the direction of B inside the solenoid can be found from the 'corkscrew rule' or the 'clenched fist rule'. The reader should verify the directions of B shown in Figure 12.1 (i) and (ii).

Experiment for B using Hall Probe

Figure 12.2 shows how the flux density B in the middle of a long solenoid can be investigated. S is a 'Slinky' (loose) coil, with its N turns uniformly spaced in a length L. P is a Hall probe in the middle of S and placed so that the flux density B is *normal* to P. As shown on p. 342, the Hall voltage produced at P is proportional to the value of B and this can be read directly in tesla (T) on the meter M.

In the experiment, the uniform spacing of S is varied by pulling out the coil more and the total length L of the coil and the value of B in the middle are measured each time. The number of turns per metre length is given by $n = N/L$, so $n \propto 1/L$ as N is constant. A graph of B against $1/L$ produces a straight line

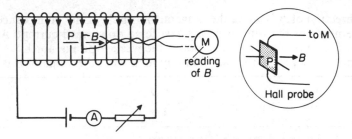

Figure 12.2 *B measured by Hall probe inside solenoid*

passing through the origin, so showing that $B \propto n$. The same circuit can be used to verify $B \propto I$, the current in the solenoid, for a given value of n.

Effect on *B* of Relative Permeability

As we have stated, the constant μ_0 in the formula for flux density B is called the permeability of free space (or vacuum) and has the value $4\pi \times 10^{-7}$ H m^{-1}. The permeability of air at normal pressure is only very slightly different from that of a vacuum. So we can consider the permeability of air to be practically $4\pi \times 10^{-7}$ H m^{-1}.

If the solenoid is wound round soft iron, so that this material is now the core of the solenoid, the permeability is increased considerably. The name 'relative permeability', symbol μ_r, is given to the number of times the permeability has increased relative to that of free space or air. So if $\mu_r = 1000$, the value of B in the solenoid is 1000 times as great as with an air core. Generally, the permeability μ of an iron core would be given by

$$\mu = \mu_r \mu_0$$

Note that μ_r is a number and has no units, unlike μ_0 and μ.

Long Straight Conductor

We now consider the magnetic field of a long straight current-carrying conductor. A submarine cable carrying messages is an example of such a conductor.

All round a straight current-carrying wire, the field pattern consists of circles concentric with the wire. Figure 12.3 (i) shows the field round one section of the conductor. Maxwell's corkscrew rule gives the field direction: If a right-handed corkscrew is turned so that the point moves along the current direction, the field direction is the same as the direction of turning.

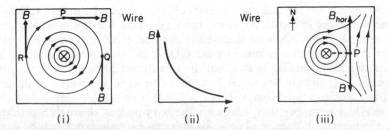

Figure 12.3 *Field due to a long straight conductor*

The direction of B is along the tangent to a circle at the point concerned. So at P due north of the wire, B points east for a downward current. At a point due east, B points south and at a point due west, B points north.

At a point distance r from an infinitely long wire, the value of B is given by

$$B = \frac{\mu_0 I}{2\pi r}$$

So for a given current, $B \propto 1/r$, Figure 12.3 (ii).

The earth's horizontal magnetic field B_{hor} is about 4×10^{-5} T and acts due north. When this cancels exactly the magnetic field of the current, a *neutral point* is obtained in the combined field of the earth and the current. Since the field due to the current must be due south, the neutral point P in Figure 12.3 (iii) is due *east* of the wire. Suppose the current is 5 A. The distance r of the neutral point from the wire is then given by

$$\frac{\mu_0 I}{2\pi r} = B_{hor} = 4 \times 10^{-5}$$

So
$$r = \frac{\mu_0 I}{2\pi \times 4 \times 10^{-5}} = \frac{4\pi \times 10^{-7} \times 5}{2\pi \times 4 \times 10^{-5}}$$

$$= 0.025 \text{ m}$$

$$= 25 \text{ mm}$$

Variation of B using a Search Coil

An apparatus suitable for finding the variation of B with distance r from a long straight wire CD is shown in Figure 12.4. Alternating current (a.c.) of the order

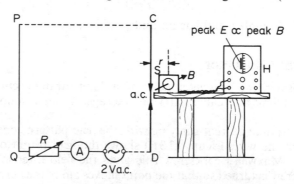

Figure 12.4 *Investigation of B due to straight conductor CD*

of 10 A, from a low voltage mains transformer, is passed through CD by using another long wire PQ at least one metre away, a rheostat R and an a.c. ammeter A. A small *search coil* S, with thousands of turns of wire, such as the coil from an output transformer, is placed near CD. It is positioned with its axis at a small distance r from CD and so that the flux from CD enters its face normally. S is joined by long twin flex to the Y-plates of an oscilloscope H and the greatest sensitivity, such as 5 mV/cm, is used.

When the a.c. supply is switched on, the varying flux through S produces an induced alternating e.m.f. E. The peak (maximum) value of E can be found by switching off the time-base and measuring the length of the line trace,

Figure 12.4. See p. 760. Now the peak value of the magnetic flux density B is proportional to the peak value of E, as shown later. So the length of the trace gives a measure of the peak value of B.

The distance r of the coil CD is then increased and the corresponding length of the trace is measured. The length of the trace plotted against $1/r$ gives a straight line graph passing through the origin. Hence $B \propto 1/r$. A similar method can be used for investigating the field B inside a solenoid.

Forces between Currents

In 1821, Ampère discovered by experiment that current-carrying conductors exert a force on each other. For example, when the currents in two long neighbouring straight conductors X and Y are in the same direction, there is a force of *attraction* between them, Figure 12.5 (i). If the currents flow in opposite directions, there is a *repulsive* force between them, Figure 12.5 (ii). Each conductor has a force on it due to the magnetic field of the other, from the law of action and reaction.

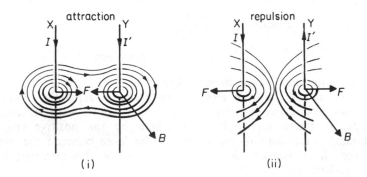

Figure 12.5 *Forces between currents*

Figure 12.5 (i) shows the resultant magnetic flux round two long straight vertical conductors X, Y in a horizontal plane when the currents are both downwards. The lines tend to pull the conductors towards each other. In Figure 12.5 (ii), the currents are in opposite directions. Here the lines tend to push the conductors apart.

Fleming's left-hand rule confirms the direction of the forces. At Y, the flux density B due to the conductor X is perpendicular to Y (the flux due to X alone consists of circles with X as centre and at Y the *tangent* to the circular line is perpendicular to Y). So, from Fleming's rule, the force F on Y in Figure 12.5 (i) is towards X. From the law of action and reaction, the force F on X is towards Y and equal to that on Y. So the conductors *attract* each other.

In Figure 12.5 (ii), the current I' in Y is *opposite* to that in Figure 12.5 (i). From Fleming's left-hand rule, the force F on Y is now away from X and so the force is *repulsive*.

Magnitude of Force, The Ampere

If two long straight conductors X and Y lie parallel and close together at a distance r apart, and carry currents I, I' respectively as in Figure 12.5 (i), then the current I is in a magnetic field of flux density B equal to $\mu_0 I'/2\pi r$ due to the current I' (p. 348). The force *per metre* length, F, on X is hence given by

$$F = BIl = BI \times 1 = \frac{\mu_0 I'}{2\pi r} \times I \times 1$$

$$\therefore F = \frac{\mu_0 II'}{2\pi r} \quad . \quad . \quad . \quad . \quad . \quad (1)$$

From the law of action and reaction, this would also be the force per metre on the other conductor Y.

The ampere can be *defined* in terms of the force between conductors.

It is *that current, which flowing in each of two infinitely long parallel straight wires of negligible cross-sectional area separated by a distance of 1 metre* in vacuo, *produces a force* between the wires *of* 2×10^{-7} *newton metre*$^{-1}$.

Taking $I = I' = 1$ A, $r = 1$ metre, $F = 2 \times 10^{-7}$ newton metre^{-1}, then, from (1),

$$2 \times 10^{-7} = \frac{\mu_0 \times 1 \times 1}{2\pi \times 1}$$

$$\therefore \mu_0 = 4\pi \times 10^{-7} \text{ henry metre}^{-1}$$

which is the value used in formulae with μ_0.

It may be noted that the electrostatic force of repulsion between the negative charges of the moving electrons in the two wires is completely neutralised by the attractive force on them by the positive charges on the stationary metal ions in the wires. So the force between the two wires is only the *electromagnetic* force, which is due to the magnetic fields of the moving electrons.

Example on Force between Conductors

A long straight conductor X carrying a current of 2 A is placed parallel to a short conductor Y of length 0·05 m carrying a current of 3 A, Figure 12.6. The two conductors are 0·10 m apart. Calculate   (i) the flux density due to X at Y,   (ii) the approximate force on Y.

(i) Due to X,

$$B = \frac{\mu_0 I}{2\pi r} = \frac{4\pi \times 10^{-7} \times 2}{2\pi \times 0.10}$$

$$= 4 \times 10^{-6} \text{ T}$$

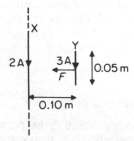

Figure 12.6 *Force between conductors*

(ii) On Y, length $l = 0.05$ m

$$\text{force } F = BIl = 4 \times 10^{-6} \times 3 \times 0.05$$
$$= 6 \times 10^{-7}\,\text{N}$$

Absolute Determination of Current, Ampere Balance

A simple laboratory form of an *ampere balance*, which measures current by measuring the force between current-carrying conductors, is shown in Figure 12.7.

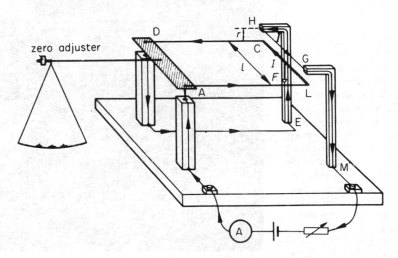

Figure 12.7 *Laboratory form of ampere balance*

With no current flowing, the zero screw is adjusted until the plane of ALCD is horizontal. The current I to be measured is then switched on so that it flows through ALCD and EHGM in series and HG repels CL downwards. The mass m necessary to restore balance is then measured, and mg is the force between the conductors since the respective distances of CL and the scale pan from the pivot are equal. The equal lengths l of the straight wires CL and HG, and their separation r, are all measured.

From equation (1) above,

$$\text{force per metre} = \frac{4\pi \times 10^{-7}\,I^2}{2\pi r}$$

$$\therefore mg = \frac{4\pi \times 10^{-7}I^2 l}{2\pi r}$$

$$\therefore I = \sqrt{\frac{mgr}{2 \times 10^{-7} l}}$$

In this expression I will be in amperes if m is in kilograms, $g = 9.8$ m s^{-2} and l and r are measured in metres.

Note that the force F on CL is still downwards if the currents are *reversed* in CL and HG. So the ampere-balance can be used to measure alternating current. Since $F \propto I^2$, the root-mean-square value is measured (p. 403).

Plate 12A *The new National Physical Laboratory moving-coil balance in its vacuum chamber with the top raised. This balance is designed to relate the electrical watt, the metre, the kilogram and the second with an accuracy of better than one part in a hundred million. The electrical watt is linked to the Planck constant, the metre to the speed of light and the second to an atomic transition*

Narrow Circular Coil

The third of our typical conductors is the narrow circular coil.

Figure 12.8 (i) shows the magnetic field pattern round a narrow vertical circular coil C carrying a current I, in the horizontal (perpendicular) plane passing through the middle of the coil. In the middle M of the coil, the field is uniform for a short distance either side. Here the field value B is given by

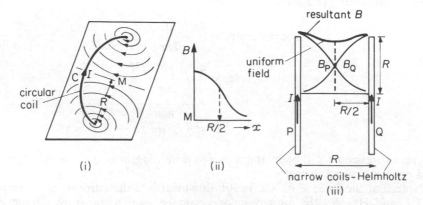

Figure 12.8 *Fields due to narrow circular coils. Helmholtz coils*

$$B = \frac{\mu_0 I}{2r}$$

where r is the radius in metres.

Figure 12.8 (ii) shows how B varies as we move from the centre of the coil along a line perpendicular to the plane of the coil. The field value decreases continuously. Helmholtz, an eminent scientist of the nineteenth century, showed that two narrow circular coils of the same radius and carrying the same current could provide a *uniform* magnetic field between them. For this purpose they are placed facing each other at a distance apart equal to their radius R. As shown in Figure 12.8 (iii), the resultant magnetic field B round a point half-way between the coils P and Q is fairly uniform for some distance on either side of the point. The flux density B of the uniform field is given approximately by

$$B = 0.72 \frac{\mu_0 N I}{R}$$

where N is the number of turns in each coil, I is the current in amperes and R is the radius in metres.

Helmholtz coils were used by Sir J. J. Thomson to obtain a uniform magnetic field of known value in a famous experiment to find the charge–mass ratio of an electron (see p. 747).

■ Magnitudes of B for Current-carrying Conductors

We conclude this chapter with proofs of the values of B used earlier for a narrow circular coil, a straight conductor and a solenoid.

Law of Biot and Savart

To calculate B for any shape of conductor, Biot and Savart gave a law which can now be stated as follows: The flux density ΔB at a point P due to a small element Δl of a conductor carrying a current is given by

$$\Delta B \propto \frac{I \Delta l \sin \alpha}{r^2} \qquad . \qquad . \qquad . \qquad . \qquad . \qquad (1)$$

where r is the distance from the point P to the element and α is the angle between the element and the line joining it to P, Figure 12.9.

The formula in (1) cannot be proved directly, as we cannot experiment with

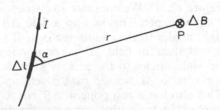

Figure 12.9 *Biot and Savart law*

an infinitesimally small conductor. We believe in its truth because the deductions for large practical conductors turn out to be true.

The constant of proportionality in equation (1) depends on the medium in which the conductor is situated. In air (or, more exactly, in a vacuum), we write

$$\Delta B = \frac{\mu_0}{4\pi} \frac{I\Delta l \sin \alpha}{r^2} \qquad . \qquad . \qquad . \qquad . \qquad . \qquad (2)$$

where
$$\mu_0 = 4\pi \times 10^{-7}$$

and its unit is *'henry per metre'* (H m^{-1}) as will be shown later.

B for Narrow Coil

The formula for the value of B at the centre of a narrow circular coil can be immediately deduced from (2). Here the radius r is constant for all the elements Δl, and the angle α is constant and equal to 90°, Figure 12.10. If the coil has

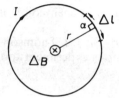

Figure 12.10 *Field of circular coil*

N turns, the length of wire in it is $2\pi rN$, and the field at its centre is therefore given, if the current is I, by

$$
\begin{aligned}
B = \int dB &= \frac{\mu_0}{4\pi} \int_0^{2\pi rN} \frac{I\, dl \sin 90°}{r^2} \\
&= \frac{\mu_0 I}{4\pi r^2} \int_0^{2\pi rN} dl = \frac{\mu_0 I}{4\pi r^2} 2\pi rN \\
&= \frac{\mu_0 NI}{2r} \qquad . \qquad . \qquad . \qquad . \qquad . \qquad . \qquad (1)
\end{aligned}
$$

B along Axis of a Narrow Circular Coil

We will now find the magnetic field at a point anywhere on the axis of a narrow circular coil (P in Figure 12.11). We consider an element Δl of the coil, at right angles to the plane of the paper. This sets up a field ΔB at P, in the plane of the paper, and at right angles to the radius vector r. If β is the angle between r and the axis of the coil, then the field ΔB has components $\Delta B \sin \beta$ along the axis, and $\Delta B \cos \beta$ at right angles to the axis. If we now consider the element $\Delta l'$ diametrically opposite to Δl, we see that it sets up a field $\Delta B'$ equal in magnitude to ΔB. This also has a component, $\Delta B' \cos \beta$, at right angles to the axis; but this component acts in the opposite direction to $\Delta B \cos \beta$ and, therefore, cancels it. By considering elements such as Δl and $\Delta l'$ all round the

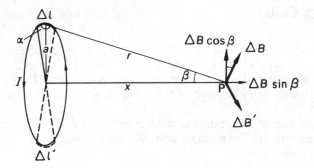

Figure 12.11 *Field on axis of flat coil*

circumference of the coil, we see that the field at P can have no component at right angles to the axis. Its value along the axis is

$$B = \int dB \sin \beta$$

From Figure 12.11, we see that the length of the radius vector r is the same for all points on the circumference of the coil, and that the angle α is also constant, being 90°. Thus, if the coil has a single turn, and carries a current I,

$$\Delta B = \frac{\mu_0 I \, \Delta l \sin \alpha}{4\pi r^2} = \frac{\mu_0 I}{4\pi r^2} \Delta l$$

And, if the coil has a radius a, then

$$B = \int dB \sin \beta = \int_0^{2\pi a} \frac{\mu_0 I}{4\pi r^2} \, dl \sin \beta$$

$$B = \frac{\mu_0 I a \sin \beta}{2r^2} \qquad \cdot \qquad \cdot \qquad \cdot \qquad \cdot \qquad (1)$$

When the coil has more than one turn, the distance r varies slightly from one turn to the next. But if the width of the coil is small compared with all its other dimensions, we may neglect it, and write,

$$B = \frac{\mu_0 N I a \sin \beta}{2r^2} \qquad \cdot \qquad \cdot \qquad \cdot \qquad \cdot \qquad (2)$$

where N is the number of turns.

Equation (2) can be put into a variety of forms, by using the facts that

$$\sin \beta = \frac{a}{r}$$

and

$$r^2 = x^2 + a^2$$

where x is the distance from P to the centre of the coil. Thus

$$B = \frac{\mu_0 N I a^2}{2r^3} = \frac{\mu_0 N I a^2}{2(x^2 + a^2)^{3/2}} \qquad \cdot \qquad \cdot \qquad \cdot \qquad (3)$$

Helmholtz Coils

The field along the axis of a single coil varies with the distance x from the coil. In order to obtain a *uniform* field, Helmholtz used two coaxial parallel coils of equal radius R, separated by a distance R. In this case, when the same current flows around each coil in the same direction, the resultant field B is uniform for some distance on either side of the point on their axis midway between the coils. See p. 352.

The magnitude of the resultant field B at the midpoint can be found from our previous formula for a single coil. We now have $a = R$ and $x = R/2$. So, for the two coils,

$$B = 2 \times \frac{\mu_0 NIR^2}{2(R^2/4 + R^2)^{3/2}} = \left(\frac{4}{5}\right)^{3/2} \times \frac{\mu_0 NI}{R}$$

$$B = 0 \cdot 72 \frac{\mu_0 NI}{R} \text{ (approx.)}$$

B on Axis of a Long Solenoid

We may regard a solenoid as a long succession of narrow coils; if it has n turns per metre, then in an element Δx of it there are $n\Delta x$ coils, Figure 12.12. At a point P on the axis of the solenoid, the field due to these is, by equation (2),

$$\Delta B = \frac{\mu_0 I a \sin \beta}{2r^2} n\Delta x$$

in the notation which we have used for the flat coil.

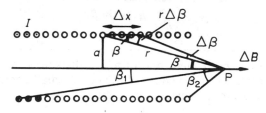

Figure 12.12 *Field on axis of solenoid*

If the element Δx subtends an angle $\Delta\beta$ at P, then, from the figure,

$$r\Delta\beta = \Delta x \sin \beta$$

so

$$\Delta x = \frac{r\Delta\beta}{\sin \beta}$$

Also,

$$a = r \sin \beta$$

Thus

$$\Delta B = \frac{\mu_0 I r \sin^2 \beta}{2r^2} n \frac{r\Delta\beta}{\sin \beta}$$

$$= \frac{\mu_0 nI}{2} \sin \beta \Delta\beta$$

If the radii of the coil, at its ends, subtend the angles β_1 and β_2 at P, then the field at P is

$$B = \int_{\beta}^{\beta_2} \frac{\mu_0 nI}{2} \sin \beta \, d\beta$$

$$= \frac{\mu_0 nI}{2} \left[-\cos \beta \right]_{\beta_1}^{\beta_2}$$

$$= \frac{\mu_0 nI}{2} (\cos \beta_1 - \cos \beta_2) \quad . \qquad . \qquad . \qquad . \qquad (1)$$

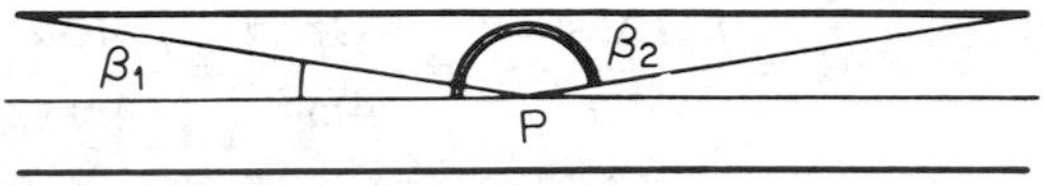

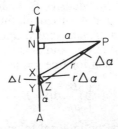

Figure 12.13 *A very long solenoid*

If the point P inside a very long solenoid—so long that we may regard it as infinite—then $\beta_1 = 0$ and $\beta_2 = \pi$, as shown in Figure 12.13. Then, by equation (1):

$$B = \frac{\mu_0 nI}{2} \left[-\cos \beta \right]_{0}^{\pi}$$

so
$$B = \mu_0 nI \qquad . \qquad . \qquad . \qquad . \qquad . \qquad (2)$$

The quantity nI is often called the 'ampere-turns per metre'.

B due to Long Straight Wire

In Figure 12.14, AC represents part of a long straight wire. P is taken as a point so near it that, from P, the wire looks infinitely long—it subtends very

Figure 12.14 *Field of a long straight wire*

nearly 180°. An element XY of this wire, of length Δl, makes an angle α with the radius vector, r, from P. It therefore contributes to the magnetic field at P an amount

$$\Delta B = \frac{\mu_0 I \Delta l \sin \alpha}{4\pi r^2} \qquad . \qquad . \qquad . \qquad . \qquad (1)$$

when the wire carries a current I. If a is the perpendicular distance, PN, from

P to the wire, then

$$PN = PX \sin \alpha \quad \text{or} \quad a = r \sin \alpha$$

so

$$r = \frac{a}{\sin \alpha} \qquad \qquad \cdot \qquad \cdot \qquad \cdot \qquad \cdot \qquad (2)$$

Also, if we draw XZ perpendicular to PY, we have

$$XZ = XY \sin \alpha = \Delta l \sin \alpha$$

If Δl subtends an angle $\Delta \alpha$ at P, then

$$XZ = r\Delta \alpha = \Delta l \sin \alpha$$

From (1)

$$\Delta B = \frac{\mu_0 I \Delta l \sin \alpha}{4\pi r^2} = \frac{\mu_0 I r \Delta \alpha}{4\pi r^2} = \frac{\mu_0 I \Delta \alpha}{4\pi r}$$

From (2),

$$\Delta B = \frac{\mu_0 I \sin \alpha \Delta \alpha}{4\pi a}$$

When the point Y is at the bottom end A of the wire, $\alpha = 0$; and when Y is at the top C of the wire, $\alpha = \pi$. Therefore the total magnetic field at P is

$$B = \frac{\mu_0}{4\pi} \int_0^\pi \frac{I \sin \alpha \Delta \alpha}{a} = \frac{\mu_0 I}{4\pi a} \left[-\cos \alpha \right]_0^\pi$$

$$\therefore B = \frac{\mu_0 I}{2\pi a} \qquad \cdot \qquad \cdot \qquad \cdot \qquad \cdot \qquad \cdot \qquad (3)$$

Equation (3) shows that the magnetic field of a long straight wire, at a point near it, is inversely proportional to the distance of the point from the wire. The result was discovered experimentally by Biot and Savart, and led to their general formula in (1) which we used to derive equation (3).

Ampère's Theorem

In the calculation of magnetic flux density B, we have used so far only the Biot and Savart law. Another law useful for calculating B is *Ampère's theorem*.

Ampère showed that if a *continuous closed line or loop* is drawn round one or more current-carrying conductors, and B is the flux density in the direction of an element dl of the loop, then for free space

$$\oint \frac{B}{\mu_0} . dl = I$$

where the symbol $\oint$ represents the integral taken completely round the closed loop and I is the total current enclosed by the loop. So we can write

$$\oint B . dl = \mu_0 I \qquad \cdot \qquad \cdot \qquad \cdot \qquad \cdot \qquad \cdot \qquad (1)$$

The proof of (1) is outside the scope of this book.

We now apply the theorem to two special cases of current-carrying conductors.

1   *Straight wire*

Figure 12.15 shows a circular loop L of radius r, drawn concentrically round a straight wire carrying a current I. The flux lines are circles and so, at every

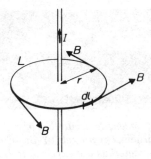

Figure 12.15 *B due to straight wire*

part of a closed line, B is directed along the tangent to the circle at that part. Further, by symmetry, B has the same value everywhere along the line.

So
$$\oint B \, . \, dl = B \oint dl = B \, . \, 2\pi r$$

since B is constant. Hence, from (1),

$$B \, . \, 2\pi r = \mu_0 I$$

and so
$$B = \frac{\mu_0 I}{2\pi r}$$

This agrees with the result derived earlier.

2 Toroid (Solenoid)

Consider the closed loop M indicated by the broken line in Figure 12.16. Again B is everywhere the same at M and is directed along the loop at every point.

$$\oint B \, . \, dl = B \oint dl = BL$$

where L is the total length of the loop M. Hence, from (1),

$$BL = \mu_0 NI$$

So
$$B = \frac{\mu_0 NI}{L} = \mu_0 nI$$

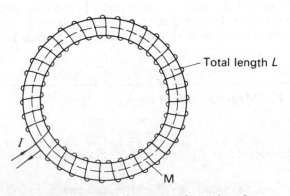

Figure 12.16 *B due to toroid or solenoid*

Plate 12B *Demonstration of magnetic levitation of one of the new high-temperature superconductors—yttrium-barium-copper oxide ($Y\text{-}Ba_2\text{-}Cu_3\text{-}O_7\text{-}x$). Discovered in 1986, the new superconducting ceramic materials are expected to lead to a technological revolution and are the subject of intensive worldwide research. The photograph shows a small, cylindrical magnet floating freely above a nitrogen-cooled, cylindrical specimen of a superconducting ceramic (made by IMI Ltd). The glowing vapour is from liquid nitrogen, which maintains the ceramic within its superconducting temperature range. Photographed at the University of Birmingham*

where N is the total number of turns, and n is the number of turns per metre. This agrees with the result previously obtained.

Superconducting Electromagnets

In theory, an electromagnet (like the solenoid on pages 345 and 356) can produce very powerful magnetic fields: all that is required is to increase the coil current I until the flux density B is as large as is required. In practice, however, the coil will dissipate power I^2R, where R is the coil's resistance, and eventually this power loss will become unacceptably large and might even cause the coil to melt.

For applications where strong magnetic fields are imperative, *superconducting electromagnets* are becoming increasingly popular. These magnets have coils made from a material which has zero resistance, so there are no internal power losses. Thus the coil current can become as large as is necessary to produce the desired field. Unfortunately, most superconducting materials only exhibit their special properties at low temperatures, and superconducting magnets usually need to be continuously cooled using liquid nitrogen.

Today, there is considerable research to develop useful room-temperature superconducting materials. Superconducting magnets are already in use in high speed, levitating trains. A very strong magnetic field holds the train in suspension above the rails, so there are no frictional forces to overcome, only wind resistance and the inertia of the train. Since there is no physical contact between the train and the rails, conventional wheel-based systems cannot be used to move the train. Instead a magnetic system is used, where a magnetic field on the train interacts with an induced field on the rails to drive the train forward.

EXERCISES 12   Magnetic fields, Forces

Multiple Choice

1 A vertical solenoid S has 200 turns in a length of 0·40 m and carries a current of 3 A in the direction shown, Figure 12A (i). The flux density in the middle in tesla (T) is about

A $1·9 \times 10^{-5}$ and downward   **B** $2·9 \times 10^{-5}$ and upward
C $1·9 \times 10^{-3}$ and upward   **D** $0·6 \times 10^{-4}$ and downward   **E** 0·19 and upward

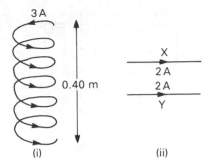

Figure 12A

2 In Figure 12A (ii), X and Y are two straight parallel conductors each carrying a current of 2 A. The force on each conductor is F newtons. When the current in each is changed to 1 A and reversed in direction, the force on each is now

A $F/4$ and unchanged in direction B $F/2$ and reversed in direction
C $F/2$ and unchanged in direction D $F/4$ and reversed in direction
E $F/8$ and unchanged in direction

3 A long vertical straight conductor (not shown) is placed at O in Figure 12B and carries a downward current of 5 A. A small straight wire X of length 0·03 m is placed along the tangent to the circle of centre O and radius 0·1 m as shown. If X carries a current of 2 A, the force on X in N is

A 9×10^{-7} B 6×10^{-7} C 4×10^{-7} D 3×10^{-7} E zero

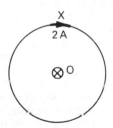

Figure 12B

Longer Questions

4 A vertical conductor X carries a downward current of 5 A.
  (a) Draw the pattern of the magnetic flux in a horizontal plane round X.
  (b) What is the flux density due to the current alone at a point P 10 cm due east of X?
  (c) If the earth's horizontal magnetic flux density has a value 4×10^{-5} T, calculate the resultant flux density at P.
    Is the resultant flux density at a point 10 cm due north of X greater *or* less than at P? Explain your answer.
5 A horizontal wire, of length 5 cm and carrying a current of 2 A, is placed in the middle of a long solenoid at right angles to its axis. The solenoid has 1000 turns per metre and carries a steady current I. Calculate I if the force on the wire is vertically downwards and equal to 10^{-4} N. Draw a diagram showing the force direction.
6 Two vertical parallel conductors X and Y are 0·12 m apart and carry currents of 2 A and 4 A respectively in a downward direction, Figure 12C (i).
  (a) Draw the resultant flux pattern between X and Y.
  (b) Ignoring the earth's magnetic field, find the distance from X of a point where the magnetic fields due to X and Y neutralise each other.

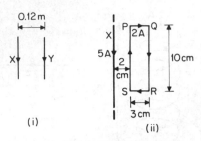

Figure 12C

7 In Figure 12C(ii), X is a very long straight conductor carrying a current of 5 A. A metal rectangle PQRS is suspended with PS 2 cm from X as shown. The dimensions of PQRS are 10 cm by 3 cm, and a current of 2 A flows in the coil. Calculate the resultant force on PQRS in magnitude and direction.

8 The magnetic flux density, B, near a long straight current-carrying conductor depends on the distance, r, from the conductor. Describe how you would investigate experimentally the relationship between B and r.

State the other factors which affect the value of B.

A long straight wire produces a field of $6.0 \, \mu T$ at a distance of 30 cm from it in vacuum. Calculate the current in the wire.

A second similar wire, carrying a current of 8.0 A, is now placed parallel to and 10 cm from the first wire. Calculate the force per unit length on the wire. The currents are in opposite directions. Is the force attractive or repulsive?

(Permeability of free space, $\mu_0 = 4\pi \times 10^{-7} \, H \, m^{-1}$.) (*L*.)

9 Two very long thin straight parallel wires each carrying a current in the same direction are separated by a distance d. With the aid of a diagram which indicates the current directions, account for the force on each wire and show on the diagram the direction of one of the forces.

Write down an expression for the magnitude of the force per unit length of wire and hence define the ampere. Why is the electrostatic force between charges ignored in the definition? ($\mu_0 = 4\pi \times 10^{-7} \, H \, m^{-1}$.) (*N*.)

10 Define the *ampere*. Write down expressions for (i) the magnitude of the flux density B at a distance of d from a very long straight conductor carrying a current I, and (ii) the mechanical force acting on a straight conductor of length l carrying a current I at right angles to a uniform magnetic field of flux density B.

Show how these two expressions may be used to deduce a formula for the force per unit length between two long straight parallel conductors *in vacuo* carrying currents I_1 and I_2 separated by a distance d.

A horizontal straight wire 5 cm long weighing $1.2 \, g \, m^{-1}$ is placed perpendicular to a uniform horizontal magnetic field of flux density 0.6 T. If the resistance of the wire is $3.8 \, \Omega \, m^{-1}$, calculate the p.d. that has to be applied between the ends of the wire to make it just self-supporting. Draw a diagram showing the direction of the field and the direction in which the current would have to flow in the wire ($g = 9.8 \, m \, s^{-2}$). (*C*.)

11 State the law of force acting on a conductor carrying an electric current in a magnetic field. Indicate the direction of the force and show how its magnitude depends on the angle between the conductor and the direction of the field.

Sketch the magnetic field due solely to two long parallel conductors carrying respectively currents of 12 A and 8 A in the same direction. If the wires are 10 cm apart, find where a third parallel wire also carrying a current must be placed so that the force experienced by it shall be zero. (*L*.)

12 (a) A long straight wire of radius a carries a steady current. Sketch a diagram showing the lines of magnetic flux density (B) near the wire and the relative directions of the current and B. Describe, with the aid of a sketch graph, how B varies along a line from the surface of the wire at right-angles to the wire.

(b) Two such identical wires R and S lie parallel in a horizontal plane, their axes being 0·10 m apart. A current of 10 A flows in R in the opposite direction to a current of 30 A in S. Neglecting the effect of the earth's magnetic flux density calculate the magnitude and state the direction of the magnetic flux density at a point P in the plane of the wires if P is   (i) midway between R and S,   (ii) 0·05 m from R and 0·15 m from S. The permeability of free space, $\mu_0 = 4\pi \times 10^{-7}\,\text{H m}^{-1}$. (N.)

13 Electromagnetic Induction

In this chapter we discuss the experiments and laws of induced e.m.f. and current due to Faraday and Lenz. We apply the laws to the straight conductor, the simple dynamo coil and the rotating disc, all moving in magnetic fields, followed by the relation between flux linkage and charge. We conclude with self and mutual induction and the variation of current and p.d. in an inductor-resistor circuit.

Plate 13A *The original Faraday apparatus*

Plate 13B *The Queen and the Duke of Edinburgh opening the Faraday Museum*

Faraday's Discovery

After Ampère and others had investigated the magnetic effect of a current Faraday tried to find its opposite. He tried to produce a current by means of a magnetic field. He began work in 1825 but did not succeed until 1831.

The apparatus with which he worked is represented in Figure 13.1. It consists of two coils of insulated wire, A, B, wound on a wooden core. One coil was connected to a sensitive meter or galvanometer, as in all his previous attempts.

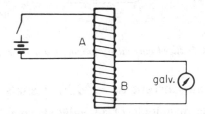

Figure 13.1 *Faraday's experiment on induction*

But when he disconnected the battery Faraday happened to notice that the galvanometer needle gave a kick. When he connected the battery back again, he noticed a kick in the opposite direction. However often he disconnected and reconnected the battery, he got the same results. The 'kicks' could hardly be all accidental—they must indicate momentary currents. Faraday had been looking for a steady current—that was why it took him six years to find it.

Conditions for Generation of Induced Current

The results of Faraday's experiments showed that a current flowed in coil B of Figure 13.1 only while the magnetic field due to coil A was *changing*—the field building up as the current in A was switched on, decaying as the current in A was switched off. And the current which flowed in B while the field was decaying was in the opposite direction to the current which flowed while the field was building up. Faraday called the current in B an *induced current*. He found that it could be made much greater by winding the two coils on an iron core, instead of a wooden one. This historic apparatus is at the Royal Institution, London.

Once he had realised that an induced current was produced only by a *change* in the magnetic field inducing it, Faraday was able to find induced currents wherever he had previously looked for them. In place of the coil A he used a magnet, and showed that as long as the coil and the magnet were at rest, there was no induced current, Figure 13.2 (i). But when he moved either the coil or the magnet an induced current flowed *as long as the motion continued*, Figure 13.2 (ii), (iii). If the current flowed one way when the north pole of the magnet was approaching the end X of the coil, it flowed the other way when the north pole was moving away from X.

Since a flow of current implies the presence of an e.m.f., Faraday's experiments showed that an e.m.f. could be induced in a coil by moving it relatively to a magnetic field, Figure 13.2 (iii). In discussing induction it is more fundamental to deal with the e.m.f. than the current, because the current depends on both the e.m.f. and the resistance.

Summarising, *relative motion* is needed between a magnet and a coil to produce induced currents. The induced current increases when the relative velocity increases, when a soft iron core is used inside the coil and when there are more turns in the coil.

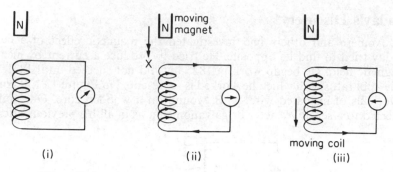

Figure 13.2 *Induced current by moving magnet or moving coil*

Direction of Induced Current or E.M.F.: Lenz's Law of Energy

Before considering the magnitude of an induced e.m.f., let us investigate its direction. To do so we must first see which way the galvanometer deflects when a current passes through it in a known direction: we can find this out with a battery and a megohm resistor, Figure 13.3 (i). We then take a coil whose direction of winding we know, and connect this to the galvanometer. In turn we push each pole of a magnet into and out of the coil; and we get the results shown in Figure 13.3 (ii), (iii), (iv) for the currents flowing in the coil.

These results were generalised into a simple rule by Lenz in 1835. He said that

---

the induced current flows always in such a direction as to oppose the change causing it.

For example, in Figure 13.3 (ii), the clockwise current flowing in the coil makes this end an S pole. So it repels the approaching S pole. In Figure 12.3 (iii), the

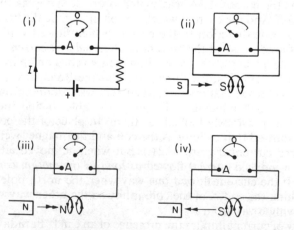

Figure 13.3 *Direction of induced currents*

induced anticlockwise current makes the end of the coil an N pole. So the approaching N pole is repelled. In Figure 13.3 (iv), the induced clockwise current in the coil now attracts the N pole moving away from it.

Lenz's law is a beautiful example of the principle of the conservation of Energy. The induced current sets up a force on the magnet, which the mover of the magnet must overcome. The work done in overcoming this force provides the electrical energy of the current. (This energy is dissipated as heat in the coil.)

If the induced current flowed in the opposite direction to that which it actually takes, then it would speed up the motion of the magnet. So the current would continuously increase the kinetic energy of the magnet. So both mechanical *and* electrical energy would be produced, without any agent having to do work. The system would be a perpetual motion machine and this is impossible. So the induced current always flows in a direction to *oppose* the motion and the electrical energy comes from the mechanical energy required to overcome the force opposing the motion. From the conservation of energy principle, the electrical energy produced = the mechanical work done.

The direction of the induced e.m.f., *E*, is the same as that of the current, as in Figure 13.4 (i). If we wished to reword Lenz's law, substituting e.m.f. for current, we would have to speak of the e.m.f.s *tending* to oppose the change ...

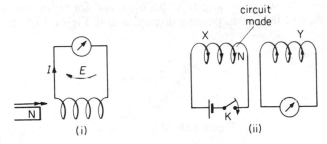

Figure 13.4 *Direction of induced e.m.f.*

etc., because there can be no opposing force unless the circuit is *closed* and a current can flow. If the terminals of a coil are *not* closed, and a flux change is made in the coil, an e.m.f. is produced between the terminals but no current flows.

In Figure 13.4 (ii), a coil X connected to a battery is placed near a coil Y. When the circuit in X is made by pressing the switch K, current in the face of the coil near Y flows anticlockwise when viewed from Y. This is similar to bringing a N-pole suddenly near Y. So the induced current in Y is anticlockwise, as shown. If the current in X is switched off, this is similar to removing a N-pole suddenly from Y. So the current in Y is now clockwise, that is, in the opposite direction to before. The induced e.m.f. in Y, which follows the direction of the current, therefore reverses when the current in X is switched on and off.

---

1   **Lenz's law follows from the Principle of the Conservation of Energy.**
2   **An induced e.m.f. is obtained when there is a CHANGE in the amount of flux linking a coil. With an open coil, this change produces an induced e.m.f. but not an induced current.**

Magnitude of E.M.F., Faraday's Law

The magnitude or size of the induced e.m.f. in a coil due to induction can be shown simply by turning the coil through 90° as shown in Figure 13.5, and noting the deflection in the meter connected to the coil. Here the plane of the coil changes from perpendicular to parallel to the magnet, so the flux linking the coil changes from maximum to zero. We find that the induced e.m.f.

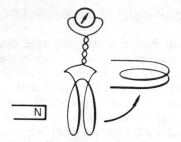

Figure 13.5 *E.m.f. induced by a turning coil*

increases with: (i) the *speed* with which we turn the coil, (ii) the *area* of the coil, (iii) the *strength of the magnetic field* (two magnets give a greater e.m.f. than one), (iv) the *number of turns* in the coil.

To generalise these results and to build up useful formulae, we use the idea of *magnetic flux*, or field lines, passing through a coil. Figure 13.6 shows a single

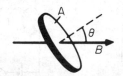

Figure 13.6 *Magnetic flux*

turn coil, of area A, whose normal makes an angle θ with a uniform field of flux density B. The component of the field at right angles to the plane of the coil is $B \cos \theta$, and we say that the magnetic flux Φ through the coil is

$$\Phi = AB \cos \theta \qquad . \qquad . \qquad . \qquad . \qquad . \qquad (1)$$

If either the strength B of the field is changed, or the coil is turned so as to change the angle θ, then the flux through the coil changes.

Results (i) to (iv) above, therefore, show that the e.m.f. induced in a coil increases with the *rate of change of the magnetic flux* through it. More accurate experiments show that the induced e.m.f. is actually *proportional* to the rate of change of flux through the coil. This result is sometimes called *Faraday's*, or *Neumann's law*: The induced e.m.f. is proportional to the *rate of change of magnetic flux linking the coil or circuit*.

The unit of magnetic flux Φ is the *weber* (Wb). So the unit of B, the flux density *or* flux per unit area, is the *weber per metre*^{2} (Wb m^{-2}) or *tesla* (T).

Flux Linkage

If a coil has more than one turn, then the flux through the whole coil is the sum of the fluxes through the individual turns. We call this the *flux linkage* through the whole coil. If the magnetic field is uniform, the flux through one turn is given, from (1), by $AB \cos \theta$. If the coil has N turns, the total flux linkage Φ is given by

$$\Phi = BAN \cos \theta \qquad . \qquad . \qquad . \qquad . \qquad . \qquad (2)$$

From Faraday's or Neumann's law, the e.m.f. induced in a coil is proportional to the rate of change of the flux linkage, Φ. Hence

$$E \propto \frac{d\Phi}{dt}$$

or
$$E = -k\frac{d\Phi}{dt} \qquad . \qquad . \qquad . \qquad . \qquad (3)$$

where k is a positive constant. The minus sign expresses Lenz's law. It means that the induced e.m.f. is in such a direction that, if the circuit is closed, the induced current *opposes* the change of flux. Note that an induced e.m.f. exists across the terminals of a coil when the flux linkage changes, even though the coil is on 'open circuit'. A current, of course, does not flow in this case.

On p. 372, it is shown that $E = -k\,d\Phi/dt$ is consistent with the expression $F = BIl$ for the force on a conductor only if $k = 1$. We may, therefore, say that

$$E = -\frac{d\Phi}{dt} \qquad . \qquad . \qquad . \qquad . \qquad (4)$$

where Φ is the flux linkage in webers, t is in seconds, and E is in volts. Equation (4) is also written as

$$E = -N\frac{d\Phi}{dt} \qquad . \qquad . \qquad . \qquad . \qquad (4A)$$

where N is the number of turns in a coil and $d\Phi/dt$ is the rate of flux change in *one* turn. We shall use equation (4), where $d\Phi/dt$ is the rate of flux change in the whole coil, and then $\Phi = BAN$ with the usual notation.

From (4), it follows that one weber is the flux linking a circuit if the induced e.m.f. is one volt when the flux is reduced to zero in one second.

1 **Lenz's law states that the induced current *opposes* the motion or change producing it.**
2 **Faraday's (Neumann's) law states that the induced e.m.f. is directly proportional to the rate of change of magnetic flux linking the circuit or coil.**

3 $\qquad\qquad\qquad\qquad E = -d\Phi/dt$

Example on e.m.f. due to Flux Change

(a) A narrow coil of 10 turns and area 4×10^{-2} m^2 is placed in a uniform magnetic field of flux density B of 10^{-2} T so that the flux links the turns normally. Calculate the average induced e.m.f. in the coil if it is removed completely from the field in 0·5 s, Figure 13.7 (i).

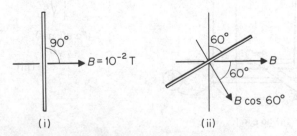

(i) (ii)

Figure 13.7 *Example*

(b) If the same coil is *rotated* about an axis through its middle so that it turns through 60° in 0·2 s in the field B, calculate the average induced e.m.f., Figure 13.7 (ii)

(a) Flux linking coil initially $= NAB = 10 \times 4 \times 10^{-2} \times 10^{-2}$

$$= 4 \times 10^{-3} \text{ Wb (Figure 13.7 (i))}$$

So average induced e.m.f. $= \dfrac{\text{flux change}}{\text{time}} = \dfrac{4 \times 10^{-3}}{0·5}$

$$= 8 \times 10^{-3} \text{ V}$$

(b) When the coil is initially perpendicular to B, flux linking coil $= NAB$, Figure 13.7 (ii). When the coil is turned through 60°, the flux density normal to the coil is now $B \cos 60°$. So

flux change through coil $= NAB - NAB \cos 60°$

$$= 4 \times 10^{-3} - 4 \times 10^{-3} \times 0·5$$

So $\qquad$ average induced e.m.f. $= \dfrac{\text{flux change}}{\text{time}} = \dfrac{2 \times 10^{-3}}{0·2}$

$$= 10^{-2} \text{ V}$$

E.M.F. Induced in Moving Straight Conductor

Generators at power stations produce high induced voltages by rotating long *straight conductors*. Figure 13.8 (i) shows a simple apparatus for demonstrating

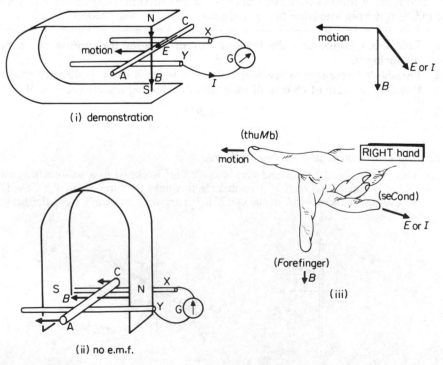

(i) demonstration

(ii) no e.m.f.

(iii)

Figure 13.8 *E.m.f. induced in moving rod. Fleming's right-hand rule*

that an e.m.f. may be induced in a straight rod or wire, when it is moved across a magnetic field. The apparatus consists of a rod AC resting on rails X and Y, and lying between the poles NS of a permanent magnet. The rails are connected to a galvanometer G.

If we move the rod to the left, so that it *cuts across* the field B of the magnet, a current I flows as shown. If we move the rod to the right, the current reverses. We notice that the current flows only while the rod is moving, and so we conclude that the motion of the rod AC induces an e.m.f. E in it.

By turning the magnet into a vertical position (Figure 13.8 (ii)) we can show that no e.m.f. is induced in the rod when it moves *parallel* to the field B. We conclude that an e.m.f. is induced in the rod only when it *cuts across* the field. And, whatever the direction of the field, no e.m.f. is induced when we slide the rod parallel to its own length. The induced e.m.f. is greatest when we move the rod at right angles, both to its own length and to the magnetic field. These results may be summarised in *Fleming's right-hand rule*:

If we place the thumb and first two fingers of the right *hand so that they are all at right angles to one another* as shown in Figure 13.8 (iii), with the *forefinger in the field direction and the thumb in the direction of motion, then the second or middle finger points in the direction of the induced current or induced e.m.f.*

Students should remember that the *right*-hand rule is used for induced current or e.m.f. but the *left*-hand rule refers to the *force* on a conductor.

To show E.M.F. ∝ Rate of Change

The variation of the magnitude of the e.m.f. in a rod with the speed of 'cutting' magnetic flux can be demonstrated with the apparatus in Figure 13.9 (i).

Here AC is a copper rod, which can be rotated by a wheel W round one pole N of a long magnet. Brush contacts at X and Y connect the rod to a galvanometer G and a series resistance R. When we turn the wheel, the rod AC cuts across the field B of the magnet, and an e.m.f. is induced in it. If we turn the wheel steadily, the galvanometer gives a steady deflection, showing that a steady current is flowing round the circuit.

To find how the current and e.m.f. depend on the speed of the rod, we keep the circuit resistance constant, and vary the rate at which we turn the wheel. We time the revolutions with a stop-watch, and find that the deflection θ is proportional to the number of revolutions per second, n, Figure 13.9 (ii). So the induced e.m.f. is directly proportional to the speed of the rod.

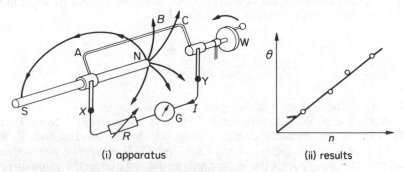

(i) apparatus   (ii) results

Figure 13.9 *Induced e.m.f. experiment*

Calculation of E.M.F. in Straight Conductor

Consider the circuit shown in Figure 13.10. PQ is a straight wire touching the two connected parallel wires QR, PS and free to move over them. All the conductors are situated in a uniform vertical magnetic field of flux density B, perpendicular to the horizontal plane of PQRS.

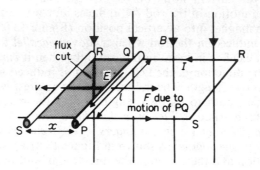

Figure 13.10 *Calculation of induced e.m.f.*

Suppose PQ, length l, moves left with uniform velocity v to a position SR. If the distance moved is x in a time t, then, as shown shaded in the diagram,

$$\text{flux cut, } \Phi = B \times \text{area PQRS} = Blx$$

So, numerically, induced e.m.f. $E = \Phi/t = Blx/t = Blv$, since velocity $v = x/t$. So

$$E = Blv$$

The induced e.m.f. E produces a current I which flows round the circuit PSRQ as shown. A *force* will now act on the wire PQ due to the current flowing and to the presence of the magnetic field. From Fleming's left-hand rule, we find that the force F acts in the *opposite* direction to the motion of the rod PQ, as shown. So, as Lenz's law states, the induced current flows in a direction so as to oppose the motion of PQ.

If the current flowing is I, and the length of PQ is l, the force on PQ is BIl. This is equal to the force moving PQ because PQ is not accelerating. From the principle of conservation of energy, work done per second by force moving PQ = electrical energy produced per second. So

$$BIl \times v = EI$$

$$\therefore Blv = E$$

as already deduced from flux changes.

It should be noted that the e.m.f. E pushes the current I from P round to Q through the external circuit PSRQ, as shown. So the end P of the rod PQ is at a *higher* potential than the end Q. If a battery was connected to the points P and Q with the same e.m.f. as the rod and in the same direction, then the positive pole of the battery would be at P and the negative pole at Q.

You should know:

When a straight conductor of length *l* moves with constant velocity *v* in a magnetic field *B*:
1   the induced e.m.f. $E = Blv$ when *l* and *v* are both 90° to *B*.
2   $E = 0$ when *l* or *v* is parallel to *B*.
3   The direction of the induced current or e.m.f. is obtained from Fleming's RIGHT hand rule—the second or middle finger is the direction.
4   If the induced e.m.f. is in the direction QP in a straight rod PQ, then P is at the higher potential.

Examples on Induced e.m.f. in Straight Conductors

1   A train travels at 30 m s^{-1} due east.

Calculate the induced e.m.f. between the ends of a horizontal axle CD of the train which is 1·5 m long, assuming the earth's magnetic field strength is 6×10^{-5} T and acts downwards at 65° to the horizontal. Which end of CD is at a higher potential? (Figure 13.11 (i).)

The induced e.m.f. along CD will be due to the *vertical* component *B* of the earth's magnetic field, since *B*, *v* and CD are at 90° to each other.

$$B = 6 \times 10^{-5} \sin 65° = 5·4 \times 10^{-5} \text{ T}$$

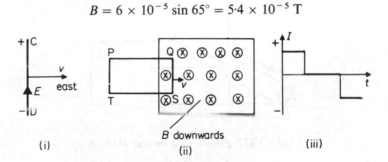

Figure 13.11 *Induced e.m.f. in straight conductors*

With the train and CD moving due east. Figure 13.11 (i),

$$\text{induced e.m.f. } E = Blv = 5·4 \times 10^{-5} \times 1·5 \times 30$$
$$= 2·4 \times 10^{-3} \text{ V}$$

Using Fleming's right-hand rule, *E* acts from D to C. So C is at the higher potential.

2   A horizontal metal frame PQST moves with uniform velocity *v* of 0·2 m s^{-1} into a uniform field *B* of 10^{-2} T acting vertically downwards, Figure 13.11 (ii). PT = 0·1 m and PQ = 0·2 m and the resistance *R* of the frame is 5 Ω. The sides QS and PT enter the field in a direction normal to the field boundary as shown.

What current flows in the metal frame when
(a) QS just enters the field,
(b) the whole frame is moving through the field,
(c) QS just moves out of the field on the other side? Draw a sketch graph showing the variation of current.

(a) The sides PQ and TS move parallel to v, so no induced e.m.f. is obtained in these sides. For the moving side QS in the field,

$$E = Blv = 10^{-2} \times 0.1 \times 0.2 = 2 \times 10^{-4} \text{ V}$$

So $\qquad I = E/R = 2 \times 10^{-4}/5 = 4 \times 10^{-5} \text{ A}$

(b) With the whole frame PQST moving through the field, the flux through PQST is constant. Since there is no flux change, no induced e.m.f. is obtained. Alternatively, the induced e.m.f. in QS and PT act in opposite directions round the frame, so their resultant e.m.f. is zero.

(c) When QS just leaves the field on the other side, the induced e.m.f. in PT moving through the field is the same value as in (a) and so the current I has the same value. But the direction of I round the frame is now *opposite* to the current in (a). So the graph of I with time t is that shown roughly in Figure 13.11 (iii).

Induced E.M.F. and Force on Moving Electrons

The e.m.f. induced in a wire moving through a magnetic field is due to the motion of electrons inside the metal, as we now explain.

When we move the wire downwards across the field B as in Figure 13.12, each electron moves downwards across the field. A downward movement of

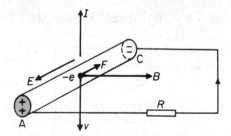

Figure 13.12 *Forces on a moving electron*

electron charge $-e$ is equivalent to an *upward* movement of positive charge or conventional current I, as shown. Applying Fleming's left-hand rule (in which the force F is at right angles to the velocity v of the wire and to B), we see that *F drives the electrons along the wire* from A to C. So if the wire is not connected in a closed circuit, electrons will pile up at C. Then the end C will gain a negative charge and A will be left with an equal positive charge. After a time the charge at C will oppose further electron movement along the wire and so the drift stops.

The charges between A and C produce an *electromotive force E*, Figure 13.12. As in a battery, A is the 'positive pole' of the wire generator and C is the 'negative pole'. So when an external resistor R is joined to A and C, the conventional current flows in it as shown.

Homopolar or Disc Generator

Another type of generator, which gives a very steady e.m.f., is illustrated in Figure 13.13 (i). It consists of a copper disc which rotates between the poles of a magnet. Connections are made from its axle X and circumference Y to a galvanometer G. We assume for simplification that the magnetic field B is uniform over the radius XY.

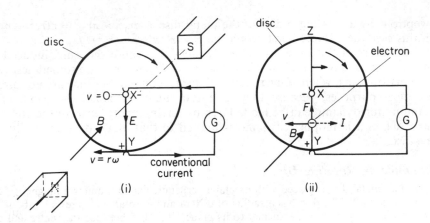

Figure 13.13 *Disc generator*

The radius XY continuously cuts the magnetic flux between the poles of the magnet. For this straight conductor, the velocity at the end X is zero and that at the other end Y is $r\omega$, where ω is the angular velocity of the disc. Since the velocity varies uniformly from X to Y,

$$\text{average velocity of XY, } v = \frac{1}{2}(0 + r\omega) = \frac{r\omega}{2}$$

Now the induced e.m.f. E in a straight conductor of length l and moving with velocity v normal to a field B is given by $E = Blv$. Since in this case $l = r$ and $v = \frac{1}{2}r\omega$, then

$$E = B \times r \times \tfrac{1}{2}r\omega = \tfrac{1}{2}Br^2\omega \qquad . \qquad . \qquad . \qquad (1)$$

As $\omega = 2\pi f$, where f is the number of revolutions per second of the disc, we can say that

$$E = B \cdot \pi r^2 \cdot f \qquad . \qquad . \qquad . \qquad . \qquad (2)$$

The direction of E is given by Fleming's *right*-hand rule. Applying the rule, we find that E acts from X to Y so that Y is at the *higher* potential, as shown.

We can understand the origin of the e.m.f. by considering an electron between X and Y, Figure 13.13 (ii). When the disc rotates, the electron moves to the left as shown. The equivalent conventional current I is then to the right. Applying Fleming's left-hand rule, we find that the force F on the electron drives it to X. So X obtains a negative charge and Y a positive charge. The radius is thus a generator with Y as its positive pole. The generator was first made by Faraday.

E.M.F. with Axle Radius

As the disc rotates clockwise, Figure 13.13 (ii), the radius XY moves to the left at the same time as the radius XZ moves to the right. If the magnetic field covered the whole disc, the induced e.m.f. in the two radii would be in *opposite* directions. So the resultant e.m.f. between the ends of the diameter YZ would be *zero*. $B \cdot \pi r^2 \cdot f$, the e.m.f. between the centre and rim of the disc, is the maximum e.m.f. which can be obtained from the dynamo.

If the disc had a radius r_1 and an axle at the centre of radius r_2, the area swept out by a rotating radius of the metal disc $= \pi r_1{}^2 - \pi r_2{}^2 = \pi(r_1{}^2 - r_2{}^2)$. In this case the induced e.m.f. would be $E = B \cdot \pi(r_1{}^2 - r_2{}^2) \cdot f$.

Generators of this kind are called *homopolar* because the e.m.f. induced in the moving conductor is always in the same direction. They are sometimes used for electroplating, where only a small voltage is required, but they are not useful for most purposes, because they give too small an e.m.f. The e.m.f. of a commutator dynamo can be made large by having many turns in the coil but the e.m.f. of a homopolar dynamo is limited to that induced in one radius of the disc.

Example on Rotating Disc

A circular metal disc is placed with its plane perpendicular to a uniform magnetic field of flux density B. The disc has a radius of 0·20 m and is rotated at 5 rev s^{-1} about an axis through its centre perpendicular to its plane. The e.m.f. between the centre and the rim of the disc is balanced by the p.d. across a 10 Ω resistor when carrying a current of 1·0 mA. Calculate B.

The induced e.m.f. $E = B \cdot \pi r^2 \cdot f = B \times \pi \times 0 \cdot 2^2 \times 5 = 0 \cdot 2\pi B$

and p.d. across 10 Ω, $V = IR = 1 \times 10^{-3} \times 10 = 10^{-2}$ V

$$\therefore 0 \cdot 2\pi B = 10^{-2}$$

$$\therefore B = 1 \cdot 6 \times 10^{-2} \text{ T}$$

The Dynamo Generator

Faraday's discovery of electromagnetic induction was the beginning of electrical engineering. Nearly all the commercial electric current used today is generated by induction, in machines which contain coils moving continuously in a magnetic field.

Figure 13.14 illustrates the principle of such a machine, which is called a *dynamo*. A coil DEFG, shown for simplicity as having only one turn, rotates on a shaft, (not shown), between the poles NS of a horseshoe magnet. The ends of the coil are connected to flat brass rings R called *slip-rings*, which are supported on the shaft by discs of insulating material, also not shown. Contact with the rings is made by small blocks of carbon H, supported on springs, and shown connected to a lamp L.

Plate 13C *This old-fashioned bicycle dynamo, used for demonstration purposes, shows the main parts of a dynamo. It is hand-cranked*

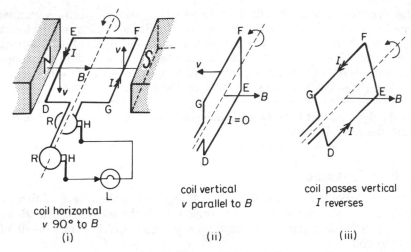

coil horizontal
v 90° to B
(i)

coil vertical
v parallel to B
(ii)

coil passes vertical
I reverses
(iii)

Figure 13.14 *Action of a simple dynamo*

As the coil rotates, the flux linking it changes, and a current I is induced in it which flows, through the carbon blocks H, to the lamp L. The magnitude (which we study shortly) and the direction of the current are not constant. So when the coil is in the position shown, the side ED is moving downwards through the lines of force, and GF is moving *upwards*. Half a revolution later, ED and GF will have interchanged their positions, and ED will be moving upwards. Consequently, applying Fleming's right-hand rule (p. 370), the current round the coil must *reverse* as ED changes from downward to upward motion, Figure 13.14 (iii). The actual direction of the current at the instant shown on the diagram is indicated by the double arrows, using Fleming's rule. By applying this rule, it can be seen that *the current reverses* every time the plane of the coil passes the vertical position. So an induced current flows continuously through the lamp L through the slip-rings joined to the two ends of the coil.

Note that when the coil is vertical, Figure 13.14 (ii), the velocity v of ED and GF are both *parallel* to the field B. So at this instant the induced current is zero.

We see shortly that the magnitude of the e.m.f. and current varies with time as shown in Figure 13.15 (i). This diagram also shows the corresponding position of DG in Figure 13.15 (ii), which should be verified by the reader.

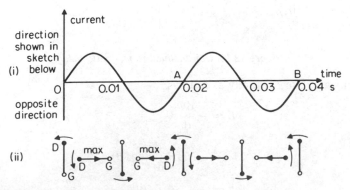

Figure 13.15 *Current generation by dynamo of Figure 13.14 plotted against time and coil position*

This type of current is called an *alternating current* (*a.c.*). A complete alternation, such as from A to B in the figure, is called a 'cycle'; and the number of cycles which the current goes through in one second is called its 'frequency'. The frequency of the current represented in the figure is that of domestic supplies in Britain—50 Hz (cycles per second). From A to B, which is one cycle, the time taken $(0·04-0·02) = 0·02$ s $= 1/50$ s. So the frequency is 50 Hz.

When the dynamo coil is horizontal, the e.m.f. is a maximum. When the coil is vertical, the e.m.f. is zero.

E.M.F. in Dynamo

We can now calculate the e.m.f. in the rotating coil. If the coil of N turns has an area A, and its normal makes an angle θ with the magnetic field B, as in Figure 13.16 (i), then the flux linkage with the coil $= NA \times$ component of B normal to coil.

So $$\Phi = NAB \cos \theta$$

Figure 13.16 (ii) shows how the flux linkage Φ varies with the angle θ starting from $\theta = 0$, when the coil is vertical (V). Since $\theta = \omega t$, then $\theta \propto t$. So the horizontal axis can also represent the time t, as indicated. When $\theta = 90°$ the coil is horizontal (H) and no flux links the coil. As the coil rotates further the flux linking the same face reverses and so Φ becomes negative as shown.

We can now find the induced e.m.f. E. If the coil turns with a steady angular velocity ω or $d\theta/dt$, then the e.m.f. induced in the coil is given by $E = -d\Phi/dt = -gradient$ of the Φ–t graph in Figure 13.16 (ii). Figure 13.16 (iii) shows the negative gradient variation found from Figure 13.16(ii). This is the variation of E with time t.

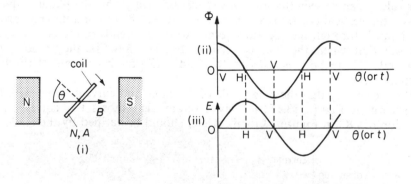

Figure 13.16 *Coil inclined to magnetic field*

We can calculate E exactly as follows.

$$E = -\frac{d\Phi}{dt}$$

$$= -NAB \frac{d}{dt}(\cos \theta)$$

$$= NAB \sin \theta \frac{d\theta}{dt} \qquad\qquad . \qquad . \qquad . \qquad . \qquad (1)$$

$d\theta/dt$ is ω, the angular velocity and $\omega = 2\pi f$ where f is the number of revolutions per second. Also, in a time t, $\theta = \omega t = 2\pi ft$. So, from (1), we can write

$$E = \omega NAB \sin \omega t \quad . \qquad . \qquad . \qquad . \qquad . \qquad (2)$$

or $$E = 2\pi fNAB \sin 2\pi ft \quad . \qquad . \qquad . \qquad . \qquad (3)$$

So the e.m.f. varies *sinusoidally* with time, that is, like a sine wave, the frequency being f cycles per second.

The maximum (peak) value or amplitude of E occurs when $\sin 2\pi ft$ reaches the value 1. If the maximum value is denoted by E_0, it follows that

$$E_0 = 2\pi fNAB$$

and $$E = E_0 \sin 2\pi ft \quad . \qquad . \qquad . \qquad . \qquad (4)$$

The e.m.f. E sends an alternating current of a similar sine equation through a resistor connected across the coil.

Dynamo E.M.F. from Energy Principles

The e.m.f. in a simple dynamo can also be found from energy principles.

At a time t, suppose the normal to the plane of the coil makes an angle θ with the field B, Figure 13.16 (i). The torque (moment of couple) acting on the coil is then given by $NABI \sin \theta$ (p. 330), where I is the current flowing if the ends of the coil are connected to an external resistor.

The work done by a torque in rotation through an angle = torque × angle of rotation (p. 119). So if the coil is rotated through a small angle $\Delta\theta$ in a time Δt,

mechanical work done per second by torque = torque × $\Delta\theta/\Delta t$

$$= NABI \sin \theta \times \omega$$

since $\omega = \Delta\theta/\Delta t$. But if E is the induced e.m.f. in the coil, the electrical energy per second generated in the coil = EI. So

$$EI = NABI \sin \theta \times \omega$$

or $$E = \omega NAB \sin \theta = \omega NAB \sin \omega t$$

This result for E agrees with the calculation using $E = -d\Phi/dt$.

A simple dynamo rotating at a constant angular velocity ω (or f rev/s) in a uniform field B has an alternating e.m.f. (a.c. voltage) given by $E = E_0 \sin \omega t$, where $E_0 =$ maximum e.m.f. $= \omega NAB = 2\pi fNAB$.

Alternators

Generators of alternating current are often called *alternators*. In all but the smallest, the magnetic field of an alternator is provided by an electromagnet called a field magnet or *field*, as shown in Figure 13.17. It has a core of cast steel, and is fed with direct current from a separate d.c. generator. The rotating coil, called the *armature*, is wound on an iron core, which is shaped so that it can turn within the pole-pieces of the field magnet. With the field magnet, the

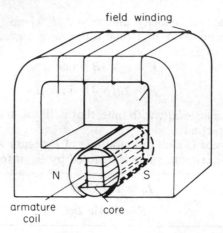

field winding

N   S

armature   core
coil

Figure 13.17 *Field magnet and armature*

armature core forms a system which is almost wholly iron, and can be strongly magnetised by a small current through the field winding. The field in which the armature turns is much stronger than if the coil had no iron core, and the e.m.f. is proportionately greater. In the small alternators used for bicycle lighting the armature is stationary, and the field is provided by permanent magnets, which rotate around it. In this way rubbing contacts, for leading the current into and out of the armature, are avoided.

When no current is being drawn from a generator, the power required to turn its armature is merely that needed to overcome friction, since no electrical energy is produced. But when a current is drawn, the power required increases, to provide the electrical power. The current, flowing through the armature winding, causes the magnetic field to set up a couple which opposes the rotation of the armature, and so demands the extra power.

Transformers

A *transformer* is a device for stepping up—or down—an alternating voltage. It has primary and secondary windings but no make-and-break, Figure 13.18. It has an iron core, which is made of thin insulated E-shaped slices of iron tightly bound so that the magnetic flux does not pass through air at all. In this way the greatest flux is obtained with a given current.

When an alternating e.m.f. E_p is connected to the primary winding, it sends an alternating current through it. This sets up an alternating flux in the core of magnitude BA, where B is the flux density and A is the cross-sectional area. This induces an alternating e.m.f. in the secondary E_s. If N_p, N_s are the number of turns in the primary and secondary coils, their linkages with the flux Φ are:

$$\Phi_p = N_p AB \quad \Phi_s = N_s AB$$

The magnitude of the e.m.f. induced in the secondary is

$$E_s = \frac{d\Phi_s}{dt} = N_s A \frac{dB}{dt}$$

The changing flux also induces a back-e.m.f. in the primary, whose

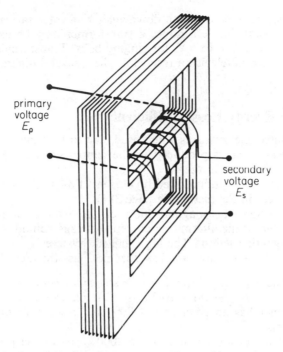

Figure 13.18 *Transformer with soft iron core (Power losses in use: (i) heat losses in coils (ii) eddy-current losses in core (reduced by laminations, p. 384), (iii) magnetic hysteresis (reduced by special iron alloys with low 'magnetic friction') (iv) magnetic flux leakage from coils.)*

magnitude is

$$E_p = \frac{d\Phi_p}{dt} = N_p A \frac{dB}{dt}$$

The voltage applied to the primary, from the source of current, is used simply in overcoming the back-e.m.f. E_p, if we neglect the resistance of the wire. Therefore, it is equal in magnitude to E_p. (This is analogous to saying, in mechanics, that action and reaction are equal and opposite.) Consequently we have

$$\frac{e.m.f. \text{ induced in secondary}}{voltage \text{ applied to primary}} = \frac{E_s}{E_p} = \frac{N_s}{N_p} \quad . \quad . \quad . \quad (1)$$

So the transformer steps voltage up or down according to its '*turns-ratio*'.

$$\frac{\textbf{secondary voltage}}{\textbf{primary voltage}} = \frac{\textbf{number of secondary turns}}{\textbf{number of primary turns}}.$$

The relation in (1) is only true when the secondary is on *open circuit*. When a load is connected to the secondary winding, a current flows in it. The power drawn from the secondary is drawn, in turn, from the supply to which the primary is connected. So now a *greater* primary current is flowing than before the secondary was loaded.

Transformers are used to step up the voltage generated at a power station from 25 000 to 400 000 volts for high-tension transmission (p. 296). After transmission they are used to step it down again to a value safer for distribution (240 volts in houses). Inside a house a transformer may be used to step the voltage down from 240 V to 6 V, for ringing bells. Transformers with several secondaries are used in television receivers, where several different voltages are required.

Transformer Energy Losses, Efficiency

Transformers have energy losses when used, due to four main causes:
(1) *Copper losses*—heat is produced in the copper coils of the primary and secondary by the current in them.
(2) *Eddy current losses*—induced (eddy) currents flow in the soft iron core due to the flux changes in the metal (see page 384).
(3) *Hysteresis losses*—the magnetisation of the iron does not follow the magnetic field due to the alternating current but lags behind it due to a form of internal 'magnetic friction'. The lag is called 'hysteresis'.
(4) *Flux leakage*—some magnetic flux does not pass through the iron core.

The transformer *efficiency* is the (*power in secondary/power in primary*) × 100% as this is (*power out/power in*) × 100%. As a simple example, suppose a 240 V mains transformer has an efficiency of 90% and is used to light normally a 12 V–36 W lamp.

The step-down transformer may have 2000 turns in the primary and 100 turns in the secondary. The turns-ratio is then $100/2000 = 1/20$, which is the ratio 12 V/240 V. Since the efficiency is 90%, the input power to the primary must be *greater* than the required 36 W for the secondary and is (100%/90%) × 36 W = $10 × 36/9 = 40$ W. The 12 V–36 W lamp has a normal secondary current $I = $ power$/V = 36$ W/12 V = 3 A. The current I in the primary = power/240 V = 40 W/240 V = 0·17 A.

If the transformer is 100% efficient, then power in secondary = power in primary. In this case power in secondary = 36 W, so current I in primary = 36 W/240 V = 0·11 A, less than the current when the efficiency was 90%.

D.C. Generators

Figure 13.19 (i) is a diagram of a *direct-current* (d.c.) generator or dynamo. Its essential difference from an alternator is that the armature winding is connected to a *commutator* instead of slip-rings.

Two commutator consists of two half-rings of copper C, D, insulated from one another, and turning with the coil. Brushes BB, with carbon tips, press against the commutator and are connected to the external circuit. The commutator is arranged so that it *reverses* the connections from the coil to the circuit at the instant when the e.m.f. reverses in the coil.

Figure 13.19 (ii) shows several positions of the coil and commutator, and the e.m.f. observed at the terminals XY. This e.m.f. varies in magnitude, but it acts always in the same way round the circuit connected to XY and so it is a varying direct e.m.f. The average value in this case can be shown to be $2/\pi$ of the maximum e.m.f. E_0, $\omega NA\beta$, of the a.c. dynamo.

In practice, as in an alternator, the armature coil is wound with insulated wire on a soft iron core, and the field magnet is energised by a current. This current is provided by the dynamo itself. The steel of the field magnet has always a small residual magnetism, so that as soon as the armature is turned an e.m.f. is induced in it. This then sends a current through the field

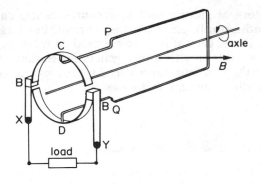

(i) principle

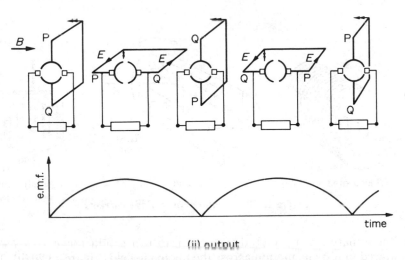

(ii) output

Figure 13.19 *D.c. generator*

winding, which increases the field and the e.m.f. The e.m.f. rapidly builds up to its working value.

Most consumers of direct current wish it to be steady, not varying as in Figure 13.19. A reasonably steady e.m.f. is given by an armature with many coils, inclined to one another, and a commutator with a correspondingly large number of segments. The coils are connected to the commutator in such a way that their e.m.f.s add round the external circuit.

Applications of Alternating and Direct Currents

Direct currents are less easy to generate than alternating currents, and alternating e.m.f.s are more convenient to step up and to step down, and to distribute over a wide area. The National Grid system in the UK, which supplies electricity to the whole country, is therefore fed with alternating current. Alternating current is just as suitable for heating as direct current, because the heating effect of a current is independent of its direction. It is also equally suitable for lighting, because filament lamps depend on the heating effect, and gas-discharge lamps—neon, sodium, mercury—run as well on alternating current as on direct.

Small motors, of the size used in vacuum-cleaners and common machine-tools, run satisfactorily on alternating current, but large ones, as a general rule, do not. Direct current is therefore used on most electric railway systems. These systems either have their own generating stations, or convert alternating current from the grid into direct current. One way of converting alternating current into direct current is to use a *rectifier*, whose principle we shall describe later.

Eddy Currents and Power Losses

The core of the armature of a dynamo is built up from thin sheets of soft iron insulated from one another by an even thinner film of oxide, as shown in Figure 13.20 (i). These are called *laminations*, and the armature is said to be laminated.

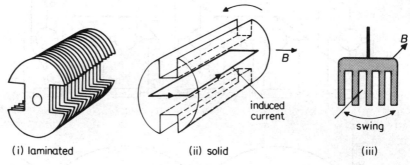

Figure 13.20 *Armature cores. Eddy currents*

If the armature were solid, then, since iron is a conductor, currents would be induced in it by its motion across the magnetic field, Figure 13.20 (ii). These currents would absorb power by opposing the rotation of the armature, and they would dissipate that power as heat, which would damage the insulation of the winding. But when the armature is laminated, these currents cannot flow, because the induced e.m.f. acts at right angles to the laminations, and therefore to the insulation between them. The magnetisation of the core, however, is not affected, because it acts *along* the laminations. So the induced currents, called *eddy currents*, are practically eliminated, while the desired e.m.f.—in the armature coil—is not.

Eddy currents, by Lenz's law, always tend to oppose the motion of a solid conductor in a magnetic field. The opposition can be shown in many ways. One of the most impressive is to make a chopper with a thick copper blade, and to try to slash it between the poles of a stronger electromagnet; then to hold it delicately and allow it to drop between them. The resistance to the motion in the case of the fast-moving chopper can be felt.

If a rectangular metal plate of aluminium is set swinging between the poles of a strong magnet, it soon comes to rest. The eddy currents circulating inside the metal oppose the motion. But if many deep slots are cut into the metal, as in a comb, the pendulum now keeps oscillating for a much longer time before coming to rest, Figure 12.20 (iii). The eddy currents are considerably reduced in this case as they cannot flow across the many air gaps formed by the slots.

Damping of Moving-Coil Meters

Sometimes eddy currents can be made use of—for example, in damping a meter. When a current is passed through the coil of an ammeter, a couple acts on the coil which sets it swinging. If the swings are opposed only by the friction of the air, they decay very slowly and are said to be naturally damped, Figure 13.21. The pointer then takes a long time to come to its final steady deflection θ.

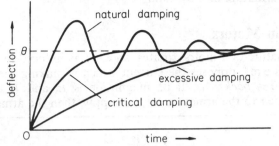

Figure 13.21 *Damping of galvanometer*

To bring the pointer more rapidly to rest, the damping must be increased. One way of increasing the damping is to wind the coil on a *metal* frame or former made of aluminium. Then, as the coil swings, the field of the permanent magnet induces eddy currents in the former, and these, by Lenz's law, oppose the motion. They therefore slow down the turning of the coil towards its eventual position, and *also stop its swings about that position*. So in the end the deflected coil comes to rest sooner than if it were not damped.

Galvanometer coils which are wound on insulating formers can be damped by short-circuiting a few of their turns, or by joining the galvanometer terminals with connecting wire so that the whole coil is short-circuited. The meter can then be carried safely from one place to another without excessive swinging of the coil, which might otherwise damage the instrument.

If the coil is overdamped, as shown in Figure 13.21, it may take almost as long to come to rest as when it is undamped. The damping which is just sufficient to prevent 'overshoot' is called 'critical' damping.

Electric Motors

If a simple direct-current dynamo, with a split-ring commutator, is connected to a battery it will run as a *motor*, Figure 13.22. Current flows round the

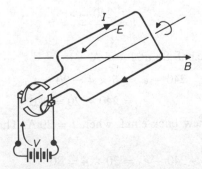

Figure 13.22 *Principle of d.c. motor*

armature coil, and the magnetic field exerts a couple on this, as in a moving-coil meter. The commutator reverses the current just as the sides of the coil are changing from upward to downward movement and vice versa. So the couple on the armature is always in the same direction, and so the shaft turns continuously.

The armature of a motor is laminated, in the same way and for the same reason, as the armature of a dynamo.

Back-e.m.f. in Motors

When the armature of a motor rotates, an e.m.f. is induced in it windings. By Lenz's law this e.m.f. opposes the current which is making the coil turn. It is therefore called a *back-e.m.f.* If its magnitude is E, and V is the potential difference applied to the armature by the supply, then the armature current is

$$I_a = \frac{V - E}{R_a} \qquad . \qquad . \qquad . \qquad . \qquad (1)$$

Here, R_a is the resistance of the armature, which is generally small—of the order of 1 ohm.

The back-e.m.f. E is proportional to the strength of the magnetic field, and to the speed of rotation of the armature. When the motor is first switched on, the back-e.m.f. is zero: it rises as the motor speeds up. In a large motor the starting current would be much too great and destroy the armature coil. To limit it, a variable resistance called a *starter resistance* is therefore put in series with the armature, and this is gradually reduced to zero by the train driver if the motor is used to drive an electric train.

When a motor is running, the back-e.m.f. E in its armature is not much less than the supply voltage V. For example, a motor running off the mains ($V = 240$ V say) might develop a back-e.m.f. $E = 230$ V. If the armature had a resistance of 1 ohm, the armature current would then be 10 A (equation (1)). When the motor was switched on, the armature current would be 240 A if no starting resistor were used.

The following example shows how the back e.m.f. is taken into account.

Example on Motor and Speed

A motor has an armature resistance of 4·0 Ω. On a 240 V supply and a light load, the motor speed is 200 rev min⁻¹ and the armature current is 5 A. Calculate the motor speed at a full load when the armature current is 20 A.

Generally,
$$I = \frac{240 - e_1}{4}$$

where e_1 is the back-e.m.f. So at $I = 5$ A,
$$240 - e_1 = 5 \times 4 = 20$$
and
$$e_1 = 240 - 20 = 220 \text{ V}$$

Suppose e_2 is the new back-e.m.f. when $I = 20$ A. Then, from the equation for I,
$$240 - e_2 = 20 \times 4 = 80$$
So
$$e_2 = 240 - 80 = 160 \text{ V}$$

Now the speed of the motor is proportional to the back-e.m.f.

So

$$\text{full load speed} = \frac{160}{220} \times 220 \text{ rev min}^{-1}$$

$$= 145 \text{ rev min}^{-1} \text{ (approx.)}$$

Back-e.m.f. and Power

If V is the supply voltage to a motor and I_a is the current, the power supplied to the motor is $I_a V$. Part of this power is dissipated in the resistance R_a of the motor coil and this is equal to $I_a^2 R$. The rest of the power is the *mechanical power* developed in the motor. As we soon see, this is the power needed to just overcome back-e.m.f.

If E_b is the back-e.m.f. in the motor, then

$$I_a = \frac{V - E_b}{R_a}$$

So

$$V = I_a R_a + E_b$$

Multiplying by I_a, then

$$I_a V = I_a^2 R + I_a E_b$$

$$\therefore I_a E_b = I_a V - I_a^2 R_a$$

So

$$I_a E_b = \textit{mechanical power} \text{ developed in motor.}$$

In the example on the motor just done, the supply voltage was 240 V and the armature current was 5 A. So the power supplied $= I_a V = 5 \times 240 = 1200$ W. The heat per second in the resistance $R_a = I_a^2 R_a = 5^2 \times 4 = 100$ W. So the mechanical power developed in the motor $= 1200 - 100 = 1100$ W.

Series- and Shunt-Wound Motors

The field winding of a motor may be connected in series or in parallel with the armature. If it is connected in series, it carries the armature current I_a, which is large, Figure 13.23. The field winding therefore has few turns of thick wire,

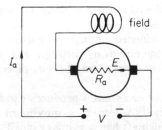

Figure 13.23 *Series-wound motor*

to keep down its resistance and so waste little power in it as heat. The few turns are enough to magnetise the iron, because the current is large.

Series motors are used where great torque is required in starting—for example, in cranes. They develop a great starting torque because the armature current flows through the field coil. At the start the armature back-e.m.f. is small, and the current is great—as great as the starting resistance will allow. The field-magnet is therefore very strongly magnetised. The torque on the armature is proportional to the field and to the armature current; since both are great at the start, the torque is very great.

If the field coil is connected in parallel with the armature, as in Figure 13.24, the motor is said to be 'shunt-wound'. The field winding has many turns of fine wire to keep down the current which it consumes.

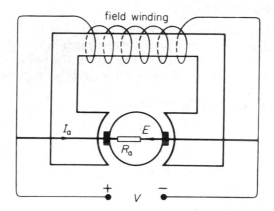

Figure 13.24 *Current and voltages in shunt-wound motor*

Shunt-wound motors are used for driving machine-tools, and in other jobs where a steady speed is required. A shunt-wound motor keeps a nearly steady speed for the following reason. If the load is increased, the speed falls a little; the back-e.m.f. then falls in proportion to the speed, and the current rises, enabling the motor to develop more power to overcome the increased load. A series motor does not keep such a steady speed as a shunt motor.

You should know:

1. An induced e.m.f. or current is produced by a *change of flux* due to relative motion between a magnet and a coil or to *flux cutting* by a straight wire.

2. Lenz's law—the induced current opposes the flux change—follows from the conservation of energy.

3. Faraday's law: the induced e.m.f. is proportional to the rate of change of the flux linking the circuit or cutting the straight conductor.

4. $E = -d\Phi/dt$. $E = Blv$ for a straight wire-direction from Fleming's *right*-hand rule.

5. Simple a.c. generator or dynamo produces e.m.f. $E = E_0 \sin 2\pi f t$ (a.c. voltage of frequency f) where E_0 = maximum e.m.f. = $2\pi f NAB$.

6. Commercial transformer: has a primary coil P and secondary coil S wound round a laminated soft-iron core. With open secondary, $E_s/E_p = N_s/N_p$. Closed secondary, (a) more current is drawn from the primary, (b) power in secondary = power in primary if 100% efficiency.

7. In a d.c. simple motor, $I = (V - E_b)/R_a$, where E_b is the back e.m.f. due to the coil cutting flux. E_b is proportional to the revs per second. Since $E_b = 0$ at the start, a *starter resistance* is needed which is gradually cut out as the speed increases.

Speedometers

Car speedometers work through the action of electromagnetic induction. A permanent magnet, which rotates with the car's wheels, is suspended inside an aluminium drum. The drum can also rotate, though it is held back by a spring so that greater rotations require greater torques on the drum. The needle of the speedometer is attached to the drum. Since aluminium is not a ferromagnetic material, we would not expect the magnetic field of the rotating magnet to directly affect the drum. However, since aluminium is a good conductor of electricity, the rotating field will induce *eddy currents* in the drum.

When we discussed eddy currents, we considered them a nuisance, since they led to power losses in the armatures of motors and dynamos. However, here we use eddy currents to our advantage. The eddy currents produce their own magnetic field, which interacts with the original magnetic field to produce a torque on the drum. So the drum (and needle) will rotate, until the torque due to the interacting fields is matched by the torque of the restraining spring. As the car moves faster, the induced currents in the drum increase. Thus the torque on the drum due to the interacting fields increases also, and the needle rotates further, indicating a higher speed.

EXERCISES 13A   Electromagnetic induction

Multiple Choice

1 In Figure 13A, the key K is depressed to close the circuit of the solenoid X. Which of the following statements **A** to **E** is correct?

   A a momentary current flows in the solenoid Y circuit,from R to Q
   B no current flows along QR   **C** the coil Y is attracted to X
   D a momentary current flows from Q to R   **E** no flux change occurs in coil Y

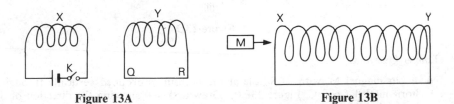

Figure 13A   Figure 13B

2 A short magnet M is moved steadily through a long solenoid from X to Y and out in Figure 13B. Which current (I)–time (t) graph from **A** to **E** would be obtained in Figure 13C (page 390)?

3 An induced e.m.f. is obtained between the ends of a horizontal steel axle X of a train moving due east. This is because

   A X points due east   **B** the earth's magnetic field has a horizontal component
   C X moves parallel to the earth's field
   D the earth has a vertical magnetic component   **E** X is a conductor

4 A 50 V d.c. motor has a coil of 0·1 Ω. At 30 rev min^{-1}, the current flowing is 5·0 A. When the current flowing is less than 5·0 A, the number of rev min^{-1} is

   A less than 30   **B** more than 30   **C** 30   **D** 10   **E** 0

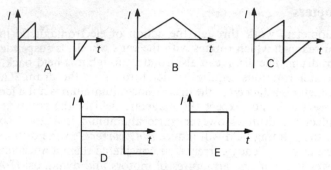

Figure 13C

5 Figure 13D (i) shows a copper disc X rotating steadily about its centre O in the uniform field between the poles N, S of a magnet. The field is 90° to the plane of the disc. The induced e.m.f. E between 0 and a point R on the rim of the disc varies with time t. Which graph in **A** to **E** shows the correct variation of E, Figure 13D (ii)?

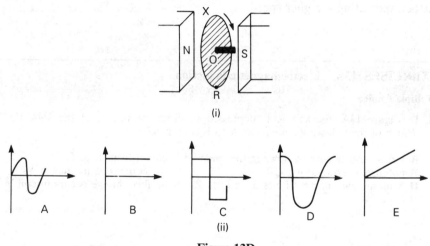

Figure 13D

Longer Questions

6 A bar magnet M, with its S pole at the bottom, is dropped vertically through a horizontal flat coil C, Figure 13E (i). Draw a sketch showing the direction of the induced current
   (a) just before M passes through C,
   (b) just after M has passed completely through C.

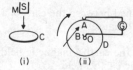

Figure 13E

7 Figure 13E (ii) shows a vertical copper disc D rotating clockwise in a uniform horizontal magnetic field B directed normally towards D. A galvanometer G is connected to contacts at O and A. The radius OA can be considered as a straight

conductor moving in the field. Copy the diagram and show in your sketch the direction of the induced current flowing through G. Has A or O the higher potential?

8 A horizontal rod PQ of length 1·5 m is perpendicular to a uniform horizontal field B of 0·1 T, Figure 13F (i). Calculate the induced e.m.f., if any, in PQ when the rod is moved through the field with a uniform velocity of 4 m s⁻¹:
   (a) in the direction of B,
   (b) perpendicular to B and upwards. Which end of PQ has the higher potential?

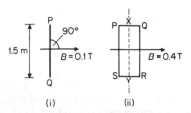

(i)   (ii)

Figure 13F

9 Figure 13F (ii) shows a vertical rectangular coil PQRS with its plane parallel to a uniform horizontal magnetic field B of 0·4 T. The coil has 5 turns, PS is 10 cm long and SR is 5 cm long. Calculate the average induced e.m.f. in the coil, if any,
   (a) when it is moved sideways in the direction of B with a velocity of 2 m s⁻¹,
   (b) when it is rotated through 90° about the vertical axis XY in 0·1 s.

10 In Figure 13F (ii) the coil PQRS is rotated about the axis XY at 50 rev s⁻¹. Calculate the maximum e.m.f. induced in the coil.
   What is the instantaneous e.m.f. in the coil when its plane is   (i) parallel to the direction of B,   (ii) 60° to B,   (iii) 90° to B?

11 Explain, using the case of a N pole of a magnet approaching a coil, why Lenz's law is a consequence of the law of the conservation of energy.
   How is Lenz's law applied to explain why an induced e.m.f. is obtained in a straight conductor cutting flux in a magnetic field?

12 State the laws of electromagnetic induction. Describe how you would demonstrate experimentally the relationship between the magnitude of the change of flux and and magnitude of the induced e.m.f.
   A long magnet is removed from the centre of a coil of 20 turns. The speed of the magnet is controlled to maintain an induced e.m.f. of 60 μV across the coil. Removing the magnet in this way takes two minutes. Calculate the change of flux through the coil.
   Explain with the aid of a diagram how you would determine the polarity of the magnet from this experiment. (L.)

13 (a) State the laws of electromagnetic induction.
   (b) A long solenoid has n turns per unit length and carries a sinusoidal current given by $I = I_0 \sin 2\pi ft$ where I_0 is the peak current and f is its frequency. A small search coil consisting of N turns of mean cross-sectional area A is positioned in the solenoid at its midpoint with its plane perpendicular to the axis of the solenoid.
      (i) The magnetic flux density, B, at the centre of the solenoid is given by $B = \mu_0 nI$, where μ_0 is a constant. Derive an expression for the flux linked with the search coil at any instant.
      (ii) Explain why an e.m.f. is induced in the search coil and derive an expression for its peak value.
   (c) For the arrangement described in (b), describe how you would attempt to verify experimentally that the magnitude of the peak induced e.m.f. is directly proportional to both the frequency and the peak value of the current in the solenoid.
   (d) An aluminium ring is suspended near to the end of a coil so that their axes coincide, Figure 13G. Explain why, in terms of the laws of electromagnetic induction, the ring is repelled when a large alternating current flows in the coil. (N.)

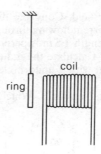

Figure 13G

14 When a current flows through a solenoid with n turns per unit length, the solenoid behaves like a magnet. Draw a diagram showing the magnetic field pattern *inside* the solenoid. When the current is switched off an e.m.f. is induced in the solenoid. Why is this? What determines how long the induced e.m.f. lasts? (*L.*)

15 Define *electromotive force* and state the *laws of electromagnetic induction*. Using the definition and the laws, derive an expression for the e.m.f. induced in a conductor moving in a magnetic field.

When a wheel with metal spokes 1·2 m long is rotated in a magnetic field of flux density 5×10^{-5} T normal to the plane of the wheel, an e.m.f. of 10^{-2} V is induced between the rim and the axle. Find the rate of rotation of the wheel. (*L.*)

16 State Lenz's law and describe how you would demonstrate it using a solenoid with two separate superimposed windings with clearly visible turns, a cell with marked polarity, and a centre-zero galvanometer. Illustrate your answer with diagrams.

A metal aircraft with a wing span of 40 m flies with a ground speed of 1000 km h^{-1} in a direction due east at constant altitude in a region of the northern hemisphere where the horizontal component of the earth's magnetic field is $1·6 \times 10^{-5}$ T and the angle of dip is 71·6°. Find the potential difference in volts that exists between the wing tips and state, with reasons, which tip is at the higher potential (*N.*)

17 State the laws relating to the electromotive force induced in a conductor which is moving in a magnetic field.

Describe the mode of action of a simple dynamo.

Find in volts the e.m.f. induced in a straight conductor of length 20 cm, on the armature of a dynamo and 10 cm from the axis when the conductor is moving in a uniform radial field of 0·5 T and the armature is rotating at 1000 r.p.m. (*L.*)

18 State Lenz's law of electromagnetic induction and describe, with explanation, an experiment which illustrates its truth.

Describe the structure of a transformer suitable for supplying 12 V from 240 V mains and explain its action. Indicate the energy losses which occur in the transformer and explain how they are reduced to a minimum.

When the primary of a transformer is connected to the a.c. mains the current in it
(a) is very small if the secondary circuit is open, but
(b) increases when the secondary circuit is closed. Explain these facts. (*L.*)

19 State the laws of electromagnetic induction and describe experiments you would perform to illustrate the factors which determine the magnitude of the induced current set up in a closed circuit.

A simple electric motor has an armature of 0·1 Ω resistance. When the motor is running on a 50 V supply the current is found to be 5 A. Explain this and show what bearing it has on the method of starting large motors. (*L.*)

20 A closed square coil consisting of a single turn of area A rotates at a constant angular speed, ω, about a horizontal axis through the midpoints of two opposite sides. The coil rotates in a uniform horizontal magnetic flux density, B, which is directed perpendicularly to the axis of rotation.
(a) Give an expression for the flux linking the coil when the normal to the plane of the coil is at an angle α to the direction of B.

(b) If at time $t = 0$ the normal to the plane of the coil is in the same direction as that of B, show that the e.m.f. E induced in the coil is given by $E = BA\omega \sin \omega t$.

(c) With the aid of a diagram, describe the positions of the coil relative to B when E is (i) a maximum (ii) zero. Explain your answer. (N.)

Self-induction

The phenomenon called *self-induction* was discovered by the American, Joseph Henry, in 1832. It is used in the *inductor*, a component widely used in communication and radio circuits.

When a current flows through a coil, it sets up a magnetic field. And that field threads the coil which produces it, Figure 13.25 (i). If the current I through

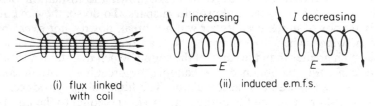

(i) flux linked with coil   (ii) induced e.m.f.s.

Figure 13.25 *Self-induction*

the coil is changed—by carrying an alternating current, for example—the flux linked with the turns of the coil changes. An e.m.f. is therefore induced in the coil. By Lenz's law the direction of the induced e.m.f. will be such as to oppose the change of current. So the e.m.f. will be against the current if it is increasing, but in the same direction if it is decreasing, Figure 13.25 (ii).

Back-e.m.f.

When an e.m.f. is induced in a circuit by a change in the current through that circuit, the e.m.f. induced is called a *back-e.m.f.* Self-induction opposes the growth of current in a coil, and so the current may increase gradually to its final value.

This effect can be demonstrated by the circuit shown in Figure 13.26 (i). Two parallel arrangements are connected to a battery B and a key K. One consists of an iron-cored coil L with many turns in series with a small lamp A_1. The other has a variable resistor R in series with a similar lamp A_2. Initially R is adjusted so that the two lamps have the same brightness in their respective circuits with steady current flowing.

With the circuit open as shown in Figure 13.26 (i), K is closed, so that B is

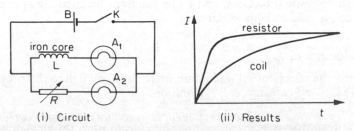

(i) Circuit   (ii) Results

Figure 13.26 *Self-induction experiment*

now connected. The lamp A_2 with R is seen to become bright almost immediately but the lamp A_1 with L increases slowly to full brightness. The induced or back-e.m.f. in the coil L *opposes* the growth of current so the glow in the lamp filament in A_1 increases slowly. The resistor R, however, has a negligible back-e.m.f. So its lamp A_2 glows fully bright as soon as K is closed, See Figure 13.26 (ii).

Induced e.m.f. across Contacts

Just as self-induction opposes the rise of an electric current when it is switched on, so also it opposes the decay of the current when it is switched off. When the circuit is broken, the current starts to fall very rapidly, so a big e.m.f. is induced, which tends to maintain the current.

This e.m.f. is often high enough to break down the insulation of the air between the switch contacts, and produce a spark. To do so, the e.m.f. must be about 350 volts or more, because air will not break down—not over any gaps, narrow or wide—when the voltage is less than that value. The e.m.f. at break may be much greater than the e.m.f. of the supply which maintained the current. A spark can easily be obtained, for example, by breaking a circuit consisting of an iron-cored coil and an accumulator. The high voltage produces a small current for a short time and the output energy (IVt) is equal to the input energy from the accumulator.

Non-inductive Coils

In some circuits containing coils, self-induction is a nuisance. To minimise their self-inductance, the coils of resistance boxes are wound so as to set up extremely small magnetic fields. As shown in Figure 13.27, the wire is doubled-back on

Figure 13.27 *Non-inductive winding*

itself before being coiled up. Every part of the coil is then travelled by the same current in opposite directions, and so its resultant magnetic field is negligible. Such a coil is said to be *non-inductive*.

Self-inductance, *L*

To discuss the effects of self-induction more fully, we define the property of a coil called its *self-inductance*. By definition,

$$self\text{-}inductance = \frac{back\text{-}e.m.f.\ induced\ in\ coil\ by\ a\ changing\ current}{rate\ of\ change\ of\ current\ through\ coil}$$

Self-inductance is denoted by the symbol L. Numerically, we may, therefore, write its definition as

$$L = \frac{E_{back}}{dI/dt}$$

or
$$E_{back} = L\frac{dI}{dt} \qquad . \qquad . \qquad . \qquad . \qquad . \qquad (1)$$

The unit of self-induction is the henry (H). *A coil has a self-inductance of 1 henry if the back-e.m.f. in it is 1 volt, when the current through it is changing at the rate of 1 ampere per second.* Equation (1) then becomes:

$$E_{back}(\text{volts}) = L(\text{henrys}) \times \frac{dI}{dt} (\text{ampere/second})$$

The iron-cored coils used for smoothing the rectified supply current to a television receiver are usually very large and have an inductance L of about 50 H or more.

An air-cored coil may have a small inductance of 0·001 H or 1 millihenry (1 mH).

resistance, R, of coil $= V/I =$ **opposition to *steady* current**

inductance, L, of coil $= E/(dI/dt) =$ **opposition to *varying* current**

L for Coil

Since the induced e.m.f. $E = d\Phi/dt = L\, dI/dt$, numerically, it follows by integration from zero that

$$\Phi = LI$$

So $L = \Phi/I$. Hence the self-inductance may be defined as the *flux linkage per unit current*. When Φ is in webers and I in amperes, then L is in henries. So if a current of 2 A produces a flux linkage of 4 Wb in a coil, the inductance $L = 4$ Wb/2 A $= 2$ H.

Earlier we saw that when a long coil of N turns and length l carries a current I, the flux density B inside the coil with an air core is given by $B = \mu_0 NI/l$, where μ_0 is the permeability of air, $4\pi \times 10^{-7}$ H m^{-1} (p. 346). With an iron core of *relative permeability* μ_r, the flux density is given by $B = \mu_r \mu_0 NI/l$. In this case

$$\text{flux linkage } \Phi = NAB = \frac{\mu_r \mu_0 N^2 AI}{l}$$

$$\therefore L = \frac{\Phi}{I} = \frac{\mu_r \mu_0 N^2 A}{l} \qquad . \qquad . \qquad . \qquad . \qquad (1)$$

This formula may be used to find the approximate value of the inductance of a coil. L is in henry when A is in metre^2, l in metres and μ_0 is $4\pi \times 10^{-7}$ henries metre^{-1}. Note that L depends on N^2, the *square* of the number of turns.

From (1), $\mu_0 = Ll/\mu_r N^2 A$. Now the unit of L is henry, H, the unit of l/A is metre^{-1}, m^{-1}, and μ_r and N^2 are numbers. So

$$\text{unit of } \mu_0 = \text{H m}^{-1}, \text{ or } \mu_0 = 4\pi \times 10^{-7} \text{ H m}^{-1}$$

Energy Stored

When the current in a coil is interrupted by breaking the circuit, a spark passes across the gap and energy is liberated in the form of heat and light. The energy has been stored in the *magnetic field of the coil*, just as the energy of a charged capacitor is stored in the electrostatic field between its plates (p. 473). When the current in the coil is first switched on, the back-e.m.f. opposes the rise of current. The current flows against the back-e.m.f. and therefore does *work* against it. When the current becomes steady, there is no back-e.m.f. and no more work done against it. The total work done in bringing the current to its final value is stored in the magnetic field of the coil. It becomes liberated when the current collapses because then the induced e.m.f. tends to maintain the current, and to do external work of some kind.

To calculate the energy stored in a coil, suppose that the current through it is rising at a rate dI/dt ampere per second. Then, if L is its self-inductance in henrys, the back-e.m.f. across it is given numerically by

$$E = L\frac{dI}{dt}$$

The total work W done against the back-e.m.f. in bringing the current from zero to a steady value I_0 is, therefore, the integral of $E \times dQ$ or $E \times I \,.\, dt$ and so

$$W = \int EI \, dt = \int_0^{I_0} LI \frac{dI}{dt} \, dt = \int_0^{I_0} LI \,.\, dI$$

So $$W = \tfrac{1}{2}LI_0{}^2$$

This is the energy stored in the magnetic field of the coil.

E.M.F. across Contacts at Break

To calculate the e.m.f. induced at break is, in general, a complicated business. But we can easily do it for one important practical circuit. To prevent sparking at the contacts of a switch in an inductive circuit, such as a relay used in telecommunications, a capacitor is often connected across the switch, Figure 13.28 (i). When the circuit is broken, the collapsing flux through the coil tends to maintain the current because the current can continue to flow for a brief time by charging the capacitor, Figure 13.28 (ii). Consequently the current does not decay as rapidly as it would without the capacitor, and the back-e.m.f. never rises as high. If the capacitance of the capacitor is great enough, the potential difference across it (and therefore across the switch) never rises high enough to cause a spark.

To find the value to which the potential difference does rise, we assume that all the energy originally stored in the magnetic field of the coil is now stored in the electrostatic field of the capacitor.

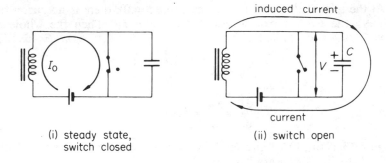

(i) steady state,   (ii) switch open
switch closed

Figure 13.28 *Prevention of sparking by capacitor*

If C is the capacitance of the capacitor in farads, and V_0 the final value of potential difference across it in volts, then the energy stored in it is $\frac{1}{2}CV_0^2$ joules (p. 253). Equating this to the original value of the energy stored in the coil, we have

$$\tfrac{1}{2}CV_0^2 = \tfrac{1}{2}LI_0^2$$

Let us suppose that a current of 1 ampere is to be broken, without sparking, in a circuit of self-inductance 1 henry and that to prevent sparking, the potential difference across the capacitor must not rise above 350 volts. The least capacitance that must be connected across the switch is, therefore, given by

$$\tfrac{1}{2}C \times 350^2 = \tfrac{1}{2} \times 1 \times 1^2$$

So
$$C = \frac{1}{350^2} = 8 \times 10^{-6}\ \mathrm{F} = 8\ \mu\mathrm{F}$$

A capacitor of capacitance $8\ \mu\mathrm{F}$, and able to withstand 350 volts, would, therefore be required.

Current in L and R Series Circuit

Consider a coil of inductance $L = 2\ \mathrm{H}$ and resistance $R = 5\ \Omega$ connected to a 10 V battery of negligible internal resistance, with a switch S in the circuit, Figure 13.29 (i).

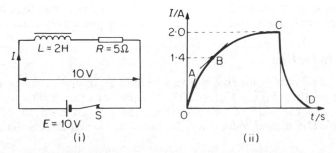

Figure 13.29 *Current variation in L, R series circuit*

When the switch is closed so that current flows, part of the 10 V is needed to maintain the current in R and the rest of the p.d. is needed to maintain the growth of the current against the back-e.m.f. E_b due to the inductance L.

1. At the instant the switch is closed (time $t = 0$), there is no current in the circuit. So there is no p.d. across R, from $V = IR$. Then the whole of the 10 V $= E_b$, the back-e.m.f. So

$$E_b = 10 = L\frac{dI}{dt} = 2\frac{dI}{dt}$$

Hence

$$\frac{dI}{dt} = \frac{10}{2} = 5 \text{ A s}^{-1}$$

In Figure 13.29 (ii), OA represents the rate of change of current with time at $t = 0$ and the line has a gradient of 5 A s^{-1}.

2. Suppose the current I rises to a value 1·4 A, which is represented by B in Figure 13.29 (ii). The p.d. across R is then given by

$$V = IR = 1·4 \times 5 = 7 \text{ V}$$

So

$$\text{back-e.m.f. } E_b = 10 - 7 = 3 \text{ V}$$

Hence

$$L\frac{dI}{dt} = 3 \quad \text{and} \quad \frac{dI}{dt} = \frac{3}{2} = 1·5 \text{ A s}^{-1}$$

So the current rise, shown by the gradient at B, decreases as time goes on.

3. When the current is finally established and is constant, there is no flux change in the coil and hence no back-e.m.f. In this case the whole of the 10 V maintains the current in R. So if I_0 is the final or steady current, $V = 10 = IR$. So

$$I_0 = \frac{V}{R} = \frac{10}{5} = 2 \text{ A}$$

This is the current value corresponding to C in Figure 13.29 (ii). The graph shows how I increases to its final value after a time t. It is an exponential graph (see below).

Generally, we see that $E = V + L\, dI/dt = IR + L\, dI/dt$. When we solve this differential equation (see *Scholarship Physics* by the author, Michael Nelkon), the result is

$$I = I_0(1 - e^{-Rt/L})$$

So the curve OC in Figure 12.30 (ii) is exponential.

When the circuit is broken, the flux in the coil decreases rapidly and the current falls quickly along CD in Figure 13.29 (ii). A spark may then be obtained across the switch due to the high voltage, as was previously explained.

Mutual Induction

We have already seen that an e.m.f. may be induced in one circuit by a changing current in another (Figure 13.1, p. 365). The phenomenon is often called *mutual induction*, and the pair of circuits which show it are said to have mutual inductance. The *mutual inductance*, M, between two circuits is defined by the equation:

$$\left. \begin{matrix} \text{e.m.f. induced in B, by} \\ \text{changing current in A} \end{matrix} \right\} = M \times \left\{ \begin{matrix} \text{rate of change of} \\ \text{current in A} \end{matrix} \right.$$

across the switch due to the high voltage, as was previously explained.

The same value of M would be obtained if we changed the current in B and observed the e.m.f. induced in A. So, from above,

$$E_B = M\,dI_A/dt \quad . \quad . \quad . \quad . \quad . \quad (1)$$

Further, since $E_B = d\Phi_B/dt$ numerically from Faraday's law, then if Φ_B is the flux change in B due to a current change I_A in A, then we can write

$$\Phi_B = MI_A \quad . \quad . \quad . \quad . \quad . \quad (2)$$

Either equation (1) or equation (2) can be used for defining or calculating M. So if 0·02 V is the induced e.m.f. in B when the current in A changes at 2 A s^{-1}, then

$$M = E_B/(dI_A/dt) = 0.02/2 = 0.01 \text{ H}$$

Also, if the flux linking a coil B is 0·04 Wb when the current in a neighbouring coil A is 2 A, then

$$M = \Phi_B/I_A = 0.04/2 = 0.02 \text{ H}$$

The greatest value of mutual inductance M occurs when the two coils A and B are wound over each other on a soft iron core, as in the case of the primary and secondary coils of a commercial (soft iron) transformer. In this case it can be shown that $M = \sqrt{L_A L_B}$, where L_A and L_B are the respective self-inductances of the separate coils A and B.

You should know:

1 **Self-inductance L of a coil opposes the rise or fall of current in the coil. The 'opposition' is due to the induced back e.m.f. due to the flux change of the varying current. High L = soft iron core, low L = air core.**
2 **$L = \Phi/I$, where Φ is the flux in the coil when I is the current. L in henrys (H) when Φ in webers (Wb) and I in A.**
   $L = E/(dI/dt)$, where E is the induced e.m.f. when the current rate of change is dI/dt. L in H when E in V and dI/dt in A s^{-1}. $E = L(dI/dt)$.
3 **Energy stored in electromagnetic field of coil $= \frac{1}{2}LI_0^2$, where I_0 is the final steady current in the coil. A spark at the break is due to the energy stored.**
4 **When a d.c. supply is connected to a coil of inductance L and resistance R, the current rises at a rate which depends on the time constant L/R. High L (50 H) slows the current appreciably**

   (a) **At $t = 0$, $dI/dt = E/L$,** (b) **final (steady) current $= E/R$.**

5 **Mutual inductance M between coils A and B $= \Phi_B/I_A$ or Φ_A/I_B.**

EXERCISES 13B Self and Mutual Induction

Multiple Choice

1 The flux linking a solenoid is 10 Wb when the steady current flowing is 2 A. If the inductance of the coil is L henries, H, then the best answer is as follows overleaf.

A $L = 5$ H and the solenoid core is air **B** $L = 5$ H and the solenoid core is iron
C $L = 0.2$ H and the solenoid core is air **D** $L = 0.02$ H and the solenoid core is air
E $L = 20$ H and the solenoid core is iron

2 In Figure 13H (i), the switch S is closed so that a current flows in the iron-core inductor L and the resistance R. When the switch is opened, a spark is obtained at the contacts. The spark is due to

A the heat produced in R **B** a slow flux change in L
C a sudden increase in the battery B e.m.f.   **D** a rapid flux change in L
E a rapid flux change in R

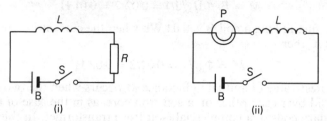

Figure 13H

3 In Figure 13H (ii), a lamp P is in series with an iron-core inductor L. When the switch S is closed, the lamp brightness rises relatively slowly to its full brightness. This is due to

A the low resistance of P   **B** the low resistance of L **C** the back-e.m.f. in L
D the high voltage of the battery B   **E** a poor connection to the lamp P

Longer Questions

4 Define *self inductance*.
A 12-V battery of negligible internal resistance is connected in series with a coil of resistance $1.0 \, \Omega$ and inductance L. When switched on the current in the circuit grows from zero. When the current is 10 A the rate of growth of the current is 500 A s^{-1}. What is the value of L? (*L.*)
5 A circuit contains an iron-cored inductor, a switch and a d.c. source arranged in series. The switch is closed and, after an interval, reopened. Explain why a spark jumps across the switch contacts.
In order to prevent sparking a capacitor is placed in parallel with the switch. The energy stored in the inductor at the instant when the circuit is broken is 2·00 J and to prevent sparking the voltage across the contacts must not exceed 400 V. Assuming there are no energy losses due to resistance in the circuit, calculate the minimum capacitance required. (*AEB.*)
6 What is meant by the statement that a solenoid has an inductance of 2 H?
A 2·0 H solenoid is connected in series with a resistor, so that the total resistance is 0·50 Ω, to a 2·0 V d.c. supply. Sketch the graph of current against time when the current is switched on. What is   (i) the final current,   (ii) the initial rate of change of current with time,   (iii) the rate of change of current with time when the current is 2·0 A?
Explain why an e.m.f. greatly in excess of 2·0 V will be produced when the current is switched off.
7 State what is meant by
(a) self induction, and
(b) mutual induction.
Describe one experiment in each case to illustrate these effects.
In the circuit shown (Figure 13I) A and B have equal ohmic resistance but A is of negligible self inductance whilst B has a high self inductance.

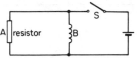

Figure 13I

Describe and explain how the currents through A and B change with time   (i) when the switch S is closed, and   (ii) when it is opened. Illustrate your answers graphically.

Describe briefly *two* applications of self inductors. (*L*.)

8 Describe the phenomena of self induction and mutual induction.

Describe the construction and explain the action of a simple form of a.c. transformer.

In an a.c. transformer in which the primary and secondary windings are perfectly coupled and in which a negligible primary current flows when there is no load in the secondary, a current of 5 A (r.m.s.) was observed to flow in the primary under an applied voltage of 100 V (r.m.s.) when the secondary was connected to resistors only. If the primary contains 100 turns and the secondary 25 000 turns, calculate

(a) the voltage

(b) the current in the secondary, stating any simplifying assumptions you make. (*O. & C.*)

9 A choke (see p. 411) of large self inductance and small resistance, a battery and a switch are connected in series. Sketch and explain a graph illustrating how the current varies with time after the switch is closed. If the self inductance and resistance of the coil are 10 H and 5 Ω respectively and the battery has an e.m.f. of 20 V and negligible resistance, what are the greatest values after the switch is closed of

(a) the current,

(b) the rate of change of current? (*N*.)

14 A.C. Circuits

A.C. circuits are needed for understanding the action and design of radio and television circuits. We start with the root-mean-square value of a.c. The single components L, C and R in a.c. circuits are then discussed, and the analogy with d.c. circuits is stressed. This is followed by series L, R and C, R circuits and the important series resonance L, C, R circuit and its application in radio reception. As we show, power in a.c. circuits depends on the phase difference between current and voltage.

Measurement of A.C.

If an alternating current (a.c.) is passed through a moving-coil meter, the pointer does not move. The coil is turned clockwise and anticlockwise at the frequency of the current—50 times per second if it is drawn from the British grid—and does not move at all. In a sensitive instrument the pointer may be seen to vibrate with a small amplitude.

The relation between current I and pointer deflection θ in a moving-coil meter is $I \propto \theta$. This is unsuitable for measuring alternating current as the deflection reverses on the negative half of the cycle. Instruments for measuring alternating currents must be so made that the pointer deflects the same way when the current flows through the instrument in either direction. As we shall see, a suitable law of deflection is $\theta \propto I^2$, a square-law deflection.

Hot-wire Instrument, Mean-square Value of Current

One type of 'square law' instrument is the hot-wire ammeter, Figure 14.1. In it the current flows through a fine resistance-wire XY, which it heats. The wire warms up to such a temperature that it loses heat—mainly by convection—at a rate equal to the average rate at which heat is developed in the wire. The rise in temperature of the wire makes it expand and sag. The sag is taken up by a second fine wire PQ, which is held taut by a spring. The wire PQ passes round

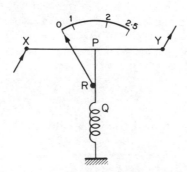

Figure 14.1 *Hot wire meter*

a pulley R attached to the pointer of the instrument, which rotates the pointer. The deflection of the pointer is roughly proportional to the average rate at which heat is developed in the wire XY. It is therefore roughly proportional to the average value of the *square* of the alternating current, and the scale is a square-law (non-uniform) one as shown.

Root-mean-square Value of Sinusoidal (Sine Wave) A.C.

Earlier we saw that an alternating current I varied sinusoidally; that is, it could be represented by the equation $I = I_m \sin \omega t$, where I_m was the peak (maximum) value of the current. In commercial practice, alternating currents are always measured and expressed in terms of their *root-mean-square* (*r.m.s.*) value.

Consider two resistors of equal resistance R, one carrying an alternating current and the other a direct current. Suppose both are dissipating the same power P, as heat. The root-mean-square (r.m.s.) value of the alternating current, I_r, is then defined as equal to the direct current, I_d. So

the root-mean-square value of an alternating current is defined as that value of steady current which would dissipate heat at the same rate in a given resistance.

Since the power dissipated by the direct current is

$$P = I_d{}^2 R$$

our definition means that, in the a.c. circuit,

$$P = I_r{}^2 R \qquad . \qquad . \qquad . \qquad . \qquad . \qquad (1)$$

Whatever the wave-form of the alternating current, if I is its value at any instant, the power which it delivers to the resistance R at that instant is $I^2 R$. Consequently, the average power P is given by

$$P = \text{average value of } (I^2 R)$$

$$= \text{average value of } (I^2) \times R$$

since R is a constant. Therefore, by equation (1), taking the average value over a cycle,

$$I_r{}^2 = \text{average value of } (I^2) \qquad . \qquad . \qquad . \qquad (2)$$

The average value of (I^2) is called the *mean-square* current. Figure 14.2 (i) shows a sinusoidal (sine variation) current I from the a.c. mains and the way its I^2 values vary. The values are *positive* on the negative half cycle. Since this graph of I^2 is a symmetrical one, the mean or average value of I^2 is $I_m{}^2/2$, where I_m is the *maximum* or *peak* value of the current. So in this case, the *root-mean-square* (r.m.s.) value, I_r, of the current is given by taking the square root of $I_r{}^2/2$ and therefore

$$I_r = \frac{I_m}{\sqrt{2}} = 0 \cdot 71 I_m \qquad . \qquad . \qquad . \qquad . \qquad (3)$$

If the r.m.s. value I_r is known, the peak value of the current I_m is calculated from

$$I_m = \sqrt{2} I_r$$

In Britain, the a.c. mains supply is 240 V (r.m.s.). So the peak or maximum

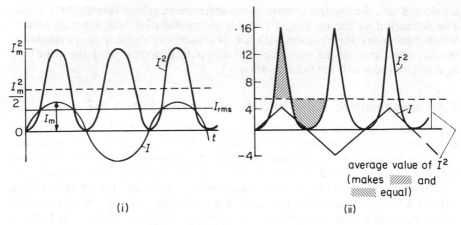

Figure 14.2 *Mean-square values*

value of the voltage is

$$V_\mathrm{m} = \sqrt{2}V_\mathrm{r} = 1{\cdot}41 \times 240 = 338 \text{ V}$$

This means that an electrical appliance is unsuitable for use on the a.c. mains if it cannot withstand a voltage of about 338 V.

Other A.C. Waveforms, Square Wave A.C.

The root-mean-square (r.m.s.) value of a varying current or voltage depends on its waveform.

Figure 14.2 (ii) shows an alternating current I which is not sinusoidal and the way its I^2 values vary with time. Unlike the sine-wave current in Figure 14.2 (i), the mean-square value is *not* half-way between the zero and the peak or maximum value. The mean-square value corresponds to the value which, for a cycle, makes the areas equal on both sides, as shown in Figure 14.2 (ii). In this case it can be seen that the mean-square value is *less* than $I_\mathrm{m}^2/2$, where I_m is the maximum value of the current. So the r.m.s. is less than $I_\mathrm{m}/\sqrt{2}$.

Figure 14.3 (i) shows one form of a *square wave* alternating current. Unlike

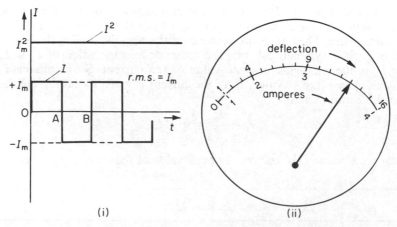

Figure 14.3 (*i*) *Square wave*   (*ii*) *Scale of a.c. ammeter*

sinusoidal a.c., this has a constant positive current I_m for half a cycle OA and a constant negative current I_m for the other half of the cycle AB. The square of the current is positive on both halves of the cycle and equal to $I_m{}^2$ throughout. So the root-mean-square value is I_m. The power delivered to a pure resistance R would therefore be $I_m{}^2 R$.

A.C. Meter and Scale

We can see that for measuring alternating current, we require a meter whose deflection measures not the current through it but the average value of the square of the current. As we have already shown, hot-wire meters have just this property.

For convenience, such meters are scaled to read amperes, not (amperes)², as in Figure 14.3 (ii). The scale reading is then proportional to the square-root of the deflection, and indicates directly the root-mean-square value of the current, I_r. An a.c. meter of the hot-wire type can be calibrated by using direct current. This follows at once from the definition of the r.m.s. value of current as the value of direct current which produces the same heat per second in a resistor.

Moving-coil meters with semiconductor diode rectifiers are widely used for measuring alternating current and voltage, as described later. They work in a different way to a hot-wire meter and give much more accurate readings of a.c. current or voltage.

A.C. and Resistor R

The mains a.c. voltage provides current and power for resistors used in electric cookers.

Figure 14.4 (i) shows the voltage V connected to a resistor R with a current I flowing and Figure 14.4 (ii) shows the waveforms of V and I. Since $I = V/R$ at all values of V, the waveform of I follows that of V. So I and V are *in phase* (in step) with each other.

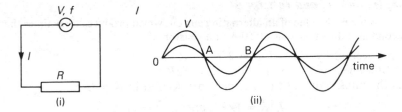

Figure 14.4 *A.C. current and voltage*

So if I_m and V_m are maximum values of I and V, then

$$I_m = \frac{V_m}{R}. \quad \text{So } \frac{I_m}{\sqrt{2}} = \frac{1}{\sqrt{2}} V_m$$

$$\therefore I_r = \frac{V_r}{R} \text{ (since r.m.s.} = 1/\sqrt{2} \times \text{peak)}$$

where I_r and V_r are r.m.s. values. So $V_r = I_r R$.

We see that the a.c. formulae for current and voltage is the same as the d.c. formulae if we use r.m.s. values *or* if we use peak (maximum) values. So 10 V peak (maximum) for an a.c. voltage connected to a 5 Ω resistor will produce a current of 2 A peak, from $I = V/R$, and a 6 V r.m.s. voltage connected to a 3 Ω resistor will produce a current of 2 A r.m.s.

Power in Resistor R

We can now find the power P in R. Since $P = IV$, we multiply the values of I and V at the same time in the two curves shown in Figure 14.4 (ii).

For the positive half cycle OA, I and V are both positive, so P is +ve. For the negative half AB of the same cycle, I and V are both negative. So their product P is again *positive*. This means that power is produced when current flows in either direction through a resistor. Figure 14.5 shows the variation of P. It has a peak value of $I_m V_m$. Since the curve is symmetrical—its shape is a sine variation—then the average power P in $R = I_m V_m/2 = (I_m/\sqrt{2}) \times (V_m/\sqrt{2}) = I_r V_r$, the product of the r.m.s. values of current and voltage. So

$$P = I_r V_r$$

for the resistor R. Since $V_r = I_r R$, we can also write

$$P = I_r^2 R = V_r^2/R$$

The formulae for a.c. power are the same as the d.c. power formulae except that *r.m.s. values* must be used for a.c. power.

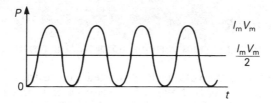

Figure 14.5 *Variation of P*

Examples on a.c. and resistor R

1 What is the peak value of an alternating current which produces three times the heat per second as a direct current of 2·0 A in a resistor R?

Heat per second by 2 A $= I^2 R = 2^2 R = 4R$
Three times heat per second $= 3 \times 4R = 12R$
If I_r is the r.m.s. value of the a.c., heat per second in $R = I_r^2 R$

So $I_r^2 R = 12R$ and $I_r^2 = 12$

So   r.m.s. value $= \sqrt{12}$

peak value $= \sqrt{2} \times$ r.m.s. $= \sqrt{2} \times \sqrt{12} = \sqrt{24} = 4·9$ A

2 An a.c. voltage of 4 V peak (maximum) is connected to a 100 Ω resistor R.
   (a) What is the phase of the current and voltage?
   (b) Calculate the current in R in mA (milliamps).
   (c) What is the power in R in mW (milliwatts)?

   (a) The current and voltage are in phase
   (b) 4 V peak $= 4$ V$/\sqrt{2}$ r.m.s.

       So $I_r = V_r/R = (4/\sqrt{2})/100 = 0·028$ A r.m.s. $= 28$ mA r.m.s.
   (c) Power $P = I_r^2 R = 0·028^2 \times 100 = 0·078$ W $= 78$ mW

A.C. with a Capacitor *C*

In many radio circuits, resistors, capacitors, and inductors or coils are present. An alternating current can flow through a resistor, but it is not obvious at first that it can flow through a capacitor. This can be demonstrated, however, by connecting a capacitor of 1 µF or more in series with a mains filament lamp of low rating such as 25 W. The lamp lights up, showing that a current is flowing through it. Direct current cannot flow through a capacitor because an insulating medium is between the plates. So with a mixture of a.c. and d.c., *only the a.c.* flows through a circuit with a capacitor.

The current flows because the capacitor plates are being continually charged, discharged, and charged the other way round by the alternating voltage of the mains, Figure 14.6 (i). So the current flows round the circuit, and can be measured by an a.c. milliammeter inserted in any of the connecting wires.

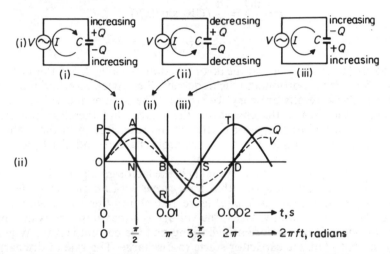

Figure 14.6 *Flow of a.c. through capacitor, frequency 50 Hz*

Figure 14.6 (ii) shows how the alternating voltage *V* varies with time, *t*. Since the charge *Q* on the capacitor plates is given at any instant by $Q = CV$, and *C* is constant, the graph of *Q* is in phase (in step) with that of *V* as shown.

The current *I* is the rate of change of *Q* with time, that is, $I = dQ/dt$. So the value of *I* at any instant is the corresponding *gradient* of the *Q–t* graph. At O, the gradient value OP is a maximum, so *I* is then a maximum. From O to A, the gradient of the *Q–t* graph decreases to zero. So *I* decreases to zero at N. From A to B, the gradient of the *Q–t* curve is *negative* and so *I* is negative from N to R. In this way we see that the *I–t* graph is PNRST.

So *I* and *V* are 90° out of phase, *with I leading V by 90°*.

If *V* is made bigger, we see that the gradient at O is bigger. So I_m increases when V_m increases. Also, if the frequency *f* of *V* is doubled, for example, so that there are now two cycles between O and B, the gradient at O becomes greater. So I_m increases when both *f* and V_m increase.

Calculation of I

To find the exact variation of I with time t, suppose the amplitude or peak of the voltage V applied to the capacitor C is V_m and its frequency is f. Then, assuming a sinusoidal voltage variation, the instantaneous voltage at any time t is

$$V = V_m \sin 2\pi ft$$

If C is the capacitance of the capacitor, then the charge Q on its plates is

$$Q = CV$$

so
$$Q = CV_m \sin 2\pi ft$$

The current, I, flowing at any instant, is equal to the rate at which charge is accumulating on the capacitor plates. Thus

$$I = \frac{dQ}{dt} = \frac{d}{dt}(CV_m \sin 2\pi ft)$$

$$= 2\pi f CV_m \cos 2\pi ft \qquad . \qquad . \qquad . \qquad . \qquad (1)$$

Equation (1) shows that the peak or maximum value I_m of the current is $2\pi f CV_m$. So I_m is proportional to the frequency, the capacitance, and the voltage amplitude. These results are easy to explain. The greater the voltage, or the capacitance, the greater the charge on the plates, and therefore the greater the current required to charge or discharge the capacitor. And the higher the frequency, the more rapidly is the capacitor charged and discharged, and therefore again the greater is the current.

A more puzzling feature of equation (1) is the factor giving the time variation of the current, $\cos 2\pi ft$. It shows that *the current varies a quarter-cycle or 90° ($\pi/2$) out of phase with the voltage*. Figure 14.6 shows this variation, and also helps to explain it physically. When the voltage is a maximum, so is the charge on the capacitor. It is therefore not charging and the current is zero. When the voltage starts to fall, the capacitor starts to discharge. The rate of discharging, or current, reaches its maximum when the capacitor is completely discharged and the voltage across it is zero. Since the current I passes its maximum a quarter-cycle ahead of the voltage V, we see that I leads V by 90° ($\pi/2$).

For a capacitor C, I leads V by 90°

Reactance of C

The *reactance* of a capacitor is its opposition in ohms to the passage of alternating current. We do not use the term 'resistance' in this case because this is the opposition to direct current.

The reactance, symbol X_C, is defined by

$$X_C = \frac{V_m}{I_m}$$

---

where V_m and I_m are the peak or maximum of the a.c. voltage and current. Since the ratio $V_m/I_m = V_r/I_r$, where V_r and I_r are the r.m.s. voltage and current respectively, we can also define reactance X_C by the ratio V_r/I_r. We shall omit the suffix r when using r.m.s. values and so

$$X_C = \frac{V}{I}$$

Here X_C is in ohms when V is in volts (r.m.s.) and I is in amperes (r.m.s.). As we have just seen, the amplitude or peak value of the current through a capacitor is given by

$$I_m = 2\pi f C V_m$$

The reactance of the capacitor is therefore

$$X_C = \frac{V_m}{I_m} = \frac{1}{2\pi f C} = \frac{V}{I}$$

X_C is in ohms when f is in Hz (cycles per second), and C in farads.

For convenience we often write $\omega = 2\pi f$. The quantity ω is called the *angular frequency* of the current and voltage. It is expressed in radians per second. Then an alternating voltage, for example, may be written as

$$V = V_m \sin \omega t$$

and the reactance of a capacitor can therefore also be written as

$$X_C = \frac{1}{\omega C}$$

Calculations of Reactance X_C

As an illustration, suppose a capacitor C of $0.1\ \mu F$ is used on the mains frequency of 50 Hz. Then the reactance is

$$X_C = \frac{1}{2\pi f C} = \frac{1}{2 \times 3.14 \times 50 \times 0.1 \times 10^{-6}}$$

$$= \frac{10^6}{2 \times 3.14 \times 50 \times 0.1} = 32\,000\ \Omega \text{ (approx.)}$$

From the formula for reactance we note that $X_C \propto 1/C$ for a given frequency. So if a 1 μF capacitor is used on the 50 Hz mains, its reactance is 10 times *less* than that of $0.1\ \mu F$, which is 32 000 Ω/10 or 3200 Ω. Figure 14.7 (ii) shows how X_C varies with C.

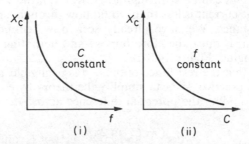

Figure 14.7 *Reactance of capacitor*

Also, since $X_C \propto 1/f$ for a given capacitor, at $f = 1000$ Hz a capacitor of 1 μF has 20 times *less* reactance than at $f = 50$ Hz. So $X_C = 32\,000\,\Omega/20 = 160\,\Omega$. See Figure 14.7 (i).

Since $X_C = V/I$, where V and I are both r.m.s. (or peak) values, we can see that

$$I = \frac{V}{X_C} \quad \text{and} \quad V = IX_C$$

These are similar formulae to d.c. circuit formulae. The difference with d.c. circuits is that we must always consider the phase difference between V and I, and V and I are 90° out of phase as we have previously shown.

Example on Capacitor Reactance

A capacitor C of 1 μF is used in a radio circuit where the frequency is 1000 Hz and the current flowing is 2 mA (r.m.s.). Calculate the voltage across C.

What current flows when an a.c. voltage of 20 V r.m.s., $f = 50$ Hz is connected to this capacitor?

(i) Reactance, $X_C = \dfrac{1}{2\pi f C} = \dfrac{1}{2\pi \times 1000 \times 10^{-6}} = 159\,\Omega$ (approx.)

$$\therefore V = IX_C = \frac{2}{1000} \times 159 = 0.32 \text{ V (approx.)}$$

(ii) When 20 V r.m.s., $f = 50$ Hz, is connected to C, the reactance of C changes. Since $X_C \propto 1/f$,

$$X_C \text{ at } f = 50 \text{ Hz is 20 times } X_C \text{ at } f = 1000 \text{ Hz}$$

So

$$X_C = 20 \times 159\,\Omega = 3180\,\Omega$$

$$\therefore I = \frac{V}{X_C} = \frac{20}{3180} = 6.3 \times 10^{-3} \text{ A r.m.s.}$$

For a capacitor C   *V lags on I by 90°*

$$X_C = \frac{1}{\omega C} = \frac{1}{2\pi f C}$$

A.C. through an Inductor

Since a coil is made from conducting wire, we have no difficulty in seeing that an alternating current can flow through it. However, if the coil has appreciable self-inductance, the current is less than would flow through a non-inductive coil of the same resistance. We have already seen how self-inductance opposes changes of current; it must therefore oppose an alternating current, which is continuously changing, Figure 14.8 (i).

Let us suppose that the resistance of the coil is negligible, a condition which can be satisfied in practice. We can simplify the theory by considering first the current, and then finding the potential difference across the coil. Suppose the current is

$$I = I_m \sin 2\pi f t \quad . \qquad . \qquad . \qquad . \qquad (1)$$

where I_m is its peak value, Figure 14.8 (ii). If L is the inductance of the coil, the

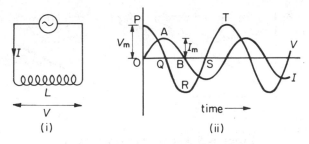

Figure 14.8 *Flow of a.c. through a coil*

changing current sets up a back-e.m.f. in the coil, of magnitude

$$E = L \frac{dI}{dt}$$

To maintain the current, the applied supply voltage must be equal to the back-e.m.f. The voltage applied to the coil must therefore be given by

$$V = L \frac{dI}{dt}$$

Since L is constant, it follows that V is proportional to dI/dt.

Figure 14.8 (ii) shows how the current I varies with time t. The values of dI/dt are the *gradients* of the I–t graph at the time concerned. At O the gradient is a maximum; so the maximum voltage of V, or V_m, occurs at O and is represented by OP as shown. From O to A, the gradient of the I–t graph decreases to zero. So the voltage V decreases from P to Q. From A to B the gradient of the I–t graph is *negative* (downward slope). So the voltage decreases along QR. We now see that *V leads I by 90° ($\pi/2$)*.

We can find the value of V_m from the equation $V = L\, dI/dt$. From (1),

$$I = I_m \sin 2\pi ft$$

So, by differentiation with respect to t,

$$V = L \frac{dI}{dt}$$

$$= L \frac{d}{dt} (I_m \sin 2\pi ft) = 2\pi f L I_m \cos 2\pi ft$$

So $\qquad V_m$ = maximum (peak) voltage = $2\pi f L I_m$

Hence the reactance of the inductor is

$$X_L = \frac{V_m}{I_m} = 2\pi f L = \frac{V}{I}$$

X_L is in ohms when f is in Hz, and L is in henrys (H). An *iron-cored* coil has a high inductance L such as 20 H. Used on the mains frequency f of 50 Hz, its reactance $X_L = 2\pi f L = 2 \times 3{\cdot}14 \times 50 \times 20 = 6280\ \Omega$. Since this type of inductor provides a high reactance, it is sometimes called a 'choke'.

Since $X_L = 2\pi f L$, it follows that $X_L \propto f$ for a given inductance, Figure 14.9 (i), and that $X_L \propto L$ for a given frequency, Figure 14.9 (ii).

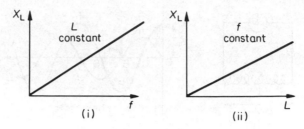

Figure 14.9 *Variation of reactance X_L*

Example on Reactance X_L

An inductor of 2 H and negligible resistance is connected to a 12 V (r.m.s.) mains supply, $f = 50$ Hz. Find the current flowing. What current flows when the inductance is changed to 6 H?

$$\text{Reactance, } X_L = 2\pi f L = 2\pi \times 50 \times 2 = 628\ \Omega$$

$$\therefore I = \frac{V}{X_L} = \frac{12}{628}\ \text{A} = 19\ \text{mA (r.m.s.) (approx.)}$$

When the inductance is increased to 6 H, its reactance X_L is increased 3 times since $X_L \propto L$ for a given frequency. So the current is reduced to 1/3 of its value. So now $I = 6$ mA (approx.).

For an inductor L, V leads on I by 90°

$$X_L = \omega L = 2\pi f L$$

Phasor Diagrams

In the Mechanics section of this book, it is shown that a quantity which varies sinusoidally with time may be represented as the projection of a rotating vector (p. 71). These quantities are called *phasors*, as the phase angle must also be represented. Alternating currents and voltages may therefore be represented as phasors. Figure 14.10 shows, on the left, the phasors representing the current

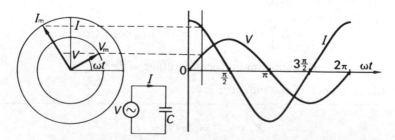

Figure 14.10 *Phasor diagram for capacitor*

through a capacitor, and the voltage across it. Since the current leads the voltage by $\pi/2$, the current vector I is displaced by 90° ahead of the voltage vector V.

Figure 14.11 shows the phasor diagram for a pure inductor. In drawing it, the voltage has been taken as $V = V_m \sin \omega t$, and the current drawn lagging $\pi/2$ behind it. This enables the diagram to be readily compared with that for

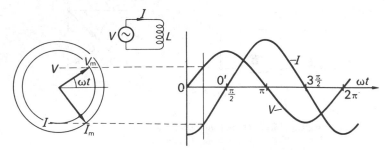

Figure 14.11 *Phasor diagram for pure inductance*

a capacitor. To show that it is essentially the same as Figure 14.8 (ii), we have only to shift the origin by $\pi/2$ to the right, from 0 to 0′.

When an alternating voltage is connected to a pure resistance R, the current I at any instant $= V/R$, where V is the voltage at that instant. So I is zero when V is zero and I is a maximum when V is a maximum. Hence I and V are in phase, Figure 14.12 (i). Since the phase angle is zero, we draw the vector or phasor I in the same direction as V, where the phasors represent either peak values or r.m.s. values.

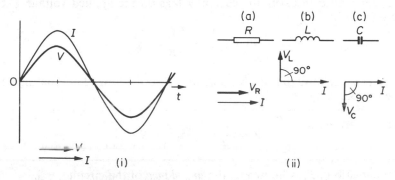

Figure 14.12 *Phasor diagrams for R, L, C*

Figure 14.12 (ii) summarises the phasor diagrams for
(a) a pure resistance R
(b) a pure inductance L and
(c) a pure capacitance C.
They should be memorised by the reader for use in all kinds of a.c. circuits.

Series Circuits

L and *R* in Series

Consider an inductor L in series with resistance R, with an alternating voltage V (r.m.s.) of frequency f connected across both components, Figure 14.13 (i).

The sum of the respective voltages V_L and V_R across L and R is equal to V. But the voltage V_L leads by 90° on the current I, and the voltage V_R is in phase with I (see above). Thus the two voltages can be drawn to scale as shown in Figure 14.13 (ii), and hence, by Pythagoras' theorem, it follows that the vector

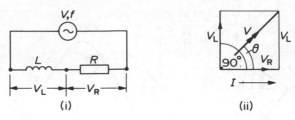

Figure 14.13 *Inductance and resistance in series*

sum V is given by

$$V^2 = V_L^2 + V_R^2$$

But $V_L = IX_L$, $V_R = IR$.

$$\therefore V^2 = I^2 X_L^2 + I^2 R^2 = I^2(X_L^2 + R^2)$$

$$\therefore I = \frac{V}{\sqrt{X_L^2 + R^2}} \qquad \cdots \qquad (1)$$

Also, from Figure 14.13 (ii), the current I lags on the applied voltage V by an angle θ given by

$$\tan \theta = \frac{V_L}{V_R} = \frac{IX_L}{IR} = \frac{X_L}{R} \qquad \cdots \qquad (2)$$

From (1), it follows that the 'opposition' Z to the flow of alternating current is given in ohms by

$$Z = \frac{V}{I} = \sqrt{X_L^2 + R^2} \qquad \cdots \qquad (2)$$

This 'opposition', Z, is known as the *impedance* of the circuit.

Example on L and R in Series

An iron-cored coil of 2 H and 50 Ω resistance is placed in series with a resistor of 450 Ω, and a 100 V, 50 Hz, a.c. supply is connected across the arrangement. Find
(a) the current flowing in the coil,
(b) its phase angle relative to the voltage supply,
(c) the voltage across the coil.

(a) The reactance $X_L = 2\pi f L = 2\pi \times 50 \times 2 = 628$ Ω.
Total resistance $R = 50 + 450 = 500$ Ω.

$$\therefore \text{ circuit impedance } Z = \sqrt{X_L^2 + R^2} = \sqrt{628^2 + 500^2} = 803 \text{ Ω}$$

$$\therefore I = \frac{V}{Z} = \frac{100}{803} \text{A} = 12 \cdot 5 \text{ mA (approx.)}$$

(b)
$$\tan \theta = \frac{X_L}{R} = \frac{628}{500} = 1 \cdot 256$$

So
$$\theta = 515 \cdot 5°$$

(c) For the coil, $X_L = 628\ \Omega$ and $R = 50\ \Omega$

So coil impedance $Z = \sqrt{X_L{}^2 + R^2} = \sqrt{628^2 + 50^2} = 630\ \Omega$

Thus voltage across coil $V = IZ = 12{\cdot}5 \times 10^{-3} \times 630$

$$= 7{\cdot}9\ V\ (\text{approx.})$$

C and *R* in Series

A similar analysis enables the impedance to be found of a capacitance *C* and resistance *R* in series, Figure 14.14 (i). In this case the voltage V_C across the capacitor lags by 90° on the current *I* (see p. 407), and the voltage V_R across the resistance is in phase with the current *I*. As the vector sum is *V*, the applied voltage, it follows by Pythagoras' theorem that

$$V^2 = V_C{}^2 + V_R{}^2 = I^2 X_C{}^2 + I^2 R^2 = I^2(X_C{}^2 + R^2)$$

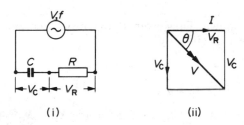

(i) (ii)

Figure 14.14 *Capacitance and resistance in series*

$$\therefore I = \frac{V}{\sqrt{X_C{}^2 + R^2}} . \qquad \qquad (1)$$

Also, from Figure 14.14 (ii), the current *I* leads on *V* by an angle θ given by

$$\tan \theta = \frac{V_C}{V_R} = \frac{IX_C}{IR} = \frac{X_C}{R} \qquad \qquad (2)$$

It follows from (1) that the impedance *Z* of the *C–R* series circuit is

$$Z = \frac{V}{I} = \sqrt{X_C{}^2 + R^2}$$

It should be noted that although the impedance formula for a *C–R* series circuit is of the same mathematical form as that for a *L–R* series circuit, the current in the *C–R* series case *leads* on the applied voltage but the current in the *L–R* series case *lags* on the applied voltage.

For *L–R* series:

> **impedance $Z = \sqrt{X_L{}^2 + R^2}$, $\tan \theta = X_L/R$**

For *C–R* series:

> **impedance $Z = \sqrt{X_C{}^2 + R^2}$, $\tan \theta = X_C/R$**

L, C, R in Series

The most general series circuit is the case of *L, C, R* in series, Figure 14.15 (i). As we see later, it is widely used in radio. The phasor diagram has V_L leading by 90° on V_R, V_C lagging by 90° on V_R, with the current I in phase with V_R, Figure 14.15 (ii). If V_L is greater than V_C, their resultant is $(V_L - V_C)$ in the

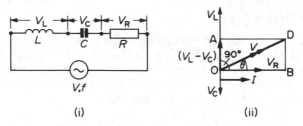

(i) (ii)

Figure 14.15 *L, C, R in series*

direction of V_L, as shown. Thus, from Pythagoras' theorem for triangle ODB, the applied voltage V is given by

$$V^2 = (V_L - V_C)^2 + V_R^2$$

But $V_L = IX_L$, $V_C = IX_C$, $V_R = IR$.

$$\therefore V^2 = (IX_L - IX_C)^2 + I^2R^2 = I^2[(X_L - X_C)^2 + R^2]$$

$$\therefore I = \frac{V}{\sqrt{(X_L - X_C)^2 + R^2}} \qquad . \qquad . \qquad . \qquad . \qquad (1)$$

Also, I lags on V by an angle θ given by

$$\tan \theta = \frac{DB}{OB} = \frac{V_L - V_C}{V_R} = \frac{IX_L - IX_C}{IR} = \frac{X_L - X_C}{R} \qquad . \qquad . \qquad (2)$$

Resonance in the *L, C, R* Series Circuit

From (1), it follows that the impedance Z of the circuit is given by

$$Z = \sqrt{(X_L - X_C)^2 + R^2}$$

The impedance varies as the frequency, f, of the applied voltage varies, because X_L and X_C both vary with frequency. Since $X_L = 2\pi fL$, then $X_L \propto f$, and so the variation of X_L with frequency is a straight line passing through the origin, Figure 14.16 (i). Also, since $X_C = 1/2\pi fC$, then $X_C \propto 1/f$, and so the variation of X_C with frequency is a curve approaching the two axes, Figure 14.16 (i). The resistance R is independent of frequency, and so it is represented by a line parallel to the frequency axis. The difference $(X_L - X_C)$ is represented by the broken lines shown in Figure 14.16 (i), and it can be seen that $(X_L - X_C)$ decreases to zero for a particular frequency f_0, and thereafter increases again. So, from $Z = \sqrt{(X_L - X_C)^2 + R^2}$, the impedance diminishes to A and then increases as the frequency f is varied.

The variation of Z with f is shown in Figure 14.16 (i), and since the current $I = V/Z$, the current varies as shown in Figure 14.16 (ii). So the current has a

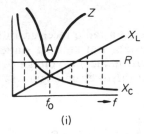 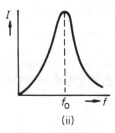

(i)   (ii)

Figure 14.16 *Resonance curves*

maximum value at the frequency f_0, and this is known as the *resonant frequency* of the circuit.

The magnitude of f_0 is given by $X_L - X_C = 0$, or $X_L = X_C$.

$$\therefore 2\pi f_0 L = \frac{1}{2\pi f_0 C} \quad \text{or} \quad 4\pi^2 LC f_0^2 = 1$$

$$\therefore f_0 = \frac{1}{2\pi\sqrt{LC}}$$

At frequencies above and below the resonant frequency, the current is less than the maximum current, see Figure 14.16 (ii), and the phenomenon is thus basically the same as the forced and resonant vibrations obtained in Sound or Mechanics.

At resonance: (1) $f_0 = 1/2\pi\sqrt{LC}$ (2) $X_L = X_C$ (3) **impedance $Z = R$**
(4) **maximum current $I = V/R$**   (5) **I and V are in phase**

Analogy with Period of Spring–Mass System

The electrical resonance of the inductor (L)–capacitance (C) system is analogous to the mechanical resonance of the spring–mass system which we met in the simple harmonic section.

The magnetic energy in an inductor carrying a current I is $\frac{1}{2}LI^2$. The kinetic energy of a mass m moving with speed v is $\frac{1}{2}mv^2$. We can think of I, the current, which is dQ/dt, as analogous to v, the speed, which is dx/dt; so *L is analogous to m*. In fact, L and m each represent a form of 'inertia'; L to a change of current in a coil and m to a change of motion.

The electric energy stored in a capacitor C is $\frac{1}{2}Q^2/C$, where Q is the charge. The molecular potential energy stored in a spring is $\frac{1}{2}kx^2$, where k is the spring constant and x is the extension. With Q analogous to x, then *1/C is analogous to k*.

In simple harmonic motion, we showed that the frequency f of the spring–mass system $= (1/2\pi)\sqrt{k/m}$. Substituting $1/C$ for k and L for m, we obtain

$$f = \frac{1}{2\pi\sqrt{LC}}$$

for electrical resonance, as we have already shown.

Tuning in Radio Receivers

The series resonance circuit is used for tuning a radio receiver. In this case the incoming waves of frequency f from a distant transmitting station induces a varying voltage in the aerial, which in turn induces a voltage V of the same frequency in a coil and capacitor circuit in the receiver, Figure 14.17 (i). When

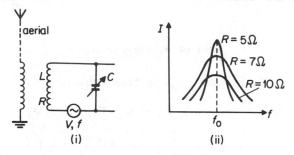

Figure 14.17 *Tuning a receiver*

the capacitance C is varied the resonant frequency is changed; and at one setting of C the resonant frequency becomes f, the frequency of the incoming waves. The maximum current is then obtained, and the station is now heard very loudly.

The sharpness of the resonance is an important matter in radio reception of transmitting stations. If it is not sharp, other transmitting stations may produce a current I or 'response' in the circuit of about the same value as the station required. Considerable 'interference' then occurs. If the resistance R in the L, C, R series is small, then the resonance is sharp, as illustrated roughly in Figure 14.17 (ii). An inductor (L, R) and a capacitor (C) in series form an L, C, R series circuit, and the resistance R is low if the inductor coil is made with the *minimum* wire needed. In this case the resonance is sharp.

Parallel Circuits

We now consider briefly the principles of a.c. parallel circuits. In d.c. parallel circuits, the currents in the individual branches are added arithmetically to find their total. In a.c. circuits, however, we add the currents by vector methods, taking into account the phase angle between them.

L, R in Parallel

In the parallel circuit in Figure 14.18 (i), the supply current I is the vector sum of I_L and I_R, Figure 14.18 (ii) shows the vector addition. I_L is 90° out of phase with I_R, since I_R is in phase with V, and I_L lags 90° behind V. So

$$I^2 = I_R{}^2 + I_L{}^2 = \left(\frac{V}{R}\right)^2 + \left(\frac{V}{X_L}\right)^2$$

Then
$$I = V\sqrt{\frac{1}{R^2} + \frac{1}{X_L{}^2}}$$

Also, from Figure 14.18 (ii), I lags behind the applied voltage V by an angle

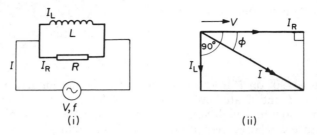

Figure 14.18 *L, R in parallel*

φ given by

$$\tan \varphi = \frac{I_L}{I_R} = \frac{R}{X_L}$$

Similar analysis shows that when an a.c. voltage V r.m.s. is applied to a parallel C, R circuit, the supply I is given by

$$I = V \sqrt{\frac{1}{R^2} + \frac{1}{X_C^2}} \quad \text{and} \quad \tan \varphi = \frac{R}{X_C}$$

where I now leads V by the angle φ.

L, C in Parallel, Coil–Capacitor Resonance

A parallel arrangement of coil (L, R) and capacitor (C) is widely used in transistor oscillators and in radio-frequency amplifier circuits. To simplify matters, let us assume that the resistance of the coil is negligible compared with its reactance. We then have effectively an inductor L in parallel with a capacitor C, Figure 14.19 (i).

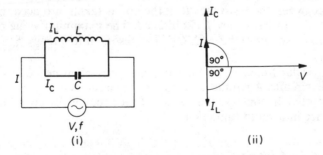

Figure 14.19 *L, C in parallel*

Figure 14.19 (ii) shows the two currents in the components. I_L lags by 90° on V but I_C leads by 90° on V. If I_C is greater than I_L at the particular frequency f, then

$$I = I_C - I_L = \frac{V}{X_C} - \frac{V}{X_L}$$

Since I leads by 90° on V in this case, we say that the circuit is 'net capacitive'.

If, however, I_L is greater than I_C, then

$$I = I_L - I_C = \frac{V}{X_L} - \frac{V}{X_C}$$

Since I lags by 90° on V in this case, the circuit is 'net inductive'.

Suppose V, L and C are kept constant and the frequency f of the supply is varied from a low to a high value. The magnitude and phase of I then varies according to the relative magnitudes of $X_L (2\pi f L)$ and $X_C (1/2\pi f C)$, as shown in Figure 14.20 (i). A special case occurs when $X_L = X_C$. Then $I_L = I_C$ and so $I = 0$, Figure 14.20 (ii). At this frequency f_0, we have $X_L = X_C$, so

$$2\pi f_0 L = \frac{1}{2\pi f_0 C}, \quad \text{or} \quad f_0 = \frac{1}{2\pi\sqrt{LC}}$$

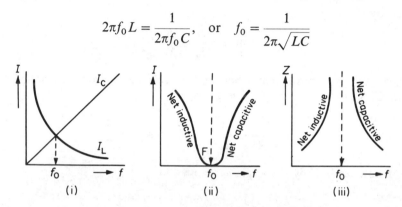

Figure 14.20 *L, C in parallel—variation of Z and I*

Figure 14.20 (ii) shows how the current I varies with the frequency f. Figure 14.20 (iii) shows how the *impedance* Z of the parallel L, C circuit varies with frequency f. Since $Z = V/I$, and I is zero at the frequency f_0, it follows that Z is infinitely high at f_0. The parallel inductor–capacitor circuit has therefore a *resonant frequency* (f_0) as far as its impedance Z is concerned.

In practice, when the resistance R of the coil is taken into account, a similar variation of Z with frequency f is obtained. The maximum value of Z is now finite and theory shows that $Z = L/CR$. The resonant frequency f_0 is practically still given by $f_0 = 1/2\pi\sqrt{LC}$.

For a particular frequency, a high impedance is often needed as a 'load' in certain radio circuits. A parallel coil–capacitor circuit is then used, tuned to the frequency wanted. In contrast, the *series L, C, R* circuit gives a maximum *current* I at resonance in a radio tuning circuit.

Power in A.C. Circuits

Resistance R. The power absorbed is usually $P = IV$. In the case of a resistance, $V = IR$, and $P = I^2 R$. The variation of power is shown in Figure 14.21 (i), where I_m = the peak (maximum) value of the current. On p. 403 we explained the reason for choosing the root-mean-square value of alternating current. So the average power absorbed in R, as we also showed earlier, is given by

$$P = I^2 R$$

where I is the r.m.s. value.

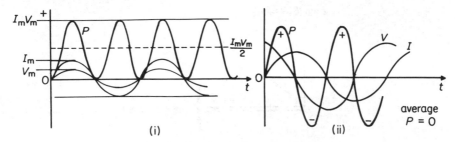

Figure 14.21 *Power in R, L and C*

The power *P* in a resistor can also be written as

$$P = IV = \frac{I_m V_m}{2}$$

as shown in Figure 14.21 (i).

Inductance L. In the case of a pure inductor, the voltage *V* across it leads by 90° on the current *I*. So if $I = I_m \sin \omega t$, then $V = V_m \sin (90° + \omega t) = V_m \cos \omega t$. Hence, at any instant,

power absorbed $= IV = I_m V_m \sin \omega t . \cos \omega t = \frac{1}{2} I_m V_m \sin 2\omega t$

The variation of power, *P*, with time *t* is shown in Figure 14.21 (ii); it is a sine curve with an average of zero. *Hence no power is absorbed in a pure inductance.* This is explained by the fact that on the first quarter of the current cycle, power is absorbed (+) in the magnetic field of the coil (see p. 396). On the next quarter-cycle the power is returned (−) to the generator, and so on.

Capacitance. With a pure capacitance, the voltage *V* across it lags by 90° on the current *I* (p. 407). So if $I = I_m \sin \omega t$,

$$V = V_m \sin (\omega t - 90°) = - V_m \cos \omega t$$

Hence, numerically,

power at an instant, $P = IV = I_m V_m \sin \omega t \cos \omega t = \frac{I_m V_m}{2} \sin 2\omega t$

So, as in the case of the inductance, *the power absorbed in a cycle is zero,* Figure 14.21 (ii). This is explained by the fact that on the first quarter of the cycle, energy is stored in the electrostatic field of the capacitor. On the next quarter the capacitor discharges, and the energy is returned to the generator.

Formulae for A.C. Power, Power Factor

It can now be seen that, if *I* is the r.m.s. value of the current in amps in a circuit containing a resistance *R* ohms, the power absorbed is $I^2 R$ watts. Care should be taken to exclude the inductances and capacitances in the circuit, as no power is absorbed in them. So if a current of 2 A r.m.s. flows in a circuit containing a coil of 2 H and resistance 10 Ω in series with a capacitor of 1 μF, the power absorbed in the circuit $= I^2 R = 2^2 \times 10 = 40$ W.

If the voltage *V* across a circuit leads by an angle θ on the current *I*, the voltage can be resolved into a component $V \cos \theta$ in phase with the current, and a voltage $V \sin \theta$ perpendicular to the current, Figure 14.22. The former

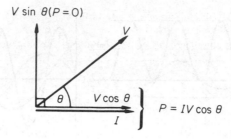

Figure 14.22 *Power absorbed*

component, $V \cos \theta$, represents that part of the voltage across the total *resistance* in the circuit, and hence the power absorbed is

$$P = IV \cos \theta$$

The component $V \sin \theta$ is that part of the applied voltage across the total inductance and capacitance. Since the power absorbed here is zero, it is sometimes called the 'wattless component' of the voltage.

The *power factor* of an a.c. circuit is defined as the ratio

$$power\ absorbed/IV$$

since IV is the maximum power which would be observed if the whole circuit was a resistance. So

$$\text{power factor} = \frac{IV \cos \theta}{IV} = \cos \theta$$

If the circuit has considerable reactance (X_L or X_C) compared to the amount of resistance (R), then the phase angle θ is nearly $90°$. So $\cos \theta$, and hence the power factor, is very small. This is because a pure inductor L and a pure capacitor C absorb no power as we have seen.

You should know:

Single components

(a) R: $I = V/R$ and I in phase with V. (b) L: $I = V/X_L$, where X_L (reactance) $= \omega L = 2\pi f L$ (Ω) and V leads I by $90°$. (c) C: $I = V/X_C$, where $X_C = 1/\omega C = 1/(2\pi f C)$ (Ω). V lags on I by $90°$.

So high L = high reactance, high C = low reactance. (d) r.m.s. value = $(1/\sqrt{2})$ × peak value or peak value = $\sqrt{2}$ × r.m.s. value.

Series components

L and R: Impedance $Z = \sqrt{R^2 + X_L{}^2}$ and V leads I by ϕ where $\tan \phi = X_L/R$.
C and R: Impedance $Z = \sqrt{R^2 + X_C{}^2}$ and V lags on I by ϕ where $\tan \phi = X_C/R$.

Series L, C, R: At resonance, $X_L = X_C$. So (a) $f_0 = 1/(2\pi LC)$
(b) $I_{max} = V/R$, in phase with V.
Power: No power absorbed by L or C, only in R. So power $= I^2 R$.

Examples on A.C. Circuits

1 A circuit consists of a capacitor of $2\,\mu F$ and a resistor of $1000\,\Omega$. An alternating e.m.f. of 12 V (r.m.s.) and frequency 50 Hz is applied. Find (i) the current flowing, (ii) the voltage across the capacitor, (iii) the phase angle between the applied e.m.f. and current, (iv) the average power supplied.

The reactance X_C of the capacitor is given by

$$X_C = \frac{1}{2\pi f C} = \frac{1}{2\pi \times 50 \times 2 \times 10^{-6}} = 1590\,\Omega \text{ (approx.)}$$

$$\therefore \text{ total impedance } Z = \sqrt{R^2 + X_C^2} = \sqrt{1000^2 + 1600^2} = 1880\,\Omega \text{ (approx.)}$$

(i) $\therefore$ current, $I = \dfrac{V}{Z} = \dfrac{12}{1880} = 6\cdot4 \times 10^{-3}$ A

(ii) Voltage across C, $V_C = IX_C = \dfrac{12}{1880} \times 1590 = 10\cdot2$ V (approx.)

(iii) The phase angle θ is given by

$$\tan\theta = \frac{X_C}{R} = \frac{1590}{1000} = 1\cdot59$$

$$\therefore \theta = 58° \text{ (approx.)}$$

(iv) Power supplied $= I^2 R = \left(\dfrac{12}{1880}\right)^2 \times 1000 = 0\cdot04$ W (approx.)

2 A capacitor of capacitance C, a coil of inductance L and resistance R, and a lamp are placed in series with an alternating voltage V. Its frequency f is varied from a low to a high value while the magnitude of V is kept constant. Describe and explain how the brightness of the lamp varies.
If $V = 0\cdot01$ V (r.m.s.) and $C = 0\cdot4\,\mu F$, $L = 0\cdot4$ H, $R = 10\,\Omega$, calculate (i) the resonant frequency, (ii) the maximum current, (iii) the voltage across C at resonance, neglecting the lamp resistance. What is the effect of reducing the resistance R to $5\,\Omega$?

When f is varied, the impedance Z of the circuit decreases to a minimum value (resonance) and then increases. Z is a minimum when $X_L = X_C$, so that $Z = R$ at resonance. Since the *current* flowing in the circuit increases to a maximum and then decreases, the brightness of the lamp increases to a maximum at resonance and then decreases.

(i) Resonant frequency $f_0 = \dfrac{1}{2\pi\sqrt{LC}} = \dfrac{1}{2\pi\sqrt{0\cdot4 \times 0\cdot4 \times 10^{-6}}}$

$$= \frac{10^3}{2\pi \times 0\cdot4} = 400 \text{ Hz (approx.)}$$

(ii) Maximum current $I = \dfrac{V}{R} = \dfrac{0\cdot01}{10} = 0\cdot001$ A (r.m.s.)

(iii) $$\text{Voltage across } C = IX_\text{C} = 0.001 \times \frac{1}{2\pi \times 400 \times 0.4 \times 10^{-6}}$$

$$= \frac{0.001 \times 10^6}{2\pi \times 400 \times 0.4} = 1 \text{ V}$$

When R is reduced to $5\,\Omega$, the maximum current I is doubled, since $I = V/R$. Also, the sharpness of resonance is considerably increased.

Power Factor Correction

As we saw on page 421, inductors and capacitors absorb no net power in an a.c. circuit, while resistors dissipate an average power of $I^2 R$ watts. An average industrial load (like a large motor, for instance) has a large inductance as well as resistance, and so has a power factor (page 421) much less than 1. While the inductance absorbs no net power, it *does* absorb instantaneous power in one half of each cycle (Figure 14.21 (ii)), before returning it to the supply in the other half. This storing and returning of power is measured in vars (volt-amps reactive), and is actually metered and charged for by the electricity companies, though at a reduced rate to resistive power (watts).

So why are vars charged for if any instantaneous reactive power taken from the supply is returned in the next half of the cycle? The problem is that reactive components do, of course, draw current from the supply, and this current produces losses in the transmission lines linking the load to the supply, so costing the electricity company money. For large inductive loads (like large motors), these losses are not small! For this reason, it is common to correct the power factor to 1 at the load, so no reactive power is drawn from the supply. This is done by connecting a large capacitor across the inductive load, which draws reactive power in anti-phase to the inductor, so effectively cancelling out the vars. Power factor correction is practised widely in industry, since being charged for vars should be avoided.

EXERCISES 14 A.C. Circuits

Multiple Choice

1 An alternating current I in amperes varies with time t in seconds as $I = 4 \sin 200\pi t$. The r.m.s. value of the current, I_r, in A and the frequency f in Hz are

A $I_r = 2, f = 100$ B $I_r = 4\sqrt{2}, f = 50$ C $I_r = 4/\sqrt{2}, f = 100$
D $I_r = 4/\sqrt{2}, f = 400$ E $I_r = 2, f = 100$

2 The resonant frequency in a series L–C circuit is 20 MHz. When the capacitance C is increased four times and L stays constant, the resonant frequency in MHz is

A 0.5 B 1.0 C 2.0 D 4.0 E 10.0

3 An a.c. voltage of 50 Hz is connected to an inductor of 2 H and negligible resistance. A current of I r.m.s. flows in the coil. When the frequency of the voltage V is changed to 400 Hz keeping the magnitude of V the same, the current is now

A $8I$ in phase with V B $4I$ and leading by 90° on V C $2I$ in phase with V
D $I/4$ and lagging 90° on V E $I/8$ and lagging 90° on V

4 A capacitor $C = 2\,\mu F$ and an inductor with $L = 10\,H$ and coil resistance $5\,\Omega$ are in series in a circuit. When an alternating current of 2 A r.m.s. flows, the power in watts in the circuit is

A 100   **B** 50   **C** 20   **D** 10   **E** 2

Longer Questions

5 Figure 14A represents alternating currents of different wave shapes, each of peak (amplitude) value 3·0 A.   (i) is a sinusoidal a.c.,   (ii) is a square wave and   (iii) is a rectangular wave. Calculate the r.m.s. value of the current in each case.

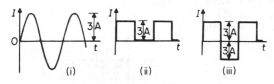

Figure 14A

6 What is the principal difference between alternating current and direct current? Calculate the peak current in a 60 W lamp working from a 240 V (r.m.s.) supply. (*L.*)

7 (a) A signal generator of negligible internal resistance is connected to a 10 kΩ resistor, R. Figure 14B (i). Figure 14B (ii) shows the variation of the potential difference V across R with time.
  (i) Find the frequency of the signal.
  (ii) Calculate the value of $V_{r.m.s.}$.
  (iii) Sketch the variation of current in R with time t for the same time interval as in Figure 14 (ii). Label the axes.

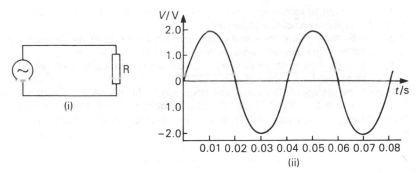

Figure 14B

  (iv) Use one or both of the graphs to determine the power dissipated in R at the instant when $t = 0·015$ s.
 (b) Another resistor of value 20 kΩ is connected in the circuit in series with R and the signal generator. Calculate the mean power dissipated in this second resistor. (*O. & C.*)

8 An inductor of inductance, L, and a capacitor of capacitance, C, are connected in series across a sinusoidal a.c. supply. The circuit has a resonant frequency of 600 Hz.
 (a) Sketch a graph to show how the impedance of the circuit varies with frequency when the frequency of the supply is increased from a low value to about 1 kHz.
 (b) By considering the reactances in the circuit, show that the resonant frequency, f, is given by

$$f = \frac{1}{2\pi\sqrt{LC}}.$$

(c) When the a.c. supply is at the resonant frequency, state the phase difference between.
(i) the current in the circuit and the applied p.d.,
(ii) the p.d. across the capacitor and the p.d. across the inductor. (*N.*)

9 If a sinusoidal current, of peak value 5 A, is passed through an a.c. ammeter the reading will be $5/\sqrt{2}$ A. Explain this.
What reading would you expect if a square-wave current, switching rapidly between $+0.5$ and -0.5 A, were passed through the instrument? (*L.*)

10 Explain what is meant by the *peak value* and *root-mean-square value* of an alternating current. Establish the relation between these quantities for a sinusoidal waveform.
What is the r.m.s. value of the alternating current which must pass through a resistor immersed in oil in a calorimeter so that the initial rate of rise of temperature of the coil is three times that produced when a direct current of 2 A passes through the resistor under the same conditions? (*N.*)

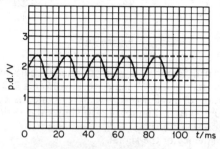

Figure 14C

11 A potential difference varying with time *t* as shown in the Figure 14C is applied to a capacitor of reactance 1·0 kΩ.
(a) Use the graph to calculate   (i) the peak value of the current in the circuit, and (ii) the capacitance of the capacitor.
(b) Sketch a graph, using the same origin and time scale as in the diagram above, to show how the current varies with time over the first 40 ms.
(c) Calculate the new peak value of the current if the frequency of the applied p.d. were increased by a factor of 10^2. (*L.*)

12 An alternating voltage of 10 V r.m.s. and frequency 50 Hz is applied to   (i) a resistor of 5 Ω,   (ii) an inductor of 2 H, and   (iii) a capacitor of 1 μF. Determine the r.m.s. current flowing in each case and draw a phasor diagram of the current and voltage for each.

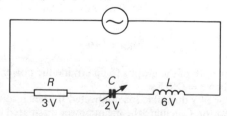

Figure 14D

13 A constant voltage source in the circuit illustrated supplies a sinusoidal alternating e.m.f., Figure 14D. The voltages marked against the other components are the peak values developed across each for a particular value of the capacitance of *C*.
(a) Determine   (i) the peak value of the applied e.m.f.,   (ii) the phase angle between the applied e.m.f. and the current.
(b) If the variable capacitor *C* is now adjusted until the voltage across *R* is a maximum, the circuit is said to be resonant. Explain why the current in the circuit has its maximum value when this is so. (*L.*)

PART THREE
Geometrical Optics, Waves, Wave Optics, Sound Waves

15 Geometrical Optics

Reflection, Refraction, Principles of Optical Fibres

*Light is an electromagnetic wave and later we shall discuss the wave theory
of light in detail. In this chapter, however, we are mainly concerned with
geometrical (ray) optics. So we start with the effect on light rays when they
meet mirrors (reflection) and when they travel from one medium such as air to
another such as glass (refraction). In particular,* total internal reflection *occurs
when rays meet the boundary between two different media at an angle of
incidence greater than the* critical angle.

The principles of optical fibres *are discussed at the end of the chapter. These
very thin strands of glass are now used in telecommunications to transmit signals.*

Light Energy and Light Beams

Light is a form of energy. We know this is the case because plants and vegetables
grow when they absorb sunlight. Also, electrons are emitted by certain metals
when light is incident on them, showing that there was some energy in the light.
This phenomenon is the basis of the *photoelectric cell* (p. 829). Substances like
wood or brick which allow no light to pass through them are called 'opaque'
substances. Unless an opaque object is perfectly black, some of the light falling
on it is reflected. A 'transparent' substance, like glass, is one which allows some
of the light energy incident on it to pass through. The rest of the energy is
absorbed and (or) reflected.

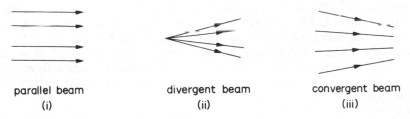

parallel beam
(i)

divergent beam
(ii)

convergent beam
(iii)

Figure 15.1 *Beams of light*

A *ray* of light is the direction along which the light energy or light waves
travel. Although rays are represented in diagrams by straight lines, in practice
a ray has a finite width. A *beam* of light is a collection of rays. A searchlight
emits a *parallel beam* of light. The rays from a point on a very distant object
like the sun are substantially parallel, Figure 15.1 (i). A lamp emits a *divergent
beam* of light; while a source of light behind a lens, as in a projection lantern,
can provide a *convergent beam*, Figure 15.1 (ii), (iii).

Reflection by Plane Mirrors, Reversibility of Light

When we see an object, rays of light enter the eye and produce the sensation
of vision. In Figure 15.2 (i), rays from a small (point) object O are reflected by

a plane mirror so that the angle of incidence i = the angle of reflection r (law of reflection). The rays enter the eye of an observer at D. We always see images in the direction *in which the rays enter the eye*. So the image of O appears to be at I, behind the mirror.

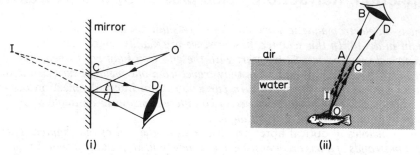

Figure 15.2 *Reflection and refraction images*

In Figure 15.2 (ii), the rays OA and OC from a point object O on a fish change their direction at the boundary with air. They travel along AB and CD. So an observer sees the image of O at I, higher in the water than O.

Light rays never change their light path. So if the ray of light CD is reversed in direction to travel along DC, it will travel along CO in the water. This is known as the *principle of reversibility of light*. We shall need to use this principle later. It follows from a general law in optics due to Fermat. This states that light travels between two points such as O and D in the minimum (or maximum) time. So the light path between O and D is the same in either direction.

Virtual and Real Images in Plane Mirrors

As was shown in Figure 15.2, an object O in front of a mirror has an image I behind the mirror. The rays reflected from the mirror do not actually pass through I, but only *appear* to do so. The image cannot be received on a screen because the image is behind the mirror, Figure 15.3 (i). This type of image is therefore called a *virtual* image. You can see that the light beam from O is a diverging beam. After reflection from the mirror it is still a diverging beam which appears to come from I.

Not only virtual images are obtained with a plane mirror. If a *convergent* beam is incident on a plane mirror M, the reflected rays pass through a point I *in front of* M, Figure 15.3 (ii). If the incident beam converges to the point O,

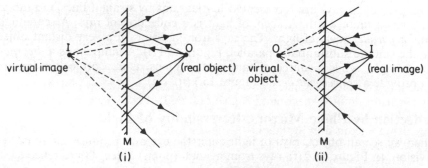

Figure 15.3 *Virtual and real image in plane mirror*

then O is called a 'virtual' object. I is called a *real* image because it can be received on a screen. Figure 15.3 (i) and (ii) should now be compared. In the former, a real object (divergent beam) gives rise to a virtual image; in the latter, a virtual object (convergent beam) gives rise to a real image. In each case the image and object are at equal distances from the mirror. A plane mirror produces an image which is the same size as the object.

Curved Mirrors, Spherical and Paraboloid

Curved mirrors are widely used as driving mirrors in cars. Make-up and dentists' mirrors are curved mirrors. The largest telescope in the world uses an enormous curved mirror to collect light from distant stars. British Telecom use large curved reflectors in suitable parts of the UK to transmit and receive radio signals, which are electromagnetic waves like light waves (p. 432).

Figure 15.4 (i) shows a *concave mirror* P. Its surface is part of a sphere of centre C. When a *narrow* parallel beam of rays from a distant object such as the sun is incident on the middle of P, all the rays are reflected to one point or *focus* F.

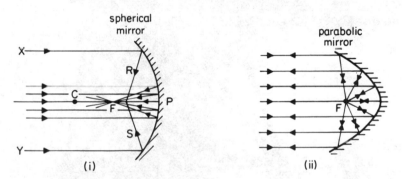

Figure 15.4 *Spherical and paraboloid reflectors*

When a *wide* beam of light XY, parallel to the principal axis, is incident on a concave spherical mirror, reflected rays such as R and S do not pass through a single point, as was the case with a narrow beam. In the same way, if a small lamp is placed at the focus F of a concave spherical mirror, those rays from the lamp which strike the mirror at points well away from the pole P will be reflected in different directions and not as a parallel beam. In this case the reflected beam diminishes in intensity as its distance from the mirror increases.

So a concave spherical mirror is useless as a searchlight mirror. For this reason a mirror whose section is the shape of a parabola (the path of a ball thrown forward into the air) is used in searchlights. A paraboloid mirror has the property of reflecting the wide beam of light from a lamp at its focus F as a perfectly parallel beam. The intensity of the reflected beam is practically undiminished as the distance from the mirror increases, Figure 15.4 (ii). For the same reason, motor headlamp reflectors and those used in torches are paraboloid in shape.

British Telecom uses aerials in the shape of a paraboloid dish to send and receive radio signals. See Plate 15A. A communications satellite high above the earth sends a parallel beam of radio signals to all parts of the dish. This is reflected to a receiver at the focus, like light waves. One aerial reflector

Plate 15A *British Telecom aerial dish at Madley, Hereford. International radio signals, received from earth satellites in geostationary orbits above the equator, are reflected by the paraboloid dish to a sensitive receiver at the focus.*

dish has a diameter of 32 m. It is steered by mechanisms to point directly at the communications satellite and so to receive maximum power from it.

Refraction at Plane Surfaces

Laws of Refraction

When a ray of light AO is incident at O on the plane surface of a glass medium, some of the light is reflected from the surface along OC in accordance with the

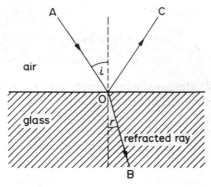

Figure 15.5 *Refraction at plane surface*

laws of reflection. The rest of the light travels along a new direction, OB, in the glass, Figure 15.5. The light is said to be 'refracted' on entering the glass. The *angle of refraction, r*, is the angle made by the refracted ray OB with the normal at O.

Snell, a Dutch professor, discovered in 1620 that the sines of the angles of incidence and refraction have a constant ratio to each other. The *laws of refraction* are:

---

1 *The incident and refracted rays, and the normal at the point of incidence, all lie in the same plane.*

2 *For two given media,* $\dfrac{\sin i}{\sin r}$ *is a constant, where i is the angle of incidence and*
r *is the angle of refraction* (**Snell's law**).

Refractive Index

The constant ratio $\sin i/\sin r$ is known as the *refractive index*, symbol n, for the two given media. As the value of n depends on the colour of the light used, it is usually given as the value for a particular yellow wavelength emitted from sodium vapour. If the medium containing the incident ray is denoted by 1, and that containing the refracted ray by 2, the refractive index can be denoted by $_1n_2$.

Scientists have drawn up tables of refractive indices when the incident ray is travelling in a vacuum and is then refracted into the medium, for example, glass or water. The values obtained are known as the *absolute* refractive indices of the media; and as a vacuum is always the first medium, the subscripts for the absolute refractive index, n, can be dropped. An average value for the magnitude of n for glass is about 1·5, n for water is about 1·33, and n for air at normal pressure is about 1·000 28. In fact the refractive index of a medium is only very slightly altered when the incident light is in air instead of a vacuum. So experiments to determine the absolute refractive index n are usually performed with the light incident from air into the medium. We can take $_{air}n_{glass}$ as equal to $_{vacuum}n_{glass}$ for most practical purposes.

Light is refracted because it has different speeds in different media. The wave theory of light, discussed later, shows that the refractive index $_1n_2$ for two given media 1 and 2 is given by

$$_1n_2 = \frac{\text{speed of light in medium 1 } (c_1)}{\text{speed of light in medium 2 } (c_2)} \qquad . \qquad . \qquad . \qquad (1)$$

This is a *definition* of refractive index which can be used instead of the ratio $\sin i / \sin r$. An alternative definition of the absolute refractive index, n, of a medium 1 is then

$$n = \frac{\text{speed of light in a vacuum, } c}{\text{speed of light in medium 1, } c_1} \qquad . \qquad . \qquad . \qquad (2)$$

In practice the velocity of light in air can replace the velocity in a vacuum in this definition.

Relations between Refractive Indices

1 Consider a ray of light, AO refracted from *glass to air* along the direction OB. The refracted ray OB is bent away from the normal, Figure 15.6. The refractive index from glass to air, $_gn_a$, is given by $\sin x / \sin y$ where x is the angle of incidence in the glass and y is the angle of refraction in the air.

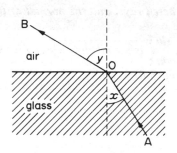

Figure 15.6 *Refraction from glass to air*

From the principle of the reversibility of light (p. 429–30), it follows that a ray travelling along BO in air is refracted along OA in the glass. The refractive index from air to glass, $_an_g$, is given by $\sin y / \sin x$. But $_gn_a = \sin x / \sin y$.

$$\therefore \; _gn_a = \frac{1}{_an_g} \qquad . \qquad . \qquad . \qquad . \qquad . \qquad (3)$$

If $_an_g$ is 1·5, then $_gn_a = 1/1·5 = 0·67$. Similarly, if the refractive index from air to water is 4/3, the refractive index from water to air is 3/4.

2 Consider a ray AO incident in air on a plane glass boundary, then refracted from the glass into a water medium, and finally emerging along a direction CD into air. *If the boundaries of the media are parallel, the emergent ray CD is parallel to the incident ray AO*, although there is a relative displacement, Figure 15.7. So the angles made with the normals by AO, CD are equal, and we shall denote them by i_a.

We can find the refractive index $_gn_w$ from glass to water from the ratio of the light speed in glass, c_g, to that in water, c_w. Assuming the speed of light in air,

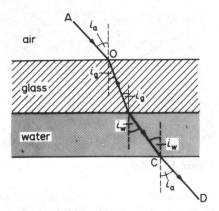

Figure 15.7 *Refraction at parallel plane surfaces*

c, is practically the same as in a vacuum,

then
$$_gc_w = \frac{c_g}{c_w} = \frac{c_g}{c} \times \frac{c}{c_w} = {}_gn_a \times {}_an_w$$

$$= {}_an_w/{}_an_g \tag{1}$$

Using ${}_an_w = 1.33$ and ${}_an_g = 1.5$, it follows that ${}_gn_w = \dfrac{1.33}{1.5} = 0.89$.

We see from the equation above that, for different media 1, 2 and 3,

$$_1n_3 = {}_1n_2 \times {}_2n_3 \quad \cdot \quad \cdot \quad \cdot \quad \cdot \quad \cdot \tag{4}$$

The order of the suffixes enables this formula to be easily memorised.

General Relation between n and $\sin i$

From Figure 15.7, $\sin i_a/\sin i_g = {}_an_g$

$$\therefore \sin i_a = {}_an_g \sin i_g \,. \quad \cdot \quad \cdot \quad \cdot \quad \cdot \tag{1}$$

Also, $\sin i_w/\sin i_a = {}_wn_a = 1/{}_an_w$

$$\therefore \sin i_a = {}_an_w \sin i_w \,. \quad \cdot \quad \cdot \quad \cdot \quad \cdot \tag{2}$$

From (1) and (2),
$$\sin i_a = {}_an_g \sin i_g = {}_an_w \sin i_w$$

If the equations are re-written in terms of the absolute refractive indices of air (n_a), glass (n_g), and water (n_w), we have

$$n_a \sin i_a = n_g \sin i_g = n_w \sin i_w$$

since $n_a = 1$. This relation shows that when a ray is refracted from one medium to another, *the boundaries being parallel,*

$$n \sin i = constant$$

where n is the absolute refractive index of a medium and i is the angle made by the ray with the normal in that medium.

This relation, used later in fibre optics, also applies to the case of light passing

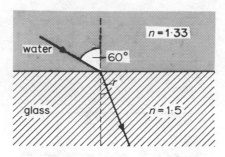

Figure 15.8 *Refraction from water to glass*

directly from one medium to another. Suppose a ray is incident on a water–glass boundary at an angle of 60°, Figure 15.8. Then, applying '$n \sin i$ is a constant', we have

$$1 \cdot 33 \sin 60° \text{ (water)} = 1 \cdot 5 \sin r \text{ (glass)} \qquad . \qquad . \qquad (3)$$

where r is the angle of refraction in the glass, and $1 \cdot 33$, $1 \cdot 5$ are the respective values of n_w and n_g. So $\sin r = 1 \cdot 33 \sin 60°/1 \cdot 5 = 0 \cdot 7679$, from which $r = 50 \cdot 1°$.

Total Internal Reflection

If a ray AO in glass is incident at a small angle α on a glass–air plane boundary, part of the incident light is reflected along OE in the glass, while the rest of the light is refracted away from the normal at an angle β into the air. The reflected ray OE is weak, but the refracted ray OL is bright, Figure 15.9 (i). This means that most of the incident light energy is transmitted, and only a little is reflected.

When the angle of incidence, α, in the glass is increased, the angle of emergence, β, is increased at the same time. At some angle of incidence C in the glass, the refracted ray OL travels along the *glass–air boundary*, making the angle of refraction 90°, Figure 15.9 (ii). The reflected ray OE is still weak in intensity, but as the angle of incidence in the glass is increased slightly the reflected ray suddenly becomes bright, and no refracted ray is seen. Figure 15.9 (iii) shows what happens. Since *all* the incident light energy is now reflected, *total internal reflection* is said to take place in the glass at O.

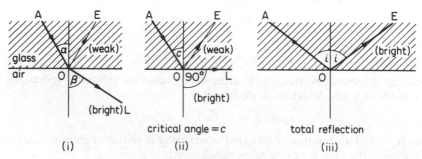

Figure 15.9 *Total internal reflection at a perfectly smooth glass surface*

Critical Angle Values

When the angle of refraction in air is 90°, a critical stage is reached at the point of incidence O. The angle of incidence *in the glass* is known as the *critical angle* for glass and air, Figure 15.9 (ii). Since '$n \sin i$ is a constant' (p. 435), we have

$$n \sin C \text{ (glass)} = 1 \times \sin 90° \text{ (air)}$$

where n is the refractive index of the glass. As $\sin 90° = 1$, then

$$n \sin C = 1$$

or, $$\sin C = \frac{1}{n} \qquad . \qquad . \qquad . \qquad . \qquad . \qquad (8)$$

Crown glass has a refractive index of about 1·51 for yellow light, and thus the critical angle for glass to air is given by $\sin C = 1/1·51 = 0·667$. Consequently $C = 41·5°$. So if the incident angle in the glass is greater than C, for example 45°, total internal reflection occurs, Figure 15.9 (iii).

The refractive index of glass for blue light is greater than that for red light (p. 433). Since $\sin C = 1/n$, we see that the critical angle for blue light is *less* than for red light.

Total reflection may also occur when light in glass ($n_g = 1·51$, say) is incident on a boundary with water ($n_w = 1·33$). Applying '$n \sin i$ is a constant' to the critical case, Figure 15.10, we have

$$n_g \sin C = n_w \sin 90°$$

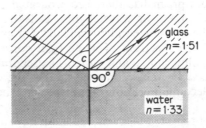

Figure 15.10 *Critical angle for water and glass*

where C is the critical angle. As $\sin 90° = 1$

$$n_g \sin C = n_w$$

$$\therefore \sin C = \frac{n_w}{n_g} = \frac{1·33}{1·51} = 0·889$$

So $$\therefore C = 63° \text{ (approx.)}$$

So if the angle of incidence in the glass exceeds 63°, total internal reflection occurs.

Note that total internal reflection can occur only when light travels from one medium to another which has a *smaller* refractive index, i.e. which is optically less dense. It cannot occur when light travels from one medium to another optically denser, for example from air to glass, or from water to glass. In this case a refracted ray is always obtained.

Dispersion

White light has a band of wavelengths of different colours. This is called the *spectrum* of white light. The longest wavelength is red light, which has a wavelength in air of about 700 nm (700×10^{-9} m or $0.7\,\mu$m). The shortest wavelength is violet, which has a wavelength in air of about 450 nm (450×10^{-9} m or $0.45\,\mu$m).

In a vacuum (and practically in air), all the colours travel at the same speed. In a medium such as glass, however, the colours travel at different speeds—red has the fastest speed and violet the slowest. According to wave theory, refraction is due to the change in speed of light when it enters a different medium. So when a ray AO of white light is incident at O on a glass prism, the colours are refracted in different directions such as OBR and OCS,

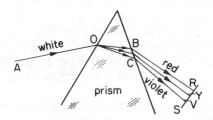

Figure 15.11 *Dispersion in a glass prism*

Figure 15.11. The glass prism has therefore separated or *dispersed* the white light into its various colours or wavelengths, as Newton first discovered in 1666. After leaving the glass, a band or spread of impure colours is formed on a white screen S. The spectrum of white light consists of (bands of) red, orange, yellow, green, blue, indigo and violet. The separation of the colours by the prism is known as *dispersion*. As we see shortly, the wavelengths in a light signal produced in telecommunications are dispersed when they travel along a glass optical fibre.

The sun and the hot tungsten filament of a lamp have a continuous spectrum of visible wavelengths. Hot gases such as hydrogen and krypton have visible wavelengths which form a *line spectrum*. The light from a laser has practically one wavelength, so it can be a *monochromatic* light source. The topic of spectra is discussed more fully in a later chapter.

You should know:

1 n (refractive index) $= \sin i/\sin r$ (i in air)

$n = c/c_m$ (c = light speed in vacuum or air, c_m = speed in medium)

$_g n_a$ (glass to air) $= 1/_a n_g$

$_1 n_2 = {}_1 n_a \times {}_a n_2 = {}_a n_2/_a n_1$.

2 *Critical angle*

   (a) total internal reflection only occurs if light travels from a dense to a *less* dense medium

   (b) $\sin C = 1/n$, from medium to air

   (c) $\sin C = n_2/n_1$, from medium 1 to *less* dense medium 2.

3 *Dispersion* = separation of colours by glass (or other material) due to their speed differences in glass. All colours travel with the same speed in a vacuum.

EXERCISES 15A   Refraction at Plane Surface, Critical Angle, Dispersion

1 A ray of light is incident at 60° at an air–glass plane surface. Find the angle of refraction in the glass (n for glass $= 1\cdot5$).

2 A ray of light is incident in water at an angle of 30° on a water–air plane surface. Find the angle of refraction in the air (n for water $= 4/3$).

3 A ray of light is incident in water at an angle of (i) 30°; (ii) 70° on a water–glass plane surface. Calculate the angle of refraction in the glass in each case ($_an_g = 1\cdot5$, $_an_w = 1\cdot33$).

4 Calculate the critical angle for (i) an air–glass surface; (ii) an air–water surface; (iii) a water–glass surface; draw diagrams in each case illustrating the total reflection of a ray incident on the surface ($_an_g = 1\cdot5$, $_an_w = 1\cdot33$).

5 State the conditions under which total reflection occurs. Show total reflection will occur for light entering normally one face of an isosceles right-angle prism of glass of $n = 1\cdot5$ but not in the case when light enters similarly a similar thin hollow prism full of water of $n = 1\cdot33$.

6 Explain the meaning of critical angle and total internal reflection. Describe fully
   (a) one natural phenomenon due to total internal reflection;
   (b) one practical application of it.
   Light from a luminous point on the lower face of a rectangular glass slab, 2·0 cm thick, strikes the upper face and the totally reflected rays outline a circle of 3·2 cm radius on the lower face. What is the refractive index of the glass? (*JMB.*)

7 Figure 15A shows a narrow parallel horizontal beam of monochromatic light from a laser directed towards the point A on a vertical wall. A semicircular glass block G is placed symmetrically across the path of the light and with its straight edge vertical. The path of the light is unchanged.
   The glass block is rotated about the centre, O, of its straight edge and the bright spot where the beam strikes the wall moves down from A to B and then disappears.

$$OA = 1\cdot50 \text{ m} \qquad AB = 1\cdot68 \text{ m}$$

   (a) Account for the disappearance of the spot of light when it reaches B.
   (b) Find the refractive index of the material of the glass block G for light from the laser.
   (c) Explain whether AB would be longer or shorter if a block of glass of higher refractive index was used. (*L.*)

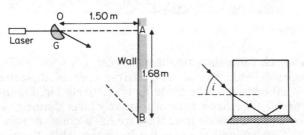

Figure 15A   Figure 15B

8 (a) For light travelling in a medium of refractive index n_1 and incident on the boundary with a medium of refractive index n_2, explain what is meant by total internal reflection and state the circumstances in which it occurs.
   (b) A cube of glass of refractive index 1·500 is placed on a horizontal surface separated from the lower face of the cube by a film of liquid, as shown in Figure 15B. A

ray of light from outside and in a vertical plane parallel to one face of the cube strikes another vertical face of the cube at an angle of incidence $i = 48° 27'$ and, after refraction, is totally reflected at the critical angle at the glass–liquid interface. Calculate (i) the critical angle at the glass–liquid interface; (ii) the angle of emergence of the ray from the cube. (*N.*)

9 Draw a sketch showing the dispersion by a glass prism when the source is (a) a hot gas such as hydrogen; (b) a hot tungsten filament; (c) the sun.

Optical Fibres in Communications

Monomode and Multimode Fibres

As we shall see shortly, light signals can travel along very fine long glass fibres roughly the same diameter as a human hair. *Optical fibres*, as they are called, have replaced the copper cables previously used in telecommunications, Plate 15B. The fibre is a very fine glass rod of diameter about $125 \, \mu m$ (125×10^{-6} m). After manufacture it has a central glass *core* surrounded by a glass coating or *cladding* of smaller refractive index than the core.

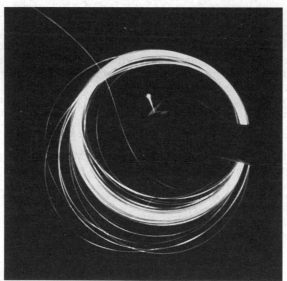

Plate 15B *Light transmitted through a coil of optical fibre*

The fibres are classified into two main types.

(a) The *monomode* fibre has a very narrow core of diameter about $5 \, \mu m$ (5×10^{-6} m) or less, so the cladding is relatively big, Figure 15.12 (i).

(b) The *multimode* fibre has a core of relatively large diameter such as $50 \, \mu m$. In one form of multimode fibre the core has a constant refractive index n_1 such as 1·52 from its centre to the boundary with the cladding, Figure 15.12 (ii). The refractive index then changes to a lower value n_2 such as 1·48 which remains constant throughout the cladding. This is called a *step-index* multimode fibre, in the sense that the refractive index 'steps' from 1·52 to 1·48 at the boundary with the cladding.

As we discuss later, to transmit light signals more efficiently a multimode fibre is made whose refractive index decreases smoothly from the middle to the

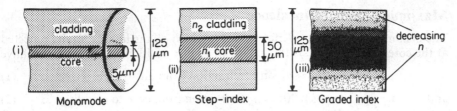

Figure 15.12 *Monomode and multimode fibres*

outer surface of the fibre, Figure 15.12 (iii). There is now no noticeable boundary between the core and the cladding. This is called a *graded index* multimode fibre.

Optical Paths in Fibres

We shall now see what happens when a light signal enters one end of an optical fibre.

Figure 15.13 shows a step-index fibre. With a large angle of incidence, a ray OA entering one end at O is refracted into the core along OP and then refracted along PQ in the cladding. At Q, the fibre surface, the ray passes into the air. In this case only a very small amount of light, due to reflection, passes along the fibre.

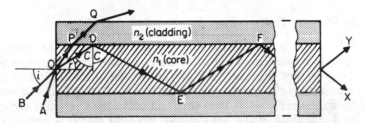

Figure 15.13 *Light path by total internal reflection—multiple reflections*

With a smaller angle of incidence, however, a ray such as BO is refracted in the core along OD and meets the boundary between the core and cladding *at their critical angle, C*. In this case, since $n \sin i$ is constant (p. 435),

$$n_1 \sin C = n_2 \sin 90° = n_2$$

where n_1 is the core refractive index and n_2 the slightly smaller cladding refractive index. If $n_1 = 1·52$ and $n_2 = 1·48$, it follows that

$$\mathbf{\sin C} = \frac{\mathbf{n_2}}{\mathbf{n_1}} = \frac{1·48}{1·52} = 0·974$$

and so $$C = 77° \text{ (approx.)}$$

The ray OD is now *totally* reflected at D along DE, where it again meets the core-cladding boundary at the critical angle. At E, therefore, it is totally reflected along EF.

In this way, by total reflection, a ray of light entering one end of a fibre can travel along the fibre by *multiple reflections* with a fairly high light intensity. At the other end of the fibre the ray emerges in a direction X (odd number of multiple reflections) or a direction Y (even number of multiple reflections).

Maximum Angle of Incidence

The maximum angle of incidence in air for which *all* the light is totally reflected at the core-cladding fibre is the angle i in Figure 15.13. To calculate i, we have

$$1 \times \sin i = n_1 \sin r \text{ (refraction from air to core)} \quad . \quad . \quad (1)$$

and $\quad n_1 \sin C = n_2 \sin 90° = n_2$ (refraction from core to cladding) . (2)

Also, $\qquad\qquad r = 90° - C$, so $\sin r = \cos C$

From (1), $\cos C = \sin i/n_1$; from (2), $\sin C = n_2/n_1$.
  Using the trigonometrical relation $\sin^2 C + \cos^2 C = 1$, then

$$\frac{n_2{}^2}{n_1{}^2} + \frac{\sin^2 i}{n_1{}^2} = 1$$

Simplifying,

$$\sin i = \pm\sqrt{n_1{}^2 - n_2{}^2}$$

With $n_1 = 1·52$ and $n_2 = 1·48$, calculation shows that $i = 20°$ (approx.). So an incident beam from air, making an angle of incidence not more than $20°$ will be transmitted along the fibre with appreciable intensity.

Losses of Power, Dispersion

When a light signal travels along fibres by multiple reflections, some light is absorbed due to impurities in the glass. Some is scattered at groups of atoms which collect together at places such as joints when fibres are joined together. Careful manufacture can reduce the power loss by absorption and scattering.

  The information received at the other end of a fibre can be in error due to *dispersion* or spreading of the light signal. No light signal is perfectly monochromatic. A narrow band of wavelengths is present in the spectrum of the light signal. As we saw when we considered dispersion in a glass prism (p. 438), the various wavelengths are refracted in different directions when the light signal enters the glass fibre and the light spreads.

  Figure 15.14 (i) shows the light paths followed by three wavelengths λ_1, λ_2 and λ_3. λ_1 meets the core-cladding at the critical angle and λ_2 and λ_3 at slightly greater angles. All the rays travel along the fibre by multiple reflections as

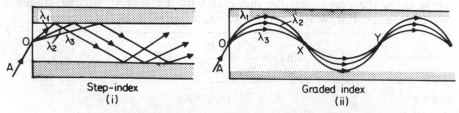

Figure 15.14 *Light paths in step-index and graded index fibres*

previously explained. But the light paths have different lengths. So the wavelengths reach the other end of the fibre *at different times*. The signal received is, therefore, faulty or distorted.

  Figure 15.14 (i) shows a step-index fibre. Its disadvantage can be considerably reduced by using a *graded index* fibre (p. 441). Figure 15.14 (ii) shows roughly what happens in this case. Each wavelength still takes a different path and at some layer in the glass, different for each, the rays are totally reflected. But unlike the step-index fibre, all the rays come to a focus at X as shown, and then again at Y, and so on. We can see this is possible because the speed is inversely-proportional to the refractive index (speed = c/n). So the wavelength λ_1 travels a longer path than λ_2 or λ_3 but at a greater speed. Fermat's principle (p. 430) states that light takes the minimum (or maximum) time to travel between points such as O and X, so the time of travel is the same whichever path is taken.

In spite of the different dispersion, then, all the wavelengths arrive at the other end of the fibre at the same time. With a step-index fibre, the overall time difference may be about 33 ns (33×10^{-9} s) per km length of fibre. Using a graded index fibre, the time difference is reduced to about 1 ns per km.

Figure 15.15 shows diagrammatically the monomode and multimode fibres, the light paths through them and their effect on an input light pulse, where I is the light intensity and t is the time.

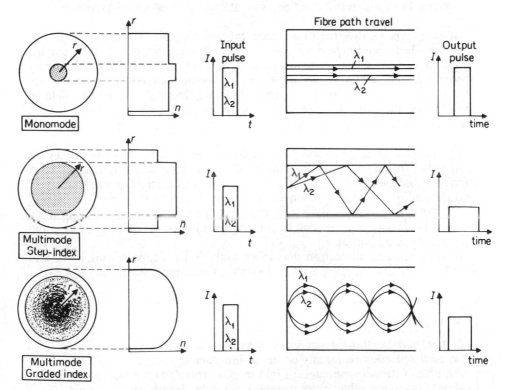

Figure 15.15 *Monomode and multimode fibres*

Light Signal Transmission, Conversion to Sound

A gallium phosphide (GaP) light-emitting diode (LED) can be used as a light source with a graded index fibre. Its light intensity is weak, however, and the absorption and scattering in a long fibre makes this an unsatisfactory source in practice.

A gallium arsenide (GaAs) semiconductor laser is a much better light source than the LED, though it is more expensive. It has a relatively high light intensity and a much narrower band of wavelengths about a mean value such as 1·3 μm, which reduces dispersion problems. With a laser light source, British Telecom prefer to use a *monomode* fibre for long-distance transmission. The monomode fibre with a very thin core of diameter 5 μm or less can now be manufactured with precision. Using a narrow band of wavelengths of mean value 1·3 μm, the light travels straight along the core with only one mode or path. See Figure 15.15.

Figure 15.16 shows in block form how sound energy is transmitted along a fibre and reconverted at the other end to sound energy. Sound information

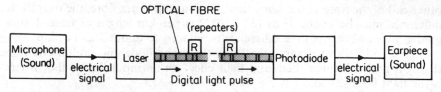

Figure 15.16 *Transmission and reception of sound information* (diagrammatic)

such as speech is converted to an electrical audio signal by a microphone and this is made to modulate light from a laser. The information is then carried along the fibre as a train of light pulses in digital form. Signal losses occur by absorption and scattering, so amplifiers or boosters called *repeaters* (R) are placed at places along the fibre cable. Between Nottingham and Sheffield in the UK, repeaters are placed every 50 km of cable. Copper cables would need many more amplifiers per 50 km length than optical fibres due to greater power losses.

At the other (receiving) end of the cable, a photodiode converts the digital light pulses into a corresponding electrical signal. Technical problems of noise carried by the incoming signal are overcome by using special types of transistors and the audio currents are reconverted to sound in the earpiece of the listener in the telephone system. By a system called *time-division multiplexing*, many thousands of telephone calls can be transmitted along one optical fibre by using light pulses in digital form.

Further telecommunication details are outside the scope of this book and should be obtained from specialist books on telecommunications.

You should know:

Optical fibre has thin glass core coated by glass of smaller refractive index. So total reflection repeatedly occurs at interface of the glasses all along the fibre. Telecom advantages: Light in glass travels faster than electrical signals in copper cable, more messages per cable length, clearer sound.

Optical Fibre Viewing Systems

Even though optical fibres have found their main application in high-speed communications (see above), they are also used to make remote viewing systems for medical and other purposes where access to the scene is restricted. The idea is that very large numbers of tiny fibres are packed together to form a flexible bundle: one end is placed in contact with the scene, the other end forms a remote image of the scene, which may be around a corner or on the other side of a small aperture. Each individual fibre transmits light along its length from the scene to the viewer. Individual fibres are incapable of forming an image: they merely collect light from a small area of the scene and transmit it to the viewer. However, since each fibre 'sees' a slightly different part of the scene, they *collectively* transmit an image from the scene to the viewer. It is important to ensure that the fibres do not change their relative positions over the length of the bundle, otherwise the image will be scrambled at the viewer's end and will be totally unintelligible.

The resolution of the image depends on how closely the fibres can be packed together in the bundle, though this is not really a problem: even in the 1960s resolutions in excess of 100 lines per mm were possible. Such systems have found a very powerful application in medicine. Small bundles of fibres need not be much larger than a hypodermic needle, and can easily be inserted without pain into a patient. Part of the bundle carries illumination from the outside, while the other part transmits the image in the way described above. Using such techniques 'keyhole' surgery can be performed, without the need for any major incisions.

EXERCISES 15B   Optical Fibres

Multiple Choice

1 When a beam of light passes through an optical fibre

   A rays are continually reflected at the outside (cladding) of the fibre
   B some of the rays are refracted from the core to the cladding
   C the bright beam coming out of the fibre is due to the high refractive index of the core
   D the bright beam coming out of the fibre is due to total internal reflection at the core-cladding interface
   E all the rays of light entering the fibre are totally reflected even at very small angles of incidence.

2 A laser is used for sending a signal along a monomode fibre because

   A the light produced is faster than from any other source of light
   B the laser has a very narrow band of wavelengths
   C the core has a low refractive index to laser light
   D the signal is clearer if the cladding has a high refractive index
   E the electrical signal can be transferred quickly using a laser.

Longer Questions

3 (a) Explain in terms of a wave model how a beam of light is refracted as it crosses an interface between two transparent media. Hence derive Snell's law of refraction in terms of the speeds of light in the media. (*see Chap.* 18.)
   (b) Describe the phenomenon of total internal reflection and explain what is meant by the critical angle. How is the critical angle related to the speeds of light in the media involved?
   (c) A portion of a straight glass rod of diameter d and refractive index n is bent into an arc of a circle of mean radius R, and a parallel beam of light is shone down it, as shown in Figure 15C.   (i) Derive an expression in terms of R and d for the angle of incidence i of the central ray C on reaching the glass-to-air surface at the circular arc.   (ii) Show that the smallest value of R which will allow *all* the Flight to pass around the arc is given by

$$R = \frac{d(n+1)}{2(n-1)}$$

(iii) Use this result to explain why glass fibres, rather than rods, are used to carry optical signals around sharp corners.

(d) A glass fibre of refractive index 1·5 and diameter 0·50 mm is bent into a semi-circular arc of mean radius 4·0 mm, and a beam of light is shone along it. (i) Show that no light escapes from the sides of the fibre. (ii) Show by a suitable calculation that if the fibre is immersed in oil of refractive index 1·4 some light will escape. (iii) Suggest an application for such a device. (*O.*)

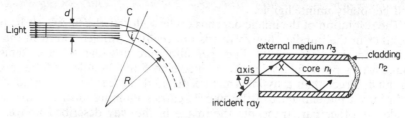

Figure 15C   **Figure 15D**

4 What do you understand by *angle of refraction* and *refractive index*?

A ray of light crosses the interface between two transparent media of refractive indices n_A and n_B. Give a formula relating the directions of the ray on the two sides of the interface. Show the angles you use in your formula on a diagram. Hence, deduce the conditions necessary for total internal reflection to take place at an interface.

In the simple 'light pipe' shown in Figure 15D, a ray of light may be transmitted (with little loss) along the core by repeated internal reflection. The diagram shows a cross-section through the diameter of the 'pipe' with a ray incident in that plane. The core, cladding and external medium have refractive indices n_1, n_2 and n_3 respectively. Show that total internal reflection takes place at X provided that the angle θ is smaller than a value θ_m given by the expression $\sin \theta_m = \sqrt{(n_1{}^2 - n_2{}^2)/n_3}$. Explain why the pipe does not work for rays for which $\theta > \theta_m$. (Reminder: $\sin^2 \theta + \cos^2 \theta = 1$.) (*C.*)

5 Draw sketches showing the different light paths through a monomode and a multimode fibre. Why is the monomode fibre preferred in telecommunications?

6 The refractive index of the core and cladding of an optical fibre are 1·6 and 1·4 respectively. Calculate:
(a) the critical angle at the interface;
(b) the (maximum) angle of incidence in the air of a ray which enters the fibre and is then incident at the critical angle on the interface.

7 A short pulse of white light is sent out at one end of an optical fibre 4 km long. (i) Calculate the time interval between the red and blue light emerging at the other end, given the speed of light in air is 3×10^8 m s^{-1} and the refractive indices of blue and red light are respectively 1·53 and 1·50. (ii) If the pulse of white light is a train of short-duration square waves of intensity against time, draw a labelled sketch of the pulse arriving at the other end of the fibre. With a telecommunication optical fibre, how is this disadvantage overcome? (iii) What is the frequency of the white light pulses at one end if the red and blue pulses at the other end are just separated?

16 Lenses and Optical Instruments

In this chapter we deal first with refraction through converging and diverging lenses and the different images obtained. We then apply the lens equation to calculate image positions and magnification. Next, we consider the astronomical telescope in normal adjustment and show that its magnifying power is the ratio of the focal lengths of its two lenses. The eye-ring and resolving power then follow, and the radio telescope is discussed. The chapter concludes with the simple microscope and its magnifying power.

Converging and Diverging Lenses

A *lens* is an object, usually made of glass, bounded by one or two spherical surfaces. Figure 16.1 (i) illustrates three types of *converging* lenses, which are thicker in the middle than at the edges. Figure 16.1 (ii) shows three types of *diverging* lenses, which are thinner in the middle than at the edges.

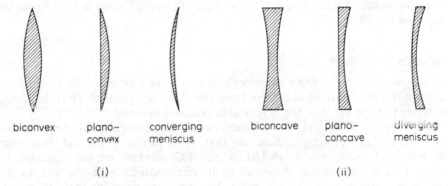

biconvex   plano-convex   converging meniscus   biconcave   plano-concave   diverging meniscus

(i)   (ii)

Figure 16.1 (i) *Converging lens*   (ii) *Diverging lens*

The *principal axis* of a lens is the line joining the centres of curvature of the two surfaces, and passes through the middle of the lens. Experiments with a ray-box show that a thin converging lens brings an incident parallel beam of rays to a *principal focus*, F, on the other side of the lens when the beam is narrow and incident close to the principal axis, Figure 16.2 (i). On account of the convergent beam obtained with it, the lens is better described as a 'converging' lens. If a similar parallel beam is incident on the other (right) side of the lens, it converges to a focus F′, which is at the same distance from the lens as F when the lens is thin. To distinguish F from F′ the latter is called the 'first principal focus'; F is known as the 'second principal focus'.

When a narrow parallel beam, close to the principal axis, is incident on a thin diverging lens, experiment shows that a beam is obtained which appears to diverge from a point F on the same side as the incident beam, Figure 16.2 (ii). F is known as the principal 'focus' of the diverging lens.

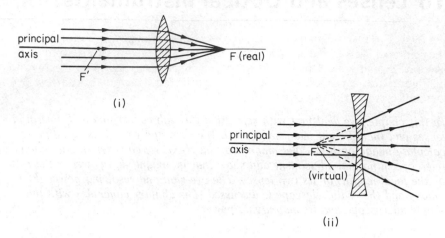

Figure 16.2 *Focus of* (i) *converging, and* (ii) *diverging lenses*

Signs of Focal Length, f

From Figure 16.2 (i), it can be seen that a converging lens has a real focus. By convention (p. 450), the focal length, f, of a *converging* lens is *positive* in sign. Since the focus of a diverging lens is virtual, the focal length of such a lens is negative in sign, Figure 16.2 (ii). These signs are needed when optical formulae are used (p. 450).

Images in Lenses

Converging lens. (i) When an object is a very long way from this lens, i.e., at infinity, the rays arriving at the lens from the object are parallel. Thus the image is formed at the focus of the lens, and is real and inverted.

(ii) Suppose an object OP is placed at O perpendicular to the principal axis of a thin converging lens, so that it is *farther* from the lens than its principal focus, Figure 16.3 (i). A ray PC incident on the midddle, C, of the lens is very slightly displaced by its refraction through the lens, as the opposite surfaces near C are parallel. We therefore consider that PC passes *straight through* the lens, and this is true for any ray incident on the middle of a thin lens.

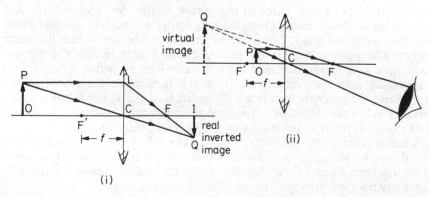

Figure 16.3 *Images in converging lenses*

A ray PL parallel to the principal axis is refracted so that it passes through the focus F. Thus the image, Q, of the top point P of the object is formed below the principal axis, and hence the whole image IQ is real and inverted. In making accurate drawings the lens should be represented by a straight line, as illustrated in Figure 16.3, as we are only concerned with thin lenses and a narrow beam incident close to the principal axis.

(iii) *Image same size as object*. When an object is placed at a distance $2f$ from the lens, the real inverted image has the same size as the object and is also a distance $2f$ from the lens on the other side. So if a converging lens has a focal length 10 cm, an object 20 cm ($2f$) from the lens forms an image of the same size at 20 cm from the lens on the other side.

If the object is *further* than 20 cm from the lens, the real inverted image moves nearer the lens and becomes *smaller* than the object. If the object is nearer than 20 cm but greater than 10 cm (f) from the lens, the image moves back and becomes bigger than the object.

(iv) The least distance between an object and a real image formed by a lens is $4f$ ($2f + 2f$). To form a real image on a screen, the distance between the object and the screen must be at least $4f$. So if a lens has a focal length 10 cm, and a screen is placed 30 cm (less than 4×10 cm) from the object, the lens cannot form an image on the screen.

(v) The image formed by a converging lens is always real and inverted until the object is placed *nearer* the lens than its focal length, Figure 16.3 (ii). In this case the rays from the top point P *diverge* after refraction through the lens, and hence the image Q is *virtual*. The whole image, IQ, is upright or the same way up as the object) and magnified, besides being virtual, and hence the converging lens can be used as a simple 'magnifying glass' (see p. 466).

Images in converging lens:
  When the object is
  1   **at distance $2f$ from lens, image is real, inverted and same size as object.**
  2   **Between $2f$ and f, image is real, inverted and bigger than object.**
  3   **Further than $2f$, the image is real, inverted and smaller than object.**
  4   **Nearer than f, image is upright, magnified and virtual (magnifying glass).**

Diverging lens. In the case of a converging lens, the image is sometimes real and sometimes virtual. In a diverging lens, the image is always virtual; in addition, the image is always upright (erect) and diminished. Figure 16.4 (i), (ii) illustrates the formation of two images. A ray PL appears to diverge from the

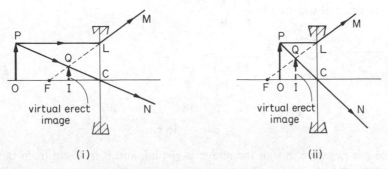

(i)   (ii)

Figure 16.4 *Images in diverging lenses*

focus F after refraction through the lens, a ray PC passes straight through the middle of the lens and emerges along CN, and hence the emergent beam from P appears to diverge from Q on the same side of the lens as the object. The image IQ is thus *virtual*.

Lens Equation and Magnification Formula

Provided a sign rule is used for the distances, the equation

$$\frac{1}{v} + \frac{1}{u} = \frac{1}{f}. \qquad \qquad (1)$$

is the relation between the object distance u from the lens, the image distance v and the focal length f. The 'real is positive' sign rule, which we shall use, is: (1) give a *plus*($+$) sign for *real* object and image distances, (2) give a *minus*($-$) sign for *virtual* object and image distances.

The sign rule also applies to focal lengths. A converging lens has a real focus. So $f = +10$ cm for a converging lens of focal length 10 cm. A diverging lens has a virtual focus. So $f = -20$ cm for a diverging lens of focal length 20 cm.

The linear (transverse) magnification m produced by a lens is defined as the ratio *height of image/height of object*. Numerically,

$$m = \frac{v}{u} \qquad \qquad (2)$$

Applications of Lens Equation and Magnification Formula

The following examples illustrate how to apply the lens equation $1/v + 1/u = 1/f$ and the magnification formula $m = v/u$. The case of a virtual object should be carefully noted.

Examples on Lenses

1 *Converging lens, real object*
An object is placed 12 cm from a converging lens of focal length 18 cm. Find the position of the image.

Since the lens is converging, $f = +18$ cm. The object is real, and therefore $u = +12$ cm. Substituting in $\frac{1}{v} + \frac{1}{u} = \frac{1}{f}$;

$$\frac{1}{v} + \frac{1}{(+12)} = \frac{1}{(+18)}$$

$$\therefore \frac{1}{v} = \frac{1}{18} - \frac{1}{12} = -\frac{1}{36}$$

$$\therefore v = -36 \text{ cm}$$

Since v is negative in sign the image is *virtual*, and it is 36 cm from the lens. See Figure 16.3 (ii).

The magnification,

$$m = \frac{v}{u} = \frac{-36}{12} = -3$$

So the object is magnified 3 times and the minus shows it is upright (magnifying glass).

2 *Converging lens, virtual object*
A beam of light, converging to a point 10 cm behind a converging lens, is incident on the lens. Find the position of the point image if the lens has a focal length of 40 cm.

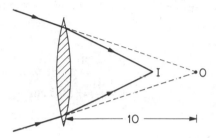

Figure 16.5 *Virtual object*

If the incident beam converges to the point O, then O is a *virtual object*, Figure 16.5. Thus $u = -10$ cm. Also, $f = +40$ cm, since the lens is converging. Substituting in $\dfrac{1}{v} + \dfrac{1}{u} = \dfrac{1}{f}$,

$$\frac{1}{v} + \frac{1}{(-10)} = \frac{1}{(+40)}$$

$$\therefore \frac{1}{v} = \frac{1}{40} + \frac{1}{10} = \frac{5}{40}$$

$$\therefore v = \frac{40}{5} = 8$$

Since v is positive in sign the image is *real*, and it is 8 cm from the lens. The images is I in Figure 16.5.

If the beam of light formed an object of finite size at O, and a real image of this object at I, then

$$\text{magnification} = \frac{v}{u} = \frac{8}{-10} = -0.8$$

So the image is smaller than the object.

Diverging lens. Suppose a beam converges to a point 10 cm behind a diverging lens of focal length 40 cm, so $f = -40$ cm. Then $u = -10$ cm (virtual object). So

$$1/v + 1/(-10) = 1/(-40)$$

Solving, $v = 40/3 = 13.3$ cm. So a real image is formed 13.3 cm behind the lens.

3 A slide projector has a converging lens of focal length 20.0 cm and is used to magnify the area of a slide, 5 cm^2, to an area of 0.8 m^2 on a screen.
Calculate the distance of the slide from the projector lens.

The ratio *area* of image/*area* of object $= 0.8 \text{ m}^2/5 \text{ cm}^2$

$$= \frac{8000 \text{ cm}^2}{5 \text{ cm}^2} = 1600$$

So linear magnification m = square root of area ratio = 40

$$\therefore \frac{v}{u} = 40, \text{ and } v = 40u$$

From the lens equation $\dfrac{1}{v} + \dfrac{1}{u} = \dfrac{1}{f}$, since u and v are both real and $+$ve,

$$\frac{1}{40u} + \frac{1}{u} = \frac{1}{+20}$$

Solving,
$$u = \frac{41}{2} = 20.5 \text{ cm}$$

The Eye, Far and Near Points

The main optical features of the eye are shown in Figure 16.6. The ciliary muscles round the eye lens alter its radii of curvature when near and far objects are viewed, so that the focal length is changed. The pupil is the opening which controls the amount of light entering the eye and becomes wider at night to allow more light through. The iris is the coloured ring round the pupil.

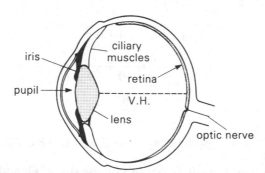

Figure 16.6 *Main optical features of the eye*

To see objects clearly, the lens should focus the light rays on to the retina. The retina has light-sensitive cells at the back of the pupil.

With normal vision, the furthest point seen is a very long way from the eye, at 'infinity'. This is called the *far point* of a normal eye. The nearest point seen distinctly by the normal eye is about 25 cm from the eye. This is called the near point of the eye.

Defects of Vision, Short Sight and Correction

With short sight, the far point of the eye is relatively near. In Figure 16.7, for example, the furthest point seen distinctly may be at A, 2 metres from the eye. The eye lens then focuses the rays from A on to the retina but the parallel

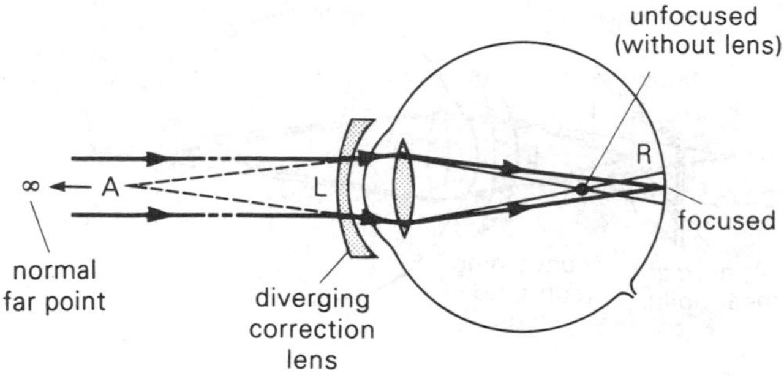

Figure 16.7 *Short sight and correction*

rays from infinity are brought to a focus in front of the retina and, therefore, form a blurred image.

To correct short sight, a suitable *diverging* (*concave*) lens is needed. Its focal length would be 2 m in this case. The parallel rays from infinity are now refracted by the lens and emerge from the lens into the eye *as if they now come from A*. So the rays from infinity are focused on the retina. Drivers with short sight need to wear spectacles with diverging lenses.

Example on Short Sight

The far point of a defective eye is 1 m. What lens is needed to correct this defect?

With this lens, at what distance from the eye is its near point, if the near point is 25 cm without the lens.

A diverging lens of focal length 1 m is needed.

Using the lens, rays from a book at the new near point must be refracted by the lens as if they come from a point 25 cm away after refraction. In this case,

object distance = new near point distance $= u$
image distance $v = -25$ cm because the image is virtual in the diverging lens
focal length $= -1$ m $= -100$ cm, since lens is diverging.

From $\qquad 1/v + 1/u = 1/f$

$$1/(-25) + 1/u = 1/-100$$

$$1/u = -1/100 + 1/25 = 3/100$$

So $\qquad u = 100/3 = 33 \cdot 3$ cm $=$ new near point distance

Long Sight and Correction

With long sight, the eye can see clearly a near point N which is further from the eye than 25 cm, the normal near point at A, Figure 16.8. So rays from N are focused on the retina but not the rays from A.

To correct long sight, a suitable converging (convex) lens is needed. The rays from A are now refracted as if they enter the eye from N. So the rays are focused by the eye on the retina, as shown.

Example on Long Sight

The near point N of a defective eye is 30 cm from the eye. If the normal near point is 25 cm from the eye, find the focal length and power of the lens needed to correct this defect.

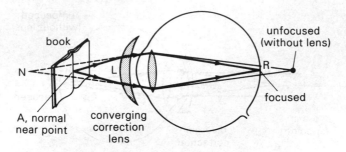

Figure 16.8 *Long sight and correction*

Here u = object distance = +25 cm,
v = image distance = −30 cm (virtual image at N)

From $1/v + 1/u = 1/f$

$1/-30 + 1/+25 = 1/f = 1/150$

So $f = +150$ cm (converging lens) and power = $1/1·5$ m = 0·67

Lens Camera

The basic principle of the lens camera is shown in Figure 16.9. Like the eye, it has a converging (convex) lens which focuses the light from an object on to a light-sensitive film at the back of the camera. A stop allows the light on to the central part of the lens only, which makes for greater clarity

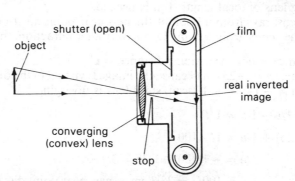

Figure 16.9 *Principle of lens camera*

of the image, and a shutter opens and closes at a speed depending on the lighting conditions.

If the lens has a focal length of 5 cm, the image of a distant object is formed clearly on the film when it is 5 cm from the lens—parallel rays are brought to the focus of the lens.

If the lens is now focused on an object which is 1 metre away, the image distance v is given by, since the object distance u = 100 cm,

$$1/v + 1/u = 1/f$$

or $1/v + 1/100 = 1/5$

So $1/v = 1/5 - 1/100 = 19/100$

$v = 100/19 = 5·3$ cm

The lens must therefore be moved further away from the film to obtain a clear image. So the lens must now be moved closer to the object after photographing a distant object.

You should know:

1 *Lens $f = +$ for converging (convex) lenses, $f = -$ for diverging (concave) lenses.*

2 *Converging (convex) lens.* **Real focus. Image is real if object *further* than f from lens. Nearer than f, image is larger than object, same way up and virtual (= magnifying glass). Image distance found from lens equation $1/v + 1/u = 1/f$. Linear magnification = image length/object length = v/u.**
Diverging (concave) lens **Virtual focus. Images of real objects are virtual.**

3 **Normal vision: Eye far point is at infinity, near point 25 cm from eye. Short sight: Far point closer than infinity. Corrected by diverging lens. Long sight: Near point farther than 25 cm. Corrected by converging lens.**

4 **Lens camera focuses light on to film at back of camera. Distant object, image on film distant f from lens. Nearer object, image on film now at greater distance than f from film and lens nearer the object.**

Colour Constancy

Colour constancy refers to the ability of humans to make sensible judgements of an object's colour under different kinds of illumination. This is quite a remarkable skill to have, since the light reaching our eyes depends not only on the actual colour of the object (its so-called *surface reflectance function*), but also on the wavelengths of the illuminating light. Yet we seem to be able to filter out the effect of the illumination, and say quite reliably that 'this object is blue', even though when viewed under certain lamps very little blue light might actually reach our eyes!

Not only would we like to understand how humans achieve colour constancy, but we would also like to give artificial computer vision systems similar capabilities. Recent research has succeeded in developing colour constancy techniques for computer vision, though it is necessary to view the scene through more than three types of sensor. This does not explain how the human system works, since the retina is generally believed to have only three types of sensor.

A recent theory suggests that *chromatic aberration* (image colouring due to dispersion in a glass lens) might hold the key to the problem. Since an imperfect lens exhibits chromatic aberration, different colours of light are focused on to different positions in the image. If this chromatic aberration is detected and processed, it provides more information about the colours in an image than we would otherwise have. The lens in the human eye appears to suffer badly from chromatic aberration, which suggests that humans might be using chromatic aberration to improve colour constancy. With a suitably poor lens, chromatic aberration is such a rich source of colour information that it is even possible to reconstruct a colour image from a black-and-white one!

EXERCISES 16A   Lenses

1 An object is placed   (i) 12 cm;   (ii) 4 cm from a converging lens of focal length 6 cm. Calculate the image position and the magnification in each case, and draw sketches illustrating the formation of the image.
2 The image obtained with a converging lens is upright and three times the length of the object. The focal length of the lens is 20 cm. Calculate the object and image distances.
3 Used as a magnifying' glass, the image of an object 4 cm from a converging lens is five times the object length. What is the focal length of the lens?
4 A slide of dimensions 2 cm by 2 cm produces a clear image of area 6400 cm^2 on a projector screen. Calculate the focal length of the projector lens if the screen is 82 cm from the slide.
5 An object placed 20 cm from a converging lens forms a magnified clear image on a screen. When the lens is moved 20 cm towards the screen, a smaller clear image is formed on the screen. Calculate the focal length of the lens.
6 (a) Using a labelled ray diagram in each case: (i) explain what is meant by *short sight*;   (ii) show how it is corrected.
   (b) A man can see clearly only objects which lie between 0·50 m and 0·18 m from his eye.
     (i) What is the power of the lens which when placed close to the eye would enable him to see distant objects clearly?
     (ii) Calculate his least distance of distinct vision when using this lens. (*N.*)

Optical Instruments

When a telescope or microscope is used to look at an object, the image we see depends on the eye. We therefore need to know some basic points about vision.

Firstly, the image formed by the eye lens L must appear on the retina R at the back of the eye if the object is to be clearly seen, Figure 16.10. Secondly, the normal eye can focus an object at infinity (the 'far point' of the normal eye). In this case the eye is relaxed or said to be 'unaccommodated'. Thirdly, the eye can see an object in greatest detail when it is placed at a certain distinct D from the eye, known as the *least distance of distinct vision*, which is about 25 cm for a normal eye. The point at a distance D from the eye is known as its 'near point'.

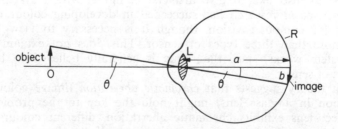

Figure 16.10 *Length of image on retina, and visual angle*

Visual Angle

Consider an object O placed some distance from the eye, and suppose θ is the angle in radians subtended by it at the eye, Figure 16.10. Since the opposite angles at L are equal, the length b of the image on the retina is given by $b = a\theta$, where a is the distance from R to L. But a is a constant; so $b \propto \theta$. We thus arrive at the important conclusion that *the length of the image formed by the eye is proportional to the* **angle** *subtended at the eye by the object.* This angle is known as the *visual angle*; the greater the visual angle, the greater is the apparent size of the object.

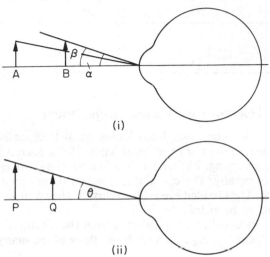

Figure 16.11 *Relation between visual angle and length of image*

Figure 16.11 (i) illustrates the case of an object moved from A to B, and viewed by the eye in both positions. At B the angle β subtended at the eye is greater than the visual angle α subtended at A. So the object appears larger at B than at A, although its physical size is the same. Figure 16.11 (ii) illustrates the case of two objects, at P, Q respectively, which subtend the same visual angle θ at the eye. The objects then appear to be of equal size, although the object at P is physically bigger than that at Q. Of course, an object is not clearly seen if it is brought closer to the eye than the near point.

Angular Magnification of Telescopes

Telescopes and microscopes are instruments designed to increase the visual angle, so that the object viewed can be made to appear much larger with their help. Before they are used, the object may subtend a small angle α at the eye; when they are used the final images should subtend an increased angle β at the eye. The *angular magnification, M,* of the instrument is defined as the ratio

$$M = \frac{\beta}{\alpha} \qquad . \qquad . \qquad . \qquad . \qquad . \qquad (1)$$

This is also popularly known as the *magnifying power* of the instrument. It should be carefully noted that we are concerned with visual angles in the theory of optical instruments, and not with the physical sizes of the object and the image obtained.

Telescopes are instruments used for looking at distant objects. High-power telescopes are used at astronomical observatories, In 1609 Galileo made a telescope through which he saw the satellites of Jupiter and the rings of Saturn. The telescope led the way for great astronomical discoveries, particularly by

Kepler. Newton also designed telescopes. He was the first to suggest the use of curved mirrors for telescopes, as we see later.

If α is the angle subtended at the unaided eye by a *distant* object, and β is the angle subtended at the eye by its final image when a telescope is used, the angular magnification M (also called the 'magnifying power') of the telescope is given by

$$M = \frac{\beta}{\alpha}$$

Astronomical Telescope in Normal Adjustment

An astronomical telescope made from lenses consists of an *objective* of long focal length and an *eyepiece* of short focal length, for a reason given on p. 459. Both lenses are converging. *The telescope is in normal adjustment when the final image is formed at infinity.* The eye is then relaxed or unaccommodated when viewing the image. The unaided eye is also relaxed when a distant object viewed can be considered to be at infinity.

The objective lens O collects parallel rays from the distant object. So it forms an image I at its focus F_0. Figure 16.12 shows three of the many non-axial rays

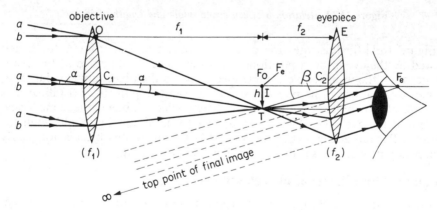

Figure 16.12 *Telescope in normal adjustment*

a from the *top* point of the object, which pass through the top point T of the image. The three rays *b* from the foot of the object would pass through the foot of I (not shown). As the final image is at infinity, I must be at the focus F_e of the eyepiece. So F_e and F_0 are at the *same* place.

To draw the final image, take one lens at a time.

(a) For O, draw a central ray *a* straight through C_1 to T, the top of the objective image below F_0. Then draw the other two rays *a* to pass through T, as shown.

(b) For E, draw a line from T to pass straight through C_2 and another line from T parallel to the principal axis to pass through F_e. The lines emerging from E are parallel; so the final image is at infinity.

(c) Now continue the rays passing through T from O so that they meet the lens E; then draw each refracted ray *parallel to* TC_2 because they must pass through the top of the image of T when produced back. Note carefully that the two lines first drawn from T to E to find the image position are construction lines and *not* actual light rays, and so should not have arrows on them.

To find the *angular magnification* M of the telescope, assume that the eye is closed to the eyepiece. Since the telescope length is very small compared with the distance of the object from either lens, we can take the angle α subtended at the unaided eye by the object as that subtended at the *objective* lens, as shown. Since I is distance f_1 from C_1, where f_1 is the focal length of O, we see that $\alpha = h/f_1$, where h is the length of I. Also, the angle β subtended at the eye when the telescope is used is given by h/f_2, where f_2 is the focal length of the eye-piece E. So

$$M = \frac{\beta}{\alpha} = \frac{h/f_2}{h/f_1}$$

$$\therefore M = \frac{f_1}{f_2} \qquad \cdot \qquad \cdot \qquad \cdot \qquad \cdot \qquad \cdot \qquad (2)$$

So the angular magnification is equal to the ratio of the focal length of the objective (f_1) to that of the eyepiece (f_2). For high angular magnification, it follows from (2) that the objective should have a *long* focal length and the eyepiece a *short* focal length. Note that the separation of the lenses is $f_1 + f_2$.

Telescope in normal adjustment = Final image at infinity
$M = f_1$ (objective)/f_2 (eyepiece)
and separation of (distance between) lenses $= f_1 + f_2$

Examples on Telescopes

1 An astronomical telescope has an objective of focal length 120 cm and an eyepiece of focal length 5 cm. If the telescope is in normal adjustment, what is
(a) the angular magnification (magnifying power);
(b) the separation of the two lenses?

(a) $$M = \frac{f_1}{f_2} = \frac{120}{5} = 24$$

(b) $$\text{separation} = f_1 + f_2 = 120 + 5 = 125 \text{ cm}$$

2 An astronomical telescope has an objective focal length of 100 cm and an eyepiece focal length of 5 cm, Figure 16.13. With the eye close to the eyepiece, an observer sees clearly the final image of a star at a distance 25 cm from the lens.
Calculate:
(a) the separation between the lenses;
(b) the angular magnification (magnifying power M.)

(a) The objective lens O forms an image I_1 of the star at its focus F_0 since parallel rays are incident on the lens. F_0 is 100 cm from O.
The eyepiece E, $f = 5$ cm, forms a magnified and virtual image of I_1 at I_2, which is 25 cm from E. Suppose u is the distance of I_1 from E.

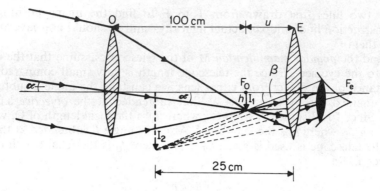

Figure 16.13 *Telescope with image at 25 cm from eye* (not to scale)

Then, for lens E, $v = -25$ cm (virtual image) and $f = +5$ cm. From $1/v + 1/u = 1/f$,

$$\frac{1}{-25} + \frac{1}{u} = \frac{1}{+5}$$

Solving

$$u = \frac{25}{6} = 4 \cdot 2 \text{ cm}$$

So   separation OE of lenses $= 100 + 4 \cdot 2 = 104 \cdot 2$ cm

(b) In Figure 16.13, final image I_2 subtends an angle β at the eye close to E. If h is the height of I_1, then

$$M = \frac{\beta}{\alpha} = \frac{h/u}{h/100} = \frac{100}{u}$$

$$= \frac{100}{4 \cdot 2} = 24$$

Eye-ring of Telescope

When an object is viewed by an optical instrument, only those rays from the object which are bounded by the perimeter or edge of the objective lens enter the instrument. The lens thus acts as a *stop* to the light from the object. With a given objective, the best position of the eye is one where it collects as much light as possible from that passing through the objective.

Figure 16.14 shows the rays from the field of view which are refracted at the *boundary* of the objective O to form an image at F_o or F_e with the telescope in normal adjustment. These rays are again refracted at the boundary of the eyepiece E to form a small image *ab*. From the ray diagram, we see that *a* is the image of A on the objective and *b* is an image of B on the objective. *So ab is the image of the objective AB in the eyepiece.*

The small circular image *ab* is called the *eye-ring*. It is the best position for the eye. Here the eye can collect the maximum amount of light entering the objective from outside so that it has a *wide field of view*. If the eye were placed closer to the eyepiece than the eye-ring the observer would have a smaller field of view.

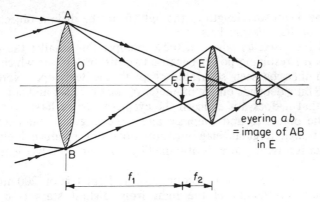

Figure 16.14 *Eye-ring position*

If the telescope is in normal adjustment, the distance u of the objective from the eyepiece E, focal length f_2, is $(f_1 + f_2)$. From the lens equation, the eye-ring distance v from E is given by

$$\frac{1}{v} + \frac{1}{+(f_1 + f_2)} = \frac{1}{+(f_2)}$$

from which

$$v = \frac{f}{f_1}(f_1 + f_2)$$

Now the objective diameter:eye-ring diameter $= \text{AB}:ab = u:v$

$$= (f_1 + f_2):\frac{f_2}{f_1}(f_1 + f_2)$$

$$- f_1/f_2$$

But the angular magnification of the telescope $= f_1/f_2$ (p. 459). So the angular magnification, M, is also given by

$$M = \frac{\text{diameter of objective}}{\text{diameter of eye-ring}} \qquad . \qquad . \qquad . \qquad (3)$$

the telescope being in normal adjustment.

The relation in (3) provides a simple way of measuring M for a telescope.

Resolving Power

If two distant objects are close together, it may not be possible to see their separate images through a telescope even though the lenses are perfect and produce high magnifying power. This is due to the phenomenon of diffraction and is explained later (p. 541). Here we can state that the *smallest* angle θ subtended at a telescope by two distant objects which can just be seen separated is given approximately by

$$\theta = \frac{1 \cdot 22\lambda}{D}$$

where λ is the mean wavelength of the light from the distant objects and D is the diameter of the *objective* lens.

θ is called the *resolving power* of the telescope. The smaller the value of θ, *the greater* is the resolving power because two distant objects which are closer together can then be seen separated through the telescope. Note that the formula for θ only depends on the diameter of the objective and *not* on its focal length, and that it does not concern the eyepiece. As we have seen, the focal lengths of the objective and eyepiece affect the angular magnification of a telescope but high angular magnification does not produce high resolving power. Higher resolving power is obtained by using an objective lens of greater diameter.

So if the objective of a telescope has a diameter of 200 mm (0·2 m), and the mean wavelength of the light from distant stars is 6×10^{-7} m, the resolving power

$$\theta = \frac{1 \cdot 22 \times 6 \times 10^{-7}}{0 \cdot 200} = 4 \times 10^{-6} \text{ rad (approx.)}$$

This means that two stars which subtend this angle at the telescope objective can just be seen separated or resolved.

Reflector Telescope

The astronomical telescope so far discussed has a lens objective and is, therefore, a *refractor* telescope. A *reflector* telescope, with a large curved mirror as its objective, was first suggested by Newton.

The construction of the Hale telescope at Mount Palomar is one of the most fascinating stories of scientific skill and invention. The major feature of the telescope is a *parabolic mirror*, 5 metres across, which is made of Pyrex, a low expansion glass. The glass itself took more than six years to grind and polish, and the front of the mirror is coated with aluminium, instead of being covered with silver, as it lasts much longer. The huge size of the mirror enables enough light from very distant stars and planets to be collected and brought to a focus for them to be photographed. Special cameras are incorporated in the instrument. This method has the advantage that plates can be exposed for hours, if necessary, to the object to be studied, enabling records to be made. It is used to obtain useful information about the building-up and breaking-down of the elements in space, to investigate astronomical theories of the universe, and to photograph planets such as Mars. The Hale telescope has been superceded by other ground-based telescopes built in Hawaii and California, where the air is particularly free of mist and other hindrances to night vision.

Besides the main parabolic mirror O, which is the telescope objective seven other mirrors are used in the 5 metre telescope. Some are plane, Figure 16.15 (i), while others are convex, Figure 16.15 (ii), and they are used to bring the light to a more convenient focus, where the image can be photographed, or magnified several hundred times by an eyepiece E for observation. The various methods of focusing the image were suggested respectively by *Newton*, *Cassegrain* and *Coudé*, the last being a combination of the other two methods, Figure 16.15 (iii).

The reflecting telescope is free from the coloured images produced by refraction at the glass lenses of the refractor telescope. This so-called 'chromatic aberration' of the lens makes the image seen indistinct. The image is also brighter than in a refractor telescope, where some loss of light occurs by

Plate 16A *Hubble Space Telescope. The vast Hubble Space Telescope (HST) seen shortly after capture by shuttle Endeavour at the start of the first HST servicing mission, SIS-61. The photo was taken as the crew manipulated the HST to allow ground controllers to study its exterior. In a subsequent series of five spacewalks, the crew replaced one of HST's cameras and its solar panels, installed a corrective optics package and completed many other tasks. Mission SIS-61 flew 1–13 December 1993.*

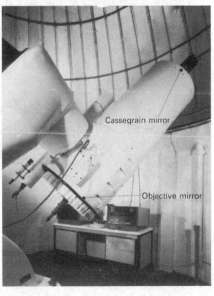

Plate 16B *Allen Reflector Telescope. This has an objective mirror of 60 cm diameter and focal length about 2 m. A Cassegrain mirror is at the top of the tube, which has a length much shorter than the refractor telescope.*

reflection at the lens' surfaces and by absorption. The large diameter of the mirror, which is the telescope objective, also produces high resolving power.

Radio waves from galaxies in outer space are detected by *radio telescopes*. These consist of a concave aerial 'dish' of metal rods which reflect the radio waves to a sensitive detector at the focus of the dish. See p. 431. The signal received is then amplified and recorded automatically. The resolving power of the telescope is increased by moving several widely-spaced dishes along rails, while pointing them skywards in the same direction. This effectively increases the diameter of the telescope objective.

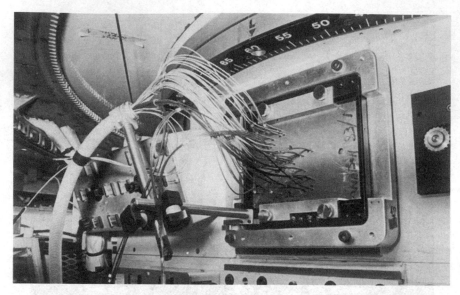

Plate 16C *Fifty optical fibres plugged into an aperture plate at the Cassegrain focus of the 4-metre Anglo-Australian Telescope. The output of the fibre bundle feeds the spectrograph slit, allowing simultaneous spectroscopy of fifty separate objects.*
(Courtesy of Peter Gray, Epping Laboratory, Anglo-Australian Observatory.)

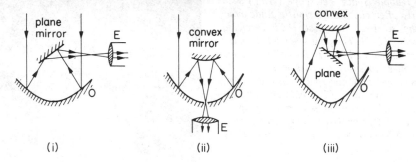

Figure 16.15 (*i*) *Newton reflector* (*ii*) *Cassegrain reflector* (*iii*) *Coudé reflector*

You should know:

1 Angular magnification of telescope in normal adjustment = angle subtended at eye by final image at infinity/angle subtended at unaided eye by object at infinity. Astronomical telescope has converging objective and eyepiece.

2 In normal adjustment, with final image at infinity, an astronomical telescope has
   (a) final image at focus of eyepiece
   (b) telescope length $= f_0 + f_e$
   (c) angular magnification $M = f_0/f_e$, so high M if long f_o and short f_e.

3 Reflector telescope has a large parabolic (concave) mirror as objective which reduces loss of light by lens telescope.

EXERCISES 16B   Telescopes

Multiple Choice

1 With an astronomical telescope in normal adjustment and used for looking at the moon
   A the final image of the moon is upright (erect)
   B the objective forms an image of the moon in front of the eyepiece focus
   C the final image is seen by a relaxed eye
   D the final image is seen at the near point of the eye
   E the angular magnification only depends on the eyepiece focal length.
2 An astronomical telescope X has an objective of focal length 100 cm and diameter 10 cm and an eyepiece of focal length 5 cm and diameter 4 cm. With the telescope in normal adjustment, the angular magnification is

   A 2·5   B 14   C 20   D 25   E 119

Longer Questions

3 A simple astronomical telescope in normal adjustment has an objective of focal length 100 cm and an eyepiece of focal length 5 cm.   (i) Where is the final image formed? (ii) Calculate the angular magnification.   (iii) How would you increase the *resolving power* of the telescope?
4 Draw a ray diagram showing how the image of a distant star is formed at the least distance of distinct vision of an observer using a simple astronomical telescope. In your sketch show the principal focus of the two lenses.
   The same telescope is now required to produce the image of the star on a photographic plate beyond the eyepiece. What adjustment is required? Draw a diagram to explain your answer.
5 What is the *eye-ring* of a telescope? Draw a ray diagram showing how the eye-ring is formed in a simple astronomical telescope and explain why this telescope has a wide field of view.
   Calculate the distance of the eye-ring from the eyepiece of a simple astronomical telescope in normal adjustment whose objective and eyepiece have focal lengths of 80 cm and 10 cm respectively.
6 Draw a sketch of a *reflector telescope* and show with a ray diagram how an observer sees the final image of a distant star.
   State   (i) the advantages of a reflector telescope over a refractor telescope; (ii) how the resolving power of the reflector telescope can be increased;   (iii) the purpose of a radio reflector telescope.
7 Explain the term *angular magnification* as related to an optical instrument. Describe, with the aid of a ray diagram, the structure and action of an astronomical telescope. With such an instrument what is the best position for the observer's eye? Why is this the best position?
   Even if the lenses in such an instrument are perfect it may not be possible to produce clear separate images of two points which are close together. Explain why this is so. Keeping the focal lengths of the lenses the same, what could be changed in order to make the separation of the images more possible? (*L.*)
8 (a) What is meant by *normal adjustment* for an astronomical telescope? Why is it used in this way?
   (b) An astronomical telescope in normal adjustment is required to have an angular magnification of 15. An objective lens of focal length 900 mm is available. Calculate the focal length of the eyepiece required and draw a ray diagram, not to scale, to show how the lenses should be arranged. The diagram should show three rays passing through the telescope from a non-axial point on a distant object. State the position of the final image, and whether or not it is inverted. (*N.*)
9 An astronomical telescope may be constructed using as objective either (a) a converging lens, or (b) a concave mirror.

Draw diagrams to illustrate the optical system of both types of telescope. Include in each diagram at least three rays reaching the instrument from an off-axial direction.

Define the magnifying power of a telescope. A telescope consists of two thin converging lenses of focal lengths 0·3 m and 0·03 m separated by 0·33 m. It is focused on the moon, which subtends an angle of 0·5° at the objective. Starting from first principles, find the angle subtended at the observer's eye by the image of the moon formed by the instrument.

Explain why one would expect this image to be coloured. Suggest how this defect might be rectified. (*O. & C.*)

Simple Microscope or Magnifying Glass

A microscope is an instrument used for viewing *near* objects. When it is in normal use, therefore, the image formed by the microscope is usually at the least distance of distinct vision, *D*, from the eye, i.e., at the near point of the eye. With the unaided eye (that is, without the instrument), the object is seen most clearly when it is placed at the near point. So the angular magnification of a microscope in *normal* use is given by

$$M = \frac{\beta}{\alpha}$$

where β is the angle subtended at the eye by the image at the near point, and α is the angle subtended at the unaided eye by the object at the near point.

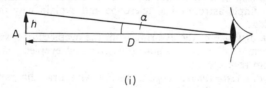

(i)

Figure 16.16 *Visual angle with unaided eye*

Suppose an object of length *h* is viewed at the near point A by the unaided eye, Figure 16.16. The visual angle, α, is then h/D in radian measure. Now suppose that a converging lens L is used as a magnifying glass to view the same object. An erect, magnified image is obtained when the object O is nearer to L than its focal length (p. 448), and the observer moves the lens until the image at I is situated at his or her near point. If the observer's eye is close to the lens at C, the distance IC is then equal to *D*, the least distance of distinct vision, Figure 16.17. Thus the new visual angle β is given by h'/D, where h' is the length of the virtual image. We can see that β is greater than α by comparing Figure 16.16 with Figure 16.17.

The angular magnification, *M*, of this simple microscope can be found in terms of *D* and the focal length *f* of the lens. From definition, $M = \beta/\alpha$.

But
$$\beta = \frac{h'}{D}, \quad \alpha = \frac{h}{D}$$

$$\therefore M = \frac{h'}{D} \bigg/ \frac{h}{D} = h'/h \qquad \qquad . \qquad . \qquad . \qquad (1)$$

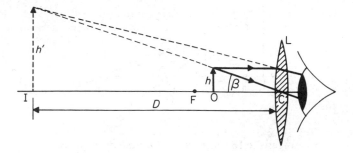

Figure 16.17 *Simple microscope or magnifying glass*

Now h'/h is the 'linear magnification' produced by the lens, and is given by $h'/h = v/u$, where v is the image distance CI and u is the object distance CO (see p. 450). Since $1/v + 1/u = 1/f$, with the usual notation, we have

$$1 + \frac{v}{u} = \frac{v}{f}, \quad \text{or} \quad \frac{v}{u} = \frac{v}{f} - 1$$

by multiplying throughout by v. Since the image is virtual, $v = \text{CI} = -D$, where D is the *numerical* value of the least distance of distinct vision,

$$\therefore \frac{v}{u} = \frac{v}{f} - 1 = -\frac{D}{f} - 1$$

$$\therefore \frac{h'}{h} = -\frac{D}{f} - 1$$

$$\therefore M = -\left(\frac{D}{f} + 1\right). \qquad \qquad (2)$$

from (1) above. So numerically, $M = \left(\frac{D}{f} + 1\right)$

If the magnifying glass has a focal length of 5 cm, $f = +5$ as it is converging; also, if the least distance of distinct vision is 25 cm, $D = 25$ numerically. Substituting in (2),

$$M = -\left(\frac{25}{5} + 1\right) = -6$$

Thus the angular magnification is 6. The position of the object O is given by substituting $v = -25$ and $f = +5$ in the lens equation $1/v + 1/u = 1/f$, from which the object distance u is found to be $+4.2$ cm.

From the formula for M in (2), it follows that a lens of *short* focal length is required for high angular magnification.

When an object OA is viewed through a converging lens acting as a *magnifying glass*, various coloured virtual images, corresponding to I_R, I_V for red and violet rays for example, are formed, Figure 16.18. The top point of each image lies on the line CA. So each image subtends the same angle at the eye close to the lens, so that the colours received by the eye will practically overlap. Thus the virtual image seen in a magnifying glass is almost free of chromatic aberration. A little colour is observed at the edges as a result of spherical

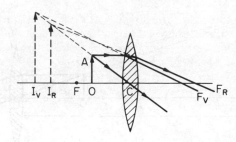

Figure 16.18 *Dispersion with magnifying glass*

aberration. A *real* image formed by a lens, however, has chromatic aberration, see page 455.

Compound Microscope

From the formula $M = -\left(\dfrac{D}{f} + 1\right)$ for the magnifying power of a single lens (p. 467), we see that M is greater numerically the smaller the focal length of the lens. As it is impracticable to decrease f beyond a certain limit, owing to the mechanical difficulties of grinding a lens of short focal length (great curvature), *two* separated lenses are used to obtain a high angular magnification. This forms a *compound* microscope. The lens nearer to the object is called the *objective*; the lens through which the final image is viewed is called the *eyepiece*. The objective and the eyepiece are both converging, and both have small focal lengths for a reason explained later.

When the microscope is used, the object O is placed at a slightly *greater* distance from the objective than its focal length. In Figure 16.19, F_o is the focus of this lens. An inverted real image is then formed at I_1 in the microscope tube,

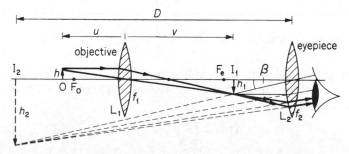

Figure 16.19 *Compound microscope in* normal *use*

and the eyepiece is adjusted so that a large virtual image is formed by it at I_2. So I_1 is *nearer* to the eyepiece than the focus F_e of this lens. It can now be seen that the eyepiece acts as a simple magnifying glass, used for viewing the image formed at I_1 by the objective.

To draw the final image I_2, we first draw construction lines from the top of I_1 to the eyepiece as shown in Figure 16.19. The actual rays, shown by heavy lines in Figure 16.19, can then be drawn as we explained for the telescope on p. 458, to which the reader should refer.

Figure 16.19 illustrates only the basic principle of a compound microscope. The single lens objective shown would produce a real image of the object which is coloured. The single lens eyepiece would produce a virtual image fairly free of colour (p. 467). In practice, both the objective and eyepiece of microscopes are made of several lenses which together reduce chromatic aberration as well as spherical aberration.

The best position for the eye is at *the image of the objective in the eyepiece* or *eye-ring*. All the rays from the object pass through this image. See p. 460. Suppose the objective is 16 cm from L_2, which has a focal length of 2 cm. The image distance, v, in L_2 is given by $\dfrac{1}{v} + \dfrac{1}{(+16)} = \dfrac{1}{(+2)}$, from which $v = 2{\cdot}3$ cm. So the eye-ring is a short distance from the eye-piece, and in practice the eye should be farther from the eyepiece than in Figure 16.19. This is arranged in commercial microscopes by having a circular opening fixed at the eye-ring distance from the eyepiece, so that the observer's eye has automatically the best position when it is placed close to the opening.

Angular Magnification with Microscope in Normal Use

When the microscope is in normal use the image at I_2 is formed at the least distance of distinct vision, D, from the eye (p. 466). Suppose that the eye is close to the eyepiece, as shown in Figure 16.19. The visual angle β subtended by the image at I_2 is then given by $\beta = h_2/D$, where h_2 is the height of the image. With the unaided eye, the object subtends a visual angle given by $\alpha = h/D$, where h is the height of the object, see Figure 16.16.

$$\therefore \text{ angular magnification, } M = \frac{\beta}{\alpha}$$

$$= \frac{h_2/D}{h/D} = \frac{h_2}{h}$$

Now $\dfrac{h_2}{h}$ can be written as $\dfrac{h_2}{h_1} \times \dfrac{h_1}{h}$, where h_1 is the length of the intermediate image formed at I_1.

$$\therefore M = \frac{h_2}{h_1} \cdot \frac{h_1}{h} \qquad . \qquad . \qquad . \qquad . \qquad . \qquad (1)$$

The ratio h_2/h_1 is the linear magnification of the 'object' at I_1 produced by the *eyepiece*. We have shown on p. 467 that the linear magnification is also given by $v/f_2 - 1$, where v is the image distance from the lens and f_2 is the focal length. Since $v = -D$ where D is the numerical value of the least distance of distinct vision, it follows that

$$\frac{h_2}{h_1} = \frac{D}{f_2} - 1 = -\left(\frac{D}{f_2} + 1\right) \qquad . \qquad . \qquad . \qquad . \qquad (2)$$

Also, the ratio h_1/h is the linear magnification of the object at O produced by the *objective* lens. So if the distance of the image I_1 from this lens is denoted by v, we have

$$\frac{h_1}{h} = \frac{v}{f_1} - 1 \qquad . \qquad . \qquad . \qquad . \qquad (3)$$

$$\therefore M = \frac{h_2}{h_1} \cdot \frac{h_1}{h} = -\left(\frac{D}{f_2} + 1\right)\left(\frac{v}{f_1} - 1\right) . \qquad . \qquad . \qquad (4)$$

It can be seen that if f_1 and f_2 are small, M is large. So the angular magnification is high if the focal lengths of the objective and the eyepiece are both *small*.

Example on Compound Microscope

A model of a compound microscope is made up of two converging lenses of 3 cm and 9 cm focal length at a fixed separation of 24 cm. Where must the object be placed so that the final image may be at infinity? What will be the magnifying power if the microscope as thus arranged is used by a person whose nearest distance of distinct vision is 25 cm? State what is the best position for the observer's eye and explain why.

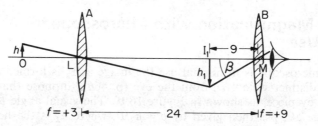

Figure 16.20 *Example on compound microscope*

(i) Suppose the objective A is 3 cm focal length, and the eyepiece B is 9 cm focal length, Figure 16.20. If the final image is at infinity, the image I_1 in the objective must be 9 cm from B, the focal length of the eyepiece. So the image distance LI_1, from the objective A $= 24 - 9 = 15$ cm. The object distance OL is thus given by

$$\frac{1}{(+15)} + \frac{1}{u} = \frac{1}{(+3)}$$

from which $\qquad\qquad u = OL = 3\tfrac{3}{4}$ cm

(ii) The angle β subtended at the observer's eye is given by $\beta = h_1/9$, where h_1 is the height of the image at I_1, Figure 16.20. Without the lenses, the object subtends an angle α at the eye given by $\alpha = h/25$, where h is the height of the object, since the least distance of distinct vision is 25 cm.

$$\therefore \text{ magnifying power } M = \frac{\beta}{\alpha} = \frac{h_1/9}{h/25} = \frac{25}{9} \times \frac{h_1}{h}$$

But $\qquad\qquad\qquad \frac{h_1}{h} = \frac{LI_1}{LO} = \frac{15}{3\tfrac{3}{4}} = 4$

$$\therefore M = \frac{25}{9} \times 4 = 11\cdot1$$

The best position of the eye is at the eye-ring, which is the image of the objective A in the eyepiece B.

Electron Microscopy

Like the telescope (p. 457), the resolving power of an optical microscope is limited by diffraction effects. It is impossible to resolve two objects which are separated by less than about half a wavelength of light, since their diffraction patterns overlap and produce a blurred image. To produce higher resolution images, it is necessary to use shorter wavelengths: this is where electron microscopy excels. On p. 858 we shall see that electrons have a wave-like side to them, just as light exhibits a particle nature in the form of the photon (p. 827). Electrons have been observed to produce diffraction effects when passed through crystal lattices or metal foils. The effective wavelength of the electron beam is reduced as the electrons move faster, to the extent that greatly accelerated electrons have an effective wavelength about 10^{-5} times less than that of light.

Electron microscopes use electrons, instead of light, to resolve images to a much finer level of detail than is possible with optical microscopes. The electrons are focused on to the specimen using a magnetic field device (sometimes referred to as an *electron lens*), and the resulting image is recorded on a fluorescent screen or a photographic plate. Today, powerful microscopes can actually resolve individual atoms in a crystal lattice!

EXERCISES 16C Microscopes

1 A converging lens of focal length 5 cm is used as a magnifying glass. If the near point of the observer is 25 cm from the eye and the lens is held close to the eye, calculate (i) the distance of the object from the lens; (ii) the angular magnification.

What is the angular magnification when the final image is formed at infinity?

2 Explain what is meant by the magnifying power of a magnifying glass.

Derive expressions for the magnifying power of a magnifying glass when the image is

(a) 25 cm from the eye and (b) at infinity.

In each case draw the appropriate ray diagram. (*N.*)

3 Figure 16A shows the paths of two rays of light from the tip of an object B through the objective O, and the eyelens E of a compound microscope. The final image is at the near point of an observer's eye when the eye is close to E. F_O and F_O' are the principal foci of O and F_E is one of the principal foci of E. The diagram is *not drawn to scale*.

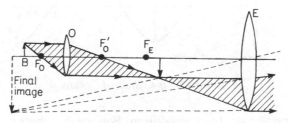

Figure 16A

(a) Explain why
   (i) the object is placed to the left of F;
   (ii) the eyepiece is adjusted so that the intermediate image is to the right of F_E.
(b) In this arrangement, the focal lengths of O and E are 10 mm and 60 mm respectively. If B is 12 mm from O and the final image is 300 mm from E, calculate the distance apart of O and E. (*N.*)

17 Oscillations and Waves

In this chapter we first study the general properties of oscillations, together with resonance and phase difference. We then consider different types of waves—longitudinal and transverse waves, and progressive and stationary or standing waves. Matter waves such as sound waves, and electromagnetic waves such as light or radio waves, are then compared and discussed, together with their speeds.

Simple Harmonic Motion

In a previous mechanics chapter we showed that simple harmonic motion (s.h.m.) occurs when the force acting on an object or system is directly proportional to its displacement x from a fixed point and is always directed towards this point. If the object moves with s.h.m., the variation of the displacement x with t is a sine relation given by

$$x = a \sin \omega t \qquad . \qquad . \qquad . \qquad . \qquad (1)$$

Here a is the greatest displacement from the mean or equilibrium position and is the *amplitude* of the motion, Figure 17.1. The constant $\omega = 2\pi f$ where f is the *frequency* of oscillation or number of cycles per second. The period T of the motion, or time to undergo one complete cycle, is equal to $1/f$, so that $\omega = 2\pi/T$.

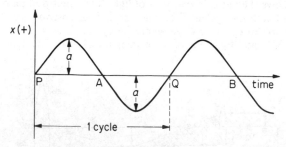

Figure 17.1 *Sine (sinusoidal) curve*

The small oscillation of a pendulum bob or a vibrating layer of air is a *mechanical oscillation*, so that x is a displacement from a mean fixed position. Later, *electrical oscillations* are considered; x may then represent the instantaneous charge on the plates of a capacitor when the charge alternates about a mean value of zero. In an *electromagnetic wave*, x may represent the component of the electric or magnetic field vectors at a particular place.

472

Energy in s.h.m.

On p. 78, it was shown that the sum of the potential and kinetic energies of a body moving with s.h.m. is *constant* and equal to the total energy in the vibration. Further, it was shown that the time averages of the potential energy (p.e.) and kinetic energy (k.e.) are equal; each is half the total energy. In any mechanical oscillation, *there is a continuous interchange or exchange of energy from p.e. to k.e. and back again.*

For vibrations to occur, therefore, an agency is needed which can have and store p.e. and another which can have and store k.e. This was the case for a mass oscillating on the end of a spring, as we saw on p. 79. The mass stores k.e. and the spring stores p.e.; and interchange occurs continuously from one to the other as the spring is compressed and released alternately. In the oscillations of a simple pendulum, the mass stores k.e. as it swings downwards from the end of an oscillation, and this is changed to p.e. as the height of the bob increases above its mean position.

Electrical Oscillations

So far we have dealt with mechanical oscillations and energy. The energy in electrical oscillations takes a different form. There are still two types of energy. One is the energy stored in the electric field, and the other that stored in the magnetic field. To obtain electrical oscillations, an inductor (a coil) is used to produce the magnetic field and a capacitor to produce the electric field. This is discussed more fully on p. 416.

Suppose the capacitor is charged and there is no current at this moment, Figure 17.2 (i). A p.d. then exists across the capacitor and an electric field is present between the plates. At this instant all the energy is stored in the electric field, and since the current is zero there is no magnetic energy. Because of the p.d. a current will begin to flow and magnetic energy will begin to be stored in the inductor. Thus there will be a change from electric to magnetic energy. The p.d. causes the transfer of energy.

One quarter of a cycle later the capacitor will be fully discharged and the

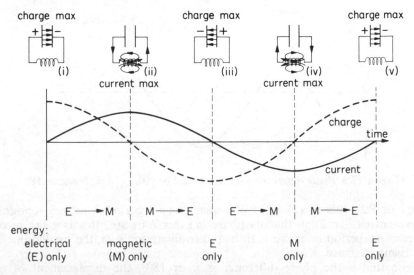

Figure 17.2 *Electrical oscillations—energy exchanges*

current will be at its greatest, so that the energy is now entirely stored in the magnetic field, Figure 17.2 (ii). The current continues to flow for a further quarter-cycle until the capacitor is fully charged in the opposite direction, when the energy is again completely stored in the electric field, Figure 17.2 (iii). The current then reverses and the processes occur in reverse order, Figure 17.2 (iv), after which the original state is restored and a complete oscillation has taken place, Figure 17.2 (v). The whole process then repeats, giving continuous oscillations.

Electrical oscillations are produced by exchange of energy between a capacitor, which stores electrical energy, and a coil or inductor, which stores magnetic energy.

Phase of Vibrations

Consider an oscillation given by $x_1 = a \sin \omega t$. Suppose a second oscillation has the same amplitude, a, and angular frequency, ω, but is out of step and reaches the end of its oscillation a fraction, β, of the period T later than the first one. The second oscillation thus *lags behind* the first by a time βT, and so its displacement x_2 is given by

$$x_2 = a \sin \omega(t - \beta T)$$
$$= a \sin (\omega t - \varphi) \quad . \qquad . \qquad . \qquad . \qquad (2)$$

where $\varphi = \omega \beta T = 2\pi \beta T/T = 2\pi \beta$. If the second oscillation *leads* the first by a time βT, the displacement is given by

$$x_2 = a \sin (\omega t + \varphi) . \qquad . \qquad . \qquad . \qquad (3)$$

φ is known as the *phase angle* of the oscillation. It represents the *phase difference* between the oscillations $x_1 = a \sin \omega t$ and $x_2 = a \sin (\omega t - \varphi)$.

Graphs of displacement against time are in Figure 17.3. Curve 1 represents $x_1 = a \sin \omega t$. Curve 2 represents $x_2 = a \sin (\omega t + \pi/2)$, so that its phase lead

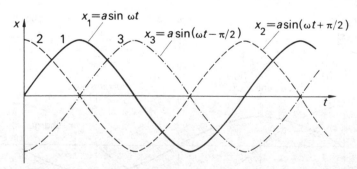

Figure 17.3 *Phase difference* (i) 1, 2 $= \pi/2$ or 90°, (ii) 2, 3 $= \pi$ or 180°

is $\pi/2$ or 90°; this is a *lead* of one quarter of a period. Curve 3 represents $x_3 = a \sin (\omega t - \pi/2)$ so that its phase *lag* is $\pi/2$ or 90°, this is a lag of one quarter of a period on curve 1. If the phase difference is 2π, the oscillations are effectively in phase.

Note that if the phase difference is π or 180°, the displacement of one oscillation reaches a positive maximum value at the same instant as the other oscillation reaches a *negative* maximum value. The two oscillations are thus sometimes said to be 'antiphase'. This is the case for curves 2 and 3 in Figure 17.3.

Damped Vibrations

In practice, the amplitude of vibration in simple harmonic motion does not remain constant but becomes progressively smaller. Such a vibration is said to be *damped*. The decrease in amplitude is due to loss of energy; for example, the amplitude of the bob of a simple pendulum diminishes slowly owing to the viscosity (friction) of the air. This is shown by curve 1 in Figure 17.4.

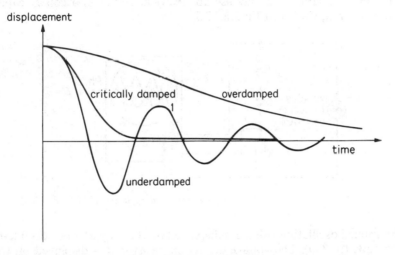

Figure 17.4 *Damped motion*

The general behaviour of mechanical systems subject to a various amounts of damping may be conveniently investigated using a coil of a ballistic galvanometer. If a resistor is connected to the terminals of a ballistic galvanometer when the coil is swinging, the induced e.m.f. due to the motion of the coil in the magnetic field of the galvanometer magnet causes a current to flow through the resistor. This current, by Lenz's law (p. 366), opposes the motion of the coil and so causes damping. The smaller the value of the resistor, the greater is the current and the degree of damping. The galvanometer coil is set swinging by discharging a capacitor through it. The time period, and the time taken for the amplitude to be reduced to a certain fraction of its original value, are then measured. The experiment can then be repeated using different values of resistor connected to the terminals.

It is found that as the damping is increased the time period increases and the oscillations die away more quickly. As the damping is increased further, there is a value of resistance which is just sufficient to prevent the coil from vibrating past its rest position. This degree of damping, called the *critical damping, reduces the motion to rest in the shortest possible time.* If the resistance is lowered further, to increase the damping, no vibrations occur but the coil takes a longer time to settle down to its rest position. The coil is now *overdamped.* Graphs showing the displacement against time for 'underdamped', 'critically damped', and 'overdamped' motion are shown in Figure 17.4.

When it is required to use a galvanometer as a current-measuring instrument, rather than ballistically to measure charge, it is generally critically damped. The return to zero is then as rapid as possible.

These results, obtained for the vibrations of a damped galvanometer coil, are quite general. All vibrating systems has a certain critical damping, which brings the motion to rest in the shortest possible time.

Forced Oscillations, Resonance

In order to keep a system, which has a degree of damping, in continuous oscillatory motion, some outside periodic force must be used. The frequency of this force is called the *forcing frequency*. In order to see how systems respond to a forcing oscillation, we may use an electrical circuit of a coil L, capacitor C and resistor R, shown in Figure 17.5.

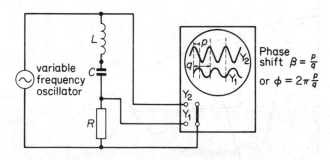

Figure 17.5 *Demonstration of oscillations*

The applied oscillating voltage is displayed on the Y_2 plates of a double-beam oscilloscope (p. 760). The voltage across the resistor R is displayed on the Y_1 plates. Since the current I through the resistor is given by $I = V/R$, the voltage across R is a measure of the current through the circuit. The frequency of the oscillator is now set to a low value and the amplitude of the Y_1 display is recorded. The frequency is then increased slightly and the amplitude again measured. By taking many such readings, a graph can be drawn of the current through the circuit as the frequency is varied. A typical result is shown in Figure 17.6 (i).

The phase difference, φ, between the Y_1 and Y_2 displays can be found by measuring the horizontal shift p between the traces, and the length q occupied by one complete waveform. φ is given by $(p/q) \times 2\pi$. A graph of the variation of phase difference between current and applied voltage can then be drawn. Figure 17.6 (ii) shows a typical curve.

You should know:

1 The current is greatest at a certain frequency f_0. This is the frequency of undamped oscillations of the system, when it is allowed to oscillate *on its own*. f_0 is called the *natural frequency* of the system. When the forcing frequency is equal to the natural frequency, *resonance* is said to occur. The largest current is then produced.

2 At resonance, the current and voltage are in phase. Well below resonance, the current leads the voltage by $\pi/2$; at very high

frequencies the current lags by $\pi/2$. **The behaviour of other resonant systems is similar.**

3 **The forced oscillations always have the same frequency as the forcing oscillations.**

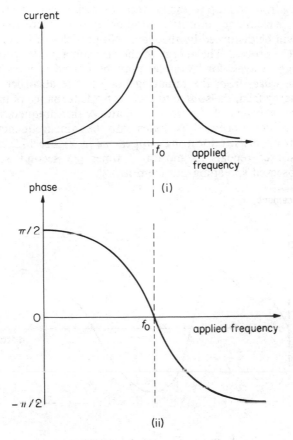

Figure 17.6 *Amplitude and phase in forced vibrations*

In addition to resonance in electrical circuits, resonance occurs in sound and in optics. This is discussed later (see p. 607). It should be noted that considerable energy is absorbed at the resonant frequency from the system supplying the external periodic force.

You should know:

Natural frequency of a system is the frequency on its own.
Forced frequency is the frequency due to an outside period force.
Resonant frequency is the frequency produced in a system when the outside periodic force has a frequency equal to the natural frequency.

Waves and Wave-motion

A *wave* allows energy to be transferred from one point to another some distance away without any particles of the medium travelling between the two points. For example, if a small weight is suspended by a string, energy to move the weight may be obtained by repeatedly shaking the other end of the string up and down through a small distance. Waves, which carry energy, then travel along the string from the top to the bottom. Likewise, *water waves* may spread along the surface from one point A to another point B, where an object floating on the water will be disturbed by the wave. No particles of water at A actually travel to B in the process. The energy in the electromagnetic spectrum, such as X-rays and light waves, for example, may be considered to be carried by *electromagnetic waves* from the radiating body to the absorber. Again, *sound waves* carry energy from the source to the ear by disturbance of the air (p. 480).

If the source or origin of the wave oscillates with a frequency f, then each point in the medium concerned oscillates with the same frequency. A snapshot of the wave profile or waveform may appear as in Figure 17.7 at a particular instant. The source repeats its motion f times per second, so a repeating *waveform* is observed spreading out from it.

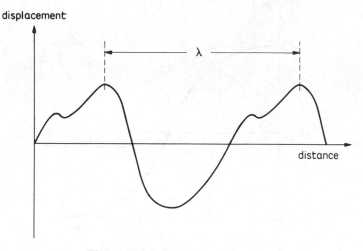

Figure 17.7 *Wave and wavelength*

The distance between corresponding points in successive waveforms, such as two successive crests or two successive troughs, is called the *wavelength*, λ. Each time the source vibrates once, the waveform moves forward a distance λ. So in one second, when f vibrations occur, the wave moves forward a distance $f\lambda$. So the speed c of the waves, which is the distance the profile moves in one second, is given by:

$$c = f\lambda$$

This relation between c, f and λ is true for all wave motion, whatever its origin, that is, it applies to sound waves, electromagnetic waves and mechanical waves.

Transverse Waves

A wave which is propagated by vibrations *perpendicular* to the direction of travel of the wave is called a *transverse* wave. Examples of transverse waves are waves on plucked strings. Electromagnetic waves, which include light waves, are also transverse waves.

The propagation of a transverse wave is illustrated in Figure 17.8. Each particle vibrates perpendicular to the direction of propagation with the same amplitude and frequency, and the wave is shown successively at $t = 0$, $T/4$, $T/2$, $3T/4$, in Figure 17.8, where T is the period.

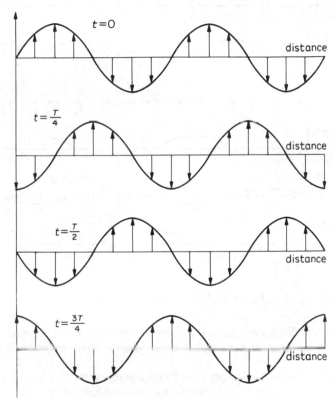

Figure 17.8 *Progressive transverse wave*

Longitudinal Waves

In contrast to a transverse wave, a *longitudinal* wave is one in which the vibrations occur in the *same* direction as the direction of travel of the wave. Figure 17.9 illustrates the propagation of a longitudinal wave. The row of dots shows the actual positions of the particles whereas the *graph* shows the *displacement* of the particles from their equilibrium positions. The positions at time $t = 0$, $t = T/4$, $t = T/2$ and $t = 3T/4$ are shown. The diagram for $t = T$ is, of course, the same as $t = 0$. With displacements to R (right) and to L (left), see graph axis, note the following. (i) The displacements of the particles cause regions of high density (*compressions* C) and of low density (*rarefactions* R) to be formed along the wave. (ii) These regions move along with the speed of the wave, as shown by the broken diagonal line. (iii) Each particle vibrates about its mean position with the same amplitude and frequency. (iv) The regions of greatest compression are one-quarter wavelength (90° phase) ahead

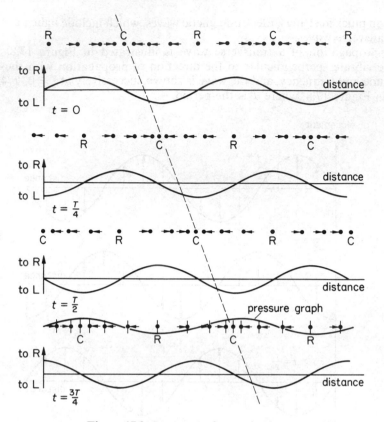

Figure 17.9 *Progressive longitudinal wave*

of the greatest displacement in the direction of the wave. Compare the *pressure graph* at $t = 3T/4$ with the displacement graph. This result is useful in sound waves.

The most common example of a longitudinal wave is a *sound* wave. This is produced by alternate compressions and rarefactions of the air.

You should know:

A *transverse* wave is one in which the direction of the oscillations is perpendicular to the direction of the wave. Light and all other electromagnetic waves are transverse.

A *longitudinal* wave is one in which the direction of the oscillations is in the same direction as the wave. Sound is a longitudinal wave.

Progressive Waves

Both the transverse and longitudinal waves described above are *progressive*. This means that the wave profile moves along with the speed of the wave. If a snapshot is taken of a progressive wave, it repeats at equal distances. The repeat distance is the *wavelength* λ. If one point is taken, and the profile is observed as it passes this point, then the profile is seen to repeat at equal intervals of time. The repeat time is the *period, T.*

The vibrations of the particles in a progressive wave are of the same amplitude and frequency. But *the phase of the vibrations changes for different points along the wave.* This can be seen by considering Figures 17.8 and 17.9. The phase difference may be demonstrated by the following experiment, in which sound waves of the order of 1000 to 2000 Hz may be used.

An audio-frequency (af) oscillator is connected to the loudspeaker L and to the Y_2 plates of a double-beam oscilloscope, Figure 17.10. A microphone M,

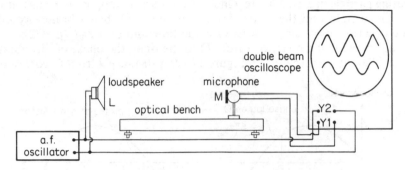

Figure 17.10 *Demonstration of phase in progressive wave*

mounted on an optical bench, is connected to the Y_1 plates. When M is moved away from or towards L, the two traces on the screen are as shown in Figure 17.11 (i) at one position. This occurs when the distance LM is equal to a whole number of wavelengths, so that the signal received by M is in phase with that sent out by L. When M is now moved further away from L through distance $\lambda/4$, where λ is the wavelength, the appearance on the screen changes to that shown in Figure 17.11 (ii). The result phase change is $\pi/2$, so that the signal now arrives a quarter of a period later. When M is moved a distance $\lambda/2$ from its 'in-phase' position, the signal arrives half a period later, a phase change of π, Figure 17.11 (iii).

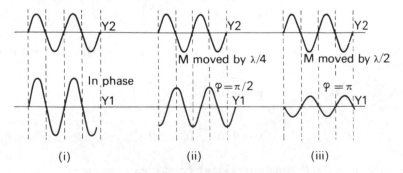

Figure 17.11 *Phase difference and wavelength*

Speed of Sound in Free Air

The speed of sound in free air can be found from this experiment. Firstly, a position of the microphone M is obtained when the two signals on the screen are in phase, as in Figure 17.10. The reading of the position of M on the optical bench is then taken. M is now moved slowly until the phase of the two signals on the screen is seen to change through $\pi/2$ to π and then to be in phase again. The shift of M is then measured. It is equal to λ, the wavelength. From several measurements the average value of λ is found, and the speed of sound is calculated from $c = f\lambda$, where f is the frequency obtained from the oscillator dial. Another method for finding the speed of sound in free air is given on p. 495.

Progressive Wave Equation

An equation can be formed to represent generally the displacement y of a vibrating particle in a medium in which a *wave* passes. Suppose the waves moves from left to right and that a particle at the origin O then vibrates according to the equation $y = a \sin \omega t$, where t is the time and $\omega = 2\pi f$ (p. 472).

At a particle P at a distance x from O to the right, the phase of the vibration will be different from that at O, Figure 17.12. A distance λ from O corresponds

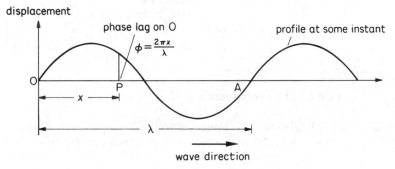

Figure 17.12 *Progressive wave equation*

to a phase difference of 2π (p. 481). So the phase difference φ at P is given by $(x/\lambda) \times 2\pi$ or $2\pi x/\lambda$. Then the displacement of any particle at a distance x from the origin is given by

$$y = a \sin(\omega t - \varphi)$$

or

$$y = a \sin\left(\omega t - \frac{2\pi x}{\lambda}\right). \qquad \qquad (4)$$

Since $\omega = 2\pi f = 2\pi c/\lambda$, where c is the speed of the wave, this equation may be written:

$$y = a \sin\left(\frac{2\pi ct}{\lambda} - \frac{2\pi x}{\lambda}\right)$$

or

$$y = a \sin \frac{2\pi}{\lambda}(ct - x). \qquad \qquad (5)$$

Also, since $\omega = 2\pi/T$, equation (4) may be written:

$$y = a \sin 2\pi\left(\frac{t}{T} - \frac{x}{\lambda}\right). \qquad \qquad (6)$$

Equations (5) and (6) represent a *plane-progressive wave*. The negative sign in the bracket indicates that, since the wave moves from left to right, the vibrations at points such as P to the right of O will *lag* on that at O. A wave travelling in the *opposite direction*, from right to left, arrives at P before O. So the vibration at P *leads* that at O. Therefore a wave travelling in the opposite direction is given by

$$y = a \sin 2\pi\left(\frac{t}{T} + \frac{x}{\lambda}\right). \qquad \cdot \qquad \cdot \qquad \cdot \qquad \cdot \qquad (7)$$

that is, the sign in the bracket is now a plus sign.

To show how the constants of a wave are calculated, suppose a wave is represented by

$$y = a \sin\left(2000\pi t - \frac{\pi x}{0\cdot17}\right)$$

where t is in seconds, y in metres. Then, comparing it with equation (5),

$$y = a \sin \frac{2\pi}{\lambda}(ct - x),$$

we have $\qquad \dfrac{2\pi c}{\lambda} = 2000\pi$ and $\dfrac{2\pi}{\lambda} = \dfrac{\pi}{0\cdot17}$

$$\therefore \lambda = 2 \times 0\cdot7 = 0\cdot34 \text{ m}$$

and $\qquad\qquad c = 1000\lambda = 1000 \times 0\cdot34$

$$= 340 \text{ m s}^{-1}$$

$$\therefore \text{ frequency, } f = \frac{c}{\lambda} = \frac{340}{0\cdot34} = 1000 \text{ Hz}$$

$$\therefore \text{ period, } T = \frac{1}{f} = \frac{1}{1000} \text{ s}$$

If two layers of the wave are 1·8 m apart, they are separated by 1·8/0·34 wavelengths, or by $5\frac{10}{34}\lambda$. Their *phase difference* for a separation λ is 2π; and hence, for a separation $10\lambda/34$, omitting 5λ from consideration, we have:

$$\text{phase difference} = \frac{10}{34} \times 2\pi = \frac{10\pi}{17} \text{ radians}$$

Since y represents the displacement of a particle as the wave travels, the *velocity* v of the particle at any instant is given by dy/dt. From equation (6),

$$v = \frac{dy}{dt} = \frac{2\pi a}{T} \cos 2\pi\left(\frac{t}{T} - \frac{x}{\lambda}\right)$$

So the graph of v against x is 90° out of phase with the graph of y against x. Some microphones used in broadcasting are 'velocity' types, in the sense that the audio current produced is proportional to the velocity of the particles of air. Other microphones, such as those used in the telephone handset, may be 'pressure' types—the audio current is here proportional to the pressure changes in the air.

You should know:

In progressive waves, the amplitude may be constant and neighbouring points are out of phase with each other. Points λ apart have a phase difference of 2π (360°).

Principle of Superposition

When two waves travel through a medium, their combined effect at any point can be found by the *principle of superposition*. This states that *the resultant displacement at any point is the sum of the separate displacements due to the two waves.*

The principle can be illustrated by means of a long stretched spring ('Slinky'). If wave pulses are produced at each end simultaneously, the two waves pass through the wire. Figure 17.13 (a) shows the stages which occur as the two pulses pass each other. In Figure 17.13 (a)(i), they are some distance apart and are approaching each other, and in Figure 17.13 (a)(ii) they are about to meet. In Figure 17.13 (a)(iii), the two pulses, each shown by broken lines, are partly overlapping. The resultant is the sum of the two curves. In Figure 17.13 (a)(iv), the two pulses exactly overlap and the greatest resultant is obtained. The last diagram shows the pulses receding from one another.

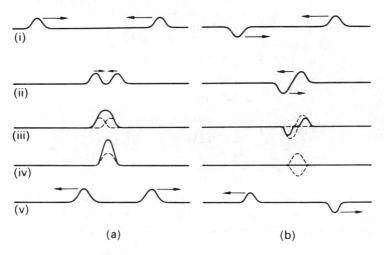

Figure 17.13 *Superposition of waves*

The diagrams in Figure 17.13 (b) show the same sequence of events (i)–(v) but the pulses are now equal and opposite. The principle of superposition is widely used in discussion of wave phenomena such as interference, as we shall see (p. 517).

Stationary or Standing Waves

We have already discussed progressive waves and their properties. Figure 17.14 shows an apparatus which produces a different kind of wave (see also p. 590). If the weights on the scale-plan are suitably adjusted, a number of *stationary vibrating loops* are seen on the string when one end is set vibrating. This time the wave-like profile on the string does *not* move along the medium, which is the string, and the wave is therefore called a *stationary* (or *standing*) wave. In Figure 17.14 a wave travelling along the string to one end is reflected here. So the stationary wave is due to the superposition of *two waves* of equal frequency and amplitude travelling in *opposite* directions along the string. This is discussed more fully later.

The motion of the string when a stationary wave is produced can be studied by using a stroboscope (strobe). This instrument gives a flashing light whose

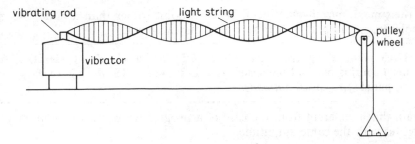

Figure 17.14 *Demonstration of stationary wave*

frequency can be varied. The apparatus is set up in a darkened room and illuminated with the strobe. When the frequency of the strobe is nearly equal to that of the string, the string can be seen moving up and down slowly. Its observed frequency is equal to the difference between the frequency of the strobe and that of the string. Progressive stages in the motion of the string can now be seen and studied, and these are illustrated in Figure 17.15.

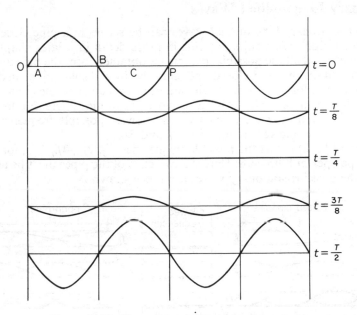

Figure 17.15 *Changes in motion of stationary wave*

Properties of Stationary or Standing Waves

The following points should be noted:

1   **There are points such as B where the displacement is permanently zero. These points are called** *nodes* **of the stationary wave.**
2   **At points between successive nodes the vibrations** *are in phase.*

This last property of the stationary wave is in sharp contrast to the progressive wave, where the phase of points near each other are all different. So when one point of a stationary wave is at its maximum displacement, *all* points are then at their maximum displacement. When a point (other than a node) has zero displacement, *all* points then have zero displacement.

3 **Each point along the wave has a *different amplitude* of vibration from neighbouring points. Points such as C in Figure 17.15 which have the greatest amplitude are called *antinodes*.**

Again this is different from the case of a progressive wave, where every point vibrates with the same amplitude.

4 **The wavelength is equal to the distance OP, Figure 17.15. So the wavelength λ is *twice* the *distance between successive nodes or successive antinodes*. The distance between successive nodes or antinodes is λ/2; the distance between a node and a neighbouring antinode is λ/4.**

These important relations between nodes and antinodes apply to all types of stationary (standing) waves and should be memorised.

Stationary Longitudinal Waves

In sound, stationary longitudinal waves can be set up in a pipe closed at one end (closed pipe). We shall study this in more detail in a later chapter. Here we may note that all the possible frequencies obtained from the pipe are subject to the condition that the closed end must be in a displacement *node* of the stationary wave formed, since the air cannot move here, and the open end must be a displacement *antinode* as the air is most free to move here. Figure 17.16 (i) shows the stationary wave formed for the lowest possible frequency f_0 and other possible or allowed frequencies $3f_0$ and $5f_0$.

Figure 17.16 (ii) shows the possible frequencies, $f_0, 2f_0, 3f_0, \ldots$, for the case of the pipe open at both ends. Here the two ends of the pipes must be antinodes and so the possible stationary waves are those shown.

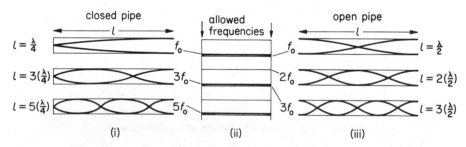

Figure 17.16 *Stationary waves in pipes*

Stationary Transverse Waves

Figure 17.17 shows the possible frequencies for stationary transverse waves produced by plucking in the middle a string fixed at both ends. Here the ends must always be displacement nodes and the middle an antinode.

Pressure in Stationary Wave

Consider the instant corresponding to curve 1 of the displacement graph of a stationary wave, Figure 17.18 (i). At the node *a*, the particles on either side

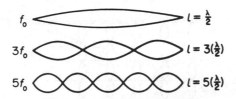

Figure 17.17 *Stationary waves in strings*

produce a compression (increase of pressure), from the direction of their displacement. At the same instant the pressure at the antinode *b* is normal and that at the node *c* is a rarefaction (decrease in pressure). Figure 17.18 (ii) shows the pressure variation along the stationary wave—the displacement nodes are the pressure antinodes. So the closed end of the pipe in Figure 17.16 (i) is a *pressure antinode*.

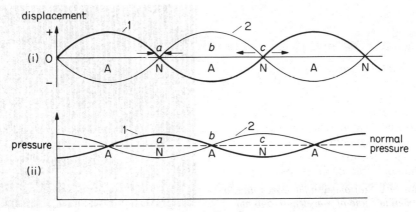

Figure 17.18 *Pressure variation due to stationary wave*

You should know:

A stationary or standing wave is one in which some points are permanently at rest (*nodes*), others between these points are vibrating with varying amplitude and the maximum amplitude is midway between the nodes (*antinodes*). Points between successive nodes are in phase with each other. (In a progressive wave, neighbouring points are out of phase with each other.)

In air or other gases, a pressure antinode occurs at a displacement node and a pressure node at a displacement antinode.

Stationary Light Waves

Light waves have extremely short wavelengths of the order of 5×10^{-7} m or 0·0005 mm. In 1890 Wiener succeeded in detecting stationary light waves. He deposited a very thin photographic film, about one-twentieth of the wavelength of light, on glass and placed it in a position XY inclined at the extremely small angle of about 4′ to a plane mirror CD. Figure 17.19 is an exaggerated sketch

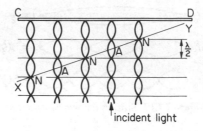

Figure 17.19 *Stationary light waves*

for clarity. When the mirror was illuminated normally by monochromatic light and the film was developed, bright and dark bands were seen. These were respectively antinodes A and nodes N of the stationary light waves formed by reflection at the mirror (see Plate 17A).

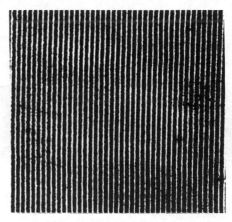

Plate 17A *Stationary light waves due to the mercury line of wavelength 546 nm*

Stationary Waves in Aerials

Stationary waves, due to oscillating electrons, are produced in *aerials* tuned to incoming radio waves, or aerials transmitting radio waves. Figure 17.20

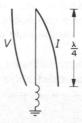

Figure 17.20 *Stationary waves in aerials*

illustrates the stationary wave obtained on a vertical metal rod acting as a 'quarter-wave' aerial. Here the electrons cannot move at the top of the rod, so this is a current (I) node. The current antinode is near the other end of the rod. The current node corresponds to a voltage (V)

antinode, as shown (compare 'displacement' and 'pressure' for the case of the closed pipe on p. 486).

Stationary Waves in Electron Orbits

Moving electrons have wave properties (see p. 858). If we consider a circular orbit of the simplest atom, the hydrogen atom, there must be a complete number of such waves in the orbit for a stable atom; otherwise some of the waves or energy would be radiated as the electron rushed round the orbit and the atom would then lose its energy. So stationary waves are formed in the orbit. Figure 17.21.

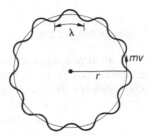

Figure 17.21 *Stationary waves in electron orbit*

So if the radius is r and there are n waves of wavelength λ, we must have

$$2\pi r = n\lambda$$

Bohr suggested that the angular momentum about the centre $= nh/2\pi$ (p. 845), where n is an integer and h is the Planck constant.

$$\therefore mv \times r = \frac{nh}{2\pi}$$

From above,
$$r = \frac{n\lambda}{2\pi} = \frac{nh}{2\pi mv}$$

$$\therefore \lambda = \frac{h}{mv}$$

So the wavelength of electrons with momentum mv is h/mv, as de Broglie first proposed (p. 860).

Stationary Wave Equation

In deriving the wave equation of a progressive wave, we used the fact that the phase changes from point to point (p. 482). In the case of a stationary wave, we may find the equation of motion by considering the *amplitude* of vibration at each point because the amplitude varies while the phase remains constant.

As we have seen, if ω is a constant, the vibration of each particle may be represented by the equation

$$y = Y \sin \omega t \qquad . \qquad . \qquad . \qquad . \qquad (8)$$

where Y is the amplitude of the vibration at the point considered. Y varies along the wave with the distance x from some origin. If we suppose the origin to be at an antinode, then the origin will have the greatest amplitude, A, say. Now the wave repeats at every distance λ, and it can be seen that the amplitudes at differing points vary sinusoidally with their particular distance x. So an equation representing the changing amplitude Y along the wave is:

$$Y = A \cos \frac{2\pi x}{\lambda} = A \cos kx \qquad . \qquad . \qquad . \qquad . \qquad (9)$$

where $k = 2\pi/\lambda$. When $x = 0$, $Y = A$; when $x = \lambda$, $Y = A$. When $x = \lambda/2$, $Y = -A$. So this equation correctly describes the variation in amplitude along the wave, as shown in Figure 17.15. The equation of motion of a stationary wave is, with equation (8),

$$y = A \cos kx \sin \omega t \qquad . \qquad . \qquad . \qquad . \qquad (10)$$

From equation (10), $y = 0$ at all times when $\cos kx = 0$. So $kx = \pi/2, 3\pi/2, 5\pi/2, \ldots$, in this case. This gives values of x corresponding to $\lambda/4, 3\lambda/4, 5\lambda/4, \ldots$ These points are *nodes* since the displacement at a node is always zero. So equation (10) gives the correct distance, $\lambda/2$, between nodes.

A stationary wave can be considered as produced by the superposition of *two progressive waves, of the same amplitude and frequency, travelling in opposite directions*, as we now show.

Mathematical Proof of Stationary Wave Properties

The properties of the stationary wave just deduced can be obtained by a mathematical treatment. Suppose $y_1 = a \sin 2x\left(\dfrac{t}{T} - \dfrac{x}{\lambda}\right)$ is a plane-progressive wave travelling in one direction along the x axis (p. 482). Then

$$y_2 = a \sin 2\pi\left(\frac{t}{T} + \frac{x}{\lambda}\right)$$

represents a wave of the same amplitude and frequency travelling in the opposite direction. So the resultant displacement, y, is given by

$$y = y_1 + y_2 = a\left[\sin 2\pi\left(\frac{t}{T} - \frac{x}{\lambda}\right) + \sin 2\pi\left(\frac{t}{T} + \frac{x}{\lambda}\right)\right]$$

from which

$$y = 2a \sin \frac{2\pi t}{T} \cdot \cos \frac{2\pi x}{\lambda} \qquad . \qquad . \qquad . \qquad . \qquad (1)$$

using the transformation of the sum of two sine functions to a product.

$$\therefore y = Y \sin \frac{2\pi t}{T} \qquad . \qquad . \qquad . \qquad . \qquad (2)$$

where

$$Y = 2a \cos \frac{2\pi x}{\lambda} \qquad . \qquad . \qquad . \qquad . \qquad (3)$$

From (2), Y is the magnitude of the *amplitude* of vibration of the various layers; and from (3) it also follows that the amplitude is a maximum and equal to $2a$

at $x = 0$, $x = \lambda/2$, $x = \lambda$, and so on. These points are, therefore, antinodes, and so consecutive antinodes are separated by a distance $\lambda/2$. The amplitude Y is zero when $x = \lambda/4$, $x = 3\lambda/4$, $x = 5\lambda/4$, and so on. These points are nodes, therefore, and they are midway between consecutive antinodes.

■ Wave Properties, Reflection

Any wave motion can be *reflected*. The reflection of light waves is discussed later.

Like light waves, sound waves are reflected from a plane surface so that the angle of incidence is equal to the angle of reflection. This can be demonstrated by placing a tube T_1 in front of a plane surface AB and blowing a whistle gently at S, Figure 17.22. Another tube T_2, directed towards N, is place on the other side of the normal NQ, and moved until a sensitive microphone, connected to

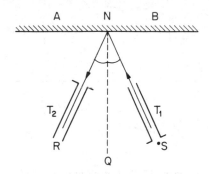

Figure 17.22 *Reflection of sound*

a cathode-ray oscilloscope, is considerably affected at R, showing that the reflected wave is in the direction NR. It will then be found that angle RNQ = angle SNQ.

It can also be shown that sound waves come to a focus when they are incident on a curved concave mirror. A surface shaped like a parabola reflects sound waves to long distances if the source of sound is placed at its focus (see also p. 431). The famous whispering gallery of St Paul's is a circular-shaped chamber whose walls repeatedly reflect sound waves round the gallery, so that a person talking quietly at one end can be heard distinctly at the other end.

Electromagnetic waves of about 21 cm wavelength from outer space are now detected by radio-telescopes. The waves are reflected by a large parabolic 'dish' to a sensitive receiver (see p. 431). A demonstration of the reflection of 3 cm electromagnetic waves is shown on p. 498.

Refraction

Waves can also be *refracted*, that is, their direction changes when they enter a new medium. This is due to the change in speed of the waves on entering a different medium. Refraction of light is discussed later.

Sound waves can be refracted as well as reflected. TYNDALL placed a watch in front of a balloon filled with carbon dioxide, which is heavier than air, and found that the sound was heard at a definite place on the other side of the balloon. The sound waves thus converged to a focus on the other side of the balloon, which therefore has the same effect on sound waves as a converging lens has on light waves (see p. 513). If the balloon is filled with hydrogen, which is lighter than air, the sound waves diverge on passing through the balloon. In this case the gas balloon acts similarly to a diverging lens when light waves are incident on it, as we see later.

The refraction of sound explains why sounds are easier to hear at night than during day-time. In the day-time, the upper layers of air are colder than the layers near the earth. Sound travels faster the higher the temperature (see p. 503), and so sound waves are refracted in a direction away from the earth. The intensity of the sound waves then diminishes. At night-time, however, the layers of air near the earth are colder than those higher up, and so sound waves are now refracted towards the earth, with a consequent increase in intensity.

For a similar reason, a distant observer O hears a sound from a source S more easily when the wind is blowing towards him than away from him, Figure 17.23. When the wind is blowing towards O, the bottom of the sound wavefront

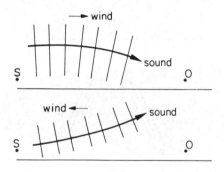

Figure 17.23 *Refraction of sound*

is moving more slowly than the upper part, and so the wavefronts turn towards the observer, who therefore hears the sound easily. When the wind is blowing in the opposite direction the reverse in the case, and the wavefronts turn upwards away from the ground and O. The sound intensity then diminishes. This phenomenon shows how sound wavefronts may change direction due to variation in wind velocity.

Radio (electromagnetic) waves are refracted in the ionosphere (a layer of electrons and ions) high above the earth when they are transmitted from one side of the earth to the other. A demonstration of the refraction of microwaves, 3 cm electromagnetic waves, is shown on p. 498.

Diffraction

Waves can also be 'diffracted'. *Diffraction* is the name given to the spreading of waves when they pass through apertures or around obstacles.

The general phenomenon of diffraction may be illustrated by using water waves in a ripple tank, with which we assume the reader is familiar. Figure 17.24 (i) shows the effect of widening the aperture and Figure 17.24 (ii) the effect of shortening the wavelength and keeping the same width of opening. In certain circumstances in diffraction, reinforcement of the waves, or complete cancellation occurs in particular directions from the aperture, as shown in Figure 17.24 (i) and (ii). These patterns are called 'diffraction bands'.

Generally, the smaller the width of the aperture in relation to the wavelength,

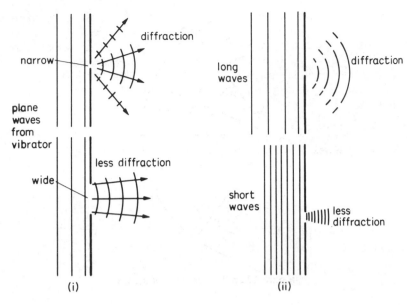

Figure 17.24 *Diffraction of waves*

the greater is the spreading or diffraction of the waves. This explains why we cannot see round corners. The wavelength of *light waves* is about 6×10^{-7} m (p. 523). This is so short that no appreciable diffraction is obtained around obstacles of normal size. With very small obstacles or narrow apertures, however, diffraction of light may be appreciable, as we see later. *Electromagnetic waves* can be diffracted, as shown on p. 498.

Sound waves are diffracted round wide openings such as doorways because their wavelength is comparable with the width of the opening. For example, the wavelength for a frequency of, say, 680 Hz is about 0·5 m and the width of a door may be about 0·8 m. Generally, the diffraction increases with longer wavelength. For this reason, the low notes of a band marching away out of sight round a corner are heard for a longer time than the high notes. Similarly, the low notes of an orchestra playing in a hall can be heard through a doorway better than the high notes by a listener outside the hall.

Interference

When two or more waves of the same frequency overlap, the phenomenon of *interference* occurs. Interference is easily demonstrated in a ripple tank. Two sources, A and B, of the same frequency are used. These produce circular waves which spread out and overlap, and the pattern seen on the water surface is shown in Figure 17.25.

The interference pattern can be explained from the principle of superposition (p. 484). If the oscillations of A and B are in phase, crests from A will arrive at the same time as crests from B at any point on the line RS. So by the principle of superposition there will be reinforcement or a large wave along RS. Along XY, however, crests from A will arrive before corresponding crests from B. In fact, every point on XY is half a wavelength, $\lambda/2$, nearest to A than to B, so that crests from A arrive at the same time as *troughs* from B. So, the resultant is *zero*. Generally, reinforcement (constructive interference) occurs at a point C

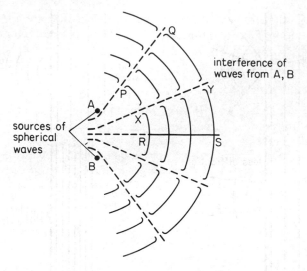

Figure 17.25 *Interference of waves*

when the path difference AC − BC = 0 *or* λ *or* 2λ, and cancellation (destructive interference) when AC − BC = λ/2 or 3λ/2 or 5λ/2.

Interference of *light waves* is discussed in detail later. An experiment to demonstrate the interference of *electromagnetic waves* (microwaves) is given on p. 517. The interference of *sound waves* can be demonstrated by connecting two loudspeakers in parallel to an audio-frequency oscillator, Figure 17.26 (i). As the ear or microphone is moved along the line MN, alternate loud (L) and soft (S) sounds are heard according to whether thhe receiver of sound is on a line of reinforcement (constructive interference) or cancellation (destructive interference) of waves. Figure 17.26 (i) indicates the positions of loud and soft sounds if the two speakers oscillate in phase. If the connections to *one* of the speakers

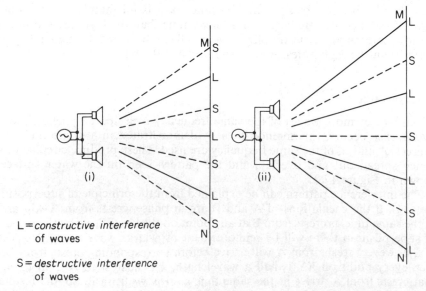

L = *constructive interference* of waves

S = *destructive interference* of waves

Figure 17.26 *Interference of sound waves*

is reversed, so that they oscillate 180° out of phase, then the pattern is altered as shown in Figure 17.26 (ii). The reader should try to account for this difference.

You should know:

All waves can be *reflected, refracted* (due to change of speed) and *diffracted* (they spread through openings comparable to their wavelength). *Interference* occurs where two waves overlap. From the *principle of superposition*, two crests arriving together at a place produce *constructive interference* but a crest and a trough produce *destructive interference*.

Example on Sound Interference

Two small loudspeakers A, B, 1·00 m apart, are connected to the same oscillator so that both emit sound waves of frequency 1700 Hz in phase, Figure 17.27. A sensitive detector, moving parallel to the line AB along PQ 2·40 m away, detects a maximum wave at P on the perpendicular bisector MP of AB and another maximum wave when it first reaches a point Q directly opposite to B.

Calculate the speed c of the sound waves in air from these measurements.

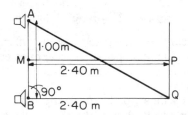

Figure 17.27 *Example*

There is constructive interference of the sound waves at P and Q. Since Q is the first maximum after P, where AP = BP, it follows that

$$AQ - BQ = \lambda \qquad . \qquad . \qquad . \qquad . \qquad . \qquad (1)$$

where λ is the wavelength of the sound waves. Now BQ = 2·40 m, AB = 1·00 m and angle ABQ = 90°. So

$$AQ = \sqrt{BQ^2 + AB^2} = \sqrt{2\cdot40^2 + 1\cdot00^2} = 2\cdot60 \text{ m}$$

From (1),
$$\lambda = 2\cdot60 - 2\cdot40 = 0\cdot20 \text{ m}$$

So
$$\text{wave speed } c = f\lambda = 1700 \times 0\cdot20 = 340 \text{ m s}^{-1}$$

Measurement of Speed of Sound

Figure 17.28 shows one method of measuring the speed of sound in free air by an interference method.

Sound waves of constant frequency, such as 1500 Hz, travel from a loudspeaker L towards a vertical board M. Here the waves are reflected and interfere with the incident waves. As explained before, the two waves travelling in opposite directions produce a *stationary wave* between the board M and L.

A small microphone, positioned in front of the board, is connected to the

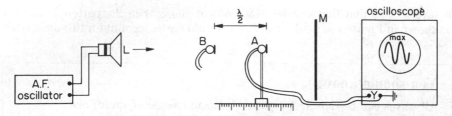

Figure 17.28 *Speed of sound in free air—interference method*

Y-plates of an oscilloscope. As the microphone is moved back from M towards L, the amplitude of the waveform seen on the screen increases to a *maximum* at one position A, as shown. This is an antinode of the stationary wave. When the microphone is moved on, the amplitude diminishes to a minimum (a node) and then increases to a maximum again at a position B, the next antinode. The distance between successive antinodes is $\lambda/2$ (p. 486). So by measuring the average distance d between successive maxima, the wavelength λ can be found. Knowing the frequency f of the note from the loudspeaker, the speed c of the sound wave can be calculated from $c = f\lambda$.

Speed of Sound by Lissajous Figures

The speed of sound can be measured by a different method using an oscilloscope. In this case, the loudspeaker L is connected to the X-plates of the oscilloscope and the microphone A to the Y-plates. The board M in Figure 17.28 is completely removed.

When A faces L, the oscilloscope beam is affected:
(a) *horizontally* by a voltage V_L due to the sound waves from L;
(b) *vertically* by a voltage V_A due to sound waves arriving at A.
These two sets of waves together produce a resultant waveform on the screen, whose geometrical form is called a *Lissajous figure*.

If V_L and V_A are exactly in phase, a straight inclined line is seen on the screen. Figure 17.29 (i). If V_L and V_A are out of phase, an ellipse is seen, Figure 17.29 (ii).

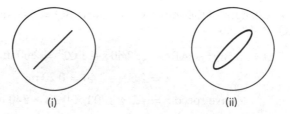

(i) (ii)

Figure 17.29 *Lissajous figures*

By moving the microphone, the average distance between successive positions when a sloping straight line in the same direction as the original appears on the screen can be measured. This distance is λ, the wavelength. The speed of sound c is then calculated from $c = f\lambda$, where f is the known frequency of the sound. This method is more accurate than the method described earlier, as the Lissajous straight line position of the microphone can be found with far greater accuracy than the maxima (or minima) positions required in Figure 17.28.

■ Wave Properties of Electromagnetic Waves

Electromagnetic waves, like all waves, can undergo reflection, refraction, interference and diffraction. In laboratory demonstrations, *microwaves* of about 3 cm wavelength may be used. These are radiated from a horn waveguide T and are received by a similar waveguide R or by a smaller *probe* X. The detected wave then produces a deflection in a connected meter. Some experiments which can be performed in a school laboratory are illustrated in Figure 17.30 (i)–(v).

Polarisation of Waves

A transverse wave due to vibrations in *one plane* is said to be *plane-polarised*. Figure 17.31 shows a plane-polarised wave due to vibrations in the vertical plane yOx and another plane-polarised wave due to vibrations in the perpendicular plane zOx. Both waves travel in the direction Ox.

Consider a horizontal rope AD attached to a fixed point D at one end, Figure 17.32 (i). Transverse waves due to vibrations in many different planes can be set up along A by holding the end A in the hand and moving it up and down in all directions perpendicular to AD, as illustrated by the arrows in the plane X. Suppose we repeat the experiment but this time we have two parallel slits B and C between A and D as shown. A wave then emerges along BC, but unlike the waves along the part AB of the rope (now shown), which are due to vibrations in many different planes, the wave along BC is due only to vibrations parallel to the slit B. This plane-polarised wave passes through the parallel slit C. But when C is turned so that it is *perpendicular* to B, as shown in Figure 17.32 (ii), *no wave* is now obtained beyond C along CD. A Slinky coil in place of the rope AD can also be used to demonstrate polarisation in the same way.

Longitudinal Waves

It is important to note that *no polarisation can be obtained with longitudinal waves*. Figure 17.32 illustrates how transverse waves can be distinguished by experiment from longitudinal waves. If the rope AD is replaced by a thick elastic cord or Slinky coil, and longitudinal waves are produced along AD, then turning the slit C round from the position shown in Figure 17.32 (i) to that shown in Figure 17.31 (ii) makes no difference to the wave—it travels through B and C undisturbed. Since sound waves are longitudinal waves, no polarisation of sound waves can be produced. As we shall see in a later chapter, because light waves are transverse waves they can be polarised. The phenomena of interference and diffraction occur both with sound and light waves, but only the phenomenon of polarisation can distinguish between waves which may be longitudinal or transverse.

Electromagnetic Waves

Figure 17.33 illustrates an experiment on polarisation carried out with *electromagnetic waves*. Here a grille of parallel metal rods is rotated between:
(a) a source T of 3 cm electromagnetic waves or microwaves;
(b) a detector, a probe P, with a meter connected to it.
When the rods are horizontal the meter reading is high, Figure 17.33 (i). So a wave travels past the grille. When the grille is turned round so that the rods are vertical, there is no deflection in the meter, Figure 17.33 (ii). Thus the wave

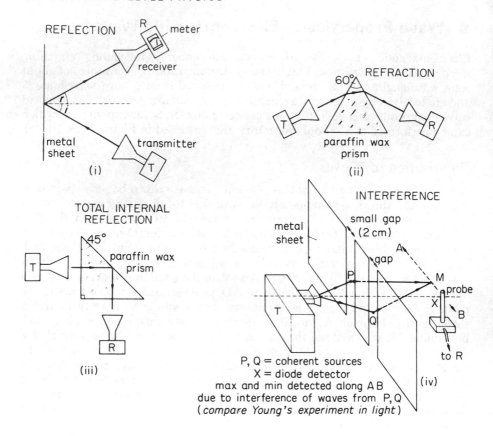

REFLECTION

R meter

receiver

metal sheet

transmitter

T

(i)

REFRACTION

60°

T

R

paraffin wax prism

(ii)

TOTAL INTERNAL REFLECTION

45°

paraffin wax prism

T

R

(iii)

INTERFERENCE

metal sheet

small gap (2 cm)

gap

A

P

M

probe

T

X

B

Q

to R

P, Q = coherent sources
X = diode detector
max and min detected along A B
due to interference of waves from P, Q
(*compare Young's experiment in light*)

(iv)

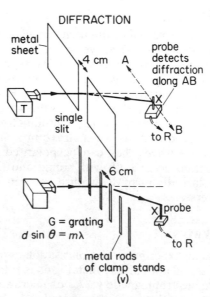

DIFFRACTION

metal sheet

4 cm A

probe detects diffraction along AB

T

X

single slit

to R

B

6 cm

T

X probe

to R

G = grating
$d \sin \theta = m\lambda$

metal rods of clamp stands

(v)

Figure 17.30 *Experiments with microwaves*: (i) *reflection*, (ii) *refraction*, (iii) *total internal reflection*, (iv) *interference*, (v) *diffraction*

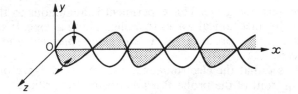

Figure 17.31 *Plane-polarised waves*

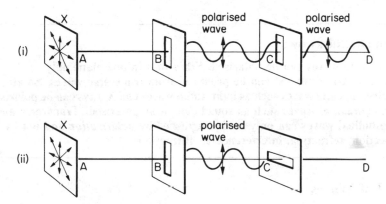

Figure 17.32 *Transverse waves and polarisation*

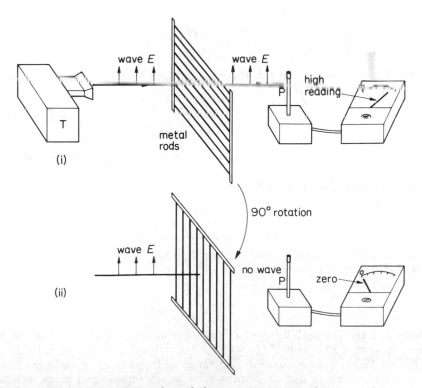

Figure 17.33 *Plane-polarised electromagnetic waves (microwaves)*

does not travel past the grille. This experiment is analogous to that illustrated in Figure 17.32 for mechanical waves travelling along a rope. It shows that the electromagnetic waves produce by T are *plane-polarised* and so they are transverse waves.

We can also see that the electromagnetic waves from T are plane-polarised by placing T in front of the probe P *without* using the grille, so that the meter joined to P indicates a high reading. If T is now rotated about its axis through 90°, the meter reading falls to zero. The probe P detects vibrations in one plane. So when T is rotated through 90° the plane-polarised waves are not detected.

You should know:

Transverse waves can be polarised. Vibrations in one plane only (plane-polarised waves) can be produced from transverse waves. So all electromagnetic waves such as light, radio waves and X-rays can be polarised.

Longitudinal waves such as sound cannot be polarised. Transverse and longitudinal waves can only be distinguished by *polarisation* and not by reflection, refraction, interference or diffraction.

Speed of Waves

We now list, for convenience, the speed c of waves of various types, some of which are considered more fully in other sections of the book:

1. *Transverse wave on string*

$$c = \sqrt{\frac{T}{\mu}} . \qquad . \qquad . \qquad . \qquad . \qquad (1)$$

where T is the tension and μ is the mass per unit length.

2. *Sound waves in gas*

$$c = \sqrt{\frac{\gamma p}{\rho}} . \qquad . \qquad . \qquad . \qquad . \qquad (2)$$

where p is the pressure, ρ is the density and γ is the ratio of the molar heat capacities of the gas.

3. *Longitudinal waves in solid*

$$c = \sqrt{\frac{E}{\rho}} . \qquad . \qquad . \qquad . \qquad . \qquad (3)$$

where E is Young modulus and ρ is the density.

4. *Electromagnetic waves*

$$c = \sqrt{\frac{1}{\mu \varepsilon}} . \qquad . \qquad . \qquad . \qquad . \qquad (4)$$

where μ is the permeability and ε is the permittivity of the medium. In free space $\mu_0 = 4\pi \times 10^{-7}$ H m^{-1}, $\varepsilon_0 = 8 \cdot 85 \times 10^{-12}$ F m^{-1}, so $c = 3 \cdot 0 \times 10^8$ m s^{-1}.

Figure 17.34 (i) shows roughly the range of frequencies in the spectrum of *matter waves*—waves due to vibrations in solids, liquids and gases including sound waves. Sound frequencies in air range from about 20 to 20 000 Hz for

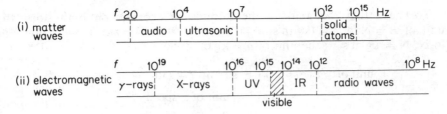

Figure 17.34 *Frequency spectrum*

those detected by the human ear but very much higher particle frequencies can be obtained in solids.

Figure 17.34 (ii) shows roughly the range of frequencies in the spectrum of *electromagnetic waves*. The various waves are discussed later. The frequency range detected by the eye is about 4×10^{14} to 7×10^{14} Hz, a range factor of about 2. The human ear, however, has a range factor of about 1000.

Speed of Sound in a Medium

When a sound wave travels in a medium, such as a gas, a liquid or a solid, the particles in the medium are subjected to varying stresses and strains (p. 479). So the speed of a sound wave is partly governed by the *modulus of elasticity*, E, of the medium, which is defined by the relation

$$E = \frac{\text{stress}}{\text{strain}} = \frac{\text{force per unit area}}{\text{change in length (or volume)/original length (or volume)}} \quad (1)$$

The speed, c, also depends on the density, ρ, of the medium, and it can be shown that

$$c = \sqrt{\frac{E}{\rho}} \quad . \qquad . \qquad . \qquad . \qquad . \qquad (2)$$

When E is in pascals or newtons per metre^2 (N m^{-2}) and ρ in kg m^{-3}, then c is in metres per second (m s^{-1}). The relation (2) was first obtained by Newton.

For a solid, E is the Young modulus of elasticity. The magnitude of E for steel is about 2×10^{11} Pa (N m^{-2}), and the density ρ of steel is 7800 kg m^{-3}. So the speed of sound in steel is given by

$$c = \sqrt{\frac{E}{\rho}} = \sqrt{\frac{2 \times 10^{11}}{7800}} = 5060 \text{ m s}^{-1}$$

For a liquid, E is the bulk modulus of elasticity. Water has a bulk modulus of $2{\cdot}04 \times 10^9$ Pa (N m^{-2}), and a density of 1000 kg m^{-3}. The calculated speed of sound in water is then given by

$$c = \sqrt{\frac{2{\cdot}04 \times 10^9}{1000}} = 1430 \text{ m s}^{-1}$$

The proof of the speed formula requires advanced mathematics, and is beyond the scope of this book.

To check that the units in the formula are correct on both sides, the unit of $E = N m^{-2} = (kg\ m\ s^{-2})\ m^{-2}$, since from $F = ma$ in mechanics for force, $N = kg\ m\ s^{-2}$; the unit of $\rho = kg\ m^{-3}$. So

$$\text{unit of } \sqrt{E/\rho} = \sqrt{kg\ m\ s^{-2}\ m^{-2}/kg\ m^{-3}} = \sqrt{m^2\ s^{-2}}$$

$$= m\ s^{-1} = \text{unit of } c, \text{ speed}$$

Speed of Sound in a Gas, Laplace's Correction

Changes in pressure and volume occur at different places in a gas when a sound wave travels along the gas. If the pressure–volume changes take place *isothermally* (constant temperature), the velocity of the sound wave is given by $c = \sqrt{p/\rho}$, where p is the pressure and ρ is the density of the gas.

For air at normal conditions, $p = 1 \cdot 01 \times 10^5$ Pa and $\rho = 1 \cdot 29$ kg m^{-3}. So

$$c = \sqrt{\frac{1 \cdot 01 \times 10^5}{1 \cdot 29}} = 280\ m\ s^{-1}$$

This theoretical value was first obtained by Newton and it is well below the experimental value of about 340 m s^{-1}.

About a century later, Laplace suggested that the pressure–volume changes at different places in the air took place quickly so no heat enters or leaves the gas while the wave travels along. In this case the speed of sound formula changes to

$$c = \sqrt{\frac{\gamma p}{\rho}} \,. \qquad . \qquad . \qquad . \qquad . \qquad (1)$$

γ for air is $1 \cdot 40$. *Laplace's correction*, as it is known, then changes the value of the speed in air at $0\ ^{\circ}$C to

$$c = \sqrt{\frac{1 \cdot 40 \times 1 \cdot 01 \times 10^5}{1 \cdot 29}} = 331\ m\ s^{-1}$$

This is in good agreement with the experimental value.

Effect of Pressure and Temperature on Speed of Sound in a Gas

Suppose that a mole of gas has a mass M and a volume V. The density is then M/V and so the speed of sound, c, is

$$c = \sqrt{\frac{\gamma p}{\rho}} = \sqrt{\frac{\gamma p V}{M}}$$

But the molar gas equation is $pV = RT$, where R is the molar gas constant and T is the absolute temprature. So

$$c = \sqrt{\frac{\gamma R T}{M}}. \qquad . \qquad . \qquad . \qquad . \qquad (1)$$

Since γ, M and R are constants for a given gas, it follows that *the speed of sound in a gas is independent of the pressure* **if the temperature is constant.**

This has been verified by experiments which showed that the speed of sound at the top of a mountain is about the same as at the bottom. It also follows from (1) that

the speed of sound is proportional to the square root of its kelvin temperature.

So if the speed in air at 16°C is 338 m s^{-1} by experiment, the speed, c, at 0°C is calculated from

$$\frac{c}{338} = \sqrt{\frac{273}{289}}$$

from which $\qquad c = 338\sqrt{\frac{273}{289}} = 329 \text{ m s}^{-1}$

Television Multipath Effects

As we saw on p. 491, all waves are subject to reflection and refraction: these two phenomena are responsible for some of the more common problems with television reception.

You may have noticed that the image on your television screen is sometimes afflicted by *ghosting*, where a second image appears horizontally displaced from the true image. This is caused by the reflection of the television signal from hills, buildings and even aircraft! The reflected signal will arrive a little later than the direct signal, and the television set will display this signal a little later in its scan across the screen. So the ghost image appears. This effect is usually avoided using a properly oriented, directional antenna pointing towards the nearest television transmitter. In this case, only the direct signal is detected, unless the reflected signal happens to arrive from exactly the same direction as the direct signal.

Refraction is responsible for a form of television interference known as *multipath fading*. The refractive index of the earth's atmosphere is not constant, but varies with height and time. So waves can travel to the receiver along many different paths, both direct and refracted in the upper atmosphere. At the receiver the interference of the multipath signals can cause the image to fade for a while. Reception problems caused by refraction are enhanced by particular atmospheric conditions, which is why weather forecasters can sometimes predict when television reception will be poor.

EXERCISES 17   Oscillations and Waves

Multiple Choice

1 Figure 17A (i) shows a progressive sound wave W at an instant when it is travelling in a direction Ox. y represents the displacement of the air at this instant. The maximum compressions (maximum pressures) are then at points

A 1, 2, 3   B 1 and 2 only   C 2, and 3 only   D 1 only   E 3 only

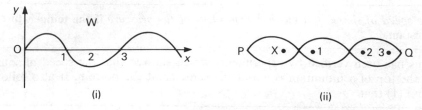

Figure 17A

2 Figure 17A (ii) shows a stationary wave between two fixed points P and Q. Which point(s) of 1, 2 and 3 are in phase with the point X?

A 1, 2 and 3   **B** 1 and 2 only   **C** 2 and 3 only
D 1 only   **E** 3 only

3 In Figure 17B (i), a radio microwave M travels towards a detector D which shows a large deflection. When a metal grill X is placed between M and D facing the transmitter as shown, no deflection is obtained at D. This is because

A X reflects the wave   **B** X refracts the wave   **C** the wave M is too weak
D the wave M is plane-polarised   **E** X diffracts the wave

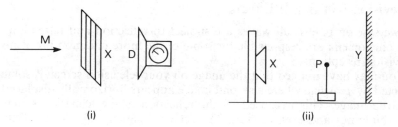

Figure 17B

4 In Figure 17B (ii), a microwave transmitter X sends a radio wave of wavelength 30 mm to a metal sheet Y. A probe P between X and Y detects an antinode. To detect two more antinodes, P must be moved towards X a distance in mm of

A 60   **B** 45   **C** 30   **D** 15   **E** 10

5 Which is the most correct statement in **A** to **E**?

A A progressive wave has a node at the beginning.
B A transverse wave cannot be reflected.
C A progressive wave cannot be refracted.
D A transverse wave can be polarised.
E A sound wave is a transverse wave.

Longer Questions

6 If the speed of sound in air is 340 metres per second, calculate   (i) the wavelength when the frequency is 256 Hz;   (ii) the frequency when the wavelength is 0·85 m.

7 A small piece of cork in a ripple tank oscillates up and down as ripples pass it. If the ripples travel at 0·20 m s^{-1}, have a wavelength of 15 mm and an amplitude of 5·0 mm, what is the maximum velocity of the cork? (*L.*)

8 A beam of electromagnetic waves of wavelength 3·0 cm is directed normally at a grid of metal rods, parallel to each other and arranged vertically about 2·0 cm apart. Behind the grid is a receiver to detect the waves. It is found that when the grid is in this position, the receiver detects a strong signal but that when the grid is rotated in a vertical plane through 90°, the detected signal strength falls to zero. What property of the wave gives rise to this effect? Account briefly in general terms for the effect described above. (*L.*)

9 State and explain the differences between progressive and stationary waves.

A progressive and a stationary simple harmonic wave each have the same frequency of 250 Hz and the same velocity of 30 m s^{-1}. Calculate (i) the phase difference between two vibrating points on the progressive wave which are 10 cm apart; (ii) the distance between nodes in the stationary wave.

10 Two identical progressive waves, travelling in opposite directions, will form a stationary wave where they superpose. Explain with the aid of diagrams how two such progressive waves combine when they meet (i) in phase; (ii) in antiphase.

Explain the terms *node* and *antinode*. How do you account for their formation?

Identify one situation that makes use of stationary waves. State whether these waves are longitudinal or transverse. (*L.*)

11 Two waves of equal frequency and amplitude travel in opposite directions in a medium.

(a) Why is the resultant wave called 'stationary'?

(b) State *two* differences between a stationary and a progressive wave.

(c) Explain briefly, using the principle of superposition, why nodes and antinodes of displacement are obtained in a stationary wave.

If the amplitudes of the two waves are 3 units and 1 unit respectively, show by the principle of superposition that the ratio of the amplitudes of the stationary wave at an antinode and node respectively is 2:1.

12 Two small loudspeakers A and B are 0·50 m apart. Figure 17C (i). Both emit sound waves of the same frequency f. A detector moves along a line CD 1·20 m from AB and parallel to it. A maximum wave is detected at C where OC is the perpendicular bisector of AB and again at D directly opposite A. Calculate f. (Speed of sound = 340 m s^{-1}.)

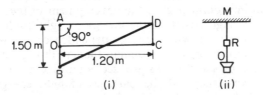

(i) (ii)

Figure 17C

13 A small source O of electromagnetic waves is placed some distance from a plane metal reflector M, Figure 17C (ii). A receiver R, moving between O and M along the line normal to the reflector, detects successive maximum and minimum readings on the meter joined to it.

(a) Explain why these readings are obtained.

(b) Calculate the frequency of the source O if the average distance between successive minima is 1·5 cm and the speed of electromagnetic waves in air = 3·0 × 10^8 m s^{-1}.

14 A plane-progressive wave is represented by the equation

$$y = 0·1 \sin(200\pi t - 20\pi x/17)$$

where y is the displacement in millimetres, t is in seconds and x is the distance from a fixed origin O in metres (m).

Find (i) the frequency of the wave; (ii) its wavelength; (iii) its speed; (iv) the phase difference in radians between a point 0·25 m from O and a point 1·10 m from O.

15 What is the principle of superposition as applied to wave motion?

Discuss as fully as you can the result of superposing two waves of equal amplitude

(a) of the same frequency travelling in opposite directions;

(b) of slightly different frequencies travelling in the same direction.

Describe how you would demonstrate the validity of your conclusions in *one* of these cases.

18 Wave Theory of Light, Reflection, Refraction

Historical

We have already mentioned that light is a form of energy which stimulates our sense of vision. About 1600 *Newton proposed that particles, or corpuscles, were emitted from a luminous object. The* corpuscular theory of light *was adopted by many scientists of the day owing to the authority of Newton, but* HUYGENS, *a famous Dutch scientist, proposed about* 1680 *that light energy travelled from one place to another by means of a* wave-motion.

If the *wave theory of light* was correct, light should bend round a corner, just as sound travels round a corner. The experimental evidence for the wave theory in Huygens's time was very small, and the theory was dropped for more than a century. In 1801, however, THOMAS YOUNG obtained evidence that light could produce wave effects such as interference, and he was among the first to see clearly the close analogy between sound and light waves. As the principles of the subject became understood, other experiments were carried out which showed that light could spread round corners, and Huygens's wave theory of light was revived. Newton's corpuscular theory was rejected since it did not agree with experimental observations. The wave theory of light has played, and is still playing, an important part in the development of the subject.

In 1905 EINSTEIN suggested that the energy in light could be carried from place to place by 'particles' whose energy depended on the wavelength of the light. This was a return to a corpuscular theory, though it was completely different from that of Newton, as we see later. Experiments showed that Einstein's theory was true, and the particles of light energy are known as 'photons' (p. 827). It is now considered that *either* the wave theory *or* the particle theory of light can be used in a problem on light, depending on the circumstances of the problem. In this section we shall consider Huygens's wave theory, which led to many important applications in light.

Wavefronts

Consider a point source of light, S, in air, and suppose that a disturbance, or wave, starts at S and travels outwards. After a time t the wave has travelled a distance ct, where c is the speed of light in air, and the light energy has thus reached the surface of a sphere of centre S and radius ct, Figure 18.1. The surface of the sphere is called the *wavefront* of the light at this instant, and every point on it is vibrating 'in step' or *in phase* with every other point. As time goes on, the wave travels farther and new wavefronts are obtained. These are all surfaces of spheres of centre S.

At points a long way from S, such as C or D, the wavefronts are parts of a sphere of very large radius, and so the wavefronts are then substantially *plane*. Light from the sun reaches the earth in plane wavefronts because the sun is so far away. Plane wavefronts are also produced by a converging lens when a point source of light is placed at its focus.

The significance of the wavefront, then, is that it shows how the light energy travels from one place in a medium to another. A *ray* is the name given to the

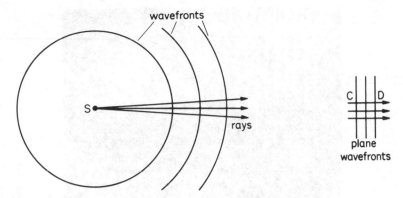

Figure 18.1 *Wavefronts and rays*

direction along which the energy travels, and so a ray of light passing through a point is *perpendicular* to the wavefront at that point. The rays diverge near S, but they are approximately parallel a long way from S, as the curved wavefronts C and D are then fairly plane, Figure 18.1.

Huygens's Construction for the New Wavefront

Suppose that the wavefront from a centre of disturbance S had reached the surface AB in a medium at some instant, Figure 18.2. To obtain the position

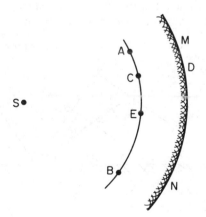

Figure 18.2 *Huygens's construction*

of the new wavefront after a further time t, Huygens said that *every point, A, . . . ,C, . . . , E, . . . , B, on AB becomes a new or 'secondary' centre of disturbance.* The wavelet from A then reaches the surface M of a sphere of radius ct and centre A, where c is the speed of light in the medium; the wavelet from C reaches the surface D of a sphere of radius ct and centre C; and so on for every point on AB. According to Huygens,

the new wavefront is the surface MN which touches all the wavelets from the secondary sources

and in the case considered, it is the surface of a sphere of centre S.

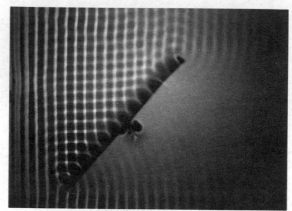

Plate 18A *Ripple tank showing reflection at plane surface*

In this simple example of drawing the new wavefront, the light travels in the same medium. Huygens's principle, however, is especially valuable for drawing the new wavefront when the light travels from one medium to another, as we soon show.

Reflection at Plane Surface

Suppose that a beam of parallel rays between HA and LC is incident on a plane mirror, and imagine a plane wavefront AB which is normal to the rays, reaching the mirror surface, Figure 18.3. At this instant the point A acts as a centre of disturbance. Suppose we require the new wavefront at a time

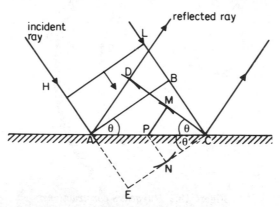

Figure 18.3 *Reflection at plane surface*

corresponding to the instant when the disturbance at B reaches C. The wavelet from A reaches the surface of a sphere of radius AD at this instant. When other points between AC on the mirror, such as P, are reached by the disturbances starting at AB, wavelets of smaller radius than AD such as PM are obtained at the instant we are considering. The new wavefront is the surface CMD which touches all the wavelets.

In the absence of the mirror, the plane wavefront AB would reach the position EC in the time considered. Thus AD = AE = BC, and PN = PM, where PN is perpendicular to EC. The triangles PMC, PNC are hence congruent, as PC is common, angles PMC, PNC are each 90°, and PN = PM.

So the angles marked θ in Figure 18.3 are all equal.

Law of reflection. We can now deduce the law of reflection for the angles of incidence and reflection. The incident wavefront AB and the reflected wavefront CD make equal angles θ with the mirror AC. Since the incident and reflected rays such as HA and HD are normal (90°) to the wavefronts, these rays also make equal angles with AC. So the angles of incidence and reflection are equal.

Point Object

Consider now a *point object* O in front of a plane mirror M, Figure 18.4. A spherical wave spreads out from O, and at some time the wavefront reaches

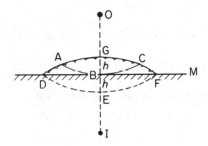

Figure 18.4 *Point object*

ABC. In the absence of the mirror, the wavefront would reach a position DEF in a time t. But every point between D and F on the mirror acts as a secondary centre of disturbance and so wavelets are reflected back into the air. At the end of the time t, a surface DGF is drawn to touch all the wavelets, as shown. DGF is part of a spherical wave which advances into the air, and since it appears to have come from a point I as centre *below* the mirror, then I is a virtual image of O.

The sphere of which DGF is part has a chord DF. Suppose the distance from B, the midpoint of the chord, to G is h. The sphere of which DEF is part has the same chord DF, and the distance from B to E is also h. It follows, from the theorem of product of intersection of chords of a circle, that $DB . BF = h(2r - h) = h(2R - h)$, where r is the radius OE and R is the radius IG. So $R = r$, or IG = OE, and so IB = OB. So the image and object are equidistant from the mirror.

Refraction at Plane Surface

Consider a beam of parallel rays between LO and PD incident on the plane surface of a water medium from air in the direction shown, and suppose that a plane wavefront has reached the position OA at a certain instant, Figure 18.5. Each point between O, A becomes a new centre of disturbance as the wavefront advances to the surface of the water, and the wavefront changes in direction when the disturbance enters the liquid because the speed is now less than in air.

Suppose that t is the time taken by the light to travel from A to D. The disturbance from O travels a distance OB, or $c_w t$, in water in a time t, where c_w is the speed of light in water. At the end of the time t, the wavefronts in the water from the other secondary centres between O, D reach the surfaces of spheres to each of which DB is a tangent. So DB is the new wavefront

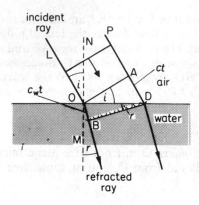

Figure 18.5 *Refraction at plane surface*

in the water, and the ray OB which is normal to the wavefront is therefore the refracted ray.

Since c is the speed of light in air, $AD = ct$. Now

$$\frac{\sin i}{\sin r} = \frac{\sin LON}{\sin BOM} = \frac{\sin AOD}{\sin ODB}$$

$$\therefore \frac{\sin i}{\sin r} = \frac{AD/OD}{OB/OD} = \frac{AD}{OB} = \frac{ct}{c_w t} = \frac{c}{c_w} \qquad . \qquad . \qquad . \qquad (1)$$

But c, c_w are constants for the given media.

$$\therefore \frac{\sin i}{\sin r} \text{ is a constant}$$

which is Snell's law of refraction (p. 433).

It can now be seen from (1) that the refractive index, n, of a medium is given by $n = \dfrac{c}{c_m}$, where c is the speed of light in air and c_m is the speed of light in the medium.

The refractive index n of crown glass is about 1·5. So $c/c_g = 1·5$ where c_g is the speed of light in glass. Now c, the speed of light in air is about 3×10^8 m s^{-1}. So

$$\frac{3 \times 10^8}{c_g} = 1·5, \text{ or } c_g = 3 \times 10^8/1·5 = 2 \times 10^8 \text{ m s}^{-1}$$

Analogy for Refracted Wavefront

An analogy may help to understand how the wavefront OA in air is refracted to BD.

Suppose a line of soldiers OA is marching in line at a steady speed along a field of well-cut grass towards the straight boundary with another field where the grass is thick and their speed will be less than before.

When the line reaches OA, the soldier at the end A will reach D in a certain time. In the same time, however, the soldier at O reaches a *smaller* distance OB in the thick grass because his speed is less than that of A. Other soldiers between O and A also move fast at first but slow down after reaching the boundary OD. To keep in line, the line of soldiers turn round from OA to BD. Then they all move in a line parallel to BD more slowly than before.

Wave Theory for Two-media Refraction

As we saw earlier, the refraction index n of a material such as glass or water is the ratio $\sin i/\sin r$ when the light travels from air (strictly, a vacuum) into the glass or water. In this case $n = 1\cdot5$ for crown glass and $n = 1\cdot33$ for water. And if c is the speed of light in air and c_g in glass, then

$$n = c/c_g \text{ for glass} = 1\cdot5$$

and

$$n = c/c_w \text{ for water} = 1\cdot33$$

From the two values of n for glass and for water, we see that the speed of light in water is greater than in glass. When light is refracted from glass to air, the light path from air to glass is reversed in direction. So from glass to air, $n = 1/1\cdot5 = c_g/c$ and from water to air, $n = 1/1\cdot33 = c_w/c$.

Suppose refraction occurs across the boundary between one medium 1 and another medium 2, for example, from water to glass. In this case the refractive index will be written $_1n_2$. If c_1 and c_2 are the respective speeds of light in medium 1 and 2, and θ_1 and θ_2 are the angles made with the normal by an incident ray in medium 1 and a refracted ray in medium 2, then (Figure 18.6)

$$_1n_2 = \frac{\sin \theta_1}{\sin \theta_2} = \frac{c_1}{c_2}$$

So for refraction from water to glass, the refracted index $_wn_g$ is given by

$$_wn_g = \frac{c_w}{c_g}$$

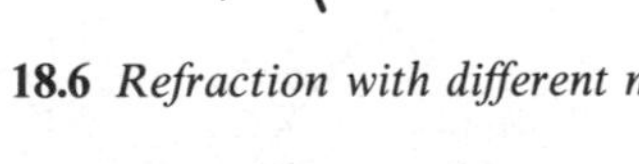

Figure 18.6 *Refraction with different media*

The refractive index of water $= 1\cdot33$ or $4/3$, so $c_w = c/(4/3) = 3c/4$. The refractive index of glass is about $1\cdot5$ or $3/2$, so $c_g = c/(3/2) = 2c/3$. So

$$_wn_g = \frac{c_w}{c_g} = \frac{3c/4}{2c/3} = \frac{9}{8} = 1\cdot13$$

Critical Angle on Wave Theory

As we saw in the optical ray section, total reflection occurs (a) when light travels from one medium to a *less dense* medium and (b) the angle of incidence in the denser medium is greater than the critical angle C between the two media. If the refractive index of the denser medium such as glass is n and the less dense medium is air, then Figure 18.7 (i),

$$\sin C = 1/n$$

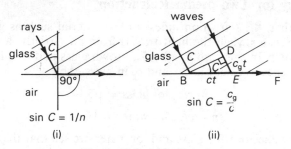

Figure 18.7 *Critical angle*

Figure 18.7 (ii) shows a wave theory explanation of the critical angle value C. When the plane wavefronts such as BD make the angle C with the interface (boundary) for glass–air, the refracted plane wavefronts in air then move in the direction BEF along the boundary. (At a slightly smaller angle than C, the critical angle, the refracted wavefronts in air move at a small angle to the boundary.) The light travels from D to B through a distance $c_g t$ in a time t. In the same time t, the light travels in air from B to E in a time ct. So in the right-angle triangle BDC,

$$\sin \text{DBE} \, (\sin C) = \text{DE/BE} = c_g/ct = c_g/c$$

So
$$\sin C = c_g/c = 1/(c/c_g) = 1/n$$

Dispersion

The dispersion (spreading) of colours produced by a medium such as glass is due to the difference in speeds of the various colours in the medium. Suppose a plane wavefront AC of white light is incident in air on a plane glass surface, Figure 18.8. In the time the light takes to travel in air from C to D, the red

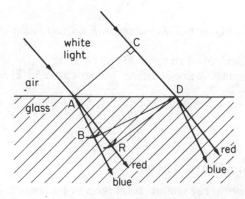

Figure 18.8 *Dispersion*

light from the centre of disturbance A reaches a position shown by the wavelet at R. The blue light from A reaches another position shown by the wavelet at B, since the speed of blue light in glass is less than that of red light, so AB is less than AR. On drawing the new wavefronts DB, DR, it can be seen that the blue wavefront BD is refracted *more* in the glass than the red wavefront DR. The refracted blue ray is AB and the refracted red ray is AR, and so dispersion occurs when white light is refracted from air into glass.

Power of a Lens

We can now consider briefly the effect of lenses on the *curvature* of wavefronts. The curvature of a spherical wavefront is defined as $1/r$, where r is the radius of the wavefront surface.

When a plane wavefront is incident on a converging lens L, a spherical wavefront, S, of radius f emerges from L, where f is the focal length of the lens, Figure 18.9 (i). This is because the light travels faster in the air at the top L than in the glass at the middle of the lens. Parallel rays, which are normal

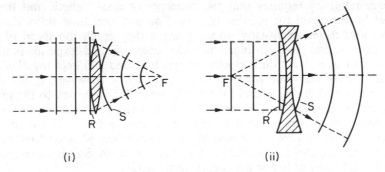

Figure 18.9 (i) *Converging lens* (ii) *Diverging lens*

to the wavefront, are thus refracted towards F, the focus of the lens. Now the curvature of a plane wavefront is zero, and the curvature of the spherical wavefront S is $1/f$. So the converging lens adds a curvature of $1/f$ to a wavefront incident on it. $1/f$ is defined as the *converging power* of the lens:

$$\text{Power } P = \frac{1}{f}$$

Figure 18.9 (ii) illustrates the effect of a *diverging* lens on a plane wavefront R. The front S emerging from the lens has a curvature opposite to S in Figure 18.9 (i), and it appears to be diverging from a point F behind the diverging lens, which is its focus. The curvature of the emerging wavefront is thus $1/f$, where f is the focal length of the lens, and the powers of the converging and diverging lens are opposite in sign.

The power of a converging lens is positive, since its focal length is positive, while the power of a diverging lens is negative. The unit of power is the *dioptre*, D, which is the power of a lens of 1 metre focal length. A lens of +8 dioptres, or +8D, is therefore a converging lens of focal length 1/8 m or 12·5 cm, and a lens of −4D is a diverging lens of 1/4 m or 25 cm focal length. Opticians classify converging and diverging lenses in dioptres, D.

> ## You should know:
>
> In Huygens's wave theory, every point on a wavefront becomes a new centre of disturbance. Refraction is due to the change in speed of light from one medium such as air to another medium such as glass or water. Light has the greatest speed c in a vacuum or air. From air to glass,
>
> $$\text{refractive index } _a n_g = c/c_g$$
>
> The critical angle for glass–air is given by $\sin C = c_g/c$

Special Relativity

One of the consequences of Einstein's theory of special relativity is that nothing can travel faster than the speed of light. Why not? Why can we not take something which is travelling near the speed of light, apply a little force, and make it go faster?

Special relativity requires that the concepts of mass, length and time be completely reassessed for moving objects. For instance, time slows down as objects move faster, eventually stopping when they reach the speed of light. This might sound very strange, but has been verified experimentally. If two identical, accurate atomic clocks are synchronised and kept together, then they tell exactly the same time for very many years. However, if one is put in an aircraft and flown around for a while, when it is returned to the ground it appears to have lost time compared with its twin. The effect is very slight, but becomes more pronounced as the object moves faster. If an astronaut were to travel at 0·9 times the speed of light, then when 30 years have elapsed on the spacecraft, 69 years would have elapsed on earth. So the astronaut will return to find most of his or her friends long dead!

Another consequence is that if an outside force acts on the rocket for 69 years, as far as the rocket is concerned it lasted only 30 years. So the force appears to act for a shorter and shorter time the closer the rocket gets to the speed of light, and the acceleration it produces is diminished. Eventually, when the object is travelling at the speed of light, any applied force will appear to act for no time at all, and the acceleration will be zero! Hence nothing can travel faster than the speed of light. This is neatly expressed by Einstein's formula $a = \dfrac{F}{m}\left(1 - \dfrac{v^2}{c^2}\right)^{3/2}$, compared with Newton's $a = \dfrac{F}{m}$. Of course, at speeds v commonly encountered in real-world scenarios the difference is very slight, and classical Newtonian mechanics is a perfectly good approximation to use. However, it *is* necessary to take relativity into account for some engineering applications. For example, both special and general relativity are applied in the Global Positioning System (GPS), a satellite system which allows users to pinpoint their location anywhere on the earth's surface to within 30 metres!

EXERCISES 18   Wave Theory

Multiple Choice

1 A point object O is placed in front of a plane mirror M. Which of the statements **A** to **E** is most correct?

A O produces plane waves reflected by M.
B Circular waves from O are reflected by M to O.
C Light waves from O are reflected towards the image of O in M.
D The curvature of circular waves from O is reversed by M.
E The image of O is produced by plane waves from O.

2 Figure 18A (i) shows plane waves refracted for air to water using Huygens's principle. *a, b, c, d, e* are lengths on the diagram. The refractive index from air to water is the ratio

  A *a/e* **B** *b/e* **C** *b/d* **D** *d/b* **E** *a/b*

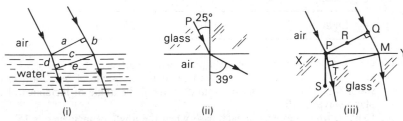

(i)   (ii)   (iii)

Figure 18A

3 Figure 18A (ii) shows a ray PQ in glass refracted into air and the angles of incidence and refraction. The speed of light in air is $3 \cdot 0 \times 10^8$ m s^{-1}. The speed of light in glass in m s^{-1} is about

  A $1 \cdot 5 \times 10^8$ **B** $2 \cdot 0 \times 10^8$ **C** $2 \cdot 2 \times 10^8$ **D** $2 \cdot 5 \times 10^8$ **E** $2 \cdot 8 \times 10^8$

4 Plane waves are incident in air on the boundary XY of a glass block, Figure 18A (iii). P, R, Q are points on the wavefront PQ which reaches the boundary.
    Which of the statements **A** to **E** is most correct?

  A P, R, Q are not coherent sources.
  B When the wave from Q reaches M, the wave from P reaches S.
  C When the wave from Q reaches M, the wave from P reaches T.
  D The time from Q to M in air is greater than from P to T in glass.
  E The wavefronts entering the glass travel faster than those in air.

Longer Questions

5 Using Huygens's principle of secondary wavelets explain, making use of a diagram, how a refracted wavefront is formed when a beam of light, travelling in glass, crosses the glass–air boundary. Show how the sines of the angles of incidence and refraction are related to the speeds of light in air and glass. (*L.*)

6 A parallel beam of monochromatic radiation travelling through glass is incident on the plane boundary between the glass and air. Using Huygens's principle draw diagrams (one in each case) showing successive positions of the wavefronts when the angle of incidence is
(a) 0°; (b) 30°; (c) 60°.
Indicate clearly and explain the constructions used. (The refractive index of glass for the radiation used is 1·5.) (*N.*)

7 A plane wavefront of monochromatic light is incident normally on one face of a glass prism, of refracting angle 30°, and is transmitted. Using Huygens's construction trace the course of the wavefront. Explain your diagram and find the angle through which the wavefront is deviated. (Refracted index of glass = 1·5.) (*N.*)

8 State *Snell's law of refraction* and define *refractive index*.
    Show how refraction of light at a plane interface can be explained on the basis of the wave theory of light.
    Light travelling through a pool of water in a parallel beam is incident on the horizontal surface. Its speed in water is $2 \cdot 2 \times 10^8$ m s^{-1}. Calculate the maximum angle which the beam can make with the vertical if light is to escape into the air where its speed is $3 \cdot 0 \times 10^8$ m s^{-1}.

At this angle in water, how will the path of the beam be affected if a thick layer of oil, of refractive index 1·5, is floated on to the surface of the water? (*O. & C.*)

9 Refraction occurs when waves pass from one medium into another of different refractive index. The table below gives data for sound waves and light waves.

Type of wave	Speed in air $(m\ s^{-1})$	Speed in water $(m\ s^{-1})$
Sound	340	1400
Light	$3\cdot00 \times 10^8$	$2\cdot25 \times 10^8$

(a) Plane waves travelling in air meet a plane water surface.
  (i) Use the data in the table above to calculate the angle of refraction in the water for light waves *and* for sound waves if the angle of incidence of the plane waves is 10°.
  (ii) Explain what happens in each case when the angle of incidence is 15°.
(b) Submarine detection helicopters are fitted with transmitters which send out sound waves. The reflected signals identify the location of the submarine. Explain why it is common for this transmitter to be suspended from the helicopter by a wire so that it is underwater. (*N.*)

10 How did Huygens explain the reflection of light on the wave theory? Using Huygens's conceptions, show that a series of light waves diverging from a point source will, after reflection at a plane mirror, appear to be diverging from a second point, and calculate its position. (*C.*)

11 (a) Explain briefly Huygens's method for constructing wavefronts.
    A parallel beam of light is projected on to the surface of a plane mirror at an angle of incidence of about 70°. Draw a diagram showing clearly how Huygens's method can be used to determine the direction of the reflected beam.
(b) What is meant by *critical angle*? Under what conditions will a wave be totally reflected on meeting a boundary between two media, both of which will allow passage of the wave?
    A beam of light travelling through a transparent medium A is incident on a plane interface into air at an angle of 20°. If the speed of light in the medium is 60% of that in air, calculate the angle of refraction in air.
    When the beam is shone through another transparent medium B and the incident angle is again 20°, it is found that the beam is just totally reflected at a plane interface with air. Calculate the speed of light in B as a percentage of the speed of light, *c*, in air. (*L.*)

12 What is Huygens's principle?
    Draw and explain diagrams which show the positions of a light wavefront at successive equal time intervals when:
(a) parallel light is reflected from a plane mirror, the angle of incidence being about 60°;
(b) monochromatic light originating from a small source in water is transmitted through the surface of the water into the air.
    Describe an experiment, and add the necessary theoretical explanation, to show that in air the wavelength of blue light is less than that of red light. (*N.*)

13 Using Huygens's concept of secondary wavelets show that a plane wave of monochromatic light incident obliquely on a plane surface separating air from glass may be refracted and proceed as a plane wave. Establish the physical significance of the refractive index of the glass.
    In what circumstances does dispersion of light occur? How is it accounted for by the wave theory?
    If the wavelength of yellow light in air is $6\cdot0 \times 10^{-7}$ m, what is its wavelength in glass of refractive index 1·5? (*N.*)

19 Interference of Light Waves

We shall first discuss coherent sources, which are necessary for the phenomenon of interference, and the use of path difference for constructive and destructive interference. We then consider the interference produced in the Young two-slit experiment, the air-wedge experiment and Newton's rings together with their applications, and conclude with the blooming of lenses for clearer images.

Coherent Sources

As we see later, light waves from a sodium lamp, for example, are due to energy changes in the sodium atoms. The emitted waves occur in bursts lasting about 10^{-8} second. The light waves produced by the different atoms are out of phase with each other, as they are emitted randomly and rapidly. We call such sources of light waves as these atoms *incoherent sources* on account of the continual change of phase.

Two sodium lamps X and Y both emit light waves of the same colour or wavelength. But owing to the random emission of light waves from their atoms, their resultant light waves are constantly out of phase. So X and Y are incoherent sources. *Coherent* sources are those which emit light waves of the same wavelength or frequency which are *always* in phase with each other or have a *constant phase difference*. As we now show, two coherent sources can together produce the phenomenon of interference.

Interference of Light Waves, Constructive Interference

Suppose two sources of light, A, B, have exactly the same wavelength and amplitude of vibration, and that their vibrations are always in phase with each other, Figure 19.1. The two sources A and B are therefore *coherent* sources.

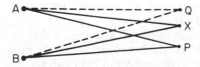

Figure 19.1 *Interference of waves*

Their combined effect at a point is obtained by adding algebraically the displacements at the point due to the sources individually. This is known as the *principle of superposition*. So their resultant effect at X, for example, is the algebraic sum of the vibrations at X due to the source A alone and the vibrations at X due to the source B alone. If X is equidistant from A and B, the vibrations at X due to the two sources are *always* in phase as (i) the distance AX travelled by the wave starting from A is equal to the distance BX travelled by the wave

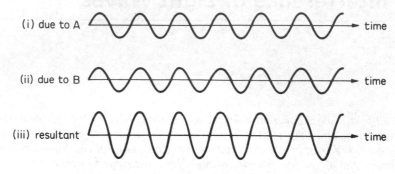

(i) due to A   time

(ii) due to B   time

(iii) resultant   time

Figure 19.2 *Vibrations at X—constructive interference*

starting from B; (ii) the sources A and B are assumed to have the same wavelength and to be always in phase with each other.

Figure 19.2 (i), (ii) illustrate the vibrations at X due to A and B, which have the same amplitude. The resultant vibration at X is obtained by adding the two curves, and has an amplitude *double* that of either curve, Figure 19.2 (iii). Now the energy of a vibrating source is proportional to the square of its amplitude (p. 79). Consequently the light energy at X is four times that due to A or B alone. A bright band of light is thus obtained at X. As A and B are coherent sources, the bright band is *permanent*. With wave crests and troughs arriving at X at the same time, we say that the bright band is due to *constructive interference* of the light waves from A and B at X.

If Q is a point such that BQ is greater than AQ by a whole number of wavelengths (Figure 19.1), the vibration at Q due to A is in phase with the vibration there due to B (see p. 474). A permanent bright band is then obtained at O.

Generally, a permanent bright band is obtained at any point Y if the *path difference*, BY − AY, is given by

$$BY - AY = n\lambda$$

where λ is the wavelength of the sources A, B, and *n* = 0, 1, 2 and so on.

We now see that permanent interference between two sources of light can only take place if they are *coherent* sources, that is they must have the same wavelength and be always in phase with each other or have a constant phase difference. This implies that the two sources of light must have the same colour. As we see later, two coherent sources of light can be produced by using a single primary source of light.

Destructive Interference

Consider now a point P in Figure 19.1 whose distance from B is half a wavelength longer than its distance from A, AP − BP = $\lambda/2$. The vibration at P due to B will then be 180° out of phase with the vibration there due to A (see p. 474), Figure 19.3 (i), (ii). The resultant effect at P is thus zero, as the displacements at any instant are equal and opposite to each other, Figure 19.3 (iii). No light is therefore seen at P. With a wave crest from A

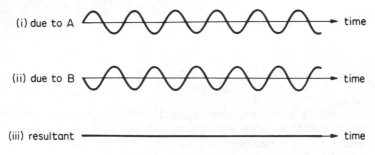

(i) due to A .... time

(ii) due to B .... time

(iii) resultant .... time

Figure 19.3 *Vibrations at P—destructive interference*

arriving at P at the same time as a wave trough from B, the permanent dark band here is said to be due to *destructive interference* of the waves from A and B.

If the path difference, AP − BP, were $3\lambda/2$ or $5\lambda/2$, instead of $\lambda/2$, a permanent dark band would again be seen at P as the vibrations there due to A and B would be 180° out of phase.

If the path difference is zero or a whole number of wavelengths, a bright band is obtained; if it is an odd number of half-wavelengths, a dark band is obtained.

From the principle of the conservation of energy, the total light energy from the sources A and B above must be equal to the light energy in all the bright bands of the interference pattern. The light energy missing from the dark bands is therefore found in the bright bands. It follows that the bright bands on a screen appear *brighter* than when the screen is uniformly illuminated by A and B without forming an interference pattern.

Optical Path, Reflection of Waves

The phase of a wave arriving at a point is affected by the medium through which it travels. For example, part of its path may be in air and part in glass. Since the speed of light is less in glass than in air, there are more waves in a given length in glass than in an equal length in air.

Suppose light travels a distance t in a medium of refractive index n. Then if λ is the wavelength in the medium, the phase difference Δ due to this path (p. 474) is

$$\Delta = \frac{2\pi t}{\lambda} \ . \qquad . \qquad . \qquad . \qquad . \qquad (1)$$

If the wave travels from a vacuum (or air) to this medium, its frequency does not alter but its wavelength and speed become smaller. Suppose λ_0 is the wavelength and c is the speed in a vacuum. Then if c_m is the speed in the medium,

$$\text{frequency} = \frac{c}{\lambda_0} = \frac{c_m}{\lambda}$$

$$\therefore \ \lambda = \frac{c_m}{c} \lambda_0 \qquad . \qquad . \qquad . \qquad . \qquad (2)$$

Substituting for λ from (2) in (1),

$$\Delta = \frac{2\pi ct}{c_m \lambda_0} = \frac{2\pi nt}{\lambda_0}. \qquad . \qquad . \qquad . \qquad (3)$$

since $n = c/c_m$.

From (1) and (3), we see that a light path of geometric length t in a medium of refractive index n produces the same phase change as a light path of length nt in a vacuum. We call 'nt', the product of the refractive index and path length, the *optical path* in the medium. In interference phenomena, we always calculate the optical paths of the coherent light rays. With the notation on p. 518, constructive interference occurs if their optical path difference is $m\lambda$.

As an illustration of optical path, suppose light travels from O to A, a distance d, in air, Figure 19.4. The optical path $= n_0 d = d$, since the refractive index n_0

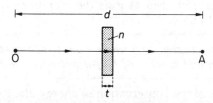

Figure 19.4 *Optical path*

of air is practically 1. Now suppose a thin slab of glass of thickness t and refractive index n is placed between O and A so that the light passes through a length t in the glass. The optical path between O and A is now

$$(d - t) + nt = d + (n - 1)t$$

since the light travels a distance $(d - t)$ in air and a distance t in glass.

Reflection of waves. Light waves may also undergo phase change by reflection at some point in their path. If the waves are reflected at a *denser* medium, for example, at an air–glass interface (boundary) after travelling in air, the reflected waves have a phase change of π or 180° compared to the incident waves, Figure 19.5 (i). This phase change also occurs with matter waves such as sound waves,

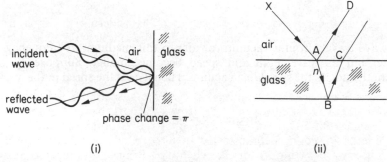

(i)   (ii)

Figure 19.5 *Phase difference and reflection*

as shown in Figure 23.2, p. 590. A phase change of 2π or $360°$ is equivalent to a path length λ

$$\therefore t = \frac{\lambda}{2}$$

To take into account reflection at a *denser* medium, then, we must add (or subtract) $\lambda/2$ to the optical path.

Figure 19.5 (ii) shows an incident ray of light XA partly refracted at A from air to glass and then reflected at B, the glass–air interface. The optical path from A to C is $n(AB + BC)$; there is no phase change by reflection at B since this occurs at an interface with the *less* dense medium, air. By contrast, a phase change equivalent to a path of $\lambda/2$ occurs when XA is reflected at A along AD, since this is reflection at a denser medium, glass.

Young's Two-slit Experiment

From our previous discussion, two conditions are essential to obtain an interference phenomenon in light: (i) two coherent sources of light must be produced; (ii) the coherent sources must be very close to each other as the wavelength of light is very small, otherwise the bright and dark patterns produced some distance away would be too close to each other and no interference pattern would then be seen.

One of the first demonstrations of the interference of light waves was given by YOUNG in 1801. He placed a source, S, of monochromatic light in front of a narrow slit C, and arranged two very narrow slits A, B, close to each other, in front of C. Young then saw bright and dark bands on either side of O on a screen T, where O is on the perpendicular bisector of AB, Figure 19.6.

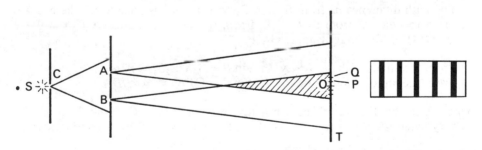

Figure 19.6 *Young's experiment—fringes are shown on right*

Young's observations can be explained by considering the light from S illuminating the two slits A, B. Since the light diverging from A has exactly the same frequency as, and is always in phase with, the light diverging from B, A and B act as *two close coherent sources*. Interference thus takes place in the shaded region, where the light beams overlap, Figure 19.6. As $AO = OB$, a bright band is obtained at O. At a point close to O, such that $BP - AP = \lambda/2$, where λ is the wavelength of the light from S, a dark band is obtained. At a point Q such that $BQ - AQ = \lambda$, a bright band is obtained; and so on for either side of O. Young demonstrated that the bands or *fringes* were due to interference by covering A or B, when the fringes disappeared. Young's experiment is an example of interference by *division of wavefront* from C at A and B. Diffraction at A and B controls the overall angular width of the bands and their intensities (see Figure 20.9, p. 547).

Separation of Fringes

We can find the separation x of the fringes in terms of d, the separation of the slits S_1, S_2, the distance D of the screen T from the slits and the wavelength λ of the monochromatic light coming from the slits Figure 19.7 (i).

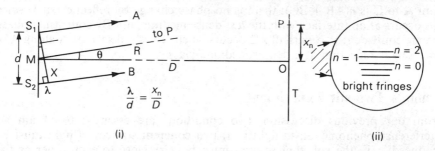

Figure 19.7 *Separation of fringes*

The centre of the fringes is at O, where the perpendicular bisector MO of S_1S_2 meets T. Here the path difference $S_2O - S_1O = 0$, since the paths are equal. On either side of O, the fringes cover only a very small distance as the wavelength of light is very small. So if P is the nth bright fringe from O, with S_1S_2 very small (0·4 mm, for example) and D very large relatively (1·0 m, for example), we can assume that S_1A and S_2B travel to P along *parallel* paths as shown. The line MR to P is also parallel to S_1A and S_2B.

The path difference $n\lambda$ from S_1 and S_2 to P is the distance S_2X, where S_2X is the perpendicular from S_1 to S_2B. From the geometry, angle $S_2S_1X = \theta =$ angle OMP in the large triangle OMP.

From the small triangle S_2S_1X, $\sin \theta = S_2X/S_2S_1 = n\lambda/d$
From the large triangle OMP, $\tan \theta = OP/OM = x_n/D$

where x_n is the distance of the nth fringe from the centre O. Now the angle θ is very small (only of the order of 0·5°). So $\tan \theta = \sin \theta$ from trigonometry.

Then
$$\frac{x_n}{D} = \frac{n\lambda}{d}$$

So
$$x_n = \frac{n\lambda D}{d}$$

For the $(n-1)$th fringe next to the nth fringe, since λ, D and d are constant,

$$n_{n-1} = \frac{(n-1)\lambda D}{d}$$

So separation of fringes, $x = x_n - x_{n-1} = \lambda D/d$
and
$$\lambda = xd/D$$

With $x = 1.2$ mm $= 1.2 \times 10^{-3}$ m, $d = 0.5$ mm $= 0.5 \times 10^{-3}$ m and $D = 1.0$ m,

$$\lambda = \frac{xd}{D} = \frac{1.2 \times 10^{-3} \times 0.5 \times 10^{-3}}{1.0} = 6 \times 10^{-7} \text{ m} = 600 \text{ nm}$$

The fringes appear on either side of the middle fringe O. O has a path difference of zero, the first fringe from O has a path-difference λ and others have path differences 2λ, 3λ and so on. Figure 19.7 (ii).

When $\theta = 0$ in Figure 19.7 (i), the diffracted wavefronts from the two slits travel in the direction MO to form the central fringe at O on the screen. When the first bright fringe is formed, the diffracted wavefronts now travel in a direction θ to MO given by $\theta = \lambda/d$. This is also the angle subtended at M by the first bright fringe and O.

Measurement of Wavelength by Young's Interference Fringes

A laboratory experiment to measure wavelength by Young's interference fringes is shown in Figure 19.8. Light from a small filament lamp is focused by a lens

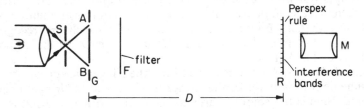

Figure 19.8 *Laboratory experiment on Young's interference fringes*

on to a narrow slit S, such as that in the collimator of a spectrometer. Two narrow slits A, B, about 0·5 millimetre apart, are placed a short distance in front of S, and the light coming from A, B is viewed in a low-powered microscope or eyepiece M about one metre away. Some coloured interference fringes are then observed by M. A red and then a blue filter, F, placed in front of the slits, produces red and then blue fringes. Observation shows that the separation x of the red fringes is more than that of the blue fringes. Now $\lambda = dx/D$, where x is the separation of the fringes, or $\lambda \propto x$. So the wavelength of red light is *longer* than that of blue light.

An approximate value of the wavelength of red or blue light can be found by placing a Perspex rule R in front of the eyepiece and moving it until the graduations are clearly seen, Figure 19.8. The average distance, x between the fringes is then measured on R. The distance d between the slits can be found by magnifying the distance by a converging lens, or by using a travelling microscope. The distance D from the slits to the Perspex rule, where the fringes are formed, is measured with a metre rule. The wavelength λ can then be calculated from $\lambda = dx/D$; it is of the order 6×10^{-7} m. Further details of the experiment can be obtained from *Advanced Level Practical Physics* by Nelkon and Ogborn (Heinemann).

Measurements can also be made using a spectrometer, with the collimator and telescope adjusted for parallel light. The narrow collimator slit is illuminated by sodium light, for example, and the double slits placed on the table. Young's fringes can be seen through the telescope after alignment. From the theory on p. 522, the angular separation of the fringes is λ/d. So by measuring the average angular separation of a number of fringes with the telescope, λ can be calculated from $\lambda = d\theta$, if d is known or measured.

The wavelengths of the extreme colours of the visible spectrum vary with the observer. This may be 4×10^{-7} m (400 nm) for violet and 7.5×10^{-7} m (750 nm) for red; an 'average' value for visible light is 5.5×10^{-7} m (550 nm), which is a wavelength in the green.

Appearance of Young's Interference Fringes

The experiment just outlined can also be used to demonstrate the following points.

1. If the source slit S is moved *nearer* the double slits the separation of the fringes is unaffected but their brightness increases. This can be seen from the formula x (separation) $= \lambda D/d$, since D and d are constant.

2. If the distance apart d of the slits is diminished, keeping S fixed, the separation of the fringes increases. This follows from $x = \lambda D/d$.

3. If the source slit S is *widened* the fringes gradually disappear. The slit S is then equivalent to a large number of narrow slits, each producing its own fringe system at different places. The bright and dark fringes of different systems therefore overlap, giving rise to uniform illumination. It can be shown that, to produce interference fringes which are recognisable, the slit width of S must be less than $\lambda D'/d$, where D' is the distance of S from the two slits A, B.

4. If one of the slits, A or B, is covered up, the fringes disappear.

5. If white light is used, the central fringe is white, and the fringes either side are *coloured*. Blue is the colour nearer to the central fringe and red is farther away. The path difference to a point O on the perpendicular bisector of the two slits A, B is zero for all colours, and so each colour produces a bright fringe here. As they overlap, a white fringe is formed. Farther away from O, in a direction parallel to the slits, the shortest visible wavelengths, blue, produce a bright fringe first.

6. The intensity of a light wave is proportional to A^2, where A is the amplitude of the wave. In constructive interference, the resultant amplitude of the wave from both slits is $2A$ if A is the amplitude of each wave. So the intensity (brightness) of the bright fringe is proportional to $(2A)^2$ or $4A^2$. The intensity of the light from one slit is proportional to A^2. So light energy is transferred from the dark fringe positions to the bright fringe positions in the Young experiment.

Examples on Young's Two-slit Experiment

1 In a Young's slits experiment, the separation between the first and fifth bright fringe is 2·5 mm when the wavelength used is $6\cdot2 \times 10^{-7}$ m. The distance from the slits to the screen is 0·80 m. Calculate the separation of the two slits.

From previous, $\qquad \lambda = \dfrac{dx}{D}$, where d is the slit separation

$$\therefore d = \frac{\lambda D}{x} = \frac{6\cdot2 \times 10^{-7} \times 0\cdot8}{2\cdot5 \times 10^{-3}/4}$$

$$= 8 \times 10^{-4} \text{ m} = 0\cdot8 \text{ mm}$$

2 In Figure 19.9, S_1 and S_2 are two coherent light sources in a Young's two-slit experiment separated by a distance 0·50 mm and O is a point equidistant from S_1 and S_2. O is on a screen A which is 0·80 m from the slits.

When a thin parallel-sided piece of glass G of thickness $3·6 \times 10^{-6}$ m is placed near S_1 as shown, the centre of the fringe system moves from O to a point P. Calculate OP if the wavelength of the monochromatic light from the two slits is $6·0 \times 10^{-7}$ m and the refractive index of the glass is 1·5.

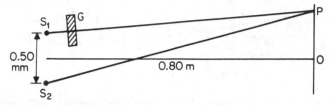

Figure 19.9 *Example on Young's two-slit experiment*

(*Analysis* (a) If P is the centre of the fringe system, the number of waves in S_1P = the number of waves in S_2P. (b) Since light travels slower in the glass G than in air, the number of wavelengths in G is *more* than in air of the same thickness as G.)
In air, $\lambda_a = 6·0 \times 10^{-7}$ m. In glass, $\lambda_g = 6·0 \times 10^{-7}/1·5 = 4 \times 10^{-7}$ m.

The *increase* in the number of waves in G when it replaces air of the same thickness

$$= \frac{3·6 \times 10^{-6}}{4 \times 10^{-7}} - \frac{3·6 \times 10^{-6}}{6 \times 10^{-7}} = 9 - 6 = 3$$

So number of bright bands from O to P = 3

Now separation of bright bands x is given by

$$x = \frac{D\lambda}{d} = \frac{0·8 \times 6 \times 10^{-7}}{0·5 \times 10^{-3}}$$

$$= 9·6 \times 10^{-4} \text{ m}$$

So $OP = 3 \times 9·6 \times 10^{-4} = 2·88 \times 10^{-3}$ m = 2·88 mm

(*Alternatively* Extra path difference due to $G = nt - t = (n - 1)t$

So extra number of wavelengths $= \dfrac{(n - 1)t}{\lambda_a}$

$$= \frac{(1·5 - 1) \times 3·6 \times 10^{-6}}{6 \times 10^{-7}} = 3$$

As shown above, $OP = 3D\lambda/d = 2·88 \times 10^{-3}$ m)

Interference in Thin Wedge Films

A very thin wedge of an air film can be formed by placing a thin piece of foil or paper between two microscope slides at one end Y, with the slides in contact at the other end X, Figure 19.10. The wedge has then a very small angle θ, as shown. When the air film is illuminated by monochromatic light from an extended source S, *straight* bright and dark fringes are seen which are parallel to the line of intersection X of the two slides.

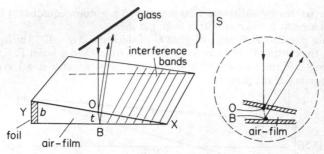

Figure 19.10 *Air-wedge fringes*

The light reflected down towards the wedge is partially reflected upwards from the *lower* surface O of the top slide, as shown inset in Figure 19.10. The rest of the light passes through the slide and some is reflected upward from the *top* surface B of the lower slide. The two trains of waves are coherent, since both have originated from the same centre of disturbance at O. So they produce an interference pattern if brought together by the eye or in an eyepiece.

The path difference is $2t$, where t is the small thickness of the air film at O. At X, where the path difference is apparently zero, we would expect a bright fringe. But a *dark* fringe is observed at X. This is due to a phase change of $180°$, equivalent to an extra path difference of $\lambda/2$, which occurs when a wave is reflected at a denser medium. See p. 520. The optical path difference between the two coherent beams is thus actually $2t + \lambda/2$. So, if the beams are brought together to interfere, a bright fringe is obtained when $2t + \lambda/2 = n\lambda$, or $2t = (n - \frac{1}{2})\lambda$. A dark fringe is obtained at a thickness t given by $2t = n\lambda$.

The bands or fringes are located at the air-wedge film, and the eye or microscope must be focused here to see them. The appearance of a fringe is the contour of all points in the air-wedge film where the optical path difference is the same. If the wedge surfaces make perfect optical contact at one edge, the fringes are straight lines parallel to the line of intersection of the surfaces. If the glass surfaces are uneven, and the contact at one edge is not regular, the fringes are not perfectly straight. A particular fringe still shows the locus of all points in the air wedge which have the same optical path difference in the air film.

In *transmitted light*, the appearance of the fringes is complementary to those seen by reflected light, from the law of conservation of energy. The bright fringes thus correspond in position to the dark fringes seen by reflected light, and the fringe where the surfaces touch is now bright instead of dark.

The air-wedge film is an example of interference by *division of amplitude*. Here part of the wave is transmitted at O and the remainder is reflected at O, so that the amplitude of the wave, which is a measure of its energy, is 'divided' into two parts. This method of producing interference is basically different from producing interference by division of wavefront (p. 521).

Thickness of Thin Foil by Air Wedge, Crystal Expansion

If there is a bright fringe at Y at the edge of the foil, Figure 19.10, the thickness b of the foil is given by $2b = (n + \frac{1}{2})\lambda$, where n is the number of bright fringes between X and Y. If there is a dark band at Y, then $2b = n\lambda$. So by counting n, the thickness b can be found. The small angle θ of the wedge is given by b/a, where a is the distance XY, and by measuring a with a travelling microscope focused on the air film, θ can be found.

Plate 19A *Interference bands in air wedges. The angle of the air wedge on the right is about $2\frac{1}{2}$ times less than that on the left, so that the separation of the bands is correspondingly greater*

*The angle θ of the wedge can also be found from the separation s of the bright bands. In Figure 19.11, B_1 and B_2 are *consecutive* bright bands. So the extra path difference in the air film from one band to the other is λ. The extra path difference is $2t_2 - 2t_1$, or $2A_2D$, where A_1D is the perpendicular from A_1 to A_2B_2. So $A_2D = \lambda/2$. Hence if s is the separation, B_1B_2, of the bands, which is equal to A_1D, then

$$\tan \theta = \frac{A_2D}{A_1D} = \frac{\lambda/2}{s} = \frac{\lambda}{2s}$$

Since θ is very small, $\tan \theta = \theta$ in radians. So

$$\theta = \frac{\lambda}{2s}$$

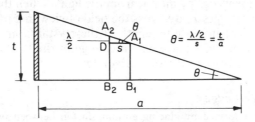

Figure 19.11 *Air-wedge theory*

As an illustration, suppose an air wedge is illuminated normally by mono-chromatic light of wavelength $5\cdot8 \times 10^{-7}$ m (580 nm) and the separation of the bright bands is $0\cdot29$ mm or $2\cdot9 \times 10^{-4}$ m. Then, from $\theta = \lambda/2s$, we have

$$\theta = \frac{5\cdot8 \times 10^{-7}}{2 \times 2\cdot9 \times 10^{-4}} = 10^{-3} \text{ rad}$$

If a *liquid wedge* is formed between the plates, the optical path difference becomes $2nt$ where the air thickness is t, n being the refractive index of the liquid. An optical path difference of λ now occurs for a change in t which is n

Plate 19B Interferometer measurement of length. *The photograph shows an interfero-meter used to measure the accuracy of block (slip) gauges, widely used for precision measurements in the engineering industry. As shown, the gauges are placed on turntables. Optical flats are used with them to form interference bands. Any error in length is determined in terms of an accurately known wavelength of two frequency-stabilised helium-neon lasers which are used to form the bands. Correction is needed for temperature changes, measured by a platinum resistance thermometer.* (Crown copyright. Courtesy of National Physical Laboratory)

times *less* than in the case of the air wedge. The spacing of the bright and dark fringes is thus n times *closer* than for air. So measurement of the relative spacing enables n to be found.

The expansivity of a crystal can be found by forming an air wedge of small angle between a fixed horizontal glass plate and the upper surface of the crystal, and illuminating the wedge by monochromatic light. When the crystal is heated a number of bright fringes, n say, cross the field of view in a microscope focused on the air wedge. The increase in length of the crystal in an upward direction is $n\lambda/2$, since a change of λ represents a change in the thickness of the film is $\lambda/2$, and the expansivity can then be calculated.

Example on Air Wedge

An air-wedge film is formed by placing aluminium foil between two glass slides at a distance of 75 mm from the line of contact of the slides. When the air wedge is illuminated normally by light of wavelength $5\cdot60 \times 10^{-7}$ m, interference fringes are produced parallel to the line of contact which have a separation of $1\cdot20$ mm. Calculate the angle of the wedge and the thickness of the foil.

If s is the separation of the bands, then

$$\text{angle of wedge, } \theta = \frac{\lambda/2}{s} = \frac{\lambda}{2s}$$

$$= \frac{5\cdot6 \times 10^{-7}}{2 \times 1\cdot2 \times 10^{-3}} = 2\cdot3 \times 10^{-4} \text{ rad}$$

If t is the thickness of the foil, then

$$\frac{t}{75 \times 10^{-3}} = \theta = 2.3 \times 10^{-4}$$

So
$$t = 75 \times 10^{-3} \times 2.3 \times 10^{-4} = 1.7 \times 10^{-5} \text{ m}$$

Newton's Rings

Newton discovered an example of interference which is known as 'Newton's rings'. In this case a lens L is placed on a sheet of plane glass H, L having a lower surface of very large radius of curvature so that a curved air wedge is

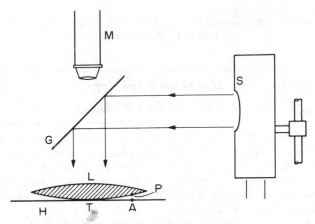

Figure 19.12 *Newton's rings experiment*

formed with H, Figure 19.12. By means of a sheet of glass G, monochromatic light from a sodium lamp S, for example, is reflected downwards towards L; and when the light reflected upwards is observed through a microscope M focused on H, a series of bright and dark rings is seen. The circles have increasing radius, and are concentric with the point of contact T of L with H, see Plate 19C.

Consider the air film PA between A on the plate and P on the lower lens surface. Some of the incident light is reflected from P to the microscope, while the remainder of the light passes straight through to A, where it is also reflected to the microscope and brought to the same focus. The two rays of light have thus a net path difference of $2t$, where $t = $ PA. The same path difference is obtained at *all points round* T which have the same distance TA from T. So if $2t = n\lambda$, where n is a whole number and λ is the wavelength, we might expect a bright *ring* with centre T. Similarly, if $2t = (n + \frac{1}{2})\lambda$, we might expect a dark ring.

When a ray is reflected from an optically *denser* medium, however, a phase change of 180° occurs in the wave, which is equivalent to an extra path difference of $\lambda/2$ (see also p. 520). The truth of this statement can be seen by the presence of the dark spot at the centre, T, of the rings. At this point there is no geometrical path difference between the rays reflected from the lower surface of the lens and H, so that they should be in phase when they are brought to a focus and should form a bright spot. The dark spot means, therefore, that one of the rays suffers a phase change of 180°. Taking the phase change into account, it follows that

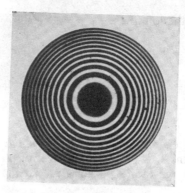

Plate 19C *Newton's rings, formed by interference of yellow light between converging lens and flat glass plate*

$$2t = n\lambda \text{ for a } \textit{dark} \text{ ring} \qquad . \qquad . \qquad . \qquad (1)$$

and
$$2t = (n + \tfrac{1}{2})\lambda \text{ for a } \textit{bright} \text{ ring} . \qquad . \qquad . \qquad (2)$$

where n is a whole number. Young verified the phase change by placing oil of sassafras between a crown and a flint glass lens. This liquid had a refractive index greater than that of crown glass and less than that of flint glass, so that light was now reflected at an optically denser medium at each lens. A *bright* spot was then observed in the middle of the Newton's rings, showing that no net phase change had now occurred.

The grinding of a lens surface can be tested by observing the appearance of the Newton's rings formed between it and a flat glass plate when monochromatic light is used. If the rings are not perfectly circular as in Plate 19C, the grinding is imperfect (see Plate 19D). As in the case of the air-wedge film, Newton's rings is an example of interference by division of amplitude (p. 526).

'Blooming' of Lenses

Whenever lenses are used, a small percentage of the incident light is reflected from each surface. In compound lens systems, as in telescopes and microscopes, this produces a background of unfocused light, so the clarity of the final image is reduced. There is also a reduction in the intensity of the image, since less light is transmitted through the lenses.

The amount of reflected light can be considerably reduced by evaporating a thin coating of a fluoride salt such as magnesium fluoride on to the surfaces, Figure 19.13. Some of the light, of average wavelength λ, is then reflected from the air–fluoride surface and the remainder penetrates the coating and is partially reflected from the fluoride–glass surface. Destructive interference occurs between the two reflected beams when there is a phase difference of 180°, or a path difference of $\lambda/2$, as the refractive index of the fluoride is less than that of glass. So if t is the required thickness of the coating and n' its refractive index, $2n't = \lambda/2$. Then $t = \lambda/4n' = 6 \times 10^{-7}/(4 \times 1\cdot38)$, assuming λ is 6×10^{-7} m and n' is $1\cdot38$, from which $t = 1\cdot1 \times 10^{-7}$ m.

For best results n' should have a value equal to about $\sqrt{n}$, where n is the refractive index of the glass lens. The intensities of the two reflected beams are then equal, and so complete interference occurs between them. No light is now

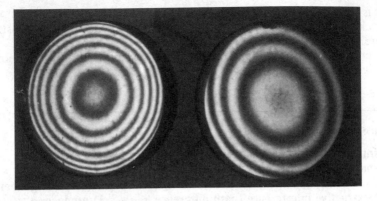

Plate 19D *Newton's rings formed by transmitted light. The central spot is bright since there is no phase change on transmission (compare the dark central spot in Newton's rings formed by reflected light). If the lens surface is imperfect the rings are distorted, as shown on the right*

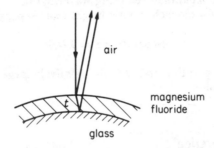

Figure 19.13 *Blooming of lens*

reflected back from the lens. In practice, since complete interference is not possible simultaneously for every wavelength of white light, an average wavelength for λ, such as green-yellow, is chosen. The lens thus appears purple, a mixture of red and blue, since these colours in white light are reflected. 'Bloomed' lenses produce a marked improvement in the clarity of the final image in optical instruments.

Vertical Soap Film Colours

An interesting experiment on thin films, due to C. V. Boys, can be performed by illuminating a vertical soap film with monochromatic light. At first the film appears uniformly coloured. As the soap drains to the bottom, however, a wedge-shaped film of liquid forms in the ring, the top of the film being thinner than the bottom. The thickness of the wedge is constant in a horizontal direction, and so horizontal bright and dark fringes are observed across the film. When the upper part of the film becomes extremely thin a *black* fringe is observed at the top (compare the dark central spot in Newton's rings experiment), and the film breaks shortly afterwards.

With white light, a succession of broad coloured fringes is first observed in the soap film. Each fringe contains colours of the spectrum, red to violet. The fringes widen as the film drains, and just before it breaks a black fringe is obtained at the top. The black fringe is due to the 180° (or π) phase change by reflection when the film is $\lambda/4$ thick and so destructive interference occurs.

The colours seen in oil films in the road are due to thin film interference of white light coming from the sky.

You should know:

1 Interference phenomena are due to overlapping waves from coherent sources.
2 Constructive interference (path difference $= n\lambda$) produces a bright band. Destructive interference (path difference $= n + \frac{1}{2}\lambda$) produces a completely dark band if the amplitudes of the two waves are equal.
3 Young's two-slit interference produces bright and dark bands when the slits are close and have coherent monochromatic light through them. With white light the bands are coloured; blue (short wavelength) is nearer the central white band than red (long wavelength).
4 Young formula for monochromatic bright band separation:

$$\text{Separation } x = \lambda D/d,$$

where d is slit separation and D is the relatively large distance of the bands from the slits.

Co-channel Interference

Interference is not peculiar to light waves, but is observed in all types of electromagnetic and other waves. It is a particularly important factor in the design of terrestrial broadcast systems for television and radio. For reliable reception it is important to have line-of-sight contact between the transmitter and the receivers: hence national broadcast systems require many transmitters, each serving a small, local area.

But what happens at the boundary between two areas, where signals from two transmitters can be detected? If both transmitters were operating at the same frequency, then the two signals would interfere and the signal reception would be considerably disturbed: this is known as *co-channel interference*. To combat this effect, it is customary to arrange for neighbouring transmitters to operate at slightly different frequencies. You may have noticed this when listening to a national radio station in the car. When the signal becomes weak (because you are losing line-of-sight contact with a particular transmitter) it is necessary to re-tune the radio slightly to pick up a stronger signal (since the neighbouring transmitter is operating at a different frequency). This also explains why national radio station frequencies are usually quoted as a range and not a single value, for example, BBC Radio 1 broadcasts on FM in the UK between 97·6 and 99·8 MHz. In practice, avoiding co-channel interference is not so straightforward because pressure on available frequency slots forces neighbouring transmitters to use similar frequencies. The problem is particularly severe in areas where more than two transmitters can be received, and unless a highly directional receiving antenna is used, co-channel interference is sometimes a problem.

EXERCISES 19   Interference

Multiple Choice

1 Figure 19A shows a Young two-slit interference experiment. S is a monochromatic source illuminating slits S_1 and S_2 a distance x apart, P is the screen distance D from the slits. The bright bands or fringes on P are separated a distance y.

   When D and x are each doubled, the separation of the fringes are

   A $8y$ **B** $4y$ **C** y **D** $y/2$ **E** $y/4$

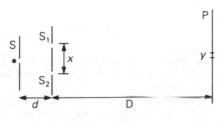

Figure 19A

2 In Figure 19A, S is a white source of light. Then:

   A The bands (fringes) on P are all white.
   B The blue bands are further apart than the red bands.
   C The light from S is not diffracted at S_1 and S_2.
   D The centre of the band system is white.
   E S_1 and S_2 are not coherent sources.

3 Which of the statements **A** to **E** is most correct?

   A Coherent sources are not needed to produce interference fringes.
   B The atoms in a tungsten filament lamp produce coherent light waves.
   C The atoms in a gas laser produce coherent light waves.
   D Two coherent light sources always produce bright and dark fringes on a screen.
   E The yellow light from a street lamp is coherent light.

Longer Questions

4 A low-flying aircraft can adversely affect the quality of television reception in the houses over which it is flying. The effects produced are caused by alternate rises and falls in the strengths of the signal received at the aerial. Suggest why these occur. (*L.*)

5 In Young's two-slit experiment using red light, state the effect of the following procedure on the appearance of the fringes.
   (a) The separation of the slits is decreased.
   (b) The screen is moved closer to the slits.
   (c) The source slit is moved closer to the two slits.
   (d) Blue light is used in place of red light.
   (e) One of the two slits is covered up.
   (f) The source slit is made wider.

6 In Young's two-slit experiment using light of wavelength $6·0 \times 10^{-7}$ m, the slits were 0·40 mm apart and the distance of the slits to the screen was 1·20 m. Find the separation of the fringes.

   What is the angle in radians subtended by a central pair of bright fringes at the slits?

7 Figure 19B (not drawn to scale) shows the apparatus used in an attempt to measure the wavelength of light using double-slit interference.
   (a) Explain how the apparatus produces double-slit interference fringes on the translucent screen.

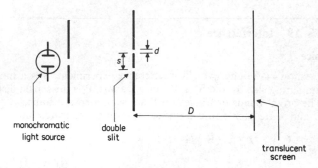

Figure 19B

(b) If measurable fringes are to be seen on the screen, then (i) the slit separation s must be small; (ii) the distance D between the slits and the screen must be large. Explain each of these conditions.

(c) The light from the monochromatic source has a wavelength of $5 \cdot 0 \times 10^{-7}$ m and it is required to produce a fringe separation of $5 \cdot 0$ mm. Suggest suitable values for s and D justifying your answer. Describe the important features of the pattern you would hope to obtain.

(d) Describe and explain any changes in the pattern which would be brought about by each of the following changes made separately: (i) one slit is covered with an opaque material; (ii) the slits are made narrower although their separation remains the same. (*AEB*.)

8 Describe, with the aid of a labelled diagram, how the wavelength of monochromatic light may be found using Young's slits. Give the theory of the experiment.

State, and give physical reasons for the features which are common to this method and to the method based on Lloyd's mirror.

In an experiment using Young's slits the distance between the centre of the interference pattern and the tenth bright fringe on either side is $3 \cdot 44$ cm and the distance between the slits and the screen is $2 \cdot 00$ m. If the wavelength of the light used is $5 \cdot 89 \times 10^{-7}$ m determine the slit separation. (*N.*)

9 (a) The thickness of a mica sheet is $4 \cdot 80$ μm. The refractive index of mica for light of wavelength 512 nm is $1 \cdot 60$. Calculate the ratio of the thickness of the sheet of mica to: (i) the wavelength of the 512 nm light in air; (ii) the wavelength of the same light in mica.

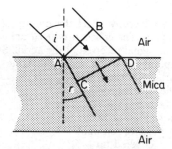

Figure 19C

(b) Figure 19C shows the positions of the slits and the screen in an optical two-slits experiment. Light of wavelength 512 nm is used, the separation of the slits is $0 \cdot 42$ mm and the perpendicular distance from the line S_1S_2 to the screen at O is 75 cm. P marks the centre of the third bright fringe from O. Calculate the distance OP.

Write down the value of the distance $PS_2 - PS_1$.

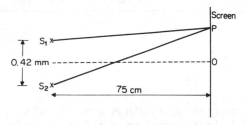

Figure 19D

The same mica sheet as was referred to in (a) above is mounted normally across the light path from S_1 without cutting into the light from S_2, see Figure 19D. Explain why the centre of the pattern on the screen moves from O to a new position O'. Calculate the distance OO'.

How would the distance OO' be affected by rotating the mica sheet about an axis parallel to S_1? (*L.*)

20 Diffraction of Light Waves

In this chapter we study the diffraction or spreading of waves through openings. We first deal with diffraction at a single slit and the variation of intensity of the image formed. This leads to the way in which the resolving power of optical and radio telescopes can be increased. The diffraction grating is then considered in detail and its use in measuring wavelengths in spectra is discussed. We conclude with a brief account of the use of diffraction in holography.

Diffraction Image

In 1665 GRIMALDI observed that the shadow of a very thin wire in a beam of light was much broader than he expected. This is due to diffraction.

Consider two points on the *same wavefront*, for example the two points A, B, on a plane wavefront, arriving at a narrow slit in a screen, Figure 20.1 and

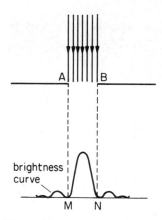

Figure 20.1 *Diffraction of light*

Plate 20A. A and B can be considered as secondary sources of light from Huygens's principle in the wave theory of light (p. 507); and as they are on the same wavefront, A and B have identical amplitudes and frequencies and are in phase with each other. So A and B are *coherent sources*. So we can expect to find an interference pattern on a screen in front of the slit, provided its width is small compared with the wavelength of light. In fact, for a short distance beyond the edges M, N, of the projection of AB, that is, in the geometrical shadow, observation shows that there are some alternate bright and dark fringes, see Figure 20.2.

Diffraction and Slit Width

Diffraction is the name given to the spreading of waves after they pass through small openings (or round small obstacles). The diffraction is appreciable when

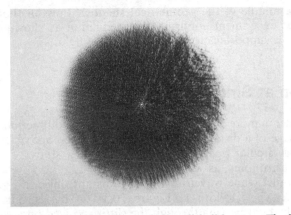

Plate 20A *Diffraction rings in the shadow of a small ball-bearing. The bright spot is at the centre of the geometrical shadow*

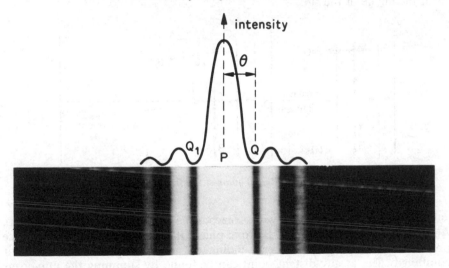

Figure 20.2 *The variation in intensity of the bright diffraction bands due to a single small rectangular aperture is shown roughly in the diagram. The central bright band, in which most of the light is concentrated, has maximum intensity in the direction of P, the middle of the band. The intensity of this band diminishes to a minimum at Q or Q_1, which are at an angle of diffraction θ to the direction of P*

the width of the opening is comparable to the wavelength of the waves and very small when the width is large compared to the wavelength. Sound has long wavelengths and can therefore diffract after passing through doorways. Light has a very short wavelength such as 6×10^{-7} m or 6×10^{-4} mm and so light waves are diffracted appreciably only through very small openings as we shall see.

If a source of white light is observed through the eyelashes a series of coloured images can be seen. These images are due to interference between sources on the same wavefront, and so the phenomenon is an example of diffraction. Another example of diffraction was deduced by POISSON at a time when the wave theory was new. Poisson considered mathematically the combined effect of the wavefronts round a circular disc illuminated by a distant small source of light, and he came to the conclusion that the light should be visible beyond the disc in the middle of the geometrical shadow. Poisson thought this was impossible; but experiment confirmed his deduction as shown in Plate 20A, and he became a supporter of the wave theory of light.

Diffraction at Single Slit

We now consider in more detail the image produced by diffraction at an opening or slit. The image of a distant star produced by a telescope objective is due to diffraction at a circular opening. On this account a study of the image has application in astronomy, as we see later.

Suppose parallel plane wavefronts from a distant object are diffracted at a rectangular slit AB of width a, Figure 20.3 (i). This is called *Fraunhofer diffraction*. If the light passing through the slit is received on a screen S a long way from AB, we may consider that parallel wavefronts have travelled to S to form the image of the slit.

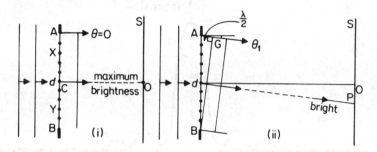

Figure 20.3 *Rectangular slit diffraction image*

Consider a plane wavefront which reaches the opening AB in Figure 20.3 (i). All points on it between A and B are in phase, that is, they are coherent. These points act as secondary centres, sending out waves beyond the slit. Their combined effect at any distant point can be found by summing the numerous waves arriving there, from the principle of superposition. The mathematical treatment is beyond the scope of this book. The general effect, however, can be seen by a simplified treatment.

Central Bright Image

Consider first a point O on the screen which lies on the normal to the slit passing through its midpoint C, Figure 20.3 (i). O is the *centre* of the diffraction image. It corresponds to a direction $\theta = 0$ through the slit, where θ is measured from the normal to AB. Now in this direction the waves from the secondary sources such as A, X, Y, B have no path difference. So all the waves arrive *in phase* at O. The centre of the diffraction image is therefore brightest.

In a direction very slightly inclined to $\theta = 0$, the waves from all the sources between A and B arrive slightly out of phase at the corresponding point P of the image near the centre, Figure 20.3 (ii). So the brightness decreases. In a particular direction θ_1, we reach the dark edge or boundary Q of the central image. As we soon show, this direction corresponds to a path difference of λ between the two sources at the edges A and B of the slit. Figure 20.4 shows

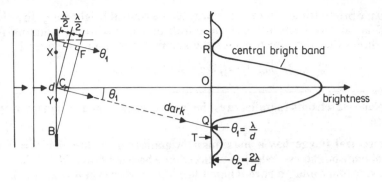

Figure 20.4 *Variation of brightness of diffraction image*

the path difference AF of the waves from A to B in this case, which totals $\left(\dfrac{\lambda}{2} + \dfrac{\lambda}{2}\right)$ or λ.

To find the resultant amplitude of the waves arriving at Q, divide the wavefront AB into two halves. The top point A of the upper half CA, and the top point C of the lower half BC, send out waves to Q which have a path difference $\lambda/2$, from above. So the resultant amplitude at Q is zero. All other pairs of corresponding points in the two halves of AB, for example, X and Y where CX = BY and the two bottom points C and B, also have a path difference $\lambda/2$. So the brightness at Q is zero. This point, then, is one *edge* of the central bright fringe on the screen. The other edge is R on the opposite side of O, where OR = OQ.

We can easily find the direction θ_1. From triangle BAF,

$$\sin \theta_1 = \frac{AF}{AB} = \frac{\lambda}{d} \quad . \qquad . \qquad . \qquad (1)$$

or if θ_1 is small, its value in radians is

$$\theta_1 = \frac{\lambda}{d} \qquad . \qquad . \qquad . \qquad (2)$$

As we discuss later, the direction $\theta_1 = \lambda/d$ becomes important in finding the resolving power of a telescope (p. 541).

Secondary Fringes

Secondary bright and dark fringes are also obtained on the screen beyond Q, as shown in the photograph in Figure 20.2. Consider, for example, a point T which lies in a direction θ_3 where the path difference of waves starting from A and B is $3\lambda/2$, Figure 20.4. We can imagine the wavefront AB divided into *three* equal parts. Now the waves from the extreme ends of the upper two parts have a path difference λ. So, as we explained for the dark fringes at Q, these two parts of the wavefront produce darkness at T. The third part produces a fringe of light at T much less bright than the central fringe, which was due to the whole wavefront between A and B. Calculation shows that the intensity at T is less than 5% of the intensity at the middle of the central bright fringe. So most of the light incident on AB is diffracted into the central bright fringe.

We can find the directions of the edge or minimum brightness of the secondary bright fringes in the same way as the central bright fringe, by dividing the diffracted wavefront at the slit into four or six, and so on, equal parts. The result shows that the directions for the successive minima are given by

$$\sin \theta_2 = 2\lambda/d \quad \sin \theta_3 = 3\lambda/d \ . \qquad . \qquad . \qquad . \qquad (3)$$

With plane wavefronts (parallel rays) incident normally on a rectangular opening of width d:

1 the central image has a maximum brightness in a direction $\theta = 0$ and a minimum brightness (edge) in a direction where $\sin \theta = \lambda/d$
2 the secondary images have minima brightness where $\sin \theta = 2\lambda/d$, $3\lambda/d$, and so on.

Rectilinear and Non-rectilinear Propagation

As we have seen, the *angular width* of the central bright fringe is 2θ, where θ is the angle between the direction of maximum intensity and the direction of minimum intensity or the edge of the bright fringe in Figure 20.4. The angle θ is given by

$$\sin \theta = \frac{\lambda}{d}$$

where d = width of slit AB. The angle θ is the angular half-width of the central fringe.

When the slit is widened and d becomes large compared with λ, then $\sin \theta$ is very small and so θ is very small. In this case the directions of the minimum and maximum intensities of the central fringe are very close to each other. Practically the whole of the light is then in a direction immediately in front of the incident direction, that is, no spreading occurs. This explains the *rectilinear propagation of light* for wide openings. When the slit width d is very small and equal to 2λ, for example, then $\sin \theta = \lambda/d = 1/2$, or $\theta = 30°$. The light waves now spread round through 30° on either side of the slit, that is, the diffraction is appreciable.

These results are true for any wave phenomenon. In the case of an electromagnetic wave of 3 cm wavelength, a slit of these dimensions produces sideways spreading. Sound waves of a particular frequency 256 Hz have a wavelength of about 1·3 m. So sound waves spread round openings such as a doorway or an open window, which have similar dimensions to their wavelengths. Light waves would not spread round these openings as they have very small wavelengths of the order 6×10^{-7} m (600 nm).

Diffraction in Telescope Objective

When a parallel beam of light from a distant object such as a star S_1 enters a telescope objective L, the lens collects light through a circular opening. So a *diffraction* image of the star is formed round its principal focus, F. This is illustrated in the exaggereated diagram of Figure 20.5.

Consider an incident plane wavefront AB from the star S_1, and suppose for a moment that the aperture is rectangular. The diffracted rays such as AG, BH normal to the wavefront are incident on the lens in a direction parallel to the principal axis LF. The optical paths AGF, BHF are equal. This is true for all

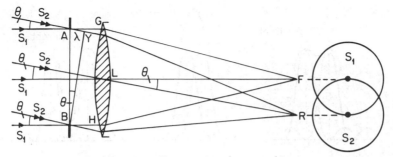

Figure 20.5 *Diffraction in telescope objective*

other diffracted rays from points between A, B which are parallel to LF, since the optical paths to an image produced by a lens are equal. The central part F of the star pattern is therefore bright.

Now consider those diffracted rays from all points between AB which enter the lens at an angle θ to the principal axis. This corresponds to a diffracted plane wavefront BY at an angle θ to AB. As explained previously, if $AY = \lambda$, then R is the dark edge of the central maximum of the diffraction pattern of the star S_1.

The angle θ corresponding to the edge R is given by

$$\sin \theta = \frac{\lambda}{D}$$

where D is the diameter of the lens aperture. This is the case where the aperture can be divided into a number of rectangular slits. For a *circular* opening such as a lens, or the concave mirror of the Palomar telescope, the formula becomes $\sin \theta = 1 \cdot 22\lambda/D$. As λ is small compared to D, then θ is small and so we may write $\theta = 1 \cdot 22\lambda/D$, where θ is in radians.

Resolving Power of Telescope

Suppose now that another distant star S_2 is at an angular distance θ from S_1, Figure 20.6. The maximum intensity of the central pattern of S_2 then falls on the minimum or edge of the central pattern of the star S_1, corresponding to R in Figure 20.6 (i). Experience shows that the two stars can then *just* be distinguished or *resolved*. Lord Rayleigh stated a criterion for the resolution of two objects, which is generally accepted: *Two objects are just resolved when the maximum intensity of the central pattern of one object falls on the first minimum or dark edge of the other.* Figure 20.6 (i) shows the two stars just resolved. The resultant intensity in the middle dips to about 0·8 of the maximum, and the eye is apparently sensitive to the change here. Figure 20.6 (ii) shows two stars S_1, S_2 unresolved, and Figure 20.6 (iii) the same stars completely resolved. See also Plate 20B.

The angular distance θ between two distant stars just resolved is thus given by $\sin \theta = \theta = 1 \cdot 22\lambda/D$, where D is the diameter of the objective. This is an expression for the *limit of resolution*, or *resolving power*, of a telescope. The resolving power increases when θ is *smaller*, as two stars closer together can then be resolved. So telescope objectives of *large diameter D* gives high resolving power.

The Yerkes Observatory has a large telescope objective diameter of about 1 metre. The angular distance θ between two stars which can just be resolved is

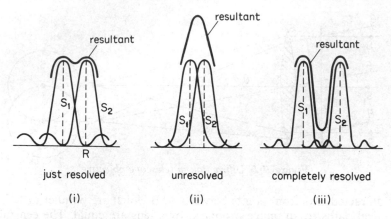

Figure 20.6 *Resolving power—Rayleigh criterion*

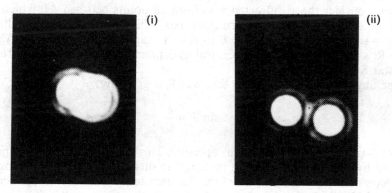

Plate 20B *Resolving*: (i) *Two sources just resolved according to Rayleigh's criterion* (ii) *Two sources completely resolved*

then given by

$$\theta = \frac{1 \cdot 22\lambda}{D} = \frac{1 \cdot 22 \times 6 \times 10^{-7}}{1} = 7 \cdot 3 \times 10^{-7} \text{ radians}$$

assuming 6×10^{-7} m for the wavelength of light. The Mount Palomar telescope has a parabolic mirror objective of aperture 5 metres. The resolving power is then five times as great as the Yerkes Observatory telescope. In addition to the advantage of high resolving power, a large aperture collects more light from the distant source and so provides an image of higher intensity or brightness.

Magnifying Power of Telescope and Resolving Power

If the width of the emergent beam from a telescope is greater than the diameter of the eye-pupil, rays from the outer edge of the objective do not enter the eye. Hence the full diameter D of the objective is not used. If the width of the emergent beam is less than the diameter of the eye-pupil, the eye itself, which has a constant aperture, may not be able to resolve the distant objects. Theoretically, the angular resolving power of the eye is $1 \cdot 22\lambda/d$, where d is the diameter of the eye-pupil. In practice an angle of 1 minute is resolved by the eye, which is more than the theoretical value.

Now the angular magnification, or magnifying power, of a telescope is the ratio β/α, where β is the angle subtended at the eye by the final image and α is the angle subtended at the objective (p. 457). To make the fullest use of the diameter D of the objective, therefore, the magnifying power should be increased to the angular ratio given by, if D is in metres,

$$\frac{\text{resolving power of eye}}{\text{resolving power of objective}} = \frac{\pi/(180 \times 60)}{1\cdot22 \times 6 \times 10^{-7}/D} = 400D \text{ (approx.)}$$

In this case the telescope is said to be in 'normal adjustment'. Any further increase in magnifying power will make the distant objects appear larger, but there will be no increase in definition or resolving power.

---

1 Smallest angle θ just resolved by a telescope objective of diameter D is $\theta = 1\cdot22\lambda/D$.
2 Resolving power increases with increased objective diameter D, *not* with increased magnifying power M.

Radio Telescope

Radio telescopes are used to investigate and map the sources of radio waves arriving at the earth from the solar system, the galaxy and the extragalactic nebulae. Basically the radio telescope consists of an aerial in the form of a paraboloid-shaped metal surface or 'dish', which can be rotated to face any part of the sky. Distant radio waves, like distant light waves, are reflected towards a focus from all parts of the paraboloid surface (p. 431). The converging waves are then passed to a sensitive receiver and after detection the signals may be recorded on a paper chart or they may be recorded on tape for feeding into a computer for analysis.

The Jodrell Bank radio telescope has a dish about 75 m in diameter. The hydrogen line emitted from interstellar space has a wavelength of about 21 cm or 0·21 m. So the angular resolution

$$= \frac{1\cdot22\lambda}{D} = \frac{1\cdot22 \times 0\cdot21}{75} = 0\cdot0034 \text{ rad} = 0\cdot2°$$

Larger telescopes can provide greater resolving power but the technical problems and cost make this unpractical. A technique using interferometry principles, in which the dishes need not be large, provides much greater resolving power. This has been developed particularly at the Mullard Radio Astronomy Observatory, Cambridge, UK.

■ Radio Interferometers

The principle of the interferometer type of radio telescope is illustrated in Figure 20.7.

Two dishes A and B, separated by a distance of say 5 km, are connected to the same receiver R, Figure 20.7 (i). Radiation from a moving source S will reach both aerials or dishes. If the path difference is a whole number of wavelengths or zero, the two signals are in phase and a maximum resultant signal is obtained (constructive interference); if the signals are in antiphase, the resultant signal is a minimum or zero (destructive interference). So as a source S moves across the sky, the resultant signal varies in intensity. The principles

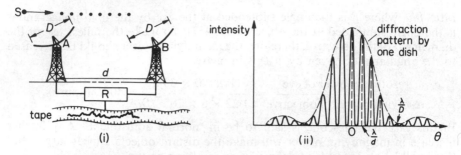

Figure 20.7 *Radio telescope—interferometer type*

concerned are similar to a Young's two-slit experiment, except that in this case we are dealing with two 'point receivers' of radiation instead of 'point emitters' of light. The mathematics of the interference is the same in both cases.

The variation in intensity with angle is shown roughly in Figure 20.7 (ii). It is similar to the variation in intensity of the interference fringes in a Young's two-slit experiment. Successive maxima or peaks are therefore obtained at angular separations equal to λ/d, where d is the separation of the two dishes. So if $\lambda = 3$ cm $= 0.03$ m and $d = 5$ km $= 5000$ m,

$$\text{angular separation} = \frac{\lambda}{d} = \frac{0.03}{5000} = 6 \times 10^{-6} \text{ rad} = 0.0003° \text{ (approx.)}$$

Further, the angular resolution θ, the angle from the central maximum to the first minimum (p. 541), is given by

$$\theta = \frac{\lambda}{2d} = 3 \times 10^{-6} \text{ rad}$$

If only one dish were used, with a diameter $D = 13$ m for example, the single-slit intensity pattern (p. 541) shows that the angular resolution is now

$$\theta = \frac{1.22\lambda}{D} = \frac{1.22 \times 0.03}{13} = 3 \times 10^{-3} \text{ rad (approx.)}$$

which is 1000 times less resolution than that obtained with the two dishes. So the angular diameter of a source, or of two sources, may be found more accurately by the interferometer method. Further, the method enables the direction of a moving source to be tracked more accurately—the fringe of the central maximum is much narrower with two dishes than with one and this helps to locate the source better.

The interferometer principle has been successfully applied to surveying regions of the sky and to mapping radio galaxies. The Mullard Radio Astronomy Observatory has a telescope consisting of eight parabolic reflectors or dishes, each about 13 m in diameter and mounted on rails about $1\frac{1}{4}$ km long, acting together as a 'grating interferometer'. The effective base line of the telescope is 5 km. The dishes can be moved along the rails to different sets of positions to map a region of the sky. The signals received from the different aerials are recorded on tape and then synthesised by means of a computer. In this way a telescope of very large aperture (5 km) can be simulated. This is called *aperture synthesis*. With this telescope extremely accurate maps have been made of radio sources in outer space. The Nobel Prize was awarded in 1974 to Sir Martin Ryle, then director of the Mullard Radio Astronomy Observatory, and to Professor A. Hewish of Cambridge for their contributions to radio astronomy, especially aperture synthesis.

Increasing Number of Slits

On p. 537 we saw that the image of a single narrow rectangular slit is a bright central or principal maximum diffraction fringe, together with subsidiary maxima diffraction fringes which are much less bright. Suppose that parallel light is incident on two more parallel close slits, and the light passing through the slits is received by a telescope focused at infinity. Since each slit produces a similar diffraction effect in the same direction, the observed diffraction pattern will have an intensity variation identical to that of a single slit. This time, however, the pattern is crossed by a number of interference fringes, which are due to interference between slits (see *Young's experiment*, p. 521). The envelope of the intensity variation of the interference fringes, shown by the broken lines in Figure 20.8, follows the *diffraction pattern variation due to a single slit*. In general, if I_s is the intensity at a point due to interference between slits and I_d that due to diffraction of a *single* slit, then the resultant intensity I is given by $I = I_d \times I_s$. So if $I_d = 0$ at any point, then $I = 0$ at this point irrespective of the value of I_s.

As more parallel equidistant slits are introduced, the intensity and sharpness of the principal maxima increase and those of the subsidiary maxima decrease.

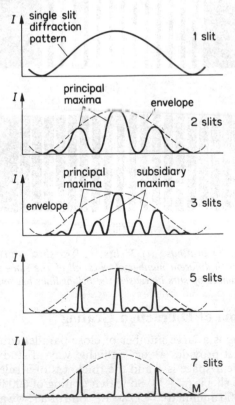

Figure 20.8 *Principle maxima with increasing slits, showing increasing sharpness and intensity*

The effect is illustrated roughly in Figure 20.8. With several hundred lines per millimetre, only a few sharp principal maxima are seen in directions discussed shortly. Their angular separation depends only on the distance between successive slits. The slit width affects the intensity of the higher-order principal maxima; the narrower the slit, the greater is the diffraction of light into the higher orders. As shown, the single-slit diffraction pattern modulates the maxima. In the direction corresponding to M (7 slits), the single-slit intensity $I_d = 0$. So, as stated above, the resultant intensity $I = 0$. This means that no light is diffracted in this direction from any of the slits. Plate 20C shows how the principal maxima diffraction images become *sharper* as the number of slits increases, which is one reason why the 'diffraction grating' was invented, as we now discuss.

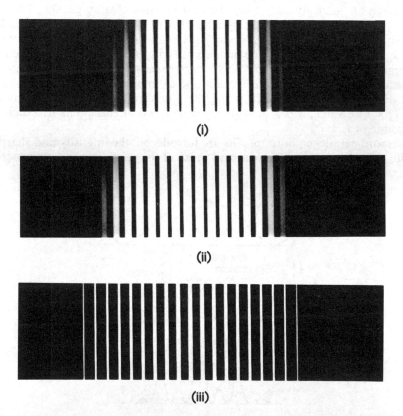

(i)

(ii)

(iii)

Plate 20C *Diffraction by gratings* (i) *3 slits;* (ii) *5 slits;* (iii) *20 slits. Diffraction patterns by gratings with different number of slits, all of the same width and separation. As the number of slits increases, the bright lines become sharper*

Principal Maxima of Diffraction Grating

A *diffraction grating* is a large number of close parallel equidistant slits, ruled on glass or metal; it provides a very valuable way of studying spectra. If the width of a slit or clear space is a and the thickness of a ruled opaque line is b, the spacing d of the slits is $(a + b)$. So with a grating of 600 lines per millimetre, the spacing $d = 1/600$ millimetre $= 17 \times 10^{-7}$ m, or a few wavelengths of visible light.

The angular positions of the principal maxima produced by a diffraction

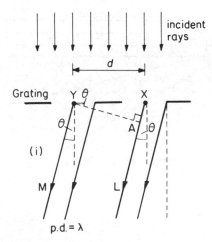

Figure 20.9 *Diffraction grating images*

grating can easily be found. Suppose X, Y are corresponding points in consecutive slits, where XY = d, and the grating is illuminated normally by monochromatic light of wavelength λ, Figure 20.9. In a direction θ, the diffracted rays XL, YM have a path difference XA of $d \sin \theta$. The diffracted rays from all other corresponding points in the two slits have a path difference of $d \sin \theta$ in the same direction. Other pairs of slits throughout the grating can be treated in the same way. So bright or principal maxima are obtained when

$$d \sin \theta = n\lambda \qquad . \qquad . \qquad . \qquad . \qquad . \qquad \text{(i)}$$

where n is a whole number, if all the diffracted parallel rays are collected by a telescope focused at infinity. The images corresponding to $n = 0, 1, 2, \ldots$ are said to be respectively of the zero, first, second ... orders respectively. The zero-order image is the image where the path difference of diffracted rays is zero. It corresponds to that seen directly opposite the incident beam on the grating. It should again be noted that all points in the slits are secondary centres on the same wavefront and therefore coherent sources.

Unlike the images obtained with glass prisms, the grating gives *two* sets of diffraction images on opposite sides of the normal where $d \sin \theta = n\lambda$. The angle between the two images of the same order is 2θ.

Figure 20.10 shows how the grating produces diffracted waves which interfere constructively in directions given by $d \sin \theta = n\lambda$. Only six slits are shown and only six corresponding secondary sources in the slits, separated by a distance d. The diffracted wavefronts correspond to a path difference of λ and 2λ respectively. They travel at an angle θ to the normal given by $d \sin \theta = \lambda$ and 2λ respectively. In practice the build-up of the diffracted wavefronts is due to the effect of many slits, for example, 300 or more per millimetre, and to the numerous secondary sources within each slit.

Diffraction Images

The *first-order* diffraction image is obtained when $n = 1$. So

$$d \sin \theta = \lambda$$

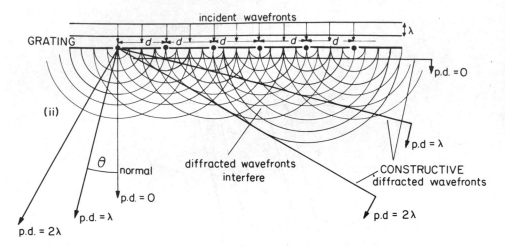

Figure 20.10 *Grating action—diffraction and interference of wavefronts*

or
$$\sin \theta = \frac{\lambda}{d}$$

If the grating has 600 lines per millimetre (600 mm^{-1}), the spacing of the slits, d, is $1/600$ mm $= 1/(600 \times 10^3)$ m. Suppose yellow light, of wavelength $\lambda = 5\cdot89 \times 10^{-7}$ m, is used to illuminate the grating. Then

$$\sin \theta = \frac{\lambda}{d} = 5\cdot89 \times 10^{-7} \times 600 \times 10^3 = 0\cdot3534$$

$$\therefore \theta = 20\cdot7°$$

The *second-order* diffraction image is obtained when $n = 2$. In this case $d \sin \theta = 2\lambda$.

$$\sin \theta = \frac{2\lambda}{d} = 2 \times 5\cdot89 \times 10^{-7} \times 600 \times 10^3 = 0\cdot7068$$

$$\therefore \theta = 45\cdot0°$$

If $n = 3$, $\sin \theta = 3\lambda/d = 1\cdot060$. Since the sine of an angle cannot be greater than 1, it is impossible to obtain a third-order image with the diffraction grating for this wavelength.

With a grating of 1200 lines per mm the diffraction images of sodium light would be given by $\sin \theta = n\lambda/d = n \times 5\cdot89 \times 10^{-7} \times 12 \times 10^5 = 0\cdot7068n$. So only $n = 1$ is possible here. As all the diffracted light is now concentrated in one image, instead of being distributed over several images, the first-order image is very bright, which is an advantage.

Missing Orders in Principal Maxima

In a diffracting grating, the order n of the principal maxima obtained is given by $\sin \theta = n\lambda/d$. As we saw previously, however, the intensity of the principal maxima will be zero if the intensity due to a *single slit* of the grating is zero, as this modulates the over-all intensity variation (see Plate 20C). Now the intensity of the single-slit pattern is zero where $\sin \theta = \lambda/a$, if a is the *width* of the slit. So the *missing order n* in the principal maxima obtained by a diffraction grating is given by

$$\sin \theta = \frac{n\lambda}{d} = \frac{\lambda}{a}$$

So

$$n = \frac{d}{a}$$

If the width a of the slit is one quarter of the separation d of the grating lines, then $n = 4$ from above. The fourth order will be missing from the principal maxima of the grating. With a particular grating, missing orders occur only if d/a is a whole number.

Measurement of Wavelength by Diffraction Grating

The wavelength of monochromatic light can be accurately measured by a diffraction grating with a spectrometer.

The collimator C and telescope T of the instrument are first adjusted for parallel light, and the grating P is then placed on the table so that its plane is perpendicular to two screws, Q, R, Figure 20.11 (i). To level the table so that the

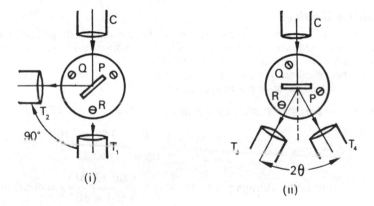

Figure 20.11 *Measurement of wavelength by diffraction grating*

plane of P is parallel to the axis of rotation of the telescope, the telescope is first placed in the position T_1 directly opposite the illuminated slit of the collimator, and then rotated exactly through 90° to a position T_2. The table is now turned until the slit is seen in T_2 by *reflection* at P, and one of the screws Q, R turned until the slit image is in the middle of the field of view. The plane of P is now parallel to the axis of rotation of the telescope. The table is then turned through 45° so that the plane of the grating is exactly perpendicular to the light from C, and the telescope is turned to a position T_3 to receive the first diffraction image, Figure 20.11 (ii). If the lines of the grating are not parallel to the axis of rotation of the telescope, the image will not be in the middle of the field of view. The third screw is then adjusted until the image is central.

The readings of the first diffraction image are observed on both sides of the normal. The angular difference is 2θ, and the wavelength is calculated from $\lambda = d \sin \theta$, where d is the spacing of the slits, obtained from the number of lines per metre of the grating. If a second-order image is obtained for a diffraction angle θ_1, then $\lambda = \frac{1}{2}d \sin \theta_1$.

Diffraction gratings with suitable values of d have been used to measure a wide range of wavelengths. For very short wavelengths such as X-rays, a grating of similar linear dimensions is provided by rows of atomic planes inside a crystal (see p. 853).

Resolving Power of Grating

The resolving power of a grating is a measure of how effectively it can separate or *resolve* two wavelengths in a given order of their spectrum. Using the Rayleigh criterion for resolving power (p. 541), it can be shown that the resolving power depends on the *total lines or width Nd* of the grating, where N is the number of lines and d is their spacing.

So a grating 5 cm wide with 3000 lines per cm has twice the resolving power of the same grating if it was masked to be only 2·5 cm wide. In both gratings the diffraction images of two close wavelengths would be formed at the same angles for a given order, since $d \sin \theta = n\lambda$ and d is the same. But the images with the 2·5 cm grating are not as sharp as with the 5 cm grating and the two close wavelengths may not be seen separated or resolved.

Note that $Nd = Nn\lambda/\sin \theta$, so the resolving power increases with the order n of the image.

Example on Gratings

A parallel beam of sodium light is incident normally on a diffraction grating. The angle between *the two first-order spectra on either side* of the normal is 27° 42′. Assuming that the wavelength of the light is $5·893 \times 10^{-7}$ m, find:
(a) the number of rulings per mm on the grating;
(b) the greatest number of bright images obtained.

(a) The first-order spectrum occurs at an angle $\theta = \frac{1}{2} \times 27°\ 42′ = 13°\ 51′$

$$\therefore d = \frac{\lambda}{\sin \theta} = \frac{5·893 \times 10^{-7}}{\sin 13°\ 51′}\ \text{m}$$

$$\therefore \text{number of rulings per metre} = \frac{1}{d} = \frac{\sin 13°\ 51′}{5·893 \times 10^{-7}} = 406\,000$$

$$= 406 \text{ per millimetre}$$

(b) From $d \sin \theta = n\lambda$, the greatest number of images is obtained from $\theta = 90°$ and so $\sin \theta = 1$. In this case,

$$d \sin \theta = d = n\lambda$$

So

$$n = \frac{d}{\lambda} = \frac{(1/4·06) \times 10^{-5}}{5·893 \times 10^{-7}}$$

$$= \frac{100}{4·06 \times 5·893} = 4·18$$

Since n is a whole number, the number of images $= 4$.

Spectra with Grating

If white light is incident normally on a diffraction grating, several coloured spectra are observed on either side of the normal, Figure 20.12 (i). The first-

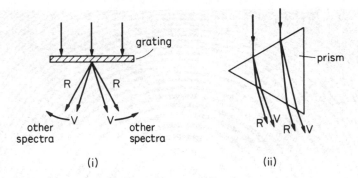

Figure 20.12 *Spectra with grating and prism*

order diffraction images are given by $d \sin \theta = \lambda$, and as violet has a shorter wavelength than red, θ is less for violet than for red. So the spectrum colours on either side of the incident white light are violet to red. In the case of a spectrum produced by dispersion in a glass prism, the colours range from red, the least deviated, to violet, Figure 20.12 (ii). Second- and higher-order spectra are obtained with a diffraction grating on opposite sides of the normal, whereas only one spectrum is obtained with a glass prism. The angular spacing of the colours is also different in the grating and the prism.

If $d \sin \theta = n_1 \lambda_1 = n_2 \lambda_2$, where n_1, n_2 are integers, then a wavelength λ_1 in the n_1-order spectrum overlaps the wavelength λ_2 in the n_2 order. The extreme violet in the visible spectrum has a wavelength about 3.8×10^{-7} m. The violet direction in the second-order spectrum would thus correspond to $d \sin \theta = 2\lambda = 7.6 \times 10^{-7}$ m, and this would not overlap the extreme colour, red, in the first-order spectrum, which has a wavelength about 7.0×10^{-7} m. In the second-order spectrum, a wavelength λ_2 would be overlapped by a wavelength λ_3 in the third order if $2\lambda_2 = 3\lambda_3$. If $\lambda_2 = 6.9 \times 10^{-7}$ m (red), then $\lambda_3 = 2\lambda_2/3 = 4.6 \times 10^{-7}$ m (blue). So overlapping of colours occurs in spectra of higher orders than the first.

Holography

When an object is photographed by a camera, only the *intensity* of the light from its different points is recorded on the photographic film to form the image. Now the intensity is a measure of the mean square value of the amplitude of the original light wave from the object. So the *phase* of the wave arriving at the film from the different points on the object is lost.

In *holography*, however, both the phase and the amplitude of the light waves are recorded on the film. The resulting photograph is called a *hologram*. As we see later, it is a speckled pattern of fine dots. Dr. D. Gabor, who laid the foundations of holography in 1948, gave this name from the Greek 'holos' meaning 'the whole', because it contains the whole information about the light wave, that is, its phase as well as its amplitude.

Gabor's method of producing a hologram and of reconstructing the original wave were difficult to put in practice in 1948 because sources which remain coherent only over very short path differences were then available. In 1962, however, the laser was invented. This gave a powerful source of light which remained coherent over long paths (see Plate 20D) and the subject of holography then developed rapidly.

Plate 20D *Newton's rings. Obtained simply by reflecting a helium–neon laser beam from the front and back surfaces of a low power lens and so demonstrating the coherency of a laser beam over long paths*

Making a Hologram

Figure 20.13 shows the basic principle of making a hologram. By means of a half-silvered mirror M, part of the coherent light from a laser is reflected towards the object O and the remainder passes straight through as shown. The photographic plate or film P is thus illuminated by:
(a) light waves scattered or diffracted from O, called the 'object beam', and
(b) a direct beam called the 'reference beam'.

The numerous points which make up the image on P are then formed by

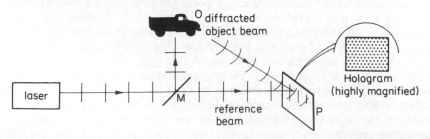

Figure 20.13 *Making a hologram*

interference between the overlapping coherent waves of the object beam and the reference beam. As shown inset diagrammatically, the hologram consists of a very large number of closely-spaced points, invisible to the naked eye but seen under a powerful microscope. The 'structure' of the hologram is like a diffraction grating; it has opaque and transparent regions very closely spaced.

In commercial broadcasting, audio-frequency currents are arranged to modulate the carrier ratio wave from the transmitter. By analogy, in making a hologram the reference beam may be considered analogous to a 'carrier wave' which is 'modulated' at the photographic film by interference with the object beam. Thus, to simplify matters, suppose the object and reference beams arriving at the plate are plane waves represented by $y = O \sin(\omega t + \Delta)$ and $y = R \sin \omega t$ respectively, where Δ is their phase difference due to the path difference. The resultant amplitude at the plate is given, from vector addition, by $(O^2 + R^2 + 2OR \cos \Delta)^{1/2}$. The third term $2OR \cos \Delta$ contains (i) the phase angle Δ and (ii) the amplitude O of the object wave modulated by the amplitude R of the reference beam. After the plate is exposed and developed, the object wave, accompanied by the others due to interference, is 'recorded' on the emulsion. It may be noted that if the object moves very slightly (of the order of a fraction of a wavelength) while the photograph is taken, the diffracted waves from the object no longer produces a hologram. Rigid mounting of the object is therefore essential.

Reconstructing the Hologram Image

Figure 20.14 shows how the object wavefront or image is 'reconstructed'. The optical arrangement is simply reversed and the hologram H is illuminated by

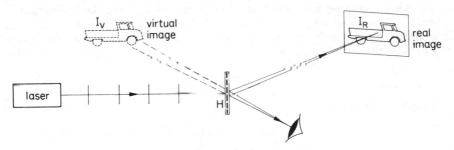

Figure 20.14 *Reconstructing hologram image*

coherent light from the laser, which was the original reference beam. Now as a general principle, if an image produced by a grating is recorded on photographic film, then the image formed by illuminating the resulting grating on the negative will be identfcal to the original grating. In a similar way, we may consider the object as a 'three-dimensional grating' and its three-dimensional image is formed when the hologram is illuminated as in Figure 20.14.

The points on the hologram H thus act as a diffraction grating. The waves diffracted through H carry the phase and amplitude of the waves originally diffracted from the object O when the hologram was made. The object wavefronts have thus been 'reconstructed'. One of the diffracted beams forms a real image I_R, as shown. Another diffracted beam forms a virtual image I_V. This image can be seen on looking through the hologram H. The hologram thus acts like a 'window' through which the image can be seen.

Plate 20E *The wonder of holography. A boy admires circular holograms at the Museum of La Villette in Paris. The hologram gives a 3-D effect of the woman's head and shoulders*

One of the most remarkable features of the hologram is that by moving the head while looking through it, one can see more of the object originally hidden from view. Thus a three-dimensional view is recorded on a two-dimensional photographic film, see Plate 20E. This is due to the fact that all parts of the object originally photographed have sent diffracted waves to the photo graphic film. Further, the 'grain' of a particular hologram is repeated regularly throughout the whole of the film. So if the hologram is cut into small pieces, although the quality is poor the whole of the image can be seen through one piece.

You should know:

1 A diffraction grating has many parallel close lines equally spaced, for example, 400 lines per millimetre apart.

2 When illuminated normally by monochromatic light, the light is beamed strongly (principal orders) in certain directions θ to the normal given by $d \sin \theta = n\lambda$ where d, the grating spacing = line thickness + clear space width, and $n = 0, 1, 2, \ldots$. With 400 lines mm^{-1}, $d = 1/400$ mm.

No image is obtained when θ is greater than $90°$. At $90°$, $\sin \theta = \sin 90° = 1$.

3 Wavelengths can be measured accurately using a diffraction grating. θ can be large and therefore accurately measured.

Applications of Holography

For many years holography provided us with little more than fascinating exhibitions. However, now applications are being rapidly developed in many varied fields. For example, holographic interferometry is in many ways superior to standard optical interferometry (Plate 19B), especially when dealing with objects with rough, diffuse reflecting surfaces.

A totally different application concerns holographic optical elements (HOEs), which can change optical wavefronts in the same way as do lenses. They can be easily combined on thin films of emulsion, and do not require the laborious grinding and polishing processes used to make conventional lenses.

HOEs can already be found in many aircraft at the heart of the *head-on display*, a device which allows the pilot to view a computer display at infinity, superimposed over his or her normal field of view. The HOE used in this application is a *holographic beam combiner*, which sits in front of the pilot's head and looks like a sheet of glass or clear plastic. The combiner is the last stage in the head-up display, and reflects the rays of the computer image towards the pilot's eyes. Obviously it must not cut down the light passing through it, since this would obscure the pilot's view of the outside world.

This is the power of holographic beam combiners: they can be made to have very high reflecting power over a narrow spectral range and very low reflectance outside this range. By matching the colour of the computer display with the reflectance of its beam combiner, the head-up display allows a virtually unimpeded view of the outside world, with the computer information clearly superimposed upon it. Politicians often use head-up displays when delivering important speeches to conferences: they can read their script while facing the audience, without having to mumble over a piece of paper!

EXERCISES 20   Diffraction

Multiple Choice

1 When a diffraction grating is illuminated normally by monochromatic light, which of the statements **A** to **E** is most correct?

   A The angle of diffraction of the first bright (principal order) image depends on the width of the line in the grating.
   B The angle of diffraction of the first bright image depends on the width of the gap between the lines.
   C Bright (principal order) images are obtained only on one side of the normal.
   D The first-order blue image has a smaller angle of diffraction than the first-order red image.
   E Any number of diffracted principal order images can be obtained.

2 Illuminated normally, a diffraction grating produces second-order bright images with an angle of 60° between them. The light is monochromatic and has a wavelength 600 nm. The spacing of the grating in mm is

   A 3.6×10^{-3} **B** 2.4×10^{-3} **C** 2.0×10^{-3} **D** 1.5×10^{-3} **E** 1.0×10^{-3}

3 In $d \sin \theta = n\lambda$ for the bright (principal order) images obtained when a diffraction grating is illuminated normally, which is the most correct statement in **A** to **E**?

   A d is the width of the grating gap between lines.
   B The images are produced by interference between secondary sources in alternate gaps.
   C The images are produced by interference between secondary sources in neighbouring gaps.
   D The images are very close together as in Young's two-slit experiment.
   E n may be $1\frac{1}{2}$ or $2\frac{1}{2}$ in calculations.

4 A diffraction grating has 500 lines per millimetre and is illuminated normally by monochromatic light of wavelength 600 nm. The total number images seen on both sides of the normal, including the central image, is

   A 10 **B** 9 **C** 8 **D** 7 **E** 5

Longer Questions

5 A diffraction grating has 400 lines per mm and is illuminated normally by monochromatic light of wavelength 600 nm (6×10^{-7} m).
   Calculate:
   (a) the grating spacing;
   (b) the angle to the normal at which the first principal order or maximum is seen;
   (c) the number of diffraction maxima obtained.
6 A diffraction grating is illuminated normally by monochromatic light of wavelength λ. Show in a diagram using waves: (i) where diffraction occurs; (ii) where interference occurs; (iii) the directions in which the first- and second-order diffraction maxima are seen, together with the corresponding path difference.
7 A plane diffraction grating is illuminated by a source which emits two spectral lines of wavelengths 420 nm (420×10^{-9} m) and 600 nm (600×10^{-9} m). Show that the third-order line of one of these wavelengths is diffracted through a greater angle than the fourth order of the other wavelength. (*L*.)
8 A spectral line of known wavelength $5 \cdot 792 \times 10^{-7}$ m emitted from a mercury vapour lamp is used to determine the spacing between the lines ruled on a plane diffraction grating. When the light is incident normally on the grating the third-order spectrum, measured using a spectrometer, occurs at an angle of $60° \ 19'$ to the normal. Calculate the grating spacing.
   Why is the value obtained using the third-order spectrum likely to be more accurate than if the first order were used? (*L*.)
9 When a plane wavefront of monochromatic light meets a diffraction grating, the diffraction pattern produced is caused by the interference of beams diffracted through the various grating slits. Explain, with the aid of a diagram, exactly where and under what conditions diffraction occurs.
   Explain why only beams diffracted in certain directions interfere constructively. (*L*.)
10 The wavelengths of the two yellow sodium lines are $589 \cdot 0$ nm and $589 \cdot 6$ nm. The eye, with the aid of the spectrometer telescope, can resolve an angle of $0 \cdot 2$ minutes. How many lines per metre must there be in a diffraction grating for the two sodium lines to be seen as just separate in the first order in the spectrometer? Assume the diffraction angle is small so that $\sin \theta \simeq \theta$ in radians. (*W*.)
11 Draw a diagram of the arrangement you would use to measure the wavelength of monochromatic light with a diffraction grating having a known number of slits per metre. List the measurements you would make.
   Explain with the aid of a diagram how the second-order fringe is produced by the grating. What limits the number of fringes produced?
   If the grating were replaced by two slits whose separation equalled the spacing of the slits in the grating, a new interference pattern would be produced. State one similarity and one difference between the diffraction grating pattern and the two-slits pattern. (*L*.)
12 In the Young's double-slit experiment *interference* occurs when *diffraction* of light takes place at each of the two slits.
   (a) Explain with the aid of a diagram the meaning of *interference* and *diffraction* in this case.
   (b) Explain how and why *sharper* and *more widely* spaced interference fringes may be obtained using a diffraction grating. (*AEB*.)
13 Give a labelled sketch and a brief description of the essential features of a spectrometer incorporating a plane diffraction grating. What part is played by:
   (a) diffraction;
   (b) interference, in the operation of the diffraction grating?

A source emitting light of two wavelengths is viewed through a grating spectrometer set at normal incidence. When the telescope is set at an angle of 20° to the incident direction, the second-order maximum for one wavelength is seen superposed on the third-order maximum for the other wavelength. The shorter wavelength is 400 nm. Calculate the longer wavelength and the number of lines per metre in the grating. At what other angles, if any, can superposition of two orders be seen using this source? (*O. & C.*)

14 What is *Huygens's theory*? Indicate with the aid of diagrams how the theory accounts for:
(a) the reflection of waves from a plane surface;
(b) the diffraction of waves by a grating.

Describe how you would demonstrate the diffraction of waves at a single slit. How does the width of the slit affect the results?

A grating has 500 lines per mm and is illuminated normally with monochromatic light of wavelength 589 nm. How many diffraction maxima may be observed? Calculate their angular positions. (*L.*)

21 Polarisation of Light Waves

In this chapter we show that light waves are transverse waves because they can be polarised. We first deal with polarisation by Polaroid, then with polarisation by reflection and by double refraction. We conclude with a brief account of applications of polarisation such as photoelasticity and saccharimetry.

Polarised Light

Polaroid is an artificial crystalline material which can be made in thin sheets. As we shall see shortly, it has the property of allowing only light waves due to vibrations in a particular plane to pass through.

Suppose two Polaroids, P and Q, are placed one behind the other in front of a window, and the light passing through P and Q is observed,

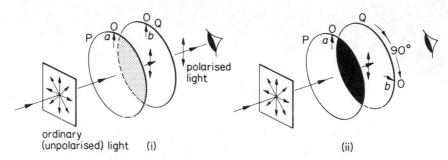

Figure 21.1 *Plane-polarisation of light by Polaroid*

Figure 21.1. When the Polaroids have their axes *a* and *b* parallel, the light passing through Q appears slightly darker, Figure 21.1 (i). If Q is now rotated slowly about the line of vision with its plane parallel to P, the light passing through Q becomes darker and darker and disappears at one stage. In this case the axes *a* and *b* are *perpendicular*, Figure 21.1 (ii). When Q is rotated further the light reappears, and becomes brightest when the axes *a*, *b* are again parallel.

This simple experiment leads to the conclusion that light waves are *transverse* waves; otherwise the light emerging from Q could never be extinguished by simply rotating the Polaroid. The experiment, in fact, is analogous to that illustrated in Figure 17.32, where transverse mechanical waves were set up along a rope and plane-polarised waves were obtained by means of a slit. Polaroid, because of its internal molecular structure, transmits only those vibrations of light in a particular plane. Consequently:

(a) plane-polarised light is obtained beyond the crystal P;

(b) no light emerges beyond Q when its axis is *perpendicular* to that of P.

Vibrations in Unpolarised and Polarised Light

Figure 21.2 (i) is an attempt to represent diagrammatically the vibrations of ordinary or unpolarised light at a point A when a ray travels in a direction AB. X is a plane perpendicular to AB, and ordinary (unpolarised) light may be imagined as due to vibrations which occur in every one of the millions of planes which pass through AB and are perpendicular to X. As represented in Figure 21.2 (ii), the amplitudes of the vibrations are all equal.

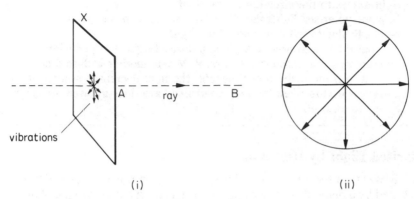

(i)   (ii)

Figure 21.2 *(i) Vibrations occur in every plane perpendicular to ray (ii) Vibrations in ordinary light*

Consider the vibrations in ordinary light when it is incident on the Polaroid P in Figure 21.1 (i). Each vibration can be resolved into two components, one in a direction parallel to *a*, the direction of easy transmission of light through P, and the other in a direction *m* perpendicular to *a*, Figure 21.3. Polaroid

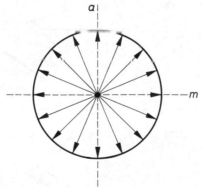

Figure 21.3 *Plane-polarised waves by selective absorption*

absorbs the light due to vibrations parallel to *m*, known as the *ordinary rays*, but allow light due to the other vibrations, known as the *extraordinary rays*, to pass through. So *plane-polarised light*, light due to vibrations in one plane, is produced near the eye, as illustrated in Figure 21.1 (i).

So Polaroid absorbs light due to vibrations in a particular direction and allows light due to vibrations in a perpendicular direction to pass through. This 'selective absorption' is also shown by certain natural crystals such as tourmaline.

Theory and experiment show that the vibrations of light are *electromagnetic* in origin. This is discussed later at the end of the chapter.

You should know:

1 Light consists of transverse waves.
2 In ordinary (unpolarised) light, the vibrations are in every plane at right angles to the direction of the light.
3 In plane-polarised light, the vibrations are only in one plane perpendicular to the direction of the light.
4 A Polaroid crystal produces plane-polarised light in a particular allowed direction through the crystal. When another Polaroid is rotated in front of the first Polaroid, the light decreases in intensity. Darkness occurs when the two Polaroids have their allowed directions at 90° to each other.

Polarised Light by Reflection

In 1808 MALUS discovered that polarised light is obtained when ordinary light is reflected by a plane sheet of glass (p. 561). The most suitable angle of incidence is about 57°, Figure 21.4. If the reflected light is viewed through a Polaroid which is slowly rotated about the line of vision, the light is practically extinguished at one position of the Polaroid. This proves that the light reflected by the glass is practically plane-polarised. Light reflected from the surface of a table becomes darker when viewed through a rotated Polaroid, showing it is partially plane-polarised.

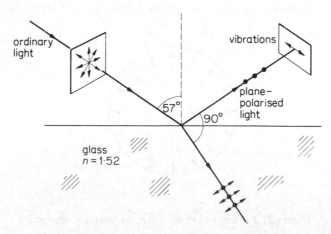

Figure 21.4 *Plane polarisation by reflection*

The production of the polarised light by the glass is explained as follows. Each of the vibrations of the incident (ordinary) light can be resolved into a component parallel to the glass surface and a component perpendicular to the surface. The light due to the components parallel to the glass is largely reflected, but the remainder of the light, due mainly to the components perpendicular to the glass, is *refracted* into the glass. So the light reflected by the glass is partially plane-polarised.

Brewster's Law, Polarisation by Pile of Plates

The particular angle of incidence i on a transparent medium when the reflected light is almost completely plane-polarised is called the *polarising angle*. BREWSTER found that, in this case, tan $i = n$, where n is the refractive index of the medium (*Brewster's law*). With crown glass of $n = 1.52$, $i = 57°$ (approx.) from tan $i = n$. Since n varies with the colour of the light, white light cannot be completely plane-polarised by reflection.

As sin i/sin $r = n$, where r is the angle of refraction, and sin i/cos i = tan i, it follows from Brewster's law that cos i = sin r, or $i + r = 90°$. So the reflected and refracted beams are at $90°$ to each other at the polarising angle, as illustrated in Figure 21.4.

The refracted beam contains light mainly due to vibrations perpendicular to that reflected and is therefore partially plane-polarised. Since refraction and reflection occur at both sides of a glass plate, the transmitted beam contains a fair percentage of plane-polarised light. A *pile of plates* increases the percentage, and thus provides a simple way of producing plane-polarised light. They are mounted inclined in a tube so that the ordinary (unpolarised) light is incident at the polarising angle, and the transmitted light is then practically plane-polarised.

---

1 Light beams reflected and refracted by glass are partially plane-polarised.
2 Maximum polarisation of the reflected beam occurs for an angle of incidence i given by tan $i = n$, refractive index of glass (Brewster's law). The reflected and refracted rays are then at $90°$ to each other.

Polarisation by Double Refraction

We have already considered two methods of producing polarised light. The first observation of polarised light, however, was made by BARTHOLINUS in 1669, who placed a crystal of Iceland spar on some words on a sheet of paper. To his surprise, two images were seen through the crystal. Bartholinus therefore gave the name of *double refraction* to the phenomenon, and experiments more than a century later showed that the crystal produced plane-polarised light when ordinary light was incident on it, see Figure 21A.

Iceland spar is a crystalline form of calcite (calcium carbonate) which cleaves in the form of a 'rhomboid' when it is lightly tapped; this is a solid whose opposite faces are parallelograms. When a beam of unpolarised light is incident one one face of the crystal, its internal molecular structure produces two beams of polarised light, E, O, whose vibrations are *perpendicular* to each other, Figure 21.5. If the incident direction AB is parallel to a plane known as the 'principal section' of the crystal, one beam O emerges parallel to AB, while the other beam E emerges in a different direction. As the crystal is rotated about the line of vision the beam E revolves round O. On account of this unusual behaviour the rays in E are called 'extraordinary' rays; the rays in O are known as 'ordinary' rays (p. 559). So two images of a word on a paper, for example, are seen when an Iceland spar crystal is placed on top of it; one image is due to the ordinary rays, while the other is due to the extraordinary rays.

With the aid of an Iceland spar crystal Malus discovered the polarisation of light by reflection (p. 560). While on a visit to Paris he looked through the crystal at the light of the sun reflected from the windows of the Palace of Luxembourg, and observed that only one image was obtained for a particular

Plate 21A *Double refraction. A ring with a spot in the centre, photographed through a crystal of Iceland spar. The two plane-polarised beams of light emerging from the crystal form two rings and two spots. The dark lines on the rings are due to the overlapping beams polarised at right angles to each other*

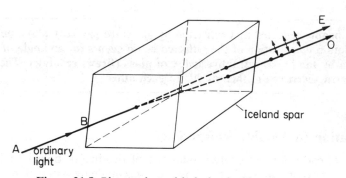

Figure 21.5 *Plane-polarised light by double refraction*

position of the crystal when it was rotated slowly. The light reflected from the windows could not therefore be ordinary (unpolarised) light, which usually gave two images, and Malus found that the reflected light was plane-polarised.

Nicol Prism

We have seen that a Polaroid produces polarised light, and that the Polaroid can be used to detect light. NICOL designed a form of Iceland spar crystal which was widely used for producing and detecting polarised light, and it is known as a *Nicol prism*. A crystal whose faces contain angles of 72° and 108° is broken into two halves along the diagonal AB, and the halves are cemented together by a layer of Canada balsam, Figure 21.6. The refractive index of the crystal for the ordinary rays is 1·66, and is 1·49 for the extraordinary rays; the refractive index of the Canada balsam is about 1·55 for both rays, since Canada balsam does not polarise light. A critical angle thus exists between the crystal and Canada balsam for the ordinary rays, but not for the extraordinary rays. So total reflection of the ordinary rays takes place at the Canada balsam if the angle of incidence is large enough, as it is with the Nicol prism. The emergent light is then due to the extraordinary rays, and is polarised.

The prism is used like a Polaroid to detect plane-polarised light, namely, the

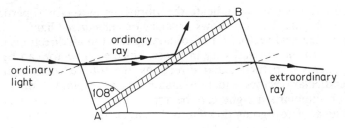

Figure 21.6 *Action of Nicol prism*

prism is held in front of the beam of light and is rotated. If the beam is plane-polarised the light seen through the Nicol prism varies in intensity, and none passes through at one position of the prism.

Differences Between Light and Sound Waves

We are now in a position to distinguish fully between light and sound waves. The physical difference, of course, is that light waves are due to varying electric and magnetic fields and are *electromagnetic waves*, while sound waves are due to vibrating layers or particles of the medium concerned and are *matter waves*. Light can travel through a vacuum, but sound cannot travel through a vacuum.

Another very important difference is that the vibrations of the particles in sound waves are in the same direction as that along which the sound travels, whereas the vibrations in light waves are perpendicular to the direction along which the light travels. Sound waves are therefore *longitudinal* waves, whereas light waves are *transverse* waves. As we have seen, sound waves can be reflected and refracted, and can give rise to interference phenomena; but no polarisation phenomena can be obtained with sound waves since they are longitudinal waves, unlike light waves.

Polarisation and Electric Vector

As we have just stated, light is a transverse electromagnetic wave and so contains electric and magnetic fields. Experiment shows that only the electric field is concerned in the blackening of a photographic film when exposed to light. So we usually define *the direction of light vibrations to be that of the electric vector E*. The 'plane of polarisation' is then defined as the plane containing the light ray and *E*.

Figure 21.7 (i) shows *plane-polarised* light due to vertical vibrations or a vector *E*. Figure 21.7 (ii) shows plane-polarised light due to horizontal vibrations;

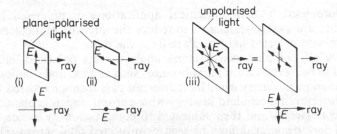

Figure 21.7 *Electric vectors in plane-polarised and unpolarised light*

in this case a dot is used to indicate the vibrations or vector E perpendicular to the paper. Figure 21.7 (iii) shows *ordinary* or *unpolarised* light.

The associated field vectors E in ordinary light act in all directions in a plane perpendicular to the ray and vary in phase. Since each vector can be resolved into components in perpendicular directions, the total effect is equivalent to two perpendicular vectors equal in magnitude as explained on p. 559. So ordinary or unpolarised light can be represented by an arrow and a dot (representing an arrow going down into the page), as shown in Figure 21.7 (iii).

Polaroid Transmission and Light Intensity

Suppose a Polaroid A produces polarised light whose electric vector vibrates in a particular direction, say AY in Figure 21.8 (i). If another Polaroid B is placed in front of A so that its 'easy' direction of transmission BX is perpendicular to AY, the overlapping areas appear black because the electric vector has no component in a perpendicular direction and so no light is transmitted.

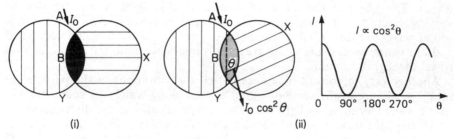

Figure 21.8 *Intensity variation by rotating Polaroid*

Suppose B is turned so that BX makes an angle θ with AY, Figure 21.8 (ii). If E_0 is the initial amplitude of the electric vector in A, the amplitude of the vector transmitted by B is the component E given by $E = E_0 \cos \theta$. The intensity is proportional to the square of the amplitude (p. 570). Since $E^2 = E_0{}^2 \cos^2 \theta$, the intensity I of the light transmitted by B is given by $I = I_0 \cos^2 \theta$, where I_0 is the light intensity incident from A on B, Figure 21.8 (iii). So if B is rotated to $\theta = 60°$, then $I = I_0 \cos^2 60° = I_0/4$, since $\cos 60° = 1/2$. So the light intensity is reduced to one quarter of that incident from A.

It should be noted that the intensity of the unpolarised light *incident on A* is $2I_0$, since, from Figure 21.8, one of the perpendicular vectors of the polarised light is absorbed by A.

Applications of Polarised Light

Polaroids are used in many practical applications of polarised light. For example, they are used in sunglasses to reduce the intensity of incident sunlight and to eliminate reflected light or glare from the road.

Photoelasticity, or photoelastic stress analysis, uses polarised light. Under mechanical stress, certain isotropic substances such as glass and celluloid become doubly refracting. Ordinary white light does not pass through crossed Polaroids (p. 558). However, if a celluloid model with a cut-out design is placed between the crossed Polaroids and then subjected to compression by a vice, coloured bright and dark fringes can now be seen or projected on a screen. The fringes spread from places where the stress is most concentrated. Now the pattern of

Plate 21B *Rolls Royce fan-blade assembly photographed in polarised light*

the fringe varies with the stress. So using a model, a study of the fringe pattern provides the engineer with valuable information on design. This is particularly useful with complicated shapes, where mathematical computation is difficult.

In *films*, it is possible to give the illusion of three dimensions or 3-D by projecting two overlapping pictures, with slightly different views, on to the same screen. Each picture has been taken by light polarised respectively in perpendicular directions. The viewer is provided with special spectacles, with perpendicular Polaroids in the respective frames. The two pictures received simultaneously by the two eyes provide a 3-D view of the scene.

Saccharimetry is the measurement of the concentration of sugars such as cane sugar in solution. Due to the molecular structure of the sugar, these solutions rotate the plane of polarisation of plane-polarised light as the light passes through. Solids such as quartz produce the same effect, which is called *optical activity*. The rotation of the plane of polarisation when the incident light is viewed may be right-handed (clockwise) or left-handed (anticlockwise).

There are various types of saccharimeter. Here we are only concerned with the basic principle of their action. A saccharimeter usually consists of a Nicol prism P, which produces plane-polarised light from an incident source S of monochromatic light and is called the *polariser*; a *tube* T containing the solution; and a Nicol A through which the emerging light is observed, called the *analyser*, Figure 21.9. Before the solution is poured into T, the analyser A is rotated until the plane-polarised light emerging from P is completely extinguished. T is then filled with the sugar solution. On looking through A, light can now be seen. A is then rotated until the light is again extinguished and the angle of rotation θ is measured.

For a solution of a given substance in a given solvent, the amount of rotation θ depends on the length of light path travelled through the liquid, the temperature of the solution and the wavelength of the light. Tables provide the rotation in degrees when the sodium D-line is used, the temperature is 20°C, and the light travels a column of length 10 cm of the liquid which contains 1 gram of the active substance per millilitre of solution. This is called

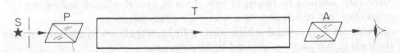

Figure 21.9 *Saccharimeter principle*

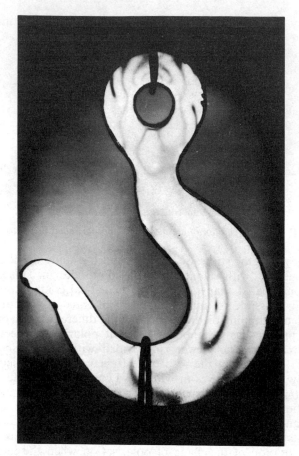

Plate 21C *Photoelasticity. The pattern in a Perspex hook between Polaroids when the hook is under stress. Similar patterns in transparent models help the engineer to investigate the stresses in mechanical structures subjected to loads or forces.* (Courtesy of Kodak Limited)

the *specific rotation* or *rotary power*. Knowing its value, the concentration of a sugar solution can be found by a polarimeter experiment and using the tables.

You should know:

1 Light waves are transverse waves like all electromagnetic waves.
2 Ordinary light is due to vibrations in all planes at 90° to the light direction. The electric vector in the light wave is taken as the light vibration.
3 Ordinary light vibrations can be resolved in two perpendicular directions normal to the light ray. This is useful when considering reflection and refraction at a water or glass surface.
4 A Polaroid is a crystal which allows light vibrations in one plane only through it. So *plane-polarised* light emerges.
5 Longitudinal waves such as sound waves cannot be polarised.

6 Ordinary light is partially plane-polarised when reflected and refracted at a water or glass surface.
Brewster's law: Maximum polarisation occurs at reflection when $\tan i = n$, where i is the angle of incidence and n is the refractive index. The reflected and refracted rays are then $90°$ to each other.

7 Polarised light through glass models helps to locate stresses in engine or building structures. The strains produce light and dark fringes. The concentration of some sugar solutions can be measured by using their rotation of plane-polarised light.

Polarimetry

Polarimetry is the science concerned with the angle of rotation of plane-polarised light. It is important in chemistry, since many chemical compounds are *optically active*; they have the power of rotating the plane of polarisation of a beam of polarised light. The phenomenon occurs when the molecular structure of the compound lacks symmetry, so that the molecule and its mirror image are not superimposable. Polarimetry has important applications in the sugar industry, since sucrose is far more optically active than many of the more common impurities, and so polarimetry can be used to measure the purity of the sugar.

Polarimetry is also used to characterise and distinguish *stereoisomers*, which are compounds with the same composition and structure, but different configurations of atoms within the molecule. Two well-known stereoisomers are tartaric and racemic acid. Tartaric acid rotates the plane of polarised light to the right, while racemic acid is optically inactive, suggesting different symmetries within the two molecules. While these acids differ in many other observable properties, it is often the case that stereoisomers can only be distinguished by polarimetry and by their reactions with other optically active substances.

EXERCISES 21   Polarisation

Multiple Choice

1 Sound waves and light waves cannot both be

   A refracted   B reflected   C made to produce an interference pattern
   D polarised   E diffracted

2 In summer, a motorist may wear Polaroid spectacles to reduce the glare of sunlight reflected from the

   A windscreen   B blue sky   C road surface   D side window   E rear window

3 Plane-polarised light of intensity I is incident normally on the face of a Polaroid P whose polarising plane, or 'allowed' light direction, is $30°$ (θ) to the plane-polarised light. Figure 21A.
   The intensity of the light emerging from P is

   A $I \cos 30°/2$ B $I \sin 30°/2$ C $I^2 \cos^2 30°/2$ D $I^2 \sin^2 30°/2$ E $I \cos^2 30°/2$

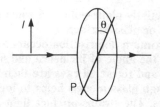

Figure 21A

Longer Questions

4 With the aid of suitable diagrams, explain the difference between transverse waves which are polarised and those which are unpolarised. State one useful application of plane polarised waves, specifying the type of wave involved (e.g. electromagnetic). (*L.*)

5 Explain what is meant by the statement that a beam of light is *plane-polarised*. Describe *one* experiment in each instance to demonstrate:
(a) polarisation by reflection;
(b) polarisation by double refraction;
(c) polarisation by scattering.
   The refractive index of diamond for sodium light is 2·417. Find the angle of incidence for which the light reflected from diamond is completely plane-polarised. (*L.*)

6 Explain the terms *linearly polarised light* and *unpolarised light*.
   A parallel beam of light is incident on the surface of a transparent medium of refractive index n at an angle of incidence i such that the reflected and refracted beams are at right angles to each other. Assuming the laws of reflection and refraction, find the relation between n and i.
   At this value of i, the Brewster angle, the reflected beam is found to be linearly polarised. Describe the apparatus and procedure you would use to find the refractive index of a material, using the Brewster angle. Has the method any advantage over other methods for measurement of n? (*O. & C.*)

7 Give an account of the action of:
(a) a single glass plate;
(b) a Nicol prism, in producing plane-polarised light. State *one* disadvantage of *each* method.
   Mention *two* practical uses of polarising devices. (*N.*)

8 Give one piece of experimental evidence in each case, with a brief account of how that evidence is obtained, that light behaves:
(a) as a wave;
(b) as an electromagnetic wave;
(c) as a stream of particles.
   When an unpolarised beam of light travelling in air is incident on a glass plate at an angle of incidence 30° the reflected light is found to be *partially linearly polarised*. Explain the meaning of the term in italics and describe how you would verify this statement experimentally. (*O. & C.*)

9 Explain what is meant by *plane-polarised* electromagnetic radiation. How may plane-polarised:
(a) light; and
(b) radio waves be produced and detected?
   Two polarising sheets initially have their polarisation directions parallel. Through what angle must one sheet be turned so that the intensity of transmitted light is reduced to a third of the original transmitted intensity?
   A vertical dipole is connected to a radio frequency generator. An identical dipole, similarly connected to the same generator, is arranged horizontally a quarter of a wavelength in front of the first so that both dipoles are normal to the horizontal line joining their midpoints. Describe as fully as you can the nature of the radiation at a distant point on this line. (*C.*)

22 Applications of Sound Waves

In the waves chapter, we discussed some properties of sound waves. Here we deal with the factors which influence the pitch, loudness and quality (timbre) of sound waves, the phenomenon of beats and its uses, and the Doppler effect and its application in sound and light.

Characteristics of Notes

Notes may be similar to or different from each other in three respects: (i) *pitch*; (ii) *loudness*; (iii) *quality*. These three quantities define or 'characterise' a note.

Pitch and Frequency

Pitch is analogous to colour in light, which depends only on the wavelength or frequency of the light wave (p. 523). Similarly, the pitch of a note depends only on the frequency of the sound vibrations. A high frequency gives rise to a high-pitched note; a low frequency produces a low-pitched note. So the high-pitched whistle of a boy may have a frequency of several thousand Hz, whereas a low-pitched hum due to a.c. mains frequency when first switched on in a television receiver may be 100 Hz. The range of sound or audio frequencies is about 15 to 20 000 Hz depending on the observer. In electronics, this is the *audio-frequency* (a.f.) range of alternating currents or voltages.

Musical or Acoustic Intervals

If a note of frequency 300 Hz, and then a note of 600 Hz, are sounded by a siren, the pitch of the higher note is recognised to be an *upper octave* of the lower note. A note of frequency 1000 Hz is recognised to be an upper octave of a note of frequency 500 Hz. So the *musical interval* between two notes is an upper octave if the ratio of their frequencies is 2:1. It can be shown that the musical interval between two notes depends on the *ratio* of their frequencies, and not on the actual frequencies.

Ultrasonics

Sound waves of higher frequency than 20 000 Hz cannot be heard by a human but a cat or dog may hear them. They are called *ultrasonics*. Since the speed of a wave = frequency × wavelength, ultrasonics have short wavelengths compared with sound waves in the audio-frequency range.

Over a hundred years ago, CURIE found that a thin plate of quartz increased or decreased in length when a battery was connected to its opposite faces. This effect is known as *piezoelectricity*. By correctly cutting the plate, the changes in length could be made to occur along the axis of the faces to which the battery was connected. When an alternating voltage of ultrasonic frequency was connected to the faces of such a quartz crystal, the faces vibrated at the same frequency. So ultrasonic waves were produced.

Ultrasonic Applications

Industrial. In recent years ultrasonics have been used for a variety of industrial purposes. They are used on board coasting vessels for depth-sounding; the time taken by the wave to reach the bottom of the sea from the surface and back is measured. Ultrasonics are also used to kill bacteria in liquids, and they are

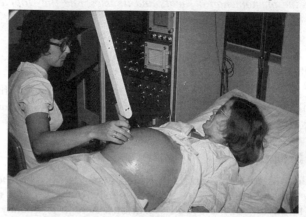

Plate 22A *Ultrasound examination of an eight-month pregnant woman. The image of the foetus is displayed on a screen and a printout can be taken. (Photograph reproduced courtesy of Charles Belinky/Science Photo Library*

used extensively to locate faults and cracks in metal castings following a method similar to that of radar. Ultrasonic waves are sent into the metal investigated, and the beam reflected from the fault is picked up on a cathode-ray tube screen together with the reflection from the other end of the metal. The position of the fault can then easily be located.

Medical. Ultrasonic techniques are used in clinical practice in hospitals. In diagnostic ultrasonics, the frequencies are very high and the waves reflected from inside the patient are amplified and displayed on the screen of a receiver. The image shows the various soft tissues and their connecting links.

Ultrasonics are now used to study the development of the foetus. The waves are reflected back from the foetus. In this way the birth can be monitored throughout the pre-natal period to check on any abnormalities, Plate 22A. In eye hospitals, ultrasonics are used to measure the eye-lens of a person with a cataract (eye) defect, which is replaced by an equivalent plastic lens.

The chief advantage of diagnostic ultrasonics is the absence of danger to the patient. Radiography, which is the use of X-rays, could be very dangerous.

Intensity and Amplitude

The *intensity* of a sound wave at a place X is defined as the energy per second per metre^2, or power per metre^2, flowing normally through an area at X. So the unit of intensity is $W\,m^{-2}$.

Suppose a *small* source, such as a small loudspeaker, produces a power of 25 W at a concert in the middle of a park. At a place B at a distance of 5 m from the speaker the sound energy passes through the surface area $4\pi r^2$ of a sphere of radius r round the speaker, where $r = 5$ m. So

$$\text{intensity at A} = \frac{25\ \text{W}}{4\pi \times 5^2\ \text{m}^2} = 0.1\ \text{W m}^{-2}\ \text{(approx.)}$$

At a distance r from the small source, we can see that

$$\textit{intensity} \propto 1/r^2$$

because the energy from the source is spread over an area $4\pi r^2$. At a distance twice as far as B, or 10 m, the intensity is $1/2^2$ of that at B or $1/4 \times 0.1\ \text{W m}^{-2} = 0.025\ \text{W m}^{-2}$

Amplitude. When the sound wave reaches B, the layer of air there vibrates with simple harmonic motion. If the mass of the layer is m and v_m is its maximum velocity we showed in the simple harmonic section that the energy of the layer was $\frac{1}{2}mv_m^2$. Since $v_m = \omega A = 2\pi f A$, where f is the frequency and A is the *amplitude* of vibration,

$$\text{energy of layer} = \tfrac{1}{2} \times m \times (2\pi f A)^2 = 2\pi^2 m f^2 A^2$$

So for a given frequency, intensity at layer $\propto A^2$. But, as we showed before, the intensity at a layer $\propto 1/r^2$, where r is the distance of the layer from a small sound source. So $A^2 \propto 1/r^2$, or $A \propto 1/r$.

Suppose the amplitude A is 4×10^{-6} m at a layer 10 m from the source. Then at a distance 20 m from the layer the amplitude is half as great or 2×10^{-6} m. So the *intensity* is proportional to $1/r^2$ at distance r from a small source but the *amplitude* of vibration of a layer is proportional to $1/r$.

---

intensity $\propto 1/r^2$

amplitude $\propto 1/r$

Loudness. The *loudness* of a sound depends on the intensity of the sound wave reaching the person concerned *and* on the person, that is, some people may hear a louder sound or a softer sound although the intensity would be the same. We say that loudness is a *subjective* quantity, unlike intensity. Loudness may be taken as proportional to the square of the amplitude, for a given person.

Power Gain, the Decibel

A rise in intensity from 1 mW m^{-2} to 100 mW m^{-2}, a ratio of 100, appears twice as loud as a rise from 1 mW m^{-2} to 10 mW m^{-2}. Now $\log_{10} 100 = 2$ and $\log_{10} 10 = 1$. So the ear appears to respond to sound intensity changes on a 'logarithmic' basis.

The *decibel*, dB, is a unit of gain or loss in power on a logarithmic basis, using base 10 logarithms. It is widely used in telecommunications and acoustics. Using logarithms, as we see soon, a wide range of power gain or loss is reduced to a more compact range by decibels.

The decibel system is defined as follows: If a power changes from a power P_1 to a power P_2, then

$$\text{number of decibels (dB) change} = 10 \log_{10}(P_2/P_1)$$

Omitting the base number 10, a power gain from 1 mW to 100 mW

$$= 10 \log (100/1) = 10 \log 100 = 10 \times 2 = 20 \text{ dB}$$

A gain from 100 mW to 200 mW $= 10 \log (200/100) = 10 \log 2 = 3 \text{ dB}$.

A *power loss* or *attenuation* from 300 mW to 100 mW

$$= 10 \log (100/300) = 10 \log (1/3) = 10 \log 1 - 10 \log 3 = -4 \cdot 8 \text{ dB}$$

since $\log 1 = 0$.

Figure 22.1 shows the gain and loss in decibels of a power P, using a power P_0 of 1 mW as a reference or zero level. Note that $dB = 0$ when the ratio $P/P_0 = 1$.

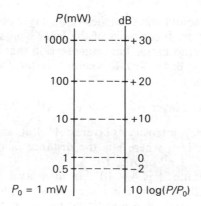

Figure 22.1 *Power and decibel range*

Threshold of Hearing and Pain

The lowest audible sound at a frequency of 1000 Hz is called the *threshold of hearing* at this frequency. This has an intensity P_0 about 10^{-12} W m^{-2}. Ordinary conversation has an intensity P about 10^{-5} W m^{-2}. This is a level in decibels dB above P which is

$$10 \log (10^{-5}/10^{-12}) = 10 \log (10^{7}) = 70 \text{ dB}$$

The *threshold of pain* is the intensity of sound which makes a painful sensation to the ear. This is about 1 W m^{-2}. Its decibel level above P_0, 10^{-12} W m^{-2},

$$= 10 \log (1/10^{-12}) = 10 \log 10^{12} = 120 \text{ dB}$$

With older people serious loss of hearing may occur at a level -20 dB or 20 dB below the threshold of hearing.

Noise Pollution

Sound which is not wanted or unpleasant to the ear is called *noise*. A large amount of noise from engines or machinery in factories can damage the ears of workers. Sound engineers make measurements to find the engines making the loudest noise and cut down the sound from them.

People living near aerodromes have noise pollution from engines in aeroplanes. So arrangements are made for fewer night flights. Traffic noise and road-building can upset people when windows and doors are left open. Double glazing of windows helps to cut down outside noise. Excessive noise from vehicle and motorcycle engines is forbidden by law.

Sound Absorption

Concert halls or large lecture rooms or broadcasting studios need to have special acoustic (sound) design. If not, sound is reflected from walls, ceilings or floors and this can disturb the sound from an orchestra or a speaker which reaches the audience.

To absorb the sound from the surfaces, suitable tiles of plastic or cork or other porous materials are used to cover the walls and other surfaces in concert halls or studios. In a large lecture room, cushions are often placed at the back of chairs. Sound waves are absorbed by the human body, so the cushions absorb the sound when the room is not full of people.

Sound investigations need special instruments such as those made for commercial use by the firm Bruel and Kjaer in Denmark. (See *Preface*.) The absorptive power of 1 m^2 of an open window has been taken as 1 unit since this is a perfect absorber. The absorptive power of a thick carpet is then about 0·5 and polished wood and glass is about 0·01.

Quality or Timbre

If the same note is sounded on the violin and then on the piano, an untrained listener can tell which instrument is being used, without seeing it. We say that the *quality* or *timbre* of the note is different in each case.

The waveform of a note is never simple harmonic in practice; the nearest approach is that obtained by sounding a tuning-fork, which produces what may be called a 'pure' note, Figure 22.2 (i). If the same note is placed on a

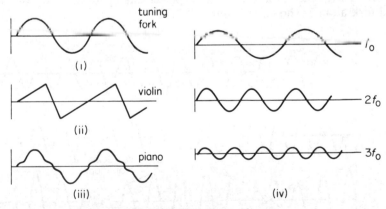

Figure 22.2 *Waveforms of notes*

violin and piano respectively, the waveforms produced might be represented by Figure 22.2 (ii), (iii), which have the same frequency and amplitude as the waveform in Figure 22.2 (i). Now curves of the shape of Figure 22.2 (ii), (iii) can be analysed mathematically into the sum of a number of *simple harmonic* curves, whose frequencies are multiples of f_0, the frequency of the original waveform; the amplitudes of these curves diminish as the frequency increases, Figure 22.2(iv), for example, might be an analysis of a curve similar to Figure 22.2 (iii), corresponding to a note on a piano. The ear is able to detect simple harmonic waves and therefore it registers the presence of notes of frequencies $2f_0$ and $3f_0$, in addition to f_0, when the note is sounded on the piano. The amplitude of the curve corresponding to f_0 is greatest, Figure 22.2 (iv), and the note of frequency f_0 is heard predominantly because the intensity is proportional to the square of the amplitude (p. 570). In the background, however, are the notes of frequencies $2f_0$, $3f_0$, which are called the *overtones*. The frequency f_0 is called the *fundamental*.

As the waveform of the same note is different when it is obtained from different instruments, it follows that the analysis of each will differ. For example, the waveform of a note of frequency f_0 from a violin may contain overtones of frequencies $2f_0$, $4f_0$, $6f_0$. The musical 'background' to the fundamental note is therefore different when it is sounded on different instruments, and so *the overtones present in a note determine its quality or timbre.*

A *harmonic* is the name given to a note whose frequency is a simple multiple of the fundamental frequency f_0. So f_0 is called the 'first harmonic'; a note of frequency $2f_0$ is called the 'second harmonic', and so on. Certain harmonics of a note may be absent from its overtones; for example, the only possible notes obtained from an organ-pipe closed at one end are f_0, $3f_0$, $5f_0$, $7f_0$, and so on (p. 601).

Beats

If two notes of nearly equal frequency are sounded together, a periodic rise and fall in loudness can be heard. This is known as the phenomenon of *beats*. The frequency of the beats is the number of intense or loud sounds heard per second.

Consider a layer of air some distance away from two pure notes of nearly equal frequency, say 48 and 56 Hz respectively, which are sounding. The variation of the displacement, y_1, of the layer due to one fork alone is shown in Figure 22.3 (i); the variation of the displacement y_2, of the layer due to the second fork alone is shown in Figure 22.3 (ii).

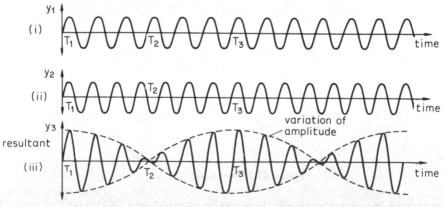

Figure 22.3 *Beats* (not to scale)

According to the principle of superposition (p. 484), the variation of the resultant displacement, y, of the layer is the algebraic sum of the two curves, which varies in amplitude in the way shown in Figure 22.3 (iii). To understand the variation of y, suppose that the displacements y_1, y_2 are in phase at some instant T_1, Figure 22.3. Since the frequency of the curve in Figure 22.3 (i) is 48 cycles per second the variation y_1 undergoes 3 complete cycles in $\frac{1}{16}$ second. In the same time, the variation y_2 undergoes $3\frac{1}{2}$ cycles, since its frequency is 56 cycles per second. Thus y_1 and y_2 are 180° out of phase with each other at this instant, and their resultant y is then a minimum at some instant T_2. So T_1T_2 represents $\frac{1}{16}$ of a second in Figure 22.3 (iii).

In $\frac{1}{8}$ of a second from T_1, y_1 has undergone 6 complete cycles and y_2 has undergone 7 complete cycles. So the two waves are in phase again at T_3 where T_1T_3 represents $\frac{1}{8}$ of a second, and their result at this instant is again a maximum, Figure 22.3 (iii). In this way it can be seen that loud sound is heard after every $\frac{1}{8}$ second, and so the beat frequency is 8 cycles per second. This is the difference between the frequencies, 48, 56, of the two notes. We show soon that *the beat frequency is always equal to the difference of the two nearly equal frequencies*. From Figure 22.3 we see that beats are a phenomenon of *interference* of two sound waves travelling in the same direction to an observer.

Beat Frequency Formula

Suppose two sounding tuning-forks have frequencies f_1, f_2 cycles per second which are close to each other. At some instant the displacement of a particular layer of air near the ear due to each fork will be a maximum to the right. The resultant displacement is then a maximum, and a loud sound or beat is heard.

After this, the vibrations of air due to each fork go out of phase, and t seconds later the displacement due to each fork is again a maximum to the right, so that a loud sound or beat is heard again. One fork has then made exactly *one cycle* more than the other. But the number of cycles made by each fork in t seconds is f_1t and f_2t respectively. Assuming f_1 is greater than f_2,

$$f_1t - f_2t = 1$$

$$\therefore f_1 - f_2 = \frac{1}{t}$$

Now 1 beat has been made in t seconds, so that $1/t$ is the number of beats per second or beat frequency.

$$\therefore f_1 - f_2 = \textit{beat frequency}$$

Uses of Beats

The phenomenon of beats can be used to measure the unknown frequency, f_1, of a note. For this purpose a note of known frequency f_2 is used to obtain beats with the unknown note, and the frequency f of the beats is obtained by counting the number made in a given time. Since f is the difference between f_2 and f_1, it follows that $f_1 = f_2 - f$, or $f_1 = f_2 + f$. Thus suppose $f_2 = 1000$ Hz, and the number of beats per second made with a tuning-fork of unknown frequency f_1 is 4. Then $f_1 = 1004$ or 996 Hz.

To decide which value of f_1 is correct, the ends of the tuning-fork prongs are loaded with a small piece of Plasticine which diminishes the frequency a little, and the two notes are sounded again. If the beat frequency is *increased*, a little thought indicates that the frequency of the note must have been originally 996 Hz. If the beats are decreased, the frequency of the note must have been originally 1004 Hz. The tuning-fork must not be overloaded, as the frequency may decrease, if it was 1004 Hz, to a frequency such as 995 Hz, in which case the significance of the beats can be wrongly interpreted.

Beats can also used to 'tune' an instrument to a tuning fork. As the instrument note approaches the note from the tuning fork, beats are heard. The instrument may be regarded as 'tuned' when the beats occur at a very slow rate.

You should know:

1 Ultrasonics are sound waves of very much higher frequences than sounds heard by humans. They are used in depth-sounding of oceans and in monitoring foetal development.

2 Sound waves: (a) pitch increases with frequency; (b) intensity (loudness) is proportional to amplitude^2—with a small source of sound, intensity $\propto 1/r^2$, so amplitude $\propto 1/r$, where r is the distance from the source; (c) timbre, or sound quality depends on the overtones produced.

3 Beats are due to the interference of two waves from sounds of close frequency. Beat frequency = frequency of loud sound heard = $f_1 - f_2$. Radio and light sources may produce beats.

Anti-sound

While sound can be both an important form of communication and a carrier of beautiful music, there are times when it can be a nuisance. Yet insulating ourselves from sound is quite a difficult problem. Often it is not practical to put a physical barrier between ourselves and the sound source; for example, the output duct of a building's air-conditioning system can often produce an extremely annoying low rumble, but we cannot physically block the sound without obstructing the desired flow of air!

For such applications engineers have recently developed a new noise-cancellation technique called *anti-sound*. The idea is to use a loudspeaker to generate sound which interferes destructively with the unwanted sound and cancels it. For our air-conditioning example, a microphone would be placed near the opening of the duct, with a loudspeaker nearby. The sound picked up by the microphone would be processed digitally to invert its phase (so that it cancels when added to the original sound) and corrected for the distance between the microphone and the loudspeaker before being played back through the loudspeaker. This produces a remarkable reduction in noise from the system, without any effect on the flow of air. Anti-sound is expected to find many applications where physical sound-insulating barriers are not practical.

EXERCISES 22A   Sound Waves

Multiple Choice

1 Two sound waves of the same frequency have respective amplitudes of 3 units and 1 unit and are travelling in opposite directions in the same straight line. At a particular place in that line, the resultant wave will vary in loudness. The ratio maximum loudness/minimum loudness is

A 9/1   B 6/1   C 9/2   D 4/1   E 1·5/1

2 A tuning fork of frequency 312 Hz is sounded with a fork of unknown frequency f Hz, 4 beats per second are heard. When a little Plasticine is added to the prongs of the fork, the beats decrease in number. Then f in Hz is

A 308 **B** 310 **C** 316 **D** 320 **E** 324

3 At a distance 20 m from a small loudspeaker, the amplitude of the sound heard is 0·012 mm. At a distance 30 m from the loudspeaker, the amplitude in mm is

A 0·036 **B** 0·024 **C** 0·018 **D** 0·010 **E** 0·008

Longer Questions

4 What is meant by:
(a) the *amplitude*; (b) the *frequency* of a wave in air? What are the corresponding characteristics of the musical sound associated with the wave? How would you account for the difference in quality between two notes of the same pitch produced by two different instruments, e.g., by a violin and by an organ pipe?
     What are 'beats'? Given a set of standard forks of frequencies 256, 264, 272, 280 and 288, and a tuning-fork whose frequency is known to be between 256 and 288, how would you determine its frequency accurately?

5 At a point 20 m from a small source of sound, the intensity is $0·5\,\mu W\,cm^{-2}$. Find a value for the rate of emission of sound energy from the source, and state the assumptions you make in your calculation.

6 Explain the origin of the beats heard when two tuning-forks of slightly different frequency are sounded together. Deduce the relation between the frequency of the beats and the difference in frequency of the forks. How would you determine which fork had the higher frequency?
     A simple pendulum set up to swing in front of the 'seconds' pendulum ($T = 2$ s) of a clock is seen to gain so that the two swing in phase at intervals of 21 s. What is the time of swing of the simple pendulum? (*L*.)

7 Describe the nature of the disturbance set up in air by a vibrating tuning-fork and show how the disturbance can be represented by a sine curve. Indicate on the curve the points of:
(a) maximum particle velocity; (b) maximum pressure.
     What characteristics of the vibration determine the pitch, intensity and quality respectively of the note? (*N*.)

8 Define *frequency* and explain the term *harmonics*. How do harmonics determine the *quality* of a musical note?
     It is much easier to hear the sound of a vibrating tuning fork if it is
(a) placed in contact with a bench, or (b) held over a *certain* length of air in a tube. Explain why this is so in both these cases and give two further examples of the phenomenon occurring in (b).
     Describe how you would measure the wavelength in the air in a tube of the note emitted by the fork. How would the value obtained be affected by changes in (i) the temperature of the air, and   (ii) the pressure of the air? (*L*.)

9 Distinguish between *longitudinal* and *transverse* wave motions, giving examples of each type. Find a relationship between the frequency, wavelength and velocity of propagation of a wave motion.
     Describe experiments to investigate quantitatively for sound waves the phenomena of (a) reflection; (b) refraction; (c) interference. (*C*.)

10 Continuous sound waves of a single frequency are emitted from two small loudspeakers A and B, fed by the same signal generator and located as shown in Figure 22A (i), which is not to scale. A small sensitive microphone placed at A is connected (via a pre-amplifier) to a cathode ray oscilloscope. The graphs in Figure 22A (ii) show the traces that appear on the screen of the c.r.o. due to A alone, then B alone, and represent the variation of the displacement at P produced by each wave separately with time.

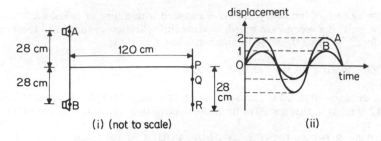

Figure 22A

(a) Calculate the relative intensities of the waves emitted by A and B. The width of the trace on the c.r.o. is 12·0 cm and the time base 'speed' is $50 \ \mu\text{s} \ \text{cm}^{-1}$. Calculate the frequency of the waves.

(b) The two waves interfere at P. Using the superposition principle, construct on the graph (Figure 22A (ii)) the resultant displacement-time curve, using the same axes as for the original waves. What is the intensity of the resultant wave at P compared with that caused by B alone?

(c) Using axes on a separate sheet, construct the resultant displacement for a point Q where the waves from A and B arrive 180° out of phase with each other. (Assume that the amplitudes of the waves arriving at Q are the same as those arriving at P.) What is the intensity of the resultant wave compared with that caused by B alone in this case?

(d) Explain why, although the displacement of the resultant wave produced at P or Q varies with time, the sound intensity does not.

(e) A maximum of sound intensity occurs when the microphone is at R. From the dimensions given in the diagram determine the largest possible value for the wavelength of the sound waves.

Doppler Effect

The whistle of a police car or a jet aeroplane appears to increase in pitch as it approaches a stationary observer; as the moving object passes the observer, the pitch changes and becomes lower. The apparent alteration in frequency was first predicted by DOPPLER in 1845. He stated that a change in frequency of the wave motion should be observed when a source of sound or light was moving, and this is known as the *Doppler effect or Doppler principle*.

The Doppler effect occurs when a source of sound or light moves *relative* to an observer. In light, the effect was observed when measurements were taken of the wavelength of the light from a moving star; they showed a marked variation. In sound, the Doppler effect can be demonstrated by placing a whistle in the end of a long piece of rubber tubing, and whirling the tube in a horizontal circle above the head while blowing the whistle. The open end of the tube acts as a moving source of sound, and an observer hears a rise and fall in pitch as the end approaches and recedes from him or her.

Everyday Examples of Doppler Effects

Police car sirens use high-frequency sound. When a police car is chasing a speeding car, the sound moving towards you appears to increase in frequency or pitch. When the car passes and moves away, the frequency appears to be lower. As we see soon, the frequency change is due to the Doppler effect. A similar change occurs when *ambulance* sirens are sounding and the ambulance is moving at high speed towards or away from you.

Plate 22B *Speed trap (Photograph reproduced by courtesy of the Metropolitan Police, London)*

In *police speed traps*, a transmitter carried in a police car or on a bridge high above a motorway, sends out radio waves towards the speeding car. The wave is reflected back and the time taken is a measure of the speed, which is recorded directly on a meter. This is similar to radar used in large airports for recording the distance and heights of incoming aeroplanes. The relatively small speed of a car compared with the speed of a radio wave (about 300 thousand kilometres per second) makes the Doppler effect negligibly small in this case. In contrast, the speed of a sound wave in air is about a third of a kilometre per second, so the Doppler effect due to a police car or ambulance siren is noticed in these cases.

A complete calculation of the apparent frequency in particular cases is given shortly, but Figure 22.4 shows why a change of wavelength, and hence

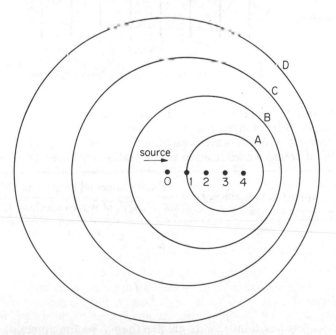

Figure 22.4 *Doppler effect: moving source, stationary observer*

frequency, occurs when a source of sound S is moving towards a stationary observer. As the source S moves at a steady speed from position O to the four equally spaced positions 1, 2, 3, 4, four circular waves are produced, A, B, C, D, which have 0, 1, 2, 3 respectively as their centres. For example, D corresponds to O and C to 1. So if the observer is on the right of the source S, he receives wavefronts which are relatively more crowded together than if S were stationary. So the frequency of S appears to increase.

When the observer is on the left of S, in which case the source is moving away from him, the wavefronts are farther apart than if S were stationary. So the observer receives correspondingly fewer waves per second. The apparent frequency is therefore lowered.

Calculation of Apparent Frequency

Suppose c is the speed of sound in air, u_s is the speed of the source of sound S, u_0 is the speed of an observer O, and f is the true frequency of the source.

(i) *Source moving towards stationary observer*. If the source S were stationary, the f waves sent out in one second towards the observer O would occupy a distance c, and the wavelength would be c/f, Figure 22.5 (i). If S moves with a

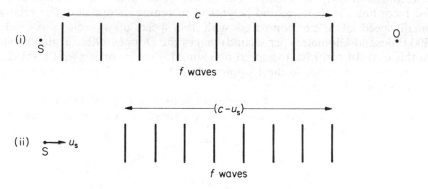

Figure 22.5 *Source moving towards stationary observer*

speed u_s towards O, however, the f waves sent out occupy a smaller distance $(c - u_s)$, because S has moved a distance u_s towards O in 1 s, Figure 22.5 (ii). So the wavelength λ' of the waves reaching O is now $(c - u_s)/f$.

The speed of sound waves relative to the stationary observer O $= c$

$$\therefore \text{ apparent frequency, } f' = \frac{\text{velocity of sound relative to O}}{\text{wavelength of waves reaching O}}$$

$$= \frac{c}{\lambda'} = \frac{c}{(c - u_s)/f}$$

$$\therefore f' = \frac{c}{c - u_s} f \qquad . \qquad . \qquad . \qquad . \qquad . \qquad (1)$$

Since $(c - u_s)$ is less than c, f' is greater than f so the apparent frequency appears to increase when a source is moving towards an observer.

(ii) *Source moving away from stationary observer.* In this case we can simply change the sign of the speed u_s in (1). So the apparent frequency f' is given by

$$\therefore f' = \frac{c}{c + u_s} \cdot f \quad . \qquad . \qquad . \qquad . \qquad . \qquad (2)$$

Since $(c + u_s)$ is greater than c, f' is less than f. So the apparent frequency decreases when a source moves away from an observer.

(iii) *Source stationary, and observer moving towards it.* Since the source is stationary, the f waves sent out by S towards the moving observer O occupies a distance c, Figure 22.6. So the wavelength of the waves reaching O is c/f, and unlike the cases considered, the wavelength is unaltered.

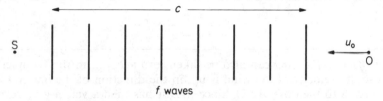

Figure 22.6 *Observer moving towards stationary source*

The speed of the sound waves relative to O is not c, however, as O is moving relative to the source. The speed of the sound waves relative to O is given by $(c + u_0)$ in this case, and so the apparent frequency f' is given by

$$f' = \frac{\text{velocity of sound relative to O}}{\text{wavelength of waves reaching O}} = \frac{c + u_0}{c/f}$$

$$\therefore f' = \frac{c + u_0}{c} \cdot f \quad . \qquad . \qquad . \qquad . \qquad . \qquad . \qquad (3)$$

Since $(c + u_0)$ is greater than c, f' is greater than f. So the apparent frequency is increased.

(iv) *Source stationary, and observer moving away from it.* In this case we can simply change the sign of u_0 in (3). So as we deduced before,

$$f' = \frac{c - u_0}{c} \cdot f \quad . \qquad . \qquad . \qquad . \qquad (4)$$

Since $(c - u_0)$ is less than c, the apparent frequency f' appears to be decreased.

Source and Observer Both Moving

If the source and the observer are both moving, the apparent frequency f' can be found from the formula

$$f' = \frac{c'}{\lambda'}$$

where c' is the velocity of the sound waves relative to the observer, and λ' is the wavelength of the waves reaching the observer. This formula can also be used to find the apparent frequency in any of the cases considered before.

Suppose that the observer has a velocity u_0, the source a velocity u_s, and that both are moving in the *same* direction. Then

$$c' = c - u_0$$

and $$\lambda' = (c - u_s)/f$$

as we deduced before.

$$\therefore f' = \frac{c'}{\lambda'} = \frac{c - u_0}{(c - u_s)/f} = \frac{c - u_0}{c - u_s} \cdot f \quad . \quad . \quad . \quad (1)$$

If the observer is moving towards the source, $c' = c + u_0$, and the apparent frequency f' is given by

$$f' = \frac{c + u_0}{c - u_s} \cdot f \quad . \quad . \quad . \quad . \quad (2)$$

The effect of the wind can also be taken into account in the Doppler effect. Suppose the velocity of the wind is u_w, in the direction of the line SO joining the source S to the observer O. Since the air has then a velocity u_w relative to the ground, and the velocity of the sound waves relative to the air is c, the velocity of the waves relative to ground is $(c + u_w)$ if the wind is blowing in the same direction as SO. All our previous expressions for f' can now be adjusted by replacing the velocity c in it by $(c + u_w)$. If the wind is blowing in the opposite direction to SO, the velocity c must be replaced by $(c - u_w)$.

Example on Doppler Principle

A car, sounding a horn producing a note of 500 Hz, approaches and then passes a stationary observer O at a steady speed of 20 m s^{-1}. Calculate the change in pitch of the note heard by O (speed of sound = 340 m s^{-1}).

Towards O. Speed of sound relative to O, $c' = 340$ m s^{-1}

Wavelength of waves reaching O, $\lambda' = (340 - 20)/500$ m

$$\therefore \text{ apparent frequency to O, } f' = \frac{c'}{\lambda'} = \frac{340}{320} \cdot 500 = 531 \text{ Hz} \quad . \quad . \quad (3)$$

Away from O. With the above notation, $c' = 340$ m s^{-1}

and $$\lambda' = (340 + 20)/500 \text{ m}$$

$$\therefore \text{ apparent frequency to O, } f'' = \frac{340}{(340 + 20)} \cdot 500 = 472 \text{ Hz}. \quad . \quad (4)$$

From (3) and (4), change in pitch $= \dfrac{f''}{f'} = 0 \cdot 9$ (approx.)

Reflection of Waves

Consider a source of sound A approaching a fixed reflector R such as a wall or bridge, for example. The reflected waves then appear to travel from R to A as if they came from the 'mirror image' A' of A in R, Figure 22.7.

As an illustration, suppose a car A approaches R with a velocity of 20 m s^{-1} when sounding a note of 1000 Hz from its horn, and that another car B behind A is travelling towards A with a velocity of 30 m s^{-1}, Figure 22.7.

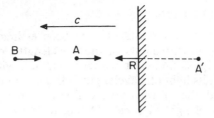

Figure 22.7 *Doppler's principle and reflection of waves*

In B, the driver hears a note from R which has an apparent frequency $f' = c'/\lambda'$, where c' is the velocity of sound relative to B and λ' is the wavelength of the waves reaching B. If the velocity of sound c is 340 m s^{-1}, then

$$c' = 340 + 30 = 370 \text{ m s}^{-1}$$

and $$\lambda' = (340 - 20)/1000 = 320/1000 \text{ m}$$

$$\therefore f' = \frac{c'}{\lambda'} = \frac{370}{320} \cdot 1000 = 1156 \text{ Hz}$$

In B, the driver also hears a note directly from A. In this case,

$$c' = 340 + 30 = 370 \text{ m s}^{-1}$$

and $$\lambda' = (340 - 20)/1000 = 360/1000 \text{ m}$$

$$\therefore f'' = \frac{c'}{\lambda'} = \frac{370}{360} \cdot 1000 = 1028 \text{ Hz}$$

With a *reflector moving* with velocity v directly towards a stationary sound source S of frequency f, the frequency f' of the waves reflected back to S is given by

$$f' = \frac{c + v}{c - v} f$$

where c is the velocity of sound in air. The frequency f_0 of the waves received by the moving reflector is $f_0 = (c + v)f/c$ and the frequency f' of the waves reflected back to S $= cf_0/(c - v) = (c + v)f/(c - v)$.

Doppler Effect in Circular Motion

Consider a sound signal of constant frequency carried by a car moving round a circular track.

At any instant, the velocity is directed along the tangent to the circle. So a distant observer O will hear a note which rises in frequency as the car approaches O in its circular path and decreases in frequency as the car goes away from O. The maximum frequency occurs when the velocity is directly towards O and the minimum frequency when the velocity is directly away from O.

If the observer O is *inside* the circle, O will hear the frequency rise when the component of the velocity is directed towards him or her. The frequency will decrease when the velocity component is directed away from O. If O is at the centre of the circle, the velocity has no component towards O. So a continuous note of the same frequency as the signal is heard in this case.

Doppler Principle in Light

The speed of distant stars and planets has been estimated by applying the Doppler principle to measurements of the wavelengths of the spectrum lines which they emit. Suppose a star or planet is moving with a velocity v away from the earth and emits light of wavelength λ. If the frequency of the vibrations is f cycles per second, then f waves are emitted in one second, where $c = f\lambda$ and c is the velocity of light *in vacuo*. Owing to the velocity v, the f waves occupy a distance $(c + v)$. So the *apparent wavelength* λ' to an observer on the earth in line with the star's motion is

$$\lambda' = \frac{c + v}{f} = \frac{c + v}{c} \cdot \lambda = \left(1 + \frac{v}{c}\right)\lambda$$

$$\therefore \lambda' - \lambda = \text{'shift' in wavelength} = \frac{v}{c}\lambda \ . \qquad . \qquad . \qquad (1)$$

and hence $\qquad \dfrac{\lambda' - \lambda}{\lambda} = \text{fractional change in wavelength} = \dfrac{v}{c} \ . \qquad . \qquad (2)$

From (1), it follows that λ' is greater than λ when the star or planet is moving away from the earth, that is, there is a 'shift' or displacement *towards the red*. The position of a particular wavelength in the spectrum of the star is compared with that obtained in the laboratory, and the difference in the wavelengths, $\lambda' - \lambda$, the 'red shift', is measured. From (1), knowing λ and c, the velocity v can be calculated.

Plate 22C *Doppler shift. Ultraviolet spectrum of the quasar 3C 273 taken in the wavelength 1300–1400 Å. The dotted line represents the centre of the spectrum with the wavelengths superimposed. The white spot next to 1200 Å is hydrogen's Lyman-alpha line at its usual wavelength while the spot near 1400 Å represents the same emission as we receive from the quasar. The difference in the two wavelengths is explained as a Doppler effect due to the expansion of the universe. According to Hubble's law of expansion, this shift puts the quasar at 2100 million light-years away from the earth. (Photograph reproduced courtesy Fred Espenak/Science Photo Library)*

If the star is moving *towards* the earth with a velocity u, the apparent wavelength λ'' is given by

$$\lambda'' = \frac{c - u}{f} = \frac{c - u}{c} \cdot \lambda = \left(1 - \frac{u}{c}\right)\lambda$$

$$\therefore \lambda - \lambda'' = \frac{u}{c}\lambda \ . \qquad . \qquad . \qquad . \qquad (3)$$

Since λ'' is less than λ, there is a displacement towards the blue in this case*.

* Equations (1)–(3) apply for velocities much less than c, otherwise relativistic corrections are required.

In measuring the speed of a star, a photograph of its spectrum is taken. The spectral lines are then compared with the same lines obtained by photographing in the laboratory an arc or spark spectrum of an element present in the star. If the lines are displaced towards the red, the star is receding from the earth; if they are displaced towards the violet, the star is approaching the earth. By this method the velocities of the stars have been found to be between about 10 km s^{-1} and 300 km s^{-1}.

The Doppler effect has also been used to measure the speed of rotation of the sun. Photographs are taken of the east and west edges of the sun; each contains absorption lines due to elements such as iron vaporised in the sun, and also some absorption lines due to oxygen in the earth's atmosphere. When the two photographs are put together so that the oxygen lines coincide, the iron lines in the two photographs are displaced relative to each other. In one case the edge of the sun approaches the earth, and in the other the opposite edge recedes from the earth. Measurements shows a rotational speed of about 2 km s^{-1}. See Worked Example 2.

Measurement of Plasma Temperature

In very hot gases or plasma, used in thermonuclear fusion experiments, the temperature is of the order of millions of degrees Celsius. At these high temperatures molecules of the glowing gas are moving away and towards the observer with very high speeds and, owing to the Doppler effect, the wavelength λ of a particular spectral line is apparently changed. One edge of the line now corresponds to an apparently increased wavelength λ_1 due to molecules moving directly away from the observer, and the other edge to an apparent decreased wavelength λ_2 due to molecules moving directly towards the observer. The line is thus seen to be *broadened*.

From our previous discussion, if v is the velocity of the molecules,

$$\lambda_1 = \frac{c+v}{c} \cdot \lambda$$

and

$$\lambda_2 = \frac{c-v}{c} \cdot \lambda$$

$$\therefore \text{ breadth of line, } \lambda_1 - \lambda_2 = \frac{2v}{c} \cdot \lambda \qquad . \qquad . \qquad . \qquad (1)$$

The breadth of the line can be measured by a diffraction grating, and as λ and c are known, the velocity v can be calculated. By the kinetic theory of gases, the velocity v of the molecules is roughly the root-mean-square velocity, or $\sqrt{3RT/M}$, where T is the absolute temperature, R is the molar gas constant and M is the mass of one mole.

Examples on Doppler Effect in Sound and Light

1 A whistle giving out 500 Hz moves away from a stationary observer in a direction towards and perpendicular to a flat wall with a velocity of 1.5 m s^{-1}. How many beats per second will be heard by the observer? (Take the speed of sound as 336 m s^{-1} and assume there is no wind.)

The observer hears a note of apparent frequency f' from the whistle directly, and a note of apparent frequency f'' from the sound waves reflected from the wall.

Now
$$f' = \frac{c'}{\lambda'}$$

where c' is the velocity of sound in air relative to the observer and λ' is the wavelength of the waves reaching the observer. Since

$$c' = 336 \text{ m s}^{-1} \quad \text{and} \quad \lambda' = \frac{336 + 1\cdot5}{500} \text{ m}$$

$$\therefore f' = \frac{336 \times 500}{337\cdot5} = 497\cdot8 \text{ Hz}$$

The note of apparent frequency f'' is due to sound waves moving towards the observer with a velocity of $1\cdot5$ m s^{-1}

$$\therefore f'' \frac{c'}{\lambda'} = \frac{336}{(336 - 1\cdot5)/500}$$

$$= \frac{336 \times 500}{334\cdot5} = 502\cdot2 \text{ Hz}$$

$$\therefore \text{ beats per second} = f'' - f' = 502\cdot2 - 497\cdot8 = 4\cdot4$$

2 In the sun's spectrum, a line of wavelength 589·00 nm differs by $7\cdot8 \times 10^{-3}$ nm when opposite edges of the sun are observed across the equatorial diameter.
Estimate the speed of rotation of the sun, assuming speed of light, $c = 3\cdot00 \times 10^8$ m s^{-1}.

One edge moves with velocity v away from the observer. From the Doppler principle, the increased wavelength is given by, if λ is the actual wavelength,

$$\lambda' = \frac{c + v}{c} \lambda$$

The opposite edge moves towards the observer with velocity v. So the decreased wavelength is given by

$$\lambda'' = \frac{c - v}{c} \lambda$$

So   change in wavelength $= \lambda' - \lambda'' = \left(\dfrac{c + v}{c} - \dfrac{c - v}{c} \right) \lambda$

$$= 2\frac{v}{c} \lambda = 7\cdot8 \times 10^{-3} \text{ nm}$$

$$\therefore v = \frac{7\cdot8 \times 10^{-3} \times 3 \times 10^8}{2 \times 589}$$

$$= 2 \times 10^3 \text{ m s}^{-1} = 2 \text{ km s}^{-1}$$

EXERCISES 22B   Doppler Effect

1 An observer travels with constant velocity of 30 m s^{-1} towards a distant source of sound, which has a frequency of 1000 Hz. Calculate the apparent frequency of the sound heard by the observer. What frequency is heard after passing the source of sound? (Assume velocity of sound = 330 m s^{-1}.)

2 The wavelength of a particular line in the emission spectrum of a distant star is measured as 600·80 nm. The true wavelength is 600·00 nm.
   (a) Is the star moving away from or towards the observer?
   (b) Calculate the speed of the star. (Velocity of light = 3·0 × 10^8 m s^{-1}.)

3 An observer, travelling with a constant velocity of 20 m s^{-1}, passes close to a stationary source of sound and notices that there is a change of frequency of 50 Hz as he passes the source. What is the frequency of the source?
   (Speed of sound in air = 340 m s^{-1}.) (L.)

4 Deduce expressions for the frequency heard by an observer:
   (a) when he is stationary and a source of sound is moving towards him;
   (b) when he is moving towards a stationary source of sound.
   Explain your reasoning carefully in each case.

   Give an example of change of frequency due to motion of source or observer from some other branch of physics, explaining either, a use which is made of it, or a deduction from it.

   A car travelling at 10 m s^{-1} sounds its horn, which has a frequency of 500 Hz, and this is heard in another car which is travelling behind the first car, in the same direction, with a velocity of 20 m s^{-1}. The sound can also be heard in the second car by reflection from a bridge head. What frequencies will the driver of the second car hear? (Speed of sound in air = 340 m s^{-1}.) (L.)

5 (a) State the conditions necessary for 'beats' to be heard and derive an expression for their frequency.
   (b) A fixed source generates sound waves which travel with a speed of 330 m s^{-1}. They are found by a distant stationary observer to have a frequency of 500 Hz. What is the wavelength of the waves? From first principles find: (i) the wavelength of the waves in the direction of the observer; and   (ii) the frequency of the sound heard if (1) the source is moving towards the stationary observer with a speed of 30 m s^{-1}, (2) the observer is moving towards the stationary source with a speed of 30 m s^{-1}, (3) both source and observer move with a speed of 30 m s^{-1} and approach one another. (N.)

6 The sun rotates about its centre. Observation of the light from the two edges at the ends of a diameter shows a Doppler shift of about 0·008 nm for a wavelength 600·00 nm.
   Estimate the angular velocity of the sun about its centre, given that its radius is 7·0 × 10^8 m and c = 3·0 × 10^8 m s^{-1}.

7 Describe: (a) the *Doppler effect*; (b) *beats*, as observed with sound waves. Derive expressions for the apparent frequency of a sound signal heard by an observer in still air   (i) from a source of frequency f moving with velocity u towards a stationary observer and (ii) from a stationary source of frequency f when the observer approaches the source with velocity v.

   An ultrasonic burglar alarm in still air transmits a signal at a frequency of 4·5 × 10^4 Hz, part of which is reflected by the burglar to a receiver alongside the transmitter.

   The burglar moves towards the transmitter at 1 m s^{-1}; calculate: (iii) the frequency of the signal received by the burglar;   (iv) the frequency detected by the receiver alongside the transmitter.

   Hence find   (v) the beat frequency between the signal reflected from the moving burglar and the original signal from the transmitter.

   The alarm is triggered by any beat frequency greater than 5 Hz; estimate (vi) the minimum velocity of approach of a burglar to activate the alarm. (Velocity of sound in air = 340 m s^{-1}.) (O. & C.)

23 Waves in Strings and Pipes

As we shall see, stationary or standing waves are formed in sounding pipes and strings. We first deal in detail with the waves produced in closed and open pipes and the range of frequencies produced in strings and the frequencies produced. We follow with closed and open pipes, and show how the speed of sound in air can be found using a pipe.

Introduction

The music from an organ, a violin or a xylophone is due to vibrations set up in these instruments. In the organ, air is blown into a pipe, which sounds its characteristic note as the air inside it vibrates. In the violin, the strings are bowed so that they oscillate; and in a xylophone a row of metallic rods are struck in the middle with a hammer, which sets them into vibration.

Before considering each of the above cases in more detail, it would be best to consider the feature common to all of them. A violin string is fixed at both ends, A, B, and waves travel along m, n to each end of the string when it is bowed and are there reflected, Figure 23.1 (i).

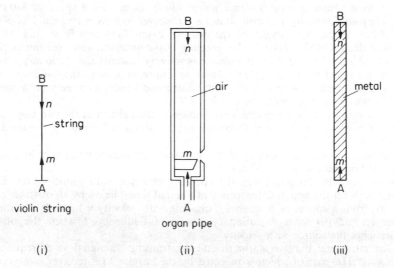

Figure 23.1 *Reflection of waves in instruments*

The vibrations of the particles of the string are therefore due to the combined or resultant effect of *two waves of the same frequency and amplitude travelling in opposite directions.* A similar effect is obtained with an organ pipe closed at one end B, Figure 23.1 (ii). If air is blown into the pipe at A, a wave travels along the direction m and is reflected at B in the opposite direction n. The air

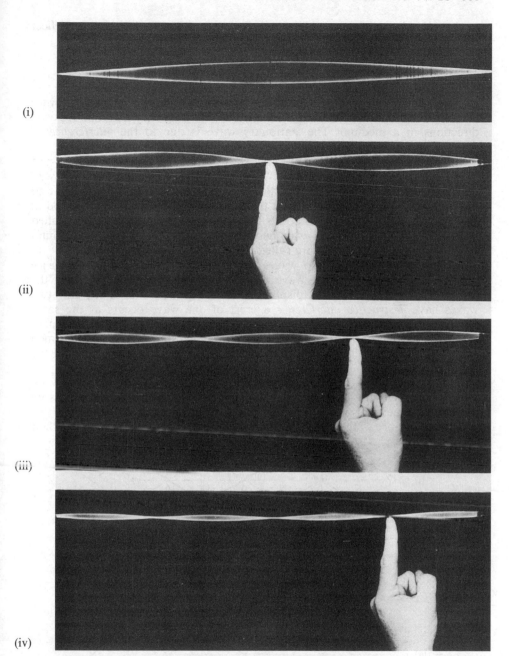

(i)

(ii)

(iii)

(iv)

Plate 23A (i)–(iv) *The photographs illustrate how 'stopping' with a light touch of a finger at different points of a vibrating cord produces successive harmonics. At the top, the mode of vibration corresponds to the fundamental frequency f_0 and the others to $2f_0$, $3f_0$, $4f_0$. As shown, a displacement node is produced at the stop in each case.*
(Courtesy of Prof. C. A. Taylor, Cardiff University)

vibrations in the pipe are therefore also due to the combined effect of two waves travelling in opposite directions. If a metal rod is fixed at its middle in a vice and stroked at one end A, a wave travels along the rod in the direction m and is reflected at the other end B in the direction n, Figure 23.1 (iii). The vibrations of the rod, which produce a high-pitched note, are due to the combined effect of two waves in opposite directions.

Stationary Waves and Wavelength

In Chapter 17 on waves, we showed that a *stationary* or *standing wave* is formed when two waves of equal amplitude and frequency travel in opposite directions in a medium. The stationary wave is due to the *interference* of the two waves.

To explain what happens, Figure 23.2 shows a plane-progressive sound wave *a* in air incident on a smooth wall W and the wave *b* reflected from the wall and travelling in the opposite direction to *a*. When the displacements due to the two waves are added together at the times shown, where *T* is the period of *a* or *b*, the resultant wave S has points N of permanent zero displacement called *nodes*. Other points A, half way between the nodes N, have a maximum amplitude of vibration; they are called *antinodes*.

At an antinode A, the two waves *a* and *b* have *constructive interference* (see p. 517). Here the two wave crests occur at the same place at the same time so the resultant displacement is big. At a node B, the two waves have *destructive interference* a crest of one wave occurs at the same place and time as a trough of the other wave, so the resultant displacement is zero. The photograph in Plate 23A shows the stationary waves formed on a vibrating cord fixed at both ends and the nodes and antinodes on the stationary waves.

Figure 23.2 shows displacement nodes N and antinodes A. As we saw on p. 479, the pressure nodes occur at displacement antinodes; the pressure antinodes occur at displacement nodes.

The importance of the nodes and antinodes in a stationary wave lies in their

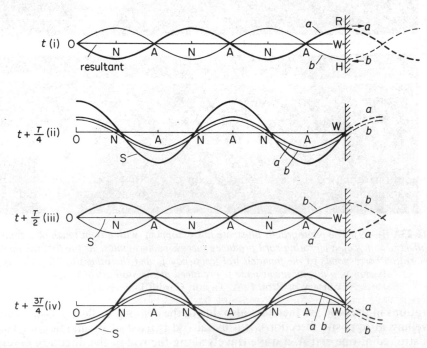

Figure 23.2 *Formation of stationary waves*

simple connection with the wavelength which we use shortly. In Chapter 17, we showed that

$$\text{the distance between consecutive nodes, } NN = \frac{\lambda}{2} \qquad . \qquad (1)$$

Where λ is the wavelength of the progressive wave which produces the stationary wave when it is reflected back, and

$$\text{the distance between consecutive antinodes, } AA = \frac{\lambda}{2} \qquad . \qquad (2)$$

and

$$\text{the distance from a node to the next antinode, } NA = \frac{\lambda}{4} \qquad (3)$$

Example on Stationary Waves

Plane sound waves of frequency 100 Hz fall normally on a smooth wall. At what distances from the wall will the air particles have: (a) maximum; (b) minimum amplitude of vibration? Give reasons for your answer. (The speed of sound in air may be taken as 340 m s^{-1}.) (L.)

A stationary wave is set up between the source and wall, due to the production of a reflected wave. The wall is a displacement node, since the air in contact with it cannot move; and other nodes are at equal distances, d, from the wall. So if wavelength is λ, the first distance d of the minimum amplitude position or node from the wall is

$$d = \frac{\lambda}{2}$$

Since
$$\lambda = \frac{c}{f} = \frac{340}{100} = 3\cdot4 \text{ m}$$

$$\therefore d = \frac{3\cdot4}{2} = 1\cdot7 \text{ m}$$

So minimum amplitude of vibration is obtained 1·7, 3·4, 5·1 m ... from the wall.
The antinodes are midway between the nodes. So maximum amplitude of vibration is obtained 0·85, 2·55, 4·25 m, ... from the wall.

Waves in Strings

If a horizontal rope is fixed at one end, and the other end is moved up and down, a wave travels along the rope. The particles of the rope are then vibrating vertically, and since the wave travels horizontally, this is an example of a *transverse* wave (see p. 478). A transverse wave is also obtained when a stretched string, such as a guitar or violin string, is plucked. Before we can study waves in strings, we need to know the speed of transverse waves travelling along them.

Speed of Transverse Waves Along a Stretched String

Suppose that a transverse wave is travelling along a thin string of length l and mass m under a constant tension T. If we assume that the string has no 'stiffness',. that is, the string is perfectly flexible, the speed c of the transverse wave along it depends only on the values of T, m, l. The speed is given by

$$c = \sqrt{\frac{T}{m/l}}$$

or
$$c = \sqrt{\frac{T}{\mu}} \quad . \quad . \quad . \quad . \quad . \quad (1)$$

where μ is the 'mass per unit length' of the string.

Using units, we can show (1) is a correct formula. T is the tension (force) in newtons N and μ is the mass per unit length in kilograms per metre or kg m^{-1}.

So the units of $\sqrt{T/\mu} = \sqrt{\text{N/kg m}^{-1}} = \sqrt{\text{Nm/kg}}$

But from $F = ma, 1\,\text{N} = 1\,\text{kg} \times 1\,\text{m s}^{-2}$.

So $\sqrt{\text{Nm/kg}} = \sqrt{\text{kg m s}^{-2} \times \text{m/kg}} = \sqrt{\text{m}^2\,\text{s}^{-2}} = \text{m s}^{-1}$
$\qquad\qquad\qquad\qquad\qquad\qquad\qquad\qquad\qquad = \text{velocity } c$

Modes of Vibration of Stretched String

If a wire is stretched between two points N, N and is plucked in the middle, a transverse wave travels along the wire and is reflected at the fixed end. A *stationary wave* is thus set up in the wire, and the simplest mode of vibration is one in which the fixed ends of the wire are nodes, N, and the middle is an antinode, A, Figure 23.3. Since the distance between consecutive nodes is $\lambda/2$,

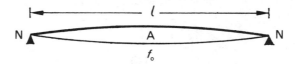

Figure 23.3 *Fundamental of stretched string*

where λ is the wavelength of the transverse wave in the wire, it follows that

$$l = \frac{\lambda}{2}$$

where l is the length of the wire. So $\lambda = 2l$. The frequency f of the vibration is then given by

$$f = \frac{c}{\lambda} = \frac{c}{2l}$$

where c is the speed of the transverse wave. But $c = \sqrt{T/\mu}$, from previous.

$$\therefore f = \frac{1}{2l}\sqrt{\frac{T}{\mu}}$$

This is the frequency of the *fundamental* or *lowest* note obtained from the string.

So if we denote the frequency by the usual symbol f_0, we have

$$f_0 = \frac{1}{2l}\sqrt{\frac{T}{\mu}} \qquad . \qquad . \qquad . \qquad . \qquad . \qquad (2)$$

Overtones of Stretched String

Plucked strings can vibrate in several ways or modes. The *first overtone* f_1 of a string plucked in the middle corresponds to a stationary wave shown in Figure 23.4 (i) which has nodes at the fixed ends and an antinode in the middle. As shown, three loops are formed. If λ_1 is the wavelength it can be seen that

$$l = \frac{3}{2}\lambda_1 \quad \text{or} \quad \mu_1 = \frac{2l}{3}$$

The frequency f_1 of the overtone is then given by

$$f_1 = \frac{c}{\lambda_1} = \frac{3c}{2l} = \frac{3}{2l}\sqrt{\frac{T}{\mu}} \qquad . \qquad . \qquad . \qquad . \qquad (3)$$

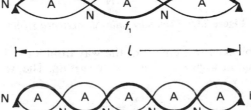

Figure 23.4 *Overtones of stretched string plucked in middle*

But the fundamental frequency, $f_0 = \frac{1}{2l}\sqrt{\frac{T}{\mu}}$, from equation (2).

$$\therefore f_1 = 3f_0$$

The second overtone f_2 of the string when plucked in the middle corresponds to a stationary wave shown in Figure 23.4 (ii). In this case $l = \frac{5}{2}\lambda_2$, where λ_2 is the wavelength

$$\therefore \lambda_2 = \frac{2l}{5}$$

$$\therefore f_2 = \frac{c}{\lambda_2} = \frac{5c}{2l}$$

where f_2 is the frequency. But $c = \sqrt{T/\mu}$

$$\therefore f_2 = \frac{5}{2l}\sqrt{\frac{T}{\mu}} = 5f_0$$

So the overtones are $3f_0$, $5f_0$, and so on, where f_0 is the fundamental frequency.

Other notes than those considered above can be obtained by touching or 'stopping' the string lightly at its midpoint, for example, so that the midpoint becomes a node in addition to those at the fixed ends. If the string is plucked one quarter of the way along it from a fixed end, the simplest stationary wave set up is that shown in Figure 23.5 (i) (see also p. 589). So the wavelength $\lambda = l$, and the frequency f is given by

$$f = \frac{c}{\lambda} = \frac{c}{l} = \frac{1}{l}\sqrt{\frac{T}{\mu}}$$

$$\therefore f = 2f_0, \text{ since } f_0 = \frac{1}{2l}\sqrt{\frac{T}{\mu}}$$

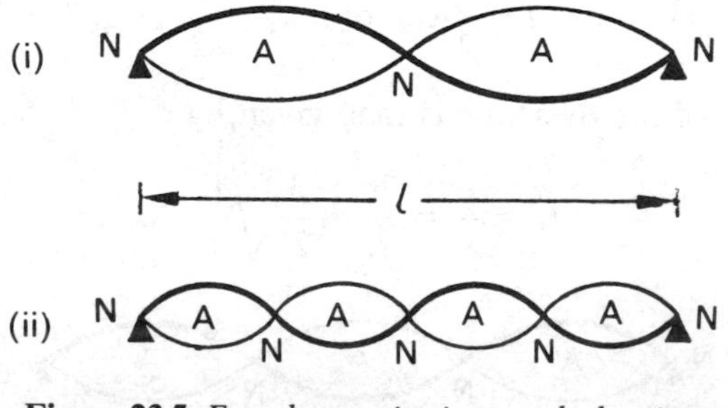

Figure 23.5 *Even harmonics in stretched string*

If the string is plucked one eighth of the way from a fixed end, a stationary wave similar to that in Figure 23.5 (ii) may be set up. The wavelength, $\lambda' = l/2$, and so the frequency

$$f' = \frac{c}{\lambda'} = \frac{2c}{l}$$

$$\therefore f' = \frac{2}{l}\sqrt{\frac{T}{\mu}} = 4f_0$$

Verification of the Laws of Vibration of a Fixed String, the Sonometer

As we have already shown (p. 592), the frequency of the fundamental of a stretched string is given by

$$f = \frac{1}{2l}\sqrt{\frac{T}{\mu}}$$

writing f for f_0. So:

1 $f \propto \dfrac{1}{l}$ *for a given tension* (*T*) *and string* (μ *constant*).

2 $f \propto \sqrt{T}$ *for a given length* (*l*) *and string* (μ *constant*).

3 $f \propto \dfrac{1}{\sqrt{\mu}}$ *for a given length* (*l*) *and tension* (*T*).

These are known as the 'laws of vibration of a fixed string'. The *sonometer* was designed to verify them.

The sonometer consists of a hollow wooden box Q, with a thin horizontal wire attached to A at one end, Figure 23.6. The wire passes over a grooved wheel H, and is kept taut by a mass M hanging down at the other end. Wooden

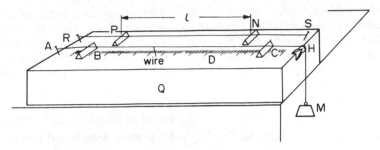

Figure 23.6 *Sonometer verification of f ∝ 1/l and* $\sqrt{T}$

bridges, B, C, can be placed beneath the wire so that a definite length of wire is obtained, and the length of wire can be varied by moving one of the bridges. The length of wire between B, C can be read from a fixed horizontal scale D, graduated in millimetres, on the box below the wire.

When the wire is plucked, the vibration passes through the bridges B, C to the sonometer box which is then set into forced vibration. The box surface is in contact with a large amount of air, and this vibrating air produces a loud sound.

(1) *To verify f ∝ 1/l for a given tenstion (T) and mass per unit length (μ)*, the mass M is kept constant so that the tension, T, in the wire AH is constant. The length, l, of the wire between B, C is varied by moving C until the note obtained by plucking BC in the middle is the same as that produced by a sounding tuning-fork of known frequency f. If the observer lacks a musical ear, the 'tuning' can be recognised by listening for beats when the wire and the tuning-fork are both sounding, as in this case the frequencies of the two notes are nearly equal (p. 574). Alternatively, a small piece of paper in the form of an inverted V can be placed on the middle of the wire, and the end of the sounding tuning-fork then placed on the sonometer box. The vibrations of the fork are transmitted through the box to the wire, which vibrates in resonance with the fork if its length is 'tuned' to the note. The paper will then vibrate considerably and may be thrown off the wire.

Different tuning-forks of known frequency f are used, and the lengths, l of the wire are observed when they are tuned to the corresponding note. A graph of f against $1/l$ is then plotted and is found to be a straight line within the limits of experimental error. So $f ∝ 1/l$ for a given tension and mass per unit length of wire.

(2) *To verify f ∝ $\sqrt{T}$ for a given length and mass per unit length*, the length BC between the bridges is kept fixed, so that the length of wire is constant, and the mass M is varied to alter the tension. To obtain a measure of the frequency f of the note produced when the wire between B, C is plucked in the middle, a second wire, fixed to R, S on the sonometer, is used. This usually has a weight (not shown) attached to one end to keep the tension constant, Figure 23.6. The wire RS has bridges P, N beneath it, and N is moved until the note from the wire between P, N is the *same* as the note from the wire between B, C. Now the tension in PN is constant as the wire is fixed to R and S. So, since frequency,

$f \propto 1/l$ for a given tension and wire, the frequency of the note from BC is proportional to $1/l$, where l is the length of PN.

By varying the mass M, the tension T in BC is varied. A graph of $1/l$ against $\sqrt{T}$ is found to be a straight line passing through the origin. So $f \propto \sqrt{T}$ for a given length of wire.

(3) *To verify $f \propto 1/\sqrt{\mu}$ for a given length and tension*, wires of different material are connected to B, C, and the same mass M and the same length BC are taken. The frequency, f, of the note obtained from BC is again found by using the second wire RS in the way already described. The mass per unit length, μ, is the mass per metre length of wire. This is $\pi r^2 \rho$ kg m^{-1}, where r is the radius of the wire in m and ρ is its density in kg m^{-3}, as $(\pi r^2 \times 1)$ m^3 is the volume of 1 m of the wire.

When $1/l$ is plotted against $1/\sqrt{\mu}$, the graph is found to be a straight line passing through the origin. So $f \propto 1/\sqrt{\mu}$ for a given length and tension.

Measurement of Frequency of A.C. Mains

The frequency of the alternating current (a.c.) mains can be measured using a sonometer wire. Using a suitable step-down transformer, a small mains alternating current is passed into the wire MP and the N, S poles of a strong magnet are placed either side of the wire as shown. Figure 23.7.

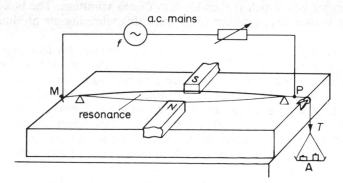

Figure 23.7 *Mains frequency by vibrating wire*

The magnetic field of the magnet is then perpendicular to the current. From Fleming's left-hand rule, a vertical force acts on the wire. So the wire vibrates up-and-down at the current frequency, which is 50 Hz if this is the mains frequency. The tension T in the wire is increased from zero by adding weights to the scale-pan A and at one stage the wire is set into *resonance* by the varying force and is seen to have a large amplitude of vibration as shown in the diagram.

The length l of wire between the bridges is now measured, and the tension T and the mass per unit length, μ, are also found. The frequency f of the alternating current is then calculated from

$$f = \frac{1}{2l}\sqrt{\frac{T}{\mu}}$$

When the tension is increased by adding more weights to A, the wire may resonate again and form three loops. As explained on p. 593, in this case $f = (3/2l) \times \sqrt{T/\mu}$, which would be the first overtone, 3×50 or 150 Hz.

Examples on Waves in Strings

1 A sonometer wire of length 76 cm is maintained under a tension of value 40 N and an alternating current is passed through the wire. A horse-shoe magnet is placed with its poles either side of the wire at its midpoint, and the resulting forces set the wire in resonant vibration. If the density of the material of the wire is 8800 kg m^{-3} and the diameter of the wire is 1 mm, what is the frequency of the alternating current?

The wire is set into resonant vibration when the frequency of the alternating current is equal to its natural frequency, f.

Now
$$f = \frac{1}{2l}\sqrt{\frac{T}{\mu}} \qquad\qquad (1)$$

where $l = 0.76$ m, $T = 40$ N, and μ = mass per metre in kg m^{-1}

Now,   mass of 1 metre = volume $\times$ density

$$= \pi r^2 \times 1 \times 8800 \text{ kg}$$

where radius r of wire $= \frac{1}{2}$mm $= 0.5 \times 10^{-3}$ m

From (1), $$f = \frac{1}{2 \times 0.76}\sqrt{\frac{40}{\pi \times 0.5^2 \times 10^{-6} \times 1 \times 8800}}$$

$$= 50 \text{ Hz}$$

2 A piano string has a length of 2.0 m and a density of 8000 kg m^{-3}. When the tension in the string produces a strain of 1%, the fundamental note obtained from the string in transverse vibration is 170 Hz. Calculate the Young modulus value for the material of the string

If E is the Young modulus, A is the cross-sectional area of the string, l is the length of the string and e is the extension due to a force (tension) T, then, from p. 144,

$$T = EA\frac{e}{l} = EA \times \frac{1}{100}$$

since the strain $e/l = 1\% = 1/100$. So

$$\text{frequency, } f = \frac{1}{2l}\sqrt{\frac{T}{\mu}} = \frac{1}{2l}\sqrt{\frac{EA}{100\,A\rho}}$$

since μ = mass per unit length = $A \times 1 \times \rho = A\rho$. So cancelling A,

$$f = \frac{1}{2l}\sqrt{\frac{E}{100\rho}}$$

Squaring, $$E = 4f^2 l^2 \times 100\rho$$

$$= 4 \times 170^2 \times 2^2 \times 100 \times 8000$$

$$= 3.7 \times 10^{11} \text{ N m}^{-2} \text{ or Pa}$$

You should know:

1 Musical instruments produce stationary (standing) waves inside them. This is due to reflection at the fixed ends of strings or at the closed or open ends of pipes. The stationary waves are formed by interference between the two waves of the same frequency travelling in opposite directions.

   The sound waves travel to the listener by a progressive wave in air.

2 In a stationary wave with node N (no displacement or maximum change) and antinode A (maximum amplitude of displacement or no pressure change),

$$NN = \lambda/2 = AA \quad \text{and} \quad NA = \lambda/4$$

3 The fundamental frequency f_0 of an instrument is the lowest note obtainable. Higher notes are overtones.

4 Strings of length l and fixed at both ends, $l = NN = \lambda/2$ if plucked gently in middle. So $\lambda = 2l$. First overtone, $f_1 = 2l/3$.

5 Speed of transverse wave along string in tension, $c = \sqrt{T/\mu}$. c in m s^{-1} when T in N and μ (mass per metre) in kg m^{-1}. $\mu = \pi r^2 \rho$, where r is string radius and ρ = string density. So

$$\text{fundamental frequency } f_0 = (1/2l)\sqrt{T/\mu}$$

6 With a 'stop' in midddle of string, and plucked one quarter from end, $NN = l/2$ and frequency $= 2f_0$.

EXERCISES 23A   Strings

Multiple Choice

1 In Figure 23A (i), R, R are paper riders on the wire XY where XR = RR = RY = 20 cm. A sounding tuning-fork placed on the box makes the riders jump off the wire. The wavelength in cm of the vibrating wire is

   A 20   **B** 35   **C** 30   **D** 40   **E** 60

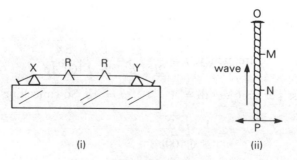

(i)   (ii)

Figure 23A

2 The fundamental frequency of a wire 80 cm long is 200 Hz. If the tension is 36 N, the mass per unit length in kg m^{-1} of the wire is about

   A 3.5×10^{-4} **B** 2.8×10^{-4} **C** 1.4×10^{-4} **D** 1.2×10^{-4} **E** 0.8×10^{-4}

3 A sonometer wire XY has a length of 100 cm and a tension of 20 N. To produce a note of twice the fundamental frequency, the wire should be plucked at a distance in cm from X of

A 10 **B** 25 **C** 50 **D** 75 **E** 90

4 A uniform heavy rope OMNP is suspended vertically from one end O. Figure 23A (ii) Transverse vibrations along the rope are produced at the bottom P. Taking into account the variation of tension along the rope, the speed of the wave in the rope is greatest near

A P **B** N **C** M **D** O **E** midway between O and P

Longer Questions

5 A taut copper wire (length 0·60 m, mass per unit length $8·0 \times 10^{-3}$ kg m^{-1}) conducts a small alternating current of frequency 50 Hz. When a strong horizontal magnetic field is applied by means of a magnet, a stationary wave with five segments is formed, as illustrated in Figure 23B.

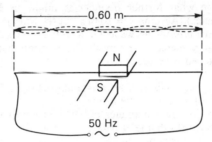

Figure 23B

(a) In which plane do the oscillations of the wire occur?
(b) Calculate: (i) the wavelength; (ii) the speed c of transverse waves on the wire; (iii) the tension T in the wire.
(c) State the effect on the stationary-wave pattern of changing the frequency of the alternating current to the following values: (i) 100 Hz; (ii) 25 Hz. (O.)

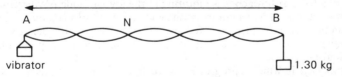

Figure 23C

6 Figure 23C illustrates one method for producing a stationary wave on a stretched wire. Two progressive waves combined to form this stationary wave. Explain how this happens. Each of the progressive waves carries energy along the wire. Discuss what happens to this energy. The point N represents one of the nodes in the stationary wave. What is a node? How is it formed? The distance AB is 0·90 m. The 1·30 kg mass holds the wire in tension and the frequency of the vibrator is 250 Hz. Calculate the mass per unit length for the wire. (L.)

7 Describe the motion of the particles of a string under constant tension and fixed at both ends when the string executes transverse vibrations of:
(a) its fundamental frequency;
(b) the first overtone (second harmonic). Illustrate your answer with suitable diagrams.
   A horizontal sonometer wire of fixed length 0·50 m and mass $4·5 \times 10^{-3}$ kg is

under a fixed tension of $1·2 \times 10^2$ N. The poles of a horse-shoe magnet are arranged to produce a horizontal transverse magnetic field at the midpoint of the wire, and an alternating sinusoidal current passes through the wire. State and explain what happens when the frequency of the current is progressively increased from 100 to 200 Hz. Support your explanation by performing a suitable calculation. Indicate how you would use such an apparatus to measure the fixed frequency of an alternating current. (N.)

8 Describe experiments to illustrate the differences between:
   (a) *transverse* waves;
   (b) *longitudinal* waves;
   (c) *progressive* waves;
   (d) *stationary* waves. To which classes belong   (i) the vibrations of a violin string;
       (ii) the sound waves emitted by the violin into the surrounding air?
   A wire whose mass per unit length is 10^{-3} kg m^{-1} is stretched by a load of 4 kg over the two bridges of a sonometer 1 m apart. If it is struck at its middle point, what will be:
   (a) the wavelength of its subsequent fundamental vibrations;
   (b) the fundamental frequency of the note emitted? If the wire were struck at a point near one bridge what further frequencies might be heard? (Do not derive standard formulae.) (Assume $g = 10$ m s^{-2}.) (O. & C.)

9 Distinguish between a *progressive* wave and a *stationary* wave. Explain in detail how you would use a sonometer to establish the relation between the fundamental frequency of a stretched wire and:
   (a) its length;
   (b) its tension. You may assume a set of standard tuning-forks and set of weights in steps of half a kilogram to be available.
   A pianoforte wire having a diameter of 0·90 mm is replaced by another wire of the same material but with diameter 0·93 mm. If the tension of the wire is the same as before, what is the percentage change in the frequency of the fundamental note? What percentage change in the tension would be necessary to restore the original frequency? (L.)

10 Describe the difference between stationary waves and progressive waves. Outline an experimental arrangement to illustrate the formation of a stationary wave in a string.
   Waves of wavelength λ, from a source S, reach a common point P by two different routes. At P the waves are found to have a phase difference $3\pi/4$ rad. Show graphically what this means. What is the minimum path difference between the two routes?
   A string fixed at both ends is vibrating in the lowest mode of vibration for which a point a quarter of its length from one end is a point of maximum vibration. The note emitted has a frequency of 100 Hz. What will be the frequency emitted when it vibrates in the next mode such that this point is again a point of maximum vibration? (L.)

11 State *two* ways in which a progressive wave differs from a stationary wave. Describe how you would set up a stationary wave on a taut string. Explain briefly how the principle of superposition accounts for the formation of this wave. Sound waves emitted from the taut string are longitudinal. How does a longitudinal wave differ from a transverse wave? How would you demonstrate by experiment that the speed of transverse waves along such a taut string is proportional to the square root of the tension in the string? (L.)

12 (a) A wave pulse of the shape shown in Figure 23D (i), is travelling along a string at a speed of 2·0 m s^{-1} towards X. The leading edge of the pulse is at a distance of 1·0 m from X, where the strong is fixed to a rigid support. The diagram shows the pulse at time $t = 0$.
    (i) Sketch, as accurately as you can, on Figure 23D (i), which is to scale, the pulse shape you would expect to see at $t = 1·0$ s.
    (ii) Draw a graph of the displacement of a point on the string at a distance of 0·50 m from the wall over the period from $t = 0$ to $t = 0·75$ s. Mark the time scale clearly.

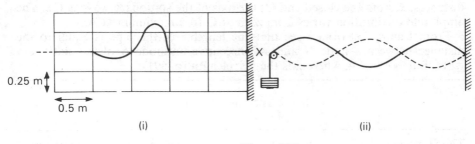

0.25 m

0.5 m

(i)   (ii)

Figure 23D

(b) The other end of the string is now tied to a weight. The string passes over a pulley and hangs vertically. The string is set into motion so that it vibrates as shown in Figure 23D (ii) at a frequency of 180 Hz.
  (i) What is the lowest possible (fundamental) frequency of vibration of the string?
  (ii) The length of the vibrating string is reduced to two-thirds by moving the pulley towards the point X. What is the new fundamental frequency of vibration? Give your reasoning. (*O. & C.*)

Waves in Pipes

Closed Pipe

A *closed* or *stopped organ pipe* consists of a metal pipe closed at one end Q, and a blast of air is blown into it at the other or open end P, Figure 23.8 (i). A sound wave then travels up the pipe to Q, and is reflected at this end

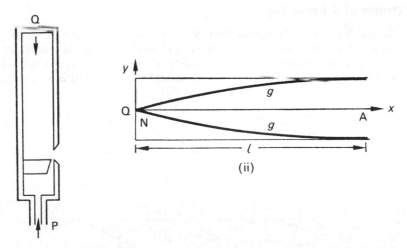

Figure 23.8 (*i*) *Closed* (*stopped*)   (*ii*) *Fundamental of closed* (*stopped*) *pipe*

down the pipe, so a *stationary wave* is formed by interference between the two waves. The end Q of the closed pipe must be a node N, since the layer in contact with Q must be permanently at rest. The open end A, where the air is free to vibrate, must be an antinode A. The simplest stationary wave in the air in the pipe is therefore represented by *g* in Figure 23.8 (ii). Here the pipe is turned horizontally to show the relative displacement, *y*, of the layers at different distances, x, from the closed end Q; the axis of the stationary wave is Qx. The amplitude of vibration varies from zero at Q to a maximum at A.

From Figure 23.8 (ii) we see that the length l of the pipe is equal to the distance between a node N and a consecutive antinode A of the stationary wave. But NA $= \lambda/4$, where λ is the wavelength (p. 591).

$$\therefore \frac{\lambda}{4} = l \quad \text{or} \quad \lambda = 4l$$

The frequency, f, of the note is given by $f = c/\lambda$, where c is the speed of sound in air.

$$\therefore f = \frac{c}{4l}$$

This is the frequency of the lowest note obtainable from the pipe, and is its *fundamental*. We denote the fundamental frequency by f_0, so

$$f_0 = \frac{c}{4l} \qquad . \qquad . \qquad . \qquad . \qquad . \qquad . \qquad (1)$$

Figure 23.8 (ii) shows the stationary wave of *displacement* of air molecules along the closed pipe. The *pressure* variation is also a stationary wave. But in contrast to Figure 23.8 (ii), the pressure node is at the open end since the air pressure here is constant and equal to the external atmospheric pressure. The pressure antinode, the position of maximum pressure, is at the closed end since here the layers of air are compressed.

Overtones of Closed Pipe

If a stronger blast of air is blown into the pipes, notes of higher frequency or overtones can be obtained which are simple multiples of the fundamental frequency f_0. Two possible cases of stationary waves are shown in Figure 23.9. In each, the closed end of the pipe is a node, and the open end is an antinode.

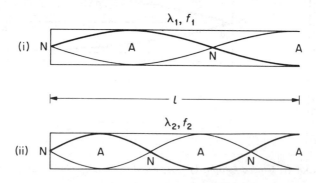

Figure 23.9 *Overtones in closed pipe*

In Figure 23.9 (i), however, the length l of the pipe is related to the wavelength λ_1 of the wave by

$$l = \frac{3}{4}\lambda_1 \qquad \therefore \lambda_1 = \frac{4l}{3}$$

The frequency f_1 of the note is given by

$$f_1 = \frac{c}{\lambda_1} = \frac{3c}{4l} \qquad . \qquad . \qquad . \qquad . \qquad . \qquad (1)$$

But from before, the fundamental $f_0 = \dfrac{c}{4l}$

$$\therefore f_1 = 3f_0 \qquad . \qquad . \qquad . \qquad . \qquad . \qquad (2)$$

As we previously explained, the stationary *pressure* wave in the air has pressure nodes at the displacement antinodes A in Figure 23.4 (i) and pressure antinodes at the displacement nodes N.

Figure 23.9 (ii) shows the overtone of higher frequency f_2. The length l of the pipe is related to the wavelength λ_2 by

$$l = \frac{5\lambda_2}{4} \quad \text{or} \quad \lambda_2 = \frac{4l}{5}$$

$$\therefore f_2 = \frac{c}{\lambda_2} = \frac{5c}{4l} \qquad . \qquad . \qquad . \qquad . \qquad . \qquad (3)$$

$$\therefore f_2 = 5f_0 \qquad . \qquad . \qquad . \qquad . \qquad . \qquad (4)$$

By drawing other sketches of stationary waves, with the closed end always a node and the open end as an antinode, it can be shown that higher frequencies can be obtained which have frequencies of $7f_0, 9f_0$, and so on. They are produced by blowing harder at the open end of the pipe. So the frequencies obtainable at a closed pipe are $f_0, 3f_0, 5f_0$, and so on, which are only odd harmonics, and so frequencies $3f_0, 5f_0$, etc. are possible *overtones*.

Open Pipe

An 'open' pipe is one which is open at both ends. When air is blown into it at P, a wave m travels to the open end Q, where it is reflected in the direction n on encountering the free air, Figure 23.10 (i). A stationary wave is therefore set

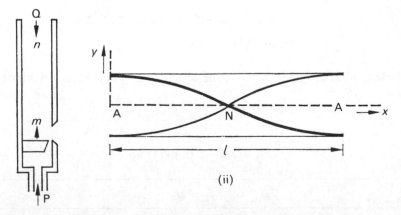

Figure 23.10 (i) *Open pipe* (ii) *Fundamental of open pipe*

up in the air in the pipe, and as the two ends of the pipe are open, they must both be *antinodes*. The simplest type of wave is therefore that shown in Figure 23.10 (ii). The x axis of the wave is drawn along the middle of the pipe, which is turned horizontal. A node N is midway between the two antinodes A, A at each end.

The length l of the pipe is the distance between consecutive antinodes. But the distance between consecutive antinodes $= \lambda/2$, where λ is the wavelength (p. 591).

$$\therefore \frac{\lambda}{2} = l \quad \text{or} \quad \lambda = 2l$$

So the frequency f_0 of the note obtained from the pipe is given by

$$f_0 = \frac{c}{\lambda} = \frac{c}{2l} \qquad . \qquad . \qquad . \qquad . \qquad . \qquad (1)$$

This is the frequency of the fundamental note of the pipe.

Overtones of Open Pipe

Notes of higher frequencies than f_0 can be obtained from the pipe by blowing harder. The stationary wave in the pipe has always an antinode A at each end, and Figure 23.11 (i) represents the case of a note of a frequency f_1 higher than the fundamental f_0.

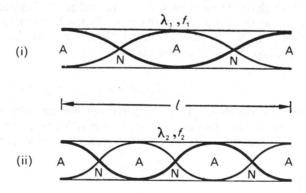

Figure 23.11 *Overtones of open pipes*

The length l of the pipe is equal to the wavelength λ_1 of the wave in this case.

So

$$f_1 = \frac{c_1}{\lambda_1} = \frac{c}{l}$$

But

$$f_0 = \frac{c}{2l}, \text{ from (1) above}$$

$$\therefore f_1 = 2f_0. \qquad . \qquad . \qquad . \qquad . \qquad . \qquad (1)$$

In Figure 23.11 (ii), the length $l = 3\lambda_2/2$, where λ_2 is the wavelength in the pipe, then so $\lambda_2 = 2l/3$. The frequency f_2 is given by

$$f_2 = \frac{c}{\lambda_2} = \frac{3c}{2l}$$

$$\therefore f_2 = 3f_0 \quad . \quad . \quad . \quad . \quad . \quad . \quad (2)$$

So the frequencies of the overtones in the open pipe are $2f_0$, $3f_0$, $4f_0$, and so on, that is, all harmonics are obtainable. The frequencies of the overtones in the closed pipe are $3f_0$, $5f_0$, $7f_0$, and so on. So the *quality* of the same note obtained from a closed and an open pipe is different (see p. 573).

Detection of Nodes and Antinodes, and Pressure Variation, in Pipes

The *nodes* and *antinodes* in a sounding pipe have been detected by suspending inside it a very thin piece of paper with lycopodium or fine sand particles on it, Figure 23.12 (ii). The particles can be heard vibrating on the paper at the antinodes, but they are still at the nodes.

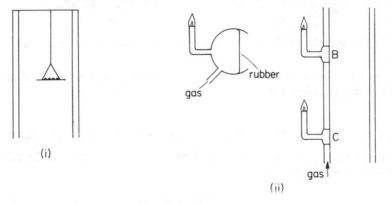

Figure 23.12 (i) *Detection of nodes and antinodes* (ii) *Detection of pressure*

The *pressure variation* in a sounding pipe has been examined by means of a sensitive flame, designed by Lord Rayleigh. The length of the flame can be made sensitive to the pressure of the gas supplied, so that if the pressure changes the length of flame is considerably affected. Several of the flames can be arranged at different parts of the pipe, with a thin rubber or mica diaphragm in the pipe, such as at B, C, Figure 23.12 (ii). At a place of maximum pressure variation or pressure antinode, which is a displacement node (p. 487), the length of flame alters accordingly. At a place of constant (normal) pressure or pressure node, which is a displacement antinode, the length of flame remains constant.

The pressure variation at different parts of a sounding pipe can also be examined by using a suitable small microphone at B, C, instead of a flame. The microphone is coupled to a cathode-ray tube and a wave of maximum amplitude is shown on the screen when the pressure variation is a maximum. At a place of constant (normal) pressure, no wave is observed on the screen.

End-correction of Pipes

The air at the open end of a pipe is free to move. So the vibrations at this end of a sounding pipe extend a little into the air outside the pipe. The antinode of the stationary wave due to any note is then a small distance e from the open end in practice, known as the *end-correction*. So the wavelength λ in the case of a closed pipe is given by $\lambda/4 = l + e$, where l is the length of the pipe, Figure 23.13 (i). In the case of an open pipe sounding its fundamental note, the wavelength λ is given by $\lambda/2 = l + e + e$, since *two* end-corrections are required, assuming the end-corrections are equal, Figure 23.13 (ii). So $\lambda = 2(l + 2e)$. See also p. 604.

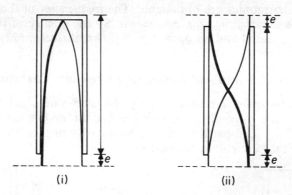

(i)   (ii)

Figure 23.13 (*i*) *Closed pipe* (*ii*) *Open pipe, both with end-corrections*

Calculation shows that $e = 0.58r$ or $0.6r$, where r is the radius of the pipe. So the wider the pipe, the greater is the end-correction.

Effect of Temperature and End-correction on Pitch of Pipes

The frequency, f_0, of the fundamental note of a closed pipe of length l and end-correction e is given by

$$f_0 = \frac{c}{\lambda} = \frac{c}{4(l + e)} \qquad\qquad\qquad (1)$$

with the usual notation, since $\lambda = 4(l + e)$. See above. Now the speed of sound, c, in air at $\theta°C$ is related to its velocity c_0 at $0°C$ by

$$\frac{c}{c_0} = \sqrt{\frac{273 + \theta}{273}} = \sqrt{1 + \frac{\theta}{273}} \qquad\qquad (2)$$

since the speed is proportional to the square root of T, the kelvin temperature. Substituting for c in (1),

$$f_0 = \frac{c_0}{4(l + e)} \sqrt{1 + \frac{\theta}{273}} \qquad\qquad (3)$$

From (3), it follows that, with a given pipe, *the frequency of the fundamental increases as the temperature increases*. Also, for a given temperature and length of pipe, the frequency decreases as e increases. Now $e = 0.6r$ where r is the radius of the pipe. So *the frequency of the note from a pipe of given length is lower the wider the pipe*, the temperature being constant. The same results hold for an open pipe.

Resonance

If a diving springboard is bent and then allowed to vibrate freely, it oscillates with a frequency which is called its *natural frequency*. When a diver on the edge of the board begins to jump up and down repeatedly, the board is forced to vibrate at the frequency of the jumps; and at first, when the amplitude is small, the board is said to be undergoing *forced vibrations*. As the diver jumps up and down to gain increasing height for her dive, the frequency of the periodic downward force reaches a stage where it is practically the same as the natural frequency of the board. The amplitude of the board then becomes very large, and the periodic force is said to have set the board in *resonance* (see also p. 476).

A mechanical system which is free to move, like a wooden bridge or the air in pipes, has a *natural frequency* of vibration, f_0, which depends on its dimensions. When a periodic force of a frequency different from f_0 is applied to the system, it vibrates with a small amplitude and undergoes *forced vibrations*. When the periodic force has a frequency equal to the natural frequency f_0 of the system, the amplitude of vibration becomes a maximum, and the system is then set into *resonance*. Figure 23.14 is a typical curve showing how the

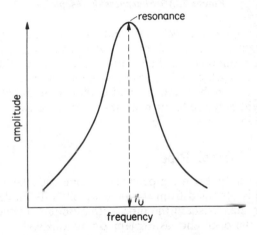

Figure 23.14 *Resonance curve*

amplitude varies with frequency. Some time ago it was reported in the newspapers that a soprano who was broadcasting had broken a glass tumbler on the table of a listener when she had reached a high note. This is an example of resonance. The glass had a natural frequency equal to that of the note sung, and was therefore set into a large amplitude of vibration sufficient to break it.

The phenomenon of resonance occurs in branches of physics other than sound and mechanics. When an electrical circuit containing a coil and capacitor is 'tuned' to receive the radio waves from a distant transmitter, the frequency of the radio waves is equal to the natural frequency of the circuit and resonance is therefore obtained. A large current then flows in the electrical circuit (p. 416).

Sharpness of Resonance

As the resonance condition is approached, the effect of the frictional or *damping* forces on the amplitude increases. Damping stops the amplitude from becoming very large at resonance. The lighter the damping, the *sharper* is the resonance, that is, the amplitude drops considerably at a frequency slightly different from the resonant frequency, as shown by the peak of the curve in Figure 23.15. A heavily damped system has a fairly flat peak in its resonance curve. Radio receivers of high quality have tuning circuits with a sharp peak in their resonance curves; unwanted frequencies then have a very small response or amplitude.

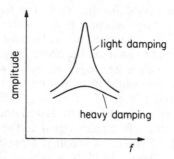

Figure 23.15 *Sharpness of resonance*

The effect of damping can be illustrated by attaching a simple pendulum carrying a very light bob, and one of the same length carrying a lead bob of equal size, to a horizontal string. The pendula are set into vibration by a third pendulum of equal length attached to the same string. It is then seen that the amplitude of the lead bob is much greater than that of the light bob. The damping of the light bob due to air resistance is much greater than for the lead bob. See also p. 475.

Resonance in a Tube or Pipe

If a person blows gently down a pipe closed at one end, the air inside vibrates freely, and a note is obtained from the pipe which is its fundamental (p. 601). A stationary wave then exists in the pipe, with a node N at the closed end and an antinode A at the open end, as explained previously.

If the prongs of a tuning-fork are held over the top of the pipe, the air inside it is set into vibration by the periodic force exerted on it by the prongs. In general, however, the vibrations are feeble, as they are *forced* vibrations. So, intensity of the sound heard is correspondingly small. But when a tuning-fork of the same frequency as the fundamental frequency of the pipe is held over the pipe, the air inside is set into *resonance* by periodic force, and the amplitude

Figure 23.16 *Resonance in closed pipe*

of the vibrations is then large. So a loud note, which has the same frequency as the fork, is heard coming from the pipe, and a stationary wave is set up with the top of the pipe acting as an antinode and the fixed end as a node, Figure 23.16. If a sounding tuning-fork is held over a pipe open at both ends, resonance occurs when the stationary wave in the pipe has antinodes at the two open ends, as shown by Figure 23.11. The frequency of the fork is then equal to the frequency of the fundamental of the open pipe. A similar case to the closed pipe, but using electrical oscillations, was discussed on p. 488.

Resonance Tube Experiment, Measurement of Velocity of Sound and 'End-correction' of Tube

If a sounding tuning-fork is held over the open end of a tube T filled with water, resonance is obtained at some position as the level of water is gradually lowered Figure 23.17 (i). We recognise when this occurs because the sound heard from the vibrating air in the tube is loudest in this case. The stationary wave set up

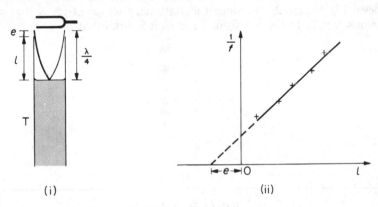

(i)   (ii)

Figure 23.17 *Resonance tube experiment*

is then as shown. If e is the end-correction of the tube (p. 606), and l is the length from the water level to the top of the tube, then

$$l + e = \frac{\lambda}{4} \qquad . \qquad . \qquad . \qquad . \qquad . \qquad (1)$$

But

$$\lambda = \frac{c}{f}$$

where f is the frequency of the fork and c is the velocity of sound in air.

$$\therefore l + e = \frac{c}{4f} \qquad . \qquad . \qquad . \qquad . \qquad . \qquad (2)$$

If different tuning-forks of known frequency f are used, and the corresponding values of l are measured when resonance occurs, it follows from equation (2) that a graph of $1/f$ against l is a straight line, Figure 23.17 (ii). Now from equation (2), the gradient of the line is $4/c$. So c can be found. Also, the negative intercept of the line on the axis of l is e, from equation (2). So the end-correction can be found.

If only one fork is available, and the tube is sufficiently long, another method

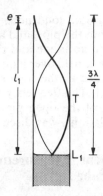

Figure 23.18 *Resonance at new water level*

for c and e can be used. In this case the level of the water is lowered further from the position in Figure 23.17 (1), until resonance is again obtained at a lower level L_1, Figure 23.18. Since the stationary wave set up is that shown and the new length to the top from L_1 is l_1, it follows that

$$l_1 + e = \frac{3\lambda}{4} \qquad . \qquad . \qquad . \qquad . \qquad . \qquad (3)$$

But

$$l + e = \frac{\lambda}{4}, \text{from (2)}$$

Subtracting,

$$l_1 - l = \frac{\lambda}{2}$$

$$\therefore \lambda = 2(l_1 - l)$$

$$\therefore c = f\lambda = 2f(l_1 - l) \qquad . \qquad . \qquad . \qquad . \qquad (4)$$

In this method for c, therefore, the end-correction e is eliminated. The magnitude of e can be found from equations (2) and (3). Thus, from (2),

$$3l + 3e = \frac{3\lambda}{4} = l_1 + e, \text{from (3)}$$

$$\therefore 2e = l_1 - 3l \quad \text{or} \quad e = \frac{l_1 - 3l}{2} \qquad (5)$$

So can be found from measurements of l_1 and l. *See Example 1.*

Speed of Sound in Air by Dust Tube Method

Figure 23.19 illustrates another way of measuring the speed of sound in air by means of stationary waves.

A measuring cylinder B is placed on its side and is arranged to lie horizontally on supports such as Plasticine. Using a ruler, the inside of the cylinder is coated lightly with lycopodium powder or cork dust along its length. A paper cone C, attached to a loudspeaker L, is fitted over the open end of B. By connecting a suitable oscillator to L, sound waves are produced which travel to the closed end of B and are reflected, so that stationary waves are formed in the air.

The frequency of the oscillator is varied. At a frequency of the order of a

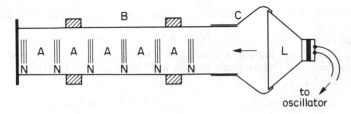

Figure 23.19 *Speed of sound in air*

kilohertz or more depending on the length of the measuring cylinder, the dust suddenly settles into regularly spaced heaps at positions along the cylinder. These are nodes, N, of the stationary wave (zero displacement positions). Midway between the nodes are antinodes, A, where the air has maximum amplitude of vibration and so little dust settles there. See Figure 23.19.

Measurement of the average distance NN between successive nodes $= \lambda/2$, where λ is the wavelength. So λ can be found. The speed in air $c = f\lambda$, where f is known, and so c can be calculated. This is an approximate method for c as:

(a) the sound waves are damped by the sides of the tube and so this is not the speed in free air (see p. 495);

(b) the distance NN cannot be measured to a high degree of accuracy. In the method outlined here, the dust and the tube must both be dry.

Examples on Waves in Pipes

1 A cylindrical pipe of length 28 cm closed at one end is found to be at resonance when a tuning fork of frequency 864 Hz is sounded near the open end. Find the mode of vibration of the air in the pipe, and deduce the value of the end-correction. (Take the speed of sound in air as 340 m s^{-1}.)

Let $\lambda =$ the wavelength of the sound in the pipe.

Then
$$\lambda = \frac{c}{f} = \frac{34\,000}{864} = 39{\cdot}35 \text{ cm}$$

If the pipe is resonating to its fundamental frequency f_0, the stationary wave in the pipe is that shown in Figure 23.13 (i) and the wavelength λ_0 is given by $\lambda_0/4 = 28$ cm. So $\lambda_0 = 112$ cm. Since $\lambda = 39{\cdot}35$ cm, the pipe cannot be sounding its resonant frequency. The first overtone of the pipe is $3f_0$, which corresponds to a wavelength λ_1 given by $3\lambda/4 = 28$ (see Figure 23.9).

$$\therefore \lambda_1 = \frac{112}{3} = 37\tfrac{1}{3} \text{ cm}$$

Consequently, allowing for the effect of an end correction, the pipe is sounding its *first overtone*.

Let $e =$ the end-correction in cm.

Then
$$28 + e = \frac{3\lambda_1}{4}$$

But, accurately,
$$\lambda_1 = \frac{c}{f} = \frac{34\,000}{864} = 39{\cdot}35$$

$$\therefore 28 + e = \tfrac{3}{4} \times 39{\cdot}35 \qquad \text{so} \qquad e = 1{\cdot}5 \text{ cm}$$

2 Explain, with diagrams, the possible states of vibration of a column of air in:
(a) an open pipe;
(b) a closed pipe.

An open pipe 30 cm long and a closed pipe 23 cm long, both of the same diameter, are each sounding their first overtone, and these are in unison. What is the end-correction of these pipes? (*L.*)

Suppose c is the speed of sound in air, and f is the frequency of the note. The wavelength, λ, is then c/f.

When the open pipe is sounding its first overtone, the length of the pipe plus end-corrections $= \lambda$.

$$\therefore \frac{c}{f} = 30 + 2e \qquad \cdot \qquad \cdot \qquad \cdot \qquad \cdot \qquad \cdot \qquad (1)$$

since there are two end-corrections.

When the closed pipe is sounding its first overtone,

$$\frac{3\lambda}{4} = 23 + e$$

$$\therefore \frac{3c}{4} = 23 + e \qquad \cdot \qquad \cdot \qquad \cdot \qquad \cdot \qquad \cdot \qquad (2)$$

From (1) and (2), it follows that

$$23 + e = \tfrac{3}{4}(30 + 2e)$$
$$\therefore 92 + 4e = 90 + 6e \qquad \text{so} \qquad e = 1 \text{ cm}$$

You should know:

1 Pipes: *Closed pipe* (closed at one end), with fundamental,
NA $= l = \lambda/4$, or $\lambda = 4l$.
So $f_0 = c/4l$, where $c =$ speed of sound in *air*.
Overtones: $3f_0$, $5f_0$, ...
Pressure antinodes at displacement nodes.
Open pipes (open at both ends), with fundamental, AA $= l = \lambda/2$.
So $f_0 = c/2l$. Overtones: $2f_0$, $3f_0$, $4f_0$...
2 *Resonance* in a closed pipe can be produced using a loudspeaker at one end and a movable plunger at the other end. With a given high frequency, the air is set into resonance at one position of the plunger. Lycopdium powder in the tube settles at the nodes and so λ can be measured.

Resonance in machines and structures

Resonance phenomena are not restricted to pipes, strings and electrical circuits. Any structure is likely to have resonant frequencies, frequencies at which it can vibrate with a large amplitude. For forced vibrations, resonance occurs when the force frequency is equal to one of the resonant frequencies of the structure.

Perhaps the most common everyday example of this is the washing machine. Have you ever noticed that as a machine slows down at the end of its spin cycle, it sometimes starts to shake for a while? This happens when the spin frequency of the drum passes through a resonant frequency of the machine's casing. In a manner analogous to a tuning fork exciting an air column (p. 608), when the drum is spinning at the case's resonant frequency, the case responds with appreciable vibration.

Such resonant vibrations can do serious damage; for example, a singer maintaining a note at a resonant frequency of a glass can cause it to shatter. More serious results could follow if soldiers did not break their march when crossing a bridge because if they march at a resonant frequency of the bridge, it is possible that the bridge could start vibrating wildly and collapse. Engineers spend much effort ensuring that the structures they design do not have significant resonant frequencies, so minimising the danger of collapse by forced vibration.

EXERCISES 23B Pipes

Multiple Choice

1 A pipe closed at one end is resonating to its first overtone. Figure 23E (i) shows roughly by arrows the magnitudes and directions of the vibrations of the air in the pipe at four equally spread points.
   Which is the correct diagram?

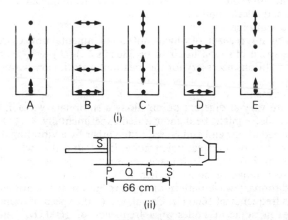

(i)

(ii)

Figure 23E

2 A tube T has a tight-fitting piston S at one end and a small loudspeaker L at the other end. Figure 23E (ii). Nodes are detected in the air at Q, R and S, where PS = 66 cm. If the frequency of the sound from L is then 800 Hz, the speed of sound in m s^{-1} is

   A 704 **B** 528 **C** 352 **D** 176 **E** 156

3 A pipe closed at one end resonates to its fundamental frequency of 400 Hz. Which frequencies from 1, 2, 3 and 4 can the pipe also resonate to?

   1 800   2 1200   3 1600   4 2000

   A 1 and 2 only **B** 2 and 3 only **C** 2 and 4 only
   D 1, 2 and 3 only **E** 1, 2, 3 and 4

Longer Questions

4 Write down in terms of wavelength, λ, the distance between (i) consecutive nodes; (ii) a node and an adjacent antinode; (iii) consecutive antinodes. Find the frequency of the fundamental of a closed pipe 15 cm long if the speed of sound in air is 340 m s^{-1}.

5 (a) (i) Explain what is meant by *displacement, amplitude* and *wavelength* for a *longitudinal* progressive wave in air.

(ii) In terms of the motion of the particles give *one* similarity and two differences between a progressive wave and a stationary wave for sound waves in air.

(b) The air in a uniform pipe, closed at one end, is vibrating in its fundamental mode. Describe, with the aid of diagrams, the variation along the length of the tube of (i) the displacement amplitude of the air molecules; (ii) the pressure amplitude.

(c) A student is performing an experiment to measure the speed of sound in air using a uniform tube 800 mm long which stands vertically and is initially full of water. The level of water in the tube can be slowly lowered. A small loudspeaker connected to a signal generator is held over the open end of the tube.

With the frequency emitted by the loudspeaker set at 600 Hz the student lowers the water level and finds resonance for the first time when the length of the air column above the water is 130 mm. He misses the second resonance and finds the third resonance when the air column is 698 mm long.

(i) Show in a sketch the displacement nodes and antinodes for the third resonance.

(ii) Calculate the speed of sound in air and the end correction for the tube.

(iii) What would be the fundamental frequency for this tube if it were open at both ends? (*N.*)

6 What are the chief characteristics of a progressive wave motion? Give your reasons for believing that sound is propagated through the atmosphere as a longitudinal wave motion, and find an expression relating the velocity, the frequency, and the wavelength.

Neglecting end effects, find the lengths of:

(a) a closed organ pipe;

(b) an open organ pipe, each of which emits a fundamental note of frequency 256 Hz. (Take the speed of sound in air to be 330 m s^{-1}.) (*O.*)

7 Explain the conditions necessary for the creation of stationary waves in air. Describe how

(a) the displacement,

(b) the pressure vary at different points along a stationary wave in air and describe how these effects might be demonstrated experimentally.

A tube is closed at one end and closed at the other by a vibrating diaphragm which may be assumed to be a displacement node. It is found that when the frequency of the diaphragm is 2000 Hz a stationary wave pattern is set up in the tube and the distance between adjacent nodes is then 8·0 cm. When the frequency is gradually reduced the stationary wave pattern disappears but another stationary wave pattern reappears at a frequency of 1600 Hz. Calculate (i) the speed of sound in air, (ii) the distance between adjacent nodes at a frequency of 1600 Hz, (iii) the length of the tube between the diaphragm and the closed end, (iv) the next lower frequency at which a stationary wave pattern will be obtained. (*N.*)

Thermal Properties of Matter

PART FOUR
Thermal Properties of Matter

24 Temperature and Thermometry

Temperature

We are interested in heat because it is the most common form of energy, and because changes of temperature have great effects on our personal comfort, and on the properties of substances, such as water, which we use every day. Temperature is a scientific quantity which corresponds to primary sensations—hotness and coldness. These sensations are not reliable enough for scientific work, because they depend on contrast—the air in a thick-walled barn or church feels cool on a summer's day, but warm on a winter's day, although a thermometer may show that it has a lower temperature in the winter. A thermometer, such as the familiar mercury-in-glass instrument (Figure 24.1), is a device whose readings depend on hotness or coldness, and which we choose to consider more reliable than our senses. We are justified in considering it more reliable because different thermometers of the same type agree with one another better than different people do.

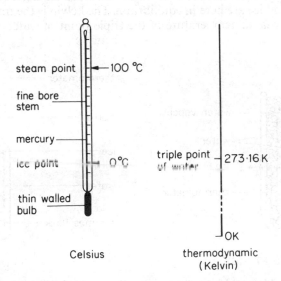

steam point ← 100 °C
fine bore stem
mercury
ice point — 0 °C
thin walled bulb

triple point of water — 273·16 K

0 K

Celsius   thermodynamic (Kelvin)

Figure 24.1 *Mercury-in-glass thermometer (left); °C and K scales*

Types of Thermometers

The temperature of an object is not a fixed number but depends on the type of thermometer used and on the temperature scale adopted, discussed shortly.

In general, thermometers use some measurable property of a substance which is sensitive to temperature change. The *constant-volume gas thermometer*, for example, uses the pressure change with temperature of a gas at constant volume. The *resistance thermometer* uses the change of electrical resistance of a pure metal with temperature, Figure 24.2. The *mercury-in-glass* thermometer depends on the change in volume of mercury with temperature relative to that of glass.

617

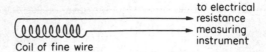

Coil of fine wire

Figure 24.2 *A resistance thermometer; the wire is usually of pure platinum*

A *thermoelectric thermometer* depends on the electromotive force change with temperature of two metals joined together.

Thermodynamic Temperature Scale

The *thermodynamic* temperature scale is the standard temperature scale adopted for scientific measurement. Thermodynamic temperature is denoted by the symbol T and is measured in *kelvins*, symbol K. The kelvin is the SI unit of temperature or of temperature change.

The thermodynamic temperature scale uses one fixed point, the *triple point of water*. This is the temperature at which saturated water vapour, pure water and melting ice are all in equilibrium. The triple point temperature is *defined* as 273·16 K. Figure 24.3 (i) shows an apparatus used for obtaining the triple point of water; distilled water (from which dissolved air has been driven out), water vapour and ice are here in equilibrium. The kelvin is the fraction 1/273·16 of the thermodynamic temperature of the triple point of water.

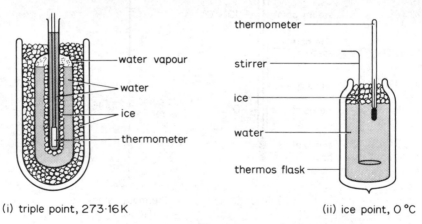

(i) triple point, 273·16 K   (ii) ice point, 0 °C

Figure 24.3 (*i*) *Triple point of water* (*ii*) *Ice point*

On the thermodynamic scale, the ice point has a temperature 273·15 K. The slight difference from the triple point is due to the difference in pressure (4·6 mm of mercury or 4·6 mmHg at the triple point and 760 mmHg at the ice point, Figure 24.3 (ii)) and to the removal of dissolved air from the distilled water used for the triple point.

Using the constant-volume gas thermometer, for example, the gas pressure p_{tr} is measured at the triple point of water, 273·16 K. If the pressure is p at an unknown temperature T on the thermodynamic scale, then, by definition,

$$T = \frac{p}{p_{tr}} \times 273 \cdot 16 \text{ K}$$

With a platinum resistance thermometer, the resistance R of the metal is measured at the unknown temperature. The temperature T_{pt} on the thermodynamic scale is then given, if R_{tr} is the resistance at the triple point, by

$$T_{pt} = \frac{R}{R_{tr}} \times 273{\cdot}16 \text{ K}$$

Celsius Temperature Scale

The *Celsius temperature*, symbol θ, is now defined by $\theta = T - 273{\cdot}15$, where T is the thermodynamic temperature. The ice point is a Celsius temperature of $0°C$ and the steam point, the temperature of steam at 760 mmHg pressure or $1{\cdot}013 \times 10^5$ Pa, is $100°C$.

It should be noted that if different types of thermometers are used to measure temperature, they only agree at the fixed points—273·16 K on the thermodynamic scale and $0°C$ and $100°C$, for example, on the Celsius scale.

The temperature change of interval of one degree Celsius, $1°C$, is exactly the same as the temperature interval 1 K in the thermodynamic scale. So '$°C$' may be replaced by 'K' in SI units, and '$°C^{-1}$' by 'K^{-1}'. Approximately, the temperature $0°C = 273$ K and $100°C = 373$ K. The absolute zero of temperature 0 K is approximately $-273°C$.

Thermometry, the Fixed Points

We have already discussed the general idea of a temperature scale. To establish such a scale we need: (i) some physical property of a substance—such as the volume of a gas or the electrical resistance of pure platinum—which increases continuously with increasing temperature, but is constant at constant temperature; (ii) fixed temperatures—fixed points—which can be accurately reproduced in the laboratory.

As we saw previously, on the thermodynamic scale the triple point of water is chosen as the fixed point and is defined as 273·16 K. The absolute zero is 0 K, the ice point is 273·15 K and the steam point is 373·15 K.

Suppose P is the chosen temperature-measuring quantity, such as a gas pressure at constant volume or the electrical resistance of a pure metal. If P_{tr} is the value of P at the triple point and P_T is the value at an unknown temperature T on the absolute thermodynamic scale, then, by definition,

$$T = \frac{P_T}{P_{tr}} \times 273{\cdot}16 \text{ K}$$

The temperatures of melting ice and pure boiling water at 760 mmHg pressure were chosen as the fixed points on the Celsius scale. Nowadays, the temperature θ on the Celsius scale is defined by $\theta = T - 273{\cdot}15$.

If P is the chosen temperature-measuring quantity, its values P_0 at the ice point and P_{100} at the steam point determine the fundamental interval, $100°C$, of the Celsius scale, $P_{100} - P_0$. And the temperature θ_P on the Celsius scale which corresponds to a value P_θ is given by

$$\theta_P = \frac{P_\theta - P_0}{P_{100} - P_0} \times 100$$

Gas Thermometry

In most accurate work, temperatures are measured by *gas thermometers*; for example, by the changes in pressure of a gas at constant volume.

At pressures of the order of one atmosphere, different gases give slightly different temperature scales, because none of them obeys the gas laws perfectly. But as the pressure is reduced, the gases approach closely to the ideal, and their temperature scales come together (see p. 715). By observing the departure of a gas from Boyle's law at moderate pressures it is possible to allow for its departure from the ideal. Temperatures measured with the gas in a constant-volume thermometer can then be converted to the values which would be given by the same thermometer if the gas were ideal.

Gas thermometer temperatures are used as *standard* temperatures. Temperatures measured with other types of thermometers are changed to a gas thermometer temperature.

Figure 24.4 shows a constant volume hydrogen thermometer. B is a bulb of platinum–iridium, holding the gas. The volume is defined by the level of the index I in the glass tube A. The pressure is adjusted by raising or lowering the mercury reservoir R. A barometer CD is fitted directly into the pressure-measuring system; if H_1 is its height, and h the difference in level between the

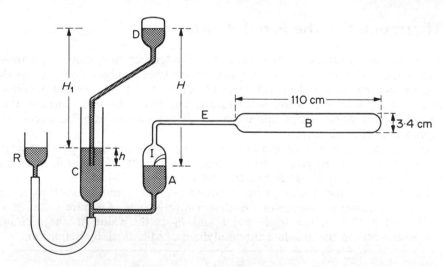

Figure 24.4 *Constant-volume hydrogen thermometer* (not to scale)

mercury surfaces in A and C, then the pressure H of the hydrogen, in mm mercury is

$$H = H_1 + h$$

H is measured with a cathetometer (a travelling telescope with vernier).

The tube E is called the 'dead-space' of the thermometer. Its diameter is made small, about 0.7 mm, so that it contains only a small fraction of the total mass of gas.

Plate 24A Gas Thermometry. *The photograph shows the whole assembly of the National Physical Laboratory constant volume gas thermometer, used to measure low temperatures in the range about 2 K to 27 K. The working parts such as the gas bulb are immersed in liquid helium in the stainless steel Dewar vessel on the left. The pressure of the gas is measured by a sensitive pressure balance top right. The value in Pa is calculated from the load on a piston divided by its cross-sectional area.* (Crown copyright. Courtesy of the National Physical Laboratory)

A gas thermometer is a large awkward instrument, demanding much skill and time, and useless for measuring changing temperatures. In practice, gas thermometers are used only for calibrating electrical thermometers—resistance thermometers and thermocouples. The readings of these, when they are used to measure unknown temperatures, can then be converted into temperatures on the ideal gas scale. Helium gas is widely used in gas thermometers.

The Celsius temperature θ on the constant volume gas thermometer scale would be calculated from

$$\theta = \frac{p_\theta - p_0}{p_{100} - p_0} \times 100°\text{C}$$

where p_θ, p_0 and p_{100} are the respective gas pressures at the unknown temperature θ, the ice point and steam point.

Examples on Temperature Measurement

1 The pressure recorded by a constant volume gas thermometer at a kelvin temperature T is $4.80 \times 10^4 \text{ N m}^{-2}$. Calculate T if the pressure at the triple point, 273.16 K, is $4.20 \times 10^4 \text{ N m}^{-2}$.

$$T = \frac{p_T}{p_{tr}} \times 273.16 \text{ K}$$

$$= \frac{4.80 \times 10^4}{4.20 \times 10^4} \times 273.16$$

$$= 312 \text{ K}$$

2 A constant mass of gas maintained at constant pressure has a volume of $200 \cdot 0$ cm^3 at the temperature of melting ice, $273 \cdot 2$ cm^3 at the temperature of water boiling under standard pressure, and $525 \cdot 1$ cm^3 at the normal boiling-point of sulphur. A platinum wire has resistance of $2 \cdot 000$, $2 \cdot 778$ and $5 \cdot 280$ Ω at the temperatures. Calculate the values of the boiling-point of sulphur given by the two sets of observations, and comment on the results.

On the gas thermometer scale, the boiling-point of sulphur is given by

$$\theta = \frac{V_\theta - V_0}{V_{100} - V_0} \times 100$$

$$= \frac{525 \cdot 1 - 200 \cdot 0}{273 \cdot 2 - 200 \cdot 0} \times 100$$

$$= 444 \cdot 1°C$$

On the platinum resistance thermometer scale, the boiling-point is given by

$$\theta_p = \frac{R_\theta - R_0}{R_{100} - R_0} \times 100$$

$$= \frac{5 \cdot 280 - 2 \cdot 000}{2 \cdot 778 - 2 \cdot 000} \times 100$$

$$= 421 \cdot 6°C$$

The temperatures recorded on the thermometers are therefore different. This is due to the fact that the variation of gas pressure with temperature at constant volume is different from the variation of the electrical resistance of platinum with temperature.

Electric Thermometers

Electrical thermometers have great advantages over other types. They are more accurate than any except gas thermometers, but are quicker in action and less cumbersome than those thermometers.

The measuring element of a *thermoelectric thermometer* is the welded junction of two fine wires. It is very small in size, and can therefore measure the temperature almost at a point. It causes very little disturbance wherever it is placed, because the wires leading from it are so thin that heat loss along them is usually negligible. It has a very small heat capacity, and can therefore follow a rapidly changing temperature. To measure such a temperature, however, the e.m.f. of the junction must be measured with a galvanometer, instead of a potentiometer, and some accuracy is then lost. The Celsius temperature θ_{th} on the thermoelectric thermometer scale would be calculated from

$$\theta_{th} = \frac{E_\theta - E_0}{E_{100} - E_0} \times 100°C$$

where E_θ, E_0 and E_{100} are the respective thermoelectric e.m.f.s at the unknown temperature θ, the ice point and the steam point.

The measuring element of a *resistance thermometer* is a spiral of fine wire. It has a greater size and heat capacity than a thermojunction, and cannot therefore measure a local or rapidly changing temperature. But, over the range from about room temperature to a few hundred degrees Celsius, it is more accurate.

The platinum resistance scale differs appreciably from the mercury-in-glass scale, for example, as the following table shows:

Mercury-in-glass	0	50	100	200	300	°C
Platinum-resistance	0	50·25	100	197	291	°C

Resistance Thermometers

Resistance thermometers are usually made of platinum. The wire is wound on two strips of mica, arranged crosswise as shown in Figure 24.5 (i). The ends of the coil are attached to a pair of leads A, for connecting them to a Wheatstone bridge. A similar pair of leads B is near to the leads from the coil, and connected in the adjacent arm of the bridge (Figure 24.5 (ii)). At the end near the coil, the pair of leads B is short-circuited. If the two pairs of leads are identical, their resistances are equal, whatever their temperature. Thus if $P = Q$ the dummy pair, B, just compensates for the pair A going to the coil; and the bridge measures the resistance of the coil alone. The Celsius temperature θ_p on the platinum resistance thermometer scale would be calculated from

$$\theta_p = \frac{R_\theta - R_0}{R_{100} - R_0} \times 100°C$$

where R_θ, R_0 and R_{100} are the respective resistances at the unknown temperature θ, the ice point and the steam point.

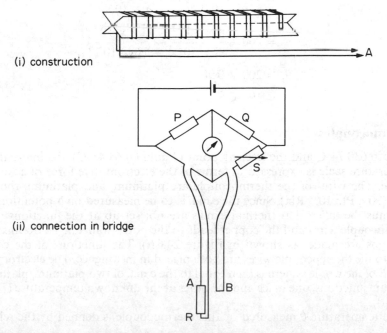

(i) construction

(ii) connection in bridge

Figure 24.5 *Platinum resistance thermometer* ($P = Q$ *and* B *compensates* A *so that* $S = R$)

The platinum resistance thermometer is used to measure temperature on the International Practical Temperature Scale between $-259\cdot34°C$ and $630\cdot74°C$. The platinum used in the coil must be strain-free and annealed pure platinum. Its purity and reliability are judged by the increase in its resistance from the ice point to the steam point. Thus if R_0 and R_{100} are the resistances of the coil at these points, then the coil is suitable if

$$\frac{R_{100}}{R_0} > 1\cdot39\,250$$

Various formulae and tables are provided to obtain the temperature over the wide temperature range.

Example on Resistance Thermometer

The resistance R_θ of a platinum wire at temperature $\theta°C$, measured on the gas scale, is given by $R_\theta = R_0(1 + a\theta + b\theta^2)$, where $a = 3\cdot800 \times 10^{-3}$ and $b = -5\cdot6 \times 10^{-7}$. What temperature will the platinum thermometer indicate when the temperature on the gas scale is $200°C$?

$$R_\theta = R_0(1 + a\theta + b\theta^2)$$
$$\therefore R_{200} = R_0(1 + 200a + 200^2 b)$$
and
$$R_{100} = R_0(1 + 100a + 100^2 b)$$

$$\therefore \theta_p = \frac{R_{200} - R_0}{R_{100} - R_0} \times 100$$

$$= \frac{R_0(1 + 200a + 200^2 b) - R_0}{R_0(1 + 100a + 100^2 b) - R_0} \times 100$$

$$= \frac{200a + 200^2 b}{a + 100b} = \frac{200(a + 200b)}{a + 100b}$$

$$= 200\,\frac{(3\cdot8 \times 10^{-3} - 11\cdot2 \times 10^{-5})}{3\cdot8 \times 10^{-3} - 5\cdot6 \times 10^{-5}}$$

$$\therefore \theta_p = 200 \times 0\cdot985$$
$$= 197°C$$

Thermocouples

Between $630\cdot74°C$ and the freezing point of gold ($1064\cdot43°C$), the international temperature scale is expressed in terms of the electromotive force of a thermocouple. The wires of the thermocouple are platinum, and platinum–rhodium alloy (90% Pt.: 10% Rh.). Since the e.m.f. is to be measured on a potentiometer, care must be taken that thermal e.m.f.s are not set up at the junctions of the thermocouple wires and the copper leads to the potentiometer. To do this three junctions are made, as shown in Figure 24.6 (i). The junctions of the copper leads to the thermocouple wires are both placed in melting ice. The electromotive force E of the whole system is then equal to the e.m.f. of two platinum/platinum–rhodium junctions, one in ice and the other at the unknown temperature (Figure 24.6 (ii)).

The temperature θ measured by this thermocouple is defined by the relation

$$E = a + b\theta + c\theta^2$$

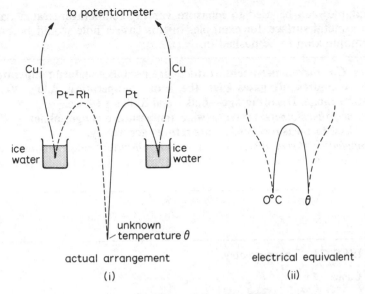

Figure 24.6 *Use of thermocouples*

where *a*, *b* and *c* are constants. The values of the constants are determined by measurements of *E* at the gold point (1064·43°C), the silver point (960·8°C), and the temperature of freezing antimony (about 630·74°C).

Other Thermocouples

Because of their convenience, thermocouples are used to measure temperatures outside their range on the international scale, when the highest accuracy is not required. The arrangement of three junctions and potentiometer may be used, but for less accurate work the potentiometer may be replaced by a galvanometer G, in the simpler arrangement of Figure 24.7 (i). The galvanometer scale may be calibrated to read directly in temperatures, the known melting-points of metals like tin and lead being used as subsidiary fixed points. For rough work, particularly at high temperatures, the cold junction may be omitted (Figure 24.6 (ii). An uncertainty of a few degrees in a thousand is often of no importance.

Owing to the small size and small heat capacity of their thermojunctions,

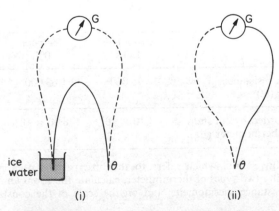

Figure 24.7 *Simple thermojunction thermometer*

thermocouples can be used to measure varying temperatures at a particular place of a metal surface, for example. In this case a hole would be bored for the thermojunction to be pushed in.

Summary: *Gas thermometer* **temperatures are used as standard temperatures. At very low pressures all gases give the same temperature value. Very wide temperature range. Disadvantage—bulky and difficult to use.**

Resistance thermometers **have a wide temperature range, about** $-200°$**C to** $1200°$**C. Used to measure liquid temperatures accurately.**

Thermoelectric thermometers **have a wide temperature range, about** $-250°$**C to** $1500°$**C. Used to measure varying temperatures as the junction has a low heat capacity.**

EXERCISES 24 Thermometry

Multiple Choice

1 A gas thermometer with helium is used as a standard thermometer because
   A helium is a rare gas
   B helium is sensitive to pressure
   C helium can be obtained in a pure state
   D at low pressures all gases approach the same ideal state
   E helium expands regularly with temperature.
2 A thermoelectric thermometer is particularly useful for measuring the temperature of
   A oil   B an aeroplane wing after a flight   C ice under pressure
   D steam at high pressure   E mercury at very low temperature.
3 The resistance of a platinum resistance thermometer is $1·20\,\Omega$ when measuring the Kelvin temperature T of oil and $1·00\,\Omega$ at the triple point of water $273·16$ K. The temperature T is

   A $(1·20 - 1·00)/1·00 \times (273·16)$ K
   B $(1·00/1·20) \times 273·16$ K
   C $(1·20/1·00) \times (273·16 - 100)°$C
   D $(1·20/1·00) \times 273·16$ K
   E $\dfrac{1·20}{1·00} \times 100 - 273·16$ K

Longer Questions

4

	Steam point 100°C	Ice point 0°C	Room temperature
Resistance of resistance thermometer (Ω)	75·000	63·000	64·992
Pressure recorded by constant volume gas thermometer (Pa)	$1·10 \times 10^5$	$8·00 \times 10^4$	$8·51 \times 10^4$

Using the above data, which refers to the observations of a particular room temperature using two types of thermometer, calculate the room temperature on the scale of the resistance thermometer and on the scale of the constant volume gas thermometer.

Why do these values differ slightly? (*L.*)

5 (a) Explain how a temperature scale is defined.
  (b) Discuss the relative merits of (i) a mercury-in-glass thermometer, (ii) a platinum resistance thermometer, (iii) a thermocouple, for measuring the temperature of an oven which is maintained at about 300°C. (N.)

6 Tabulate various physical properties used for measuring temperature. Indicate the temperature range for which each is suitable.

Discuss the fact that the numerical value of a temperature expressed on the scale of the platinum resistance thermometer is not the same as its value on the gas scale except at the fixed points.

If the resistance of a platinum thermometer is $1.500 \, \Omega$ at 0°C, $2.060 \, \Omega$ at 100°C and $1.788 \, \Omega$ at 50°C on the gas scale, what is the difference between the numerical values of the latter temperature on the two scales? (N.)

7 The table below gives data for two thermometers at three different temperatures (the ice point, the steam point and room temperature).

Type of thermometer	Property	Value of property		
		ice point	steam point	room temp.
Gas	Pressure in mmHg	760	1040	795
Thermistor	Current in mA	12·0	54·0	15·0

  (a) Calculate the temperature of the room according to each thermometer.
  (b) State why the thermometers disagree in their value for room temperature.
  (c) Explain why a gas thermometer is seldom used for temperature measurement in the laboratory. (AEB.)

8 The resistance of the element of a platinum resistance thermometer is $2.00 \, \Omega$ at the ice point and $2.73 \, \Omega$ at the steam point. What temperature on the platinum resistance scale would correspond to a resistance value of $8.43 \, \Omega$?

Measured on the gas scale, the same temperature corresponded to a value of 1020°C. Explain the discrepancy. (L.)

9 Explain what is meant by a change in temperature of 1°C on the scale of a platinum resistance thermometer.

Draw and label a diagram of a platinum resistance thermometer together with a circuit in which it is used.

Give two advantages of this thermometer and explain why, in its normal form, it is unsuited for measurement of varying temperatures.

The resistance R_θ of platinum varies with the temperature θ°C as measured by a constant-volume gas thermometer according to the equation

$$R_\theta = R_0(1 + 8000\alpha\theta - \alpha\theta^2)$$

where α is a constant. Calculate the temperature on the platinum scale corresponding to 400°C on this gas scale. (N.)

25 Heat and Energy (Heat Capacity, Latent Heat)

Heat and Energy

Heat and Temperature

If we run hot water into a lukewarm bath, we make it hotter; if we run in cold water, we make it cooler. The hot water, we say, gives *heat* to the cooler bath-water; but the cold water takes heat from the warmer bath-water. The quantity of heat which we can get from hot water depends on both the mass of water and on its temperature: a bucket-full at 80°C will warm the bath more than a cup-full at 100°C.

Roughly speaking, temperature is analogous to electrical potential in electricity, and heat is analogous to quantity of electricity. We can detect temperature changes, and whenever the temperature of a body rises, that body has gained heat. The converse is not always true; when a body is melting or boiling, it is absorbing heat from the flame beneath it to change its solid or liquid state but its temperature is not rising.

Thermal Equilibrium, The Zeroth Law

When two bodies are in thermal contact and there is no net flow of heat between them, they are said to be in *thermal equilibrium*. Experimentally, it is found that when two bodies A and B are each in thermal equilibrium with a third body C, then A and B are also in thermal equilibrium with each other. This is called the 'zeroth law of thermodynamics'—the number 'zero' was used because this law logically precedes the 'first' and 'second' laws of thermodynamics and, in fact, is assumed in the two laws.

The zeroth law leads to the conclusion that temperature is a well-defined physical quantity. For example, suppose we wish to determine whether two bodies A and B are in thermal equilibrium. In practice, we do this by bringing each in turn into contact with a third body, a thermometer T say. Experimentally, then, we bring A and T into thermal equilibrium, and B and T into thermal equilibrium. If the temperature reading is the same in the two cases, we deduce, but only with the help of the zeroth law, that A and B are in thermal equilibrium. The law thus enables temperature to be defined as that property of a body which decides whether or not it is in thermal equilibrium with another body.

Heat and Energy

The idea of heat as a form of energy was developed particularly by BENJAMIN THOMPSON (1953–1814). He was an American who, after adventures in Europe, became a Count of the Holy Roman Empire, and war minister of Bavaria. He is now generally known as Count Rumford.

While supervising his arsenal, he noticed the great amount of heat which was liberated in the boring of cannon. The idea common at the time was that this heat was a fluid, pressed out of the chips of metal as they were bored out of the barrel. To measure the heat produced, Rumford used a blunt borer, and surrounded it and the end of the cannon with a wooden box, which was filled with water. From the mass of water, and the rate at which its temperature rose, he showed that the amount of heat liberated was in no way connected with the mass of metal bored away, and concluded that it depended only on the *work done* against friction. It followed that *heat was a form of energy.*

Rumford published the results of his experiments in 1798. No similar experiments were made until 1840, when Joule began his study of heat and other forms of energy. Joule measured the work done, and the heat produced, when water was churned in an apparatus he designed for the experiment. He also measured the work done and heat produced when oil was churned, when air was compressed, when water was forced through fine tubes, and when cast iron bevel wheels were rotated one against the other. Always, within the limits of experimental error, he found that the heat liberated was *proportional* to the mechanical work done, and that the ratio of the two was the same in all types of experiment.

In other experiments, Joule measured the heat liberated by an electric current in flowing through a resistance; at the same time he measured the work done in driving the dynamo which generated the current. He obtained about the same ratio for work done to heat liberated as in his direct experiments. This work linked the ideas of heat, mechanical, and electrical energy. He also showed that the heat produced by a current is related to the chemical energy used up.

We can thus say:

(a) an object gains *energy* when its temperature rises;

(b) *energy* (heat) passes from a warm to a cold object if they are placed in contact.

The metric unit of work or energy is the *joule*, J. Since experiment shows that heat is a form of energy, *the joule is the scientific unit of heat.* 'Heat per second' is expressed in 'joules per second' or *watts*, W.

The Conservation of Energy

As a result of all his experiments, Joule developed the idea that energy in any one form could be converted into any other. There might be a loss of useful energy in the process—for example, some of the heat from the furnace of a steam-engine is lost up the chimney, and some more down the exhaust—but no energy is destroyed. The work done by the engine, added to the heat lost as described and the heat developed as friction, is equal to the heat provided by the fuel burnt. The idea underlying this statement is called the *principle of the conservation of energy*. It means that, if we start with a given amount of energy in any one form, we convert it in turn into all other forms. We may not always be able to convert it completely, but if we keep an accurate balance-sheet we shall find that the *total* amount of energy, expressed in any one form—say, heat or work—is always the same, and is equal to the original amount.

The principle of the conservation of energy is often expressed concisely in mathematical form by the equation

$$\Delta Q = \Delta U + \Delta W \qquad . \qquad . \qquad . \qquad . \qquad (1)$$

In (1), ΔQ is the *quantity of heat or energy* given to a system, ΔU is the consequent rise in *internal energy* of the system and ΔW is the *external work* done by the system. As we see later, the rise in internal energy, ΔU, of a system is shown by a temperature rise. If the system also expands, it does external work, ΔW, against the external forces. An alternative mathematical equation to (1) is

$$\Delta U = \Delta Q + \Delta W \quad . \qquad . \qquad . \qquad . \qquad . \qquad (2)$$

Here ΔU is the change in internal energy of a system when the system is given a quantity of heat ΔQ and, in addition, work ΔW is done *on* the system. The principle of conservation of energy will be applied later to many cases of energy changes.

The conservation of energy applies to living organisms—plants and animals—as well as to inanimate systems. For example, we may put a man or a mouse into a box or a room, give him a treadmill to work, and feed him. His food is his fuel; if we burn a sample of it, we can measure its chemical energy, in heat units. And if we now add up the heat value of the work which the man does, and the heat which his body gives off, we find that their total is equal to the chemical energy of the food which the man eats. Because food is the source of man's energy, food value may be expressed in *kilojoules*. A man needs about 120 000 kilojoules per day.

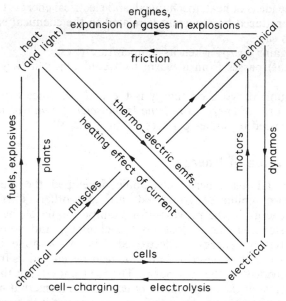

Figure 25.1 *Forms of energy and their transfers*

All the energy by which we live comes from the sun. The sun's ultraviolet rays are absorbed in the green matter of plants, and make them grow; the animals eat the plants, and we eat them—we are all vegetarians at one remove. The plants and trees of an earlier age decayed, were buried, and turned into coal. Even water-power comes from the sun—we would have no lakes if the sun did not evaporate the water and provide the rainfall which fills the lakes. The relationship between the principal forms of energy are summarised in Figure 25.1. It shows how different kinds of energy can be transferred from one form to another.

Heat Capacity, Specific Heat Capacity

We shall now deal with the topic of the *heat capacity* of objects, which needs calculations and measurements.

A large iron tank and an aluminium kettle have a different *heat capacity* which depends on their respective masses and on the metal used. By definition, the heat capacity of a body, is the quantity of heat required to raise its temperature by 1 degree. It is expressed in joules per kelvin ($J\,K^{-1}$) in SI units.

Information of the heat capacity of *1 kilogram* of iron or aluminium would help to find the heat capacity of the iron tank or aluminium kettle if their mass is known.

The *specific heat capacity* of a substance is the heat required to raise the temperature of 1 kg of it through 1 degree; it is the heat capacity per kg of the substance. Specific heat capacities are expressed in *joules per kilogram per kelvin* and the symbol is *c*. The specific heat capacity of water, c_w, is about $4200\,J\,kg^{-1}\,K^{-1}$, or $4\cdot2\,kJ\,kg^{-1}\,K^{-1}$, where 1 kJ = 1 kilojoule = 1000 J. The megajoule, MJ, is a larger unit than the kilojoule used in the gas or electrical industry. $1\,MJ = 10^6\,J = 1000\,kJ$.

From the definition of specific heat capacity, it follows that, for a particular object,

---

heat capacity, C = mass × specific heat capacity

The specific heat capacity of copper, for example, is about $400\,J\,kg^{-1}\,K^{-1}$. Hence the heat capacity of 5 kg of copper = $5 \times 400 = 2000\,J\,K^{-1} = 2\,kJ\,K^{-1}$. If the copper temperature rises by 10°C, then heat gained = $5 \times 400 \times 10 = 20\,000\,J$.

Generally, then, the heat *Q* gained (or lost) by an object is given by

---

$$Q = mc\,\Delta\theta$$

---

where *m* is the mass of the object, *c* its specific heat capacity and $\Delta\theta$ its temperature change.

Temperature Changes

From $Q = mc\Delta\theta$, the temperature change θ of an object of mass *m* which loses or gains a given quantity of heat *Q* is given by

$$\Delta\theta = \frac{Q}{mc}$$

where *c* is the specific heat capacity of the object. So for a given loss of heat *Q* to the surroundings, the temperature fall $\Delta\theta$ of a small mass of warm water in a room is greater than that of a large mass of water at the same temperature. We shall see later that the rate at which a hot solid or liquid cools depends on the nature and area of its surface, in addition to its temperature, mass and specific heat capacity.

When a thermoelectric thermometer is used, the thermometer junction is placed in contact with the object whose temperature is required. The junction has a small thermal capacity, *mc*, since its mass is small. So the temperature of the junction quickly reaches the temperature of the object, which is an advantage of the thermometer.

Measurement of Specific Heat Capacity

Specific Heat Capacity of Solid by Electrical Method

The specific heat capacity of a metal can be found by an *electrical method*. Figure 25.2 shows a simple form of laboratory heater apparatus. M is a thick solid block of metal such as aluminium, with an electric heater element H completely inside a deep hole bored into the metal and a thermometer T inside another deep hole. Both H and T must make good thermal contact with the block. An insulating jacket J is placed round the metal.

Plate 25A *Laboratory set-up for measuring specific heat capacity*

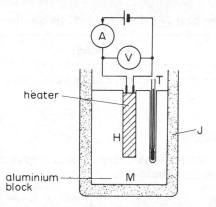

Figure 25.2 *Electrical method for specific heat capacity*

In an experiment, suppose the voltmeter reads 12 V, the ammeter A reads 4·0 A and the block, mass 1·0 kg, rises by 16°C in 5 min or 300 s. Then (p. 295)

$$\text{electrical heat supplied} = IVt = 12 \times 4 \times 300 = 14400 \text{ J}$$

If c is the specific heat capacity of the metal, then, assuming negligible heat losses,

$$Q = mc\Delta\theta = 1 \times c \times 16 = 14400$$

$$\therefore c = 900 \text{ J kg}^{-1} \text{ K}^{-1}$$

If $\Delta\theta$ is the temperature rise corrected for heat losses (p. 636), then, generally,

$$IVt = mc\Delta\theta$$

The electrical energy supplied, $\Delta Q = IVt$. From the principle of conservation of energy (p. 629), this is equal to the rise in internal energy ΔU of the metal $(mc\Delta\theta)$ plus the heat losses h by cooling *plus* the external work ΔW done against external atmospheric pressure by the metal when it expands on warming. Since metals expand very slightly in volume on warming, ΔW can be neglected in this experiment. So $IVt = mc\Delta\theta + h$.

If $\Delta\theta$ is not corrected for heat losses, we see from $IVt = mc\Delta\theta$ that the result for c is too high.

Specific Heat Capacity of a Liquid by Continuous Flow Method

Callendar and Barnes devised an electrical method for the specific heat capacity of a liquid in which only steady temperatures are measured. They used platinum resistance thermometers, which are more accurate than mercury ones but take more time to read. In the measurement of steady temperatures, however, this is no drawback. As we shall see shortly *the heat capacity of the apparatus is not required*, which is a great advantage of the method.

Figure 25.3 shows Callendar and Barnes' apparatus used to measure the specific heat capacity of water. Water from the constant-head tank K flows through the glass tube U, and can be collected as it flows on. It is heated by the spiral resistance wire R, which carried a steady electric current I. Its temperature, as it enters and leaves, is measured by the thermometers T_1 and T_2. (In a simplified laboratory experiment, these may be mercury thermometers.) Surrounding the apparatus is a glass jacket G, which is evacuated, so that heat cannot escape from the water by conduction or convection.

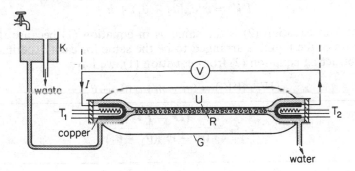

Figure 25.3 *Callendar and Barnes' apparatus (contracted several times in length relative to diameters)*

When the apparatus is running, it settles down eventually to a steady state, in which the heat supplied by the current is all carried away by the water. *None is then taken in warming the apparatus, because every part of it is at a constant temperature.* The mass of water m, which flows out of the tube in t seconds, is then measured. If the water enters at a temperature θ_1 and leaves at θ_2, then if c_w is its mean specific heat capacity,

$$\text{heat gained by water} = Q = mc_w(\theta_2 - \theta_1) \text{ joules}$$

The energy which liberates this heat is electrical. To find it, the current I, and the potential difference across the wire V, are measured with a potentiometer. If I and V are in amperes and volts respectively, then, in t seconds, energy supplied to the wire = IVt joules.

If we ignore any heat losses and any external work, then, from the principle of conservation of energy,

$$mc_w(\theta_2 - \theta_1) = IVt$$

$$\therefore c_w = \frac{IVt}{m(\theta_2 - \theta_1)}$$

Elimination of Heat Losses

To get the highest accuracy from this experiment, the small heat losses due to radiation, and conduction along the glass, must be allowed for. These are determined by the temperatures θ_1 and θ_2. For a given pair of values of θ_1 and θ_2, and constant-temperature surroundings (not shown), let the heat lost per second be h. Then if m now represents the mass of liquid flowing *per second*,

$$\text{power supplied by heating coil} = mc_w(\theta_2 - \theta_1) + h$$

$$\therefore IV = mc_w(\theta_2 - \theta_1) + h \quad . \quad . \quad (1)$$

To allow for the loss h, the rate of flow of water is changed, to about half or twice its previous value. The current and voltage are then adjusted to bring θ_2 back to its original value, for example, if the rate of flow of the water is reduced, then I and V are reduced. The inflow temperature θ_1 is fixed by the temperature of the water in the tank. If I', V', are the new values of I, V and m' is the new mass of water flowing per second, then:

$$I'V' = m'c_w(\theta_2 - \theta_1) + h \quad . \quad . \quad . \quad (2)$$

Note that h in equation (2) is the same as in equation (1) because the *mean* temperature of the liquid is arranged to be the same for each experiment.

On subtracting equation (2) from equation (1), we have

$$(IV - I'V') = (m - m')c_w(\theta_2 - \theta_1)$$

$$\therefore c_w = \frac{(IV - I'V')}{(m - m')(\theta_2 - \theta_1)} \quad . \quad . \quad . \quad (3)$$

When the temperature rise, $\theta_2 - \theta_1$, is made small, for example, $\theta_1 = 20 \cdot 0°C$, $\theta_2 = 22 \cdot 0°C$, then c_w may be considered as the specific heat capacity at $21 \cdot 0°C$, the mean temperature. If the inlet water temperature is now raised to say $\theta_1 = 40 \cdot 0°C$ and θ_2 is then $42 \cdot 0°C$, c_w is now the specific heat capacity at $41 \cdot 0°C$. In this way it was found that *the specific heat capacity of water varies with temperature*. The continuous flow method can be used to find the variation in specific heat capacity of any liquid in the same way.

The table below shows the relative variation of the specific heat capacity of water, taking the value at 15°C as 1·0000 in magnitude.

Specific heat capacity of water

Temperature (°C)	5	15	25	40	70	100
c_w	1·0047	1·0000	0·9980	0·9973	1·0000	1·0057

Example on Constant Flow

Water flows at the rate of 0.1500 kg min^{-1} through a tube and is heated by a heater dissipating 25.2 W. The inflow and outflow water temperatures are $15.2°C$ and $17.4°C$ respectively. When the rate of flow is increased to 0.2318 kg min^{-1} and the rate of heating to 37.8 W, the inflow and outflow temperatures are unaltered. Find   (i) the specific heat capacity of water,   (ii) the rate of loss of heat from the tube.

Suppose c_w is the specific heat of water in J kg^{-1} K^{-1} and h is the heat lost in J s^{-1}. Then, since 1 W = 1 J per second,

$$25.2 = \frac{0.1500}{60} c_w (17.4 - 15.2) + h \qquad . \qquad . \qquad . \qquad (1)$$

and

$$37.8 = \frac{0.2318}{60} c_w (17.4 - 15.2) + h \qquad . \qquad . \qquad . \qquad (2)$$

Subtracting (1) from (2),

$$37.8 - 25.2 = \frac{0.2318 - 0.1500}{60} c_w (17.4 - 15.2)$$

$$\therefore c_w = \frac{12.6 \times 60}{0.0818 \times 2.2} = 4200 \text{ J kg}^{-1} \text{ K}^{-1}$$

Substituting for c_w in (1),

$$h = 25.2 - \frac{0.15}{60} \times 4200 \times 2.2 = 2.1 \text{ J s}^{-1} = 2.1 \text{ W}$$

Method of Mixtures

A common way of measuring specific heat capacities of solids (or liquids) is the method of mixtures.

As an illustration of a specific heat capacity determination, suppose a metal of mass 0.2 kg at $100°C$ is dropped into 0.08 kg of water at $15°C$ contained in a calorimeter of mass 0.12 kg and specific heat capacity 400 J kg^{-1} K^{-1}. The final temperature reached is $35°C$. Then, assuming negligible heat losses,

$$\text{heat capacity of calorimeter} = 0.12 \times 400 = 48 \text{ J K}^{-1}$$

$$\text{heat capacity of water} = 0.08 \times 4200 = 336 \text{ J K}^{-1}$$

$$\therefore \text{ heat gained by water + cal.} = (336 + 48) \times (35 - 15) \text{ J}$$

$$\text{and heat lost by hot metal} = 0.2 \times c \times (100 - 35) \text{ J}$$

$$\therefore 0.2 \times c \times 65 = 384 \times 20$$

$$\therefore c = \frac{384 \times 20}{0.2 \times 65} = 590 \text{ J kg}^{-1} \text{ K}^{-1} \text{ (approx.)}$$

Heat Losses

In a calorimetric experiment, some heat is always lost by leakage. Leakage of heat cannot be prevented, as leakage of electricity can, by insulation, because even the best insulator of heat still has appreciable conductivity (p. 690).

When convection is prevented, gases are the best thermal insulators. Hence calorimeters are often surrounded with a shield S and the heat loss due to conduction is made small by packing S with insulating material or by supporting the calorimeter on an insulating ring, or on threads. The loss by radiation is small at small excess temperatures over the surroundings. In some simple calorimetric experiments the final temperature of the mixture is reached quickly, so that the time for leakage is small. The total loss of heat is therefore negligible in laboratory experiments on the specific heats of metals, but not on the specific heat capacities of bad conductors, such as rubber, which give up their heat slowly. Here a 'cooling correction' must be added to the observed final temperature. When great accuracy is required, the loss of heat by leakage is always taken into account.

Newton's Law of Cooling

Newton was the first person to investigate the heat lost by a body in air. He found that *the rate of loss of heat is proportional to the excess temperature over the surroundings*. This result, called *Newton's law of cooling*, is approximately true in still air only for a temperature excess of about 20 K or 30 K; but it is true for all excess temperatures in conditions of forced convection of the air, i.e. in a draught. With natural convection Dulong and Petit found that the rate of loss of heat was proportional to $\theta^{5/4}$, where θ is the excess temperature, and this appears to be true for higher excess temperatures, such as from 50 K to 300 K.

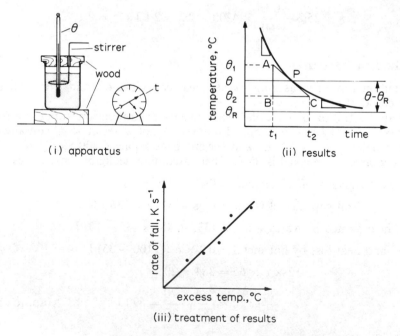

(i) apparatus

(ii) results

(iii) treatment of results

Figure 25.4 *Newton's law of cooling*

To demonstrate Newton's law of cooling, we plot a temperature (θ)–time (t) cooling curve for hot water in a calorimeter placed in a draught (Figure 25.4 (i)). If θ_R is the room temperature, then the excess temperature of the water is ($\theta - \theta_R$). At various temperatures, such as θ in Figure 25.4 (ii), we draw tangents such as APC to the curve. The slope of the tangent, in degrees per second, gives us the rate of fall of temperature, when the water is at the temperature θ:

$$\text{rate of fall} = \frac{AB}{BC} = \frac{\theta_1 - \theta_2}{t_2 - t_1}$$

We then plot these rates against the excess temperature, $\theta - \theta_R$, as in Figure 25.4 (iii), and find a straight line passing through the origin. Since the heat lost per second by the water and calorimeter is proportional to the rate of fall of the temperature, Newton's law is thus verified.

Heat Loss and Temperature Fall

Besides the excess temperature, the rate of heat loss depends on the exposed area of the calorimeter, and on the nature of its surface: a dull surface loses heat a little faster than a shiny one, because it is a better radiator (p. 723). This can be shown by doing a cooling experiment twice, with equal masses of water, but once with the calorimeter polished, and once after it has been blackened in a candle-flame. In general, for any body with a uniform surface at a uniform temperature θ, we may write, if Newton's law is true,

$$\text{heat loss/second} = \frac{dQ}{dt} = kS(\theta - \theta_R)$$

where S is the area of the body's surface, θ_R is the temperature of its surroundings, k is a constant depending on the nature of the surface, and Q denotes the heat lost from the body.

When a body loses heat Q, its temperature θ falls; if m is its mass, and c its specific heat capacity, then its rate of fall of temperature, $d\theta/dt$, is given by

$$\frac{dQ}{dt} = -mc\frac{d\theta}{dt}$$

Now the mass of a body is proportional to its volume. The rate of heat loss, however, is proportional to the surface area of the body. The rate of fall of temperature is therefore proportional to the ratio of surface to volume of the body

For bodies of similar shape, the ratio of surface to volume is inversely proportional to any linear dimension. If the bodies have surfaces of similar nature, therefore, the rate of fall of temperature is inversely proportional to the linear dimension; a small body cools faster than a large one. This is a fact of daily experience: a small coal which falls out of the fire can be picked up sooner than a large one; a tiny baby should be more thoroughly wrapped up than a grown man. In calorimetry by the method of mixtures, the fact that a small body cools faster than a large one means that the larger the specimen, the less serious is the heat loss in transferring it from its heating place to the calorimeter. It also means that the larger the scale of the whole apparatus, the less serious are the errors due to loss of heat from the calorimeter.

Specific Latent Heat

Evaporation or Vaporisation, Electrical Method for Specific Latent Heat

The *specific latent heat of evaporation or vaporisation* of a liquid is the heat required to convert unit mass of it, at its boiling point, into vapour *at the same temperature.* It is expressed in joules per kilogram (J kg^{-1}), or, with high values, in kJ kg^{-1}.

A modern electrical method for the specific latent heat of evaporation of water is illustrated in Figure 25.5. The liquid is heated in a vessel U by the heating coil R. As shown, the vapour escapes through H at the top of U into a surrounding jacket J and then passes down the tube T. Here the vapour is condensed by cold water flowing through the jacket K. When the apparatus has reached its *steady state*, the liquid is at its boiling-point, and heat supplied by the coil is used in evaporating the liquid, and in offsetting the losses. The heat losses are considerably reduced by the use of the vapour jacket. The liquid coming from the condenser is then collected for a measured time, and weighed.

If I and V are the current through the coil, and the potential difference across

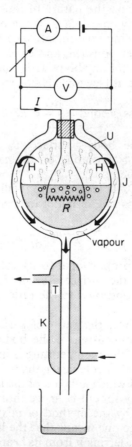

Figure 25.5 *Electrical method for specific latent heat of evaporation*

it, the electrical power supplied is IV. And if h is the heat lost from the vessel per second, and m the mass of liquid collected per second, then

$$IV = ml + h \qquad . \qquad . \qquad . \qquad . \qquad . \qquad (1)$$

The heat losses h are determined by the temperature of the vessel, which is fixed at the boiling-point of the liquid. So they can be eliminated by a second experiment with a different rate of evaporation, as in the case of the continuous flow method, p. 633. If I', V' are the new current and potential difference, and if m' is the mass per second evaporated, then

$$I'V' = m'l + h$$

Hence by subtraction from equation (1)

$$l = \frac{(IV - I'V')}{(m - m')}$$

It may be noted that a much higher power supply is needed for determining the specific latent heat of evaporation of water than for alcohol, as water has a much higher value of l. The specific latent heat of evaporation of water is about $l = 2260 \text{ kJ kg}^{-1}$ or $2 \cdot 26 \times 10^6 \text{ J kg}^{-1}$.

Method of Mixtures

To find the specific latent heat of evaporation of water by mixtures, we pass steam into a calorimeter with water (Figure 25.6). On its way the steam passes

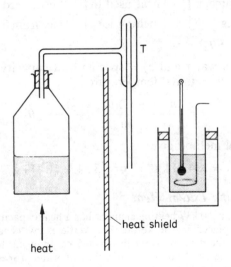

Figure 25.6 *Specific latent heat of evaporation of water by mixtures*

through a vessel, T in the figure, which traps any water carried over by the steam and is called a steam-trap. The mass m of condensed steam is found by weighing. If θ_1 and θ_2 are the initial and final temperatures of the water, the specific latent heat l is given by:

$$\left.\begin{array}{l}\text{heat given by steam}\\\text{condensing}\end{array}\right\} + \left\{\begin{array}{l}\text{heat given by condensed}\\\text{water cooling from}\\100°C \text{ to } \theta_2\end{array}\right\} = \left\{\begin{array}{l}\text{heat taken by}\\\text{calorimeter and}\\\text{water}\end{array}\right.$$

$$ml \qquad + \qquad mc_w(100 - \theta_2) \qquad = (m_1c_w + C)(\theta_2 - \theta_1)$$

where m_1 is the mass of water in the calorimeter, c_w is the specific heat capacity of water and C is the heat capacity (mass × specific heat capacity) of the metal calorimeter. Hence

$$l = \frac{(m_1 c_w + C)(\theta_2 - \theta_1)}{m} - c_w(100 - \theta_2)$$

The result for l is only approximate as the steam-trap is not efficient.

Fusion

The specific latent heat of fusion of a solid is the heat required to convert unit mass of it, at its melting-point, into liquid at the same temperature. It is expressed in joules per kilogram (J kg^{-1}). High values can be more conveniently expressed in kJ kg^{-1}.

Ice is one of the substances whose specific latent heat of fusion we are likely to have to measure. To do so, place warm water, at a temperature of θ_1 a few degrees above room temperature, inside a calorimeter. Then add small lumps of ice, dried by blotting paper, until the temperature reaches a value θ_2 as much below room temperature as θ_1 was above. Weigh the mixture, to find the mass m of ice which has been added. Then the specific latent heat l is given by:

$$\left.\begin{array}{l}\text{heat given by calorimeter}\\\text{and water in cooling}\end{array}\right\} = \left\{\begin{array}{l}\text{heat used in}\\\text{melting ice}\end{array}\right\} + \left\{\begin{array}{l}\text{heat used in warming melted}\\\text{ice from } 0°C \text{ to } \theta_2\end{array}\right.$$

$$\therefore \quad (m_1 c_w + C)(\theta_1 - \theta_2) \quad = \quad ml \quad + \quad mc_w(\theta_2 - 0)$$

where m_1 = mass of water and c_w = specific heat capacity, C = heat capacity of calorimeter, and θ_1 = initial temperature.

Hence
$$l = \frac{(m_1 c_w + C)(\theta_1 - \theta_2)}{m} - c_w \theta_2$$

A modern electrical method gives

$$l = 334 \text{ kJ kg}^{-1} \quad \text{or} \quad 3\cdot 34 \times 10^5 \text{ J kg}^{-1}$$

Example on Specific Latent Heat

An electric kettle with a 2·0 kW heating element has a heat capacity of 400 J K^{-1}. 1·0 kg of water at 20°C is placed in the kettle. The kettle is switched on and it is found that 13 minutes later the mass of water in it is 0·5 kg. Ignoring heat losses calculate a value for the specific latent heat of vaporisation of water. (Specific heat capacity of water = $4\cdot 2 \times 10^3$ J kg^{-1} K^{-1}.)

$$\text{Total heat supplied} = 200 \times 13 \times 60 = 1\cdot 56 \times 10^6 \text{ J}$$

$$\text{Heat used for kettle} = C\Delta\theta = 400 \times (100 - 20) = 32\,000 = 0\cdot 032 \times 10^6 \text{ J}$$

Heat used to raise temperature of 1 kg of water from 20°C to 100°C

$$= mc\Delta\theta = 1 \times 4200 \times (100 - 20) = 0\cdot 336 \times 10^6 \text{ J}$$

So total heat to change water at 100°C to steam at 100°C

$$= 1\cdot 56 \times 10^6 - (0\cdot 032 \times 10^6 + 0\cdot 336 \times 10^6) \text{ J}$$

$$= 1\cdot 192 \times 10^6 \text{ J}$$

Since mass of water changed to steam = $1\cdot 0 - 0\cdot 5 = 0\cdot 5$ kg, then

$$l = \frac{1\cdot 192 \times 10^6}{0\cdot 5} = 2\cdot 38 \times 10^6 \text{ J kg}^{-1}$$

The Weather

When a weather forecaster predicts the next day's temperatures, he or she often suggests that it will be 'a few degrees warmer in the cities'. Why is this? The answer lies largely in the theory of heat capacities and latent heat. First, the heat capacity of building and paving materials (like brick and tarmac) is generally larger than that of soil. So cities store more heat during the day than the countryside does, helping to keep up the temperature at night. Another important factor lies in the drainage of cities. Because surface water is rapidly removed in modern cities, there is less evaporation after rain, which would otherwise cool the city by the latent heat principle. The final significant factor is cooling by wind (the *wind chill factor* so often referred to by weather forecasters), which is disrupted by the tall buildings found in cities. In relation to these warming mechanisms, heat loss from buildings and cars is relatively insignificant!

Heat capacity can also be used to explain why the coldest months of the year (January and February in the UK) follow significantly later than the shortest days (which occur at the end of December in the UK). At first sight, because the UK receives minimal heating by the sun during December, it is reasonable to expect this to be the coldest part of the year. However, the soil and atmosphere have a heat capacity, and therefore take time to cool after their heating from the sun reaches its minimum. Similarly, the hottest part of the year follows the longest day by a month or so, since the earth takes time to warm up.

EXERCISES 25   Specific Heat, Latent Heat, Energy

Multiple Choice

1 Metal pellets of mass 0.1 kg are fired into an iron plate of large mass m and specific heat capacity c at a temperature of 15°C. 25% of the kinetic energy of the pellets is converted to thermal energy and the plate temperature rises to 16°C. The average speed of the pellets before hitting the plate was

   A $\sqrt{20mc}$ B $\sqrt{80mc}$ C $4mc$ D $\sqrt{40mc}$ E $0.4mc$

2 A solar furnace has a concave mirror of collecting area 0.8 m^2. The average thermal radiation from the sun reaching the earth is about 750 W m^{-2}. A small mass 0.5 kg, specific heat capacity 2000 J kg^{-1} K^{-1}, is heated by the furnace from 10°C to 40°C. The time taken in seconds for the heating is

   A 100   B 60   C 50   D 30   E 20

3 A small mass x kg of metal, specific heat capacity c in J kg^{-1} K^{-1}, at 60°C is placed on a large mass y kg of ice at 0°C. If l is the specific latent heat of ice in J kg^{-1}, the mass of ice in kg melted by the metal is

   A $cx/60l$ B y/xc C $60xc/l$ D $yl/60xc$ E $60xc/y$

Longer Questions

4 An electrical heater of 2 kW is used to heat 0.5 kg of water in a kettle of heat capacity 400 J K^{-1}. The initial water temperature is 20°C.
   Neglecting heat losses: (i) how long will it take to heat the water to its boiling point, 100°C?   (ii) starting from 20°C, what mass of water is boiled away in 5 min? (Assume for water, specific heat capacity = 4200 J kg^{-1} K^{-1} and specific latent heat of vaporisation = 2×10^6 J kg^{-1}.)

5 (a) In an espresso coffee machine, steam at 100°C is passed into milk to heat it. Calculate:
  (i) the energy required to heat 150 g of milk from room temperature (20°C) to 80°C;
  (ii) the mass of steam condensed.
  (b) A student measures the temperature of the hot coffee as it cools. The results are given below:

Time (min)	0	2	4	6	8
Temp (°C)	78	66	56	48	41

A friend suggests that the rate of cooling is exponential.
  (i) Show quantitatively whether this suggestion is valid.
  (ii) Estimate the temperature of the coffee after a total of 12 min.
(Specific heat capacity of milk = $4 \cdot 0 \text{ kJ kg}^{-1} \text{ K}^{-1}$, specific heat capacity of water = $4 \cdot 2 \text{ kJ kg}^{-1} \text{ K}^{-1}$, specific latent heat of steam = $2 \cdot 2 \text{ MJ kg}^{-1}$.) (*O. & C.*)

6 In the absence of bearing friction, a winding engine would raise a cage weighing 1000 kg at 10 m s^{-1}, but this is reduced by friction to 9 m s^{-1}. How much oil, initially at 20°C, is required per second to keep the temperature of the bearings down to 70°C? (Specific heat capacity of oil = $2100 \text{ J kg}^{-1} \text{ K}^{-1}$; $g = 9 \cdot 81 \text{ m s}^{-1}$.) (*O. & C.*)

7 What is meant by the specific latent heat of vaporisation of a liquid? Explain how latent heat of vaporisation can be regarded as molecular potential energy.
  Calculate the potential energy per molecule released when 18 g of steam condenses to water at 100°C. (Specific latent heat of vaporisation of water = $2 \cdot 26 \times 10^6 \text{ J kg}^{-1}$. Mass of 1 mole of water = 18 g. Number of molecules in a mole of molecules = $6 \cdot 02 \times 10^{23}$.) (*L.*)

8 Give an account of an electrical method of finding the specific latent heat of vaporisation of a liquid boiling at about 60°C. Point out any cases of inaccuracy and explain how to reduce their effect.
  Ice at 0°C is added to 200 g of water initially at 70°C in a vacuum flask. When 50 g of ice has been added and has all melted the temperature of the flask and contents is 40°C. When a further 80 g of ice has been added and has all melted the temperature of the whole becomes 10°C. Calculate the specific latent heat of fusion of ice, neglecting any heat lost to the surroundings.
  In the above experiment the flask is well shaken before taking each temperature reading. Why is this necessary? (*C.*)

9 Figure 25A represents a laboratory apparatus for determining the specific latent heat of evaporation of water. Draw a diagram of a suitable electric circuit in which the heater coil could be incorporated, and describe briefly how the experiment is carried out and how the result is calculated from the observations made.

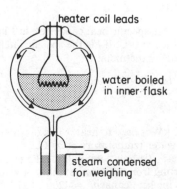

Figure 25A

A pupil performing the experiment finds that, when the heat supply is 16 W, it takes 30 minutes for the temperature of the water to rise from 20°C to 100°C, and that the rate of evaporation is very slow even at the latter temperature. Estimate an upper limit to the value of the heat capacity of the inner flask and its contents. Calculate the mass of water collected after 30 minutes of steady boiling when the power supply is 60 W. (Take specific latent heat of evaporation of water to be 2.26×10^6 J kg^{-1}.) (*O.*)

10 Give an account of a method of determining the specific latent heat of evaporation of water, pointing out the ways in which the method you describe achieves, or fails to achieve, high accuracy.

A 600 W electric heater is used to raise the temperature of a certain mass of water from room temperature to 80°C. Alternatively, by passing steam from a boiler into the same initial mass of water at the same initial temperature the same temperature rise is obtained in the same time. If 16 g of water were being evaporated every minute in the boiler, find the specific latent heat of steam, assuming that there were no heat losses. (*O. & C.*)

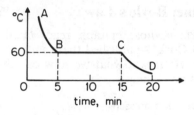

Figure 25B

11 Describe, giving the necessary theory, how the specific latent heat of a liquid may be determined by a method which involves a constant rate of evaporation. What are the advantages of this method over a method of mixtures?

In calorimetry experiments, a cooling correction has to be applied for heat losses. Explain how this is done in the experiment you describe.

Figure 25B shows a cooling curve for a substance which, starting as a liquid, eventually solidifies. Explain the shape of the curve and use the following data to obtain a value for the specific latent heat of fusion of the substance: room temperature = 20°C, slope of tangent to curve when temperature is 70°C = 10 K min^{-1}, specific heat capacity of liquid = 2.0×10^3 J kg^{-1} K^{-1}, mass of liquid = 1.50×10^{-2} kg. (You may assume that Newton's law of cooling holds.) (*L.*)

26 Gas Laws, Basics of Thermodynamics

The Gas Laws

In this chapter we shall first discuss the relationship between the temperature, pressure and volume of a gas. Unlike the case of a solid or liquid this can be expressed in very simple laws, called the Gas Laws, and reduced to a simple equation called the Ideal Gas Equation. We shall then deal with the relation between heat, internal energy of a gas and work done by a gas in basic thermodynamics.

Pressure and Volume: Boyle's Law

In 1660 ROBERT BOYLE—whose epitaph reads 'Father of Chemistry, and Nephew of the Earl of Cork'—published the results of his experiments on the 'natural spring of air'. He meant what we now call the relationship between the pressure of air and its volume.

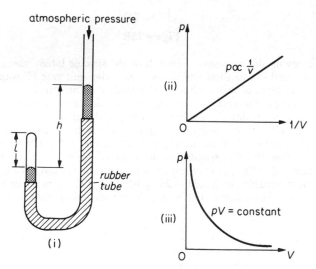

Figure 26.1 *Boyle's law apparatus and results*

We can repeat Boyle's experiment with the apparatus shown in Figure 26.1 (i), which contains dry air above mercury in a closed uniform tube. We set the open limb of the tube at various heights above and below the closed limb and measure the difference in level, h, of the mercury. When the mercury in the open limb is *below* that in the closed, we reckon h as negative. At each value of h we measure the corresponding length l of the air column in the closed limb. To find the pressure of the air we add the difference in level h to the height of the barometer, H; their sum gives the pressure p of the air in the closed limb:

$$p = g\rho(H + h)$$

where g is the acceleration of gravity and ρ is the density of mercury.

If A is the area of cross-section of the closed limb, the volume of the trapped air is

$$V = lA$$

To interpret our measurements we may either plot $H + h$, which is a measure of p, against $1/l$, or make a table of the product $(H + h)l$. We find that the plot is a straight line, and therefore

$$(H + h) \propto \frac{1}{l} \qquad \cdots \qquad \cdots \qquad (1)$$

Alternatively, we find

$$(H + h)l = \text{constant} \cdot \qquad \cdots \qquad \cdots \qquad (2)$$

which means the same as (1).

Since g, ρ, and A are constants, the relationships (1) and (2) give

$$p \propto \frac{1}{V}$$

or $\qquad\qquad\qquad\qquad$ **$pV =$ constant**

We shall see shortly that the pressure of a gas depends on its temperature as well as its volume. To express the results of the above experiments, we say that

the pressure of a given mass of gas, at constant temperature, is inversely proportional to its volume, or $pV =$ constant.

This is *Boyle's law*. See Figure 26.1 (ii), (iii).

Examples on Boyle's Law

1 A faulty barometer tube has some air at the top above the mercury. When the length of the air column is 250 mm, the reading of the mercury above the outside level is 750 mm. When the length of the air column is decreased to 200 mm, by depressing the barometer tube further into the mercury, the reading of the mercury above the outside level becomes 746 mm. Calculate the the atmospheric pressure.

Initial volume V_1 of air is proportional to the length 250 mm. Initial air pressure $p_1 = (A - 750)$ mmHg (mm of mercury), where A is the atmospheric pressure. Also, new volume V_2 of air is proportional to length 200 mm, and new pressure of air, $p_2 = (A - 746)$ mmHg.

From Boyle's law, $\qquad\qquad\qquad p_1V_1 = p_2V_2$

So $\qquad\qquad\qquad 250 \times (A - 750) = 200 \times (A - 746)$

Thus $\qquad\qquad\qquad 5(A - 750) = 4(A - 746)$

So $\qquad\qquad\qquad 5A - 4A = A = 3750 - 2984 = 766 \text{ mmHg}$

2 A vacuum pump has a cylinder of volume v and is connected to a vessel of volume V to pump out air from the vessel. The initial pressure in the vessel is p. Calculate the reduced pressure after n strokes of the pump.

What number of strokes is needed to reduce the pressure from $2 \cdot 0 \times 10^5$ Pa to $1 \cdot 0 \times 10^3$ Pa if the closed vessel has a volume of 200 cm^3 and the pump cylinder is 20 cm^3?

On the first part of the stroke when the pump piston goes down, the air in the vessel expands from V to $(V + v)$. So, from Boyle's law, the reduced pressure p_1 is given by

$$p_1(V + v) = pV$$

So

$$p_1 = p\left(\frac{V}{V + v}\right) \qquad . \qquad . \qquad . \qquad . \qquad (1)$$

On the next part of the stroke, when the piston rises, the air drawn in is pushed out into the atmosphere by the pump, leaving a pressure p_1 in the vessel. After the second stroke, the reduced pressure p_2 is given, from (1), by

$$p_2 = p_1\left(\frac{V}{V + v}\right) = p\left(\frac{V}{V + v}\right)^2$$

So after n strokes, the reduced pressure p_n is given by

$$p_n = p\left(\frac{V}{V + v}\right)^n \qquad . \qquad . \qquad . \qquad . \qquad (2)$$

Calculation Since $V = 200 \text{ cm}^3$, $v = 20 \text{ cm}^3$, $p = 2 \cdot 0 \times 10^5 \text{ Pa}$, $p_n = 1 \cdot 0 \times 10^3 \text{ Pa}$, from (2),

$$1 \cdot 0 \times 10^3 = 2 \cdot 0 \times 10^5 \left(\frac{200}{200 + 20}\right)^n$$

So

$$\frac{1}{200} = \left(\frac{200}{220}\right)^n = \left(\frac{10}{11}\right)^n$$

Taking logs to base 10, $\log(1/200) = n(\log 10 - \log 11)$

$$\therefore -2 \cdot 301 = n(1 - 1 \cdot 041)$$

$$\therefore n = \frac{2 \cdot 301}{0 \cdot 041} = 56 \text{ strokes}$$

Volume and Temperature: Charles's Law

Measurements of the change in volume of a gas with temperature, at constant pressure, can be made with the apparatus shown in Figure 26.2.

Dry air is trapped by mercury in the closed limb C of the tube AC; a scale engraved upon C enables us to measure the length of the air column, l. The tube is surrounded by water-bath W, which we can heat by passing in steam. After making the temperature uniform by stirring, we level the mercury in the limbs A and C, by pouring mercury in at A, or running it off at B. The air in C is then always at atmospheric (constant) pressure. We measure the length l and plot it against the temperature, θ (Figure 26.3).

If A is the constant cross-section of the tube, the volume of the trapped air is

$$V = lA$$

The cross-section A, and the distance between the divisions on which we read l, both increase with the temperature θ. But their increases are very small compared with the expansion of the gas. So we may say that the volume of the gas is proportional to the scale-reading of l. The graph then shows that the volume of the gas, at constant pressure, increases *uniformly* with its temperature as measured on the mercury-in-glass scale. A similar result is obtained with twice the mass of gas, as shown in Figure 26.3.

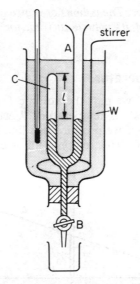

Figure 26.2 *Charles's law experiment*

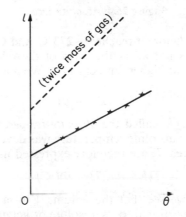

Figure 26.3 *Results of experiment*

Expansitivity of Gas (Constant Pressure)

The rate at which the volume of a gas increases with temperature can be defined by a quantity called its *expansivity at constant pressure*, α_p.

$$\alpha_p = \frac{\text{volume at } \theta°\text{C} - \text{volume at } 0°\text{C}}{\text{volume at } 0°\text{C}} \times \frac{1}{\theta}$$

Thus, if V is the volume at $\theta°$C, and V_0 the volume at $0°$C, then

$$\alpha_p = \frac{V - V_0}{V_0\theta}$$

or $$V = V_0(1 + \alpha_p\theta)$$

Charles found that α_p had the same numerical value, $\frac{1}{273}$, for all gases. This so-called Charles's law states: *The volume of a given mass of any gas, at constant pressure, increases by $\frac{1}{273}$ of its volume at 0°C, for every degree Celsius rise in temperature.*

Absolute (Kelvin) Temperature

Charles's law shows that, if we plot the volume V of a given mass of any gas at constant pressure against its temperature θ, we get a straight line graph A as shown in Figure 26.4. If we produce this line backwards, it will meet the temperature axis at $-273°C$. This temperature is called the *absolute zero*. If a

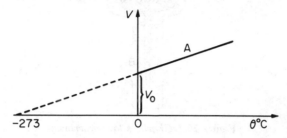

Figure 26.4 *Absolute zero*

gas is cooled, it liquefies before it reaches $-273°C$, and Charles's law no longer holds; but that fact does not affect the form of the relationship between the volume and temperature at higher temperatures. We may express this relationship by writing

$$V \propto (273 + \theta)$$

The quantity $(273 + \theta)$ is called the *absolute temperature* of the gas, and is denoted by T. The idea of absolute temperature was developed by Lord Kelvin, and absolute temperatures, T, are therefore expressed in kelvin (K). So

$$T(K) = (273 + \theta)(°C)$$

From Charles's law, we see that the volume V of a given mass of gas at constant pressure is proportional to its absolute or kelvin temperature T, since

---

$$V \propto (273 + \theta) \propto T$$

So, if a given mass of gas has a volume V_1 at $\theta_1°C$, and is heated at constant pressure to $\theta_2°C$, its new volume is given by

$$\frac{V_1}{V_2} = \frac{273 + \theta_1}{273 + \theta_2} = \frac{T_1}{T_2}$$

Pressure and Temperature

The effect of temperature on the *pressure* of a gas, at constant volume, can be investigated with the apparatus shown in Figure 26.5 (i). The bulb B contains air, which can be brought to any temperature θ by heating the water in the surrounding bath W. When the temperature is steady, the mercury in the closed limb of the tube is brought to a fixed level D, so that the volume of the air is fixed. The difference in level, h of the mercury in the open and closed limbs is then added to the height of the barometer, H, to give the pressure p of the gas

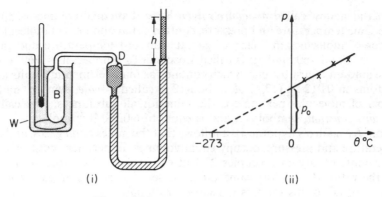

Figure 26.5 *Pressure and temperature*

in mm of mercury. If p, $(h + H)$, is plotted against the temperature, the plot is a straight line passing through $-273°C$ approximately. This is a similar result to the case of volume V at constant pressure plotted against $\theta°C$. Figure 26.5 (ii).

We may therefore say that, at constant volume, the pressure of a given mass of gas is proportional to its absolute or kelvin temperature T, since

$$p \propto (273 + \theta) \propto T$$

We can now summarise our results.

Gas Laws: At constant pressure, $V \propto T$ (kelvin) or $V_1/V_2 = T_1/T_2$
At constant volume, $p \propto T$ (kelvin) or $p_1/p_2 = T_1/T_2$
At constant temperature, $p \propto 1/V$ or $pV = $ constant (Boyle)

Ideal Gas Equation, the Gas Constant

In practice, real gases obey the gas laws only at moderate pressures and at temperatures well above the temperature at which the gas would liquefy. To simplify matters, we shall deal with an *ideal gas*, which may be defined as one which obeys Boyle's law, $pV = $ constant at constant temperature, and the constant volume gas law, $p \propto T$.

For an ideal gas, one equation connects the pressure p, volume V and the kelvin temperature T. This is

$$\frac{pV}{T} = \text{constant} \qquad . \qquad . \qquad . \qquad . \qquad . \qquad (1)$$

We can see this equation fits all the gas laws because, from (1),　(a) $pV = $ constant when T is constant,　(b) $p/T = $ constant when V is constant,　(c) $V/T = $ constant when p is constant. As we soon see, the size of the constant in (1) depends on the amount of gas used and the nature of the gas.

Avogadro's Hypothesis: Molar Gas Constant

Avogadro, with one simple-looking idea, illuminated chemistry as Newton illuminated mechanics. In 1811 he suggested that chemically active gases, such as oxygen, existed not as single atoms, but as pairs: he proposed to distinguish between an atom, O, and a molecule, O_2. In 1814 Avogadro also put forward another idea, now called *Avogadro's hypothesis*: that equal volumes of all gases, at the same temperature and pressure, contain equal numbers of molecules. The number of molecules in 1 cm³ of gas at standard temperature and pressure (s.t.p.—0°C and 760 mmHg) is called Loschmidt's number; it is $2 \cdot 69 \times 10^{19}$.

The amount of a substance which contains as many elementary units as there are atoms in 0·012 kg (12 g) of carbon-12 is called a *mole*, symbol 'mol'. The number of molecules per mole is the same for all substances. It is called the *Avogadro constant*, symbol N_A, and is equal to $6 \cdot 02 \times 10^{23}$ mol^{-1}.

From Avogadro's hypothesis, it follows that the moles of *all* gases, at the same temperature and pressure, occupy equal volumes. Experiment confirms this; at s.t.p. 1 mole of any gas occupies 22·4 litres. So if V is the volume of 1 mole, then the ratio pV/T is the same for all gases. So the molar gas constant R, which is equal to this ratio, is the *same for all gases*.

At s.t.p.
$$V = 22 \cdot 4 \text{ litres} = 22 \cdot 4 \times 10^{-3} \text{ m}^3$$

$$p = 760 \text{ mmHg} = 1 \cdot 013 \times 10^5 \text{ Pa}$$

$$T = 273 \text{ K}$$

$$\therefore R = \frac{pV}{T} = \frac{1 \cdot 013 \times 10^5 \times 22 \cdot 4 \times 10^{-3}}{273}$$

$$= 8 \cdot 31 \text{ J K}^{-1} \text{ mol}^{-1}$$

General Gas Equation

At the same temperature and pressure, the volume of two moles of a gas is twice the volume of one mole. So for two moles, $pV/T = 2R$, or $pV = 2RT$. For n moles, the gas equation is

$$pV = nRT \qquad . \qquad . \qquad . \qquad . \qquad . \qquad (2)$$

This equation is widely used in problems involving gases.

If the mass of a gas is m, and M is the mass of 1 mole of the gas, then the number of moles $n = m/M$. For example, for 4 g of hydrogen whose molar mass is 2 g, $n = 4 \text{ g}/2 \text{ g} = 2$ and the gas equation is therefore $pV = 2RT$. So with a mass m,

$$pV = nRT = \frac{m}{M} RT \qquad . \qquad . \qquad . \qquad (3)$$

The density ρ of a gas $= m/V$. So from (3),

$$p = \frac{m}{V} \frac{RT}{M} = \rho \frac{RT}{M} \qquad . \qquad . \qquad . \qquad (4)$$

At constant temperature, since RT and M are all constants, we see that the pressure p of a gas is proportional to its density ρ.

Applying the Gas Equation

We can apply the equation to calculate the mass of a gas. Suppose oxygen gas, contained in a cylinder of volume V of 1×10^{-2} m³ has a temperature T of 300 K and a pressure p_1 of $2 \cdot 5 \times 10^5$ Pa. After some of the oxygen is used at constant temperature, the pressure falls to $1 \cdot 3 \times 10^5$ Pa, p_2.

From $pV = nRT$, the number of moles in the cylinder initially is given by

$$n_1 = \frac{p_1 V}{RT}$$

and finally by

$$n_2 = \frac{p_2 V}{RT}$$

So the number of moles used $= n_1 - n_2$

$$= \frac{(p_1 - p_2)V}{RT}$$

$$= \frac{(2 \cdot 5 - 1 \cdot 3) \times 10^5 \times 1 \times 10^{-2}}{8 \cdot 3 \times 300} = 0 \cdot 48$$

using the value $R = 8 \cdot 3 \, \text{J mol}^{-1} \, \text{K}^{-1}$.

But   molar mass of oxygen $= 32 \, \text{g} = 32 \times 10^{-3} \, \text{kg}$

$\therefore$ mass of oxygen used $= 0 \cdot 48 \times 32 \times 10^{-3} \, \text{kg} = 0 \cdot 015 \, \text{kg}$ (approx.)

Connected Gas Containers

As we have seen, the number of moles, n, in a gas can be found from the relation

$$n = \frac{pV}{RT}$$

In a closed system of connected gas containers, some gas may flow out of one container when its temperature rises, to other containers. The *total* number of moles of gas in all the containers remains constant, however, no matter what changes take place in individual containers. So, in a closed system, the sum of the values of pV/RT for all the containers is constant. This is used in the following example.

Example on Connected Gas Containers

Two gas containers with volumes of $100 \, \text{cm}^3$ and $1000 \, \text{cm}^3$ respectively are connected by a tube of negligible volume, and contain air at a pressure of $1000 \, \text{mmHg}$. If the temperature of both vessels is originally $0°C$, how much air will pass through the connecting tube when the temperature of the smaller is raised to $100°C$? Give your answer in cm^3 measured at $0°C$ and $760 \, \text{mmHg}$.

The *total* number of moles of air in the two containers remains constant, although some air is transferred from the hot to the cold container on heating.
From $n = pV/RT$,

$$\text{initially, total moles of air} = \frac{pV}{RT} = \frac{1000 \times 1100}{R \times 273}$$

$$\text{and finally, total moles of air} = \frac{p \times 100}{R \times 373} + \frac{p \times 1000}{R \times 273}$$

where p is the new pressure in both containers after heating.

Equating the two numbers of moles, which are unchanged by any transfer,

$$\frac{p \times 100}{R \times 373} + \frac{p \times 1000}{R \times 273} = \frac{1000 \times 1100}{R \times 273}$$

Cancelling R and simplifying, then

$$p = \frac{11\,000 \times 373}{4003} = 1025 \text{ mmHg}$$

As the question requires, we now have to convert the initial volume of 100 cm³ of air in the smaller container at 0°C and 1000 mmHg to 0°C and 760 mmHg, and the final volume of 100 cm³ at 100°C and 1025 mmHg to 0°C and 760 mmHg.

$$\text{Initial volume at 0°C and 760 mmHg} = 100 \times \frac{1000}{760} = 131 \cdot 6 \text{ cm}^3$$

$$\text{Final volume at 0°C and 760 mmHg} = 100 \times \frac{1025}{760} \times \frac{273}{373} = 98 \cdot 7 \text{ cm}^3$$

$\therefore$ volume of air at 0°C and 760 mmHg flowing out = 33 cm³ (approx.)

Mixture of Gases: Dalton's Law

Figure 26.6 shows an apparatus with which we can study the pressure of a mixture of gases. A is a bulb, of volume V_1, containing air at atmospheric pressure, p_1. C is another bulb, of volume V_2, containing carbon dioxide at a pressure p_2. The pressure p_2 is measured on the manometer M; in millimetres of mercury it is

$$p_2 = h + H$$

where H is the height of the barometer. (In the same units, the air pressure, $p_1 = H$.)

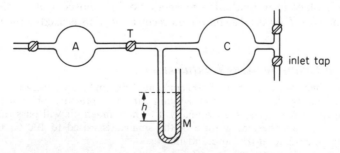

Figure 26.6 *Apparatus for demonstrating Dalton's law of partial pressure*

When the bulbs are connected by opening the tap T, the gases mix, and reach the same pressure, p; this pressure is given by the new height of the manometer. Its value is found to be given by

$$p = p_1 \frac{V_1}{V_1 + V_2} + p_2 \frac{V_2}{V_1 + V_2}$$

Now the quantity $p_1 V/(V_1 + V_2)$ is the pressure which the air originally in A would have, if it expanded to occupy A and C; for, if we denote this pressure by p', then $p'(V_1 + V_2) = p_1 V_1$. Similarly $p_2 V_2 / (V_1 + V_2)$ is the pressure which the carbon dioxide originally in C would have, if it expanded to occupy A and C. Thus the total pressure of the mixture is the sum of the pressures which the individual gases exert, when they have expanded to fill the vessel containing the mixture.

The pressure of an individual gas in a mixture is called its *partial pressure*: it is the pressure which would be observed if that gas alone occupied the volume of the mixture, and had the same temperature as the mixture. The experiment described shows that **the pressure of a mixture of gases is the sum of the partial pressures of its parts**. This statement was first made by Dalton, in 1801, and is called *Dalton's law of partial pressures*.

So if air and water vapour have a total pressure of $1·01 \times 10^5$ Pa in a container, and the water vapour alone has a pressure of $0·01 \times 10^5$ Pa, then the air pressure in the mixture has a pressure $= 1·01 \times 10^5 - 0·01 \times 10^5 = 1·00 \times 10^5$ Pa. The volume occupied by either the air or the water vapour is the whole volume of the container.

Unsaturated and Saturated Vapours

An *unsaturated* vapour is a vapour which is not in contact with its own liquid in a closed space. So a mixture of air and water vapour in a flask, without any water present, is a mixture of air and unsaturated vapour.

A *saturated* vapour is a vapour in contact with, or in equilibrium with, its own liquid. So if a flask contains air and some water at 20°C, the water vapour above the water is saturated vapour at 20°C. The mixture, then, is air and saturated water vapour. Values of saturated vapour pressure (s.v.p.) vary from a very low value at 0°C to 760 mmHg ($1·013 \times 10^5$ Pa) at 100°C. If the external atmospheric pressure is 760 mmHg, water will boil at 100°C. The bubbles inside the water then have sufficient vapour pressure just to overcome the outside atmospheric pressure and burst open at the surface.

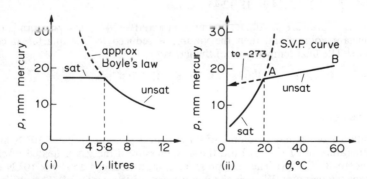

Figure 26.7 *Effect of volume and temperature on pressure of water vapour*

Gas Laws for Vapours

Saturated vapours do not obey Boyle's law: *their pressure is independent of their volume*. Unsaturated vapours obey Boyle's law roughly, as they also obey roughly Charles's law, Figure 26.7 (i). Vapours, saturated and unsaturated, are gases because they spread throughout their vessels; but we find it convenient to distinguish them by name from gases such as air, which obey Charles's and Boyle's laws closely.

Figure 26.7 (ii) shows the effect of heating a saturated vapour. More and more of the liquid evaporates, and the pressure rises very rapidly. As soon as all the liquid has evaporated, however, the vapour becomes unsaturated, and its pressure rises more readily along a straight line AB. Well away from the saturated state, the unsaturated vapour obeys the relation $p \propto T$ at constant volume, which is a gas law.

EXERCISES 26A   Gas Laws

Multiple Choice

1 Which of the graphs of pV against T, the kelvin temperature, in **A** to **E** is correct for an ideal gas in Figure 26A?

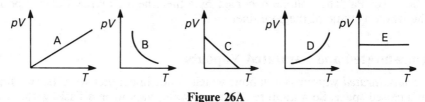

Figure 26A

2 At 27°C and 1.0×10^5 Pa pressure, air has a density of 0.09 kg m^{-3}. At 87°C and 2.0×10^5 Pa pressure, the air density in kg m^{-3} is

   A 0·45   **B** 0·40   **C** 0·35   **D** 0·15   **E** 0·10

3 The gas constant k for 2 moles of hydrogen, 4 g, is 16·6 in the equation $pV = kT$ where p is the gas pressure in Pa, V is the volume in m^3 and T is the kelvin temperature. The gas constant for 3 moles of nitrogen, 84 g, with the same units as hydrogen, is

   A 66·4   **B** 33·2   **C** 24·9   **D** 12·45   **E** 6·6

4 A gas has a pressure p, volume V and temperature 300 K. The pressure is doubled at constant volume and then the volume is reduced to one quarter at constant pressure. The final temperature of the gas is

   A 600 K   **B** 450 K   **C** 300 K   **D** 250 K   **E** 150 K

Longer Questions

5 A container of gas has a volume of 0.10 m^3 at a pressure of $2.0 \times 10^5 \text{ N m}^{-2}$ and a temperature of 27°C.   (i) Find the new pressure if the gas is heated at constant volume to 87°C.   (ii) The gas pressure is now reduced to $1.0 \times 10^5 \text{ N m}^{-2}$ at constant temperature. What is the new volume of the gas?   (iii) The gas is cooled to -73°C at constant pressure. Find the new volume of the gas.

6 Using a vertical (y) axis to represent pressure, p, and a horizontal axis to represent volume, V, illustrate by rough sketches the changes which take place in (i), (ii) and (iii) in question 5.

7 A cylinder of gas has a mass of 10·0 kg and a pressure of 8·0 atmospheres at 27°C. When some gas is used in a cold room at -3°C, the gas remaining in the cylinder at this temperature has a pressure of 6·4 atmospheres. Calculate the mass of gas used.

8 Two glass bulbs of equal volume are joined by a narrow tube and are filled with a gas at s.t.p. When one bulb is kept in melting ice and the other is placed in a hot bath, the new pressure is 877·6 mm mercury. Calculate the temperature of the bath. (*L.*)

9 What volume of liquid oxygen (density 1140 kg m^{-3}) may be made by liquefying completely the contents of a cylinder of gaseous oxygen containing 100 litres of oxygen at 120 atmospheres pressure and 20°C? Assume that oxygen behaves as an ideal gas in this region of pressure and temperature.
(1 atmosphere $= 1 \cdot 01 \times 10^5 \text{ N m}^{-2}$; molar gas constant $= 8 \cdot 31 \text{ J mol}^{-1} \text{ K}^{-1}$; relative molecular mass of oxygen $= 32 \cdot 0$.) (*O. & C.*)

10 Explain the following observations:
(a) when pumping up a bicycle tyre the pump barrel gets warm;
(b) when a gas at high pressure in a container is suddenly released, the container cools.

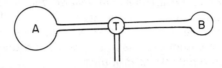

Figure 26B

Two bulbs, A of volume 100 cm^3 and B of volume 50 cm^3, are connected to a three way tap T which enables them to be filled with gas or evacuated. The volume of the tubes may be neglected, Figure 26B.
(c) Initially bulb A is filled with an ideal gas at 10°C to a pressure of $3 \cdot 0 \times 10^5$ Pa. Bulb B is filled with an ideal gas at 100°C to a pressure of $1 \cdot 0 \times 10^5$ Pa. The two bulbs are connected with A maintained at 10°C and B at 100°C. Calculate the pressure at equilibrium.
(d) Bulb A, filled at 10°C to a pressure of $3 \cdot 0 \times 10^5$ Pa, is connected to a vacuum pump with a cylinder of volume 20 cm^3. Calculate the pressure in A after one inlet stroke of the pump. The air in the pump is now expelled into the atmosphere. Calculate the pressure in A after the second inlet stroke. Calculate the number of strokes of the pump to reduce the pressure in A to $1 \cdot 0 \times 10^3$ Pa. The whole system is maintained at 10°C throughout the process. (*O. & C.*)

Basic Thermodynamics

We now deal with the relation between work and heat energy. This is a branch of *thermodynamics*, a subject widely used by engineers in their studies and researches into engines, for example. Here we are concerned only with the main principles.

Work Done by Gas

Consider some gas, at a pressure p, in a cylinder fitted with a piston (Figure 26.8).
If the piston has an area A, the force on it is

$$f = pA$$

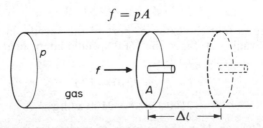

Figure 26.8 *Work done by gas in expansion*

If we allow the piston to move outwards a distance Δl, the gas will expand, and its pressure will fall. But by making the distance very short, we can make the fall in pressure so small that we may consider the pressure constant. The force f is then constant, and the work done is

$$\Delta W = f . \Delta l = pA . \Delta l$$

The product $A . \Delta l$ is the increase in volume, ΔV, of the gas, so that

$$\Delta W = p . \Delta V \qquad . \qquad . \qquad . \qquad . \qquad (1)$$

The product of pressure and volume, in general, therefore represents work. If the pressure p is in Pa or N m^{-2}, and the area A is in m^2, the force f is in newtons. And if the movement Δl is in m, the work $f . \Delta l$ is in newton $\times$ metre or *joules* (J). The increase of volume, ΔV, is in m^3. So the product of pressure in Pa, and volume in m^3, represents work in joules.

From (1),   **work done = pressure $\times$ volume change**

So if the volume of a gas at constant pressure of 10^5 Pa expands by 0·01 m^3, then

$$\text{work done} = 10^5 \times 0·01 = 1000 \text{ J}$$

Latent Heat and Internal Energy, Molecular Potential Energy

The volume of 1 g of steam at 100°C and 760 mmHg pressure is 1672 cm^3. Therefore when 1 g of water turns into steam, it expands by 1671 cm^3; in doing so, it does work against the atmospheric pressure. The heat equivalent of this work is that part of the latent heat which must be supplied to the water to make it overcome atmospheric pressure as it evaporates; it is called the 'external latent heat'. The rest of the specific latent heat—the internal part—is the equivalent of the work done in separating the molecules against their mutual attractions so that the liquid changes to vapour. This amount of energy, then, is a measure of the *increase in potential energy* of the molecules of water in the gaseous state over that in the liquid state, at the same temperature.

The work done, W, in the expansion of 1 g from water to steam is the product of the atmospheric pressure p and the increase in volume ΔV:

$$W = p . \Delta V$$

Normal atmospheric pressure corresponds to a pressure $p = 1·013 \times 10^5$ Pa,

So $W = p . \Delta V = 1·013 \times 10^5 \times 1671 \times 10^{-6} = 170 \text{ J}$

The external specific latent heat is therefore

$$l_{ex} = 1·013 \times 10^5 \times 1671 \times 10^{-6}$$
$$= 170 \text{ J g}^{-1} = 170 \text{ kJ kg}^{-1}$$

This result shows that the external part of the specific latent heat is much less than the internal part. Since the total specific latent heat of vaporisation l is 2270 J g^{-1}, the internal part is

$$l_{in} = l - l_{ex} = 2270 - 170$$
$$= 2100 \text{ J g}^{-1} = 2100 \text{ kJ kg}^{-1}$$

For 1 mole of water, 18 g, the number of molecules is about 6×10^{23}, the Avogadro constant. So the gain in potential energy of a molecule of water when changing from liquid at 100°C to vapour at 100°C is

$$\frac{18 \times 2100}{6 \times 10^{23}} = 6{\cdot}3 \times 10^{-20} \text{ J molecule}^{-1}$$

Internal Energy of Gas

The *internal energy* of an ideal gas is the kinetic energy of thermal motion of its molecules. As we see later, the magnitude depends on the temperature of the gas and on the number of atoms in its molecule.

The thermal motion of the molecules is a random motion—it is often called the thermal 'agitation' of the molecules. We must appreciate that the energy of the gas which we call internal energy is quite independent of any motion of the gas in bulk. When a cylinder of oxygen is being carried by an express train, its kinetic energy as a whole is greater than when it is standing on the platform; but the random motion of the molecules relative to the cylinder is unchanged—and so is the temperature of the gas.

The same is true of a liquid. In a water-churning experiment to convert mechanical energy into heat, baffles must be used to prevent the water from acquiring any mass-motion—all the work done must be converted into random motion, if it is to appear as heat. Likewise, the internal energy of a solid is the kinetic energy of the vibration of its atoms about their mean positions. Throwing a lump of metal through the air does not raise its temperature, but hitting it with a hammer does.

We shall use the symbol U for the internal energy. Although its absolute magnitude is not known, we are mainly concerned with *changes* in internal energy, denoted by ΔU. We can often calculate ΔU, as we soon show.

First Law of Thermodynamics

The first law of thermodynamics states that the *total energy in a closed system is constant*, that is, the energy is conserved in any transfer of energy from one form to another. This law is also known as the principle of conservation of energy (p. 629).

When we warm a gas so that it expands, the heat ΔQ we give to it appears partly as an increase ΔU to its *internal energy*—and hence its temperature—and partly as the energy needed for the *external work* done, ΔW. Thus from the first law of thermodynamics, we may write

$$\Delta Q = \Delta U + \Delta W \quad . \qquad . \qquad . \qquad . \qquad . \qquad (1)$$

If the expansion of the gas occurs and no friction forces are present, $\Delta W = p.\Delta V$. So, from (1),

$$\Delta Q = \Delta U + p.\Delta V \quad . \qquad . \qquad . \qquad . \qquad . \qquad (2)$$

Equation (2), then, is a mathematical statement of the first law of thermodynamics applied to the case of energy changes associated with a gas.

In the relation $\Delta Q = \Delta U + \Delta W$, ΔW is the external work done *by* the gas. We can express the relation in a different way if heat ΔQ is given to the gas and ΔW is the external work done *on* the gas when it is compressed, for example. In this case, the increase in the internal energy of the gas,

ΔU, is given by

$$\Delta U = \Delta Q + \Delta W \quad . \qquad . \qquad . \qquad . \qquad . \qquad (3)$$

The work done on the gas is $p \cdot \Delta V$ and so in this case

$$\Delta U = \Delta Q + p \cdot \Delta V \quad . \qquad . \qquad . \qquad . \qquad (4)$$

In electrical heating of a wire, ΔW = work done on the electrons as they flow = energy supplied. From (3), when a lamp filament is *first* switched on, $\Delta W = \Delta U$. So the temperature rises. The filament reaches a constant temperature a short time later. So $\Delta U = 0$ and $\Delta W = -\Delta W = -\Delta Q$ = heat lost by radiation.

Internal Energy Changes, Ideal Gas

If the volume of a gas is kept *constant* when it is warmed, then no external work is done. So *all* the heat supplied goes in raising the internal energy of the gas. In this special case, then, $\Delta U = \Delta Q$, the heat supplied. The change in internal energy of an ideal gas is proportional to its temperature change ΔT, no matter how the change has occurred. *The internal energy of an ideal gas is independent of its volume*—it depends only on its temperature.

We can define an ideal gas as one which obeys Boyle's law exactly and whose internal energy is independent of its volume. No such gas exists, but at room temperature, and under moderate pressures, many gases approach the ideal closely enough for most purposes.

Suppose a gas in a vessel is thermally insulated and the gas expands. In this case *no heat* enters or leaves the system. The work done by the gas is then taken from its internal energy, with the result that the gas *cools*. If the gas is compressed instead of expanding, the work done on the gas increases the internal energy. So this time the gas temperature *rises*. In either case the change in internal energy ΔU is proportional to its temperature change ΔT.

Work Done from p–V Graphs

The work done by or on a gas can be found from a graph of pressure p against volume V when the work is done. So p–V graphs of gases are studied in the design of engines.

In Figure 26.9 (i), a gas has pressure, volume and temperature represented by

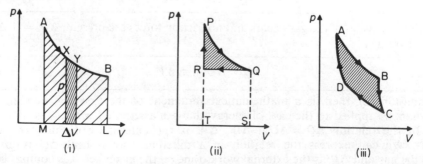

Figure 26.9 *Work done from p–V graphs*

the point A and then does work and expands to a new volume and a new pressure and temperature represented by the point B. For a *very* small volume change V from X to Y, the small amount of work done W = pressure × volume change, from our result on page 656.

So $$W = p \cdot \Delta V$$

$$= area \text{ of shaded strip below XY}$$

From A to B, then, we can say that total work done = area below AB, ABLM

So **work done = area between p–V graph and volume-axis**

In Figure 26.9 (ii), a gas expands from P to Q, then it is compressed from Q to R while its pressure is kept constant (QR is a horizontal line), and finally the gas is compressed back to P while its volume is kept constant (RP is a vertical line). Now from P to Q, the work done *by* the gas is the area PQST. From Q to R, the done *on* the gas is the area QRTS. No work is done along PR (volume constant). So, by subtracting the two areas, as shown shaded,

---

net work done by gas = area enclosed, PQR

Figure 26.9 (iii) shows a gas taken round a so-called *cycle* of changes ABCD, from A to and back to A. As before,

net work done by gas = area ABCD

Isothermal Changes

In a car, motorcycle or aeroplane engine, gases expands and are compressed, and are heated and cooled, in ways more complicated than those already described. We now consider some chief ways in which such changes take place.

When a gas expands or is compressed at *constant temperature*, the gas is said to undergo an *isothermal expansion or compression*.

For a mole of gas, we have already seen that the pressure p, the volume V and the absolute temperature T are related by $pV = RT$. So if the temperature is constant,

$$pV = \text{constant}$$

the value of the constant depending on T and R. So the curve of pressure and volume represents a Boyle's law relation. Such a curve is called an *isothermal* for the given mass of gas. Figure 26.10 shows a family of isothermals for 1 g of air, each isothermal being labelled with the particular constant temperature. At the higher temperatures, the pressure and volume values are higher from $pV = RT$.

Kinetic Theory in Isothermal Change

When a gas expands, it does work—for example, in driving a piston. The molecules of gas bombard the piston, and if the piston moves they give up

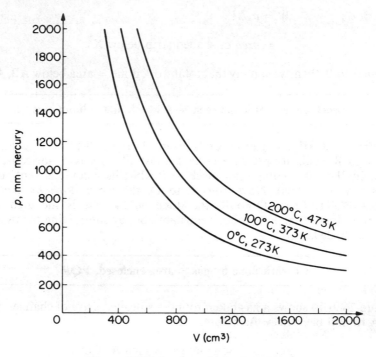

Figure 26.10 *Isothermals for 1 g of air*

some of their kinetic energy to it. When a molecule bounces off the *moving* piston, it does so with a velocity less than that with which it struck. The change in velocity is small, because the piston moves much more slowly than the molecule; but there are many molecules striking the piston at any instant, and their total loss of kinetic energy is equal to the work done in driving the piston forward. The work done by a gas expanding, therefore, is done at the expense of its internal energy. The temperature of the gas will consequently *fall* during expansion, unless heat is supplied to it.

Conversely, if a gas is compressed, its temperature rises. The molecules now rebound from the forward-moving piston with a velocity greater than their incident velocity. The total increase in kinetic energy of all the molecules is equal to the work done in moving the piston. In an isothermal compression or expansion, the gas must be held in a thin-walled, highly conducting vessel, surrounded by a constant temperature bath. And the expansion must take place slowly, so that heat can pass into the gas to maintain its temperature at every instant during the expansion.

External Work Done in Expansion

The heat taken in when a gas expands isothermally is the heat equivalent of the mechanical work done, because there is no change in internal energy of an ideal gas when its temperature is constant. As we have seen earlier, if the volume of the gas increases by a small amount ΔV, at the pressure p, then the work done is

$$\Delta W = p\Delta V$$

and in an expansion from V_1 to V_2, the work done is equal to the area between the isothermal curve and the volume-axis from V_1 to V_2.

Now let us consider an isothermal compression. When a gas is compressed, work is done on it by the compressing agent. To keep its temperature constant, therefore, heat must be withdrawn from the gas, to prevent the work done from increasing its internal energy. The gas must again be held in a thin well-conducting vessel, surrounded by a constant-temperature bath; and it must be compressed slowly.

Adiabatic Change

Let us now consider a change of volume in which the conditions are opposite to isothermal; no heat is allowed to enter or leave the gas.

An expansion or contraction in which *no heat* enters or leaves the gas is called an *adiabatic expansion or contraction*.

In an adiabatic expansion, the external work is done wholly at the expense of the internal energy of the gas, and the gas therefore cools. In an adiabatic compression, all the work done on the gas by the compressing agent appears as an increase in its internal energy and therefore as a rise in its temperature.

Heat and Mechanical Work in Engines

Finally, we consider what happens when an engine transfers heat to mechanical energy when it is working.

If one metal surface is rubbed against another, practically the whole of the mechanical work done against friction is transformed into heat. So the conversion from mechanical energy to heat can be almost 100% efficient. As we now show, however, a practical machine or *engine* which converts heat into mechanical energy, the opposite process, can never be 100% efficient.

If the engine operates in a *cycle*, so that it returns to its original state after a series of operations, then it can be made to work continuously.

Consider a gas taken through a cycle of pressure (p)–volume (V) changes. Two examples of a cycle are shown in Figure 26.11 (i) and (ii). Starting from the state represented by A, the gas expands from A to B at constant kelvin temperature T_1, Figure 26.11 (i). Here the gas takes in a quantity of heat Q_1. The gas then expands from B to C when no heat is taken in. The work done is then taken from the internal energy of the gas, so its temperature falls at C to a value T_2.

The mechanical work done *by* the gas along ABC is represented by the *area* ABCXYA between the curves and the V-axis. To restore the gas to its initial state at A the gas must now be compressed along CD and DA. In this case the work done *on* the gas is represented by the area CDAYXC. The gas now gives up a quantity of heat Q_2, whereas it takes in a quantity of heat Q_1 along ABC while doing work.

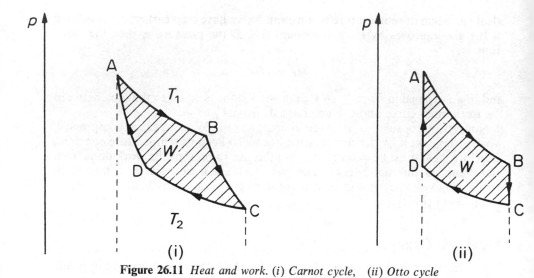

Figure 26.11 *Heat and work.* (*i*) *Carnot cycle,* (*ii*) *Otto cycle*

Since the gas returns to its initial state at A, there is no change in the internal energy of the gas at the end of a cycle. From the first law of thermodynamics, then, the net work done in a cycle = heat gained by gas = $Q_1 - Q_2$. The *net* work done W is represented by the area ABCD in both Figures 26.11 (i) and (ii).

The efficiency E of an engine is defined by

$$E = \frac{\text{work obtained}}{\text{energy supplied}} \times 100\%$$

$$= \frac{Q_1 - Q_2}{Q_1} \times 100\%$$

$$= \left(1 - \frac{Q_2}{Q_1}\right) \times 100\%$$

It follows that an engine can never be 100% efficient, that is, all the heat supplied can *never* be transferred or converted into mechanical energy during a complete cycle. This is one statement of the *second law of thermodynamics*.

Figure 26.11 (i) represents a *Carnot cycle*. Figure 26.11 (ii) represents an *Otto cycle*. Here no heat is taken in along AB and CD as the gas expands and contracts, and BC and DA are constant volume pressure changs. The Otto cycle is used in motorbikes and other engines.

With the help of an engine, refrigerator pumps use the Carnot cycle in reverse. Heat is then taken from a 'cold' object and given up to a 'warm' object. In this way water at room temperature is gradually frozen to ice in a refrigerator and the heat given up is removed to the room by metal rods outside the refrigerator.

Further information on engines must be found from specialist books.

You should know:

1 Work done by or on a gas = pressure × volume change, $p\Delta V$.
2 When water is boiling, some of the latent heat is used in doing work against the external pressure when the liquid expands to form vapour. The vapour molecules gain molecular potential energy which can be calculated using the specific latent heat value.
3 First law of thermodynamics = law of conservation of energy. It can be written:
$\Delta U = \Delta Q + \Delta W(\Delta W$ is done *on* gas) or $\Delta Q = \Delta U + \Delta W$ (work done *by* gas).
4 ΔU = internal energy change. For an ideal gas, $\Delta U \propto \Delta T$.
At isothermal (constant temperature) change, $\Delta U = 0$.
5 Work done by gas $\propto$ *area* between $p - V$ curve and V-axis.
In a complete cycle, work done $\propto$ area enclosed.
6 An engine has efficiency = $(Q_1 - Q_2)/Q_1 \times 100\%$, where Q_1 is energy taken in at high temperature and Q_2 at lower temperature.
Second law of thermodynamics: in cycles of heat and work interchanges, the efficiency can never be 100%.

Refrigerators and the Second Law

A consequence of the second law of thermodynamics (p. 662) is that it is impossible for heat to flow from a cooler body to a hotter one without putting in some work or energy. This is why refrigerators need connecting to the mains. Even though the aim is simply to transfer heat from the interior of the refrigerator to the exterior, which is a perfectly acceptable energy transfer by the first law of thermodynamics (p. 657), this cannot be accomplished without the input of some extra work. So the overall refrigerator cycle involves taking a quantity of heat ΔQ from the interior of the refrigerator and a quantity of electrical energy ΔW from the mains, and then giving out a total amount of heat $\Delta Q + \Delta W$ to the environment. Of course, we would ideally like to dispense with the ΔW term, but this is impossible by the second law of thermodynamics. This is why leaving the refrigerator door open on a hot day to cool a room does not work. Even though an amount of heat ΔQ is taken from the room per cycle, an even greater amount $\Delta Q + \Delta W$ is returned to it, so the room just gets hotter!

EXERCISES 26B Basic Thermodynamics

Multiple Choice

1 The first law of thermodynamics can be written as $\Delta U = \Delta Q + \Delta W$ for an ideal gas. Which statement in **A** to **E** is most correct?
A ΔU is always zero when no heat enters or leaves the gas.
B ΔW is the work done by the gas in this written law.
C ΔU is zero when heat is supplied and the temperature stays constant.
D $\Delta Q = -W$ when the temperature increases very slowly.
E The law is due to the properties of gases.

2 A gas in a cylinder has a pressure of $2 \cdot 0 \times 10^5$ Pa. Figure 26C (i). The piston has an area $3 \cdot 0 \times 10^{-3}$ m^2 and is pulled out slowly a distance of 10 mm. If the cylinder was insulated, the change in internal energy of the gas in J is

A $+10$ **B** $+6$ **C** -6 **D** $-0 \cdot 6$ **E** $-0 \cdot 06$

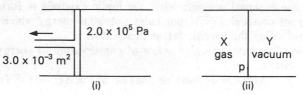

Figure 26C

3 When two metal blocks X and Y of iron and steel and different mass are in thermal equilibrium with each other, then
1 each has the same amount of internal energy
2 their temperatures are the same
3 there is no transfer of energy between them
A 1, 2, 3 are all correct **B** only 1, 2
C 2/3 **D** 1 only **E** 2 only

4 A container has one half filled with an ideal gas X separated by a plate P from the other half which contains a vacuum. Y, Figure 26C (ii). When P is removed, X moves into Y. Which statement from A to E is most correct?
A X decreases in temperature.
B X increases in internal energy.
C No work is done by X.
D X doubles in pressure.
E The container expands.

Longer Questions

5 A domestic kettle is marked 250 V, $2 \cdot 3$ kW and the manufacturer claims that it will heat a pint of water to boiling point in 94 s.
(a) Test this claim by calculation and state any simplifying assumptions you make.
(b) If the kettle is left switched on after it boils, how long will it take to boil away half a pint of water measured from when it first boils?
(c) Estimate the work done against an atmospheric pressure of 100 kPa when 1 cm^3 of water evaporates at 100°C, producing 1600 cm^3 of steam. Express this as a percentage of the total energy required to evaporate 1 cm^3 of water at 100°C.
(Specific heat capacity of water $= 4 \cdot 2 \times 10^3$ J kg^{-1} K^{-1}, specific latent heat of evaporation of water $= 2 \cdot 3 \times 10^6$ J kg^{-1}, density of water $= 1 \cdot 0$ g cm^{-3}, 1 pint $= 570$ cm^3. (*N*.)

6 The cylinder in Figure 26D holds a volume $V_1 = 1000$ cm^3 of air at an initial pressure $p_1 = 1 \cdot 10 \times 10^5$ Pa and temperature $T_1 = 300$ K. Assume that air behaves like ideal gas.
Figure 26E shows a sequence of changes imposed on the air in the cylinder.
(a) AB—the air heated to 375 K at constant pressure. Calculate the new volume, V_2.
(b) BC—the air is compressed isothermally to volume V_1. Calculate the new pressure, p_2.
(c) CA—the air cools at constant volume to pressure p_1. State how a value for the work done on the air during the full sequence of changes may be found from the graph in Figure 26E. (*L*.)

7 A gas has a volume of $0 \cdot 02$ m^3 at a pressure of 2×10^5 Pa and temperature of 27°C. It is heated at constant pressure until its volume increases to $0 \cdot 03$ m^3.
Calculate: (i) the external work done; (ii) the new temperature of the gas; (iii) the increase in internal energy of the gas if its mass is 16 g, its molar heat capacity at constant volume is $0 \cdot 8$ J mol^{-1} K^{-1} and its molar mass is 32 g.

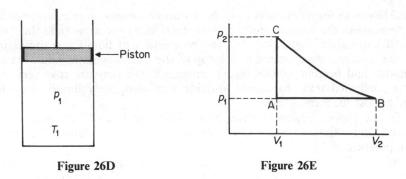

Figure 26D   **Figure 26E**

Real Gases, Critical Temperature

An ideal gas is one which obeys Boyle's law (pV = constant at constant temperature) and whose internal energy is independent of its volume. No such gas exists but at room temperature and moderate pressures, many gases approach the ideal closely for most purposes.

We now consider briefly the behaviour of *real gases*. We shall then appreciate better the relationship between liquid vapour and gas, and we shall then understand how gases such as oxygen and air, for example, can be liquefied.

Experiments on Carbon Dioxide, Critical Temperature

About 1870, THOMAS ANDREWS carried out an important experiment on carbon dioxide. He used a capillary tube to measure the volume V of the gas above a mercury column and applied varying pressures p to the gas while the temperature was kept constant. Andrews' results for the pressure (p)–volume (V) curves at various constant temperatures, called *isothermals*, are shown in Figure 26.12.

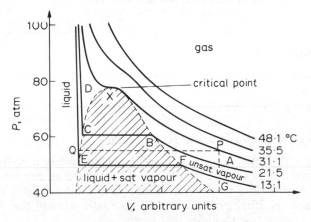

Figure 26.12 *Andrews' isothermals for carbon dioxide*

Let us consider the one for 21·5°C, ABCD. Andrews noticed that, when the pressure reached the value corresponding to B, a meniscus appeared above the mercury in the capillary containing the carbon dioxide—liquid had begun to form. From B to C, he found no change in pressure, but simply a decrease in the volume of the carbon dioxide. At the same time the meniscus moved upwards, suggesting that the proportion of liquid was increasing. At C the meniscus disappeared at the top of the tube, suggesting that the carbon dioxide had become wholly liquid. Beyond C the pressure rose very rapidly; this confirmed that the carbon dioxide was completely liquid, since liquids are almost incompressible.

So the vapour begins to change to a liquid at B and there is a mixture of liquid and vapour along BC. A similar result was obtained for the lower temperature of 13·1°C.

At 31.1°C, however, there is a significant change. There is no horizontal line such as BC or DF in the $p-V$ changes at this temperature. At that temperature, Andrews observed no meniscus; he concluded that it was the *critical temperature*. The isothermals for temperatures above 31·3°C never become horizontal, and show no breaks such as B or F.

At temperatures above the critical, no change from gas to liquid can be seen.

The isothermal for 48·1°C agrees fairly well with Boyle's law; even when the gas is highly compressed its behaviour is not far from that of an ideal gas.

Liquefaction of Gases

We can now see the importance of Andrews' experiments. A gas *above its critical temperature* cannot be liquefied by pressure. Early attempts to liquefy gases such as air, by compression without cooling, failed; and the gases were wrongly called 'permanent' gases. We still, for convenience, refer to a gas as a vapour when it is below its critical temperature, and as a gas when it is above it. The critical temperature of nitrogen is $-147°C$ (126 K). So nitrogen must be cooled below $-147°C$ to liquefy it by pressure. Air has a critical temperature of $-190°C$ (183 K) and oxygen a critical temperature of $-118°C$ (153 K).

Today, liquid nitrogen has many uses in industry. Liquid oxygen supplied in containers is stored in hospitals for medical use.

You should know:

A gas must be brought below its critical temperature in order to liquefy it by pressure.

Van der Waals' Equation for Real Gases

We conclude with a brief account of a gas equation widely used for real gases.

In deriving the ideal gas equation $pV = RT$ from the kinetic theory of gases, a number of assumptions need to be made. These are listed on p. 669. Van der Waals modified the ideal gas equation to take into account that two of these assumptions may not be valid for real gases:

1 *The volume of the molecules may not be negligible in relation to the volume V occupied by the gas.*
2 *The attractive forces between the molecules may not be negligible.*

In the bulk of the gas, the resultant force of attraction between a particular molecule and those all round it is zero when averaged over a period. Molecules which strike the wall of the containing vessel, however, are retarded by an unbalanced force due to molecules behind them. So the observed pressure p of a gas is *less* than the pressure in the ideal gas, when the attractive forces due to molecules is zero.

Van der Waals derived an expression for this pressure 'defect'. He considered that it was proportional to the product of the number of molecules per second striking unit area of the wall and the number per unit volume behind them, since this is a measure of the force of attraction. For a given volume of gas, both these numbers are proportional to the *density* of the gas. Consequently the pressure defect, p_1 say, is proportional to $\rho \times \rho$ or ρ^2. For a fixed mass of gas, $\rho \propto 1/V$, where V is the volume. So $p_1 = a/V^2$, where a is a constant for the particular gas. Taking into account the attractive forces between the molecules, it follows that, if p is the observed pressure,

the gas pressure in the bulk of the gas $= p + a/V^2$

The attraction of the walls on the molecules arriving there is to increase their velocity from v say to $v + \Delta v$. Immediately after rebounding from the walls, however, the force of attraction decreases the velocity to v again. Thus the attraction of the walls has no net effect on the momentum change due to collision. Likewise, the increase in momentum of the walls due to their attraction by the molecules arriving is lost after the molecules rebound.

Molecules have a particular diameter or volume because repulsive forces occur when they approach very closely and they cannot be compressed indefinitely. So volume of the space inside a container occupied by the molecules is not V but $(V - b)$, where b is a factor depending on the actual volume of the molecules. The magnitude of b is not the actual volume of the molecules, as if they were swept into one corner of the space, since they are in constant motion. b has been estimated to be about four times the actual volume.

So *van der Waals' equation* for real gases is:

$$\left(p + \frac{a}{V^2}\right)(V - b) = RT$$

At high pressures, when the number of molecules is high and they are closely packed together, the volume fraction b and pressure 'defect' a/V^2 both become important.

Conversely, at low pressures, where the molecules are relatively few and far apart on the average, a gas behaves like an ideal gas and obeys the equation $pV = RT$. This is why the temperature in gas thermometers is calculated as the value it would have had at infinitely low gas pressure.

Isothermals of Real Gas

A graph of pressure p against volume V at constant temperature is called an *isothermal* or *isotherm*. Figure 26.13 (i) shows some isothermals for an ideal gas, which obeys the perfect law $pV = RT$. Figure 26.13 (ii) shows a number of isothermals for a gas which obeys van der Waals' equation, $(p + a/V^2)(V - b) = RT$.

At high temperatures the isothermals are similar. As the temperature is lowered, however, the isothermals in Figure 26.13 (ii) change in shape. One curve has a point of inflexion at C, which corresponds to the *critical point* of

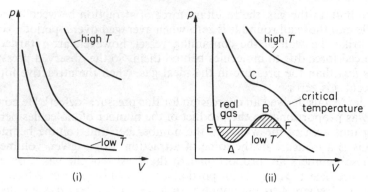

Figure 26.13 *Isothermals for ideal and van der Waals gases*

a real gas. The isothermals thus approximate to those obtained by Andrews in his experiments on actual gases such as carbon dioxide, described previously.

Below this temperature, however, isothermals such as EABF are obtained by using van der Waals' equation. These are unlike the isothermals obtained with real gases, because in the region AB the pressure increases with the volume, which is impossible. However, an actual isothermal in this region corresponds to a straight line EF, as shown. Here the liquid and vapour are in equilibrium and the line EF is drawn to make the shaded areas above and below it equal. Thus van der Waals' equation roughly fits the isothermal of actual gases above the critical temperature but below the critical temperature it must be modified considerably. Many other gas equations have been suggested for real gases but quantitative agreement is generally poor.

EXERCISES 26C   Real Gases, Critical Temperature

1 (a) Under what conditions would you expect the behaviour of a real gas to differ from that of an ideal gas? State and explain which assumptions of the kinetic theory are invalid under these conditions.

   (b) What is meant by the critical temperature of a gas? Sketch graphs, using the same axes, to show the variation of pressure with volume for a fixed mass of carbon dioxide under isothermal conditions: (i) well above the critical temperature;   (ii) at the critical temperature;   (iii) below the critical temperature.

   (c) For the graph of (b) (iii) describe the changes which occur as the pressure is increased for the range drawn. (*N.*)

2 Describe experiments in which the relation between the pressure and volume of a gas has been investigated at constant temperature over a wide range of pressure. Sketch the form of the isothermal curved obtained.

   Explain briefly how far van der Waals' equation accounts for the form of these isothermals. (*L.*)

3 What are the conditions under which the equation $pV = RT$ gives a reasonable description of the relationship between the pressure p, the volume V and the temperature T of a real gas?

   Sketch p–V isothermals for the gas–liquid states and indicate the region in which $pV = RT$ applies. Indicate the state of the substance in the various regions of the p–V diagram. Mark and explain the significance of the critical isothermal.

   Discuss a way in which the equation $pV = RT$ may be modified so that it can be applied more generally. Explain and justify on a molecular basis the additional terms introduced. Discuss the success of this modification. (*L.*)

27 Kinetic Theory of Gases

In this chapter we deal with the kinetic theory of gases and show how it can explain all the gas laws of an ideal gas. We shall also see that the molecules have a range of speeds, as Maxwell first showed.

Gas Pressure, Assumptions

In the kinetic theory of gases, we explain the behaviour of gases by considering the motion of their molecules. We suppose that the pressure of a gas is due to the molecules bombarding the walls of its container. Whenever a molecule bounces off a wall, its momentum at right-angles to the wall is reversed; the force which it exerts on the wall is equal to the rate of change of its momentum. The average force exerted by the gas on the whole of its container is the average rate at which the momentum of its molecules is changed by collision with the walls. Since pressure is force per unit area, to find the pressure of the gas we must find this force, and then divide it by the area of the walls.

The following assumptions are made to simplify the calculation:

(a) **The attraction between the molecules is negligible.**
(b) **The volume of the molecules is negligible compared with the volume occupied by the gas.**
(c) **The molecules are like perfectly elastic spheres and have random motion.**
(d) **The time during a collision is negligible compared with the time between collisions.**

Calculation of Pressure

Consider for convenience a cube of side l containing N molecules of gas each of mass m, Figure 27.1. A typical molecule will have a velocity c at any instant and this will have components of u, v, w respectively in the directions of the three perpendicular axes Ox, Oy, Oz as shown. So $c^2 = u^2 + v^2 + w^2$.

Consider the force exerted on the face X of the cube due to the component u. Just before impact, the momentum of the molecule due to u is mu. After an elastic impact, the momentum is $-mu$, since the momentum reverses. So

$$\text{momentum change on impact} = mu - (-mu) = 2mu$$

The time taken for the molecule to move across the cube to the opposite face and back to X is $2l/u$.

This is the time to make one impact. So the number of impacts per second n' on a given face $= 1 \div 2l/u = u/2l$. So at face X

$$\text{momentum change per second} = n' \times \text{one momentum change}$$

$$= \frac{u}{2l} \times 2mu = \frac{mu^2}{l}$$

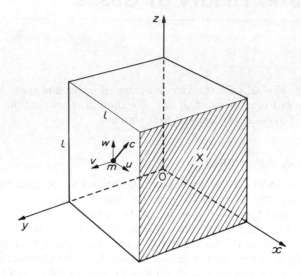

Figure 27.1 *Calculation of gas pressure*

$$\therefore \text{ force on } X = \frac{mu^2}{l}$$

$$\therefore \text{ pressure on } X = \frac{\text{force}}{\text{area}} = \frac{mu^2}{l \times l^2} = \frac{mu^2}{l^3} \qquad (1)$$

We now take account of the N molecules in the cube. Each has a different velocity and hence a component of different size in the direction Ox. Let the symbol $\overline{u^2}$ represent the average or mean value of all the squares of the components in the Ox direction, that is, for the different N molecules,

$$\overline{u^2} = \frac{u_1{}^2 + u_2{}^2 + u_3{}^2 + \cdots + u_N{}^2}{N}$$

So, from (1), the average pressure on X due to all the component velocities is given by

$$p = \frac{Nm\overline{u^2}}{l^3} \qquad (2)$$

Now with a large number of molecules of varying speed in random motion, the mean square of the component speed in any one of the three axes in the *same*.

$$\therefore \overline{u^2} = \overline{v^2} = \overline{w^2}$$

But, for each molecule, $c^2 = u^2 + v^2 + w^2$, so that the mean square $\overline{c^2}$ is given by $\overline{c^2} = \overline{u^2} + \overline{v^2} + \overline{w^2}$.

$$\therefore \overline{u^2} = \tfrac{1}{3}\overline{c^2}$$

So, from (2), with N molecules, the gas pressure is

$$p = \tfrac{1}{3}\frac{Mm\overline{c^2}}{l^3}$$

The *number of molecules per unit volume, $n = N/l^3$*. So we may write

$$p = \tfrac{1}{3}nm\overline{c^2}* \qquad (3)$$

If n is in molecules per metre^3, m in kilograms and c in metres per second, then the pressure p is in *newtons per metre*^{2} ($N m^{-2}$) or *pascals* (Pa).

In our calculation, we assumed that molecules of a gas do not collide with other molecules as they move to-and-fro across the cube. If, however, we assume that their collisions are perfectly elastic, both the kinetic energy and the momentum are conserved in them. The average momentum with which all the molecules strike the walls is then not changed by their collisions with one another; what one loses, another gains. The important effect of collisions between molecules is to distribute their individual speeds; on the average, the fast ones lose speed to the slow. We suppose, then, that different molecules have different speeds, and that the speeds of individual molecules vary with time, as they make collisions with one another; but we also suppose that the average speed of all the molecules is constant. These assumptions are justified by the fact that the kinetic theory leads to conclusions which agree with experiment.

Root-mean-square (r.m.s.) Speed

In equation (1) the factor nm is the product of the number of molecules per unit volume and the mass of one molecule. It is therefore the total mass of the gas per unit volume, its density ρ. So the equation gives

$$p = \tfrac{1}{3}\rho\overline{c^2}* \qquad . \qquad . \qquad . \qquad . \qquad . \qquad (4)$$

or

$$\overline{c^2} = \frac{3p}{\rho} \qquad . \qquad . \qquad . \qquad . \qquad (5)$$

If we substitute known values of p and ρ in equation (3), we can find $\overline{c^2}$. For hydrogen at s.t.p., $\rho = 0{\cdot}09 \text{ kg m}^{-3}$ and $p = 1{\cdot}013 \times 10^5$ Pa.

$$\therefore \overline{c^2} = \frac{3p}{\rho} = \frac{3 \times 1{\cdot}013 \times 10^5}{9 \times 10^{-2}}$$

$$= 3{\cdot}37 \times 10^6 \text{ m}^2 \text{ s}^{-1}$$

The square root of $\overline{c^2}$ is called the *root-mean-square speed*; it is of the same magnitude as the average speed, but not quite equal to it. See p. 678. Its value is

$$\sqrt{\overline{c^2}} = \sqrt{3{\cdot}37} \times 10^3 = 1840 \text{ m s}^{-1} \text{ (approx.)}$$

$$= 1{\cdot}84 \text{ km s}^{-1}$$

From (3), note that $\qquad \sqrt{\overline{c^2}} = \sqrt{\dfrac{3p}{\rho}} = \textbf{r.m.s. speed}$

Molecular speeds were first calculated in this way by Joule in 1848; they turn out to have a magnitude which is high, but reasonable. The value is reasonable because it has the same order of magnitude as the speed of sound ($1{\cdot}30 \text{ km s}^{-1}$ in hydrogen at $0°C$). The speed of sound is the speed with which the molecules of a gas pass on a disturbance from one to another, and this we expect to be of the same magnitude as the speeds of their natural motion.

* $\overline{c^2}$ may also be printed as $\langle c^2 \rangle$.

Root-mean-square Speed and Mean Speed

It should be carefully noted that the pressure p of the gas depends on the 'mean square' of the speed. This is because:

(a) the momentum change at a wall is proportional to u, as previously explained;

(b) the number of impacts per second on a given face is proportional to u.

So the rate of change of momentum is proportional to $u \times u$ or to u^2. Further, the mean square speed is *not* equal to the square of the average speed. As an example, let us suppose that the speeds of six molecules are, 1, 2, 3, 4, 5, 6 units. Their mean speed $\bar{c}$ is given by

$$\bar{c} = \frac{1 + 2 + 3 + 4 + 5 + 6}{6} = \frac{21}{6} = 3 \cdot 5$$

and its square is

$$(\bar{c})^2 = 3 \cdot 5^2 = 12 \cdot 25$$

Their mean square speed, however, is

$$\overline{c^2} = \frac{1^2 + 2^2 + 3^2 + 4^2 + 5^5 + 6^2}{6} = \frac{91}{6} = 15 \cdot 2$$

So the root-mean-square speed, $\sqrt{\overline{c^2}} = \sqrt{15 \cdot 2} = 3 \cdot 9$, which is about 12% different from the mean speed $\bar{c}$ in this simple case.

Introduction to Temperature in Kinetic Theory

Consider a volume V of gas, containing N molecules. The number of molecules per unit volume $n = N/V$. So the pressure of the gas, by equation (3) is

$$p = \tfrac{1}{3}nm\overline{c^2} = \tfrac{1}{3}\frac{N}{V}m\overline{c^2}$$

$$\therefore pV = \tfrac{1}{3}Nm\overline{c^2} \quad . \qquad . \qquad . \qquad . \qquad . \qquad (6)$$

But the ideal gas equation for 1 mole is

$$pV = RT$$

We can therefore make the kinetic theory consistent with the observed behaviour of a gas, if we write

$$\tfrac{1}{3}Nm\overline{c^2} = RT \quad . \qquad . \qquad . \qquad . \qquad . \qquad (7)$$

Essentially, we are here *assuming* that the mean square speed of the molecules, $\overline{c^2}$, is proportional to the kelvin temperature T of the gas. This is a reasonable assumption, because we have learnt that heat is a form of energy; and the translational kinetic energy of a molecule, due to its random motion within its container, is proportional to the square of its speed. When we heat a gas, we expect to speed up its molecules. So we write

$$pV = \tfrac{1}{3}Nm\overline{c^2} = RT \quad . \qquad . \qquad . \qquad . \qquad (8)$$

This is the equation for 1 mole of a gas. So N is the number of molecules in 1 mole, which is the Avogadro number, $6 \cdot 02 \times 10^{23}$. For a gas with n moles, we must write nRT in place of RT and remember that the number of molecules N is *not* the Avogadro number but n times that number.

Variation of Speed

Since N molecules each have a mass m in the mole of gas considered, the molar mass M is Nm. So, from (7),

$$\tfrac{1}{3}M\overline{c^2} = RT$$

$$\therefore \text{ r.m.s. speed, } \sqrt{\overline{c^2}} = \sqrt{\frac{3RT}{M}}.$$

So

1   The r.m.s. speed or velocity of the molecules of a *given gas* $\propto \sqrt{T}$.
2   The r.m.s. speed of the molecules of different gases at the *same* temperature $\propto 1/\sqrt{M}$, so gases of higher molecular mass have smaller r.m.s. speeds.

To illustrate the numerical changes, hydrogen of relative molecular mass about 2 has a r.m.s. speed at s.t.p. (273 K and 1.013×10^5 Pa pressure) of roughly 1840 m s^{-1}. At 100°C or 373 K and the same pressure, the r.m.s. speed c_r is given by

$$\frac{c_r}{1840} = \sqrt{\frac{373}{273}}$$

or

$$c_r = 1840 \times \sqrt{\frac{373}{273}} = 1930 \text{ m s}^{-1} \text{ (approx.)}$$

Oxygen has a relative molecular mass of about 32. From (4) above, $c_r \propto 1/\sqrt{M}$. So at s.t.p. the r.m.s. speed of oxygen molecules, by comparison with the r.m.s. speed of hydrogen molecules, is given by

$$c_r = \sqrt{\frac{2}{32}}$$

or

$$\text{or } c_r = 1840 \times \sqrt{\frac{2}{32}} = 460 \text{ m s}^{-1}$$

Boltzmann Constant, Mean Energy of Molecule

The kinetic energy of a molecule moving at an instant with a speed c is $\tfrac{1}{2}mc^2$; the average kinetic energy of translation of the random motion of the molecule of a gas is therefore $\tfrac{1}{2}m\overline{c^2}$. To relate this to the temperature, we put equation (7) into the form

$$RT = \tfrac{1}{3}Nm\overline{c^2} = \tfrac{2}{3}N(\tfrac{1}{2}m\overline{c^2})$$

so

$$\tfrac{1}{2}m\overline{c^2} = \tfrac{3}{2}\frac{R}{N}T \qquad . \qquad . \qquad . \qquad . \qquad . \qquad . \qquad (9)$$

Thus, *the average kinetic energy of translation of a molecule is proportional to the absolute temperature of the gas.*

The ratio R/N in equation (9) is a universal constant, since $R =$ molar gas constant, $8 \cdot 31 \text{ J mol}^{-1} \text{K}^{-1}$ for all gases, and $N = N_A$, the Avogadro constant, $6 \cdot 02 \times 10^{23} \text{ mol}^{-1}$ for all gases. So

$$\frac{R}{N_A} = k$$

The constant k, the gas constant per molecule, is called the *Boltzmann constant*. In terms of k equation (9) becomes

$$\tfrac{1}{2}m\overline{c^2} = \tfrac{3}{2}kT \qquad . \qquad . \qquad . \qquad . \qquad (10)$$

The Boltzmann constant is usually given in joules per degree, since it relates energy to temperature: $k = \tfrac{1}{2}m\overline{c^2}/\tfrac{3}{2}T$. Its value is $k = 1 \cdot 38 \times 10^{-23} \text{ J K}^{-1}$. Since $R = N_A k$, we can write $pV = N_A kT$.

Diffusion: Graham's Law

When a gas passes through a porous plug, a cotton-wool wad, for example, it is said to 'diffuse'. Diffusion differs from the flow of a gas through a wide tube because it is not a motion of the gas in bulk, but is a result of the motion of its individual molecules.

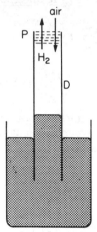

Figure 27.2 *Graham's apparatus for diffusion*

Figure 27.2 shows an apparatus devised by THOMAS GRAHAM (1805–69) to compare the rates of diffusion of different gases. D is a glass tube, closed with a plug P of plaster of Paris. It is first filled with mercury, and inverted over mercury in a bowl. Hydrogen is then passed into it until the mercury levels are the same on each side; the hydrogen is then at atmospheric pressure. The volume of hydrogen, V_H, is proportional to the length of the tube above the mercury. The apparatus is now left; hydrogen diffuses out through P, and air diffuses in. Ultimately no hydrogen remains in the tube D. The tube is then adjusted until the level of mercury is again the same on each side, so that the air within it is at atmospheric pressure. The volume of air, V_A, is proportional to the new length of the tube above the mercury.

The volumes V_A and V_H are, respectively, the volumes of air and hydrogen which diffused through the plug in the same time. Therefore the rates of diffusion of the gases air and hydrogen are proportional to the volumes V_A and V_H:

$$\frac{\text{rate of diffusion of air}}{\text{rate of diffusion of hydrogen}} = \frac{V_A}{V_H}$$

Graham found in his experiments that the volumes were inversely proportional to the square roots of the densities of the gases, ρ:

$$\frac{V_A}{V_H} = \sqrt{\frac{\rho_H}{\rho_A}}$$

So
$$\frac{\text{rate of diffusion of air}}{\text{rate of diffusuion of hydrogen}} = \sqrt{\frac{\rho_H}{\rho_A}}$$

In general:
$$\text{rate of diffusion} \propto \frac{1}{\sqrt{\rho}}$$

and in words

---

the rate of diffusion of a gas is inversely proportional to the square root of its density. This is Graham's law.

Graham's law of diffusion is readily explained by the kinetic theory. At the same kelvin temperature T, the mean kinetic energies of the molecules of different gases are equal, since

$$\tfrac{1}{2}m\overline{c^2} = \tfrac{3}{2}kT$$

and k is the Boltzmann constant. So if the subscripts A and H denote air and hydrogen respectively,

$$\tfrac{1}{2}m_A\overline{c_A^2} = \tfrac{1}{2}m_H\overline{c_H^2}$$

and
$$\frac{\overline{c_A^2}}{\overline{c_H^2}} = \frac{m_H}{m_A}$$

At a given temperature and pressure, the density of a gas, ρ, is proportional to the mass of its molecule, m, since equal volumes contain equal numbers of molecules.

So
$$\frac{m_H\rho_H}{m_A\rho_A}$$

and
$$\frac{\overline{c_A^2}}{\overline{c_{H_2}^2}} = \frac{\rho_H}{\rho_A}$$

$$\therefore \frac{\sqrt{\overline{c_A^2}}}{\sqrt{\overline{c_H^2}}} = \frac{\sqrt{\rho_H}}{\sqrt{\rho_A}} \quad\quad \cdot\quad\cdot\quad\cdot\quad\cdot\quad\cdot\quad\cdot \quad (1)$$

The average speed of the molecules of a gas is roughly equal to—and strictly proportional to—the square root of its mean square speed. Equation (1) therefore shows that the average molecular speeds are inversely proportional to the square roots of the densities of the gases. And so it explains why the rates of diffusion—which depend on the molecular speeds—are also inversely proportional to the square roots of the densities.

Example on Kinetic Theory

Helium gas occupies a volume of 0.04 m^3 at a pressure of 2×10^5 Pa and temperature 300 K.

Calculate: (i) the mass of helium; (ii) the r.m.s. speed of its molecules; (iii) the r.m.s. speed at 432 K when the gas is heated at constant pressure to this temperature; (iv) the r.m.s speed of hydrogen molecules at 432 K. (Relative molecular mass of helium and hydrogen = 4 and 2 respectively, molar gas constant = 8.3 J mol^{-1} K^{-1}.)

(i) *Mass*   For n mols, $pV = nRT$

So
$$n = \frac{pV}{RT} = \frac{2 \times 10^5 \times 0.04}{8.3 \times 300}$$

Hence   mass of helium = 3.2×4 g = 12.8 g

(ii) *r.m.s. speed* Pressure $p = 2 \times 10^5$ Pa

$$\text{density } \rho = \frac{\text{mass}}{\text{volume}} = \frac{12.8 \times 10^{-3} \text{ kg}}{0.04 \text{ m}^3} = 0.32 \text{ kg m}^{-3}$$

So   r.m.s. speed $= \sqrt{\dfrac{3p}{\rho}}$

$$= \sqrt{\frac{3 \times 2 \times 10^5}{0.32}} = 1369 \text{ m s}^{-1}$$

(iii) *Temperature 432 K*   Since r.m.s. speed $\propto \sqrt{T}$, the new value c_r at 432 K is given by

$$\frac{c_r}{1369} = \sqrt{\frac{432}{300}} = \sqrt{1.44} = 1.2$$

So $c_r = 1.2 \times 1369 = 1643$ m s^{-1}

(iv) *Hydrogen*   One mole of hydrogen has a mass of 2 g and one mole of helium has a mass of 4 g. So ratio of molar masses = $2:4 = 1:2$.

But r.m.s. speed at a given temperature $\propto 1/\sqrt{M}$, where M is the molar mass. So at 432 K,

$$\text{r.m.s. speed of hydrogen molecules} = \sqrt{2} \times 1643$$

$$= 2324 \text{ m s}^{-1}$$

Distribution of Molecular Speeds

So far we have used the 'root-mean-square' speed and the 'mean' speed of the large number of molecules in a given mass of gas. The actual distribution of the speeds among the numerous molecules can be investigated by an apparatus whose principle is illustrated in Figure 27.3 (i) and from which all the air is evacuated.

A furnace F maintains a sample of molten metal at a constant high temperature T. Molecules from the vapour emerge from an opening O in the furnace and pass through a narrow collimator slit C which produces a parallel beam of molecules. This beam is incident on a wheel A, which rotates with the same angular speed as another wheel B on the same axle and is distant l from B. The wheel A has a narrow slit, A_1, and above it a wide slit YX through which all the vapour molecules may pass. The wheel B has a narrow slit B_1 in it which is displaced from the slit A_1 by an angle θ when both wheels are viewed end-on.

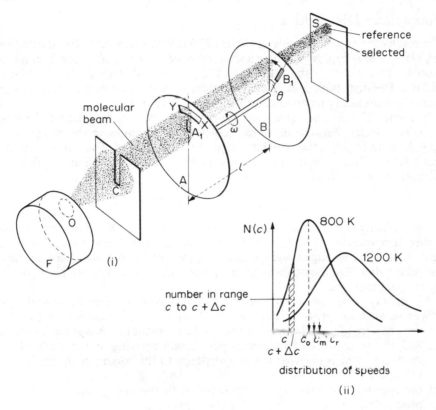

Figure 27.3 *Distribution of molecular speeds*

With B_1 originally displaced by an angle θ from A_1, both wheels are rotated at the same high angular speed ω. Only molecules passing through A_1, and which cross the distance l in the same time as the wheel B takes to turn through an angle θ, will pass through the slit B_1. These molecules have a velocity v given by

$$\frac{l}{v} = \frac{\theta}{\omega}$$

so

$$v = \frac{\omega l}{\theta}$$

The molecules passing through A_1 and B_1 are thus a 'velocity selected' beam.

The slit YX, however, is wide enough to allow molecules of practically all speeds to pass through itself and through B_1.

These molecules form an 'unselected' or reference beam. The 'velocity selected' and reference beams are incident on a surface S cooled by liquid nitrogen and the ratio of the intensities of the two beams is a measure of the fraction of all the molecules which have velocities close to v in magnitude. Thus by rotating the wheels at different speeds, measurements can be made of the distribution of velocities among the molecules. Further, by varying the temperature T of the furnace, the molecular distribution can be found at different temperatures.

Maxwellian Distribution

The results are shown roughly in Figure 27.3 (ii). The quantity $N(c)$ plotted on the vertical axis represents the number of molecules ΔN in a small range of speeds c to $c + \Delta c$, so that $\Delta N = N(c)$. Δc = area of strip shaded in Figure 28.3 (ii). The distribution of velocities agrees with the Maxwellian distribution derived theoretically from advanced kinetic theory of gases.

In Figure 27.3 (ii), the value c_0 at the maximum of a curve is called the *most probable velocity*, because more molecules have velocities in the range c_0 to $c_0 + \Delta c$ than in any other similar range, Δc, of velocities. The value c_m is the mean velocity and c_r is the root-mean-square. For a Maxwellian distribution, calculation shows that

$$c_0 : c_m : c_r = 1 \cdot 00 : 1 \cdot 13 : 1 \cdot 23$$

The velocity distribution curves at 800 K and 1200 K show that at the higher temperature 1200 K, the distribution curve flattens more round its peak value. So at higher temperatures, more molecules have speeds near the peak value. The mean, root-mean-square and peak values are all higher at higher temperatures.

As already seen, the *mean-square velocity* is concerned in large-scale gas properties such as 'pressure' and 'specific heat capacity'. This is because:

(a) the pressure of a gas is proportional to the momentum change per molecule and to the number of molecules per second arriving at the walls of the container, which together are proportional to the square of the individual velocities;

(b) the specific heat capacity is proportional to the energy gained, which is proportional to $\frac{1}{2}Mc^2$ where M is the mass of gas.

On the other hand, the *mean velocity* is concerned in gas properties such as (1) 'diffusion' through porous partitions, since this rate of diffusion is proportional to the mean velocity, and (2) 'viscosity', or frictional forces in a gas, since there is a transfer of momentum from fast to slow moving gas layers in gas flow through pipes, for example.

Thermal Agitations and Internal Energy

The random motion of the molecules of a gas, whose kinetic energy depends upon the temperature, is often called the *thermal agitation* of the molecules. And the kinetic energy of the thermal agitation is called the *internal energy* of the gas, which we discussed earlier on p. 657.

The internal energy of a gas depends on the number of atoms in its molecules. A gas whose molecules consist of single atoms is said to be monatomic: for example, chemically inert gases and metallic vapours, Hg, Na, He, Ne, A. A gas with two atoms to the molecule is said to be diatomic: O_2, H_2, Cl_2, CO.

And a gas with more than two atoms to the molecule is said to be polyatomic: H_2O, O_3, H_2S, CO_2, CH_4. The molecules of a monatomic gas we may regard as points, but those of a diatomic gas we must regard as 'dumb-bells', and those of a polyatomic gas as more complicated structures (Figure 27.4). A molecule which extends appreciably in space—a diatomic or polyatomic molecule—has an appreciable moment of inertia. It will therefore have kinetic energy of rotation, as well as of translation. A monatomic molecule, however, must have a much smaller moment of inertia than a diatomic or polyatomic. Its kinetic energy of rotation can therefore be neglected and so a helium molecule, for example, has only translational kinetic energy.

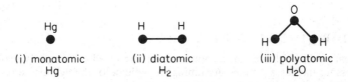

(i) monatomic　　(ii) diatomic　　(iii) polyatomic
Hg　　　　　　　　H_2　　　　　　　H_2O

Figure 27.4 *Types of gas molecule*

Diffusion Processes in Integrated Circuit Fabrication

Diffusion (p. 674) is a very important process in making integrated circuits used in the electronics industry. To construct tiny electronic devices such as diodes and transistors on wafers of silicon, it is necessary to create controlled n- and p-type regions in the silicon; this is usually done using a process known as *solid-state diffusion.*

Here the silicon is exposed to a vapour of donor or acceptor impurities (see Chapter 31) at a temperature in excess of 900°C. The atoms in the vapour are in random motion, and so there is a statistical chance that impurities will enter the silicon lattice, where few impurities atoms are present compared with the impurity-rich vapour. Before diffusion, a photolithographic masking technique is used to put down a thin layer of silicon dioxide where impurities are *not* required; the presence of the silicon dioxide greatly slows down the diffusion of the impurities in these areas.

In this way thousands of tiny electronic devices can be created on a small piece of silicon, so producing an *integrated circuit.* Solid-state diffusion is popular because it is gentle, causing little damage to the semiconductor's crystal lattice. However, it is not always as successful in compound semiconductor materials (such as gallium arsenide) as it is in silicon. The alternative technique, *ion implantation,* allows a more precisely controlled amount of impurity to be introduced, but at the expense of damage to the crystal lattice. After implantation the semiconductor must be annealed to repair this damage.

EXERCISES 27　Kinetic Theory

Multiple Choice

1 An ideal gas has a temperature T in kelvin and a r.m.s. value c_r of molecular speed. At a temperature $2T$, the r.m.s. value is

  A $2c_r$　B $\sqrt{2c_r}$　C $2\sqrt{2c_r}$　D $4c_r$　E $4\sqrt{2c_r}$

2 In the kinetic theory for an ideal gas, which of the statements **A** to **E** is *not* a correct assumption?
   A The duration of a collision is negligible compared with the time between collisions.
   B The volume of the molecules is negligible compared with the volume in which they move.
   C The molecules have negligible attraction for each other.
   D The molecules have negligible momentum change on collision with the container walls.
   E There is no total kinetic energy change of the molecules by collision with each other or with the walls of the container.

3 A container at room temperature has one mole of hydrogen, 0·002 kg, and one mole of oxygen, 0·032 kg. The ratio of the pressure of hydrogen to that of oxygen on the container walls is

   A 16:1 **B** 4:1 **C** 1:1 **D** 1:4 **E** 1:16

4 An ideal gas in a cylinder has a root-mean-square speed u. The gas is heated so that its pressure increases 9 times. Assuming the volume of the cylinder stays constant, the new root-mean-square speed of the gas molecules is

   A $9u$ **B** $6u$ **C** $\sqrt{3u}/2$ **D** $3u$ **E** $3u/2$

Longer Questions (If required, assume $R = 8·3 \, \text{J mol}^{-1} \, \text{K}^{-1}$.)

5 Show that the relation $p = \frac{1}{3}\rho \overline{c^2}$ is dimensionally correct, where p is the pressure of a gas of density ρ and $\overline{c^2}$ is the mean square velocity of all its molecules.
   Write down:
   (a) two assumptions made in deriving this relation from a simple kinetic theory;
   (b) the meaning of (i) mean velocity and   (ii) mean square velocity.

6 Calculate the root-mean-square speed at 0°C of (i) hydrogen molecules and (ii) oxygen molecules, assuming 1 mole of a gas occupies a volume of $2 \times 10^{-2} \, \text{m}^3$ at 0°C and $10^5 \, \text{N m}^{-2}$ pressure. (Relative molecular masses of hydrogen and oxygen = 2 and 32 respectively.)

7 Assuming helium molecules have a root-mean-square speed of $900 \, \text{m s}^{-1}$ at 27°C and $10^5 \, \text{N m}^{-2}$ pressure, calculate the root-mean-square speed at (i) 127°C and $10^5 \, \text{N m}^{-2}$ pressure,   (ii) 27°C and $2 \times 10^5 \, \text{N m}^{-2}$ pressure.

8 Using the kinetic theory, show that (i) the pressure of an ideal gas is doubled when its volume is halved at constant temperature,   (ii) the pressure of an ideal gas decreases when it expands in a thermally insulated vessel.

9 (a) Explain briefly what is meant by the term *ideal gas*.
   (b) A volume of 0·23 m³ contains nitrogen at a pressure of $0·50 \times 10^5 \, \text{Pa}$ and temperature 300 K. Assuming that the gas behaves ideally, calculate the amount in mol of nitrogen present.
   (c) Calculate the root-mean-square speed of nitrogen molecules at a temperature of 300 K. (Molar mass of nitrogen = $0·028 \, \text{kg mol}^{-1}$.) (*O. & C.*)

10 Figure 27A shows curves (not to scale) relating pressure, p, and volume, V, for a fixed mass of an ideal monatomic gas at 300 K and 500 K. The gas is in a container fitted with a piston which can move with negligible friction.
   (a) Give the equation of state for n moles of an ideal gas, defining the symbols used.
   Show by calculation that:
      (i) the number of moles of gas in the container is $2·01 \times 10^{-2}$;
      (ii) the volume of the gas at B on the graph is $1·67 \times 10^{-3} \, \text{m}^3$.
      (Molar gas constant, $R = 8·31 \, \text{J mol}^{-1} \, \text{K}^{-1}$.)
   (b) The kinetic theory gives the equation $p = \frac{1}{3}\rho \overline{c^2}$ where ρ is the density of the gas.
      (i) Explain what is meant by $\overline{c^2}$.
      (ii) Use the equation to derive an expression for the total internal energy of one mole of an ideal monatomic gas at kelvin temperature T.
      Calculate the total internal energy of the gas in the container at point A on the graph.

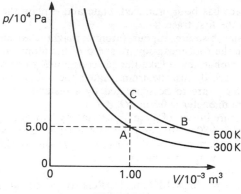

Figure 27A

(c) State the first law of thermodynamics as applied to a fixed mass of an ideal gas when heat energy is supplied to it so that its temperature rises and it is allowed to expand. Define any symbols used.

(d) Explain how the first law of thermodynamics applies to the changes represented on the graph by (i) A to C and (ii) A to B. Calculate the heat energy absorbed in each case. (*N.*)

11 Write down *four* assumptions about the properties and behaviour of molecules that are made in the kinetic theory in order to define an ideal gas. On the basis of this theory derive an expression for the pressure by an ideal gas.

Use the kinetic theory to explain why hydrogen molecules diffuse out of a porous container into the atmosphere even when the pressure of the hydrogen is equal to the atmospheric pressure outside the container.

Air at 273 K and $1 \cdot 01 \times 10^5$ N m^{-2} pressure contains $2 \cdot 70 \times 10^{25}$ molecules per cubic metre. How many molecules per cubic metre will there be at a place where the temperature is 223 K and the pressure is $1 \cdot 33 \times 10^{-4}$ N m^{-2}? (*L.*)

12 The kinetic theory of gases leads to the equation $p = \frac{1}{3}\rho \overline{c^2}$, where p is the *pressure*, ρ is the *density* and $\overline{c^2}$ is the *mean square molecular speed*. Explain the meaning of the terms in italics and list the simplifying assumptions necessary to derive this result. Discuss how this equation is related to Boyle's law.

Air may be taken to consist of 80% nitrogen molecules and 20% oxygen molecules of relative molecular masses 28 and 32 respectively. Calculate:

(a) the ratio of the root-mean-square speed of nitrogen molecules to that of oxygen molecules in air;

(b) the ratio of the partial pressures of nitrogen and oxygen molecules in air;

(c) the ratio of the root-mean-square speed of nitrogen molecules in air at 10°C to that at 100°C. (*O. & C.*)

13 Define the term *mean square speed* as applied to the molecules of a gas. Explain why the pressure exerted by a gas is proportional to the mean square speed of its molecules.

Figure 27B shows apparatus designed to measure speeds of molecules. Atoms of vapour of a heavy metal emerge from oven O into an evacuated space. The atoms pass through fixed slit S' in a well-defined beam and enter radially through slit S'' in the curved surface of a cylindrical drum D. When the drum is stationary the atoms strike the inner surface of the drum, giving a well-defined trace T. The drum is then set into rotation about its axis C and maintained at a high constant speed

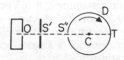

Figure 27B

until a second trace has been produced. State and explain the ways in which this trace differs from the first trace.

Oven O contains bismuth vapour (atomic weight 208) and is maintained at 1500°C. Calculate the root-mean-square speed of the atoms in the oven, assuming the vapour to be monatomic. (Take the gas constant R to be $8\cdot314$ J mol^{-1} K.)

At what angular speed must the drum rotate if the traces for atoms having speeds of 400 and 800 m s^{-1} are to be separated by a distance of 10 mm on the drum surface? The drum diameter is 0·5 m. (*O. & C.*)

14 Calculate the pressure in mm of mercury exerted by hydrogen gas if the number of molecules per cm^3 is $6\cdot80 \times 10^{15}$ and the root-mean-square speed of the molecules is $1\cdot90 \times 10^3$ m s^{-1}. Comment on the effect of a pressure of this magnitude:

(a) above the mercury in a barometer tube;

(b) in a cathode tube.

(Avogadro constant = $6\cdot02 \times 10^{23}$ mol^{-1}. Relative molecular mass of hydrogen = 2·02.) (*N.*)

15 What do you understand by the term *ideal gas*? Describe a molecular model of an ideal gas and derive the expression $p = \frac{1}{3}\rho c^2$ for such a gas. What is the reasoning which leads to the assertion that the temperature for an ideal monatomic gas is proportional to the mean kinetic energy of its molecules?

The Doppler broadening of a spectral line is proportional to the r.m.s. speed of the atoms emitting light. Which source would have less Doppler broadening, a mercury lamp at 300 K or a krypton lamp at 77 K? (Take the mass numbers of Hg and Kr to be 200 and 84 respectively.)

What causes the behaviour of real gases to differ from that of an ideal gas? Explain qualitatively why the behaviour of all gases at very low pressures approximates to that of an ideal gas. (*O. & C.*)

16 (a) The kinetic theory of gases predicts that the root-mean-square (r.m.s) speed of the molecules of an ideal gas is given by the expression $(3p/\rho)^{1/2}$, where p is the pressure and ρ is the density of the gas.

The graphs in Figure 27C (i) show how the pressure of oxygen gas depends upon its density at two different constant temperatures, T and 300 K. (i) Use the graph to calculate a value for the r.m.s. speed of the oxygen molecules at 300 K. Explain your working. (ii) Is the temperature T higher or lower

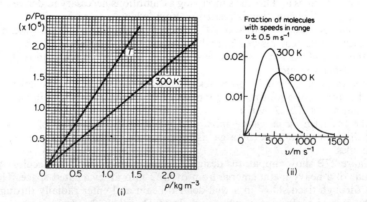

Figure 27C

than 300 K? Explain your reasoning. (iii) The graphs above are based upon experimental results. What conclusion can you draw from them about the behaviour of oxygen? (iv) Outline a simple experimental procedure for investigating how the pressure of a known mass of air varies as its density changes at room temperature.

Include a labelled diagram of the apparatus used.

(b) The graphs in Figure 27C (ii) show how the speeds of the molecules in an ideal gas are distributed at two temperatures. Use them to help you answer the following questions.   (i) In what two main ways does the temperature appear to affect the distribution of speeds?   (ii) What is the value of v for which the fraction of molecules with speeds as in the range $v \pm 0.5 \, \mathrm{m \, s^{-1}}$ is a maximum at a temperature of 300 K? How does this value compare with the r.m.s. speed calculated in (a) (i) above? (*L.*)

28 Transfer of Heat: Conduction and Radiation

Conduction

Heat can be transferred from one place to another by conduction, convection or radiation. Transfer by convection was discussed in an earlier chapter. In this chapter we deal mainly with conduction. We shall see that the conduction of heat obeys similar laws to the conduction of electricity and we shall consider good and bad conductors and conductors in series. Heat radiation will be shown to be infrared radiation, which is in the electromagnetic spectrum.

If we put a poker into the fire, and hold on to it, then heat reaches us along the metal. We say the heat is *conducted*. We find that some substances—metals— are good conductors, and others—such as wood or glass—are not. Good conductors such as a metal bar feel cold to touch on a cold day, because they rapidly conduct away the body's heat.

Temperature Distribution along a Conductor

In order to study conduction in more detail consider Figure 28.1 (i), which shows a metal bar AB whose ends have been soldered into the walls of two metal tanks, H, C. H contains boiling water, and C contains ice-water. Heat flows along the bar from A to B, and *when conditions are steady* the temperature θ of the bar is measured at points along its length. The measurements may be qmade with thermojunctions, not shown in the figure, which have been soldered to the rod. The curve in the upper part of the figure shows how the temperature falls along the bar, less and less steeply from the hot end to the cold. So the *temperature gradient* decreases from the hot end to the cold.

The Figure 28.1 (ii) shows how the temperature varies along the bar, if the bar is *well lagged* with a bad conductor, such as asbestos wool. It now falls *uniformly* from the hot to the cold end, so the temperature gradient along the bar is *constant*.

The difference between the temperature distributions is due to the fact that, when the bar is unlagged, heat escapes from its sides, by convection in the surrounding air, Figure 28.1 (i). So the heat flowing past D per second is less than that entering the bar at A by the amount which escapes from the surface AD. The arrows in the figure represent the heat escaping per second from the surface of the bar, and the heat flowing per second along its length. The heat flowing per second along the length decreases from the hot end to the cold. But when the bar is lagged, the heat escaping from its sides is negligible, and the flow per second is now constant along the length of the bar, Figure 28.1 (ii).

We therefore see that the *temperature gradient* along a bar is greatest where the heat flow through it is greatest. We also see that the temperature gradient is uniform only when there is a negligible loss of heat from the sides of the bar.

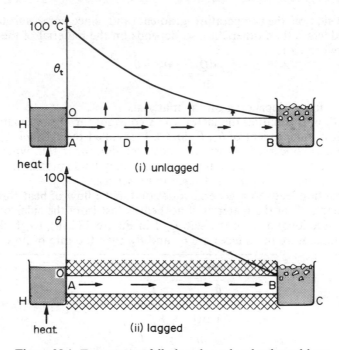

Figure 28.1 *Temperature fall along lagged and unlagged bars*

Thermal Conductivity

Consider a very large thick bar, of which AB in Figure 28.2 (i) is a part, and along which heat is flowing steadily. We suppose that the loss of heat from the sides of the bar is made negligible by lagging. XY is a slice of the bar, of thickness l, whose faces are at temperatures θ_2 and θ_1. Then the *temperature gradient* over the slice is

$$\frac{\theta_2 - \theta_1}{l}$$

We now consider an element *abcd* of the slice of unit cross-sectional area, and we denote by dQ/dt the heating flowing through it *per second*. The value of

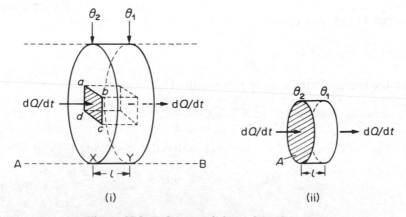

(i)   (ii)

Figure 28.2 *Definition of thermal conductivity*

dQ/dt depends on the temperature gradient, and, since some substances are better conductors than others, it also depends on the material of the bar.

We therefore write

$$\frac{dQ}{dt} = k\frac{\theta_2 - \theta_1}{l}$$

where k is a factor depending on the material.

To a fair approximation the factor k is a constant for a given material; that is to say, it is independent of θ_2, θ_1 and l. It is called the *thermal conductivity* of the material concerned. From the above relation for dQ/dt, when the heat flow is *normal* to an area inside the material, k may be defined as *the heat flow per second per unit area per unit temperature gradient*.

This definition leads to a general equation for the flow of heat through any parallel-sided slab of the material, if no heat is lost from the sides of the slab. If the cross-sectional area of the slab is A in Figure 28.2 (ii), its thickness is l, and the temperature of its faces are θ_1 and θ_2, then the rate of flow of heat is

$$\frac{dQ}{dt} = \frac{kA(\theta_2 - \theta_1)}{l} \qquad (1)$$

So

$$\frac{1}{A}\frac{dQ}{dt} = k\frac{\theta_2 - \theta_1}{l} \qquad (2)$$

heat flow per metre² per second = conductivity × temperature gradient . (2a)

U Values

Instead of using thermal conductivity k, building materials such as brick or glass use *U values* for their heat conduction property.

The U value for a material is the *heat conducted per second per unit area per unit temperature difference of its opposite sides*. So the unit of U value is $W\,m^{-2}\,K^{-1}$. Suppose the glass window in a room has a U value $5\ W\,m^{-2}\,K^{-1}$, an area $4\ m^2$ and a temperature difference of $10°C$ or $10\ K$ between the inside and outside of the window. Then

heat per second conducted through window $= 5 \times 4 \times 10 = 200\ W$

Generally, heat per second conducted through a uniform material $= U \times A \times (\theta_2 - \theta_1)$ where U is the U value, A is the area and $(\theta_2 - \theta_1)$ is the temperature difference across the opposite sides. We can write this relation as

$$\frac{dQ}{dt} = UA(\theta_2 - \theta_1) \qquad (3)$$

Using the thermal conductivity relation in (1),

$$\frac{dQ}{dt} = kA\frac{(\theta_2 - \theta_1)}{l} \qquad (4)$$

where l is the thickness of the uniform material. Comparing (3) and (4), then

$$U = \frac{k}{l} \quad \text{or} \quad k = U \times l$$

So with glass of U value $5 \text{ W m}^{-2}\text{ K}^{-1}$ and thickness 3 mm or 3×10^{-3} m, the thermal conductivity of the glass is $k = 5 \times 3 \times 10^{-3} = 15 \times 10^{-3}$ or $0.015 \text{ W m}^{-1}\text{ K}^{-1}$. Note the different units for k ($\text{W m}^{-1}\text{ K}^{-1}$) and the U ($\text{W m}^{-2}\text{ K}^{-1}$) value.

Heat Conductors in Series

To reduce heat losses from a house, insulating material is placed between the cavity brick walls and the top floor has lagging material placed above it. So the combined U value of two materials in series is needed to find how much heat is conducted through them. If U_1 and U_2 are the respective U values of the materials, their combined U value is given by $1/U = 1/U_1 + 1/U_2$, as we now show.

Suppose θ_2 and θ_1 are the temperatures outside the two materials X and Y and θ is the temperature of their junction. Figure 28.3. If l_1 and l_2 are the thicknesses and A is the uniform area, then, for X,

$$dQ/dt = U_1 A(\theta_2 - \theta)$$

So
$$\theta_2 - \theta = \frac{1}{U_1 A}\frac{dQ}{dt}$$

Similarly for Y, $\theta - \theta_1 = \dfrac{1}{U_2 A}\dfrac{dQ}{dt}$

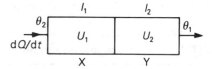

Figure 28.3 *Heat conductors in series*

Adding, θ cancels, and $\theta_2 - \theta_1 = \left(\dfrac{1}{U_1} + \dfrac{1}{U_2}\right)\dfrac{1}{A}\dfrac{dQ}{dt}$ (1)

If U is the combined U value, $\theta_2 - \theta_1 = \dfrac{1}{U A}\dfrac{dQ}{dt}$ (2)

From (1) and (2), $\dfrac{1}{U} = \dfrac{1}{U_1} + \dfrac{1}{U_2}$

Example on U Values

A roof of a house has a U value $3.0 \text{ W m}^{-2}\text{ K}^{-1}$ when unlagged and a U value $0.5 \text{ W m}^{-2}\text{ K}^{-1}$ when lagged. Calculate the U value of the lagging material and its thermal conductivity if its thickness is 20 mm or 20×10^{-3} m.

From $1/U = 1/U_1 + 1/U_2$

$1/0.5 = 1/3.0 + 1/U_2$

Solving, $U_2 = 0.6 \text{ W m}^{-2}\text{ K}^{-1}$

So $k = U \times l = 0.6 \times 20 \times 10^{-3} = 12 \times 10^{-3} = 0.012 \text{ W m}^{-1}\text{ K}^{-1}$

Lagged and Unlagged Bars

We continue with the thermal conductivity basic formula in equation (1) discussed earlier on p. 686. This was:

$$\frac{dQ}{dt} = kA \frac{\theta_2 - \theta_1}{l}$$

In terms of the calculus, it may be re-written

$$\frac{dQ}{dt} = -kA \frac{d\theta}{dl} \qquad \qquad (1)$$

the temperature gradient being negative since θ diminishes as l increases.

If a bar is *lagged* perfectly, as in Figure 28.1 (ii) on p. 685, then the heat per second, dQ/dt, flowing through every cross-section from the hot to the cold end is constant since no heat escapes through the sides. Hence, from (1), the temperature gradient, $d\theta/dl$, is constant along the bar. This is illustrated in Figure 28.1 (ii); the temperature variation with distance along the bar is a straight line.

If the bar is *unlagged*, as in Figure 28.1 (i), then heat is lost from the sides of the bar. In this case the heat per second, dQ/dt, flowing through each section decreases from the hot to the cold end. Hence, from (1), the temperature gradient, $d\theta/dl$, decreases with distance along the bar. This is shown by Figure 28.1 (i); the gradient at a point of the curve decreases with distance from the hot end of the bar.

From (1), we can see that the heat flow per second in thermal conduction is always given by

---

$dQ/dt = kA \times$ temperature gradient $\qquad \qquad (2)$

Units and Magnitude of Conductivity

Equation (2) above helps us to find the unit of thermal conductivity. We have

$$k = \frac{dQ/dt}{A \times \text{temperature gradient}}$$

The unit of dQ/dt is $J\,s^{-1}$ or W (1 watt $= 1$ joule s^{-1}), the unit of A is m^2, the unit of temperature gradient is $K\,m^{-1}$. So the unit of k is

$$\frac{W}{m^2\,K\,m^{-1}} = W\,m^{-1}\,K^{-1}$$

The thermal conductivity of copper, cardboard, water and air are roughly 380, 0·2, 0·6 and 0·03 $W\,m^{-1}\,K^{-1}$ respectively. To a rough approximation we may say that the conductivities of metals are about 1000 times as great as those of other solids, and of liquids; and they are about 10 000 times as great as those of gases such as air, which is a bad conductor or good insulator.

Examples on Conduction

1 Calculate the quantity of heat conducted through 2 m^2 of a brick wall 12 cm thick in 1 hour if the temperature on one side is 8°C and on the other side is 28°C. (Thermal conductivity of brick $= 0{\cdot}13 \text{ W m}^{-1}\text{ K}^{-1}$.)

$$\text{Temperature gradient} = \frac{28 - 8}{12 \times 10^{-2}} \text{ K m}^{-1} \text{ and } t = 3600 \text{ s}$$

$$\therefore Q = kAt \times \text{temperature gradient}$$

$$= 0{\cdot}13 \times 2 \times 3600 \times \frac{28 - 8}{12 \times 10^{-2}}$$

$$= 156\,000 \text{ J}$$

2 Figure 28.4 (i) shows a lagged bar XY of non-uniform cross-section. One end X is kept at 100°C and the other end at 0°C.

Describe and explain how the temperature varies from X to Y in the steady state.

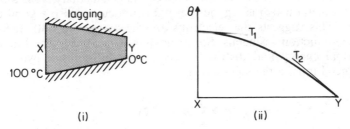

(i)　　　　　　　　　　　(ii)

Figure 28.4 *Examples on conduction*

Figure 28.3 (ii) shows how the temperature θ varies from the hot end X to the cold end Y. Since the bar is lagged, the heat per second, dQ/dt, through each section of the bar is the same. Now

$$\text{temperature gradient, } g = \frac{1}{A} \cdot \frac{dQ}{dt}$$

So

$$g \propto \frac{1}{A}$$

So at X, where A is greatest, g is smallest, as shown by slope of the tangent T_1 to the temperature curve in Figure 28.3 (ii). At Y, where A is smallest, g is greatest as shown by the slope of the tangent T_2.

Thermal and Electrical Conductivity

We can make a useful analogy between thermal conductivity and electrical conductivity.

The electric current I flowing along a conductor $= dQ/dt$, since ΔQ is the quantity of charge passing a given section in a time Δt. Also, $I = V/R$, where V is the potential difference between the ends of the conductor and R is its resistance (p. 270). Now $R = \rho l/A$, where ρ is the resistivity of the material, l is the length and A is its cross-sectional area (p. 278). So

$$I = \frac{dQ}{dt} = \frac{V}{\rho l/A}$$

So

$$\frac{dQ}{dt} = \frac{1}{\rho} A \frac{V}{l}$$

The quantity V/l is the *potential gradient* along the conductor. So

$$\frac{dQ}{dt} = \frac{1}{\rho} A \times \text{potential gradient} . \qquad . \qquad . \qquad . \qquad (1)$$

For heat conduction,

$$\frac{dQ}{dt} = kA \times \text{temperature gradient} \qquad . \qquad . \qquad . \qquad (2)$$

Comparing (1) with (2), we see that $1/\rho$ is analogous to k. The inverse of resistivity is defined as the electrical *conductivity*, symbol σ. So thermal conductivity k is analogous to electrical conductivity σ.

Wiedemann and Franz discovered a law which states that, at a given temperature, the *ratio of the thermal to electrical conductivity is the same for all metals*. So a metal which is a good thermal conductor is also a good electrical conductor. This suggests that electrons are the carriers in both thermal and electrical conduction in metals. So on heating a metal bar the free electrons gain thermal energy and distribute this energy by collision with the fixed positive metal ions in the solid lattice.

Effect of Thin Layer of Bad Conductor

Figure 28.5 shows a lagged copper bar AB, whose ends are pressed against metal tanks at 100° and 0°C, but are separated from them by layers of dirt. The length of the bar is 10 cm or 0·1 m, and the dirt layers are 0·1 mm or

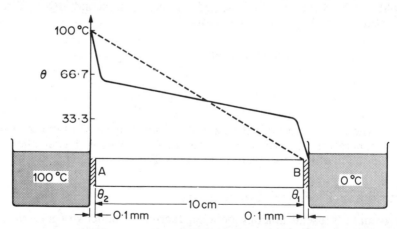

Figure 28.5 *Temperature gradients in good and bad conductors*

$0\cdot1 \times 10^{-3}$ m thick. Assuming that the conductivity of dirt is 1/1000 that of copper, let us find the temperature of each end A and B of the bar.

Suppose k = thermal conductivity of copper

A = cross-section of copper

θ_2, θ_1 = temperature of hot and cold ends

Since the bar is lagged, the heat flow per second, dQ/dt, is constant through the two dirt layers at A and B and through the bar. So using the dQ/dt

conductivity formula for the two layers and the bar,

$$\frac{dQ}{dt} = \frac{k}{1000} A \frac{100 - \theta_2}{0 \cdot 1 \times 10^{-3}} = kA \frac{\theta_2 - \theta_1}{0 \cdot 1} = \frac{k}{1000} A \frac{\theta_1 - 0}{0 \cdot 1 \times 10^{-3}}$$

Dividing through by kA, these equations give

$$\frac{100 - \theta_2}{0 \cdot 1} = \frac{\theta_2 - \theta_1}{0 \cdot 1} = \frac{\theta_1}{0 \cdot 1}$$

or
$$100 - \theta_2 = \theta_2 - \theta_1 = \theta_1$$

from which
$$\theta_2 = 66 \cdot 7 °C, \ \theta_1 = 33 \cdot 3 °C$$

So the total temperature drop, 100°C, is divided equally over the two thin layers of dirt and the long copper bar. The heavy lines in the figure show the temperature distribution; the broken line shows what it would be if there were no dirt.

Good and Bad Conductors

This numerical example shows what a great effect a thin layer of a bad conductor may have on thermal conditions; 0·1 mm of dirt causes as great a temperature fall as 10 cm of copper. We can generalise this result with the help of equation (2a) on p. 686:

heat flow/m^2 s = conductivity $\times$ temperature gradient

The equation shows that, if the heat flow is uniform, the temperature gradient is *inversely* proportional to the conductivity. So if the conductivity of dirt is 1/1000 that of copper, the temperature gradient in it is 1000 times that in copper. So 1 mm of dirt sets up the same temperature fall as 1 m of copper. In general terms we express this result by saying that the dirt prevents a good thermal contact, or that it provides a bad one. The reader who has already studied electricity will see an obvious analogy here. We can say that a dirt layer has a high thermal resistance, and so causes a great temperature drop.

Boiler plates are made of steel, not copper, although copper is about eight times as good a conductor of heat. The material of the plates makes no noticeable difference to the heat flow from the furnace outside the boiler to the water inside it, because there is always a layer of gas between the flame and the boiler plates. This layer may be very thin, but its conductivity is about 1/10 000 that of steel; if the plate is a centimetre thick, and the gas-film 1/1000 centimetre, then the temperature drop across the film is ten times that across the plate. So the rate at which heat flows into the boiler is determined mainly by the gas and not on the kind of metal used for the boiler plates. To save fuel, the gas is cleaned regularly from the bottom of the boiler.

If the water in the boiler deposits scale on the plates, the rate of heat flow is further reduced. Scale is a bad conductor, and, though it may not be as bad a conductor as gas, it can build up a much thicker layer. Scale must therefore be prevented from forming, if possible; and if not, it must be removed from time to time.

Heat Insulators

Badly conducting materials are often called *insulators*. The importance of building houses from insulating materials hardly needs to be pointed out.

Window-glass is a ten-times better conductor than brick, and it is also much thinner. A room with large windows therefore needs more heating in winter than one with small windows. Wood is as bad a conductor (or as good an insulator) as brick, but is also thinner. Wooden houses therefore have double walls, with an air-space between them. Air is an excellent insulator, and the walls stop much convection. Air is used as the insulator between glass plates in double-glazing windows. In polar climates, wooden huts must not be built with steel bolts going right through them; otherwise the inside ends of the bolts grow icicles from the moisture from the explorer's breath.

Examples on Conduction

1 A cavity wall is made of bricks 0·1 m thick with an air-space 0·1 m thick between them. (i) Assuming the thermal conductivity of brick is 20 times that of air, calculate the thickness of brick which conducts the same quantity of heat per second per unit area as 0·1 m of air. (ii) If the thermal conductivity of brick is 0·5 W m^{-1} K^{-1}, calculate the rate of heat conducted per unit area through the cavity wall when the outside surfaces of the brick walls are respectively 19°C and 4°C.

(i) Suppose θ_1, θ_2 are the respective temperatures at the ends of a brick of thickness l_B and thermal conductivity k_B, and at the ends of air of thickness l_A and thermal conductivity k_A. Then, with the usual notation,

$$\frac{dQ}{dt} = k_B A \frac{\theta_1 - \theta_2}{l_B} = k_A A \frac{\theta_1 - \theta_2}{l_A}$$

So

$$\frac{k_B}{l_B} = \frac{k_A}{l_A}$$

Then

$$l_B = l_A \times \frac{k_B}{k_A} = 0·1 \text{ m} \times 20 = 2 \text{ m}$$

Note that the brick length, 2 m, is 20 times greater than the air thickness, 0·1 m, because the brick conductivity is 20 times greater than that of air. This length of brick is 'thermally equivalent' to the air thickness or length.

(ii) Since the two bricks and the air are thermally in series, Figure 28.6 (i), we can replace the thickness of 0·1 m of air by the thermally equivalent 2 m of brick and add the three thicknesses of brick. So

$$\text{total brick thickness} = 0·1 + 2 + 0·1 = 2·2 \text{ m}$$

Then

$$\frac{1}{A} \cdot \frac{dQ}{dt} = k_B \frac{\theta_1 - \theta_2}{l_B} = 0·5 \frac{19 - 4}{2·2}$$

$$= 3·4 \text{ W m}^{-2}$$

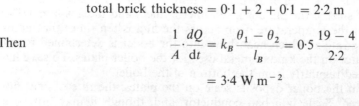

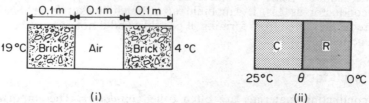

(i)   (ii)

Figure 28.6 *Series conductors*

2   A sheet of rubber and a sheet of cardboard, each 2 mm thick, are pressed together and their outer faces are maintained respectively at 0°C and 25°C. If the thermal conductivities of rubber and cardboard are respectively 0·13 and 0·05 W m⁻¹ K⁻¹, find the quantity of heat which flows in 1 hour across a piece of the composite sheet of area 100 cm².

We must first find the temperature, θ°C, of the junction of the rubber R and cardboard C, Figure 28.6 (ii). The temperature gradient across the rubber $= (\theta - 0)/2 \times 10^{-3}$; the temperature gradient across the cardboard $= (25 - \theta)/2 \times 10^{-3}$.

$$\therefore \text{d}Q/\text{d}t \text{ per m}^2 \text{ across rubber} = 0{\cdot}13 \times (\theta - 0)/2 \times 10^{-3}$$

$$\text{and d}Q/\text{d}t \text{ per m}^2 \text{ across cardboard} = 0{\cdot}05 \times (25 - \theta)/2 \times 10^{-3}$$

But in the steady state, the quantities of heat above are the same.

$$\therefore \frac{0{\cdot}13(\theta - 0)}{2 \times 10^{-3}} = \frac{0{\cdot}05(25 - \theta)}{2 \times 10^{-3}}$$

$$\therefore 13\theta = 125 - 5\theta$$

$$\therefore \theta = \frac{125}{18} = 7°C$$

Now area $= 100 \text{ cm}^2 = 100 \times 10^{-4} \text{ m}^2$. So, using the rubber alone, Q through area in 1 hour (3600 seconds)

$$= \frac{0{\cdot}13 \times 100 \times 10^{-4} \times 7 \times 3600}{2 \times 10^{-3}} = 16\,380 \text{ J}$$

Thermal Conductivity Measurements

We shall not be concerned with the details of measuring the thermal conductivity of good and bad conductors. For a *good conductor* such as a metal, a lagged, long, thick bar of uniform cross-section is used. It is heated at one end by an electrical heater and after a time the temperature of all the sections of the bar reaches a steady state. Two thermometers, placed in holes drilled into the middle of the bar at a distance of 10 cm, for example, are used to measure the temperature gradient along the bar in the steady state. The area A of a bar section is found by using calipers to measure the bar diameter. The thermal conductivity k is calculated from $\text{d}Q/\text{d}t = kA \times$ temperature gradient.

To measure k for a *bad conductor* such as cardboard or glass, a large area of the material is needed which is thin. In this case a reasonable heat flow is obtained through the material as the area A is large and the temperature gradient across the cardboard, for example, is high since it is thin. The cardboard is placed between two thick horizontal metal plates and the top plate is heated electrically. The temperature difference across the cardboard is measured by thermometers in the metal plates.

Thermal Conduction in Solids

Metals. Metals are good thermal conductors and good electrical conductors. Wiedemann and Franz showed that, at a given temperature, *the ratio of thermal to electrical conductivity is the same for all metals* (p. 690).

Since electrons are the carriers in electrical conduction, it is considered that electrons transport thermal energy through metals. So on heating a metal bar, the free electrons gain thermal energy and distribute this energy by collision with the fixed positive metal ions in the solid lattice.

Poor conductors. These have no free electrons. The transport of thermal energy through solids such as crystals is mainly due to waves. They are produced by lattice vibrations due to the thermal motion of the atoms. The waves are scattered by the atoms or by defects such as dislocations or impurity atoms and so distribute thermal energy to the solid.

The energy and momentum of the waves can also be considered carried by particles (p. 861). These particles are called *phonons*. They travel with the speed of sound.

EXERCISES 28A   Conduction

Multiple Choice

1 Figure 28A (i) shows two lagged metal bars X and Y of the same length 50 cm and the same cross-section area, with a common junction. The ends of X and Y are respectively 100°C and 0°C. Their thermal conductivities are respectively in the ratio $k_X : k_Y = 2 : 1$.

  Which of the graphs **A** to **E** is correct?

2 The SI unit of thermal conductivity is

  A $W\,m^{-2}\,K^{-1}$ **B** $W\,m^{-1}\,K^{-1}$ **C** $W\,m^{-1}\,K^{-2}$ **D** $W\,m^{-2}\,K^{-2}$ **E** $W\,m\,K^{-1}$

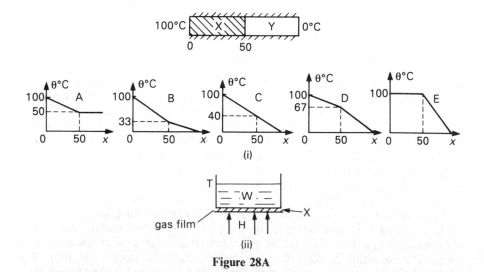

(i)

(ii)

Figure 28A

3 Figure 28A (ii) shows water W in a large tank T with burners H below the steel base X to boil the water. A thin film of gas is formed above the flames below X. To make the water boil much more quickly, which of the answers **A** and **E** is most correct?

  A A thicker steel base should be used.
  B Replace the steel base by a copper one.
  C Use a different gas for the burners.
  D Use a thinner steel base.
  E Clean the gas film from below the base.

4 Figure 28B shows a section of a double-glazing unit for the windows of a room, where G represents the glass and d is the thickness of air between the glass used. The room temperature is 20°C and the outside R is 0°C.

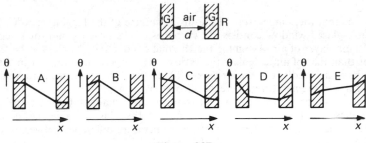

Figure 28B

Which graph from **A** to **E** best shows how the temperature θ from the room varies with the distance x along the unit to the outside?

Longer Questions

5 A closed metal vessel contains water (i) at 30°C and then (ii) at 75°C. The vessel has a surface area of 0·5 m² and a uniform thickness of 4 mm. If the outside temperature is 15°C, calculate the heat loss per minute by conduction in each case. (Thermal conductivity of metal = 400 W m^{-1} K^{-1}.)

6 A metal cylinder, containing water at 60°C, has a thickness of 4 mm and thermal conductivity 400 W m^{-1} K^{-1}. It is lagged by felt of thickness 2 mm and thermal conductivity 0·002 W m^{-1} K^{-1}. The room temperature is 10°C.

Using the relation $dQ/dt = kAg$ to find the temperature gradient g for the metal and for the felt, show that:
(a) the temperature θ of the metal–felt interface is practically 60°C;
(b) the rate of loss of heat by conduction is practically unaffected when the metal cylinder is replaced by a metal of smaller thermal conductivity, 100 W m^{-1} K^{-1}, and the same thickness.

7 An iron bar 0·10 m long and a copper bar 1·2 m long are joined together as shown in Figure 28C. The hot end of the iron bar is kept at 60°C while the cold end of the copper bar is kept at 0°C by the mixture of ice and water. The apparatus is lagged. At thermal equilibrium the temperature at the junction of the metals is T_j. Both bars have a diameter of 0·16 m.

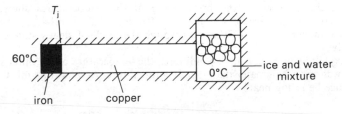

Figure 28C

Write down expressions for the temperature gradients across each of the bars.

Using these expressions, write down equations for the power transfers through each of the bars.

Explain why is it reasonable to assume that the power transfer through each conductor is the same. Using the above assumption, calculate T_j and the power flow.

Describe how you would check your calculations by experiment.

(Thermal conductivity of copper = 390 W m^{-1} K^{-1}, thermal conductivity of iron = 75 W m^{-1} K^{-1}.) (L.)

8 A copper hot-water cylinder of length 1·0 m and radius 0·20 m is lagged by 2·0 cm of material of thermal conductivity 0·40 W m^{-1} K^{-1}. Estimate the temperature of the outer surface of the lagging, assuming heat loss is through the sides only, if heat has to be supplied at a rate of 0·25 kW to maintain the water at a steady temperature of 60°C.

Assume that the temperature of the inside surface of the lagging is 60°C. (*L.*)

9 A double-glazed window consists of two panes of glass each 4 mm thick separated by a 10 mm layer of air. Assuming the thermal conductivity of glass to be 50 times greater than that of air, calculate the ratios: (i) temperature gradient in the glass, to temperature gradient in the air gap;   (ii) temperature difference across one pane of the glass to temperature difference across the air gap.

Sketch a graph showing how the temperature changes between the surface of the glass in the room and the surface of the glass outside, i.e. across the double-glazed window, if there is a large temperature difference between the room and the outside.

Explain why, in practice, the value of the ratio calculated in (i) is too high. (*L.*)

10 What is the relation between the *U value* of glass and its *thermal conductivity*?

An uninsulated roof has a U value of 1·9. Its U value when lagged is 0·4.

Calculate the U value of the lagging material and its thermal conductivity if it is 2 cm thick.

11 The U value of four construction components are given below.

Component	U value (W m^{-2} K^{-1})
Single-glazed window	5·6
Double-glazed window	3·2
Uninsulated roof	1·9
Well-insulated roof	0·4

What do you understand by the *U value* of a component?

A house has windows of area 24 m^2 and roof of area 60 m^2. The occupier heats the house for 3000 hours per year to a temperature which on average is 14 K above that of the air outside. Calculate the energy lost per year through: (i) the single-glazed windows;   (ii) the uninsulated roof, expressing your answers in kWh.

If electricity costs 5·5p per unit, calculate the annual savings the occupier could make by: (iii) installing double-glazing;   (iv) insulating the roof.

If double-glazing costs £3000 and roof insulation costs £100, which if either, of the two energy-saving steps would you advise the occupier to take? (*L.*)

12 (a) The diagram (Figure 28D(i)) shows a section of a house wall one brick thick, the surfaces of the brick being at the temperatures shown. If the thermal conductivity of the brick is 0·6 W m^{-1} K^{-1}, what is the rate of heat flow per unit area (W m^{-2}) through the bricks if steady-state conditions apply?

Explain why, under these conditions, the temperature of the outer surface of the wall must be greater than the air temperature. At what rate must the outer surface be losing heat?

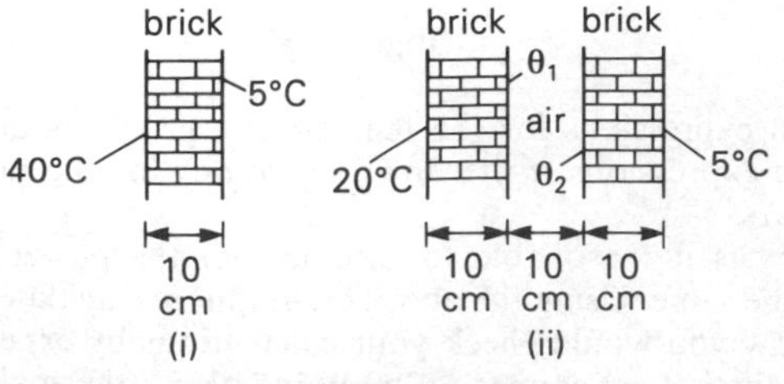

Figure 28D

(b) The diagram (Figure 28D (ii)) shows a section of a cavity wall made of brick, air and brick. The thermal conductivity of air is $0.02 \text{ W m}^{-1} \text{ K}^{-1}$. Explain why, when steady-state conditions apply, the rate of heat flow across each layer is the same. Assuming this to be the case, draw a sketch graph to show how the temperature changes between the brick surface at 20°C and that of 5°C.

13 Two equations which relate to electrical conduction, $I = V/R$ and $R = \rho l/A$, can be combined to give a single equation:

$$I = \frac{1}{\rho} A \frac{V}{l}$$

This last equation can be written in the form

$$\frac{dQ}{dt} = \sigma A \frac{V}{l}$$

What quantity does the symbol σ represent? A similar equation describes thermal conduction.

$$\frac{dQ}{dt} = kA \frac{\Delta T}{\Delta x}$$

In the cases of both thermal and electrical conduction, something can be considered as flowing. In each of these cases, identify the quantity which is flowing and give a suitable unit.

Identify, for both types of conduction, the quantity which may be thought of as 'driving the flow'. State another way in which these equations illustrate similarities between electrical and thermal conduction.

Good electrical conductors are good thermal conductors: this suggests that electrons play a part in both mechanisms. But diamond is a good thermal conductor despite it being an electrical insulator. Suggest a mechanism for thermal conduction in diamond. (L.)

Radiation

All heat comes to us, directly or indirectly, from the sun. The heat travels through 150 million km of space, mostly empty, and travels in straight lines, as does the light: the shade of a tree coincides with its shadow. Because heat and light travel with the same speed, they are both cut off at the same instant in an eclipse. Since light is propagated by waves of some kind we conclude that the heat from the sun is propagated by similar waves, and we say it is 'radiated'.

Measurements have been made which give the amount of radiant energy approaching the earth from the sun, called the *solar constant*. At the upper limit of our atmosphere, it is about 1340 W m^{-2}.

At the surface of the earth it is always less than this value because of absorption in the atmosphere. Even on a cloudless day it is less, because the ozone in the upper atmosphere absorbs much of the ultraviolet.

Experiment shows that more radiation is obtained from a dull black body than from a transparent or polished one. Black bodies are also better absorbers of radiation than polished or transparent ones, which either allow radiation to pass through themselves, or reflect it away from themselves. If we hold a piece of white card, with a patch of black drawing ink on it, in front of the fire, the black patch soon comes to feel warmer than its white surround.

Reflection and Refraction

If we focus the sun's light on our skin with a converging lens or a concave mirror, we feel heat at the focal spot. The heat from the sun has therefore been reflected or refracted in the same way as the light.

To show the refraction of heat apart from the refraction of light is more difficult. It was first done by the astronomer HERSCHEL in 1800. Herschel passed a beam of sunlight through a prism, as shown diagrammatically in Figure 28.7,

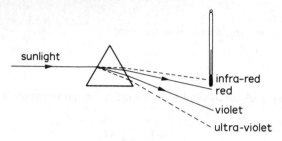

Figure 28.7 *Infrared and ultraviolet (diagrammatic)*

and explored the spectrum with a sensitive thermometer, whose bulb he had blackened. He found that in the visible part of the spectrum the mercury rose, showing that the light energy which it absorbed was converted into heat. But the mercury rose *more* when he carried the bulk into the darkened portion a little beyond the red of the visible spectrum. The sun's rays therefore carried energy which was not light energy.

Ultraviolet and Infrared

The radiant energy which Herschel found beyond the red is now called *infrared* radiation, because it is less refracted than the red. Radiant energy is also found beyond the violet and it is called *ultraviolet* radiation, because it is refracted more than the violet.

Ultraviolet radiation is absorbed by the human skin and causes sunburn. More importantly, it stimulates the formation of vitamin D, which is necessary for the assimilation of calcium and the prevention of rickets. It is also absorbed by green plants; in them it assists water to combine with carbon dioxide to form carbohydrates. This process is called *photosynthesis*. Ultraviolet radiation causes the emission of electrons from metals, as in photoelectric cells; and it produces a latent image on a photographic emulsion. It is harmful to the eyes.

Ultraviolet radiation is strongly absorbed by glass—spectacle-wearers do not get sunburn round the eyes—but enough of it gets through to affect a photographic film. It is transmitted with little absorption by quartz.

Infrared radiation is transmitted by rock-salt, but most of it is absorbed by glass. A little near the visible red passes easily through glass—if it did not, Herschel would not have discovered it. When infrared radiation falls on the skin, it gives the sensation of warmth. It is what we usually have in mind when we speak of 'heat radiation', and it is the main component of the radiation from a hot body; but it is in no essential way different from the other components, visible and ultraviolet radiation, as we shall now see.

Wavelengths of Radiation

In the section on optics, we show how the wavelength of light can be measured with a *diffraction grating*—a series of fine, close lines ruled on a glass. The wavelength ranges from 4.0×10^{-7} m for the violet, to 7.5×10^{-7} m for the red. Wavelengths are often given in units such as the micron (μm) or the nanometre (nm).

$$1 \text{ μm} = 10^{-6} \text{ m} \quad 1 \text{ nm} = 10^{-9} \text{ m}$$

We denote wavelength by the symbol λ; its value for visible light ranges from 400 nm to 750 nm, or 0.4 μm to 0.75 μm, and for infrared radiation from 750 nm to about 10^5 nm, or 0.75 μm to about 100 μm.

We now consider that X-rays and radio waves also have the same nature as light, and so do the γ-rays from radioactive substances. For reasons which we cannot here discuss, we consider all these waves to be due to oscillating electric and magnetic fields. Figure 28.8 shows roughly the range of their wavelengths: it is a diagram of the *electromagnetic spectrum*.

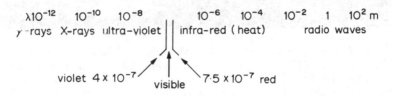

Figure 28.8 *The electromagnetic spectrum*

Infrared Photography, Thermal Imaging

Like other waves in the electromagnetic spectrum, reflection, refraction, diffraction, interference and polarisation can be obtained with infrared waves. Clear photographs can be taken at night using infrared lenses and detectors. Pilots in aeroplanes can detect enemy aircraft in this way, and photograph cities from the thermal radiation emitted from the ground.

Thermal imaging, using sensitive infrared detectors, can help to find people trapped below buildings which have collapsed. Plate 28A shows photographs taken at night.

Planck's Quantum Theory of Radiation

Measurements on the energy radiation from a white-hot source showed how the energy was distributed or shared between the different wavelengths. The wave theory, which assumed that the energy spread out from the source as waves, was not able to explain the distribution of energy among the wavelengths.

In 1895, however, Planck showed that a 'particle' theory, not a wave theory, could account for the distribution of energy. He assumed that the energy was carried in tiny separate units or *quanta* of energy, and that one unit or *quantum* of energy was equal to hf, where f was the frequency of the particular radiation and h was a constant now called the *Planck constant*. On the quantum theory, the quanta are separate (discrete) packets like grains of sand; there are

A

B

Plate 28A *Infrared photography or thermal imaging. Photos A and B are of the same scene; a Boeing 747 aircraft being loaded at night. A is a normal (visible wavelengths) photo; it shows relatively little detail. B, however, is a clear photo. It uses only infrared wavelengths from the scene and is produced with infrared lenses and plates. In this photo, black is hot and white is cold. As can be seen, the aircraft has not been parked for very long as its tyres and engines are still warm. Note also the hot parts of the light beacon above the aircraft.* (Courtesy of Barr and Stroud Limited)

amounts hf, $2hf$, $3hf$, and so on but no fractional units. On the wave theory, however, the energy can be any amount, including fractional values.

In 1905, Einstein showed that light could also be considered to consist of energy units equal to hf, where f was the frequency of the particular light wave. As we see later in the book, this 'particle' theory of light explains the experimental results in photoelectricity, which could not be explained using the wave theory of light.

Examples on Radiation

1 The tungsten filament of an 60 W electric lamp has a length of 0·5 m and a diameter of 6×10^{-5} m. The radiation energy from the lamp is 80% of its power value.

Calculate the average power per metre^2 radiated from the filament surface area.

$$\text{Power radiated as heat} = 80\% \text{ of } 60 \text{ W} = 48 \text{ W}$$

$$\text{Surface area of filament} = 2\pi rl = 2\pi \times 3 \times 10^{-5} \times 0{\cdot}5 = 9{\cdot}4 \times 10^{-5} \text{ m}^2$$

$$\text{So power radiated per metre}^2 = \frac{48}{9{\cdot}4 \times 10^{-5}} = 5{\cdot}1 \times 10^5 \text{ W m}^{-2}$$

Note: The power radiated is high because the lamp filament temperature is high, approximately 2000 K.

2 The solar constant, the energy per second arriving at the earth's surface, E, from the sun, S, is about 1400 W m^{-2}. Figure 28.9. Estimate the sun's temperature if its radius is 7×10^8 m, the sun's distance from the earth is $1 \cdot 5 \times 10^{11}$ m, and the sun's radiation $= kAT^4$, where $k = 6 \times 10^{-8}$, A is the surface $\times$ area of the spherical sun and T is its kelvin temperature.

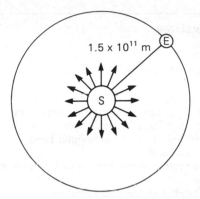

Figure 28.9 *Radiation from the sun*

Radiation from sun's surface $= kAT^4 = 6 \times 10^{-8} \times 4\pi \times (7 \times 10^8)^2 \times T^2$ since surface area A of a sphere is $4\pi r^2$ where r is the radius.

At a distance $1 \cdot 5 \times 10^{11}$ m, all the radiation falls on the surface of a sphere of this radius. So radiation on earth's surface per metre^2

$$= \frac{\text{radiation from sun}}{4\pi \times (1 \cdot 5 \times 10^{11})^2} = 1400$$

So $$\frac{6 \times 10^{-8} \times 4\pi \times (7 \times 10^8)^2 \times T^4}{4\pi \times (1 \cdot 5 \times 10^{11})^2} = 1400$$

Simplifying and solving for T, we find T is about 6000 K for the sun's surface.
If required, further details on radiation can be found on p. 721.

Greenhouses and the Greenhouse Effect

The so-called *greenhouse effect* describes how excessive CO_2 (carbon dioxide) emissions might cause long-term warming of the earth. However, the way in which *real* greenhouses work is very different from the way in which CO_2 allegedly causes global warming. Exposed plants and shrubs absorb radiated heat from the sun, but are cooled by breezes and convective air currents. In a greenhouse these cooling airflows are virtually eliminated, which is why the contents stay so warm.

Global warming, however, clearly does not work like this, since increased CO_2 in the atmosphere is not going to eliminate the planet's airflows! The undesirable property of CO_2 is that it is transparent to shortwavelength but not longwavelength radiation. So shortwavelength radiation from the sun passes through the CO_2 in the atmosphere and is absorbed by the earth, warming it up. The earth re-emits this radiation as black-body radiation consistent with its increased temperature. However, the re-emitted radiation has a relatively *long* wavelength, and so it is trapped by the CO_2 in the atmosphere. So there is an overall process by which radiation from the sun passes through the atmosphere, is absorbed and re-emitted by the earth as longwavelength radiation, and then trapped by the CO_2 in the atmosphere. This is the theory of global warming.

EXERCISES 28B   Radiation

Multiple Choice

1 Which of the following statements is NOT true?
Thermal radiation can
A be reflected   B be refracted   C be diffracted
D only travel through a vacuum   E produce interference.

2 Radiation of frequency 10^{14} Hz is in the
A radio band   B infrared band   C ultraviolet band
D visible band   E X-ray band.

3 When the filament of a bright electric lamp is switched off, which of the statement A to E is most correct?
A The light travels slower than the heat radiation.
B The light travels faster than the heat radiation.
C The light and heat travel at the same speed.
D The amount of heat radiation obtained depends on the filament volume.
E The amount of light obtained depends on the filament volume.

4 A solar furnace has a concave mirror of collecting area 0.6 m^2. A solid X of mass 0.5 kg, specific heat capacity 700 J kg K^{-1} and temperature $18°C$, is placed at the focus of the mirror.
    What temperature is reached by X in 0.5 minutes if the average solar energy reaching the earth is 1400 W m^{-2}?
    State the assumptions made in your calculation.

If required, further questions on radiation are on p. 734.

29 Further Topics in Thermal Properties[1]

Gases, Thermodynamics, Radiation

In this chapter we deal in more detail with (1) heat capacities of gases, internal energy of a gas and mean free path on kinetic theory; (2) the laws about real gases and critical temperature; (3) thermodynamics; (4) radiation.

Heat Capacities of Gases

Heat Capacities at Constant Volume and Constant Pressure

When we warm a gas, we may let it expand or not, as we please. If we do not let it expand—if we warm it in a closed vessel—then it does no external work, and all the heat we give it goes to increase its internal energy. *The heat required to warm one mole of gas through one degree, when its volume is kept constant, is called the molar heat capacity of the gas at constant volume.* It is denoted by C_V and is generally expressed in $J\ mol^{-1}\ K^{-1}$.

We can also warm a gas while keeping its pressure constant, and define the corresponding heat capacity. *The molar heat capacity of a gas at constant pressure is the heat required to warm one mole of it by one degree, when its pressure is kept constant.* It is denoted by C_p, and is expressed in the same units as C_V.

Any number of heat capacities can be defined for a gas, according to the mass and the conditions imposed upon its pressure and volume. For unit mass, 1 kg, of a gas, the heat capacities at constant pressure c_p, and at constant volume c_V, are called the *principal specific heat capacities*.

Molar Heat Capacities: their Difference

Figure 29.1 shows how we can find a relationship between the molar heat capacities of a gas. We first consider 1 mole of the gas warmed through 1 K at

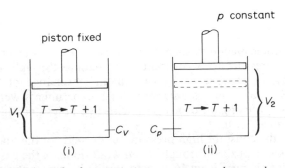

Figure 29.1 *Molar heat capacity at constant volume and pressure*

---

[1] Students limited in time for A level study may omit this chapter at a first reading, depending on the syllabus. Consult your teacher.

constant volume, (i). The heat required is C_V joules, and goes wholly to increase the internal energy U.

We next consider 1 mol warmed through 1 K at constant pressure, (ii). It expands from V_1 to V_2, and does an amount of external work given by

$$W = p \times \text{volume change} = p(V_2 - V_1)$$

Further, since the temperature rise of the gas is 1 K, and the internal energy of the gas is independent of volume, the rise in internal energy is C_V, the molar heat capacity at constant volume. Hence, from $\Delta Q = \Delta U + p.\Delta V$, the total amount of heat required to warm the gas at constant pressure is therefore

$$C_p = C_V + p(V_2 - V_1) \qquad . \qquad . \qquad . \qquad . \qquad (1)$$

We can simplify the last term of this expression by using the equation of state for one mole:

$$pV = RT$$

where T is the absolute temperature of the gas, and R is the molar gas constant. If T_1 is the absolute temperature before warming, then

$$pV_1 = RT_1 \qquad . \qquad . \qquad . \qquad . \qquad . \qquad (2)$$

The absolute temperature after warming is $T_1 + 1$; therefore

$$pV_2 = R(T_1 + 1) \qquad . \qquad . \qquad . \qquad . \qquad . \qquad (3)$$

and on substracting (2) from (3) we find

$$p(V_2 - V_1) = R$$

Equation (1) now gives $\qquad C_p = C_V + R$

or $\qquad\qquad\qquad C_p - C_V = R \qquad . \qquad . \qquad . \qquad . \qquad . \qquad (4)$

On p. 632, the specific heat capacity of a metal is measured at constant atmospheric pressure. The volume expansion of a metal at constant pressure is very small compared to that of a gas. So the external work done is very small. Hence it follows that there is not much difference between the specific heat capacities of a metal at constant pressure and constant volume.

Enthalpy

In engines and other machines such as refrigerator units, a gas may pass from a constant high pressure to a constant low pressure through a fine opening such as a needle or throttle valve, without any exchange of heat with the surroundings. So the change is adiabatic. Such experiments were first carried out many years ago using a porous plug with fine holes and this is known as a *Joule–Kelvin experiment*.

Suppose p_1 is the constant high pressure on one side of the fine opening and a volume V_1 of gas is pushed through. Then work done *on* the gas $= p_1 V_1$, since the volume of gas changes from V_1 to zero on the high-pressure side and work done $= p.\Delta V$.

If p_2 is the constant low pressure on the other side of the opening and the gas expands to a volume V_2, then work done *by* the gas in expanding from zero to $V_2 = p_2 V_2$.

So net work done by gas $= p_2 V_2 - p_1 V_1$.

If the process takes place adiabatically, then the work done by the gas = decrease in internal energy of gas = $U_1 - U_2$, where U_1 is the initial and U_2 is the final internal energy. So

$$U_1 - U_2 = p_2 V_2 - p_1 V_1$$

and

$$U_1 + p_1 V_1 = U_2 + p_2 V_2$$

The quantity $(U + pV)$ for a gas, which remained constant as the gas passed through the fine opening is called its *enthalpy* or *total heat*, symbol H.

So

$$\Delta H = \Delta U + p \Delta V$$

The enthalpy is constant in a throttle process as we have seen and is therefore an important quantity in mechanical engineering and industrial chemistry.

With most gases the internal energy decreases and the gas cools when passing through a fine opening from a high- to a low-pressure side. Exceptions are hydrogen and helium, which increase in temperature on account of the way their values of pV change with pressure (see p. 714). With an ideal gas, $p_2 V_2 = p_1 V_1$ since pV is independent of pressure, and $U_2 = U_1$. So no temperature change occurs in this case. In a refrigerator unit, when the coolant liquid is pumped from the high-pressure to the low-pressure side through a valve, the liquid evaporates and produces a drop in temperature.

The *latent heat of evaporation* of water $= \Delta U + p . \Delta V$ where ΔU is the internal energy gain and $p . \Delta V$ is the external work done (page 655). So the specific latent heat of evaporation is the *enthalpy change per kg* from water to steam. The specific latent heat of fusion of ice is the enthalpy change per kg from ice to water.

Example on Work done by Gas and Internal Energy Change

At 27°C and a pressure of $1 \cdot 0 \times 10^5$ Pa, an ideal gas has a volume of $0 \cdot 04$ m^3. It is heated at constant pressure until its volume increases to $0 \cdot 05$ m^3.

Find (i) the external work done; (ii) the new temperature of the gas; (iii) the change in internal energy of the gas if its mass is 45 g, its molar mass is 28 g and its molar heat capacity at constant volume is $0 \cdot 6$ J mol^{-1} K^{-1}; (iv) the total heat given to the gas.

(i)   External work done $= p \times$ volume change

$$= 1 \cdot 0 \times 10^5 \times (0 \cdot 05 - 0 \cdot 04)$$

$$= 1000 \text{ J}$$

(ii) From the gas law $V \propto T$ (constant pressure), the new kelvin temperature T_2 is given by

$$\frac{V_2}{V_1} = \frac{T_2}{T_1}$$

So

$$\frac{0 \cdot 05}{0 \cdot 04} = \frac{T_2}{(273 + 27)} = \frac{T_2}{300}$$

Then

$$T_2 = 300 \times 0 \cdot 05 / 0 \cdot 04 = 375 \text{ K}$$

(iii)   Number of moles of gas, $n = 45/28$. So

$$\text{internal energy rise} = nC_V(T_2 - T_1)$$

$$= \frac{45}{28} \times 0.6 \times (375 - 300)$$

$$= 72 \text{ J}$$

(iv)   Total heat given to gas, $\Delta Q = \Delta U + \Delta W$

$$= 72 + 1000$$

$$= 1072 \text{ J}$$

EXERCISES 29A   Heat Capacities of Gases, Internal Energy, Work

1 For hydrogen, the molar heat capacities at constant volume and constant pressure are respectively $20.5 \text{ J mol}^{-1} \text{ K}^{-1}$ and $28.8 \text{ J mol}^{-1} \text{ K}^{-1}$. What does this mean? (i) Which heat capacity is related to internal energy assuming hydrogen is an ideal gas? Explain your answer. (ii) From the values given, calculate the molar gas constant. (iii) Is C_p always greater than C_V? Explain your answer.

2 From the values of C_V and C_p given in question 1, calculate: (i) the heat needed to raise the temperature of 8 g of hydrogen from 10°C to 15°C at constant pressure; (ii) the increase in internal energy of the gas; (iii) the external work done. (Molar mass of hydrogen = 2 g.)

3 (a) The specific heat capacities of air are $1040 \text{ J kg}^{-1} \text{ K}^{-1}$ measured at constant pressure and $740 \text{ J kg}^{-1} \text{ K}^{-1}$ measured at constant volume. Explain briefly why the values are different.

   (b) A room of volume 180 m^3 contains air at a temperature of 16°C having a density of 1.13 kg m^{-3}. During the course of the day, the temperature rises to 21°C. Calculate an approximate value for the amount of energy transferred to the air during the day. Assume that air can escape from the room but no fresh air enters. Explain your reasoning. (L.)

4 Why is the energy needed to raise the temperature of a given mass of gas by a certain amount greater if the pressure is kept constant than if the volume is kept constant? (L.)

5 Explain why the values of the specific heat capacities of a gas when measured at constant pressure and at constant volume respectively are different. Derive an expression for the difference, for an ideal gas, in terms of its relative molecular mass M and the molar gas constant R.

   Given that the volume of a gas at s.t.p. is $2.24 \times 10^{-2} \text{ m}^3 \text{ mol}^{-1}$ and that standard pressure is $1.01 \times 10^5 \text{ N m}^{-2}$, calculate a value for the molar gas constant R and use it to find the difference between the quantities of heat required to raise the temperature of 0.01 kg of oxygen from 0°C to 10°C when:

   (a) the pressure;

   (b) the volume of the gas is kept constant.

   (Relative molecular mass of oxygen = 32.) (O. & C.)

6 Explain why the specific heat capacity of a gas is greater if it is allowed to expand while being heated than if the volume is kept constant. Discuss whether it is possible for the specific heat capacity of a gas to be zero.

   When 1 g of water at 100°C is converted into steam at the same temperature, 2264 J must be supplied. How much of this energy is used in forcing back the atmosphere? Explain what happens to the remainder of the energy. (1 g of water at 100°C occupies 1 cm^3. 1 g of steam at 100°C and 760 mmHg occupies 1601 cm^3. Density of mercury = $13\,600 \text{ kg m}^{-3}$.) (C.)

Kinetic Theory of Gases

Thermal Agitations and Internal Energy

This subject has been discussed on p. 678.

Figure 29.2 shows a monatomic molecule whose velocity c has been resolved into its components u, v, w along the x, y, z axes:

$$c^2 = u^2 + v^2 + w^2$$

The x, y, z axes are called the molecules' degrees of freedom: they are three directions such that the motion of the molecule along any one is independent of its motion along the others.

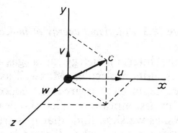

Figure 29.2 *Components of velocity*

If we average the speed c, and the components u, v, w, over all the molecules in a gas, we have

$$\overline{c^2} = \overline{u^2} + \overline{v^2} + \overline{w^2}$$

so

$$\overline{u^2} = \overline{v^2} = \overline{w^2} = \tfrac{1}{3}\overline{c^2}$$

The average kinetic energy of a molecule of the gas is given by (see p. 673)

$$\tfrac{1}{2}m\overline{c^2} = \tfrac{3}{2}kT$$

Therefore the average kinetic energy of a monatomic molecule, in each degree of freedom, is

$$\tfrac{1}{2}m\overline{u^2} = \tfrac{1}{2}m\overline{v^2} = \tfrac{1}{2}m\overline{w^2} = \tfrac{1}{2}kT$$

So the molecule has kinetic energy $\tfrac{1}{2}kT$ per degree of freedom.

Rotational Energy

Let us now consider a diatomic or polyatomic gas. When two of its molecules collide, they will, in general, tend to rotate, as well as to deflect each other. In some collisions, energy will be transferred from the translations of the molecules to their rotations; in others, from the rotations to the translations. We may assume, then, that the internal energy of the gas is shared between the rotations and translations of its molecules.

To discuss the kinetic energy of rotation, we must first extend the idea of degrees of freedom to it. A diatomic molecule can have kinetic energy of rotation about any axis at right-angles to its own. Its motion about any such axis can be resolved into motions about two such axes at right-angles to each other, Figure 29.3 (i). Motions about these axes are independent of each other, and a diatomic molecule therefore has two degrees of rotational freedom. A polyatomic molecule, unless it happens to consist of molecules all in a straight line, has no axis about which its moment of inertia is negligible. It can therefore have kinetic energy of rotation about three mutually perpendicular axes, Figure 29.3 (ii). It has three degrees of rotational freedom.

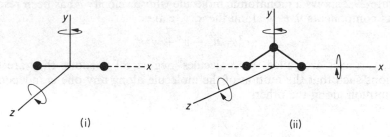

(i)   (ii)

Figure 29.3 *Rotational energy of molecules*

We have seen that the internal energy of a gas is shared between the translations and rotations of its molecules. Maxwell assumed that the average kinetic energy of a molecule, in each degree of freedom, *rotational as well as translational*, was $\frac{1}{2}kT$. This assumption is called the *principle of equipartition of energy*; experiment shows, as we shall find, that it is true at room temperature and above. At very low temperatures, when the gas is near liquefaction, it fails. At ordinary temperatures, then, we have:

---

average k.e. of monatomic molecule $= \dfrac{3}{2}kT$ **(trans.);**

average k.e. of diatomic molecule $= \dfrac{3}{2}kT$ **(trans.)** $+ \dfrac{2}{2}kT$ **(rot.)** $= \dfrac{5}{2}kT$;

average k.e. of polyatomic molecule $= \dfrac{3}{2}kT$ **(trans.)** $+ \dfrac{3}{2}kT$ **(rot.)** $= \dfrac{6}{2}kT.$

Internal Energy of a Gas

From the average kinetic energy of its molecules, we can find the internal energy of a mole of gas. Its internal energy, U, is the total kinetic energy of its molecules' random motions; so

$$U = N_A \times \text{average k.e. of molecule}$$

since there are N_A molecules mole where N_A is the Avogadro constant. For a monatomic gas, therefore,

$$U = \frac{3}{2} N_A kT \text{ (monatomic)}$$

The constant k is the gas constant per molecule; the product $N_A k$ is therefore the gas constant R for one mole. So the internal energy per mole is

---

$$U = \frac{3}{2} RT \text{ (monatomic)} \qquad . \qquad . \qquad . \qquad (1)$$

Similarly, for 1 mole of a diatomic gas,

$$U = \frac{5}{2} N_A kT = \frac{5}{2} RT \qquad . \qquad . \qquad . \qquad . \qquad (2)$$

Similar relations to (1) and (2) also apply to the internal energy of unit mass (1 kg) of a gas. In this case $U = 3rT/2$ (monatomic) and $U = 5rT/2$ (diatomic), where r is the gas constant per unit mass.

Molar Heat Capacities

We have seen that the internal energy U of a gas, at a given temperature, depends on the number of atoms in its molecule. For 1 mole of a monatomic gas, its value is, from equation (1),

$$U = \frac{3}{2} RT \qquad . \qquad . \qquad . \qquad . \qquad (3)$$

The molar heat capacity at constant volume is the heat required to increase the internal energy of 1 mole of the gas through 1 K. So, from (1),

$$C_V = \frac{3}{2} R$$

The molar heat capacity of a monatomic gas at constant pressure is therefore (p. 704):

$$C_p = C_V + R = \frac{3}{2} R + R = \frac{5}{2} R$$

Ratio of Molar Heat Capacities

Let us now divide C_p by C_V. This is called the ratio of the molar heat capacities, and is denoted by γ.

For a monatomic gas, its value is

$$\gamma = \frac{C_p}{C_V} = \frac{\frac{5}{2}R}{\frac{3}{2}R} = \frac{5}{3} = 1\cdot667$$

For a diatomic molecule, from equation (2), $U = \frac{5}{2} RT$.

So

$$C_V = \frac{5}{2} R$$

and

$$C_p = C_V + R = \frac{7}{2} R$$

Hence
$$\gamma = \frac{C_p}{C_V} = \frac{7}{5} = 1\cdot40$$

The ratio of the molar heat capacities of a gas thus gives us a measure of the number of atoms in its molecule, at least when the number is less than three. This ratio is easy to obtain by measuring the speed of sound in the gas. The poor agreement between the observed and theoretical values of γ for some of the polyatomic gases shows that, in its application to such gases, the theory is oversimple.

Mean Free Path

Although gas molecules move with high speeds such as several hundred metres per second, it still takes some time for two gases, initially separated, to diffuse into each other. This is due to numerous *collisions* which a molecule makes with other molecules surrounding it. Although the distance moved between successive collisions is not constant, we can consider the molecules of a gas under given conditions to have some average *mean free path*, λ.

Molecules have size. As an approximation, suppose a molecule in a gas is represented by a rigid sphere of diameter σ, and let us assume that a molecule moves a distance l while other molecules remain stationary. Then molecules such as A and B, situated at a distance σ from C, are those with which C *just* makes a collision, Figure 29.4. Consequently C makes a collision with all those molecules whose centres lie inside a volume $\pi\sigma^2 l$.

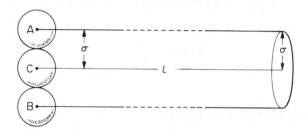

Figure 29.4 *Mean free path of molecule*

Suppose n is the number of molecules per unit volume in the gas. Then the number of collisions = the number of molecules in a volume $\pi\sigma^2 l$.

$$\therefore \lambda = \frac{\text{distance moved}}{\text{number of collisions}} = \frac{l}{n\pi\sigma^2 l}$$

$$\therefore \lambda = \frac{1}{n\pi\sigma^2} \qquad\qquad . \qquad . \qquad . \qquad . \qquad (1)$$

Since the molecules are not stationary as we have assumed but moving about in all directions, the chance of a collision by a molecule is greater. Taking this into account, the mean free path can be shown to be $\sqrt{2}$ less than that given in (1). Thus

$$\lambda = \frac{1}{\sqrt{2}n\pi\sigma^2} \qquad\qquad . \qquad . \qquad . \qquad . \qquad (2)$$

The relation in (2) shows that the mean free path λ decreases in inverse proportion to the number of molecules per unit volume, n, or the pressure of the gas.

Calculation of Mean Free Path

Suppose that, for hydrogen, $\sigma = 3 \times 10^{-10}$ m approximately. At a standard pressure of 760 mmHg, a mole of the gas (2 g) has 6×10^{23} molecules approximately (the Avogadro constant) and occupies a volume of about 22·4 litres or $22\cdot4 \times 10^{-3}$ m³. Then, from (2),

$$\lambda = \frac{22\cdot4 \times 10^{-3}}{\sqrt{2} \times 6 \times 10^{23} \times \pi \times 9 \times 10^{-20}}$$

$$= 9 \times 10^{-8} \text{ m (approx.)}$$

Thus the mean free path of hydrogen molecules at standard atmospheric pressure is about 10^{-7} m. Since the mean speed, the average distance travelled per second, of the molecule is about 2×10^3 m s^{-1}, we see that a molecule makes about 2×10^{10} collisions per second.

At extremely low pressures, the mean free path becomes comparable with the dimensions of the container and no meaning is then attached to the term 'mean free path' of the molecules.

Viscosity of Gas

The viscosity of a gas is due to a transfer of momentum by molecules moving between fast and slow gas layers. By kinetic theory, the coefficient of viscosity η of a gas can be shown to be given by the formula

$$\eta = \frac{1}{3} mn\bar{c}\lambda \qquad . \qquad . \qquad . \qquad . \qquad . \qquad (3)$$

where n is the number of molecules per unit volume, m is the mass of a molecule and $\bar{c}$ is the mean velocity and λ is the mean free path. But, from (2), $\lambda \propto 1/n$. It follows from (3) that η is *independent* of the pressure. This surprising result was shown to be true experimentally for normal pressures by Maxwell. He found that the damping of oscillations of a suspended horizontal disc was independent of the pressure of the gas when this was varied. The viscosity experiment is considered as one piece of sound evidence for the kinetic theory of gases, which leads to the relation in (3).

EXERCISES 29B   Kinetic Theory

1 The kinetic energy of translation of 1 mole of an ideal monatomic gas is given by $3RT/2$. Show that the molar heat capacities of the gas at constant volume and constant pressure respectively are $3R/2$ and $5R/2$.

2 A diatomic gas has three translational degrees of freedom and two vibrational degrees of freedom. Show that the molar heat capacity at constant volume is $5R/2$ and that the ratio of the two molar heat capacities is 1·4.

3 2 moles of oxygen at 27°C are compressed isothermally from a volume of 0·40 m³ to 0·10 m³. Assume $R = 8\cdot3$ J mol^{-1} K^{-1}, calculate the work done on the gas. What amount of heat is given up by the gas in this change? Explain your answer.

4 Argon has a molecular diameter about 3×10^{-10} m. The mean free path of the molecules is given by $\lambda = 1/(\sqrt{2}\pi\sigma^2 n)$, where λ is the mean free path, σ is the molecular diameter and n is the number of molecules per unit volume.

Estimate the mean free path in one mole of argon at standard pressure when its volume is 22.4×10^{-3} m^3, assuming the Avogadro constant $= 6 \times 10^{23}$ mol^{-1}.

5 List briefly the main assumptions on which the kinetic theory for an ideal gas is based.

In what follows, the ideal gas equation $p = \frac{1}{3}nm\langle c^2 \rangle$ may be assumed. *Do not* deduce it.

A gas consists of 5 molecules having speeds (in m s^{-1}) of 100, 600, 700, 800 and 1000. Find $\langle c^2 \rangle$.

Write down an expression, containing only terms from the above equation, which is proportional to the temperature of an ideal gas. If the speed of each molecule is doubled, by what factor is the temperature changed?

Show how Avogadro's law may be predicted from the above.

Show that the molar heat capacity at constant volume, for an ideal gas, is $\frac{3}{2}R$.

Deduce, with careful explanation, an expression for the mean free path λ between molecular collisions in a gas. How does λ vary with: (i) pressure, when the temperature is kept constant; (ii) temperature, then the pressure is kept constant? (*W.*)

6 (a) List the assumptions which are made in establishing the simple kinetic model of an ideal gas. Indicate which of these assumptions may not be justified for a real gas.

(b) Van der Waals' equation of state for one mole of a real gas is written as

$$\left(p + \frac{a}{V^2}\right)(V - b) = RT$$

(i) Explain the significance of the terms a/V^2 and b in this equation. (ii) With the aid of suitable sketch graphs relating p and V, compare the predictions of this equation with those of the ideal gas equation $pV = RT$. (*See next section*)

(c) Show that the mean free path path λ of a molecule in a gas is given approximately by the formula

$$\lambda = \frac{1}{n\pi\sigma^2}$$

where n is the number of molecules per unit volume and σ is the molecular diameter.

(d) Helium at 273 K (assumed an ideal gas) is introduced into a previously evacuated tube of length 0.40 m to such a pressure p that helium atoms, on average, traverse the length of the tube without colliding with each other.

(Take the volume of one mole of helium atoms at a temperature of 273 K and a pressure of 10^5 Pa to be 22×10^{-3} m^3, the atomic diameter σ of helium atoms to be 2.6×10^{-10} m, and the Avogadro constant N_A to be 6.0×10^{23} mol^{-1}.)

Calculate: (i) the number of helium atoms per unit volume under these conditions; (ii) the number of helium atoms per unit volume at 273 K temperature and 10^5 Pa pressure; (iii) the pressure p of the helium. (*O.*)

Real Gases, Critical Phenomena

Andrews' Experiments on Carbon Dioxide

Earlier we briefly discussed Andrews' important experiments on carbon dioxide which led to the idea of *critical temperature*.

Figure 29.5 shows his apparatus. In the glass tube A, he trapped carbon dioxide above the pellet of mercury X. To do this, he started with the tube open at both ends and passed dry gas through it for a long time. Then he sealed

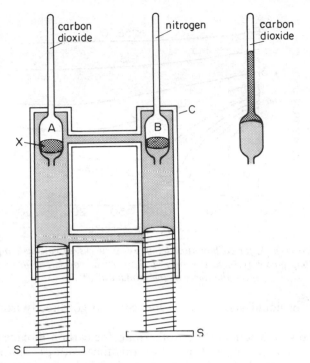

Figure 29.5 *Andrews' apparatus for isotherms of carbon dioxide at high pressures*

the end of the capillary. He introduced the mercury pellet by warming the tube, and allowing it to cool with the open end dripping into mercury. Similarly, he trapped nitrogen in the tube B.

Andrews then fitted the tubes into the copper casing C, which contained water. By turning the screws S, he forced water into the lower parts of the tubes A and B, and drove the mercury upwards. The wide parts of the tubes were under the same pressure inside and out, and so were under no stress. The capillary extensions were strong enough to withstand hundreds of atmospheres. Andrews actually obtained 108 atmospheres.

When the screws S were turned far into the casing, the gases were forced into the capillaries, as shown on the right of the figure, and greatly compressed. From the known volumes of the wide parts of the tubes, and the calibrations of the capillaries, Andrews determined the volumes of the gases. He estimated the pressure from the compression of the nitrogen, assuming that it obeyed Boyle's law. Air was also used in place of nitrogen.

For work above and below room temperature, Andrews surrounded the capillary part of A with a water bath, which he kept at a constant temperature between about 10°C and 50°C.

p–V Curves of Carbon Dioxide, Critical Temperature

Andrews' results for the pressure (p)–volume (V) curves at various constant temperatures, called *isothermals*, are shown in Figure 26.12 on page 665 to which the reader should refer.

Andrews' experiments also helped to illustrate the continuity between the liquid and gaseous states of a substance. Thus carbon dioxide is a gas above 31·1°C. Carbon dioxide can exist below 31·1°C as a liquid or as a vapour or

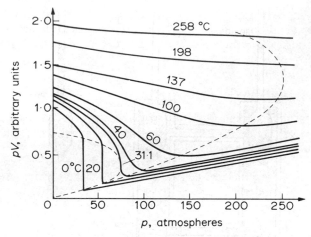

Figure 29.6 *Isothermals for carbon dioxide, as pV–p curves, at various temperatures. The small dotted loop passes through the ends of the vertical parts. The large dotted loop is the locus of the minima of pV*

as a mixture of liquid and vapour. One can also go directly from vapour to liquid.

Figure 29.6 shows some of the isothermals for carbon dioxide obtained by Andrews over a wide temperature range, this time with pV plotted against p. Corrections were made for the departure of nitrogen from Boyle's law. The vertical parts of the isothermals below the critical temperature of 31·1°C correspond to the change from the gaseous to the liquid state or to saturated vapour. If the gas obeyed Boyle's law, $pV =$ constant at constant temperature, the graph of pV against p would be a straight horizontal line. At the highest temperature shown, this is nearly true.

Departures from Boyle's Law at High Pressure

In 1847 Regnault measured the volume of various gases at pressures of several atmospheres. He found that, to halve the volume of the gas, he did not quite have to double the pressure on it. The product pV, therefore, instead of being constant, decreased slightly with the pressure. He found one exception to this

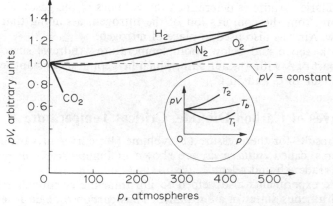

Figure 29.7 *Isothermals for various gases, at room temperature and high pressure*

rule: hydrogen. By compressing the gases further, Regnault found the variation of pV with p at constant temperature, and obtained results which are represented by the early parts of the curves in Figure 29.7. The complete curves in the figure show some of the results obtained by Amagat in 1892.

Real Gas Laws, Boyle Temperature

The results for real gases in Figure 29.6 can be expressed by the general relation

$$pV = A + Bp + Cp^2 + \cdots$$

where A, B, C are coefficients called 'virial coefficients' which are functions of temperature. At moderate pressure the gas obeys the relation

$$pV = A + Bp$$

to a good approximation. At very low pressures, $pV = A = RT$, that is, all real gases obeys the same gas law. In these conditions the volume of the molecules themselves is negligible and the attractive forces between them are negligible, so the law, in fact, is that for an ideal gas. For this reason, gas thermometer measurements are extrapolated to zero pressure in accurate work (p. 620). From $pV = A + Bp + Cp^2 + \cdots$, it follows that the coefficient B is the *gradient* at $p = 0$ of the pV graph. Figure 29.7 shows that B is negative for some gases such as carbon dioxide or oxygen and positive for others such as hydrogen. From the gradient at $p = 0$ in Figure 29.7, we see that the value of B becomes less negative as the temperature rises.

At one particular temperature T_b called the *Boyle temperature*, pV = constant for moderate pressures, to a good approximation, Figure 29.7, inset. For an ideal gas, the curve would be a straight line parallel to the axis of p, as shown in Figure 29.7. Below the Boyle temperature, the pV against p curves have a minimum, as at T_1. Above the Boyle temperature, the pV values increase, as at T_2. Hydrogen has a Boyle temperature of $-167°C$; for many other gases the Boyle temperature is above room temperature.

At high pressures and high temperatures, then, the laws for real gases depart considerably from those for real gases. At very low pressures, however, real gases obey the ideal gas laws.

Internal Energy and Volume

In our simple account of the kinetic theory of gases, we assumed that the molecules of a gas do not attract one another. If they did, any molecule approaching the boundary of the gas would be pulled towards the body of it, as is a molecule of water approaching the surface. The attractions of the molecules would thus reduce the pressure of the gas.

Since the molecules of a substate are presumably the same whether it is liquid or gas, the molecules of a gas must attract one another. But except for brief instants when they collide, the molecules of a gas are much further apart than those of a liquid. In 1 cubic centimetre of gas at s.t.p. there are 2.69×10^{19} molecules, and in 1 cubic centimetre of water there are 3.33×10^{22}; there are a thousand times as many molecules in the liquid, and so the molecules in the gas are ten times further apart. We may therefore expect that the mutual attraction of the molecules of a gas, for most purposes, can be neglected, as experiment, in fact, shows.

The experiment consists in allowing a gas to expand without doing external work; that is, to expand into a vacuum. Then, if the molecules attract one another, work is done against their attractions as they move further apart. But if the molecular attractions are negligible, the work done is also negligible. If any work is done against the molecular attractions, it will be done at the expense of the molecular kinetic energies. So as the molecules move apart, the internal energy of the gas, and therefore its temperature, will fall.

The expansion of a gas into a vacuum is called a 'free expansion'. If a gas does not cool when it makes a free expansion, then the mutual attractions of its molecules are negligible. Joule and Kelvin showed that most gases, in expanding from high pressure to low, do lose a little of their internal energy. The loss represents work done against the molecular attractions, which are therefore not quite negligible.

If the internal energy of a gas is independent of its volume, it is determined only by the temperature of the gas. The simple expression for the pressure, $p = \frac{1}{3}\rho c^2$, then holds and the gas obeys Boyle's law. Such a gas is called an *ideal* gas. All gases, when far from liquefaction, behave for most practical purposes as though they were ideal.

Real Gas and van der Waals' Equation

As stated on p. 715, the relation of the product pV to p for a real gas can be expressed by

$$pV = A + Bp + Cp^2 + \cdots$$

where $A, B, C, \ldots$ are coefficients decreasing in magnitude. The most important coefficients are A and B. So, at moderate pressures, $pV = A + Bp$. Further, when a real gas has a very low pressure, that is, p approaches zero, a real gas has the properties of an ideal gas. So $pV = A = RT$. At moderate pressures, then, we can write

$$pV = A + Bp = RT + Bp$$

or
$$p(V - B) = RT$$

From this relation we see that B represents a volume. Comparing this with the van der Waals' equation (p. 667), in which the term $(V - b)$ occurs, then B is roughly equal to the volume of the gas molecules themselves. Suppose 1 mole of a particular gas has a value B of 3×10^{-5} m^3. Since the Avogadro constant is about 6×10^{23} mol^{-1}, then roughly

$$\text{volume of 1 molecule} = \frac{3 \times 10^{-5}}{6 \times 10^{23}} = 5 \times 10^{-29} \text{ m}^3$$

The linear size of a molecule is of the order of the cube root of its volume. So an estimated size $= (5 \times 10^{-29})^{1/3} = 4 \times 10^{-10}$ m.

EXERCISES 29C   Real Gases, Critical Temperature

1 (a) Under what conditions would you expect the behaviour of a real gas to differ from that of an ideal gas? State and explain which assumptions of the kinetic theory are invalid under these conditions.

(b) What is meant by the critical temperature of a gas? Sketch graphs, using the same axes, to show the variation of pressure with volume for a fixed mass of carbon dioxide under isothermal conditions: (i) well above the critical temperature;   (ii) at the critical temperature;   (iii) below the critical temperature.

(c) For the graph of (b) (iii) describe the changes which occur as the pressure is increased for the range drawn. (N.)

2 Oxygen can be liquefied by first allowing the gas to expand very rapidly (to reduce its temperature to below 154 K) and then applying sufficient external pressure to it.

(a) Explain why a sudden expansion can lead to cooling.

(b) What is the significance of the temperature 154 K? (L.)

3 Define *critical temperature*.

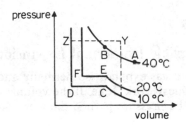

Figure 29A

Figure 29A shows the form of the relationship between the pressure and volume for isothermal changes in volume of carbon dioxide at three different temperatures. Explain qualitatively, using kinetic theory:

(a) the rise in pressure from A to B;

(b) the constancy of the pressure from C to D;

(c) why the pressure at E and F is greater than the pressure at C and D. Explain the energy interchanges which would take place in going from A to B and from C to D.

If the carbon dioxide were in a glass container, so that its condition could be seen, what would be observed as it was taken: (i) along the path CD: (ii) along the path indicated from 10°C to Y and then to Z? (L.)

4 (a) Draw a fully labelled diagram of the apparatus used in Andrews' experiments to investigate the relationship between the pressure p and the volume V for a fixed mass of a real substance over a range of pressures to produces a change of state, and for several different temperatures. Explain carefully how the pressure was varied and how corresponding values of p and V were determined.

(b) Sketch a graph of the results obtained with this apparatus with pV as ordinate and p as abscissa for three temperatures; one above, one below and one equal to the critical temperature. Indicate on your graph the corresponding isothermals you would expect for an ideal gas.

(c) (i) Which assumptions of the kinetic energy of an ideal gas need to be modified for a real gas?   (ii) Show how these modifications lead to an equation of state for a real gas different from that for an ideal gas. (N.)

5 Describe, with a diagram, the essential features of an experiment to the study the departure of a real gas from ideal gas behaviour.

Give freehand, labelled sketches of the graphs you would expect to obtain on plotting:

(a) pressure P against volume V;

(b) PV against P for such a gas at its critical temperature and at one temperature above and one below the critical temperature.

Explain van der Waals' attempt to produce an equation of state which would describe the behaviour of real gases.

Show that van der Waals' equation is consistent with the statement that all gases approach ideal gas behaviour at low pressures. (O. & C.)

6 (a) With pressure p as ordinate and volume V as abscissa, draw sketch graphs of $p-V$ isothermals for a real gas indicating: (i) the region in which the gas approximately obeys Boyle's law, (ii) the critical isothermal; (iii) the region in which the gas liquefies, and gas and liquid are in equilibrium. Justify your answers in each case.

  (b) At high pressure, the equation of state is modified from that of an ideal gas so that it becomes $pV = A + Bp$. Explain: (i) how A depends on the temperature; (ii) why B is related to the size of the gas molecules.

    In a particular experiment on 1 mole of oxygen, the value of B was found to be $20.2 \times 10^{-6} \, \text{m}^3$. Estimate the volume of the oxygen molecule. The Avogadro constant $N_A = 6.03 \times 10^{23} \, \text{mol}^{-1}$. (N.)

Thermodynamics

Work Done in Reversible Isothermal Expansion

The heat taken in when a gas expands isothermally and reversibly is the heat equivalent of the mechanical work done. If the volume of the gas increases by a small amount ΔV, at the pressure p, then the work done is (p. 655)

$$\Delta W = p\Delta V$$

In an expansion from V_1 to V_2, therefore, the work done is

$$W = \int dW = \int_{V_1}^{V_2} p \, dV$$

Assuming we have 1 mole of gas, $p = \dfrac{RT}{V}$

so
$$W = \int_{V_1}^{V_2} p \, dV = \int_{V_1}^{V_2} RT \frac{dV}{V} = RT \ln\left(\frac{V_2}{V_1}\right)$$

The heat required, Q, is therefore

$$Q = W = RT \ln\left(\frac{V_2}{V_1}\right)$$

Equation of Reversible Adiabatic

Consider one mole of the gas, and suppose that its volume expands from V to $V + \Delta V$, and that an amount of heat ΔQ is supplied to it. In general, the internal energy of the gas will increase by an amount ΔU, and the gas will do an amount of external work equal to $p\Delta V$, where p is its pressure. The heat supplied is equal to the increase in internal energy, plus the external work done:

$$\Delta Q = \Delta U + p\Delta V \quad . \quad . \quad . \quad . \quad (1)$$

The increase in internal energy represents a temperature rise, ΔT. We have seen already that the internal energy is independent of the volume, and is related to the temperature by the molar heat capacity at constant volume, C_V (p. 704). Therefore

$$\Delta U = C_V \Delta T$$

Equation (1) becomes

$$\Delta Q = C_V \Delta T + p\Delta V . \qquad . \qquad . \qquad . \qquad . \qquad (2)$$

Equation (2) is the fundamental equation for any change in state of one mole of an ideal gas.

For a reversible isothermal change, $\Delta T = 0$, and $\Delta Q = p\Delta V$.
For a reversible adiabatic change, $\Delta Q = 0$ and therefore

$$C_V \Delta T + p\Delta V = 0 . \qquad . \qquad . \qquad . \qquad . \qquad (3)$$

To eliminate ΔT we use the general equation, relating pressure, volume and temperature:

$$pV = RT$$

where R is the molar gas constant. Since both pressure and volume may change, when we differentiate this to find ΔT we must write

$$p\Delta V + V\Delta p = R\Delta T . \qquad . \qquad . \qquad . \qquad (4)$$

Substituting in (4) for $R(= C_p - C_V)$ and for $\Delta T(= -p.\Delta V/C_V$ from (3)), and then simplifying, we obtain

$$C_V V\Delta p + C_p p\Delta V = 0$$

or

$$V\Delta p + \gamma p\Delta V = 0 \left(\text{where } \gamma = \frac{C_p}{C_V} \right)$$

Integrating,

$$\int \frac{dp}{p} + \gamma \int \frac{dV}{V} = 0$$

or

$$\ln p + \gamma \ln V = A$$

where A is a constant.

Therefore,

$$pV^\gamma = \textbf{constant} \qquad . \qquad . \qquad . \qquad . \qquad (5)$$

This is the equation of a reversible adiabatic; the value of the constant can be found from the initial pressure and volume of the gas.

Carnot Cycle Efficiency

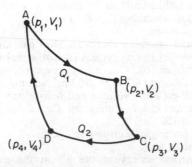

Figure 29.8 *Carnot cycle*

As we saw in an earlier chapter, the Carnot cycle ABCD consists of two reversible isothermal changes in the gas concerned, AB and CD in Figure 29.8, and two reversible adiabatic changes, BC and DA.

The heat Q_1 taken in along AB at a temperature T_1 = external work done.

So
$$Q_1 = RT_1 \ln (V_2/V_1) \qquad \qquad (1)$$

Similarly heat Q_2 given back at T_2 along CD is

$$Q_2 = RT_2 \ln (V_3/V_4) \qquad \qquad (2)$$

From the adiabatics for BC and AD, pV^γ = constant. So

$$p_1 V_1 \cdot V_1^{\gamma-1} = p_4 V_4 \cdot V_4^{\gamma-1}$$

and
$$p_2 V_2 \cdot V_2^{\gamma-1} = p_3 V_3 \cdot V_3^{\gamma-1}$$

Since $p_1 V_1 = p_2 V_2$ (isothermal AB) and $p_4 V_4 = p_3 V_3$ (isothermal CD), by division we have

$$\left(\frac{V_1}{V_2}\right)^{\gamma-1} = \left(\frac{V_4}{V_3}\right)^{\gamma-1}$$

So
$$V_1/V_2 = V_4/V_3 \qquad \qquad (3)$$

Now $$\text{efficiency} = \frac{\textbf{net work done}}{\textbf{heat taken in}}$$

$$= \frac{Q_1 - Q_2}{Q_1} = \frac{T_1 - T_2}{T_1} \qquad \qquad (4)$$

using equations (1) and (2) for Q_1 and Q_2, and cancelling, by using equation (3), the ratio of the volumes.

From equation (4), we see that it is impossible to obtain 100% efficiency when a gas is taken through a cycle. The maximum efficiency is obtained for reversible changes, which is an ideal situation.

In a *refrigerator* or *heat pump*, heat is transferred from a cold body or 'sink' at T_2 to the hotter body or 'source' at T_1. In this case a motor does work by driving a reversible Carnot cycle 'backwards', taking in heat from the 'sink' and pumping it to the 'source'.

EXERCISES 29D   Thermodynamics

1 1 mole of air expands isothermally at 27°C from a volume of 0·01 m³ to 0·04 m³. Calculate the work done, assuming $R = 8\cdot3$ J mol^{-1} K^{-1}. What heat is taken in by the gas? Explain your answer.

2 Using 4 moles of air and assuming this is an ideal gas, an engine goes through the following reversible changes in one cycle: (i) isothermal expansion from a volume of 0·02 m³ to 0·05 m³, at 100°C;   (ii) at constant volume, cooling to 27°C; (iii) isothermal compression at 27°C to 0·02 m³ volume;   (iv) at constant volume, compression to original pressure, volume and temperature in (i).

Show these four changes in a p–V diagram. Calculate:
(a) the heat taken in at 100°C;
(b) the work done by the engine during one cycle;
(c) the efficiency of the engine;
(d) the change in internal energy of the air at the end of the cycle. Assume $R = 8\cdot3$ J mol^{-1} K^{-1}.

3 A Carnot cycle uses an ideal gas as a working substance in reversible changes. (i) What is meant by *reversible*? (ii) Draw a sketch of a Carnot cycle and state what happens along each of the four stages. (iii) What is the work done in one cycle? (v) Write down an expression for the efficiency of the cycle in terms of temperature. How can the efficiency be increased?

4 Working between temperatures of 27°C and 127°C, an engine does 40 000 J of work in a Carnot cycle. Find the amount of heat taken in at the higher temperature, assuming no heat losses.

5 In a refrigerator unit, a Carnot engine using an ideal gas is driven in reverse by a 1 kW electric motor of efficiency 80% to freeze water at 0°C. Assuming the working temperatures of the engine are 20°C and 0°C respectively, calculate the mass of water frozen in 5 minutes, assuming no heat losses in the unit and $l = 3.4 \times 10^5 \, \text{J} \, \text{kg}^{-1}$. (*Hint* $Q_2/W = Q_2/(Q_1 - Q_2) = T_2/(T_1 - T_2)$, where Q_2 is heat abstracted from water, Q_1 is the heat given up and W is work done by motor.)

Radiation

Earlier we discussed briefly the reflection of radiation from black and silvery surfaces and the refraction of radiation through a prism of rocksalt. Heat radiation has *infrared* wavelengths in the electromagnetic spectrum. We now consider radiation in greater detail.

Detection of Heat (Infrared) Radiation

A thermometer with a blackened bulb is a sluggish and insensitive detector of radiant heat. More satisfactory detectors, however, are electrical. One kind consists of a long thin strip of blackened platinum foil arranged in a component zigzag on which the radiation falls (Figure 29.9).

The foil is connected in a Wheatstone bridge, to measure its electrical resistance. When the strip is heated by the radiation, its resistance increases, and the increase is measured on the bridge. The instrument was devised by LANGLEY in 1881; it is called a bolometer, *bole* being Greek for a ray.

Figure 29.9 *Bolometer strip*

The other, more common, type of radiation detector is called a *thermopile*. Its action depends on the electromotive force which appears between the junctions of two different metals, when one junction is hot and the other cold. The modern thermopile consists of many junctions between the fine wires, as shown diagrammatically in Figure 29.10, the wires are of silver and bismuth, 0·1 mm or less in diameter. Their junctions are attached to thin discs of tin, about 0·2 mm thick, and about 1 mm square. One set of discs is blackened and mounted behind a slit, through which radiation can fall on them; the junctions attached to them become the *hot* junctions of the thermopile. The other, cold, junctions are shielded from the radiation to be measured; the discs attached to them help to keep them cool, by increasing their surface area.

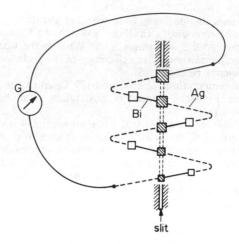

Figure 29.10 *Thermopile*

When radiation falls on the blackened discs of a thermopile, it warms the junctions attached to them, and sets up an e.m.f. This e.m.f. can be measured with a potentiometer, or, for less accurate work, it can be used to deflect a galvanometer, G, connected directly to the ends of the thermopile (Figure 29.10).

The Infrared Spectrometer

Infrared spectra are important in the study of molecular structure. They are observed with an infrared spectrometer, whose principle is shown in Figure 29.11. Since glass is opaque to the infrared, the radiation is focused by concave mirrors instead of lenses; the mirrors are plated with copper or gold on their front surfaces. The sources of light is a Nernst filament, a metal filament coated with alkaline-earth oxides, and heated electrically. The radiation from such a filament is rich in infrared.

Plate 29A *Researching Greenhouse Effect. A technician checks the trace of an infrared spectrometer to determine the carbon dioxide content of a sample of air, taken at the top of a nearby tower. Any water vapour is removed before the sample is passed into the spectrometer. Repeated observations over a period of years indicate that the carbon dioxide content of the atmosphere is increasing. Carbon dioxide is the main 'greenhouse gas', thought to increase global warming. (Photograph reproduced courtesy Hank Morgan/ Science Photo Library*

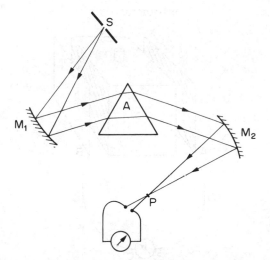

Figure 29.11 *Infrared spectrometer*

The slit S of the spectrometer is at the focus of one mirror which acts as a collimator. After passing through the rock-salt prism, A, the radiation is focused on to the thermopile P by the mirror M_2, which replaces the telescope of an optical spectrometer. Rotating the prism brings different wavelengths on to the slit of the thermopile; the position of the prism is calibrated in wavelengths with the help of a grating.

To a fair approximation, the deflection of the galvanometer is proportional to the radiant power carried in the narrow band of wavelengths which fall on the thermopile. If an absorbing body, such as a solution of an organic compound, is placed between the source and the slit, it weakens the radiation passing through the spectrometer, in the wavelengths which it absorbs. These wavelengths are therefore shown by a fall in galvanometer deflection.

Radiation and Absorption

We have already pointed out that black surfaces are good absorbers and radiators of heat, and that polished surfaces are bad absorbers and radiators. This can be demonstrated by the apparatus in Figure 29.12, called a *Leslie cube*. It is a cubic metal tank whose sides have a variety of finishes: dull black, dull white, highly polished. It contains boiling water, and therefore has a constant

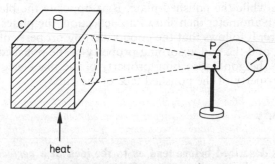

heat

Figure 29.12 *Comparing radiators*

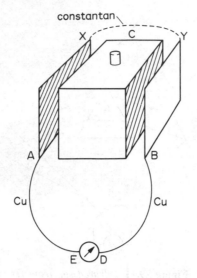

Figure 29.13 *Comparing absorbers*

temperature. Facing it is a thermopile, P, which is fitted with a blackened conical mouth-piece.

Provided the face of the cube occupies the whole field of view of the cone, its distance from the thermopile does not matter. The galvanometer deflection is greatest when the thermopile is facing the dull black surface of the cube, and least when it is facing the highly polished surface. The highly polished surface is therefore the worst radiator of all, and the dull black is the best.

A Leslie cube can also be used in an experiment to compare the absorbing properties of surfaces, Figure 29.13. The cube, C, full of boiling water, is placed between two copper plates, A, B, of which A is blackened and B is polished. The temperature difference between A and B is measured by making each of them one element in a thermojunction: they are joined by a constantan wire, XY, and connected to a galvanometer, by copper wires, AE, DB. If A is hotter than B, the junction, X, is hotter than the junction, Y, and a current flows through the galvanometer in one direction. If B is hotter than A, the current is reversed.

The most suitable type of Leslie cube is one which has two opposite faces similar—say grey—and the other two opposite faces very dissimilar—one black, one polished. At first the plates A, B are set opposite similar faces. The blackened plate, A, then becomes the hotter, showing that it is the better absorber.

The cube is now turned so that the blackened plate, A, is opposite the polished face of the cube, while the polished plate, B, is opposite the blackened face of the cube. The galvanometer then shows no deflection; the plates thus reach the same temperature. It follows that the good radiating property of the blackened face of the cube, and the bad absorbing property of the polished plate, are just compensated by the good absorbing property of the blackened plate, and the bad radiating property of the polished face of the cube.

The Black Body

The experiments described before lead us to the idea of a *perfectly black body*; one which absorbs all the radiation that falls upon it, and reflects and transmits none. The experiments also lead us to suppose that such a body would be the best possible radiator.

A good black body can be made simply by punching a small hole in the lid of a closed empty tin. The hole looks almost black, although the shining tin is a good reflector. The hole looks black because the light which enters through it is reflected many times round the walls of the tin, before it passes through the hole again, Figure 29.14. At each reflection, about 80% of the light energy is reflected, and 20% is absorbed. After two reflections, 64% of the original light goes on to be reflected a third time; 36% has been absorbed. After ten reflections, the fraction of the original energy which has been absorbed is 0.8^{10}, or 0.1. So the closed tin with a hole in it is a very good absorber of radiation.

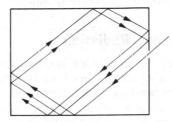

Figure 29.14 *Multiple reflections make a black body*

Any space which is almost wholly enclosed approximates to a black body. And, since a good absorber is also a good radiator, an almost closed space is the best radiator we can find.

A form of black body used in radiation measurements is shown in Figure 29.15. It consists of a porcelain sphere, S, with a small hole in it. The inside is blackened with soot to make it as good a radiator and as bad a reflector as possible. (The effect of multiple reflections is then to convert the body from nearly black to very nearly black indeed.) The sphere is surrounded by a high-temperature bath of, for example, molten salt (the melting-point of common salt is 800°C).

Figure 29.15 *A black body*

Quality of Radiation

The deepest parts of a coal or wood fire are black bodies. Anyone who has looked into a fire knows that the deepest parts of it look brightest—they are radiating most power. In the hottest part, no detail of the coals or wood can be seen. So the radiation from an almost enclosed space is uniform; its character does not vary with the nature of the surface of the space. This is so because the radiation coming out from any area is made up partly of the radiation emitted by that area, and partly of the radiation from other areas, reflected at the area in question. And if the hole in the body is small, the radiations from every area inside it are well mixed by reflection before they can escape. So the intensity and quality of the radiation escaping thus does not depend on the particular surface from which it escapes.

When we speak of the *quality* of radiation we mean the relative intensities of the different wavelengths in it; the proportion of red to blue, for example. The quality of the radiation from a perfectly black body depends *only on its temperature*. When the body is made hotter, its radiation becomes not only more intense, but also more nearly white; the proportion of blue to red in it increases. Because its quality is determined only by its temperature, black body radiation is sometimes called 'temperature radiation'.

Properties of Temperature Radiation

The quality of the radiation from a black body was investigated by LUMMER and PRINGSHEIM in 1899. They used a black body represented by B in Figure 29.16 and measured its temperature with a thermopile; they took it to 2000°C. To measure the intensities of the various wavelengths, Lummer and Pringsheim used an infrared spectrometer and a linear bolometer (p. 721) consisting of a single platinum strip.

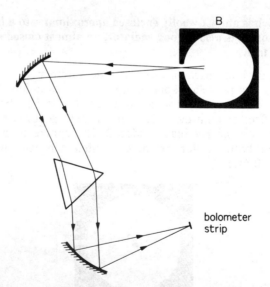

Figure 29.16 *Lummer and Pringsheim's apparatus for study of black body radiation (diagrammatic)*

The results of the experiments are shown in Figure 29.17. Each curve gives the relative intensities of the different wavelengths, for a given temperature of the body. The curves show that, as the temperature rises, the intensity of every wavelength increases, but the intensities of the shorter wavelengths increase more rapidly. Thus the radiation becomes, as we have already observed, less red, that is to say, more nearly white. The curve for sunlight has its peak at about 5×10^{-7} m in the visible green; from the position of this peak we conclude that the surface temperature of the sun is about 6000 K. Stars which

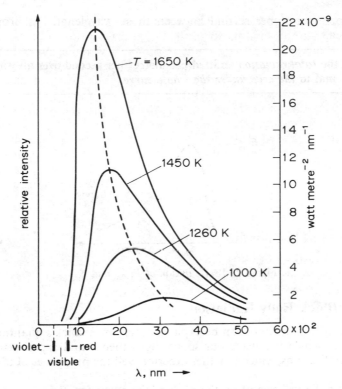

Figure 29.17 *Distribution of intensity in black body radiation*

are hotter than the sun, such as Sirius and Vega, look blue, not white, because the peaks of their radiation curves lie further towards the visible blue than does the peak of sunlight.

The actual intensities of the radiations are shown on the right of the graph in Figure 29.17. To speak of the intensity of a single wavelength is meaningless. The slit of the spectrometer always gathers a *band of wavelengths*—the narrower the slit the narrower the band—and we always speak of the intensity of a given band. We express it as follows:

$$\text{energy radiated m}^{-2}\text{ s}^{-1}, \text{ in band } \lambda \text{ to } \lambda + \Delta\lambda = E_\lambda \Delta\lambda \qquad (1)$$

The quantity E_λ is called *emissive power* of a black body for the wavelength λ and at the given temperature; its definition follows from equation (1):

$$E_\lambda = \frac{\text{energy radiated m}^{-2}\text{ s}^{-1}, \text{ in band } \lambda \text{ to } \lambda + \Delta\lambda}{\text{bandwidth}, \Delta\lambda}$$

The expression 'energy per second' can be replaced by the word 'power', whose unit is the watt. Thus

$$E_\lambda = \frac{\text{power radiated m}^{-2} \text{ in band } \lambda \text{ to } \lambda + \Delta\lambda}{\Delta\lambda}$$

In the figure, E_λ is expressed in watts per m^2 per nanometre (10^{-9} m).

The quantity $E_\lambda\Delta\lambda$ in equation (1) is the *area* beneath the radiation curve between the wavelengths λ and $\lambda + \Delta\lambda$ (Figure 29.18). Thus the energy radiated per metre^2 per second between those wavelengths is proportional to that area.

Similarly, the *total radiation* emitted per metre^2 per second over all wavelengths is proportional to *the area under the whole curve*.

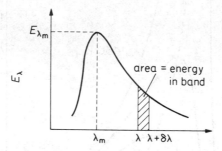

Figure 29.18 *Definition of E, λ_m and E_{λ_m}*

Laws of Black Body Radiation

The curves of Figure 29.17 can be explained only by Planck's quantum theory of radiation, which is outside our scope. Both theory and experiment lead to three generalisations, which together describe well the properties of black body radiation:

(i) If λ_m is the wavelength of the peak of the curve for T (in K), then

$$\lambda_m T = \text{constant} \qquad . \qquad . \qquad . \qquad . \qquad (2)$$

The value of the constant is 2.9×10^{-3} m K. In Figure 29.17 the dotted line is the locus of the peaks of the curves for different temperatures.

The relationship in (2) is sometimes called *Wien's displacement law*.

(ii) If E_{λ_m} is the height of the peak of the curve for the temperature T, then

$$E_{\lambda_m} \propto T^5 \qquad . \qquad . \qquad . \qquad . \qquad (3)$$

(iii) The curve showing the variation of E_λ with λ at constant temperature T in Figure 29.17 obeys the *Planck formula*

$$E_\lambda = \frac{c_1}{\lambda^5(e^{c_2/\lambda T} - 1)}$$

where c_1 and c_2 are constants.

Stefan's Law

If E is the *total* energy radiated per metre^2 per second at a temperature T, which is represented by the *total area* under the particular $E_\lambda - \lambda$ curve, then

$$E = \sigma T^4$$

where σ is a constant. This result is called *Stefan's law, and the constant σ is called the Stefan constant*. Its value is

$$\sigma = 5.7 \times 10^{-8} \text{ W m}^{-2} \text{ K}^{-4}$$

So in Figure 29.17, which shows four E_λ–λ graphs at different temperatures T, the total area below the graphs should be proportional to the corresponding value of T^4.

The energy per second or *power* P radiated by an area A of a black body radiator is thus given by

$$P = A\sigma T^4$$

where P is in watts. This is illustrated in the example which follows.

Example on Stefan's Law of Radiation

The tungsten filament of an electric lamp has a length of 0·5 m and a diameter 6×10^{-5} m. The power rating of the lamp is 60 W. Assuming the radiation from the filament is equivalent to 80% that of a perfect black body radiator at the same temperature, estimate the steady temperature of the filament. (Stefan constant = $5·7 \times 10^{-8}$ W m^{-2} K^{-4}.)

When the temperature is steady,

power radiated from filament = power received = 60 W

$$\therefore 0·8 \times 5·7 \times 10^{-8} \times 2\pi \times 3 \times 10^{-5} \times 0·5 \times T^4 = 60$$

since surface area of cylindrical wire is $2\pi rh$ with the usual notation.

$$\therefore T = \left(\frac{60}{0·4 \times 5·7 \times 10^{-8} \times 2\pi \times 3 \times 10^{-5}}\right)^{1/4}$$

$$= 1933 \text{ K}$$

Hot Object in Enclosure

Consider a black body at a temperature of T_0 where T_0 is the temperature of the room or enclosure containing the body. Since the body is in temperature equilibrium, the energy per second it radiates must equal the energy per second it absorbs (see also Prévost's theory of exchanges, p. 731). If A is the surface area of the body, then, from Stefan's law,

energy per second radiated = $\sigma A T_0^4$

So the *energy per second absorbed from the surroundings or enclosre* = $\sigma A T_0^4$.

Now suppose the black body X is heated electrically by a heater of power W watts and finally reaches a constant temperature T. In this case, from Prévost's theory,

energy per second from heater, W = net energy per second radiated by X

The net energy per second radiated by X = $\sigma A T^4 - \sigma A T_0^4$. So

$$W = \sigma A T^4 - \sigma A T_0^4 = \sigma A(T^4 - T_0^4)$$

Examples on Radiation

1 A metal sphere with a black surface and radius 30 mm, is cooled to $-73°C$ (200 K) and placed inside an enclosure at a temperature of $27°C$ (300 K). Calculate the initial rate of temperature rise of the sphere, assuming the sphere is a black body. (Assume

density of metal = 8000 kg m^{-3} specific heat capacity of metal = 400 J kg^{-1} K^{-1}, and Stefan constant = 5·7 × 10^{-8} W m^{-2} K^{-4}.)

Energy per second radiated by sphere = $\sigma A(T^4 - T_0^4)$

where A is the surface area $4\pi r^2$ of the sphere of radius r, $T = 200$ K and $T_0 = 300$ K. Since the temperature of the surroundings is greater than that of the sphere, the energy per second, Q, *gained* from the surroundings is given by

$$Q = \sigma . 4\pi r^2 . (300^4 - 200^4)$$

The mass m of the sphere = volume × density = $\frac{4}{3}\pi r^3 \rho$, where ρ is the density. If c is the specific heat capacity of the metal, and θ is the initial rise per second of its temperature, then

$$Q = mc\theta = \frac{4}{3}\pi r^3 \rho c\theta = \sigma 4\pi r^2 (300^4 - 200^4)$$

Dividing by $4\pi r^2$, and then simplifying,

$$\theta = \frac{\sigma(300^4 - 200^4) \times 3}{r\rho c}$$

$$= \frac{5·7 \times 10^{-8} \times (300^4 - 200^4) \times 3}{30 \times 10^{-3} \times 8000 \times 400}$$

$$= 0·012 \text{ K s}^{-1} \text{ (approx.)}$$

2   Estimate the temperature T_e of the earth, assuming it is in radiative equilibrium with the sun. (Assume radius of sun, $r_s = 7 \times 10^8$ m, temperature of solar surface = 6000 K, distance of earth from sun, $R = 1·5 \times 10^{11}$ m.)

Power radiated from sun = σ × surface area × T^4

$$= \sigma \times 4\pi r_s^2 \times T_s^4$$

Power received by earth = $\dfrac{\pi r_e^2}{4\pi R^2}$ × power radiated by sun

since πr_e^2 is the effective area of the earth on which the sun's radiation is incident *normally*, Figure 29.19, and $4\pi R^2$ is the total area on which the sun's radiation falls at a distance R from the sun where the earth is situated.

Now power radiated by earth = $\sigma . 4\pi r_e^2 . T_e^4$

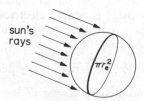

sun's
rays

πr_e^2

Figure 29.19 *Example*

Assuming radiative equilibrium,

power radiated by earth = power received by earth

$$\therefore \sigma . 4\pi r_e^2 . T_e^4 = \sigma . 4\pi r_s^2 . T_s^4 \times \frac{\pi r_e^2}{4\pi R^2}$$

Cancelling r_e^2 and simplifying, then

$$T_e^4 = T_s^4 \times \left(\frac{r_s^2}{4R^2}\right)$$

$$\therefore T_e = T_s \times \left(\frac{r_s}{2R}\right)^{1/2}$$

$$= 6000 \times \left(\frac{7 \times 10^8}{2 \times 1\cdot5 \times 10^{11}}\right)^{1/2}$$

$$= 209 \text{ K}$$

Note that the calculation is approximate, for example, the earth and the sun are not perfect black body radiators and the earth receives heat from its interior.

Prévost's Theory of Exchanges

In 1792 Prévost applied the idea of dynamic equilibrium to radiation. He asserted that a body radiates heat at a rate which depends on the nature of its surface and its temperature, and that it absorbs heat at a rate depending on the nature of its surface and the temperature of its surroundings. *When the temperature of a body is constant, the body is losing heat by radiation, and gaining it by absorption, at equal rates.*

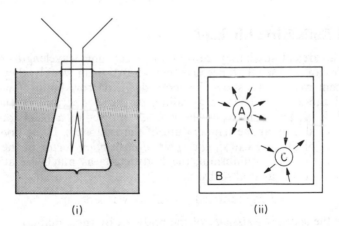

Figure 29.20 *Illustrating Prévost's theory of exchanges*

Figure 29.20 shows a simple experiment to demonstrate Prévost's theory. A high vacuum, electric lamp is carefully put into a can of water (Figure 29.20 (i)). The temperature of the lamp's filament is found by measuring its resistance. We find that, whatever the temperature of the water, the filament comes to that temperature, if we leave it long enough. When the water is cooler than the filament, the filament cools down; when the water is hotter, the filament warms up.

In the abstract language of theoretical physics, Prévost's theory is easy enough to discuss. If a hot body A (Figure 29.20 (ii)) is placed in an evacuated enclosure B, at lower temperature than A, then A cools until it reaches the temperature of B. If a body C, cooler than B, is put in B, then C warms up to the temperature of B. We conclude that radiation from B falls on C, and therefore also on A, even though A is at a higher temperature. Thus A and C each come to equilibrium at the temperature of B when each is absorbing and emitting radiation at equal rates.

Now let us suppose that, after it has reached equilibrium with B, one of the bodies, say C, is transferred from B to a cooler evacuated enclosure D (Figure 29.21 (i)). It loses heat and cools to the temperature of D. Therefore it is radiating heat. But if C is transferred from B to a warmer enclosure F, then C gains heat and warms up to the temperature of F (Figure 29.21 (ii). It seems unreasonable to suppose that C stops radiating when it is transferred to F. It is more reasonable to suppose that it goes on radiating but, while it is cooler than F, it absorbs more than it radiates.

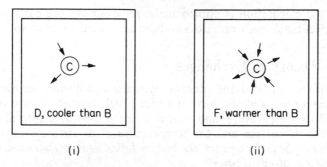

(i)   (ii)

Figure 29.21 *Illustrating Prévost's theory*

Spectral Emissivity: Kirchhoff's Law

Most bodies are coloured; they transmit or reflect some wavelengths better than others. We can now see that, because they absorb them weakly, they must also radiate them weakly. To show this, consider a narrow band of wavelengths between λ and $\lambda + \Delta\lambda$. The energy falling per m^2 per second on the body, in this band, is $E_\lambda \Delta\lambda$ where E_λ is the emissive power of a black body in the neighbourhood of λ, at the temperature of the enclosure. If the body absorbs a fraction a_λ of this, we call a_λ the *spectral absorption factor* of the body, for the wavelength λ. In equilibrium, the body emits as much radiation in the neighbourhood of λ as it absorbs. So

$$\text{energy radiated} = a_\lambda E_\lambda \Delta\lambda \text{ watts per } m^2$$

We define the *spectral emissivity* of the body e_λ, by the equation

$$e_\lambda = \frac{\text{energy radiated by body in range } \lambda \text{ to } \lambda + \Delta\lambda}{\text{energy radiated in same range, by black body at same temperature}}$$

$$= \frac{\text{energy radiated by body in range } \lambda \text{ to } \lambda + \Delta\lambda}{E_\lambda \Delta\lambda} = \frac{a_\lambda E_\lambda \Delta\lambda}{E_\lambda \Delta\lambda}$$

Thus $$e_\lambda = a_\lambda \ . \qquad . \qquad . \qquad . \qquad . \qquad (1)$$

Equation (1) expresses a law due to Kirchhoff:
The spectral emissivity of a body for a given wavelength is equal to its spectral absorption factor for the same wavelength.

Kirchhoff's law is not easy to demonstrate by experiment. If a plate when cold shows a red pattern on a blue background, it glows blue on a red ground when heated in a furnace. Not all such plates do this, because the spectral

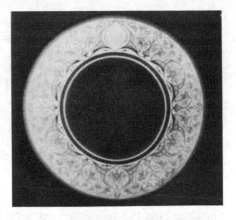

Plate 29B *Photographs showing how a decorated plate appears (left) by reflected light at room temperature and (right) by its own emitted light when incandescent at about 1100 K*

emissivity of many coloured pigments vary with their temperature. However, Plate 29A shows two photographs of a decorated plate, one taken by reflected light at room temperature (left), the other by its own light when heated to a temperature of about 1100 K (right).

Absorption by Gases

An experiment which shows that, if a body radiates a given wavelength strongly, it also absorbs that wavelength strongly, can be made with sodium vapour (see p. 842).

The process of absorption by sodium vapour—or any other gas—is not, however, the same as the process of absorption by a solid. When a solid absorbs radiation, it turns it into heat—into the random kinetic energy of its molecules. It then re-radiates it in all wavelengths, but mostly in very long ones, because the solid is cool. When a vapour absorbs light of its characteristic wavelength, however, its atoms are excited; they then re-radiate the absorbed energy, in the same wavelength (589·3 nm for sodium). But they re-radiate it in all directions, and therefore less of it passes on in the original direction than before

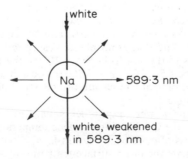

Figure 29.22 *Absorption by sodium vapour*

(Figure 29.22). So the yellow component of the original beam is weakened, but the yellow light radiated sideways by the sodium is strengthened. The sideways strengthening is hard to detect, but it was shown by WOOD in 1906. He used mercury vapour instead of sodium. See *Fraunhofer lines*, p. 843.

Pyrometry, Infrared Imagery

It is common to describe an object as glowing 'red hot', or even 'white hot', but what exactly does this mean? In fact, as we have seen, every object radiates energy at a frequency dependent on its temperature. This so-called *black body radiation* increases in frequency the hotter the object gets. At room temperature the radiation is in the infrared region of the spectrum, and so invisible to the human eye (we can still see the object because it reflects higher frequency, visible radiation from external light sources). As the temperature increases, however, the black body radiation moves into the visible part of the spectrum, and the object appears to glow red. At even higher temperatures the radiation increases further in frequency and the object turns yellow and even white.

We can use this black body radiation to measure temperature, known as *pyrometry*. The idea is to measure the spectrum of the radiation emitted by the object and use this to infer its temperature, using the black body radiation laws on p. 728: it is important to compensate for any reflected light from external light sources. Pyrometry is a very important thermometry technique, and is especially useful for measuring very high temperatures where other detectors would simply melt! If we had a sensor tuned to the infrared part of the spectrum, then we could detect the black body radiation of cool objects as well; in effect, we could see the objects without any external light sources, or see them in the dark! This is the basis for infrared (or thermal) imagery, which is now widely used for night-time surveillance and other applications (see p. 700 for an illustration).

EXERCISES 29E Radiation

1 Explain why a body at 1000 K is 'red hot' whereas at 2000 K it is 'white hot'. (*C.*)
2 The silica cylinder of a radiant wall heater is 0·6 m long and has a radius of 5 mm. If it is rated at 1·5 kW, estimate its temperature when operating. State *two* assumptions you have made in making your estimate. (The Stefan constant, $\sigma = 6 \times 10^{-8}$ W m^{-2} K^{-4}.) (*L.*)
3 (a) (i) What is meant by a *black body*? (ii) Sketch curves to show how the distribution of energy with wavelength in the radiation from a black body varies with temperature. (iii) With the aid of a diagram, describe a black body source suitable for investigations over the temperature range 0°C to 1000°C.
   (b) The solar power received at normal incidence on the surface of the earth is $1·4 \times 10^3$ W m^{-2}. Assume that the earth and the sun are uniform spheres and act as black bodies, and that atmospheric absorption can be neglected.
      (Take the radius of the sun to be $6·5 \times 10^8$ m, the radius of the earth's orbit around the sun to be $1·5 \times 10^{11}$ m, the Stefan constant σ to be $5·7 \times 10^{-8}$ W m^{-2} K^{-4}, and the speed of light c in vacuum to be $3·0 \times 10^8$ m s^{-1}. The surface area of a sphere of radius r is $4\pi r^2$.)
      (i) Show that the power output per square metre of the sun's surface is 75×10^6 W m^{-2}. (ii) Calculate the surface temperature of the sun. (iii) Estimate the rate of loss of mass of the sun in kg s^{-1} due to conversion of mass to energy. (iv) Estimate the mean surface temperature of the earth, assuming that it intercepts solar energy as a disc and re-radiates it uniformly from the whole of its spherical surface. (*O.*)
Figure 29B shows how E_λ, the energy radiated per unit area per second per unit wavelength interval, varies with wavelength λ for radiation from the sun's surface.
   Calculate the wavelengths λ_{max} at which the corresponding curves peak for:
   (a) radiation in the sun's core where the temperature is approximately 15×10^6 K;

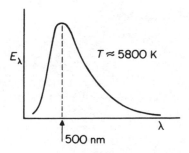

Figure 29D

(b) radiation in interstellar space which corresponds to a temperature of approximately 2·7 K.

(You may use the relation $\lambda_{max} \times T = \text{constant.}$)

Name the part of the electromagnetic spectrum to which the calculated wavelength belongs in each case. (*L.*)

4 As the temperature of a black body rises, what changes take place in:

(a) the total energy radiated from it;

(b) the energy distribution among the wavelengths radiated?

Illustrate (b) by suitable graphs and explain how the information required in (a) could be obtained from these graphs.

Use your graphs to explain how the appearance of the body changes as its temperature rises and discuss whether or not it is possible for a black body to radiate white light.

The element of an electric fire, with an output of 1·0 W, is a cylinder 25 cm long and 1·5 cm in diameter. Calculate its temperature when in use, if it behaves as a black body. (The Stefan constant $= 5\cdot7 \times 10^{-8}$ W m^{-2} K^{-4}.) (*L.*)

5 How can the temperature of a furnace be determined from observations on the radiation emitted?

Calculate the apparent temperature of the sun from the following information: sun's radius: $7\cdot04 \times 10^5$ km, distance from earth: $14\cdot72 \times 10^7$ km, solar constant: 1400 W m^{-2}, Stefan constant: $5\cdot7 \times 10^{-8}$ W m^{-2} K^{-4}. (*N.*)

6 Explain what is meant by:

(a) a *black body*;

(h) *black body radiation*.

State *Stefan's law* and draw a diagram to show how the energy is distributed against wavelength in the spectrum of a black body for two different temperatures. Indicate which temperature is the higher.

A roof measures 20 m × 50 m and is blackened. If the temperature of the sun's surface is 600 K, Stefan constant $= 5\cdot72 \times 10^{-8}$ W m^{-2} K^{-4}, the radius of the sun is $7\cdot8 \times 10^8$ m and the distance of the sun from the earth is $1\cdot5 \times 10^{11}$ m, calculate how much solar energy is incident on the roof per minute, assuming that half is lost in passing through the earth's atmosphere, the roof being normal to the sun's rays. (*O. & C.*)

PART FIVE

Electrons, Electronics, Atomic Physics

30 Electrons: motion in fields, The Cathode-Ray Oscilloscope

In this chapter, we first discuss how the charge and mass of the electron were measured and the properties of the electron. We then deal with the parabolic path of electrons in a uniform electric field and the circular and spiral path in uniform magnetic fields. The cathode-ray oscilloscope and its uses are described in the final section.

Basic Unit of Charge: The Electron

As we discussed earlier, matter has a particle nature—it consists of extremely small but separate particles which are atoms or molecules. Helmholtz, an eminent German scientist, stated about a century ago that charges were built up of basic units, that is, electricity is 'granular' and not continuous. He was led to this conclusion from the laws of electrolysis.

Particle Nature of Electricity

In electrolysis, we assume that the carriers of current through an acid or salt solution are ions, which may be positively or negatively charged.

Experiment shows that 96 500 coulombs (the Faraday constant) is the quantity of charge required to deposit one mole of a monovalent element. The number of ions carrying this charge is about 6.02×10^{23}, the Avogadro constant, for a monatomic element. So the charge on each of the ions is given by $96\,500/6.02 \times 10^{23}$ or 1.6×10^{-19} C. If 1.6×10^{-19} C is denoted by the symbol e, the charge on any ion is then e, $2e$ or $3e$, etc., depending on its valency. So e is a basic unit of charge.

All charges, whether produced in electrostatics, current electricity or any other method, are multiples of the basic unit e. Evidence that this is the case was obtained by MILLIKAN. In 1909, he designed an experiment to measure accurately the unit e. His method will be given later. Firstly, however, we describe a more direct experiment based on the Millikan method which can be carried out in the school laboratory.

Oil-drop Experiment for Electron Charge

Figure 30.1 (i) shows the principle of the experiment, A variable p.d. V can be applied to two circular plates P and Q separated by a small distance d. The upper plate P has a small hole in the middle and by means of a spray above it, fine oil drops can pass through to the space between P and Q.

In a darkened room, the drops are illuminated by a light beam and are then seen by using a low-power microscope M to focus them. M has a vertical fine line in the eyepiece graduated in millimetres, so that the motion of the drop can be observed and measured, Figure 30.1 (ii).

A particular oil drop is then focused so that it is clearly seen. The spray produces a charge q on the drop by friction and by altering the p.d. V between P and Q, the weight mg of the drop can be exactly balanced by an upward

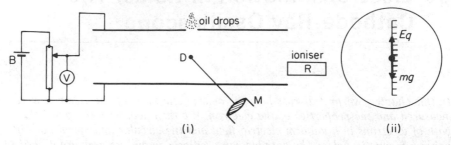

Figure 30.1 *Oil-drop experiment*

force Eq, where E is the electric field strength between P and Q. Since $E = V/d$, then

$$Eq = \frac{Vq}{d} = mg$$

and so

$$q = \frac{mgd}{V} \qquad . \qquad . \qquad . \qquad . \qquad (1)$$

Strictly, mg is the weight of the drop less the upthrust on the drop due to the surrounding air. The upthrust is equal to the weight of air displaced by the drop and is taken into account in the accurate Millikan experiment described shortly. Since the air density is very small compared to the density of oil, we may neglect the upthrust in this experiment.

Results of Experiment

The charge q on the *same* oil drop can be increased by ionising the air between the plates, so that the drop picks up more charges from the ions. This is done by using an α-particle source from apparatus placed at R in Figure 30.1 (i). Each time a new charge is obtained by the drop, the p.d. V is altered until the drop becomes stationary.

When the charge q is varied, suppose the values for V are 150 V, 190 V, 250 V and 750 V. From equation (1), since mgd is constant, then $q \propto 1/V$. The values of $1/V$ are given roughly in the table below.

V	150 V	190 V	250 V	750 V	
$1/V$	6·6	5·2	4·0	1·3	$\times 10^{-3}$
Ratios for $1/V$ or q	5	4	3	1	

Using the lowest value of $1/V$, which is 1/750 or 1·3, as the 'unit' of charge, the different charges on the drop were practically whole numbers 5, 4 and 3 in terms of this unit 1. This leads to the conclusion that a charge is a *multiple* of a basic or fundamental charge.

Calculation of Charge

From equation (1), the actual value of a charge is given by $q = mgd/V$. The values of g, d and V are known. The unknown value of m, the mass of the oil drop, can be found by allowing the drop to fall under gravity and measuring its *terminal velocity* v, using the graduated eyepiece.

In this case Stokes' law for the upward frictional (viscosity) force on a sphere of radius a is $F = 6\pi\eta av$, where η is the coefficient of viscosity of air. So, at the terminal velocity (zero acceleration),

$$F = mg = 6\pi\eta av$$

Now m = volume $V \times$ density ρ of oil $= 4\pi a^3\rho/3$, since the volume of a sphere is $4\pi a^3/3$. So

$$\frac{4\pi a^3\rho g}{3} = 6\pi\eta av$$

Simplifying,

$$a = \sqrt{\frac{9\eta v}{2\rho g}}$$

Knowing η, v, ρ and g, the radius a of the drop can be calculated from the formula. The mass m of the drop $= 4\pi a^3\rho/3$ and can now be found. So the charge q can then be calculated from $q = mgd/V$.

Millikan's accurate experiment, which we now briefly describe, shows that the basic or smallest value of charge is about $1·6 \times 10^{-19}$ C, which is the charge e on an electron. All charges are made up of units of e.

Millikan's Experiment

In his experiments Millikan used two horizontal plates A, B about 20 cm in diameter and 1·5 cm apart, with a small hole H in the centre of the upper plate, Figure 30.2. He used a fine spray to 'atomise' the oil and create tiny drops above H. Occasionally one would find its way through H, and would be observed in a low-power microscope by reflected light when the chamber was brightly illuminated. The drop was seen as a pin-point of light, and its downward velocity was measured by timing its fall through a known distance by means of a scale in the eyepiece. The field was applied by connecting a battery of several thousand volts across the plates A, B, and its strength F was known, since $E = V/d$, where V is the p.d. between the plates and d is their distance apart. Millikan found that the fraction between the drops when they were formed by the spray created electric charge, but to give a drop an increased charge an X-ray tube was operated near the chamber to ionise the air. In elementary laboratory experiments, increased charges may be obtained by using a radioactive source to ionise the air.

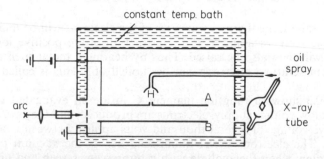

Figure 30.2 *Principle of Millikan's experiment*

Results

Millikan first measured the *terminal velocity* v_1 of an oil-drop through air. He then charged the oil-drop and applied an electric field to oppose gravity. The drop now moved with a different terminal velocity, v_2, which was again measured.

Millikan obtained a formula for the charge q on a drop which showed that $q \propto (v_1 - v_2)$. (See *Scholarship Physics* by the author.) In his experiment, Millikan found that, when an oil-drop gained charges, the *velocity change*, $v_1 - v_2$, varied by integral multiples of a basic unit. Now the charge q is proportional to $(v_1 - v_2)$ for a particular drop. Consequently *the charge is a multiple of some basic unit.* So a particular charge consists of 'grains' or multiples of this basic unit. From the formula, the charge q on the drop can be calculated. Millikan found, working with hundreds of drops, that the charge q was always a simple multiple of a basic unit, about $1 \cdot 6 \times 10^{-19}$ C, which was the *lowest* charge found. Earlier Sir J. J. Thomson had discovered the existence of the electron, a particle inside atoms which carried a negative charge $-e$ (p. 752). Millikan concluded that the charge e was $1 \cdot 6 \times 10^{-19}$ C numerically.

Thermionic Emission of Electrons, Heated Cathodes

Nowadays a *hot cathode* is used to produce a supply of electrons. This may consist of a fine tungsten wire or filament, which is heated to a high temperature when a low voltage source of 6 V is connected to it. Metals contain free electrons, moving about rather like the molecules in a gas. If the temperature of

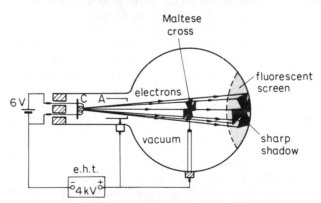

Figure 30.3 *Electrons travel in straight lines*

the metal is raised, the thermal velocities of the electrons will be increased. The chance of electrons escaping from the attraction of the positive ions, fixed in the lattice, will then also be raised. Thus by heating a metal such as tungsten to a high temperature, electrons can be 'boiled off'. This is called *thermionic emission.*

Figure 30.3 shows a tungsten filament C inside an evacuated tube. When heated by a low voltage supply, electrons are produced, and they are accelerated by a positive voltage of several thousand volts applied between C and a metal cylinder A. The electrons travel unimpeded across the vacuum past A, and produce a glow when they collide with a fluorescent screen and give up their energy.

Properties of Cathode Rays or Electrons

Before they were known to be tiny particles carrying a small charge e, beams of electrons were called *cathode rays* because they came from the cathode inside the tube producing them.

Fast-moving electrons emitted from a heated cathode C produce a sharp shadow of a Maltese cross on the fluorescent screen, as shown in Figure 30.3. Thus the electrons travel in straight lines. They also produce heat when incident on a metal—a fine piece of platinum glows, for example.

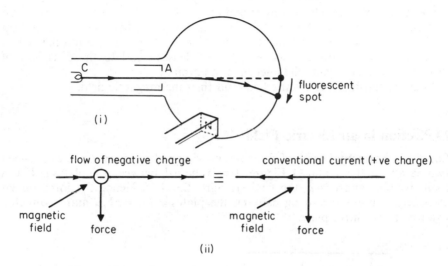

Figure 30.4 *Deflection shows electrons are negatively charged*

When a magnet is brought near to the electron beam, the glow on the fluorescent screen moves, Figure 30.4 (i). If Fleming's left-hand rule is applied to the motion to deduce the conventional current (+ve charge) direction, the middle finger points in a direction *opposite* to the electron flow, Figure 30.4 (ii). This shows that electrons are particles which carry a *negative* charge.

This is confirmed by collecting electrons inside a *Perrin tube*, shown in Figure 30.5. The electrons are deflected by the magnet S until they pass into a metal cylinder called a 'Faraday cage' (see p. 219). The cylinder is connected

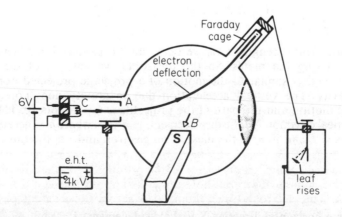

Figure 30.5 *Perrin tube. Direct method for testing electron charge*

to the plate of an electroscope, which has been negatively charged using an ebonite rod and fur. As soon as the electrons are deflected into the cage the leaf rises further, showing that an extra *negative* charge has been collected by the cage.

Electron Motion in Electric and Magnetic Fields

Gravitational fields such as those round the earth and the sun affect the motion of masses in these fields. Similarly, electric and magnetic fields affect the motion of charged particles which enter these fields. A study of the motion has led to many useful applications, such as 'electron lenses', and to the discovery of unknown particles inside the nucleus of an atom.

First we consider the electric field and then the magnetic field.

Deflection in an Electric Field

Suppose a horizontal beam of electrons, moving with velocity v, passes between two parallel plates P and Q, Figure 30.6. If the p.d. between the plates is V and their distance apart is d, the field strength $E = V/d$. Hence the force on an electron of charge e moving between the plates $= Ee = eV/d$ and is directed towards the positive plate.

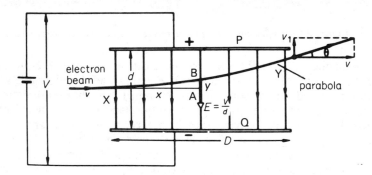

Figure 30.6 *Deflection in electric field*

Since the field strength E is vertical, no horizontal force acts on the electron entering the plates. So the horizontal velocity, v, of the beam is unaffected. This is similar to the motion of a projectile projected horizontally under gravity. The vertical acceleration due to gravity does not affect the horizontal motion and the path of the projectile is a parabola (p. 11).

In a vertical direction AB the displacement, $y = \frac{1}{2}at^2$, where $a =$ acceleration $=$ force/mass $= Ee/m_e$ if m_e is the mass of an electron and t is the time.

$$\therefore y = \frac{1}{2}\frac{Ee}{m_e}t^2 \qquad \qquad (1)$$

In a horizontal direction XA, the displacement, $x = vt$. $\qquad$ (2)

Eliminating t between (1) and (2), we obtain

$$y = \frac{1}{2}\left(\frac{Ee}{m_e}\right)\frac{x^2}{v^2} = \frac{Ee}{2m_e v^2}x^2$$

Between the plates, E and v stay constant. Since e *and* m_e are constant, we can write $y = kx^2$ for the motion, where k is a constant equal to $Ee/2m_e v^2$. This is the equation for a parabola. **So the path is a parabola between the plates.**

When the electron just passes the plates at Y, $x = D$. The value of y is then $y = EeD^2/2m_e v^2$. Outside the plates or field, the beam then moves *in a straight line*, as shown in Figure 30.6. The time for which the electron is between the plates is D/v. So the component of the velocity v_1, gained in the direction of the field during this time, is given by

$$v_1 = \text{acceleration} \times \text{time} = \frac{Ee}{m_e} \times \frac{D}{v}$$

Then the angle θ at which the beam comes out from the field is given by:

$$\tan\theta = \frac{v_1}{v} = \frac{EeD}{m_e v}\cdot\frac{1}{v} = \frac{EeD}{m_e v^2}$$

The energy of the electron is increased by an amount of $\frac{1}{2}mv_1^2$ as it passes through the plates, since the energy due to the horizontal motion is unaltered. This energy increase is equal to *force* × *distance*, where the force $= Ee$ and the distance in the direction of the force $=$ the vertical displacement of the electron in the field direction.

When an electron moves horizontally into a uniform vertical electric field it describes a parabolic path. The horizontal motion of the electron is not affected by the field. (Compare masses in gravitational field.) A charge gains energy when it moves in the direction of an electric field.

Example on Electron Motion in Electric Field

A beam of electrons, moving with a velocity of $1\cdot0 \times 10^7$ m s^{-1}, enters midway between two horizontal parallel plates P, Q in a direction parallel to the plates, Figure 30.7. P and Q are 5 cm long, 2 cm apart, and have a p.d. V applied between them. Calculate V if the beam is deflected downwards so that it just grazes the edge of the lower plate Q. (Assume $e/m_e = 1\cdot8 \times 10^{11}$ C kg^{-1}.)

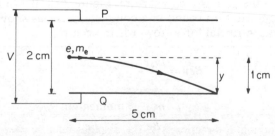

Figure 30.7 *Example*

Electric field strength between plates, $E = \dfrac{V}{d} = \dfrac{V}{2 \times 10^{-2}}$

Downward acceleration on electron, $a = \dfrac{\text{force}}{\text{mass}} = \dfrac{Ee}{m_e}$

So vertical distance, $y = \tfrac{1}{2}at^2 = \tfrac{1}{2}\dfrac{Ee}{m_e}t^2$

But $y = 1 \text{ cm} = 10^{-2} \text{ m}$ and $t = \dfrac{5 \times 10^{-2}}{1 \times 10^7} = 5 \times 10^{-9}$ s, since the horizontal velocity is not affected by the vertical electric field and remains constant.

So $\qquad 10^{-2} = \dfrac{1}{2} \times \dfrac{V}{2 \times 10^{-2}} \times 1 \cdot 8 \times 10^{11} \times (5 \times 10^{-9})^2$

Simplifying, $\qquad\qquad\qquad V = 89 \text{ V (approx.)}$

Deflection in a Magnetic Field

Consider an electron beam, moving with a speed v, which enters a *uniform magnetic field* of magnitude B acting *perpendicular* to the direction of motion, Figure 30.8. The force F on an electron is then Bev. The direction of the force is perpendicular to both B and v. Consequently, unlike the electric force, the magnetic force cannot change the *energy* of the electron. It **deflects the electron but does not change its speed or kinetic energy.**

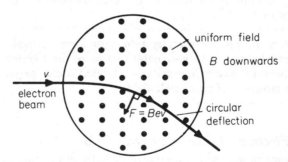

Figure 30.8 *Circular motion in uniform magnetic field*

The force Bev is always normal to the path of the beam. If the field is uniform the force is constant in magnitude and the beam then travels in a *circle* of radius r. If Bev is small, the path is a circular arc, Figure 30.8. If Bev is large enough, the electron beam can turn round in a complete circle, as shown in Plate 30B. Since Bev is the centripetal force (towards the centre),

$$Bev = \dfrac{m_e v^2}{r}$$

$$\therefore r = \dfrac{m_e v}{Be} = \dfrac{\text{momentum}}{Be}$$

If the velocity v of the electron decreases continuously due to collisions, for example, its momentum decreases. So, from the relation for r above, the radius of its path decreases and the electron will thus tend to spiral instead of moving in a circular path of constant radius.

Example on Magnetic Field Deflection

Protons, with a charge–mass ratio of $1.0 \times 10^8 \, \text{C kg}^{-1}$, are rotated in a circular orbit of radius r when they enter a uniform magnetic field of 0.5 T. Show that the number of revolutions per second, f, is independent of r and calculate f.

Suppose m_p is the proton mass and e is the charge. Then, with the usual notation,

$$Bev = \frac{m_p v^2}{r}$$

So
$$v = \frac{Ber}{m_p} = r\omega$$

where ω is the angular velocity. Since r cancels,

$$\omega = \frac{Be}{m_p} = 2\pi f$$

Then
$$f = \frac{Be}{2\pi m_p}$$

This result for f shows that it is independent of r. Also,

$$f = \frac{0.5 \times 1 \times 10^8}{2 \times 3.14} = 8 \times 10^6 \, \text{rev s}^{-1}$$

Specific Charge of Electron by Magnetic Deflection

The 'charge per unit mass' or *specific charge* of an electron (e/m_e) can be measured in the school laboratory by a teltron tube designed for this purpose, Figure 30.9.

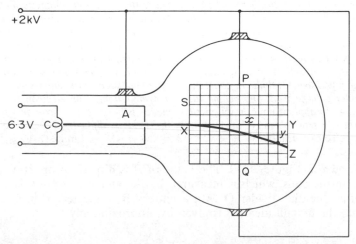

Figure 30.9 *Deflection of electron beam (Helmholtz coils not shown)*

The evacuated tube has (i) a hot cathode C which produces an electron beam, (ii) an accelerating anode A, (iii) a vertical screen S coated with luminous paint, between two plates P, Q not used in this experiment. The screen S has horizontal and vertical lines equally spaced and is slightly inclined to the electron beam incident on it. The beam then produces a luminous glow and its path can be seen.

Two Helmholtz coils H, H connected in series are on opposite sides of the tube. Plate 30A shows the coils and the tube in a photograph. The Helmholtz coils produce a *uniform* magnetic field B perpendicular to the beam over the region of S, so that the beam is deflected in a circular path. The circuit for the

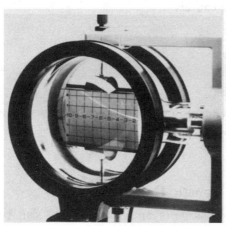

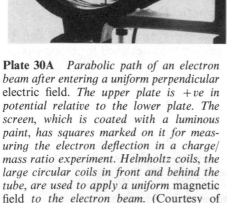

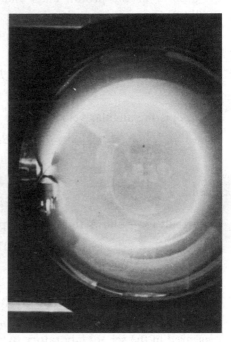

Plate 30A *Parabolic path of an electron beam after entering a uniform perpendicular electric field. The upper plate is +ve in potential relative to the lower plate. The screen, which is coated with a luminous paint, has squares marked on it for measuring the electron deflection in a charge/ mass ratio experiment. Helmholtz coils, the large circular coils in front and behind the tube, are used to apply a uniform magnetic field to the electron beam.* (Courtesy of Teltron Limited)

Plate 30B *Circular deflection of electron beam in a uniform magnetic field due to Helmholtz coils (not visible). Faster electrons produce the outer diffuse beam.*

Unlike the electric deflection tube (Plate 30A), this tube has a small amount of helium gas inside it. The gas molecules are ionised after collision with high-speed electrons and emit light. The electron track is thus made visible as a fine straight beam before the field is applied. (Courtesy of Teltron Limited)

coils is shown in Figure 30.10. The value of B is directly proportional to the current I in the coils, which is measured by the ammeter A, and is varied by means of a potential divider D with a battery B connected to it. Calculation shows that the magnitude of B is given by, approximately,

$$B = 0.72 \frac{\mu_0 N I}{R}$$

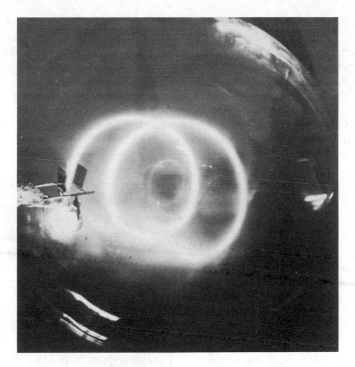

Plate 30C *Spiral electron path, produced when an electron beam enters a* magnetic field *at an angle to the field. The component of the velocity normal to the field produces the circular motion; the component parallel to the field produces the translational motion. Initially the beam passes through the two plates shown, which can be used to apply an electric field.* (Courtesy of Leybold-Heraeus GMB and Co)

where N is the number of turns in each coil, I is the current in amperes and R is the radius in metres of the coils (p. 353)

Experiment. In an experiment, the cathode C is heated by a 6·3 V supply. The anode A has a high potential V relative to C of say 2 kV or 2000 V. The plates P, Q are joined together and kept at the same potential as A; this eliminates electric fields beyond A, which would alter the speed of the electron beam. After passing A, the beam produces a luminous horizontal line XY on S.

A current is now passed into the Helmholtz coils, H, H. The electron beam is then deflected along a circular path XZ seen on S, Figure 30.9. By altering the

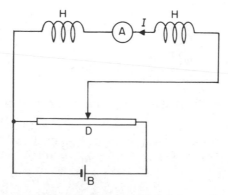

Figure 30.10 *Circuit for Helmholtz coils*

current in the coils, the magnetic field B can be varied to produce a path XZ of suitable radius of curvature r. The horizontal and vertical distances x, y from X of a convenient point on XZ are then read from the graduations on S. From the geometry of the circle, r is given by

$$r = \frac{x^2 + y^2}{2y}$$

Theory. If v is the velocity of the electrons at the anode A, then, assuming zero velocity at C, their gain in kinetic energy $= \frac{1}{2}m_e v^2$. So if V is the anode potential,

$$\frac{1}{2}m_e v^2 = eV \qquad . \qquad . \qquad . \qquad . \qquad . \qquad (1)$$

The radius r of the circular path XZ of the beam is given by

$$Bev = \frac{m_e v^2}{r} \qquad . \qquad . \qquad . \qquad . \qquad (2)$$

From (1) $$v^2 = 2V\left(\frac{e}{m_e}\right)$$

From (2) $$v = Br\left(\frac{e}{m_e}\right)$$

So $$B^2 r^2 \left(\frac{e}{m_e}\right)^2 = v^2 = 2V\left(\frac{e}{m_e}\right)$$

Cancelling e/m_e, $$\frac{e}{m_e} = \frac{2V}{B^2 r^2} \qquad . \qquad . \qquad . \qquad . \qquad . \qquad (3)$$

Knowing V, r and B, then e/m_e can be calculated.

The main errors in the experiment are: (i) the difficulty of measuring the radius r with accuracy; (ii) B may not be uniform over the whole region of S; (iii) error in the voltmeter measuring V. An error in r or in B would produce double the percentage error in e/m_e since r^2 and B^2 occur in (3) above.

Mass of Electron

Modern determinations show that

$$\frac{e}{m_e} = 1 \cdot 76 \times 10^{11} \, \text{C kg}^{-1}$$

or $$\frac{m_e}{e} = \frac{1}{1 \cdot 76} \times 10^{-11} \, \text{kg C}^{-1} \qquad . \qquad . \qquad . \qquad . \qquad (1)$$

Now from electrolysis the mass–charge ratio for a hydrogen ion is $1 \cdot 04 \times 10^{-8} \, \text{kg C}^{-1}$.

$$\therefore \frac{m_H}{e} = 1 \cdot 04 \times 10^{-8} \, \text{kg C}^{-1}$$

assuming the hydrogen ion carries a charge e numerically equal to that on an electron, m_H being the mass of the hydrogen ion. Hence, with (1),

$$\frac{m_e}{m_H} = \frac{1}{1 \cdot 76 \times 1 \cdot 04 \times 10^3} = \frac{1}{1830}$$

So the mass of an electron is about two-thousandths that of the hydrogen atom.

Helical Path of Electrons

Consider an electron P of charge e, mass m_e entering a uniform magnetic field B at a small angle θ with velocity v, Figure 30.11. The component $v \cos \theta$ parallel to B produces a *translational* (linear) motion, since no electromagnetic force acts in this case. The component $v \sin \theta$ normal to B, however, produces a *circular* motion. So the electron path is a helix (spiral). See Plate 30C, page 749.

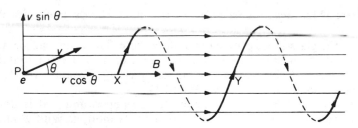

Figure 30.11 *Helical electron path*

The pitch XY of the helix, the distance between neighbouring turns, is given by $XY = v \cos \theta \cdot T$, where T is the period of rotation of the electron. For circular motion (see p. 746),

$$Bev \sin \theta = \frac{m_e (v \sin \theta)^2}{r}$$

so

$$\frac{r}{v \sin \theta} = \frac{m_e}{Be}$$

Hence

$$T = \frac{2\pi r}{v \sin \theta} = \frac{2\pi m_e}{Be}$$

So

$$XY = v \cos \theta \cdot T = \frac{2\pi m_e v \cos \theta}{Be}$$

If $v = 10^6 \text{ m s}^{-1}$, $e/m_e = 1 \cdot 8 \times 10^{11} \text{ C kg}^{-1}$, $\theta = 10°$, $B = 2 \times 10^{-4}$ T,

then

$$XY = \frac{2\pi \times 10^6 \times \cos 10°}{1 \cdot 8 \times 10^{11} \times 2 \times 10^{-4}} = 0 \cdot 17 \text{ m}$$

Thomson's Experiment for e/m

In 1897, Sir J. J. THOMSON devised an experiment for measuring the ratio *charge/mass* or e/m_e for an electron, called its *specific charge*.

Thomson's apparatus is shown simplified in Figure 30.12. C and A are the cathode and anode respectively, and narrow slits are cut in opposite plates at A so that the electrons passing through are limited to a narrow beam. The beam

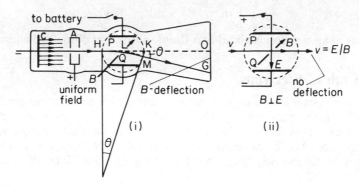

Figure 30.12 *Thomson's determination of e/m_e for electron* (not to scale)

then strikes the glass at O, producing a glow there. The beam can be deflected *electrostatically* by a high voltage battery connected to the horizontal plates P, Q, or *magnetically* by means of a current passing through two Helmholtz coils on either side of the tube near P and Q.

The *magnetic field B* is perpendicular to the paper, and if it is uniform a constant force acts on the electrons normal to its motion. With B alone, the particles move along the arc HM of a circle of radius r, Figure 30.12 (i). When they leave the field, the particles move in a straight line MG and strike the glass at G.

With the usual notation

$$\text{force } F = Bev = \frac{m_e v^2}{r}$$

where e is the charge on an electron and m_e is its mass.

$$\therefore \frac{e}{m_e} = \frac{v}{rB} \qquad \cdot \qquad \cdot \qquad \cdot \qquad \cdot \qquad \cdot \qquad (1)$$

To find the radius r, from Figure 30.12 (i), $\tan \theta = OG/OL = HK/r$

$$\therefore r = \frac{HK \cdot OL}{OG}$$

L is about the middle of the magnetic field coils.

The *velocity v* of the electron beam was found by varying the p.d. or electric field between P and Q until the beam returned to O from its deflected position G, Figure 30.12 (ii). The beam now travels *undeflected* through both the electric field E and the magnetic field B. The two fields are sometimes described as 'crossed' fields, because E is vertical and B is horizontal and perpendicular to E.

Since the downward forcd $Bev =$ upward force Ee,

$$v = \frac{E}{B}$$

If E is in V m^{-1} and B is in T, then v is in m s^{-1}. Thomson found that v was considerably less than the velocity of light, 3×10^8 m s^{-1}, so that electrons are not electromagnetic waves.

On substituting for v, r and B in (1), the ratio charge/mass (e/m_e) for an electron was obtained.

Until Sir J. J. THOMSON's experiment, it was believed that the hydrogen atom was the lightest particle in existence. As explained on p. 751, comparison with the mass–charge ratio of a hydrogen ion showed that the electron was about 1/2000th of the mass of the hydrogen atom.

Examples on Electron Motion in Fields

1 An electron beam passes undeflected with uniform velocity v through two parallel plates, when a magnetic field of 0·01 T is applied perpendicular to an electric field between the plates produced by a p.d. of 100 V. The separation of the plates is 5 mm. Calculate v.

For no deflection,

$$Bev = Ee$$

So
$$v = \frac{E}{B} = \frac{V/d}{B} = \frac{V}{dB}$$

$$= \frac{100}{5 \times 10^{-3} \times 0·01} = 2 \times 10^6 \text{ m s}^{-1}$$

2 Describe and give the theory of a method to determine e, the electronic charge. Why is it considered that all electric charges are multiples of e?

An electron having 450 eV of energy moves at right angles to a uniform magnetic field of flux density $1·50 \times 10^{-3}$ T. Show that the path of the electron is a circle and find its radius. Assume that the specific charge of the electron is $1·76 \times 10^{11}$ C kg^{-1}. (L.)

With the usual notation, the velocity v of the electron is given by

$$\tfrac{1}{2}m_e v^2 = eV, \text{ where } V \text{ is } 450 \text{ V}$$

$$\therefore v = \sqrt{\frac{2eV}{m_e}} \qquad \qquad . \qquad . \qquad . \qquad . \qquad (1)$$

The path of the electron is a circle because the force Bev is constant and always normal to the electron path. Its radius r is given by

$$Bev = \frac{m_e v^2}{r}$$

Cancelling v,
$$\therefore r = \frac{m_e}{e} \cdot \frac{v}{B} = \frac{m_e}{e} \cdot \frac{1}{B}\sqrt{\frac{2eV}{m_e}}, \text{ from (1)}$$

$$\therefore r = \frac{1}{B}\sqrt{2\frac{m_e V}{e}}$$

Now $e/m_e = 1.76 \times 10^{11}$ C kg^{-1}, $V = 450$ V, $B = 1.5 \times 10^{-3}$ T

$$\therefore r = \frac{1}{1.5 \times 10^{-3}} \sqrt{\frac{2 \times 450}{1.76 \times 10^{11}}}$$

$$= 4.8 \times 10^{-2} \text{ m}$$

3 A charged oil drop of mass 6×10^{-15} kg falls vertically in air with a steady velocity between two long parallel vertical plates 5 mm apart. When a p.d. of 3000 V is applied between the plates, the drop now falls with a steady velocity at an angle 58° to the vertical. Calculate the charge Q on the drop. (Assume $g = 10$ m s^{-2}.)

The drop falls with steady velocity due to the viscosity of the air. Neglecting the upthrust, if v_1 is the vertical velocity

downward force on drop = upward viscous force

or
$$mg = kv_1$$

Similarly, when the electric field of strength E is applied,

horizontal force on drop $= EQ = kv_2$

where v_2 is the horizontal velocity.

By vector addition, if θ is the angle to the vertical,

$$\tan \theta = \frac{v_2}{v_1} = \frac{EQ}{mg}$$

Now
$$E = \frac{V}{d} = \frac{3000}{5 \times 10^{-3}} = 6 \times 10^5 \text{ V m}^{-1}$$

$$\therefore \tan 58° = \frac{EQ}{mg} = \frac{6 \times 10^5 Q}{6 \times 10^{-15} \times 10} = 10^{19} Q$$

$$\therefore Q = \frac{\tan 58°}{10^{19}} = 1.6 \times 10^{-19} \text{ C}$$

You should know:

1 Electrons and other charges, moving with uniform velocity, have a *circular* path after entering a uniform magnetic field perpendicular to their initial direction.
  (a) The circle radius r is given by $mv^2/r = Bev$
  (b) No energy is given to the electrons moving round the circle because the force on them is always 90° to their motion.
  (c) The time for one circular orbit (period), is independent of the velocity.
  (d) Moving initially at an angle to B, electrons move in a helical path.
2 Electrons and other charges, moving initially with uniform velocity parallel to an electric field, have a *parabolic* path in the field (compare a ball thrown forward into the gravitational field).
  The electrons gain energy in the direction of E.
3 Moving initially with a velocity v, electrons passing through perpendicular magnetic and electric fields are not deflected if $v = E/B$. So v can be found.

EXERCISES 30A Electrons and Charges in Fields

Multiple Choice

1 Electrons moving with a constant velocity enter a uniform magnetic field B in a direction perpendicular to B. The electron path in the field is
A a helix **B** a straight line parallel to B
C a straight line perpendicular to B **D** a circle **E** a parabola.

2 In Figure 30A (i), P, Q are plates with a uniform electric field between them. An electron beam, moving parallel to P, is deflected a distance x as shown after leaving the field. The length of the plates is y. If the length is increased to $3y$ and the separation of P and Q and the electric field strength are kept constant, the new deflection on leaving the field is

A $3x$ **B** $6x^2$ **C** $9x$ **D** $9x^2$ **E** $12x$

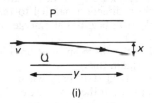   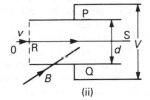

(i)   (ii)

Figure 30A

3 In Figure 30A (ii), parallel metal plates separated a distance d have a p.d. V between them. A uniform magnetic field B is applied perpendicular to the electric field. A charge enters the fields with a velocity v as shown along OR. It will emerge at S undeflected if B has a value

A Vvd **B** Vv/d **C** d/Vv **D** V/vd **E** d/Vd

4 A charge moves in a circular orbit in a uniform magnetic field. Which of the statements **A** to **E** is correct?
A The force on the charge is parallel to the field.
B The radius of the circle is proportional to the charge
C The radius of the circle is proportional to the field value B.
D The momentum of the charge is independent of the circle radius.
E The period in the orbit is independent of the speed of the charge.

5 A charged particle enters a uniform magnetic field perpendicular to its initial direction travelling in air. The path of the particle is seen to follow the path in Figure 30B.
   Which of statements 1–3 is/are correct?
1 The magnetic field strength may have been increased while the particle was travelling in air:
2 The particle lost energy by ionising the air.
3 The particle lost charge by ionising the air.
A 1, 2, 3 are correct **B** 1, 2 only are correct
C 2, 3 only are correct **D** 1 only **E** 3 only

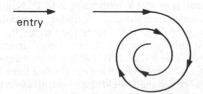

Figure 30B

Longer Questions

6 Electrons are accelerated from rest by a p.d. of 100 V. What is their final velocity? The electron beam now enters normally a uniform electric field of intensity 10^5 V m^{-1}. Calculate the flux density B of a uniform magnetic field applied perpendicular to the electric field if the path of the beam is unchanged from its original direction. Draw a sketch showing the electron beam and the two fields. (Assume $e/m_e = 1.8 \times 10^{11}$ C kg^{-1}.)

7 A beam of protons is accelerated from rest through a potential difference of 2000 V and then enters a uniform magnetic field which is perpendicular to the direction of the proton beam. If the flux density is 0.2 T calculate the radius of the path which the beam describes. (Proton mass $= 1.7 \times 10^{-27}$ kg. Electronic charge $= -1.6 \times 10^{-19}$ C.) (L.)

Two plane metal plates 4.0 cm long are held horizontally 3.0 cm apart in a vacuum, one being vertically above the other. The upper plate is at a potential of 300 V and the lower is earthed. Electrons having a velocity of 1.0×10^7 m s^{-1} are injected horizontally midway between the plates and in a direction parallel to the 4.0 cm edge. Calculate the vertical deflection of the electron beam as it emerges from the plates. (e/m for electron $= 1.8 \times 10^{11}$ C kg^{-1}.) (N.)

8 (a) A charged oil drop falls at constant speed in the Millikan oil drop experiment when there is no p.d. between the plates. Explain this.

(b) Such an oil drop, of mass 4.0×10^{-15} kg, is held stationary when an electric field is applied between the two horizontal plates. If the drop carries six electric charges each of value 1.6×10^{-19} C, calculate the value of the electric field strength. (L.)

9 An electron with a velocity of 10^7 m s^{-1} enters a region of uniform magnetic flux density of 0.10 T, the angle between the direction of the field and the initial path of the electron being 25°. By resolving the velocity of the electron find the axial distance between two turns of the helical path. Assume that the motion occurs in a vacuum and illustrate the path with a diagram. ($e/m = 1.8 \times 10^{11}$ C kg^{-1}.) (N.)

10 An electron charge $-e$ and mass m is initially projected with speed v at right angles to a uniform field of flux density B. Show that the electron moves in a circular path and derive an expression for the radius of the circle. Show also that the time taken to describe one complete circle is independent of the speed of the electron.

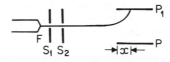

Figure 30C

Electrons are emitted with negligible speed, *in vacuo*, from a filament F. They are accelerated by a potential difference of 1200 V applied between the plates S_1 and F, as shown in Figure 30C. The electrons are collimated into a narrow horizontal beam by passing through holes in S_1 and in a second plate S_2 which is at the same potential as S_1. The electron beam subsequently enters the space between the two large parallel horizontal plates P_1 and P which are 0.02 m apart. The point of entry is midway between the plates. The mean potential of P_1 and P is equal to that of S_1 but P_1 is at a positive potential of 150 V with respect to P. Neglecting the effect of gravity and of non-uniform fields near the plate boundaries, calculate the distance x travelled by the electrons between P_1 and P before they strike P_1. (O. & C.)

11 (a) In an evaluated tube electrons are accelerated from rest through a potential difference of 3600 V and then travel in a narrow beam through a field free space before entering a uniform magnetic field the flux lines of which are perpendicular to the beam. In the magnetic field the electrons describe a circular arc of radius 0.10 m. Calculate (i) the speed of the electrons entering the magnetic field, (ii) the magnitude of the magnetic flux density.

(b) If an electron described a complete revolution in a magnetic field how much energy would it acquire? ($e/m = 1.8 \times 10^{11}$ C kg^{-1}.) (N.)

12 Define *magnetic flux density* (B). Describe how the variation in B along the axis both inside and outside a long solenoid carrying a current may be investigated. Sketch a graph showing the results you would expect to obtain.

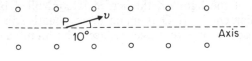

Figure 30D

Figure 30D represents a long solenoid in which a steady current is flowing. An electron is emitted at P with an initial velocity v in the direction shown. By considering the components of v: (i) perpendicular to the axis; (ii) parallel to the axis, deduce the path the electron will follow. If the value of v is 2.0×10^6 m s^{-1} and the flux density at P is 3.0×10^{-4} T, at what distance along the axis will the electron next cross it? (The specific charge of an electron, e/m_e, may be assumed to be 1.8×10^{11} C kg^{-1}.) (L.)

13 Show that if a free electron moves at right angles to a magnetic field the path is a circle. Show also that the electron suffers no force if it moves parallel to the field. Point out how the steps in your proof are related to fundamental definitions.

If the path of the electron is a circle, prove that the time for a complete revolution is independent of the speed of the electron.

In the ionosphere electrons execute 1.4×10^6 revolutions in a second. Find the strength of the magnetic flux density B in this region. (Mass of an electron = 9.1×10^{-31} kg; electronic charge = 1.6×10^{-19} C.) (C.)

Cathode-Ray Oscilloscope (C.R.O.)

We now discuss an important type of electron tube, widely used, called a *cathode-ray oscilloscope*. It is called a cathode-ray oscilloscope because it traces a required wave-form with a beam of electrons, and beams of electrons were originally called cathode rays.

A cathode-ray oscilloscope is essentially an electrostatic instrument. It consists of a highly evacuated glass tube, T in Figure 30.13, one end of which opens out to form a screen S which is internally coated with zinc sulphide or other fluorescent material. A hot filament F, at the other end of the tube, emits electrons. These are then attracted by the cylinders A_1 and A_2, which have increasing positive potentials with respect to the filament. Many of the electrons, however, shoot through the cylinders and strike the screen; where they do so, the zinc sulphide fluoresces in a green spot. On their way to the

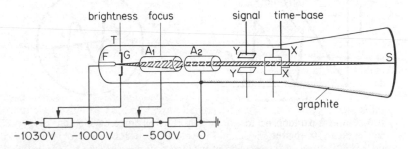

Figure 30.13 *A cathode-ray oscilloscope tube*

screen, the electrons pass through two pairs of metal plates, XX and YY, called the *deflecting plates*. The inner walls of the tube are coated with graphite, a conductor, which is connected to the final anode A_2. This makes the space between A_2 and S an equipotential volume so that the speed of the electrons is maintained from A_2 to S.

The *brightness* of the light on the screen is controlled by an electrode G in front of the filament F. If G is made more negative in potential relative to F, the increased repulsive force reduces the number of electrons per second passing G.

Electrode Potentials

In practice, the screen S, the tube and A_2 are all earthed to avoid danger due to high voltages. Further, touching the outside of S with earthed fingers does not then alter the electrostatic field inside the tube and affect the deflection of the beam. So, as indicated, the filament F is at a high negative potential relative to earth and is therefore dangerous to touch. The filament, electrode G and accelerating electrodes A_1, A_2 are often called the 'gun assembly' of a tube. Figure 30.13 illustrates a potential divider arrangement for the simple cathode-ray oscilloscope shown. The anode A_2, tube and screen are earthed; A_1 has a varying voltage for the focus control as explained soon; and the brightness control G has a varying negative potential relative to the potential of F, which is about -1000 V.

Deflection; Time-base

If a battery were connected between the Y-plates, so as to make the upper one positive, the electrons in the beam would be attracted towards that plate, and the beam would be deflected upwards. In the same way, the beam can be deflected horizontally by a potential difference applied between the X-plates.

When the oscilloscope is in use, the varying voltage to be examined is applied between the Y-plates. If that were all, then the spot would be simply drawn out into a vertical line. To trace the *wave-form* of the varying voltage, the X-plates are used to provide a *time-axis*. A special type of circuit generates a potential difference which rises steadily to a certain value, as shown in Figure 30.14 (i), and then falls rapidly to zero. It can be made to go through these changes tens, hundreds, thousands or millions of times per second. This *time-base voltage*, a saw-tooth trace, is applied between the X-plates, so that the spot

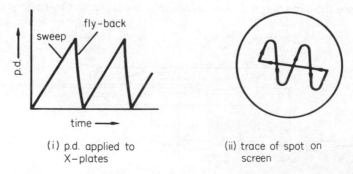

(i) p.d. applied to
    X-plates

(ii) trace of spot on
     screen

Figure 30.14 *Action of a C.R.O. time-base*

is swept steadily to the right, and then flies swiftly back and starts out again. The horizontal motion provides the *time-base* of the oscillograph. On it is superimposed the vertical motion produced by the Y-plates. Then as shown in Figure 30.14 (ii), the wave-form of the voltage to be examined can be displayed on the screen.

In a *double beam oscilloscope*, the two Y-plates are joined to terminals labelled Y_1 and Y_2 respectively. An earthed plate between the plates splits the beam into two halves. One half can be deflected by an input voltage connected to Y_1 and the other half can be deflected by another input voltage connected to Y_2. With a common time-base applied to the X-plates, two traces can be obtained on the screen and two different waveforms, from Y_1 and Y_2 respectively, can then be compared.

Focusing

To give a clear trace on the screen, the electron beam must be focused to a sharp spot. This the purpose of the cylinders A_1 and A_2, called the first and second anodes.

Figure 30.15 shows the equipotentials of the field between them, when their difference of potential is 500 volts. Electrons entering the field from the filament experience forces from low potential to high at right angles to the equipotentials. They have, however, considerable momentum, because

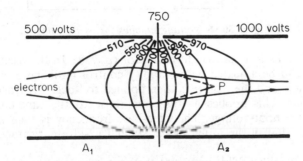

Figure 30.15 *Focusing in an oscilloscope tube by electron lens*

they have been accelerated by a potential difference of about 500 volts, and are travelling fast. Consequently the field merely deflects them, and, because of its cylindrical symmetry, it converges the beam towards the point P. Before they can reach this point, however, they enter the second cylinder. Here the potential rises from the axis, and the electrons are deflected outwards. However, they are now travelling faster than when they were in the first cylinder, because the potential is everywhere higher. Consequently their momentum is greater, and they are less deflected than before. The second cylinder, therefore, diverges the beam less than the first cylinder converged it, and so the beam emerges from the second anode still somewhat convergent. By adjusting the potential of the first anode, the beam can be *focused* on the screen, to give a spot a millimetre or less in diameter.

Electron-focusing devices are called *electron lenses*, or electron-optical systems. For example, the action of the anodes A_1 and A_2 is roughly analogous to that of a pair of glass lenses on a beam of light, the first glass lens being converging, and the second diverging, but weaker.

C.R.O. Supplies

The necessary voltage supplies or circuits for a cathode-ray oscillograph or oscilloscope are shown in block form in Figure 30.16.

1 The *power pack* supplies e.h.t.—very high d.c. voltage for the tube electrodes such as the electron lens; h.t.—rectified and smoothed d.c. voltage for the amplifiers, for example; and l.t.—low a.c. voltage such as 6·3 V for the valve heaters.

2 *Tube controls*. The brightness and focus, and correction of astigmatism, are controlled by varying the voltage of the appropriate tube electrode.

3 *Y- or Input-Amplifier*. This is a variable gain linear amplifier which amplifies the signal applied to the Y-plates.

4 *Y-shift*. This is in the Y-amplifier circuit and shifts the signal trace up or down on the screen.

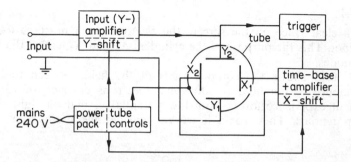

Figure 30.16 *Voltage supplies for oscilloscope*

5 *Time-base*. This is a 'sawtooth' oscillator (p. 758). In the oscillator circuit, the feedback can be adjusted with stability control until the oscillator is just not free running, and the trigger is then applied to 'lock' the trace.

6 *Trigger unit*. This applies a triggering pulse to the time-base oscillator from the Y- or input-voltage. The time-base frequency is then synchronised ('locked') with that of the input signal, so that a stationary trace is obtained on the screen.

7 *Time-base amplifier*. This amplifies the time-base voltage and applies it to the X-plates. The trace can be made to expand or contract horizontally by varying the amplification.

8 *X-shift*. This shifts the time-base horizontally to the left or right.

Oscilloscope controls are shown in Plate 30D. The *voltage sensitivity* of the Y-plates is the 'voltage per cm (or mm)' vertical deflection of the electron beam and the *time-base* is the 'time per cm' horizontal deflection of the electron beam due to the X-plates, so that voltage and time can be measured.

Uses of Oscilloscope

In addition to displaying waveforms, the oscilloscope can be used for measurement of voltage, frequency and phase, and act as a clock.

1 *A.C. voltage*

An uknown a.c. voltage, whose peak value is required, is connected to the Y-plates. With the time-base switched off, the vertical line on the screen is centred and its length then measured, Figure 30.17 (i). This is proportional to twice the amplitude or peak voltage, V_0. By measuring the length corresponding to a known a.c. voltage V, then V_0 can be found by proportion.

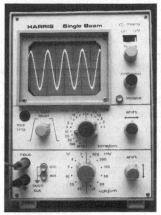

Plate 30D *The supply of electricity to homes and industry is alternating current, in order to overcome the problems of energy loss in transmitting large currents at low voltages. If the p.d. of the mains electricity supply is displayed on an oscilloscope screen, the trace looks like this.* (Photograph reproduced courtesy Peter Gould)

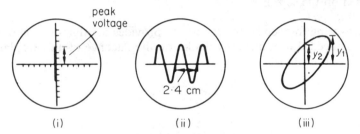

Figure 30.17 *Uses of oscilloscope*

Alternatively, using the same gain, the waveforms of the unknown and known voltages, V_0 and V, can be displayed on the screen. The ratio V_0/V is then obtained from measurement of the respective peak-to-peak heights.

2 *Measurement and comparison of frequencies*

If a calibrated time-base is available, frequency measurements can be made. In Figure 30.17 (ii), for example, the trace shown is that of an alternating waveform with the time-base switched to the '5 ms/cm' scale. This means that the time taken for the spot to move 1 cm horizontally across the screen is 5 milliseconds. The horizontal distance on the screen for one cycle is 2·4 cm. This corresponds to a time of $5 \times 2\cdot4$ ms or $12\cdot0$ ms $= 12 \times 10^{-3}$ seconds, which is the period T.

$$\therefore \text{ frequency} = \frac{1}{T} = \frac{1}{12 \times 10^{-3}} = 83 \text{ Hz}$$

If a comparison of frequencies f_1, f_2 is required, then the corresponding horizontal distances on the screen for one or more cycles are measured. Suppose these are d_1, d_2 respectively. Then, since $f \propto 1/T$,

$$\frac{f_1}{f_2} = \frac{T_2}{T_1} = \frac{d_2}{d_1}$$

3 Measurement of phase

The use of a double beam oscilloscope to measure phase difference between two a.c. voltages is given on p. 481. With the time-base switched off, in the single beam tube one input can be joined to the X-plates and the other to the Y-plates. We consider only the case when the frequencies of the two signals are the same. An ellipse will then be seen generally on the screen, as shown in Figure 31.17 (iii). This is a Lissajous figure (see p. 86).

The trace is centred, and peak vertical displacement y_2 at the middle O, and the peak vertical displacement y_1 of the ellipse, are then both measured. Suppose the x-displacement is given by $x = a \sin \omega t$, where a is the amplitude in the x-direction, and the y-displacement by $y = y_1 \sin (\omega t + \varphi)$, where y_1 is the amplitude in the y-direction and φ is the phase angle. When $x = 0$, $\sin \omega t = 0$, so that $\omega t = 0$. In this case, $y = y_2 = y_1 \sin \varphi$. Hence $\sin \varphi = y_2/y_1$, from which φ can be found. See also p. 86.

The frequencies f_x and f_y of two a.c. voltages can also be compared by connection to the X- and Y-inputs respectively of the oscilloscope. The display on the screen will be a Lissajous figure. Generally, if a horizontal line makes n_x intersections with the Lissajous figure and a vertical line makes n_y oscillations, then $f_x : f_y = n_y : n_x$. So if the frequency ratio $f_x : f_y = 2:1$, a figure of eight is obtained on the screen.

4 Use as a clock

In measuring frequency, the time-base provides a value of a time interval. The time is needed in radar for calculating the distance of an aeroplane from a station or aerodrome.

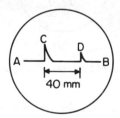

Figure 30.18 C.R.O. as clock

Figure 30.18 illustrates the principle. Using a time-base of 100 microseconds per millimetre ($10 \ \mu s \ mm^{-1}$), AB is the horizontal trace. A radar signal sent at C to a distant plant arrives back at D, 40 mm from C. So the time for the signal to travel to the plane and back $= 40 \times 10 \times 10^{-6} \ s = 4 \times 10^{-4} \ s$. Now a radio signal travels through space with a speed of 300 000 km s^{-1} (3×10^8 m s^{-1}). So

$$\text{to-and fro distance of plane} = 4 \times 10^{-4} \times 300 \, 000 = 120 \ km$$

and $\quad\quad\quad\quad\quad$ distance of plane $= \frac{1}{2} \times 120 = 60 \ km$

Electron-beam Welding

A high-power version of the cathode ray oscilloscope (without the screen!) forms the basis of electron-beam welding equipment. Electrons emitted from a cathode are accelerated through a voltage of up to 200 kV, and then focused into a narrow beam by a magnetic field device (a so-called *electron lens*). The beam then passes through a set of deflection coils, which can be used either to position the beam or to cause the beam to move in some sort of pattern. Finally the beam strikes the workpiece to be welded, where the power density can exceed 10 kW mm^{-2}. The whole process is carried out in a vacuum chamber, so that the beam is not dispersed by collisions with air molecules.

The electron-beam technique has many advantages over simpler welding techniques (such as gas welding and electric arc welding), though it is a far more expensive process. The workpiece gets so hot that a cavity is formed all the way through the piece, allowing *keyhole* welding to take place. When the electron beam moves, the cavity is filled with molten metal, so that the weld penetrates all the way through the workpiece, producing an extremely strong join. Not only can the beam penetrate very thick pieces of metal, but the process is also very efficient, with about 75% of the input electrical energy being converted into heat. Since the beam is highly focused, only a small amount of metal actually gets hot, and so distortion of the workpiece is negligible. In fact, electron-beam welding can be used to join together finished, precision parts, such as gear clusters cleanly.

These advantages often outweigh the capital expenditure on the equipment, and so electron-beam welding is now widely used in the car industry, as well as in the more high-technology aerospace and nuclear industries.

EXERCISES 30B Oscilloscope

Multiple Choice

1 Figure 30E shows an alternating voltage trace on an oscilloscope screen. The Y-sensitivity is 2 V cm^{-1} and the time-base is 0.1 ms mm^{-1}. The amplitude and frequency of the voltage is

A 6 V, 2·5 kHz **B** 3 V, 5·0 kHz **C** 3 V, 1·5 kHz **D** 6 V, 5·0 kHz **E** 2 V, 2 kHz

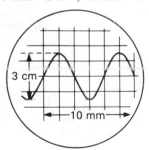

3 cm

←—10 mm—→

Figure 30E

Longer Questions

2 An oscilloscope is used to measure the time it takes to send a pulse of charge along a 200-m length of coaxial cable and back again. Figure 30F shows the appearance of the oscilloscope screen, A indicating the original pulse and B the same pulse after reflection.

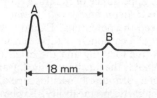

A

B

18 mm

Figure 30F

If the time base speed is set 10 mm μs^{-1}, calculate the speed of the pulse along the cable. (*L*.)

3 A cathode-ray oscilloscope has its Y-sensitivity set to 10 V cm^{-1}. A sinusoidal input is suitably applied to give a steady trace with the time base set so that the electron beam takes 0·01 s so traverse the screen. If the trace seen has a total peak to peak height of 4·0 cm and contains 2 complete cycles, what is the r.m.s. voltage and frequency of the input signal? (*L*.)

4 When a sine-form voltage of frequency 1250 Hz is applied to the Y-plates of a cathode-ray oscilloscope the trace on the tube is as shown in Figure 30G (i).

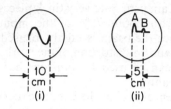

Figure 30G

If a radar transmitter sends out short pulses, and at the same time gives a voltage to the Y-plates of the oscilloscope, with the time-base setting unchanged, the deflection A is produced as shown in Figure 30G (ii). An object reflects the radar pulse which, when received at the transmitter and amplified, gives the deflection B. What is the distance of the object from the transmitter? (Speed of radar waves = 3 × 10^8 m s^{-1}.) (*L*.)

5 (a) Draw a clear labelled diagram showing the essential features of a single beam cathode-ray oscilloscope tube. Explain, without giving circuit details, how the brightness and focusing of the electron beam are controlled.

(b) What is the *time-base* in an oscilloscope? Sketch a graph showing the variation of time-base voltage with time.

(c) How would you use an oscilloscope, the time-base of which is not calibrated, to measure the frequency of a sinusoidal potential difference which is of the order of 50 Hz?

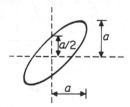

Figure 30H

(d) With the time-base disconnected, two alternating potential differences of the same frequency are applied to the X- and Y-plates respectively, the gains being equal. Figure 30H shows the appearance of the trace on the screen. The potentials may be represented by $x = a \sin \omega t$ and $y = a \sin (\omega t + \varphi)$. What is the value of φ, the phase angle between the potentials? Explain your reasoning. (*L*.)

6 (a) (i) Describe the layout of a cathode-ray tube that utilises electrical deflection and focusing. Explain how the focusing and the brightness of the spot on the screen are controlled. (ii) What is the function of the linear time-base in an oscilloscope? (iii) Why is the tube of a cathode-ray oscilloscope always mounted inside a metal screen of very high magnetic permeability? Explain how the screen achieves its purpose.

(b) Explain in detail how you would use an oscilloscope: (i) to observe the waveform of the musical note emitted by a clarinet;   (ii) to check the frequency calibration of an audio sine-wave oscillator at any one chosen frequency (you may assume that the frequency of the mains supply is 50 Hz exactly);   (iii) to measure the r.m.s. value of a sinusoidal alternating current of about 0·2 A flowing in a 5·0 Ω resistor. (O.)

31 Semiconductors, Diodes, Transistor Telecommunications

In this chapter, we consider semiconductors and their applications in junction diodes and transistors. We begin with the semiconductor electron and hole carriers and their action in diodes and transistors. Rectifiers, switching circuits and amplifiers are then described. A brief account of telecommunications concludes the chapter.

Semiconductors, Electron and Hole Charge Carriers

Semiconductors are a class of materials which have a resistivity about ten million times higher than that of a good conductor such as copper. Silicon and germanium are examples of semiconductor elements widely used in the electronics industry.

Silicon and germanium atoms are tetravalent. They have four electrons in their outermost shell, called *valence electrons*. One valence electron is shared with each of four surrounding atoms in a tetrahedral arrangement, forming 'covalent bonds' which maintain the crystalline solid structure. Figure 31.1 (i) is a two-dimensional diagram of the structure, showing four silicon atoms, each having four valence electrons round them.

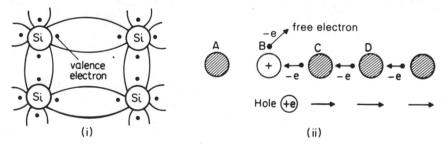

Figure 31.1 *Electrons and holes*

At 0 K, all the valence electrons are firmly bound to the nucleus of their particular atoms. At room temperature, however, the thermal energy of a valence electron may become greater than the energy binding to its nucleus. The covalent bond is then broken. The electron leaves the atom, B say, and becomes a *free electron*. This leaves B with a vacancy or *hole*, Figure 31.1 (ii). Since B now has a net positive charge, an electron in a neighbouring atom C may then be attracted. So the hole appears to move to C. So C now has a positive charge and therefore attracts an electron from D. This leaves D with a hole.

In this way we can see that, due to the movement of valence electrons from atom to atom, *holes* spread throughout the semiconductor. Since an electron carries a negative charge $-e$, a hole, moving in the *opposite* direction to the electron, is equivalent to a *positive* charge $+e$. So

---

moving holes are equivalent to moving positive charges $+e$

Summarising: In semiconductors, then, there are *two* kinds of charge carriers: a free electron $(-e)$ and a hole $(+e)$. In contrast, a metal such as pure copper, a good conductor, has only one kind of charge carrier, the free electron. In semiconductors, the escape of a valence (bound) electron from an atom produces *electron–hole pairs* of charge carriers.

Currents in Semiconductors

The free electrons and holes move about randomly in different directions inside the semiconductor. But when a battery is connected to the semiconductor, the $+$ve charges or holes drift through the metal in opposite directions to the $-$ve charges or electrons. The small current I obtained is carried by both holes and electrons, Figure 31.2 (i).

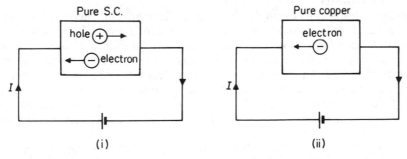

Figure 31.2 *Current in pure (intrinsic) semiconductor and pure copper*

In a pure metal such as copper, the current I is given by $I = nevA$, where n is the number of electrons per m^3, A is the cross-sectional area in m^2 and v is the drift velocity of the electrons in m s^{-1} (see p. 268), Figure 31.2 (ii). In a pure semiconductor, called an *intrinsic semiconductor*, there are equal numbers n of electrons and holes. The drift velocity v_n of the electrons or negative charges is actually greater than that v_p of the holes or positive charges. Taking both charges carriers into account, the current I is given by

---

$$I = nev_n A + nev_p A = neA(v_n + v_p)$$

In a pure (intrinsic) semiconductor:
1   there are two kinds of charge carriers, holes $(+e)$ and electrons $(-e)$
2   the number of holes $=$ the number of electrons
3   the drift velocity of the electrons is greater than that of the holes.

Effect of Temperature Rise

As we have seen, the charge carriers in a metal such as copper are only free electrons. As the temperature of the metal rises, the amplitude of vibration of the atoms increases and more 'collisions' with atoms are then made by the drifting electrons. So the resistance of a pure metal increases with temperature rise (p. 268).

In the case of a semiconductor, however, the increase in thermal energy of the valence electrons due to temperature rise enables more of them to break the covalent bonds and become free electrons. Thus *more electron–hole pairs* are produced which can act as carriers of current. Hence, in contrast to a pure metal, the electrical resistance of a pure semiconductor *decreases* with temperature rise. This is one way of distinguishing between a pure metal and a pure semiconductor. Note that the pure or *intrinsic* semiconductor always has equal numbers of electrons and holes, whatever its temperature.

N-Semiconductors

A pure or intrinsic semiconductor has charge carriers which are thermally generated. These are relatively few in number. By 'doping' a semiconductor with a tiny amount of impurity such as one part in a million, forming a so-called *extrinsic* or impure semiconductor, an enormous increase can be made in the number of charge carriers. The impure semiconductor is widely used in the electronics industry.

Arsenic atoms, for example, have five electrons in their outermost or valence band. When an atom of arsenic is added to a silicon crystal, the atom settles in a lattice site with four of its electrons shared with neighbouring silicon atoms. The fifth electron may thus become free to wander through the crystal. Since an impurity atom may provide one free electron, an enormous increase occurs in the number of charge carriers. For example, 1 milligram of arsenic has about 8×10^{18} atoms and so provides this large number of free electrons in addition to the relatively small number of thermally-generated electrons.

Since there are a great number of negative (electron) charge carriers, the impure semiconductor is called an 'n-type semiconductor' or *n-semiconductor*, where 'n' represents the negative charge on an electron. Thus the *majority carriers* in an n-semiconductor are electrons. Positive charges or holes are also present in the n-semiconductor, Figure 31.3 (i). These are thermally generated, as previously explained, and since they are relatively few they are called the *minority carriers*. The impurity (arsenic) atoms are called *donors* because they donate electrons as carriers.

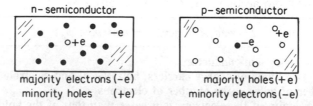

Figure 31.3 *N- and P-semiconductors*

The conductivity σ of an n-semiconductor increases with temperature from very low temperatures to about 250 K as more electron–hole pairs are formed. But from 250 K to about 650 K, σ decreases with temperature rise. In this temperature range the increase in electron–hole pairs is more than counteracted by the increase in the metal lattice vibration, which oppose the motion of the carriers.

P-Semiconductors

P-semiconductors are made by adding foreign atoms which are *trivalent* to pure germanium or silicon. Examples are boron or indium. In this case the reverse happens to that previously described. Each trivalent atom at a lattice site attracts an electron from a neighbouring atom, thereby completing the four valence bonds and *forming a hole* in the neighbouring atom. In this way an enormous increase occurs in the number of holes. So in a p-semiconductor, the majority carriers are holes or positive charges. The minority carriers are electrons, negative charges, which are thermally generated, Figure 31.3 (ii). The impurity atoms are called *acceptors* in this case because each 'accepts' an electron when the atom is introduced into the crystal.

Summarising: In an n-semiconductor, conduction is due mainly to negative charges or electrons, with positive charges (holes) as minority carriers. In a p-semiconductor, conduction is due mainly to positive charges or holes, with negative charges (electrons) as minority carriers.

P-N Junction

By a special manufacturing process, p- and n-semiconductors can be melted so that a boundary or *junction* is formed between them. This junction is extremely thin and of the order 10^{-3} mm. It is called a *p-n junction*, Figure 31.4 (i).

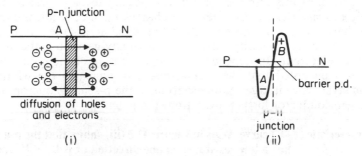

Figure 31.4 *n-p junction and barrier p.d.*

When a scent bottle is opened, the high concentration of scent molecules in the bottle causes the molecules to diffuse into the air. In the same way, the high concentration of holes (positive charges) on one sides of a p-n junction, and the high concentration of electrons on the other side, causes the two carriers to diffuse respectively to the other side of the junction, as shown diagramatically in Figure 31.4 (i). The electrons which move to the p-semiconductor side recombine with holes there. These holes therefore disappear and an excess negative charge *A* appears on this side, as shown in Figure 34.4 (ii).

In a similar way, an excess positive charge *B* builds up in the n-semiconductor when holes diffuse across the junction. Together with the negative charge *A* on the p-side, an e.m.f. or p.d. is produced which *opposes* more diffusion of charges across the junction. This is called a *barrier p.d.* and when the flow ceases it has a magnitude of a few tenths of a volt. The narrow region or layer at the p-n junction which contains the negative and positive charges is called the *depletion layer*. The width of the depletion layer is of the order 10^{-3} mm.

Junction Diode as Rectifier

When a battery B, with an e.m.f. greater than the barrier p.d., is joined with its positive pole to the p-semiconductor, P, and its negative pole to the n-semiconductor, N, p-charges (holes) are urged across the p-n junction from P to N and n-charges (electrons) from N to P, Figure 31.5 (i).

We can understand the movement if we consider the +ve pole of the battery to repel +ve charges (holes) in the p-semiconductor and the −ve pole to repel the −ve charges (electrons) in the n-semiconductor. So an appreciable current OA is obtained. The p-n junction is now said to be *forward-biased*, and when the applied p.d. is increased, the current increases along the curve OA.

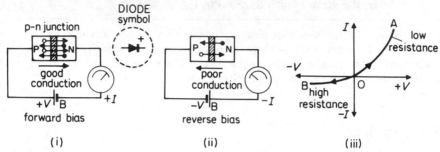

Figure 31.5 *Junction diode characteristics*

When the poles of the battery are reversed, only a very small current flows, Figure 31.5 (ii). In this case the p-n junction is said to be *reverse-biased*. This time only the *minority carriers*, negative charges in the p-semiconductor and positive charges in the n-semiconductors, are urged across the p-n junction by the battery. Since the minority carriers are thermally-generated, the magnitude of the reverse current OB depends on the *temperature* of the semiconductors. It may also be noted that the reverse-bias p.d. increases the width of the depletion layer, since it urges more electrons in the p-semiconductor and holes in the n-semiconductor further away from the junction.

The characteristic (I–V) curve AOB in Figure 31.5 (iii) shows that the p-n junction acts as a *rectifier*. It has a low resistance in one direction of p.d., $+V$, and a high resistance in the opposite p.d. direction, $-V$.

It is called a *junction diode*. In the diode symbol in Figure 31.5, the low resistance is from left to right (towards the triangle point) and the high resistance is in the opposite direction. The junction diode has several advantages, for example, it needs only a low voltage battery B to work; it does not need time to warm up; it is not bulky; and it is cheap to manufacture in large numbers.

Full Wave Rectifiers, Bridge Circuit

Figure 31.6 (i) shows how four diodes, two D1 and two D2, can be arranged in a so-called *bridge circuit* to measure alternating current using a moving-coil meter M.

On the *positive* half of the a.c. cycle of V, suppose the p.d. has its +ve side connected to R. The current then flows through the low-resistance path RSQP back to V. On the *negative* half of the same a.c. cycle, the +ve side of the p.d. is now joined to P. So the current now flows through the low-resistance path PSQR back to V.

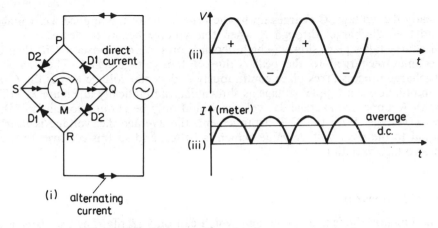

Figure 31.6 *Bridge rectifier circuit—a.c. measurement*

So on *both* halves of the cycle, the current flows the same way through the meter M. Figure 31.6 (ii) shows the alternating p.d. V and Figure 31.6 (iii) the varying current through M, which has an average positive or direct current (d.c.) value. So the meter registers a deflection, and the scale is calibrated in r.m.s. values of current or p.d. The r.m.s. value $= 1\cdot1 \times$ average value in meter M.

Full Wave A.C. Voltage Rectification by Bridge Circuit

In a.c. mains transistor receivers, diodes are used to rectify the alternating mains voltage and to produce steady or d.c. voltage needed for the circuits in the receiver.

Figure 31.7 shows a *bridge circuit* which produces full-wave rectification. Four diodes are used, and the action is similar to that described for measuring

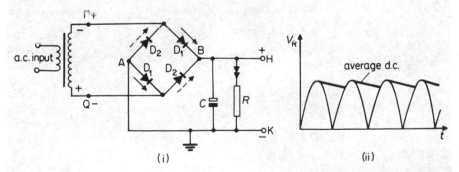

Figure 31.7 *Bridge rectifier circuit—d.c. voltage from a.c. mains*

a.c. with a moving-coil meter. Using a transformer, the mains a.c. voltage between P and Q is applied to the circuit.

On one half of a cycle, when P is +ve relative to Q, only the diodes D_1 conduct. On the other half of the same cycle, only the diodes D_2 conduct. In both cases the current goes through the resistor R in the *same* direction. Figure 31.7 (ii) shows the varying voltage across R; it is a varying d.c. voltage, so rectification is obtained.

The large (electrolytic) capacitor C is used to produce a large and fairly steady d.c. voltage. C charges up to the *peak* value of the applied a.c. voltage and then discharges through R when this voltage begins to decrease. If the discharge takes place slowly (which depends on the time-constant CR) then C becomes recharged to the peak value of the voltage again. The charge–discharge process takes place continuously and so the voltage across the C–R combination is a 'ripple' voltage as shown diagrammatically in Figure 31.7 (ii). With R alone, the average d.c. voltage would only be about two-thirds of the peak voltage. With the C–R arrangement, the average d.c. voltage is nearly equal to the peak voltage value, which is better, and so this removes (filters) the voltage variation.

The Transistor

The junction diode is a component which can only rectify. The *transistor* is a more useful component; it is a *current amplifier*. A transistor is made from three layers of p- and n-semiconductors. They are called respectively the *emitter* (E), *base* (B) and *collector* (C). Figure 31.8 (i) illustrates a *p-n-p transistor*, with electrodes connected to each of the three layers. In a *n-p-n transistor*, the emitter is n-type, the base is p-type and the collector is n-type, Figure 31.8 (ii). The base is deliberately made very thin in manufacture. So the transistor is a *three-terminal* device. The junction transistor is called a *bipolar transistor* because its action is due to two charge carriers, the electron ($-$) and the hole ($+$).

Figure 31.8 shows the circuit symbols for n-p-n and p-n-p transistors. The arrows show the directions of conventional current ($+$ve charge or hole

Plate 31A *The first transistor, invented 1947. It consisted of a triangular piece of silicon in contact with a germanium slab, and gold contacts.* (Photograph reproduced courtesy AT&T (UK) Ltd)

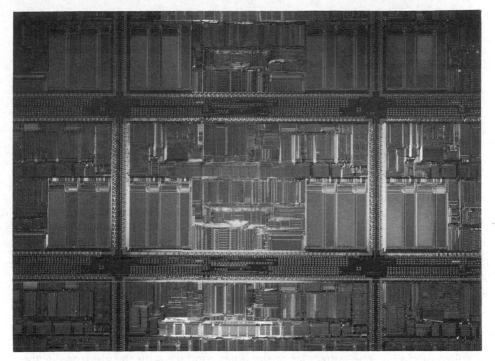

Plates 31B *The INMOS T900 transistor microprocessor, a chip about the size of a finger nail, which has more than 3 million working transistors.* (Photograph reproduced courtesy INMOS Ltd)

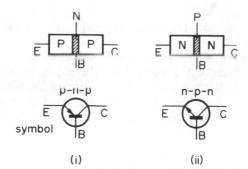

Figure 31.8 *Transistors and symbols*

movement) between the emitter E and base B, so electrons would flow in the opposite direction. In an actual transistor, the collector terminal is displaced more than the others for recognition or has a dot near it.

Current Flow in Transistors

Figure 31.9 (i) shows battiers correctly connected to a p-n-p transistor. The emitter-base is forward-biased; the collector-base is *reverse*-biased; and the base

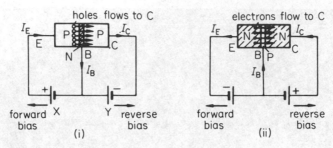

Figure 31.9 *Transistor action*

is common. This is called the *common-base* (C-B) mode of using a transistor. Note carefully the polarities of the two batteries. The positive pole of the supply voltage X is joined to the emitter E but the *negative* pole of the supply voltage Y is joined to the collector C. In the case of a n-p-n transistor, the negative pole of one battery is joined to the emitter and the positive pole of the other is joined to the collector, Figure 31.9 (ii). In this way the emitter-base is forward-biased and the collector-base is reverse-biased.

Consider Figure 31.9 (i). Here the emitter-base is forward biased by X, so that positive charges or holes flow across the junction from E to the base B. The base is so thin, however, that the great majority of the holes are pulled across the base to the collector by the battery Y. So a current I_C flows in the collector circuit. The remainder of the holes combine with the electrons in the n-base, and this is balanced by electron flow in the base circuit. So a small current I_B is obtained here. From Kirchhoff's first law, it follows that always, if I_E is the emitter current,

$$I_E = I_C + I_B$$

Typical values for a.f. amplifier transistors are: $I_E = 1 \cdot 0$ mA, $I_C = 0 \cdot 98$ mA, $I_B = 0 \cdot 02$ mA.

Although the action of n-p-n transistors are similar in principle to p-n-p transistors the carriers of the current in the n-p-n transistors are mainly electrons but holes in the p-n-p transistor. Electrons are more speedy carriers than holes (p. 767). So n-p-n transistors are used in high-frequency and computer circuits, where the carriers are required to respond very quickly to signals.

Common-Emitter (C-E) Characteristics

In general, the transistor has an *input circuit*, for the voltage or current to be changed, and an *output circuit* for the amplified voltage or current, for example. In Figure 31.9, the emitter-base (E-B) is the input circuit and the collector-base (C-B) is the output circuit. This is called the *common-base* mode of using the transistor because the base is common to the input and outut circuits. The transistor can be used in the *common-collector* or *common-emitter* mode, as we see later. Firstly, however, we need the characteristics of the common-emitter circuit, a circuit widely used for amplification. The characteristics help to estimate the performance of the common-emitter amplifier.

Figure 31.10 shows a circuit for obtaining the characteristics of a n-p-n transistor in the common-emitter mode. X and Y may be batteries of 1·5 V and 4·5 V respectively, connected to rheostats P and Q of 1 kΩ and 5 kΩ which act

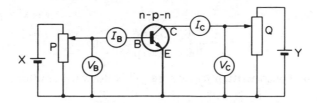

Figure 31.10 *Common-emitter characteristics*

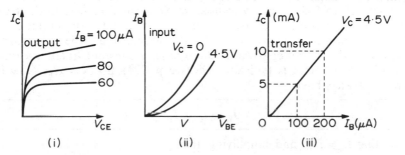

Figure 31.11 *Output, input and transfer characteristics*

as potential dividers. This enables the base-emitter p.d., V_{BE} or V_B, and the collector-emitter p.d., V_{CE} or V_C to be varied. The p.d. is measured by high resistance voltmeters, capable of measuring p.d. in steps such as 50 mV. The meter for base current, I_B, should be a microammeter and for the collector current, I_C, a milliammeter. Typical results are shown in Figure 31.11 (i), (ii) and (iii). The input circuit is the *base-emitter* circuit. The output circuit is the *collector-emitter* circuit.

Output characteristic ($I_C - V_C$, with I_B constant). The 'knee' of the curves shown in Figure 31.11 (i) correspond to a low p.d. of the order of 0·2 V. For higher p.d. the output current I_C varies *linearly* with V_{CE} or V_C for a given base current. The *linear* part of the characteristic is used in a.f. amplifier circuits, so that the output voltage variation is undistorted.

The *output resistance* r_0 is defined as $\Delta V_C/\Delta I_C$, where the changes take place on the straight part of the characteristic. r_0 is an a.c. resistance; it is the effective resistance in the output circuit for an a.c. signal input. It should be distinguished from the d.c. resistance, V_C/I_C, which is not required.

The small gradient of the straight part of the characteristic shows that r_0 is high. For example, suppose $\Delta V_C = 2V$ and $\Delta I_C = 0·02$ mA $= 2 \times 10^{-5}$ A. Then $r_0 = 2/(2 \times 10^{-5}) = 100\,000\ \Omega$. If a varying resistance load is used in the output or collector circuit, the high value of r_0 relative to the load shows that the output current is fairly constant. So the output voltage is proportional to the load resistance.

Input characteristic ($I_B - V_B$, V_C constant). The *input resistance* r_i is defined as the ratio $\Delta V_B/\Delta I_B$. As the input characteristic in Figure 31.11 (ii) is non-linear, then r_i varies. At any point of the curve, r_i is equal to the gradient of the tangent to the curve and is of the order of kilohms.

Transfer characteristic ($I_C - I_B$, V_C constant). The output current I_C varies fairly linearly with the input current I_B, Figure 31.11 (iii). The current *transfer ratio* β, or *current gain*, is defined as the ratio $\Delta I_C/\Delta I_B$ under a.c. signal conditions. It should be distinguished from the d.c. current gain, I_C/I_B. From

Figure 31.11 (iii),

$$\beta = \frac{(10 - 5)\ \text{mA}}{(200 - 100)\ \mu\text{A}} = 50$$

β, current gain, is also written h_{fe}, where 'fe' is 'forward emitter'.

Current Amplification in C-E Mode

In general, the magnitude of β, $\Delta I_C / \Delta I_B$, for the common-emitter circuit is high, from about 20 to 500 for many transistors. So a small change in the base or input current can produce a large change in the collector or output current.

We can obtain a rough value for β by assuming that when electrons are emitted from the n-emitter towards the p-base, a constant fraction α reaches the n-collector where α is typically 0·98 (see p. 774). So $I_C = \alpha I_E$. Now from before, we always have

$$I_E = I_C + I_B$$

Substituting $I_E = I_C / \alpha$ and simplifying, then

$$I_C = \left(\frac{\alpha}{1 - \alpha}\right) I_B$$

$$\therefore \Delta I_C = \left(\frac{\alpha}{1 - \alpha}\right) \Delta I_B$$

$$\therefore \frac{\Delta I_C}{\Delta I_B} = \beta = \frac{\alpha}{1 - \alpha} = \frac{0·98}{0·02} = 49$$

In practice, the current gain β of a transistor may be much higher.

Voltage Amplification and Power Gain

As we have just seen, the transistor in the common-emitter mode is a current amplifier. To change the output (a.c.) current to a voltage V_o, a resistance load R must be used in the collector or output circuit. Figure 31.12 shows also diagrammatically the base bias (steady voltage) b-b necessary for no distortion of V_o, the amplified a.c. output voltage.

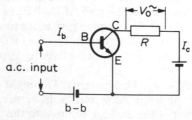

Figure 31.12 *Voltage amplification and power gain (diagrammatic)*

We can illustrate the voltage amplification in the output circuit by supposing that $R = 5\ \text{k}\Omega$, the a.c. input resistance $r_i = 1\ \text{k}\Omega$, the input (a.c.) voltage is 5 mV or 0·005 V peak value, and the current gain $\beta = 50$.

The peak a.c. current I_B flowing in the base circuit is, from $I = V/R$,

$$I_b = \frac{0 \cdot 005 \text{ V}}{1000 \text{ }\Omega} = 5 \times 10^{-6} \text{ A}$$

$$\therefore I_c = \beta I_b = 50 \times 5 \times 10^{-6} = 2 \cdot 5 \times 10^{-4} \text{ A}$$

$$\therefore V_o = I_c R = 2 \cdot 5 \times 10^{-4} \times 5000$$

$$= 1 \cdot 25 \text{ V peak}$$

$$\therefore \text{ voltage amplification } A_V = \frac{V_o}{V_i} = \frac{1 \cdot 25}{0 \cdot 005} = 250$$

Also, power gain = current gain × voltage gain = $50 \times 250 = 12\,500$

Note that the source of power gain is the battery or voltage supply.

C-E Amplifier Circuit

Figure 31.13 shows a n-p-n transistor in a simple C-E amplifier circuit. It uses one battery supply, V_{CC}. A load, R_L, is placed in the collector or output circuit. A resistor R provides the necessary bias, V_{BE}, for the base-emitter circuit for no output distortion. The base-emitter is then forward-biased but the collector-base is reverse-biased, that is, the potential of B is positive relative to E but negative relative to C.

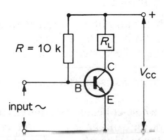

Figure 31.13 *Simple amplifier circuit*

Generally, with no distortion in the output voltage, the a.c. voltage amplification or gain in the simple common-emitter a.c. amplifier is given by

$$-\beta R_L / r_i$$

where β = a.c. current gain $\Delta I_c / \Delta I_b$, R_L is the resistance load and r_i = a.c. input resistance, $\Delta V_B / \Delta I_B$. The minus shows the 180° phase change of the a.c. input voltage when amplified, shown in Figure 31.13 and explained later. Using our values for the circuit in Figure 31.12, $\beta = 50$, $R_L = 5000 \text{ }\Omega$, r_i $(r_{be}) = 1000 \text{ }\Omega$, and voltage gain = $50 \times 5000/1000 = 250$.

The common-emitter circuit is very sensitive to temperature changes. So if no arrangement is made for *temperature stabilisation*, the output would become distorted when the temperature changed. Silicon transistors are much less sensitive to temperature change than germanium transistors and so are used more widely.

In practice, Figure 31.13 is unsuitable as an amplifier circuit since there is no arrangement for temperature stabilisation. A more practical C-E a.f. amplifier circuit is shown in Figure 31.14. Its principal features are: (i) a potential divider arrangement, R_1, R_2, which provides the necessary base-bias;

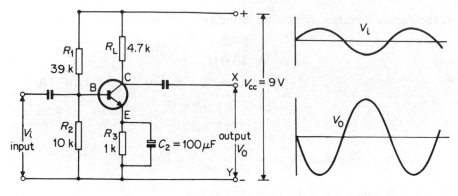

Figure 31.14 *Amplifier circuit and phase change*

(ii) a load R_L which produces the output across X, Y; (iii) a capacitor which stops the d.c. component in the input signal entering the circuit; (iv) a large capacitor C_2 across a resistor R_3, which prevents undesirable feed-back of the amplified signal to the base-emitter circuit; (v) an emitter resistance R_3, which stabilises the circuit for excessive temperature rise. So if the collector current rises, the current through R_3 increases. This lowers the p.d. between E and B, so that the collector current is automatically lowered.

C-E Amplifier Phase Change

Figure 31.15 (i) shows a basic amplifier circuit with a n-p-n transistor, a base resistance R_B of 50 kΩ to limit the base current to a suitable value, a load resistance R_L of 4 kΩ in the collector circuit, and a supply voltage V_{CC} of 9 V.

The *input circuit* is the whole base-emitter circuit. So the input voltage V_i is connected to this circuit.

The *output circuit* is the whole collector-emitter circuit. Note carefully that the output voltage V_o is always the voltage *between the collector C and the emitter E terminals* or earth. This the voltage which is passed to another transistor circuit if this is required. So you should remember that $V_o = V_{CE}$, as shown in Figure 31.15 (i).

The potential difference or 'potential drop' across the resistance $R_L = I_C R_L$.

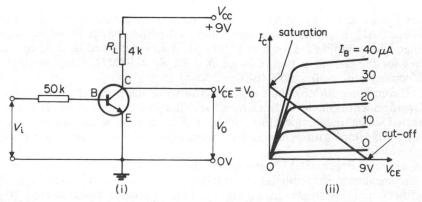

Figure 31.15 *C-E amplifier. Saturation and cut-off*

Since V_{CC} is the supply voltage, the output voltage V_o is

$$V_o = V_{CE} = V_{CC} - I_C R_L \qquad . \qquad . \qquad . \qquad . \qquad (1)$$

Suppose $I_C = 2$ mA, $R_L = 4$ kΩ $= 4000$ Ω, then potential drop $= 2 \times 10^{-3} \times 4 \times 10^3 = 8$ V. If $V_{CC} = 9$ V, then $V_o = 9 - 8 = 1$ V.

From equation (1), we also see that when I_C increases because the input voltage V_i increases, the output voltage V_o *decreases*. Similarly, if I_C decreases because V_i decreases, then V_o *increases*. So the input and output voltages are 180° out of phase or in antiphase in an amplifier.

Example on A.C. Amplifier

Figure 31.16 shows a simple form of silicon common-emitter amplifier. When the collector-emitter voltage V_{CE} is between $+1$ V and $+9$ V, the collector current is about 100 times the base current and the base-emitter voltage V_{BE} is about 0·7 V.

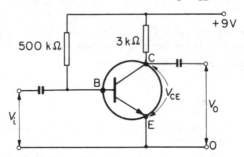

Figure 31.16 *Common-emitter amplifier*

Calculate: (i) the base current I_B and the voltage V_{CE} in the circuit; (ii) the voltage gain if the input (base-emitter) a.c. resistance is 2000 Ω, (iii) the largest (limiting) peak value of the input a.c. voltage if V_{CE} varies between 2·2 V and 8·2 V for linear amplification.

(i) I_B. Since $V_{BE} = 0·7$ V, p.d. across 500 kΩ resistor $= 9 - 0·7 - 8·3$ V

So
$$I_B = \frac{8·3}{5 \times 10^5} = 1·66 \times 10^{-5} \text{ A} = 17 \text{ μA (approx.)}$$

V_{CE}. So $I_C = 100 I_B = 100 \times 1·66 \times 10^{-5} \text{ A} = 1·66 \times 10^{-3} \text{ A}$

Then p.d. across 3 kΩ $= I_c R = 1·66 \times 10^{-3} \times 3000 = 4·98$ V

So
$$V_{CE} = 9 \text{ V} - 4·98 \text{ V} = 4 \text{ V (approx.)}$$

(ii) *Voltage gain.* If V_i is input, a.c. base current is

$$I_b = \frac{V_i}{R_b} = \frac{V_i}{2000}$$

So a.c. output $I_c = 100 I_b = 100 \times \frac{V_i}{2000} = \frac{V_i}{20}$

Then a.c. output $V_o = I_c R = \frac{V_i}{20} \times 3000 = 150 V_i$

So
$$\text{voltage gain} = \frac{V_o}{V_i} = 150$$

D.C. Amplifier

When the transistor is used as a switch in computer circuits, discussed soon, the input voltage V_i is a d.c. voltage. Suppose V_i is 2·0 V in Figure 31.15 (i) on p. 778, and the amplification factor or d.c. gain is 50 for the transistor. Neglecting the relatively small base-emitter d.c. resistance compared with the large base resistor R_B of 50 kΩ,

$$\text{base current, } I_B = V_i/R_B = 2\cdot0/(50 \times 10^3) = 4 \times 10^{-5} \text{ A} \qquad (1)$$

$$\text{So collector current, } I_C = 50 \times I_B = 50 \times 4 \times 10^{-5} = 2 \times 10^{-3} \text{ A} . \qquad (2)$$

To find the output voltage V_o or V_{CE}, we next find the potential drop V across the load resistance R_L of 4 kΩ. Then

$$V = I_C R_L = 2 \times 10^{-3} \times 4000 = 8 \ V \qquad . \qquad . \qquad . \qquad (3)$$

Since the supply voltage V_{CC} is 9 V, $V_o = 9 - 8 = 1$ V. So an input V_i from 0 to 2 V produces an output switch from 9 V (when $V_i = 0$) to 1 V. In practice, the base-emitter voltage is also taken into account.

Saturation and Cut-off

The *maximum* or *saturation current* I_S in the output or collector circuit will be obtained when the output voltage is zero.

A potential drop of about 9 V across the load resistance R_L of 4 kΩ will make V_o or V_{CE} zero because the supply voltage V_{CC} is 9 V. So

$$\text{saturation current } I_S = V/R = 9/4000 = 2\cdot25 \times 10^{-3} \text{ A} = 2\cdot25 \text{ mA}$$

Although the base current I_B may rise, the collector current remains at its saturation value.

If no current flows in the collector circuit, that is, $I_C = 0$, then we have *cut-off*. In this case there is no potential drop across R_L. So $V_o = V_{CE} = 9$ V.

At cut-off, $I_B = 0$. Figure 31.15 (ii) illustrates saturation and cut-off values. The small collector current at $I_B = 0$ is due to minority carriers in the transistor.

Example on D.C. Transistor Amplification

In Figure 31.17, the transistor is just saturated with an input d.c. voltage V_i. If the gain is 100, calculate V_i. Neglect the base-emitter resistance and p.d.

Since the transistor is just saturated, the output voltage $V_{CE} = 0$ as shown.

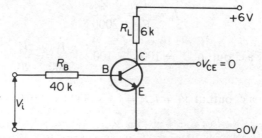

Figure 31.17 *Calculation on transistor with d.c. voltage input*

So potential drop across $R_L = 6$ V.

$$\therefore I_C = 6/6000 = 1/1000 = 1 \text{ mA}$$

$$\therefore I_B = I_C/100 = 1 \text{ mA}/100 = 1 \times 10^{-5} \text{ A}$$

Now $\qquad\qquad I_B = V_i/R_B$

$$\therefore V_i = I_B \times R_B = 1 \times 10^{-5} \times 40\,000 = 0.4 \text{ V}$$

Transistor Switch

In addition to its use as a current amplifier, the transistor can be used as a *switch* in computer circuits. Millions of switching operations are needed daily in working computers, so swift switches are required. On this account n-p-n transistors are preferred. Here the charge carriers are mainly electrons, which have a much greater speed for a given voltage than holes or p-charges.

The basic circuit, shown in Figure 31.18 (i), consists of a n-p-n transistor connected in the common-emitter way, with a resistance R in the output or collector circuit as we have already discussed.

A typical output voltage (V_o)–input voltage (V_i) characteristic of the circuit is shown in Figure 31.18 (ii). At very low input voltages the output voltage is practically $+6$ V, the supply voltage V_{cc} for the circuit, since the collector current is then low and there is a low potential drop I_cR across R. At input voltages of more than a fraction of a volt, however, the collector current is high, the potential drop across R is then high, and so the output voltage is nearly zero, as shown. In this way the transistor can switch voltages used in computer circuits from high to low values, or from '1' to '0' in logic language.

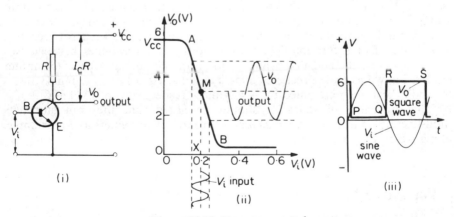

Figure 31.18 *Transistor switch*

Sine Wave Input, Amplifier Use

This type of transistor can be used:
(a) to change a sine wave to a square wave;
(b) to amplify a.c. voltages.

Suppose the input V_i is a *sine wave* voltage of peak value 6 V, Figure 31.18 (iii). For a large part of the $+$ve half cycle, V_i will be greater than $+0.4$ V. So the output voltage V_o will be practically zero along PQ. When V_i is less than $+0.4$ V and negative on the $-$ve half of the cycle, V_o will be practically a constant high voltage RS as shown. So the output V_o is roughly a *square wave* voltage.

To use the transistor as an *amplifier*, the output voltage V_o must have the same waveform as the input a.c. voltage V_i. This time the straight inclined line AB of the characteristic in Figure 31.18 (ii) must be used. So, as shown:

(a) the base bias should correspond to the midpoint M of AB, a bias equal to OX or about 0·2 V;

(b) the input voltage to be amplified must have a peak value not greater than 0·1 V, otherwise the output waveform is distorted during part of the input cycle.

In this and other a.c. amplification, it can be seen that the input voltage V_i is 180° out of phase, or antiphase, to the output voltage V_o.

States of Transistor

We can explain the characteristic curve in Figure 31.18 (ii) by noting that, if I_c is the collector current flowing for a particular input voltage, the output voltage V_o, or V_{CE}, is less than the supply voltage V_{CC} by the potential drop across R, which is $I_c R$. Thus

$$V_o = V_{CC} - I_c R \qquad \qquad (1)$$

In general, I_c depends on the base current I_b and this is controlled by the base-emitter or input voltage V_i.

Suppose V_i is very low or practically zero. Then I_c is practically zero, and the transistor is said to be 'cutoff'. From equation (1), we can see that the output voltage V_o is then practically equal to V_{CC} or *high*, as shown in Figure 31.18 (ii). Conversely, suppose V_i is high so that the transistor is 'saturated', so I_C is very high. The p.d. across R is then large and so the output voltage is practically zero from equation (1), as shown in Figure 31.18 (ii).

So depending on the input voltage, the transistor can switch between two states—cutoff or saturation. The output voltage then switches between two levels, $+V_{CC}$ (high) and practically 0 (low). In the special type of computer circuits known as *logic circuits* or *logic gates*, discussed later, the binary digits '1' and '0' can be represented by $+V_{CC}$ and 0 respectively, or by 0 and $+V_{CC}$, by this switching of states. It should be noted that the transistor acts 'non-linearly' in this case, whereas it acts 'linearly' in amplifiers as, for example, along AB in Figure 31.18 (ii).

You should know:

1 Pure semiconductors have resistances many millions of times greater than a good conductor such as copper.
  In semiconductors, the charge carriers are electrons ($-e$) and holes ($+e$). In a pure semiconductor there are equal numbers of electrons and holes.
  Electrons move faster than holes. So $I = nA(v_n + v_p)e$.
  Copper has one carrier, the electron and $I = nAve$.

2 Pure semiconductors *decrease* in resistance with temperature rise; more electrons break covalent bonds, forming electron–hole pairs.
  Copper *increases* in resistance with temperature rise; the lattice vibrations increase in amplitude.

3 **p-semiconductor is a pure semiconductor doped with a small amount of a trivalent element (e.g. indium). It has a large majority of holes and a small minority of electrons. It is now a fairly good conductor. n-semiconductor is a pure semiconductor doped with a small amount of a pentavalent element (e.g. phosphorus). It has a large majority of electrons and a small minority of holes. It is a fairly good conductor.**
4 **The junction diode has a p-n barrier. It can be used in a bridge rectifier.**
5 **The n-p-n transistor is a three-terminal device which is bipolar (i.e. with two carriers, electrons and holes). The terminals are joined to the emitter (n), base (p), collector (n).**
6 **With the *common-emitter* amplifier, the base current I_b is the input current and I_c, the collector current, is the output current. The amplification factor $\beta = \Delta I_c / \Delta I_b$. A load (resistor) is needed in the collector circuit for voltage gain, which is $-\beta R_L / r_i$, where R_L is load resistor, r_i is a.c. r_{be}.**
7 **When a transistor is used as a power amplifier, the increased power comes from the power supply. In digital circuits the transistor switches between high and low values.**

Very Large Scale Integration (VLSI)

A key factor behind the success of the electrons revolution has been VLSI, or *Very Large Scale Integration*. Transistors on their own are not especially powerful, acting as a single switch or a crude amplifier.

To build more useful circuits typically requires many thousands of transistors; for example, a modern memory circuit providing four million bits of storage needs more than four million transistors. It would not be practical to construct such circuits if each transistor came in its own little package taking up, say, 20 mm^2 of area; our memory circuit would then occupy 80 m^2, which is hardly practical!

In fact, such circuits are available on wafers of silicon about the size of a fingernail. This is the power of VLSI. Many transistors and other devices can be constructed and interconnected on a very small piece of silicon (the so-called *silicon chip*), which, once designed, is easy and cheap to mass-produce. A circuit constructed in this way is called an *integrated circuit*. The more dense the integration, the better the product, since it takes up less space, operates faster and requires less silicon. A lot of research is being carried out to produce ever smaller devices, and to find alternative semiconductors to silicon which operate faster, like gallium arsenide. One of the limiting factors of VLSI is the need to provide connections with external circuits. Even though the actual silicon chip might be only the size of a fingernail, it is usually placed in a much larger plastic package so that the right number of connecting pins can be positioned around it for input, output and power supply.

EXERCISES 31A   Semiconductors, Diodes, Transistors

Multiple Choice

1 Which of the following statements is correct?
  A Pure silicon is used in radio circuits.
  B Pure copper decreases in resistance as its temperature rises.

C Pure silicon and pure copper have similar charge carriers.
D Pure copper has the same charge carrier even when its temperature rises.
E Pure silicon changes its charge carriers as its temperature rises.

2 Figure 31A shows an a.c. supply V, diodes D1 and D2, and three resistors of 100 Ω. Which trace in **A** to **E** would appear on the oscilloscope?

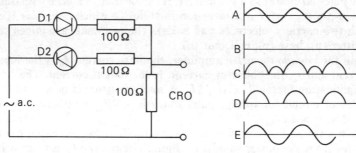

Figure 31A

3 In Figure 31B, the transistor circuit has a thermistor T whose resistance decreases with temperature rise. When the temperature increases, which statements in 1, 2 and 3 is/are correct?
1 Lamp L becomes brighter.
2 Collector current decreases.
3 Base current decreases.
A 1, 2, 3 all correct   B 1, 2 only are correct
C 2, 3 only are correct   D 1 only   E 3 only

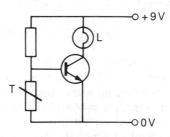

Figure 31B

Longer Questions

4 (a) Why does the electrical conductivity of an intrinsic semiconductor increase as the temperature rises?
   (b) Introducing certain impurities into semiconducting material also increases its electrical conductivity. Describe briefly *either* an *n*-type *or* a *p*-type semiconductor and explain why this increase in conductivity occurs. (*L*.)

5 A semiconductor diode and a resistor of constant resistance are connected in some way inside a box having two external terminals, as shown in Figure 31C. When a potential difference V of 1·0 V is applied across the terminals the ammeter reads 25 mA. If the same potential difference is applied in the reverse direction the ammeter reads 50 mA.

What is the most likely arrangement of the diode and the resistor? Explain your deduction. Calculate the resistance of the resistor and the forward resistance of the diode. (*L*.)

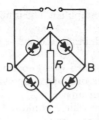

Figure 31C

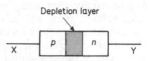

Depletion layer

Figure 31D

6 The circuit in Figure 31D shows four junction diodes and a resistor R connected to a sinusoidally alternating supply. Sketch graphs showing the variation with time over two cycles of the supply of:
(a) the potential of C with respect to A,;
(b) the potential of B with respect to A;
(c) the potential of B with respect to D. (*L*.)
7 Figure 31E shows a transistor which should operate the relay when the switch is closed. If the relay works at a current of 8 mA, calculate the value of R_B which will just make the relay operate. The direct current gain of the transistor is 80, and assume the base-emitter resistance is negligible here.

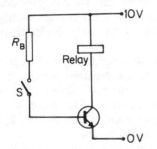

Figure 31E

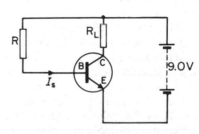

Figure 31F

8 In the transistor circuit shown in Figure 31F, R has a resistance of 150 kΩ, R_L has a resistance of 750 Ω, and the direct current gain of the transistor is 80.
Assume that there is a negligible potential difference between B and E, calculate:
(a) the base current I_B;
(b) the potential difference between the collector and emitter. (*AEB*.)
9 Figure 31G shows an arrangement for investigating the characteristics of a transistor circuit. The input voltage V_i is varied using the potentiometer, P. The corresponding output voltage V_o is shown graphically in Figure 31H.

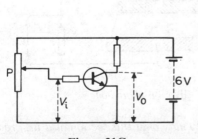

Figure 31G

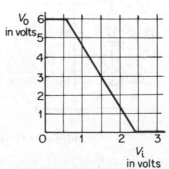

Figure 31H

The circuit is to be used as an alternating voltage amplifier. The input voltage must first be fixed at a suitable value by adjusting P.

(a) Suggest the most suitable value for this fixed input voltage, explaining your answer.

(b) A sinusoidally alternating voltage of amplitude 0·5 V is superimposed on this fixed voltage. What will be the amplitude of the output voltage variations? Will the output variations be sinusoidal? Justify your answers.

(c) Sketch one complete cycle of the output voltage which would be obtained if the amplitude of the superimposed sinusoidal voltage were increased to 1·5 V. (*AE.*)

Telecommunications[1]

After discussing how circuits work with electronic components such as diodes and transistors, we now briefly deal with some components used in carrying electrical signals along cables in telecommunications.

Optical Fibre Telecommunications

On p. 443, we saw that a *laser* is needed in optical fibre telecommunications. It provides a modulated light signal from electrical signals produced by a microphone at one end of the fibre cable and the light travels swiftly to the other end of the cable where it is demodulated (changed) to electrical signals and then changed to sound.

A *light emitting diode* (LED) can also provide a light signal, though it is not so intense as the laser. An LED consists of a forward-biased p-n diode made of gallium arsenide semiconductor material. As in normal diodes, electrons then drift into the p-region and holes into the n-region. The recombination of electron–hole pairs produces excess energy which is emitted as light. The intensity of the light is proportional to the input current.

LEDs are available in a range of colours, such as red, orange, yellow, green. A series resistor R is needed to limit the current through an LED and prevent damage, Figure 31.19 (i). For example, with colour red, an operating current of 20 mA, and a forward voltage of 2 V at 20 mA, a 400 Ω resistor would be needed with a 10 V supply, as the reader should check.

Basically, the *semiconductor laser* is an LED operating at high current. The p- and n-regions are made from gallium arsenide and gallium aluminium arsenide. Stimulated emission is produced by making the faces of the semi-

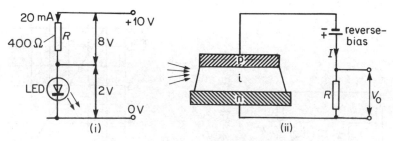

Figure 31.19 (i) *LED* (ii) *p-i-n photodiode*

[1] *Students limited in time for A-level study may omit this section at a first reading depending on the syllabus. Consult your teacher.*

conductor optically flat so that some light is reflected back repeatedly as with the gas laser (p. 843).

The semiconductor laser is superior to the LED for telecommunications. The light is not only much greater in intensity but the spectral spread is typically 1 to 2 nm compared to a much greater spread using an LED. So low dispersion is produced in a fibre cable at high bit rates, which is an advantage.

Photodiode. At the output end of optical fibre communications, a photodiode is needed to change the arriving digital light pulses to an output voltage signal.

The photodiode uses the principle that when light falls on an p-n junction, the photo energy hf can produce more electron–hole pairs in a crystal lattice. For example, visible light of wavelength 550 nm has photon energy of about 2·3 eV. This is greater than the ionisation energy of silicon, 1·1 eV.

Figure 31.19 (ii) shows the principle of a *p-i-n photodiode*. It has an intrinsic (pure) silicon layer *i* between the p-layer and n-layer. Effectively this increases the size of the depletion layer. Many more electron–hole pairs are then generated compared to the normal p-n photodiode and so the current I, and the response to the illumination, are then greater.

As shown, the diode is *reverse-biased*. The output voltage V_o across the resistor R increases with the intensity of illumination.

Radio Telecommunications, Aerials

We now turn to radio telecommunication and the topic of *aerials*. In a radio transmitter, the *aerial* has the same function as the sounding board in a violin or as the resonance tube used with a tuning-fork. It helps the radio oscillator, which generates radio waves, to radiate its energy into space. If a tuning-fork is sounding when held in the hand, only a weak sound is heard. If, however, the sounding tuning-fork is moved over the top of a resonance tube of length $\lambda/4$, where λ is the wavelength of the sound in air, a loud sound is heard coming from the tube. Waves of large amplitude, stationary or standing waves, are now set up in the air in the tube (p. 608).

In the same way, the aerial of a transmitter is coupled to the oscillator or radio source, and adjusted so that standing electrical waves are set up along it. The radiofrequency waves from the oscillator are then most strongly radiated into space.

Quarter- and Half-Wave Aerials, Standing Waves

Figure 31.20 (i) shows a long vertical wire PQ with an r.f. oscillator Q near the bottom earthed end. A *current node* is set up by Q at P, where electrons cannot move, and a *current antinode* at Q where electrons are free to move. The electrons have greatest pressure at P, which is therefore a *voltage antinode*, and least pressure at Q which is a *voltage node*. Figure 31.20 (ii) shows the analogy with air molecules in a closed pipe at resonance. The displacement curve is similar to the current $\lambda/4$ curve I and the pressure curve to the voltage curve V. The aerial and pipe have a length $\lambda/4$, where λ is the wavelength. So PQ is a *quarter-wave* $(\lambda/4)$ aerial. Medium-wave aerials are generally of the quarter-wave type. For 200 m wavelength, the aerial height is 50 m.

On short wavelengths such as television wavelengths, a *half-wave* or *dipole* aerial is used, Figure 31.21 (i). Here the ends P, P′ of the serial are current nodes (voltage antinodes) while the middle M is a current antinode (voltage node). In sound, this corresponds to a pipe closed at both ends and excited in the

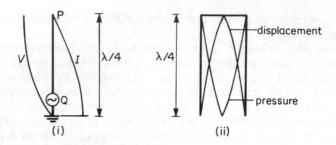

Figure 31.20 *Quarter-wave aerial*

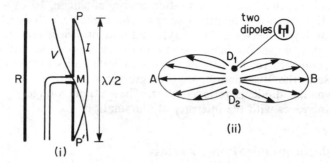

Figure 31.21 *Half-wave (dipole) aerials*

middle. A reflector R behind the dipole increases the radiation in a required direction.

Figure 31.21 (ii) shows roughly how the radiation intensity varies round a two-dipole or H-aerial, D_1, D_2. In this so-called *polar diagram*, the lengths of the arrows lines are proportional to the intensity in the direction concerned. So the radiation is mainly in the direction AB. More dipole arrays can provide greater intensity in a required direction.

Microwaves (very short wavelengths of the order of a few cm) can be transmitted from small dipoles placed at the focus of paraboloid dishes (p. 431).

Receiving aerials are similar to transmitting aerials. Dipole aerials are used in medium-wave and television reception, and small dipoles for microwave reception with paraboloid dishes.

To avoid interference at receivers by neighbouring transmitters, the two radio signals are sent out polarised in different directions, horizontal and vertical. The receiving aerial must be correctly positioned, horizontally or vertically to receive the transmitter required.

Amplitude Modulation

The energy radiated from an aerial is practically zero when the frequencies are below about 15 kHz, which is in the audio-frequency (a.f.) or sound range. So speech and music cannot be radiated directly.

Aerials radiate most strongly only high frequencies, which are those known as *radio-frequency* (r.f.) waves. They may range from very high frequencies such as 100 MHz (10^8 Hz), 3 cm waves, to 200 kHz (2×10^5 Hz), 1500 m waves. For r.f. wave calculations, a middle value of 1 MHz (10^6) may often be taken.

Practical methods of radiating a.f. (sound) energy consists basically of carrying it on an r.f. wave, which is called the *carrier wave*. The carrier r.f. wave

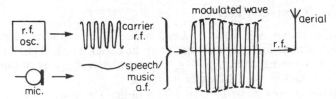

Figure 31.22 *Amplitude modulation*

is then said to be *modulated* by the a.f. wave. Figure 31.22 shows an *amplitude modulated* wave radiated from an aerial. It is an r.f. wave whose amplitude varies at the same frequency as the a.f. voltage generated by speech or music by the studio microphone. If A is the amplitude of the r.f. carrier wave and B is that of the a.f. wave carried, then B/A is defined as the 'depth of modulation'. With 50% modulation, $B = A/2$.

Sidebands, Bandwidth

Analysis of the modulated wave for a sine wave carrier, with a modulating a.f. sine wave, shows that it consists of three different r.f. waves of constant amplitude, each of which can be detected separately.

If f_c is the carrier r.f. and f_m is the much smaller modulating a.f., the three waves have respective frequencies of (i) f_c, (ii) ($f_c + f_m$), called the *upper sideband frequency*, and (iii) ($f_c - f_m$), called the *lower sideband frequency*. If $f_c = 1$ MHz and the highest modulating a.f. is 10 kHz or 0·01 MHz, then

$$\text{upper side frequency} = f_c + f_m = 1\!\cdot\!01 \text{ MHz}$$

and

$$\text{lower side frequency} = f_c - f_m = 0\!\cdot\!99 \text{ MHz}$$

In practice, the carrier is modulated by all frequencies in the a.f. range, from high to low. This is shown diagrammatically in Figure 31.23 (i). The bandwidth of the sidebands transmitted is 20 kHz in this case, from 1·01 to 0·99 MHz.

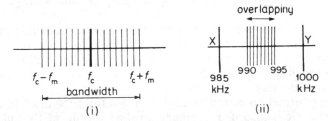

Figure 31.23 (i) *Sideband frequencies*   (ii) *Overlapping*

Figure 31.23 (ii) shows the frequencies X and Y of two transmitting stations which are less than 20 kHz apart. The carrier X has a frequency of 985 kHz (0·985 MHz) and the carrier Y has a frequency of 1000 kHz (1·0 MHz). In this case some of the upper sidebands of X have the same frequencies as some of the lower sidebands of Y, as shown, and overlapping occurs. So *interference* may occur from one transmitter when reception is required from the other. Since, in practice, two broadcasting stations may be separated by about 9 kHz, interference takes place between two powerful stations. It can be heard on an unselective radio receiver.

AM Radio Receiver (TRF)

Figure 31.24 shows the basic system needed for a radio receiver when amplitude-modulated (AM) waves arrive at the aerial. The coil (L, R)–capacitor (C) circuit is a tuning circuit, p. 418. Here a high voltage is obtained across C when it is tuned to the r.f. carrier wave.

The voltage is now passed to an r.f. amplifier for more amplification, and then to a suitable diode detector circuit. A diode conducts only one half, say the positive half, of the modulated wave. So the output voltage is a positive r.f. voltage with an average amplitude value which follows the a.f. variation. The diode thus 'detects' the a.f. voltage carried by the modulated wave.

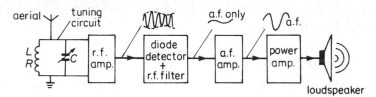

Figure 31.24 *Tuned radiofrequency (trf) receiver*

An electrical filter circuit cuts off the r.f. voltage, leaving only the a.f. voltage to pass on to an a.f. amplifier. This is finally passed to an a.f. power amplifier, so that maximum sound energy is obtained from the loudspeaker.

Optical Storage Systems

A photodiode (p. 787) can be found at the heart of every compact disc (CD) player, of which millions now exist in households and industries around the world. The compact disc is just one example of an *optical storage system*; other types of optical disc, which can be used for recording as well as playback, are becoming increasingly popular as data storage devices on computer systems.

Optical discs benefit from a very high information capacity and rapid access to any particular item of data, but where they really triumph over alternative storage devices (such as magnetic discs and tapes) is in the total lack of contact between the reading head and the disc: this reduces wear, prolonging the life of the disc considerably.

So how is information stored on and retrieved from the disc? In the case of the CD, the music is recorded in binary form as a series of 1's and 0's, which appear as a track of tiny 'pits' on the disc's surface. The disc is read by shining a laser beam onto the surface and measuring the reflected light using a photodiode. The pits change the reflecting power of the disc's surface, and so the pattern of 1's and 0's may be obtained simply by measuring the reflected light, without any physical contact between the pick-up and the disc.

To reproduce hi-fidelity music, the sound wave must be sampled to an accuracy of 16 bits (giving 2^{16} different sound levels) at a rate of 44·1 kHz; after error-correcting bits are added, the overall data-rate stored on the disc is over 4 million bits per second. Given that a CD can hold over 70 minutes of music, this means that there are over $70 \times 60 \times 4 \times 10^6 = 1·68 \times 10^{10}$ bits on each disc, a phenomenal amount of information on such a small area.

EXERCISES 31B Telecommunications

Multiple Choice

1 Figure 31I(i) shows a half-wave aerial TB. Which statement in **A** to **E** is correct?
   A The top T is a current antinode.
   B A progressive radio wave is set up along TB.
   C If the radio wavelength transmitted is 50·0 m, the length of TB is 12·5 m.
   D The radio wave transmitted from this aerial is mainly in the direction of P.
   E The aerial cannot be used for radio reception.

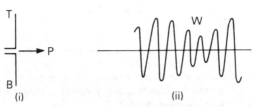

(i)   (ii)

Figure 31I

2 Figure 31I(ii) shows an amplitude-modulated wave W. Which statement **A** to **E** is correct?
   A The wave is an audio-frequency one.
   B The amplitude is modulated by the radio frequencies of the carrier.
   C The wave is not a radio-frequency wave.
   D The wave has a constant single frequency.
   E Modulation is due to the audio-frequencies transmitted.

3 A 2·0 MHz radio station sends out an upper sideband of 2·01 MHz and a lower sideband of 1·99 MHz. The bandwidth is

   A 2·00 MHz **B** 1·995 MHz **C** 200 kHz **D** 20 kHz **E** 2 kHz

4 Amplitude-modulated radio waves are received by a tuned radio-frequency (trf) receiver. The receiver has a suitable detector circuit in order to
   A Amplify the carrier waves
   B Amplify the audio-frequencies carried
   C Rectify the carrier waves
   D Detect the carrier waves
   E Transfer the audio-frequencies to the radio-frequency amplifier

Longer Questions

5 (a) What do you understand by the term *amplitude-modulated carrier wave*? Draw a labelled block diagram showing the basic elements of a simple amplitude-modulated radio transmitter for the broadcasting of audio signals developed by a microphone.
   (b) Speech signals in the frequency range 300 Hz to 3400 Hz are used to amplitude-modulate a carrier wave of frequency 200·0 kHz. Determine: (i) the bandwidth of the resultant modulated signals; (ii) the frequency range of the lower sideband; (iii) the frequency range of the upper sideband.
   Figure 31J shows the range of speech signals and the carrier wave. Copy the diagram and use it to represent your answers to (i), (ii) and (iii).
   (c) Long distance intercontinental telecommunications may use free-space electro-magnetic wave propagation linking ground station to ground station. Use a diagram to show how this is achieved for: (i) microwave signals; (ii) radio signals. (Apparatus details are NOT required.)
   Estimate the transit times for a U.K.–U.S.A. telephone signal link using: (i) satellite communication; (ii) surface wave propagation. Assume that the

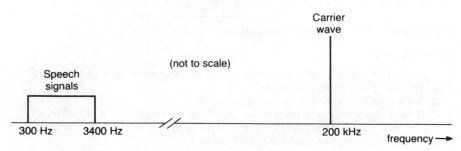

Figure 31J

satellite is 36 000 km equidistant from each ground station and that the great circle distance between the two ground stations is 8400 km. (Speed of electro-magnetic waves in free space $= 3.00 \times 10^8$ m s^{-1}.) (*L.*)

6 (a) A sinusoidal radio-frequency carrier wave of frequency f_c is to be *amplitude modulated* by a sinusoidal audio-frequency signal of frequency f_s.

  (i) Explain the term *amplitude modulated*.

  (ii) State the frequencies of the radio-frequency waves which are present in the transmitted signal.

  (iii) State the frequencies of the radio-frequency.

  (iv) Hence, or otherwise, explain what is meant by *bandwidth*.

 (b) Figure 31K shows a block diagram of the various elements in a simple amplitude modulated radio receiver.

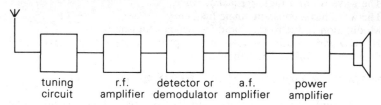

Figure 31K

  (i) Draw a circuit diagram of a simple tuning circuit.

  (ii) Why is it desirable to amplify the r.f. signal before de-modulation?

  (iii) List the elements shown in Figure 31K for which the output power is greater than the input power. What is the source of this increased power?

  (iv) The frequency of the audio modulating signal is 5 kHz. What are the minimum possible bandwidths of (1) the r.f. amplifier, (2) the a.f. amplifier?

  (*C.*)

32 Analogue and Digital Electronics

Analogue Electronics

In analogue electronics we deal with circuits which process d.c. or a.c. signals, such as those in radio or television or in oscillators. In such circuits we have a continuously varying signal. The main circuit used is the operational amplifier (opamp) *and this is the 'building block' we use in this section. We shall see that the opamp can be used for many different types of useful amplifiers, in oscillators, in switching devices and as an integrator.*

Amplifiers, Voltage Gain

As we have seen, a single transistor can be used as an amplifier. Most practical amplifiers, however, use a combination of transistors. In this section the amplifiers have many transistors and are called *operational amplifiers* or *opamps* for short.

We shall deal later with the *voltage gain* of opamp circuits. If the input voltage to a circuit is V_{in} and the output voltage is V_{out}, the voltage gain is defined as:

$$voltage\ gain = \frac{V_{out}}{V_{in}}$$

The input and output voltages can be a.c. or d.c. and may be measured with a cathode-ray oscilloscope.

Opamps

Opamps are amplifiers built from many transistors and other circuit components in an 'integrated circuit' inside a chip of semiconductor material. They can amplify a voltage from zero frequency (d.c.) to very high frequencies. In practice the range of frequencies is often controlled by external components, especially capacitors.

Figure 32.1 shows the symbol for the opamp. The opamp has two inputs and one output.

The opamp is a *differential amplifier*, that is, the output is directly proportional to the *difference* in voltage between the two inputs of the opamp. The two inputs are the 'non-inverting input' (+) and the 'inverting input' (−). The function of the inputs will be explained soon.

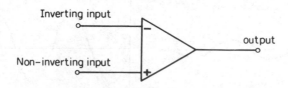

Figure 32.1 *Opamp symbol*

Function of Opamp

Figure 32.2 shows a simplified circuit of an opamp. The input voltage V_{in} is connected to the non-inverting (+) input and earth or ground voltage, 0 V, the inverting input (−) is connected to earth, and the output voltage V_{out} is

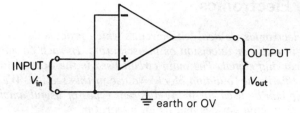

Figure 32.2 *Non-inverting amplifier*

between the output terminal and earth. The opamp now acts as a *non-inverting amplifier*. This means that the output is the exact, amplified copy of the input as shown in Figure 32.3. So the output voltage V_{out} is *in phase* with the input voltage V_{in} in a non-inverting amplifier.

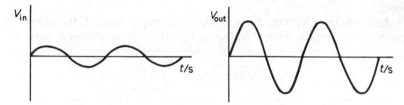

Figure 32.3 *Output voltage in phase with input voltage (non-inverting amplifier)*

If, however, the amplifier is connected as shown in Figure 32.4, it is said to be connected as an inverting amplifier. Here the input is between the inverting input (−) and earth and the non-inverting (+) input is earthed. In this case the output is an exactly opposite amplified copy of the input, as shown in Figure 32.5. So V_{out} is 180° out of phase or in antiphase with V_{in} in an inverting amplifier.

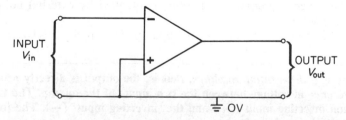

Figure 32.4 *Inverting amplifier*

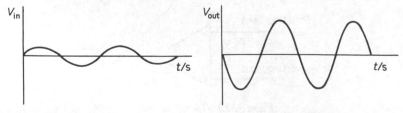

Figure 32.5 V_{out} *in antiphase to* V_{in} *(inverting amplifier)*

If an opamp were to be connected as shown in Figures 32.2 and 32.4, the voltage gain (A) V_{out}/V_{in} is about 10^5. In most practical uses of this opamp, however, the gain is limited by *negative feedback*, which we now discuss.

Negative Feedback

'Negative feedback' is said to occur when a little of the output signal is fed back to the *inverting* input. As we saw in Figure 32.5, the output is 180° out of phase with the input. So the feedback *reduces* the signal that the amplifier has to amplify and therefore reduces the gain. The amount of the output fed back is controlled by a *feedback resistor* R_f, as shown in Figure 32.6. Resistor R_i is an *input resistor* whose purpose will be explained later when we calculate the circuit voltage gain.

As we see soon, there are several advantages in using negative feedback.

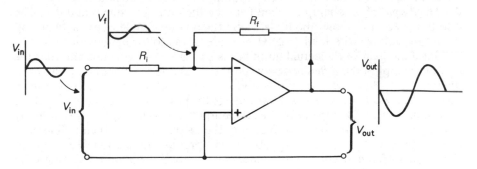

Figure 32.6 *Negative feedback—input and output voltages*

Frequency Response of Opamp Circuit

The gain of an opamp depends on frequency. A graph of gain against frequency is shown in Figure 32.7 for a typical opamp. With no feedback, or *open-loop gain* A_{OL}, the gain drops rapidly from 10^5 at low frequencies to 1 at very high

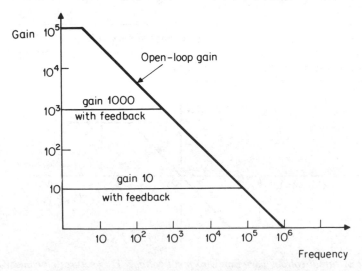

Figure 32.7 *Frequency response of opamp circuit*

frequencies. The situation is clearly not ideal since amplifiers generally require the gain to be constant at all frequencies. However, as the negative feedback is increased and the overall gain becomes lower, the gain remains constant over a wider range of frequencies and so the frequency response of the amplifier is improved. With feedback, the gain is called *closed-loop gain* or A_{CL}.

Some Opamp Characteristics

In order to calculate the gain of any given opamp circuit, it is necessary to know a little more about the characteristics of an opamp. These are:

1 There is a *very high impedance* between the + or − input, and ground about 2 MΩ or more. The high input impedance means that practically no current (in practice very, very little current, called the *bias current*) will flow into or out of the inputs when a voltage is applied. So there is no effect on the input circuit when connected to the opamp, which is an advantage.

2 The *output* of an opamp is taken between the output terminal and 0 V. The output impedance is low, about 100 Ω. So the opamp can deliver practically all its output voltage to a load of 2 kΩ or more in the following circuit.

3 Since the gain of the actual amplifier is so high, and remembering that the amplifier amplies the *difference* between the − and the + inputs, the slightest difference in voltage between the − and + inputs will make the output voltage go to its highest value, which is the voltage of the power supply. This is called the *saturation* value (V_S) of the output as the output can go no higher. If the supply voltage is 15 V and the open-loop gain of the amplifier is 10^5, then a difference in voltage of 15 V/10^5, which is 150 μV, will produce saturation. We see later that with such small voltage differences (V_{in}) between the non-inverting (V_+) and inverting (V_-) inputs, the output voltage (V_{out}) can swing either way between + 15 V ($+ V_s$) or − 15 V ($- V_s$) (Figure 32.8). The output is $+ V_s$ when V_+ is greater than V_-, and $- V_s$ when V_- is greater than V_+.

So for our amplifier to be of any use, the − input must be at virtually the same voltage as the + input since the two inputs have only a very small voltage (such as 150 μV) between them. Now in the inverting amplifier circuit the + input is connected to earth or ground which is also 0 V, so the − input must be always virtually at the same earth voltage. The − input is therefore known as a *virtual earth*.

Saturation occurs when output voltage reaches supply voltage, + 15 V or − 15 V. **Very small voltage difference between inverting and non-inverting terminals produces saturation because the gain is high.**

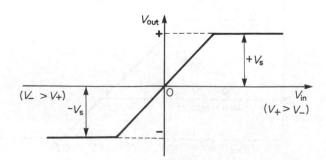

Figure 32.8 *Output voltage variation with input voltage.* V_S = *saturation value* V_+ = *non-inverting voltage,* V_- = *inverting voltage*

Gain of Inverting Amplifier

We are now in a position to calculate the gain with negative feedback.

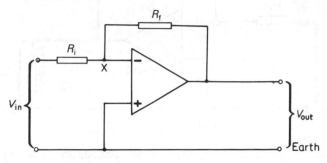

Figure 32.9 *Gain with negative feedback*

Referring to Figure 32.9, since the − input is at earth potential (virtual earth) the current through R_i will be V_{in}/R_i.

The current through R_f will be V_{out}/R_f.

But since the input impedance of the amplifier is very high, no current can flow into the − input. The sum of the currents at junction X must be equal to zero (Kirchhoff's law). So whatever current goes into junction X through R_1 must leave the junction through R_f.

So

$$V_{in}/R_i + V_{out}/R_f = 0$$

$$\therefore V_{in}/R_i = -V_{out}/R_f$$

$$\therefore V_{out}/V_{in} = -R_f/R_i = \text{gain}$$

---

so gain of this inverting amplifier $= -R_f/R_i$

The minus sign indicates that the output is *inverted*. Notice that the gain of the amplifier is independent of the open-loop gain of the opamp. Since the gain depends only on the *external* components R_i and R_f, the gain is not affected by any changes that may take place inside the opamp, such as a change in gain due to temperature change. So the negative feedback provides stability.

---

1 **Amplifiers with almost infinitely variable gain can be produced using one standard opamp circuit.**
2 **The use of negative feedback improves the range of frequencies that the amplifier will amplify and improves stability.**

A Practical Inverting Amplifier

The simplified circuit shown earlier in Figure 32.9 for an inverting amplifier needs some small modification. A resistor, R_3 needs to be connected between the + input and 0 V, Figure 32.10. R_3 should be equal to the sum of R_1 and R_2 in *parallel* in order to keep the inputs balanced as far as possible. In this circuit, the power supply connections $+V_S$ and $-V_S$ are shown although these connections will be omitted in following circuits.

The input impedance of this circuit is determined by the value of R_1. Since the − input is virtual earth, the input is effectively connected across the parallel

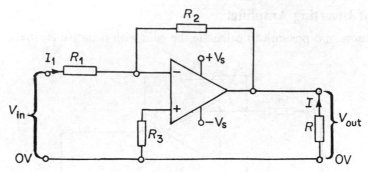

Figure 32.10 *Inverting amplifier*

combination of R_1 and the (very large) input impedance of the opamp itself. The advantage of this circuit is that the input impedance can be varied to suit the preceeding stage.

Output of Opamp

The *output* is taken between the 0 V and the output terminal of the opamp circuit. Suppose R is the opamp resistance load (or the input resistance of the next circuit) and I is the current in R (Figure 32.10). Then $V_{out} = -IR$. So gain $= V_{out}/V_{in} = -IR/I_1R_1 = -R_2/R_1$ (p. 797), where I_1 is the current in R_1. So $I/I_1 = R_2/R$. The opamp can typically deliver a current of about 5 mA which means that the minimum load R is about 1 kΩ.

Since V_- is a virtual earth, *power gain* $= V_{out}^2/R \div V_{in}^2/R_1 = $ (voltage gain)$^2 \times$ $(R_1/R) = (R_2/R_1)^2 \times (R_1/R)$.

Example on Saturation in Opamp Circuit

An inverting opamp circuit has a voltage amplification of 100 and a supply of ±9 V. If an input a.c. voltage of 50 Hz has a peak value of 0·6 V, sketch the output voltage waveform.

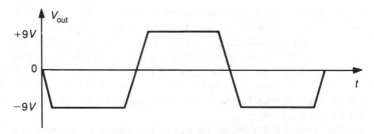

Figure 32.11 *Saturation of opamp*

The inverted output voltage *saturates* at -9 V and $+9$ V. Since the voltage amplification is 100, the input voltage is then respectively $+0·09$ V and $-0·09$ V. The 'clipped' waveform is shown roughly in Figure 32.11. The time t of saturation from zero first occurs when $0·09 = 0·6 \sin \omega t$, where $\omega = 2\pi f = 100\pi$, from which t can be calculated.

Opamp as a Summing Amplifier

One of the most frequent uses of an opamp is in audio pre-amplifiers and mixers. An opamp can be used to add any number of signals or voltages and the circuit is called a *summing amplifier*. Figure 32.12 shows a simple summing amplifier to add three voltages, V_1, V_2, V_3, using an inverting amplifier.

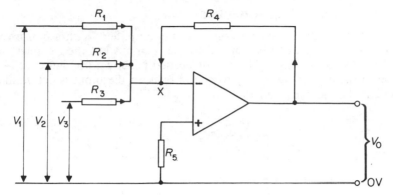

Figure 32.12 *Opamp as summing amplifier*

Point X is a virtual earth and therefore the input currents into point X are:

$$V_1/R_1, \ V_2/R_2, \ V_3/R_3$$

and the feedback current $= V_o R_4$

So $$V_1/R_1 + V_2/R_2 + V_3/R_3 + V_o/R_4 = 0$$

since all the currents at a junction must add to zero.

$$\therefore \ -V_o = \frac{R_4}{R_1} \times V_1 + \frac{R_4}{R_2} \times V_2 + \frac{R_4}{R_3} \times V_3 \ . \qquad . \qquad . \qquad (1)$$

So the output voltage, V_o, is the sum of the three inputs with each input multiplied by a factor R_4/R, where R is the corresponding input resistance.

This result applies to a.c. or d.c. voltages. It is especially useful for a microphone mixer since we do not want one microphone to affect another. It might also be necessary to change the gain of the amplifier for different inputs, in which case R_1, R_2 and R_3 can be varied.

Example on Summing Amplifier

What is the output voltage of the opamp circuit of Figure 32.13?

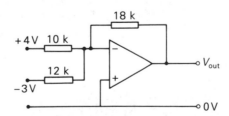

Figure 32.13 *Example on summing amplifier*

This is a summing amplifier and therefore the output is given by:

$$-V_o = \frac{R_4}{R_1} \times V_1 + \frac{R_4}{R_2} \times V_2$$

$$\therefore \ -V_o = \frac{18 \times 10^3 \times 4}{10 \times 10^3} + \frac{18 \times 10^3 \times (-3)}{12 \times 10^3}$$

$$= 7 \cdot 2 - 4 \cdot 5 = 2 \cdot 7 \text{ V}$$

So $$V_o = -2 \cdot 7 \text{ V} \ (V_o \text{ is antiphase to } +4 \text{ V input})$$

Gain of Non-Inverting Amplifier

We now discuss the circuit of a non-inverting amplifier. With a non-inverting amplifier, the input has to be applied to the + input but the feedback has to be applied to the − input. The circuit needed is shown in Figure 32.14. The fraction of the output signal V_{out} to be fed back to the input is determined by the potential divider R_1 and R_2.

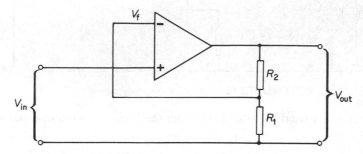

Figure 32.14 *Gain of non-inverting amplifier*

The fraction of V_{out} sent to the inverting (−) input is V_f, where

$$V_f = \frac{V_{out} \times R_1}{(R_1 + R_2)} \quad . \qquad . \qquad . \qquad . \qquad . \qquad (1)$$

The voltage *difference* between the two inputs = V_T, where

$$V_T = V_{in} - V_f \quad . \qquad . \qquad . \qquad . \qquad . \qquad (2)$$

V_T is the voltage actually amplified by the opamp.

Now
$$V_{out} = A_o \times V_T \quad . \qquad . \qquad . \qquad . \qquad . \qquad (3)$$

where A_o is the open-loop gain of the opamp (see p. 796). Substituting for V_T in (2) using (3), then, since $V_T = V_{out}/A_o$,

$$V_{in} - V_f = V_{out}/A_o$$

and
$$V_f = V_{in} - V_{out}/A_o \quad . \qquad . \qquad . \qquad . \qquad (4)$$

Substituting for V_f in (1) using (4),

$$V_{in} - \frac{V_{out}}{A_o} = \frac{V_{out} \times R_1}{(R_1 + R_2)}$$

$$\therefore V_{in} = V_{out}\left(\frac{R_1}{R_1 + R_2} + \frac{1}{A_o}\right)$$

Since A_o is typically about 10^5 and $1/A_o$ is negligible,

$$\therefore \text{gain} = V_{out}/V_{in} = (R_1 + R_2)/R_1$$

So gain of this amplifier $= 1 + (R_2/R_1)$

s with the inverting amplifier, the gain depends only on the values of R_2 and
and is independent of the open-loop gain of the opamp.

For example: If $R_1 = 1\text{ k}\Omega$ and $R_2 = 100\text{ k}\Omega$

$$\text{gain} = 1 + 100/1 = 101$$

If $R_1 = 10\text{ k}\Omega$ and $R_2 = 10\text{ k}\Omega$

$$\text{gain} = 1 + 10/10 = 2$$

Practical Opamps

Most practical opamps are integrated circuits. They are usually made as 8-pin, dual-in-line packages. The pin connections have become standard so that opamps can be readily interchanged with one another. The 741 type opamp, first manufactured in the late 1960s, is still perhaps the best known, tried and tested opamp circuit. More modern replacements are available now with better specifications. Figure 32.15 shows the pin connections for the 8-pin, opamp package. The only variation between types is in pin 8. This pin is used for a variety of functions or if not required is not connected. The function of each pin will be explained in the following section.

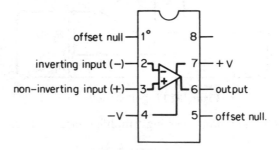

Figure 32.15 *Opamp connections*

An opamp is slightly unusual in its requirements for a power supply. It requires a *dual*, balanced voltage supply such as $+15$ V, 0 V, -15 V $(=30$ V supply) as shown below.

```
——————— +V (+15 V)   +V is positive with respect to the centre 0 V line and
——————— 0 (0 V)       has a maximum of about 15 V for most opamps.
——————— -V (-15 V)    -V is negative with respect to the centre 0 V line and
                       must have the same magnitude as the +V.
```

0 V is connected to earth or ground. The potential difference between $-V$ and $+V$ will be $2V$. By convention, once the power supply has been established, the connections from the opamp to the power supply are not shown in the circuit diagram but are assumed to be made. Figure 32.10 shows an opamp connected as an inverting amplifier with power supply connections $+V_S$ and $-V_S$.

Offset Voltage

In practice, due to manufacturing tolerances, even when there is no signal applied to the input the internal components of the opamp may supply a small differential voltage to the inputs. Even if this voltage is small, the gain of the opamp results in a large offset voltage being present at the output, which should be at exactly 0 V under no signal conditions. In many circuits, this offset voltage at the output does not matter but for d.c. circuits it must be removed and this is done by using the two 'offset null' terminals of the opamp as shown in Figure 32.16. The 10 kΩ potentiometer is adjusted until the output is at exactly 0 V with no signal present at the inputs.

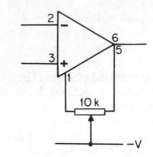

Figure 32.16 *Offset voltage*

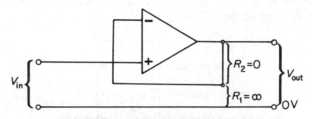

Figure 32.17 *Voltage follower*

The advantages of the non-inverting amplifier circuit discussed on p. 800 are that:

(a) the output is not an inverted copy of the input;
(b) since the + input is not a virtual earth, the input impedance of this circuit is high, typically 50 MΩ.

A further application of this circuit is shown in Figure 33.17. Here the output is connected directly to the inverting input so that the feedback is 100%. The gain is now 1 ($R_2 = 0$ and $R_1 =$ infinity in the expression gain $= 1 + (R_2/R_1)$, derived previously). So V_{out} follows V_{in} exactly. This is why the circuit is called a *voltage follower*.

At first sight this does not appear to be useful. But:

(a) the input impedance is very high so that practically no current is taken from the circuit connected to the input;
(b) the output impedance is low so that a current can be supplied to the following circuit or stage. So the voltage follower is used as a *buffer* between a high impedance (low current) circuit and a low impedance (high current) circuit.

Voltage follower: output directly connected to input so 100% feedback and V_{out} follows V_{in} exactly. Can act as a buffer connecting high and low impedance circuits.

Voltage Follower Uses

Figure 32.18 (i) shows how a voltage follower can be used to provide a very high impedance to a low impedance moving-coil voltmeter when measuring the voltage V across a capacitor C. The output voltage (V_o) measured by the voltmeter = the input voltage (V_i) across C. Then $Q = CV$ accurately. If the voltmeter was placed directly across C, the discharge of C through the meter would give inaccurate results.

Figure 32.18 (ii) shows a voltage follower used to provide a high impedance to a low impedance moving-coil voltmeter when measuring the p.d. across a resistor R carrying a very small current of the order of nanometers (10^{-9} A). No current is then taken from R in measuring its p.d. V, so $I = V/R$ accurately. A moving-coil meter in series (or in parallel) with R would give inaccurate measurements with such a small current.

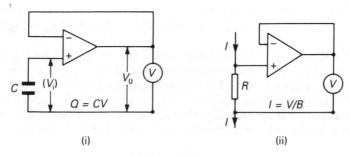

(i)   (ii)

Figure 32.18

Opamp as Voltage Comparator, Switching Circuit

Without any feedback, an opamp can be used as a *voltage comparator*. Figure 32.19 illustrates the principle. R_1 and R_2 form a potential divider between $-V$ and $+V$. The relative values of R_1 and R_2 will determine the voltage at the non-inverting input (Y) of the opamp. The voltage at (Y) is a reference voltage.

If the inverting input $(-)$ of the opamp is at an even slightly higher voltage than the non-inverting input $(+)$, then the output will be the maximum negative $(-V)$. As soon as the voltage of the inverting input $(-)$ falls below that of the non-inverting input $(+)$, or the voltage of the $+$ input rises above that of the $-$ input, the output immediately switches to the maximum positive $(+V)$.

A known reference voltage is applied to the $+$ input. If the voltage on the $-$ input is below the reference voltage, the output is $+V$. If the voltage

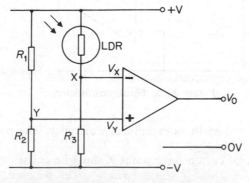

Figure 32.19 *Switching circuit with LDR*

on the − input is above the reference, then the output is −V. This is especially useful in circuits that are required to detect a changing level of voltage and switch at a specific level.

In Figure 32.19 another potential divider is formed by the light dependent resistor (LDR) and R_3. As the light level becomes lower, the resistance of the LDR increases and the voltage on the inverting input falls until it is below the reference voltage at Y set by the potential divider R_1 and R_2. At this point the output, which had been at −V, switches rapidly to +V. This switching could be used to operate a relay. The whole circuit might then be used to switch on a street light.

The circuit in Figure 32.19 is called a 'voltage comparator' because it compares the voltage V_X at X with the voltage V_Y at Y. If V_X is higher than V_Y, then a high negative output is obtained. If V_X is less than V_Y, a high positive output is obtained.

The light level at which switching occurs can be altered by adjusting the reference voltage. The LDR can also be replaced by a *thermistor* to detect a predetermined temperature and perhaps sound an alarm if a fire occurs and the temperature rises above a pre-set value. A thermistor is a resistance which varies with temperature; usually the resistance decreases with increasing temperature. So the voltage V_X relative to V_Y will alter as the temperature rises or falls.

Switching circuit: V_X (− input) from potential divider with LDR (lighting control) or thermistor (temperature control). V_Y fixed by another potential divider. If LDR or thermistor resistance changes, output switches between $-V_S$ and $+V_S$, the supply voltage, and a relay operates.

Opamp as Integrator

In certain circumstances, for example an analogue computer, a circuit is needed that will provide an output that is the integral of the input voltage. A basic integrator circuit is shown in Figure 32.20.

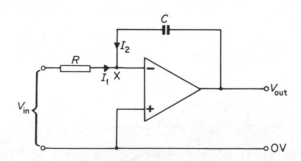

Figure 32.20 *Opamp as integrator*

Point X is a virtual earth as explained previously. The p.d. across R is V_{in} and the p.d. across C is V_{out}.

As before, the total current into point X must be equal to the total current leaving point X.

The current into point $X = I_1 = V_{in}/R$. I_2 is the current out of point X and is equal to the charging current for capacitor C.

So
$$I_2 = \frac{dQ}{dt} = \frac{C\,dV_{\text{out}}}{dt}$$

But since $\quad\quad\quad I_1 + I_2 = 0$

then $\quad\quad\quad V_{\text{in}}/R = -C\,dV_{\text{out}}/dt$ . . . (1)

If we integrate this expression:

$$(1/RC) \int V_{\text{in}}\,dt = -\int dV_{\text{out}}$$

which gives:

$$V_{\text{out}} = -(1/RC) \int V_{\text{in}}\,dt \quad . \quad . \quad . \quad (2)$$

Thus the output voltage is proportional to the *integral* of the input voltage.

If the input is a pulse as shown in Figure 32.21 (i), the output will be as shown in Figure 32.21 (ii). We can see this from equation (1). Here V_{in} is constant, Figure 33.21 (i), and so

$$dV_{\text{out}}/dt = -V_{\text{in}}/CR = -k,$$

where k is a constant. So V_{out} varies *linearly* with time t and has a negative gradient as shown.

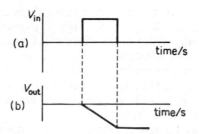

Figure 32.21 (i) *Input pulse* (ii) *Integral of input*

From equation (2), one example of the use of the circuit would be to integrate an input proportional to the speed of a car, so as to give an output proportional to the distance covered.

When using an opamp as an integrator, any initial variation of the input from zero will cause an output which will be integrated. The output will thus tend to drift slowly towards saturation even before the signal to be integrated has been connected. It is therefore necessary to use the offset-null adjustment described earlier to remove any small output offset.

Positive Feedback, Square Wave Oscillator or Astable Multivibrator

As we have seen, negative feedback *reduces* the differential voltage at the input. But if some of the output is taken back to the + (non-inverting) input, any change in the output will tend to *increase* the differential voltage at the input, since the output voltage will be in phase with the input voltage (p. 794). The output voltage will then quickly reach the saturation voltage, $+V_S$, which is the supply voltage. If the opamp circuit is designed to make the output switch continually from $+V_S$ to $-V_S$, and from $-V_S$ to $+V_S$ an *oscillating* output voltage is obtained. Figure 32.22 shows a circuit with positive feedback which produces square wave oscillations and acts as an *astable multivibrator*.

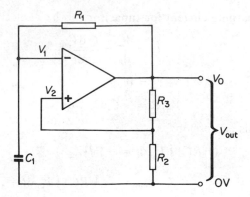

Figure 32.22 *Oscillatory circuit—astable multibrator*

To explain the action of this circuit, suppose the capacitor C_1 is initially uncharged and that the output voltage V_o has its maximum positive value $(+V)$ due to a small differential voltage at the inputs.

A fraction (V_2) of V_o is fed back to the non-inverting input:

$$V_2 = (V_o \times R_2)/(R_2 + R_3)$$

But V_o is also fed back into the inverting input through R_1. So C_1, which is initially uncharged, starts to charge up through R_1 towards $+V$ and the voltage V_1 then rises exponentially with time, as shown in Figure 32.23 (i). After a time which depends on the time constant $C_1 \times R_1$, V_1 reaches a value higher than V_2; whereupon the output of the opamp switches over so that the output is $-V$, as shown in Figure 32.23 (ii). The positive feedback encourages the opamp to switch quickly since V_2 then drops, making V_2 much lower than V_1 and so forcing V_o to go negative even more quickly.

C_1 now discharges and starts to charge in the opposite direction until V_1 becomes lower than V_2. The opamp then switches back again so that V_o becomes positive $(+V)$ again. The cycle is then repeated. Figure 32.23 (i) shows how the voltage V_1 varies with time and Figure 32.23 (ii) shows how the output

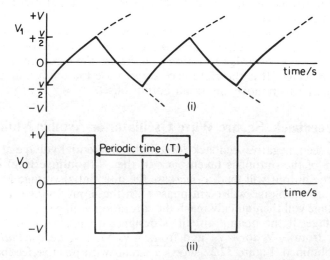

Figure 32.23 (*i*) *Variation of V_1 with time* (*ii*) *Variation of V_o—square wave*

*voltage V_o varies with time. The periodic time of the multivibrator is given by:

$$T = 2C_1R_1 \ln (1 + 2R_3/R_2)$$

the proof of which is beyond the scope of this book.

A Practical Astable Multivibrator

The circuit in Figure 32.24 gives component values to construct an astable multivibrator from an opamp. Its frequency is about 270 Hz.

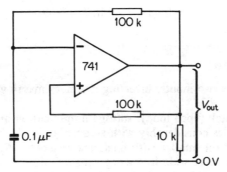

Figure 32.24 *Astable multivibrator*

Oscillator, square-wave or astable multivibrator: (1) Negative feedback from $C_1 - R_1$ provides charge–discharge voltage. (2) Positive feedback from potential divider provides switching. (3) Period T depends on C_1R_1 and feedback resistance ratio.

Example on C-R Switching

In the circuit shown in Figure 32.25, at time $t = 0$ the capacitor $C = 1\ \mu F$ is uncharged and the output voltage (V_{out}) is $+14$ V. At a later time the voltmeter connected to the output shows that V_{out} has changed to -14 V. Calculate the time at which V_{out} changes from $+14$ V to -14 V.

Initially, $$V_{out} = +14 \text{ V}$$

$$\therefore V_p = \frac{10 \times 10^3}{20 \times 10^3} \times 14 \text{ V} = 7 \text{ V}$$

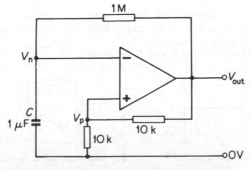

Figure 32.25 *Example on switching*

The voltage V_n is the voltage across the capacitor C of 1 μF, which is in series with the 1 MΩ across the voltage V_{out} initially at 14 V. The time-constant of the circuit $= CR = 1 \times 10^{-6} \times 10^6 = 1$ s.

So
$$V_n = V_{out}(1 - e^{-t/CR}) = 14(1 - e^{-t})$$

When V_n is just greater than V_p, or 7 V, V_{out} will switch to -14 V. So time t when this happens is given by

$$7 = 14(1 - e^{-t})$$

from which
$$-7 = -14\,e^{-t}$$

So
$$t = \ln(14/7) = \ln 2 = 0{\cdot}69 \text{ s}$$

You should know:

1 The opamp has two inputs, inverting and non-inverting, and one output V_o.
   Without feedback (open loop), voltage amplification or gain $A(V_o/V_i)$ is high but varies considerably with frequency.

2 Inverting use: gain smaller with feedback resistance R_f but advantage is constant gain over wide frequency range. Gain $= -R_f/R_i$, where R_i is input resistance (the minus shows inversion). Remember input terminal = virtual earth.
   Summing amplifier: voltages in parallel can all be added after connection to the inverting input, using gain $= -R_f/R_i$ for each voltage.

3 Non-inverting use: input between non-inverting ($+$) terminal and 0 V.
   Feedback by using potential divider across output.
   Gain $= 1 + (R_2/R_1)$, where R_2 is resistor between output and inverting ($-$) input terminal.
   Voltage follower: output directly connected to input so 100% feedback. Provides a very high input impedance and a very high output impedance, so can be used as a buffer between low and high current circuits.

4 Oscillator (or multivibrator): Uses (a) positive feedback and (b) capacitors to provide a continually switching output.

EXERCISES 32A Opamp Circuits

1 In the opamp shown in Figure 32A, R and S are resistances.
   Write down: (i) the gain; (ii) the phase difference between the input and output voltages. Why is X a 'virtual earth'?
   Explain why the currents in R and S are equal.

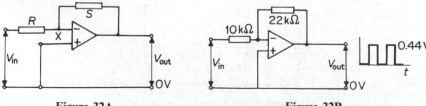

Figure 32A   Figure 32B

2 In Figure 32B, the output voltage V_{out} is a square-wave voltage of amplitude 0·44 V. Calculate the amplitude of the input voltage and draw a sketch of it.

3 Figure 32C shows a summing amplifier. One of the input voltages V_1 is 0·55 V and the other is V_2. Calculate V_2 if the output voltage is 0·9 V.

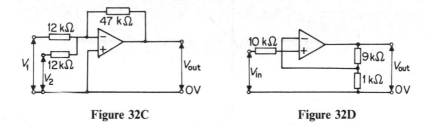

Figure 32C   **Figure 32D**

4 Calculate the gain of the circuit in 32D and draw sketches showing roughly the input and output voltages if the input voltage is a square wave voltage varying from 0 to 0·5 V.

5 Figure 32E shows an opamp circuit whose power supply is +15 V and −15 V. What is the gain of the circuit? Prove the formula used.

   Sine waves of frequency 1 kHz and amplitude:
   (a) 100 mV, and then
   (b) 1 V are applied to the input. In each case sketch the input and output waveforms, and comment on them.

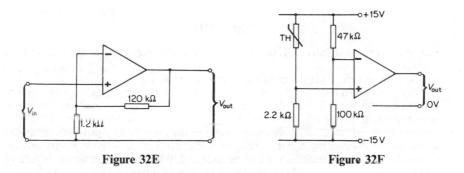

Figure 32E   **Figure 32F**

6 The circuit in Figure 32F shows the opamp used as a voltage comparator. The resistance of the thermistor shown decreases with increasing temperature. With the resistor cold, the output voltage is at maximum −15 V, the power supply being +15 V and −15 V.
   (a) What is the voltage at the inverting input of the opamp?
   (b) What voltage would the non-inverting input have to reach for the output to switch to maximum positive +15 V?
   (c) What would be the resistance of the thermistor at this point?

7 The transfer characteristic in Figure 32G (on page 810) shows the relationship between the output voltage and the input voltage for an amplifier.
   (i) State whether this characteristic refers to an inverting or a non-inverting amplifier and explain your answer.
   (ii) What is the voltage gain of the amplifier?
   (iii) Sketch a graph, giving voltage values on the axes, to show how the characteristic of the amplifier would change if the power supplies to the amplifier were reduced to +10 V and −10 V and the voltage gain was doubled. (*N.*)

8 An input voltage of the form shown in Figure 32H (i) is amplified by an inverting opamp circuit connected to supply rails of ±9 V. Sketch the output voltage waveform when the closed loop gain of the opamp circuit is 100.

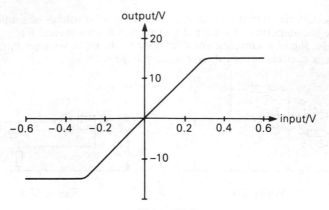

Figure 32G

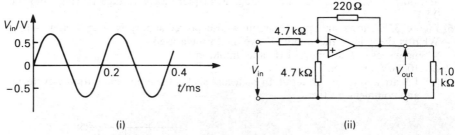

(i)   (ii)

Figure 32H

For the circuit in Figure 32H (ii) V_{in} is a sinusoidal signal of amplitude 15 mV. Calculate:
(i) the voltage gain;   (ii) the power gain of the circuit.
For (i) you may quote any formula you use and in (ii) you may take the power gain to be the mean power dissipated in the load resistor divided by the mean power drawn from the signal source. (*L.*)

9 Figure 32I shows an ideal operational amplifier (opamp) in an open-loop configuration. The non-inverting input is held at a d.c. voltage of $+4.0$ V whilst an a.c. sinusoidal voltage of amplitude 5·0 V, frequency 250 Hz, is applied to the inverting input. The output voltage of the opamp has values of either $+15$ V or -15 V.
(a) Explain what is meant by an open-loop configuration.
(b) What factors determine the output voltages of an ideal opamp in an open-loop configuration?
(c) Draw on squared paper (using the same axes) graphs showing the variation with time over two complete cycles of:
(i) the a.c. input voltage; (ii) the output voltage of the opamp.
Use scales of 1 cm ≡ 2·5 V and 2 cm ≡ 1 ms for both graphs.
(d) From your graphs estimate the ratio $T_1 : T_2$, where T_1 and T_2 are the times for which the output voltage is respectively positive and negative. (*O.*)

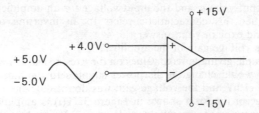

Figure 32I

10 Explain the relationship between the r.m.s. current in a resistor in an a.c. circuit and the power dissipated in it. On one set of labelled axes, sketch graphs showing the instantaneous and mean power dissipated in a resistor carrying a current $I = I_0 \sin \omega t$.

A transformer can be arranged to give a larger output voltage than the input voltage applied to the primary, yet it is not usually regarded as an amplifier. Explain the distinction between the step-up transformer and the amplifier.

Explain what is meant by *feedback* in operational amplifier circuits. Discuss the part played by positive and negative feedback in operational amplifier circuits. Give one example of each. Draw suitable circuit diagrams and identify the feedback components in each case.

The circuit in Figure 32J shows an ideal operational amplifier and its associated components. The input voltage V_{in} is a sinusoidal signal of amplitude 0·14 V.
(a) Calculate values for:
  (i) the voltage gain;
  (ii) the power gain of the circuit.
The power gain is defined as the ratio of the power dissipated in the load resistor (here 1·0 kΩ) to the power absorbed by the input circuit.
(b) On one set of axes, sketch graphs of the input and output voltages against time. Put a scale on the voltage axis. (*O. & C.*)

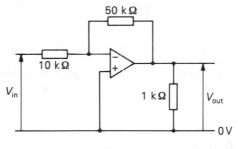

Figure 32J

11 Figure 32K shows an inverting amplifier operated from a ±15 V supply.
(a) Calculate the voltage gain of the amplifier.
(b) A sinusoidal p.d. of 0·50 V r.m.s. and frequency 1·0 kHz is applied to the input.
  (i) Using the same set of axes showing suitable voltage and time scales, sketch the variation with time of the input p.d. and of the output p.d.
  (ii) On a separate set of labelled axes, sketch the variation with time of the output p.d. of the input p.d. were 2·0 V r.m.s. and the frequency remained unchanged. (*N.*)

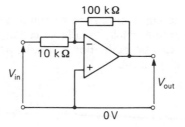

Figure 32K

12 (a) A coulombmeter made from the components shown in Figure 32L (on page 812) is used to measure the charge on a conducting sphere.
  (i) If the charge on the sphere is 1·6 μC, calculate the reading on the voltmeter when the charged sphere touches the terminal P. Explain, from first principles, each step in your calculation and state any assumptions you make.

(ii) It is suggested that if a meter of resistance 1 kΩ is used for the voltmeter in your coulombmeter, the capacitor will rapidly discharge. Explain whether you agree or disagree with this suggestion.

(iii) The sphere was charged from a +5000 V d.c. supply. Discuss whether all its charge would be measured by the coulombmeter.

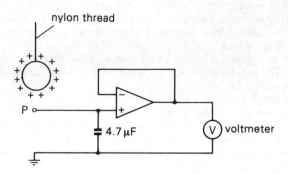

Figure 32L

(b) Give a practical application for a power amplifier. Explain why it is necessary to use it in the situation you give.

What problem is likely to occur when using power amplifiers? How, in the design of an amplifying system, is this problem limited? (*L.*)

Digital Electronics

Unlike analogue electronics, which deals with continuous signals, digital electronics is concerned with electrical signals that can be either ON or OFF. No other situation, for example 'HALF ON', can possibly be recognised.

In digital circuits, the 'ON' or '1' state can be represented as a high voltage (with respect to the grounded negative of the power supply). Thus in a typical circuit, '1' or 'ON' would be represented by a voltage of +5 V or 'HIGH' and a '0' or 'OFF' would be represented by a voltage of 0 V or 'LOW'. In this way, digital signals or numbers in binary can be recorded in a circuit without confusion.

Digital circuits are designed to process digital signals and there are a number of standard 'building blocks' available called *logic gates*, from which complex digital circuits can be built. For example, as shown later, they can be used in a simple burglar alarm.

Sensors

Electronic circuits which control lighting and temperature, for example, need *sensors* connected to the logic gates used.

A *light dependent resistor*, LDR, is a sensor whose resistance varies with the light intensity. Figure 32.26 (i) shows the circuit symbol for an LDR. In the dark, its resistance may be as high as 2 MΩ. In daylight, its resistance falls to about 1 Ω. The LDR is made from a suitable semiconductor sensitive to light.

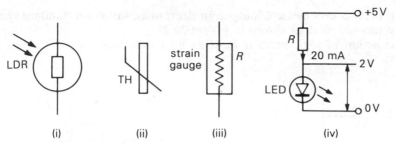

Figure 32.26 *Circuit symbols for* (i) *LDR* (ii) *thermistor* (iii) *strain gauge* (iv) *LED*

A *thermistor*, TH, is a sensor for temperature changes. Made from a semiconductor, its resistance may vary from 50 kΩ at high temperatures such as 100°C to 200 Ω at low temperatures. Figure 32.26 (ii) shows a circuit symbol for a thermistor.

A *strain gauge* (Figure 32.26(iii)) is a resistance wire sensitive to strains produced by stresses, for example, when testing aeroplane surfaces under stress. Since resistance $R = \rho l/A$, a strain produces a change in l and A and this can be linked with the electrical resistance change which is measured.

The *light-emitting diode*, LED, can indicate voltage levels and is used to display results at athletics meetings and football matches, for example. Made from suitable semiconductor material, and with the diode in a conducting direction, electrons and holes move across the p-n junction to combine with holes and electrons. Some of the energy produced is transferred to light. In one type of LED the maximum current is 20 mA when the p.d. across the LED is 2 V. A series resistor R is therefore essential when the supply voltage is 5 V, for example. In Figure 32.26 (iv), which shows the LED circuit symbol,

$$R = \frac{(5-2)}{0\cdot02} = 600\ \Omega, \text{ from } R = V/I$$

Logic Gates

Logic gates are generally made in the form of an integrated circuit. Several gates can then be made in one package, called a 'dual-in-line' (dil) package, Figure 32.27. Each gate will have one or more *inputs* and one *output*. A power supply is needed, which is often 5 volts.

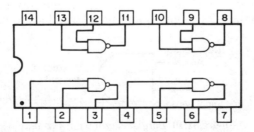

Figure 32.27 *Integrated circuit. 2-input* NAND *gate*

NOT Gate or INVERTER

The NOT gate is the simplest of all logic gates as it has only one input and one output. It changes the input A so that a '1' input becomes a '0' output at Q and a '0' input becomes a '1' output. In electronics, American standard symbols are commonly used, as shown in Figure 32.28.

The action of logic gates is most easily represented by means of a *truth table*.

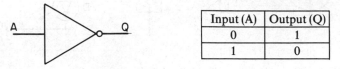

Input (A)	Output (Q)
0	1
1	0

Figure 32.28 *NOT gate*

The truth table for a NOT gate is also shown in Figure 32.28. The function of the NOT (INVERTER) gate can be represented briefly by $\bar{A}$(A bar) = Q or in words, 'not A equals Q'.

AND **Gate**

An AND gate can have any number of inputs but only one output. For simplicity, we deal with a two-input AND gate, Figure 32.29. An AND gate gives a high output (1) if input A AND input B are both high (1); otherwise the output is low (0). See Truth Table.

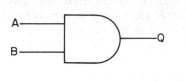

Inputs		Outputs
A	B	Q
0	0	0
0	1	0
1	0	0
1	1	1

Figure 32.29 *AND gate*

As we stated earlier, a burglar alarm might be required to sound if it is dark AND a door is opened. A simple electronic circuit could give a high voltage in the dark and a switch on the door could be arranged to give a high voltage when it is opened. These signals are then fed to an AND gate and the output used to sound an alarm, as shown in Figure 32.30. If either input A or B is 0, the output Q is low (0) and the alarm will not sound. If input A and input B are both 1, the output Q is high (1) and the alarm does sound.

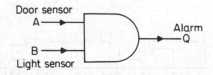

Figure 32.30 *Burglar alarm with AND gate*

NAND (NOT AND) **Gate**

Electronically, it is easy to make a gate which is equivalent to an AND gate followed by a NOT gate. The combination is given its own symbol and called a NAND (NOT AND) gate. Notice how the symbol is the same as for the AND gate but with a small circle at the output. This small circle is always taken to indicate a NOT or INVERTER operation. The truth table for the AND gate is shown in Figure 32.31.

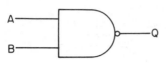

Inputs		Output
A	B	Q
0	0	1
0	1	1
1	0	1
1	1	0

Figure 32.31 *NAND gate*

As well as being a very useful gate in itself, the NAND gate is widely used because all other gates can be constructed using only NAND gates. For example, if a NOT gate is needed, it can be made from a NAND gate in two ways:
(a)   One input such as A is permanently tied (connected) high, so that A is 1 in the truth table, Figure 32.32 (i). In this case the only lines possible are lines 3 and 4, so if B is high, the output Q is low (0) and if B is low, Q is high (1).

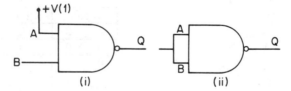

Figure 32.32 *NOT gate from NAND gate*

(b)   Both inputs may be tied (connected) together so that inputs A and B must both be high or low together, Figure 32.32 (ii). This means that only lines 1 and 4 of the truth table are possible and so, if A and B are both high, Q is low; and if A and B are both low, Q is high. However, this method is not good electronic practice and should not be used.

OR Gate

An OR gate can have any number of inputs but for simplicity we shall consider a two-input OR gate. An OR gate gives a high output (1) if either input A or B, Figure 32.33, OR both are high (1), otherwise the output is low (0). The truth table shows the function of the OR gate.
An OR gate could be used if a gardener wanted an alarm to sound when the air temperature in his greenhouse fell below 0°C OR rose above 30°C. Two

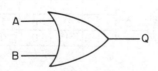

Inputs		Output
A	B	Q
0	0	0
0	1	1
1	0	1
1	1	1

Figure 32.33 *OR gate*

sensors such as thermistors could be arranged so that either sensor gave a high input (1) to an OR gate, as, for example, in Figure 32.34. The output would then trigger an alarm.

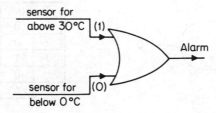

Figure 32.34 *Application of OR gate*

NOT OR or NOR Gate

A NOR gate is equivalent to an OR gate followed by a NOT gate (INVERTER), that is, all the outputs of the OR gate are inverted (changed from 0 to 1 or 1 to 0). The truth table for the NOR gate is therefore opposite to the OR gate as you should verify, Figure 32.35.

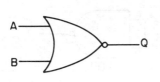

Inputs		Output
A	**B**	**Q**
0	0	1
0	1	0
1	0	0
1	1	0

Figure 32.35 *NOR gate*

Logic Gates Using NAND Gates

NAND gates can be used to construct all other logic gates such as OR, AND, NOR and NOT. Figure 32.36 shows three NAND gates that are equivalent to an OR gate. Figure 32.37 shows four NAND gates which are equivalent to a NOR gate. This can be seen from the truth tables, where Q_1, Q_2, Q are the outputs for the OR gate and Q_1, Q_2, Q_3, Q are the outputs for the NOR gate.

NOR gates can be used in a similar way to construct all other gates, AND, NAND, OR, NOT.

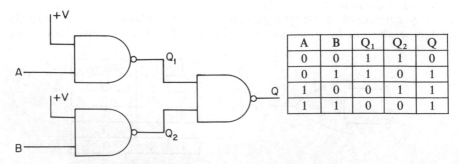

A	B	Q_1	Q_2	Q
0	0	1	1	0
0	1	1	0	1
1	0	0	1	1
1	1	0	0	1

Figure 32.36 *OR gate using NAND gates*

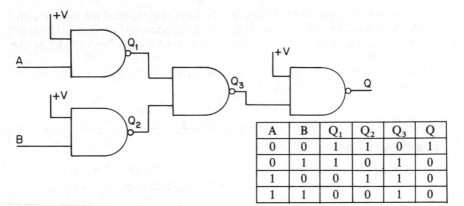

A	B	Q_1	Q_2	Q_3	Q
0	0	1	1	0	1
0	1	1	0	1	0
1	0	0	1	1	0
1	1	0	0	1	0

Figure 32.37 NOR gate from NAND gates

Logic gates: NOT (input 0, output 1; input 1, output 0).
AND (output 1 if inputs 1 *and* 1). NAND (output 0 if inputs 1 *and* 1).
OR (output 1 if inputs A *or* B *or* both 1). NOR (output 0 if A *or* B *or* both 1).
NAND gate inputs connected together = NOT gate.
All gates can be made from all NAND or all NOR gates.

Central Heating Control Using Logic Gates

All the logic gates described can be connected together in various combinations to perform an endless variety of functions. These may range from simple control of traffic lights to arithmetic and data processing. One example is the use of logic gates in the control of a central heating system. The boiler might need to be on if: (i) the timeswitch is on AND the room thermostat is on OR; (ii) there is a frost outside.

The logic combination for this would be as in Figure 32.38.

The truth table shows how the output Q of the AND gate is combined with

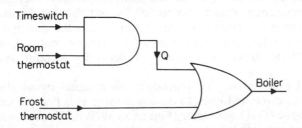

Time switch	Room Thermo.	Q	Frost Thermo.	Boiler
0	0	0	0	0
0	0	0	1	1
0	1	0	0	0
1	1	1	0	1
1	1	1	1	1

Figure 32.38 *Central heating control using* AND *plus* OR *gates*

the input to the OR gate to make the boiler work correctly. This circuit could be redrawn using three NAND gates in place of the AND and OR gates. The truth table shows the outputs Q_1, Q_2 and the output Q to the boiler. The reader should verify this.

The truth tables in Figure 32.38 contain mainly those lines showing the conditions for the boiler to be 'ON'.

You should know:

Logic circuits (combinatorial logic): Simple gates (NOT, AND, OR, ...) can be connected together to produce useful logic circuits. The outputs depend only on the inputs used.

Converting between Analogue and Digital Signals

Most natural signals are inherently *analogue*—they vary smoothly within some range, and are not restricted to a few specific values. Typical examples of analogue signals are speech and music waveforms. However, storing or transmitting analogue signals is tricky; they are easily corrupted by the smallest amount of noise, and need high quality electronic devices to handle them without distortion. For these reasons analogue signals are often converted to *digital* form for storage or transmission.

Analogue to Digital (A-D) Conversion. The principle is best illustrated by a simple example. We start with an analogue signal, for example a speech signal PQ detected by a microphone, Figure 32.39 (i). The analogue signal is first *sampled* at a sampling frequency f_s, as illustrated in Figure 32.39 (i). In this example the samples are taken every millisecond, so $f_s = 1$ kHz. This produces a sequence of voltage levels or numbers 4·4, 6·4, 7·0 ... 6·2. In Figure 32.39 (ii) we see the action of the analogue-to-digital converter. Each number is converted to 3-bit digital form, rounding to the nearest integer: this rounding process is known as *quantisation*. The sequence of 3-bit digital numbers can then be stored (perhaps on a CD or digital audio tape) or transmitted (usually over an optical fibre or metallic conductor) as a stream of 1s and 0s—see Figure 32.39 (iii). The pulse stream is far more tolerant to noise than the analogue signal. Unless the interfering noise is so intense that a 1 is confused with a 0, or vice versa, the signal will be stored or transmitted perfectly, without *any* degradation.

Digital to Analogue (D-A) Conversion. To reproduce the original signal, the digital pulse stream must be converted back to analogue form. It is first passed through a digital-to-analogue (D-A) converter (Figure 32.39 (iv)), which translates each 3-bit number into an analogue level and holds it constant for $1/f_s$ seconds, 1 ms in this case. This stepped signal would sound very harsh if played through a loudspeaker, and so it is first passed through a low-pass filter to smooth it. The smoothed signal, seen in Figure 32.39 (v), closely resembles the original waveform.

Sampling frequency. Quantisation. There are two main issues which affect the performance of the overall system: the choice of sampling frequency f_s, and the effects of quantisation. If the original analogue signal contains components of a maximum frequency f_m, then *Nyquist's theorem* states that the signal can be accurately represented only if it is sampled at a frequency $f_s \geq 2f_m$. An

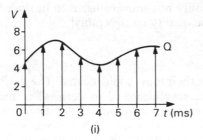

Time	Voltage	Quantise	Binary
0	4.4	4	100
1	6.4	6	110
2	7.0	7	111
3	5.0	5	101
4	4.3	4	100
5	5.2	5	101
6	6.0	6	110
7	6.2	6	110

(ii)

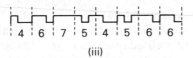

(iii)

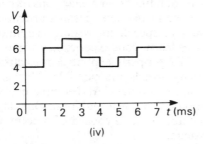

(iv)

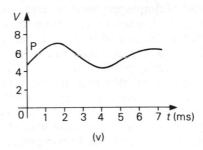

(v)

Figure 32.39 (i) *Sampled waveform* (ii) *Analogue-to-digital (A-D) conversion* (iii) *Digital pulse stream* (iv) *Digital-to-analogue (D-A) conversion* (v) *smoothed signal*

intelligible speech signal contains components up to about 3·5 kHz, so it is necessary to sample the waveform at 7 kHz at least for accurate digital transmission or storage. In fact, most modern telecommunications systems employ a sampling frequency of about 8 kHz. Music signals contain higher frequency components, such as cymbal crashes. These are audible up to about 20 kHz, so it would be necessary to sample music signals at at least 40 kHz. Compact disc (CD) systems sample at about 44 kHz. If the sampling frequency is too low ($f_s < 2f_m$), then a type of distortion known as *aliasing* distortion is present in the reconstructed analogue signal. To avoid this, it is usual to employ a low-pass *anti-aliasing* filter at the input, which removes all components of the input signal of frequency greater than $\frac{1}{2}f_s$, so ensuring that $f_s \geq 2f_m$.

Quantisation limits the accuracy of the reconstructed analogue signal. If greater accuracy is required, then it is necessary to use more bits for each sample, at the expense of greater storage or transmission. Going back to our simple example in Figure 32.39, the reconstructed signal in Figure 32.39 (iv) is limited to one of 8 discrete values: 0, 1, 2 ...7. If we had used 4 bits instead of 3 bits for each sample, then we could have had 16 possible values (0, 0·5, 1·0 ... 7·5), giving greater resolution. CD systems use 16 bits for each sample, giving 65 536 possible values and much more resolution than in Figure 32.39 (iv). If 8-bit sampling was used instead, then it would be possible to store twice as much music on each CD. But you probably would not want to listen to it, since the poor resolution would degrade the music quality considerably!

Computing and Neural Networks

Computers represent digital electronics at their most sophisticated. The flexible, powerful computers available today have revolutionised modern life in many different ways. Over the years improvements to computer hardware have generally served to increase the speed of the computer while decreasing its physical size (the computing power of an early 1980s mainframe machine, which would typically occupy a suite of specially ventilated rooms, is now commonly exceeded by compact, desktop computers).

However, the fundamental purpose of computers, that they should operate a user-defined program of precise tasks, has changed little since the 1960s. The problem is that while highly effective on numerically-intensive tasks (which humans typically find difficult), they are not well adapted to more intelligent tasks like speech and vision recognition (which humans find straightforward!).

For this reason, a new type of computer, modelled loosely on the human brain, was proposed. *Artificial Neural Networks* (ANNs) comprise a large number of simple processing elements (corresponding to single neurons or nerve cells in the human brain) connected together in massive, parallel arrays. An ANN may be trained to perform tasks such as speech and vision recognition, and, like the human brain, has the ability to learn from experience. Moreover, with the highly parallel structure of the ANN properly exploited in electronic hardware, very high information processing speeds are possible. This gives ANNs a great advantage over conventional computers. ANNs have enjoyed considerable success when applied to computer speech recognition. One day it might be possible to dispense with the cumbersome keyboard of a computer altogether, and just talk to computers instead!

EXERCISES 32B Digital Electronics

(The following symbols are used in the Exercise)

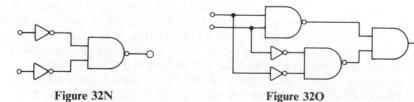

AND ⟦D⟧– NOT –▷o– OR ⟦D⟩– NAND ⟦D⟧o– NOR ⟦D⟩o–

Figure 32M

Figure 32N	**Figure 32O**

Multiple Choice

1 Which single gate from **A** to **E** is equivalent to the three gates in Figure 32N?

    A NOT **B** OR **C** AND **D** NAND **E** NOR

2 Which single gate from **A** to **E** is equivalent to the five gates in Figure 32O?

    A NOR **B** AND **C** EOR **D** NAND **E** OR

3 Figure 32P (i) shows three NOR gates with inputs 1 and 0. Which line in the table in Figure 32P (ii) shows the correct outputs P, Q, R?

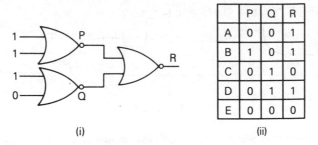

	P	Q	R
A	0	0	1
B	1	0	1
C	0	1	0
D	0	1	1
E	0	0	0

(i) (ii)

Figure 32P

4

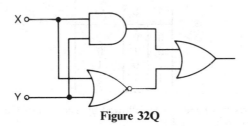

Figure 32Q

Figure 32Q shows inputs X and Y to AND and NOR gates and an output Z from an OR gate. When Z is high (1), which statements may be true in 1, 2, 3?
1 The output from the AND gate is low (0).
2 The inputs X and Y are both high (1).
3 The inputs X and Y are both low (0).
A 1, 2, 3 are all correct   B 1 and 3 only
C 2 and 3 only   D 1 only   E 2 only

Longer Questions

5 Figure 32R shows inputs X, Y, Z to a system of logic gates. Draw a truth table showing the outputs P and Q and the final output R.
   Draw a NAND gate system equivalent to all the NOR gates.

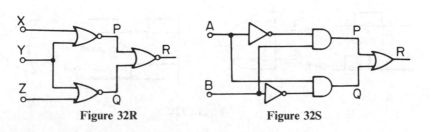

Figure 32R   **Figure 32S**

6 Figure 32S shows a system of logic gates. Draw a truth table showing the outputs P, Q and R.

7 Construct a truth table for the arrangement of NAND gates shown in Figure 32T.

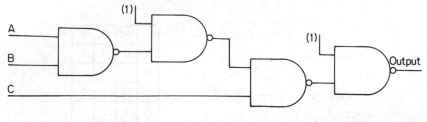

Figure 32T

8 Design a circuit that will give the truth table shown below.

Inputs		Output
A	B	Q
0	0	1
0	1	0
1	0	1
1	1	1

9 Figure 32U shows two NAND gates followed by a NOR gate. Construct a truth table showing the outputs P, Q and R.

The system is equivalent to one gate G with inputs X, Y, Z and output R. What is G?

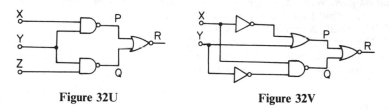

Figure 32U   **Figure 32V**

10 Figure 32V shows a system of logic gates. Construct a truth table showing the outputs P, Q and R.

From the truth table, what inputs X and Y will produce a high output (1) at R?

11 The waveforms of inputs A and B to a logic gate are shown in Figure 32W (i) and (ii). In each case copy the inputs and show beneath them the output waveform at Q.

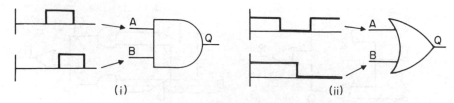

(i)   (ii)

Figure 32W

12 In a car, a red warning light comes on when the ignition is switched on if the door is not closed properly, or the seat belt is not fastened, or both door and seat belt are not properly secured.

(a) Construct the truth table, showing the door and seat belt inputs to a logic gate system as high (1) if each if properly secured and low (0) if not properly secured, the ignition input as high (1) if switched on and low (0) if off, and the red warning light, the output, as high (1) if on and low (0) if off.

(b) Then draw a suitable logic gate system for operating the red warning light.

13 A big saw for cutting wood at a timber store is designed to operate by switching on an electric motor only when a safety guard is lowered completely round the saw. In this case a green light comes on when the motor is switched on. If the safety guard is not lowered, a red light comes on when the machine is switched on.

(a) Construct a truth table for a logic gate system showing the motor switched on as high (1) input and off as low (0) input, the safety guard input as high (1) when lowered and low (0) when not lowered, and the outputs high (1) for switching on the lights and low (0) for not switching.

(b) Draw a logic gate system showing how the green and red lights can be operated as required.

14 (a) An aircraft door is locked by two bolts A and B each of which operates a sensor giving a logic 1 when the bolt is fully inserted. A further sensor C gives a logic 0 output when the door is shut.

A circuit using logic gates is required to give logic output 1 only when both bolts ar inserted and the door is shut.

By completing its truth table, show that the circuit of Figure 32X meets this requirement.

(b) Draw a circuit using two-input NOR gates *only* that performs the same logic function. (*O.*)

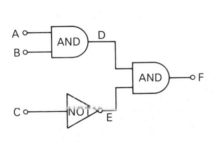

A	B	C	D	E	F
0	0	0			
0	0	1			
0	1	0			
0	1	1			
1	0	0			
1	0	1			
1	1	0			
1	1	1			

Figure 32X

15 (a) The circuit in Figure 32Y (i) shows a simple two-way mechanical switch S which can be used to make either output A or output B change from +5 volts to 0 volts. As it is operated, the contacts of this switch will bounce rather than give a clean switching action.

(i) Explain the purpose of the two 10 kΩ resistors.

(ii) Sketch a graph to show how the output voltage from terminal A will change as the switch is moved from position Y to X and then back to Y.

(iii) Explain the features of your graph.

(b) The circuit in Figure 32Y (ii) shows how two NAND gates can be connected to the circuit at points A and B to ensure a clean switching action.

With the aid of a truth table, explain how the circuit with the two NAND gates operates. The truth table should give the state of the outputs from each of the NAND gates when they switch it

(i) at Y, (ii) between X and Y, (iii) at X, (iv) between X and Y, and (v) back at Y.

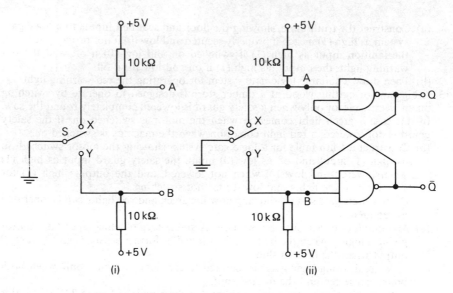

Figure 32Y

Sketch graphs to show how the output voltage from the NAND gates changes with time now that the circuit gives a clean switching action.

(c) Give *two* examples in which mechanical switches provide an input into an electronic system. In each case explain why it is important to have a clean switching action. (*N.*)

33 Photoelectricity, Energy Levels, X-Rays, Wave–Particle Duality

In this chapter, we first deal with photoelectricity and the Einstein photon theory. We then discuss the energy levels in the atom and their application to spectra, X-rays and their properties, the de Broglie formula and the wave–particle duality.

Photoelectricity: Particle Nature of Waves

Photoelectricity

In 1888 Hallwachs discovered that an insulated zinc plate, negatively charged, lost its charge if exposed to ultraviolet light. Later investigators such as Lenard and others showed that electrons were liberated from a zinc plate when exposed to ultraviolet light. So light gives energy to the electrons in the surface atoms of the metal, and enables them to break through the surface. This is called the *photoelectric effect*.

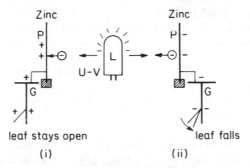

Figure 33.1 *Photoelectric demonstration*

Figure 33.1 (i) shows diagrammatically a simple demonstration of the photoelectric effect. The surface of a zinc plate P was rubbed with emery paper until the surface was clean and bright. P was then insulated and connected to the cap of a gold-leaf electroscope G, as shown, and given a *positive* charge by induction. Some of the charge spread to the leaf which then opened.

In a dark room, P was exposed to ultraviolet radiation from a small map L placed near it. The leaf stayed open. However, when the experiment was repeated with the plate P charged *negatively*, the leaf slowly collapsed, Figure 33.1 (ii).

The results are explained as follows. Electrons are usually emitted by the plate P when exposed to ultraviolet light. When P is positively-charged, any electrons (negative charges) would be attracted back to P. When P is negatively charged, however, the electrons emitted by P are now *repelled away from the plate*. So P loses negative charge and the leaf slowly falls.

Velocity or Kinetic Energy of Photoelectron

In 1902 Lenard found that the *velocity* or *kinetic energy* of the electron emitted from an illuminated metal was independent of the intensity of the particular incident monochromatic light. It appeared to vary only with the *wavelength* or *frequency* of the incident light.

Further, for a given metal, no electrons were emitted when it was illuminated by light of wavelength *longer* than a particular wavelength, called the *threshold wavelength*, no matter how great was the intensity of the light beam. But as soon as the metal was illuminated by light whose wavelength was *lower* than the threshold wavelength, electrons were emitted. Even though the light beam was made extremely weak in intensity, it was estimated that the electrons were emitted about 10^{-9} second after exposure to the light, that is, practically simultaneously with exposure to the weak light.

Classical or Wave Theory

On the wave theory of light, the so-called classical theory, these results are very surprising. If we assume light is sent out in waves from a source, the greater the intensity of the light the greater will be the energy per second reaching the illuminated plate. So the classical theory can explain why the number of electrons emitted increases as the light intensity increases.

But it cannot explain the result that the velocity or kinetic energy of the emitted electrons is independent of the intensity of the incident light beam. According to the classical theory, the greater the intensity of the beam, the greater should be the kinetic energy of the emitted electrons because the energy per second reaching the plate increases with the intensity of the light.

Further, for the classical theory electrons should always be emitted by light of *any* wavelength if the incident light beam is strong enough. Experiment, however, shows that however intense the light beam, *no* electrons are emitted if the wavelength is lower than the threshold value.

Experiments show that: (1) the maximum energy of the emitted electrons is *not* proportional to the light intensity, and that (2) electrons are *not* emitted for all wavelengths, which contradicts the classical (wave) theory of light.

Quantum Theory of Radiation, Planck Constant

In 1902 Planck had shown that the experimental observations in black body (heat) radiation could be explained on the basis that the energy from the body was emitted in separate 'packets' of energy. Each packet was called a *quantum* of energy and the amount of energy E carried was equal to hf, where f is the *frequency* of the radiation and h was a constant called the *Planck constant*. So

---

$$E = hf \qquad \qquad \qquad \qquad (1)$$

This is the *quantum theory of radiation*. Until Planck's quantum theory, it was considered that radiation was emitted continuously and not in separate packets of energy. Since $h = E/f$, the unit of h is joule seconds or J s. Measurements of radiation showed that, approximately, $h = 6.63 \times 10^{-34}$ J s.

The quantum of energy carried by radiation of wavelength 3×10^{-6} m can be calculated from $E = hf$. Since $f = c/\lambda$, where c is the speed of electromagnetic waves, about 3×10^8 m s^{-1} in a vacuum or air,

$$f = \frac{3 \times 10^8}{3 \times 10^{-6}} = 10^{14} \text{ Hz}$$

So $\qquad E = hf = 6\cdot63 \times 10^{-34} \times 10^{14} = 6\cdot6 \times 10^{-20}$ J (approx.)

Work Function

The *least* or *minimum* amount of work or energy necessary to take a free electron out of a metal against the attractive forces of surrounding positive ions is called the *work function* of the metal, symbol w_0. The work function is related to thermionic emission since this phenomenon is also concerned with electrons breaking free from the metal.

The work functions of caesium, sodium and beryllium are respectively about 1·9 eV, 2·0 eV and 3·9 eV. 1 eV (electron-volt) is a unit of energy equal to 1 electron charge $e \times 1$ volt, which is (p. 222)

$$1\cdot6 \times 10^{-19} \text{ C} \times 1 \text{ V} = 1\cdot6 \times 10^{-19} \text{ J}$$

So the work function of sodium, $w_0 = 2 \text{ eV} = 2 \times 1\cdot6 \times 10^{-19} \text{ J} = 3\cdot2 \times 10^{-19}$ J.

Einstein's Particle (Photon) Theory

In 1905 Einstein suggested that the experimental results in photoelectricity could be explained by applying a quantum theory of light. He assumed that light of frequency f contains packets or quanta of energy hf. On this basis, light consists of *particles*, and these are called *photons*. The *number* of photons per unit area of cross-section of the beam of light per second is proportional to its intensity. But the *energy* of a photon is proportional to its frequency and is *independent* of the light intensity.

Let us apply the theory to the metal sodium. Its work function w_0 is 2·0 eV or $3\cdot2 \times 10^{-19}$ J. According to the theory, if the quantum energy in the incident light is $3\cdot2 \times 10^{-19}$ J, then electrons are just liberated from the metal. The particular frequency f_0 is called the *threshold frequency* of the metal. The *threshold wavelength* $\lambda_0 = c/f_0$.

We can now calculate f_0 and λ_0 for the metal sodium. Since $E = hf_0 = w_0$,

$$f_0 = \frac{w_0}{h} = \frac{3\cdot2 \times 10^{-19}}{6\cdot6 \times 10^{-34}} = 4\cdot8 \times 10^{14} \text{ Hz}$$

So $\qquad \lambda_0 = \frac{c}{f} = \frac{3 \times 10^8}{4\cdot8 \times 10^{14}} = 6\cdot2 \times 10^{-7}$ m

So electrons are not liberated from sodium if the incident light has a frequency *less* than $4\cdot8 \times 10^{14}$ Hz or a wavelength *longer* than $6\cdot2 \times 10^{-7}$ m. This agrees with experiment.

Einstein's Photoelectric Equation

Using Einstein's theory, a simple result is obtained for the maximum energy of the electrons liberated from an illuminated metal.

If the quantum of energy hf in the light incident on a metal is say

4.2×10^{-19} J, and the work function w_0 of the metal is 3.2×10^{-19} J, the *maximum* energy of the liberated electrons is $(4.2 \times 10^{-19} - 3.2 \times 10^{-19})$ or 1.0×10^{-19} J because w_0 is the least energy to liberate electrons from the metal. The electrons with maximum energy come from the metal surface, as we have already stated. Owing to collisions, others below the surface emerge with a smaller energy.

We now see that the maximum kinetic energy, E_{max} or $\frac{1}{2}m_e v_m^2$, of the emitted electrons (photoelectrons) is given generally by

$$E_{max} = hf - w_0 \qquad . \qquad . \qquad . \qquad . \qquad . \qquad (1)$$

So
$$hf = E_{max} + w_0 \qquad . \qquad . \qquad . \qquad . \qquad . \qquad (2)$$

Equation (1) or (2) is called *Einstein's photoelectric equation.*

Example on Maximum Energy and Threshold Frequency

Sodium has a work function of 2·0 eV. Calculate the maximum energy and speed of the emitted electrons when sodium is illuminated by radiation of wavelength 150 nm.

What is the least frequency of radiation (threshold frequency) for which electrons are emitted? (Assume $h = 6.6 \times 10^{-34}$ J s, $e = -1.6 \times 10^{-19}$ C, $m_e = 9.1 \times 10^{-31}$ kg, $c = 3 \times 10^8$ m s^{-1}.)

From (1) incident photon energy $= hf = h\dfrac{c}{\lambda} = \dfrac{6.6 \times 10^{-34} \times 3 \times 10^8}{150 \times 10^{-9}}$

$$= 13.2 \times 10^{-19} \text{ J}$$

So maximum kinetic energy $= hf - w_0 = 13.2 \times 10^{-19} \text{ J} - 2 \times 1.6 \times 10^{-19} \text{ J}$

$$= 10 \times 10^{-19} = 10^{-18} \text{ J} \qquad . \qquad . \qquad . \qquad (3)$$

Thus $\frac{1}{2}m_e v_m^2 = 10^{-18}$

So
$$v_m = \sqrt{\frac{2 \times 10^{-18}}{9.1 \times 10^{-31}}} = \sqrt{\frac{20}{9.1}} \times 10^6$$

$$= 1.5 \times 10^6 \text{ m s}^{-1} \qquad . \qquad . \qquad . \qquad (4)$$

From (2) the threshold frequency is given by $hf_0 = w_0$. So

$$f_0 = \frac{w_0}{h} = \frac{2 \times 1.6 \times 10^{-19}}{6.6 \times 10^{-34}}$$

$$= 4.8 \times 10^{14} \text{ Hz} \qquad . \qquad . \qquad . \qquad . \qquad (5)$$

Measuring Maximum Kinetic Energy, Stopping Potential

The maximum kinetic energy of the liberated photoelectrons can be found by a method analogous to finding the kinetic energy of a ball moving horizontally. In Figure 33.2 (i), the moving ball is made to move up a smooth inclined plane PQ each time the slope of the plane is increased from zero. At one inclination the ball will *just* reach the top of the plane. In this case the kinetic energy, $\frac{1}{2}mv^2$, at the bottom P is just equal to the potential energy mgh at Q. Knowing the height h of Q above P, the kinetic energy is equal to mgh.

Figure 33.2 (ii) shows an analogous electrical experiment using a potential 'hill' whose gradient can be varied. Here a varying p.d. V is applied between the plates A and C inside an evacuated glass tube, and the potential of A is *negative* relative to C. This means that when photoelectrons are liberated from C, they are acted on by a retarding force.

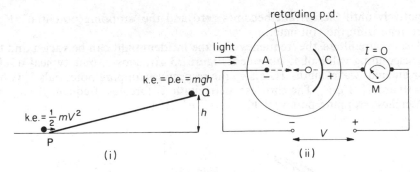

Figure 33.2 *Stopping potential*

When C is illuminated by a suitable beam, the photoelectrons liberated have a varying kinetic energy as we have previously explained. If the negative p.d. V between A and C is very small, many electrons can reach A from C. As V is increased negatively, however, the current I recorded on the meter M decreases because fewer electrons have energies sufficient to overcome the retarding force.

At a particular negative value V_s the current I becomes zero. This is the value of the negative p.d. which just stops the electrons with maximum energy from reaching A. V_s is called the *stopping potential.* Since an electron would lose an amount of energy given by *charge × p.d.* in moving from C to A, we see that

$$E_{max} = eV_s$$

So eV_s is a measure of the maximum energy of the emitted photoelectrons.

Verifying Einstein's Equation, Measurement of *h*

Figure 33.3 illustrates the basic features of laboratory apparatus for investigating photoelectricity. It contains (1) a photoelectric cell X, which has inside it a photosensitive metal C of large area, the cathode, and a collector A of the electrons, in a vacuum, (2) a potential divider arrangement Y for varying the p.d. V between the anode A and cathode C and (3) a d.c. amplifier for measuring a small current.

As shown, A is made *negative* in potential relative to C. The photoelectrons emitted from C then experience a retarding p.d. The p.d. V is increased

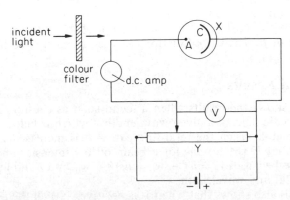

Figure 33.3 *Photoelectricity experiment:* $eV_s = hf - w_0$

negatively until the current becomes zero and the 'stopping potential', V_s, is then read from the voltmeter.

Using pure filters, the frequency f of the incident light can be varied and the corresponding value of V_s obtained. Figure 33.4 (i) shows some typical results. For different wavelengths $\lambda_1, \lambda_2, \lambda_3$, the respective stopping potentials V_s (when $I = 0$) are V_1, V_2, V_3. The shortest wavelength λ_1 (greatest frequency) required the highest stopping potential V_1.

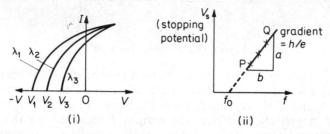

Figure 33.4 *Results of experiment*

The values of V_s are now plotted against the corresponding frequency f of the incident light. Figure 33.4 (ii) shows the result. A straight line graph PQ is obtained. Now from Einstein's photoelectric equation, $eV_s = hf - w_0$. So

$$V_s = \frac{h}{e}f - \frac{w_0}{e}$$

So from Einstein's theory, the gradient a/b of the line PQ is h/e, or

$$h = e \times \frac{a}{b}$$

Knowing e, the electron charge, h can be calculated. Careful measurement first carried out by Millikan gave a result for h of 6.26×10^{-34} J s, which was very close to the value of h found from experiments on black-body radiation. This confirmed Einstein's photoelectric theory that light can be considered to consist of particles (photons) with energy hf.

The line PQ in Figure 33.4 (ii) also enables the threshold frequency f_0 and the work function w_0 to be found. From $eV_s = hf - w_0$, we have $0 = hf - w_0$ when $V_s = 0$. So $f = w_0/h = f_0$. So the intercept of PQ with the frequency axis gives the value of f_0. The work function can then be calculated from $w_0 = hf_0$.

Experimental Results

Figure 33.5 (i) shows the results when a metal such as caesium is illuminated by monochromatic light of a given wavelength λ which is below the threshold value for the metal. When the retarding p.d. $-V$ is increased negatively, the stopping potential V_s is the *same* for a beam of low intensity and one of high intensity. This is because $eV_s = hf - w_0 = hc/\lambda - w_0$, and λ and w_0 are constant for a given metal and wavelength.

Figure 33.5 (i) also shows that when V is positive, (i) all the photoelectrons are now collected so that the current is constant, and (ii) a beam of high

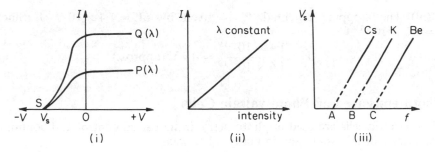

Figure 33.5 *Photoelectricity results*

intensity produces more electrons than one of low intensity. If Q has twice the intensity of P, the current I is twice as much.

Figure 33.5 (ii) shows how the current I varies with intensity for a given wavelength λ below the threshold value. The number of electrons emitted is proportional to the intensity and a straight line graph is obtained.

Figure 33.5 (iii) shows the results when the stopping potential V_s is plotted against the frequency f for three different metals—caesium (Cs), potassium (K) and beryllium (Be). The respective threshold frequencies A, B, C are different because the metals have different work functions; but the *slope of the three straight line graphs is the same* since the slope is given by h/e (see p. 830) and h and e are constants.

Example on Stopping Potential

Caesium has a work function of 1·9 electronvolts. Find: (i) its threshold wavelength; (ii) the maximum energy of the liberated electrons when the metal is illuminated by light of wavelength $4\cdot5 \times 10^{-7}$ m; (iii) the stopping p.d. (1 electronvolt $= 1\cdot6 \times 10^{-19}$ J, $h = 6\cdot6 \times 10^{-34}$ J s, $c = 3\cdot0 \times 10^{8}$ m s^{-1}).

(i) The threshold frequency f_0 is given by $hf_0 = w_0 = 1\cdot9 \times 1\cdot6 \times 10^{-19}$ J.

Now threshold wavelength, $\lambda_0 = c/f_0$

$$\therefore \lambda_0 = \frac{c}{w_0/h} = \frac{ch}{w_0}$$

$$= \frac{3 \times 10^8 \times 6\cdot6 \times 10^{-34}}{1\cdot9 \times 1\cdot6 \times 10^{-19}}$$

$$= 6\cdot5 \times 10^{-7} \text{ m}$$

(ii) Maximum energy of liberated electrons $= hf - w_0$, where f is the frequency of the incident light. But $f = c/\lambda$.

$$\therefore \text{max. energy} = \frac{hc}{\lambda} - w_0$$

$$= \frac{6\cdot6 \times 10^{-34} \times 3 \times 10^8}{4\cdot5 \times 10^{-7}} - 1\cdot9 \times 1\cdot6 \times 10^{-19}$$

$$= 1\cdot4 \times 10^{-19} \text{ J}$$

(iii) The stopping potential V_s is given by $eV_s = 1.4 \times 10^{-19}$ J. Since $e = 1.6 \times 10^{-19}$ C,

$$\therefore V = \frac{1.4 \times 10^{-19}}{1.6 \times 10^{-19}} = 0.9 \text{ V (approx.)}$$

Photo-emissive and Photo-voltaic Cells

Photoelectric cells are used in photometry, in industrial control and counting operations, in television, and in many other ways.

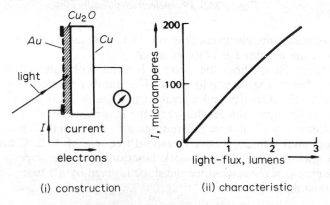

(i) construction   (ii) characteristic

Figure 33.6 *A photo-voltaic cell*

Photoelectric cells of the kind we describe on p. 829 are called *photo-emissive* cells, because in them light causes electrons to be emitted. Another type of cell is *photo-voltaic*, because it generates an e.m.f. and can therefore provide a current without a battery. One form of such a cell consists of a copper disc, oxidised on one face (Cu_2O/Cu), as shown in Figure 33.6 (i). Over the exposed surface of the oxide a film of gold (Au) is deposited, by evaporation in a vacuum; the film is so thin that light can pass through it. When it does so it generates an e.m.f. in a way which we cannot describe here.

Photo-voltaic cells are sensitive to visible light. Figure 33.6 (ii) shows the current from such a cell, through a galvanometer of resistance about 100 Ω, varies with the light-flux falling upon it. The current is not quite proportional to the flux. Photo-voltaic cells are obviously convenient for photographic exposure meters, for measuring illumination in factories, and so on, but as measuring instruments they are less accurate than photo-emissive cells.

Solar batteries are photo-voltaic cells in series. Large panels of suitable semiconductor material are attached to satellites and form solar batteries which work when sunlight falls on them. Electrical apparatus inside the satellite is connected to the batteries.

Photo-conductive Cells

A photo-conductive cell is one whose resistance changes when it is illuminated. A common form consists of a pair of interlocking comb-like electrodes made of gold (Au) deposited on glass (Figure 33.7 (i)). Over these a thin film of cadmium sulphide (CdS) is deposited. In effect, the material forms a large number of strips, electrically in parallel. The resistance between the terminals, XY, falls from about 10 MΩ in the dark to about 100 Ω in bright light.

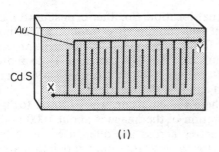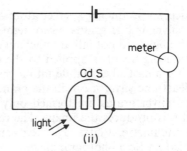

Figure 33.7 (*i*) *A selenium cell* (*ii*) *Circuit for CdS lightmeter*

This is the basis of a *light-dependent resistor* (LDR). As explained previously on p. 803, an LDR can be used in a switching circuit for street-lighting, for example, and as a light-sensitive cell for use in exposure meters for cameras, as shown in Figure 33.7 (ii).

You should know:

1   Light and all other electromagnetic radiation of frequency f have photons of energy $E = hf$, where h is the Planck constant.
2   The *work function* w_0 of a metal is the least energy needed to just liberate an electron from it. The *threshold frequency* f_0 is the lowest radiation frequency to just liberate an electron from the surface of a metal. So $hf_0 = w_0 = hc/\lambda_0$, where λ_0 is the threshold wavelength.
3   The wave theory of light cannot explain the photoelectric effect.
4   Einstein's photoelectric equation: $KE_{max} = hf - w_0$, where KE_{max} is the maximum kinetic energy of the emitted electrons from the surface. Other electrons are emitted with less energy than the maximum.
5   When a beam of light emits electrons from the surface of a metal, (a) the number of electrons emitted per second is proportional to the beam intensity, (b) the KE_{max} of the emitted electrons does *not* depend on the intensity but on the light wavelength or frequency for a given metal.
6   The *stopping potential* V_s is the least negative potential to stop all the electrons emitted from a cathode in a photoelectric cell.

$$eV_s = hf - w_0$$

So a graph of V_s against f is a straight line of gradient h/e and intercept w_0/e.

Photography and Photoelectricity

Photography is still one of the most powerful ways of recording images, even though modern electronics has provided us with many other possibilities (such as CCD chips). It is interesting to note, though, that photography is very much a photoelectric process. A photographic emulsion contains millions of grains of silver halide, each about half to one micrometre in diameter. A photon incident on one of these grains can free an electron from a single silver ion. This electron is free to move until it is trapped by an imperfection in the crystal lattice, so producing a silver atom at that location. So, after exposure to light, the silver halide grains where light fell will contain silver atoms, while those where light did not fall are unchanged. In the subsequent development process, a reducing agent is applied to the emulsion. This has the capacity to reduce each grain of silver halide into a grain of silver, though it is necessary to have at least one silver atom in the grain to catalyse the reaction.

So silver grains are formed only where the emulsion was exposed to light and this produces the image. The resolution of the image is about 1000 grains per millimetre. Plates over ten centimetres across are often used in certain telescopes. So a photographic plate can store a very large amount of information.

EXERCISES 33A Photoelectricity

Multiple Choice

1 Which of the statements **A** to **E** is correct?
The threshold wavelength λ_0 of a metal surface:
A increases with the frequency of the light
B decreases with the frequency of the light
C increases with the light intensity
D is given by $hc/\lambda_0 = w_0$ (w_0 is work function)
E is given by $h\lambda_0/c = w_0$.

2 Light of wavelength 450 nm is shone on to the surface of a metal of work function $3\cdot2 \times 10^{-19}$ J. The maximum energy of the emitted electrons in 10^{-19} J is

   A 0·8 **B** 1·2 **C** 2·4 **D** 2·8 **E** 3·2

3 Figure 33A (i) shows a typical graph of current I against p.d. V for a photoelectric cell when a beam of monochromatic light illuminates the cathode and the photo-electrons are collected by the anode. V is varied from $+$ to $-$ values.
   Which of the changes **A** to **E** is correct when the light is replaced by another monochromatic light of greater intensity and greater wavelength?
A PQ unchanged, OH increases.
B PQ decreases, OH decreases.
C PQ increases, OH decreases.
D PQ increases, OH increases.
E PQ decreases, OH increases.

4 In Figure 33A (ii), photoelectrons are emitted by C when illuminated by a beam L of monochromatic light and collected by A. 1, 2, 3 are three statements:
   The maximum energy of the emitted electron is:
   1 proportional to the intensity of L
   2 increases when V is greater
   3 increases when the inverse of the wavelength increases.
A 1, 2, 3 are correct **B** 1, 2 only are correct
C 2, 3 only **D** 2 only **E** 3 only

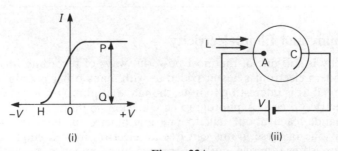

(i)   (ii)

Figure 33A

Longer Questions

5 A photoelectric cell is illuminated with monochromatic light. Figure 33B shows how the current through the cell varies with the potential difference across the cell.

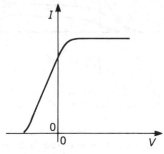

Figure 33B

Draw a diagram to show the circuit you would use to obtain the values needed to plot the graph.

How would you find the stopping potential from the graph?

Explain one feature of the graph which you would expect to change if there were an increase in: (a) the wavelength of the light illuminating the cell; (b) the intensity of the light illuminating the cell. (L.)

6 (a) When electromagnetic radiation falls on a metal surface, electrons may be emitted. This is the photoelectric effect.

  (i) State Einstein's photoelectric equation, explaining the meaning of each term.

  (ii) Explain why, for a particular metal, electrons are emitted only when the frequency of the incident radiation is greater than a certain value.

  (iii) Explain why the maximum speed of the emitted electrons is independent of the intensity of the incident radiation.

 (b) A source emits monochromatic light of frequency 5.5×10^{14} Hz at a rate of 0.10 W. Of the photons given out, 0.15% fall on the cathode of a photocell which gives a current of $6.0 \, \mu A$ in an external circuit. You may assume this current consists of all the photoelectrons emitted.

  Calculate: (i) the energy of a photon; (ii) the number of photons leaving the source per second; (iii) the percentage of the photons falling on the cathode which produce photoelectrons. (N.)

7 When light is incident on a metal plate electrons are emitted only when the frequency of the light exceeds a certain value. How has this been explained?

  The maximum kinetic energy of the electrons emitted from a metallic surface is 1.6×10^{-19} J when the frequency of the incident radiation is 7.5×10^{14} Hz. Calculate the minimum frequency of radiation for which electrons will be emitted. Assume that Planck constant $= 6.6 \times 10^{-34}$ J s. (N.)

8 When light of frequency 5.4×10^{14} Hz is shone on to a metal surface the maximum energy of the electrons emitted is 1.2×10^{-19} J. If the same surface is illuminated with light of frequency 6.6×10^{14} Hz the maximum energy of the electrons emitted is 2.0×10^{-19}. Use this data to calculate a value for the Planck constant. (L.)

9 Explain the terms *photoelectric threshold frequency* and *work function* used in connection with the photoelectric effect.

  A photoelectric cell consists of a conductor plate coated with photo-emissive material and a metal ring as current collector, mounted *in vacuo* in a transparent envelope. Graphs are plotted showing the relation between the collector current I and the potential V of the collector with respect to the emitter, for the following cases (Figure 33C).

  A. Emissive material illuminated with monochromatic light of wavelength λ.

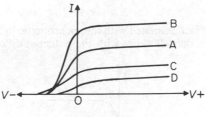

Figure 33C

B. As *A*, but the intensity of the incident light has been changed. *C.* As *A*, but the wavelength of the light has been changed. *D.* As *A*, but the emissive material is different.

Explain the general form of curve *A*. In what way do curves *B, C, D*:   (i) resemble *A*;   (ii) differ from *A*? What explanation of the differences can be offered? (*O. & C.*)

10 Describe and explain one experiment in which light exhibits a wave-like character and one experiment which illustrates the existence of photons.

Light of frequency 5.0×10^{14} Hz liberates electrons with energy 2.31×10^{-19} J from a certain metallic surface. What is the wavelength of ultra-violet light which liberates electrons of energy 8.93×10^{-19} J from the same surface? (Take the velocity of light to be 3.0×10^8 m s^{-1}, and Planck constant (*h*) to be 6.62×10^{-34} J s.) (*L.*)

11 (a) The Einstein equation for the photoelectric effect can be written

$$hf = \tfrac{1}{2}mv^2 + \phi$$

(i) Explain the meaning of each term in the equation and how it relates to Einstein's explanation of the photoelectric effect;   (ii) What does Einstein's explanation imply about the nature of light?

(b) Figure 33D shows the apparatus used in an experiment to investigate the photoelectric effect. Light falls on the photo-sensitive surface of a photocell, and the kinetic energy of the fastest moving emitted electrons can be measured by increasing the voltage provided by the battery until no current flows. The

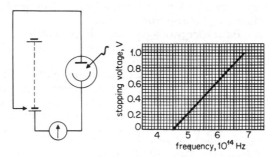

Figure 33D

results obtained for monochromatic light of various frequencies are shown graphically below.

(i) The graph indicates that there is a frequency below which no electrons are emitted. Why is this?   (ii) When the frequency is 5.5×10^{14} Hz, the required stopping voltage is 0·43 V. Given that the electronic charge is -1.6×10^{-19} C, estimate the kinetic energy of the fastest moving electrons emitted by light of this frequency. Justify your answer.   (iii) Use the graph to obtain a value for the Planck constant, explaining carefully how you obtain your result.   (iv) State and explain the effect, if any, of increasing the intensity of the light on the experimental observations and on the graph. (*AEB.*)

12 Describe an experiment to measure the Planck constant using the photoelectric effect.

A strip of clean magnesium ribbon, surrounded by a cylinder of copper gauze maintained at a positive potential of 6·0 V with respect to the magnesium, is connected to the input of an amplifier which measures the p.d. across the resistor R as shown in the Figure 33E.

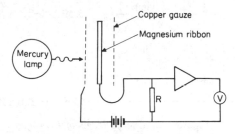

Figure 33E

When the magnesium is illuminated with mercury light of wavelength 254 nm the amplifier detects a current flowing in R. State the origin of this current.

Calculate the energies of photons of mercury light of wavelengths 254 nm and 546 nm.

Using these values and the data at the end of the question, describe and explain in detail what you would observe when each of the following experiments is carried out separately.

(a) The polarity of the 6·0 V battery is reversed.
(b) The mercury lamp is moved farther away from the magnesium ribbon.
(c) A filter selects mercury light of wavelength 546 nm in place of that of wavelength 254 nm.
(d) The e.m.f. of the battery is increased to 10 V.
(e) The magnesium ribbon is replaced by a strip of copper. (The photoelectric work function of magnesium = $4·5 \times 10^{-19}$ J (2·8 eV). The photoelectric work function of copper = $8·1 \times 10^{-19}$ J (5·05 eV) Planck constant = $6·6 \times 10^{-34}$ J s.) (O. & C.)

■ Quantisation of Energy, Energy Levels

In 1911, Bohr proposed that an atom has a number of separated energy values or *energy levels*. These levels are characteristic of the particular atom so that a hydrogen atom, for example, has different energy levels to a nitrogen atom.

An important point in the theory, given later, is that an atom *cannot* have any energies between these levels. We say that the energy is not continuous but 'quantised' in definite amounts. In an analogous way, a building has different floors from the ground floor upwards but no floors or levels between.

The energy of an atom is the energy of its electrons, which are moving round the positively-charged nucleus. These electrons therefore 'occupy' one of these separated energy levels.

The energy level values are expressed in terms of a number n, which can only be a whole number such as $n = 1, 2, 3$ and so on. n is called the *quantum number* for the particular energy level, as we now see for the hydrogen atom.

Hydrogen Energy Levels

Figure 33.8 shows roughly some of the energy levels of the hydrogen atom in electronvolts (eV) and in joules (J). The quantum numbers $n = 1, 2, \ldots, \infty$ on

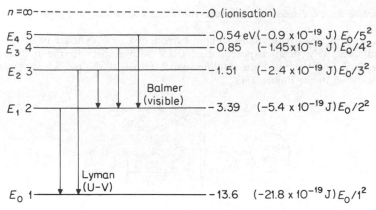

Figure 33.8 *Energy levels of hydrogen*

Plate 33A *Semiconductor laser* (courtesy Philips Research Laboratories)

the left correspond to values used to calculate the energy values. The *ground state*, $n = 1$, is -13.6 eV (-21.8×10^{-19} J) and this is the lowest energy level, where the hydrogen atom is most stable. The higher energy levels are all calculated from the formula -13.6 eV$/n^2$, where $n = 1, 2, 3, 4, 5 \ldots$ respectively, as illustrated in the diagram.

When an atom is raised from the ground state to any of its higher energy levels, it is said to be in an *excited state*. The *excitation energy* needed to raise a hydrogen atom from the ground state E_0 or -13.6 eV to its energy level E_2 or -1.51 eV

$$= E_2 - E_0 = (-1.51) - (-13.6) = 12.09 \text{ eV}$$

The energy levels *increase* to the value 0, which is the *ionisation level*. When the electron or atom is given an amount of energy equal to the *ionisation energy*, the electron is just able to become a 'free' electron. From the energy level values for hydrogen, the energy from the ground state (-13.6 eV) to the ionisation level (0) is 13.6 eV, which is the ionisation energy. If the electron is given a greater amount of energy than the ionisation energy of 13.6 eV or 21.8×10^{-19} J, such as 22.8×10^{-19} J, the excess energy of 1.0×10^{-19} J is then the kinetic energy of the free electron outside the atom. In general, the free electron can have a continuous range of energies. Inside the atom, however, it can have only one of the energy level values characteristic of the atom.

Electromagnetic Radiation and Energy Change

When an atom is excited from the ground state to a higher energy level, it becomes unstable and falls back to one of the *lower* energy levels. This energy change occurs for any system. It becomes unstable at a higher energy than its normal energy value and tends to fall back to its stable state. So the excited atom may reach, for example, a level corresponding to $n = 2$ in Figure 33.8 and then fall back to the ground state where $n = 1$.

Bohr proposed that the decrease in energy when the fall occurs is released in the form of *electromagnetic radiation*. Now from quantum theory, the energy in radiation of frequency f in hf, where h is the Planck constant (p. 826). So if the atom falls from the first energy level E_1 to the ground energy level E_0 after it was excited, the radiation emitted is given by

$$E_1 - E_0 = hf = \frac{hc}{\lambda}$$

where c is the speed of electromagnetic waves in a vacuum or air and λ is the wavelength of the radiation.

Emission Spectra of Hydrogen

We can now calculate the wavelength of the emitted radiation when the hydrogen atom is excited from its ground state ($n = 1$) where its energy level E_0 is $-21\cdot8 \times 10^{-19}$ J to the higher level ($n = 2$) of energy $E_1 - 5\cdot4 \times 10^{-19}$ J, and then falls back to the ground state.

Since $E_1 - E_0 = hf = hc/\lambda$, then, using standard values,

$$\lambda = \frac{hc}{E_1 - E_0} = \frac{6\cdot6 \times 10^{-34} \times 3 \times 10^8}{(-5\cdot4 \times 10^{-19}) - (-21\cdot8 \times 10^{-19})}$$

$$= \frac{6\cdot6 \times 3}{16\cdot4} \times 10^{-34+8+19}$$

$$= 1\cdot2 \times 10^{-7} \text{ m}$$

This wavelength is in the *ultraviolet spectrum*.

Suppose we now calculate the wavelength of the radiation emitted from an energy level $-2\cdot4 \times 10^{-19}$ J ($n = 3$) to $-5\cdot4 \times 10^{-19}$ J ($n = 2$). This is an energy change E of $3\cdot0 \times 10^{-19}$ J. Now from $\lambda = hc/E$ used above, we see that $\lambda \propto 1/E$ since h and c are constants. This means that the *smaller* the value of the energy change E, the *greater* is the wavelength. We have just shown that $\lambda = 1\cdot2 \times 10^{-7}$ m when $E = 16\cdot4 \times 10^{-19}$ J. So, by proportion, for $E = 3\cdot0 \times 10^{-19}$ J, λ is given by

$$\frac{\lambda}{1\cdot2 \times 10^{-7} \text{ m}} = \frac{16\cdot4 \times 10^{-19}}{3\cdot0 \times 10^{-19}}$$

So

$$\lambda = \frac{1\cdot2 \times 10^{-7} \times 16\cdot4}{3\cdot0} = 6\cdot6 \times 10^{-7} \text{ m}$$

This wavelength is in the *visible spectrum*.

Excitation and Ionisation Potentials

The energy required to raise an atom from its ground state to an excited state is called the *excitation energy* of the atom. If the energy is eV, where e is the electron charge, V is known as the *excitation potential* of the atom (p. 839).

The ground state energy level $= -13 \cdot 6 \, \text{eV}$ and the energy level $E_4 = -0 \cdot 85 \, \text{eV}$ for hydrogen. So the excitation energy from the ground state to $E_4 = (-0 \cdot 85) - (-13 \cdot 6) = 12 \cdot 75 \, \text{eV}$. The excitation potential is then $12 \cdot 75 \, \text{V}$.

If the atom is in its ground state with energy E_0, and absorbs an amount of energy eV which *just* removes an electron completely from the atom, then V is said to be the *ionisation potential* of the atom. The potential energy of the atom is here denoted by E_∞, as the ejected electron is so far away from the attractive influence of the nucleus as to be, in effect, at infinity. E_∞ is taken as the 'zero' energy of the atom, and the lower energy levels are thus negative. The ionisation potential V is given by $E_\infty - E_0 = eV$, or by $-E_0 = eV$. In the case of the hydrogen atom, $E_0 = -13 \cdot 6 \, \text{eV}$. So the ionisation energy $= 13 \cdot 6 \, \text{eV}$ and the ionisation potential $V = 13 \cdot 6 \, \text{V}$.

Emission Line Spectrum

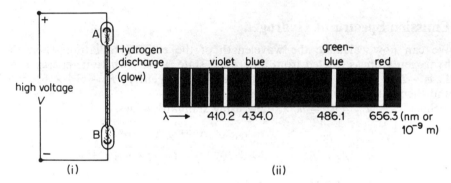

Figure 33.9 *Emission line spectrum*

Gases such as hydrogen, neon or carbon dioxide can be placed inside a narrow discharge tube at low pressure, as shown in Figure 33.9 (i). When the metal electrodes A and B at the ends of the tube are connected to a high voltage such as 1000 V, a discharge or light column is obtained between A and B. When the light from hydrogen or neon gas is examined using a diffraction grating, the *emission spectrum* is seen to consist of well-defined separated lines, such as those shown in Figure 33.9 (ii). This type of emission spectrum is called a *line spectrum*.

As they move through the discharge in the gas, some electrons obtain sufficient energy to excite atoms to a higher energy level. When the atoms fall to a lower energy level, the excess energy is emitted as electromagnetic radiation, as previously explained. The visible line spectrum of hydrogen shows the change in energy. Each line has a particular frequency or wavelength given by energy change $= hf$. **The fact that the lines are separated is *experimental evidence* for the existence of separate or 'quantised' energy levels in the atom.**

Types of Emission Spectra

Emission spectra are classified into *line*, *band* or *continuous* spectra. Line spectra are obtained from *atoms* in gases such as hydrogen or neon at

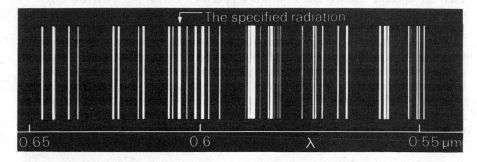

Plate 33B *Line spectrum (atoms). Part of the spectrum of the gas krypton-86 in the range 0·65 μm (650 nm) to 0·55 μm (550 nm). The wavelength in a vacuum of the specified radiation shown is used in a definition of the metre.* (Crown copyright. Courtesy of National Physical Laboratory)

low pressure in a discharge tube. Plate 33B shows the line spectrum of the gas krypton.

Gases such as carbon dioxide in a discharge tube produce a *band spectrum*. Each band consists of a series of lines very close together at the sharp edge or head of the band and farther apart at the other end or tail, Figure 33.10. Band spectra are essentially due to *molecules*. The different band heads in a band system are due to small allowed discrete energy changes in the *vibrational* energy of the molecule. The fine lines in a given band are due to still smaller allowed discrete energy changes in the *rotational* energy of the molecule.

Figure 33.10 *Diagrammatic representation of band spectra (molecules)*

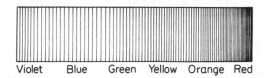

Violet   Blue   Green   Yellow   Orange   Red

Figure 33.11 *Continuous spectrum from a solid or liquid glowing white hot*

The spectrum of the sun is an example of a *continuous spectrum*, and, in general, these spectra are obtained from *solids and liquids*, Figure 33.11. In these states of matter the atoms and molecules are close together, and the energy changes in a particular atom are influenced by neighbouring atoms to such an extent that radiations of all different wavelengths are emitted. In a gas the atoms are comparatively far apart, and each atom is uninfluenced by any other. The gas therefore emits radiations of wavelengths which result from energy changes in the atom due solely to the high temperature of the gas, and the line spectrum is obtained. When the temperature of a gas is decreased and pressure applied so that the liquid state is approached, the line spectrum of the gas is observed to broaden out considerably.

Absorption Spectra

The spectra just discussed are classified as *emission spectra*. There is another class of spectra known as *absorption spectra*, which we shall now briefly consider.

If light from a hot source having a continuous spectrum is examined after it has passed through a sodium flame at a lower temperature, the spectrum is found to be crossed by a dark line; this dark line is in the position corresponding to the bright line emission spectrum obtained with the sodium flame alone. The continuous spectrum with the dark line is naturally characteristic of the absorbing substance, in this case sodium, and it is known as an *absorption spectrum*. An absorption spectrum is obtained when red glass is placed in front of sunlight, as it allows only a narrow band of red rays to be transmitted.

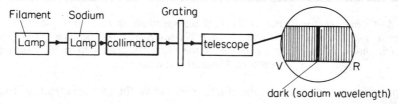

Figure 33.12 *Absorption spectrum of sodium (diagrammatic)*

Figure 33.12 shows how the absorption spectrum of sodium can be obtained. An over-run tungsten filament emits a continuous spectrum of light which is incident on a glowing sodium vapour lamp. The sodium lamp emits only visible yellow light. The light from the lamp passes through the hot sodium to a diffraction grating and the resulting spectrum is seen through a spectrometer telescope.

A darkened line is seen in the yellow part of the continuous spectrum. Also, when the sodium lamp is removed, the dark line becomes bright yellow. The dark line is the absorption spectrum of sodium. It has the same wavelength as the emission spectrum of sodium. Generally, an absorption spectrum is characteristic of the absorbing substance, so unknown elements can be identified from their absorption spectrum.

We now explain absorption spectra.

Absorption Spectra of Sun, Fraunhofer Lines

Atoms can absorb energy in a number of ways. In a flame, inelastic collisions with energetic molecules can raise atoms to higher energy levels. In a discharge tube, inelastic collisions with bombarding electrons can raise atoms to higher energy levels.

An atom can also absorb energy from a photon. If the photon energy $E = hf$ is just sufficient to excite an atom to one of its higher energy levels, the photon will be absorbed. When it returns to the ground state the excited atom emits the same wavelength as the photon but equally in all directions. So the intensity of the radiation in the direction of the incident photon is reduced. A *dark* line is thus seen whose wavelength is that of the absorbed photon.

Absorption of photons explains the dark lines in the sun's visible spectrum, first observed by Fraunhofer. The sun emits a continuous spectrum of photons. Vaporised elements in the outer or cooler parts of the sun's atmosphere absorb those photons which have the same frequency or energy to excite them to higher energy levels. The sun's spectrum is now darker at wavelengths *characteristic of the elements in the sun's atmosphere*. Since absorption spectra are always characteristic of the absorbing elements, these elements can be identified from their absorption spectra. In this way we know that elements such as iron exist in the sun.

In the sun's atmosphere, hot gases such as hydrogen are already excited to a higher energy level by collision with atoms. The absorbed photons are thus able to excite the atoms to still higher energy levels, which are in the sun's visible spectrum. So the Fraunhofer (dark) lines are present in the visible spectrum of the sun.

Normally, however, hydrogen is in the *ground* state, corresponding to the energy level -13.6 eV (p. 838). The next higher energy level is about -3.4 eV. This is a relatively high energy 'jump' of 10.2 eV and corresponds to a photon in the ultraviolet. Visible light has photons of energy less than 3.4 eV (see p. 827). So photons in visible light are not able to absorbed by hydrogen atoms in their normal or ground state. Consequently hydrogen is *transparent* to visible light. So if a wide band of wavelengths is passed through hydrogen, only those wavelengths in the ultraviolet would produce absorption lines.

Spontaneous and Stimulated Emission

As we have seen, an atom may undergo a transition between energy states if it emits or absorbs a photon of the appropriate energy. When an atom in the ground state, for example, absorbs a photon and makes a transition to a higher energy state or level, the photon is said to have 'stimulated' the absorption. The atom cannot increase its energy 'spontaneously', that is, in the absence of a photon.

We might expect that the atom in its higher energy state would emit a photon spontaneously, that is, the probability of emission would be independent of the number of photons present in the environment of the atom. In 1917 Einstein proved, however, that this is not the case. The probability per unit time that an atom will decay to a lower energy state and emit a photon is the sum of two terms. One is a spontaneous emission term. The other is a 'stimulated emission' term, which is proportional to the number of photons of the relevant energy *already present in the environment of the atoms*. Further, the photon produced by stimulated emission is always *in phase* with the stimulating photon.

The Laser

This idea led to the invention of the *laser*. (The word 'laser' is short for *l*ight *a*mplification by *s*timulated *e*mission of *r*adiation.)

Figure 33.13 shows the basic features of a gas laser. A helium–neon gas mixture (carbon dioxide gas is also used) is contained inside a long quartz tube

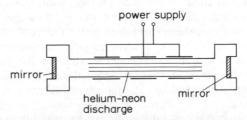

Figure 33.13 *Gas laser*

with optically plane mirrors at each end. A powerful radio-frequency generator is used to produce a discharge in the gas so that the helium atoms are excited or 'pumped up' to a higher energy level. By collision with these atoms, the neon atoms are excited to a higher energy level so that a 'population inversion' is obtained, that is, there are a much greater number of atoms in the higher energy level than in the ground state. Any stray photon, with energy equal to the difference in energies of the higher level and ground state, will stimulate many of the large population of excited atoms to fall from the higher energy level to the ground state. Repeated reflection at the silvered mirrors at the ends of the tube will multiply the number of photons rapidly. So strong stimulated emission is obtained and an enormously powerful burst of light or pulse of light is produced. The light pulse from a laser is:

(i) *monochromatic* light because all the photons have the same energy $(E - E_0)$ above the ground state and hence the same frequency $(E - E_0)/h$, where h is the Planck constant;

(ii) *coherent* light because all the photons or waves are in phase;

(iii) *intense* light because all the emitted waves are in phase or coherent. If the photons or waves were out of phase or incoherent, the resultant intensity would be the sum of the individual intensities and this is proportional to $n \times a^2$, where n is the number of waves and a is the amplitude of each wave. Since the waves are in phase, however, their total amplitude is na. Hence the resultant intensity is proportional to $(na)^2$ or n^2a^2. The intensity is greater by factor n than that obtained from incoherent waves, such as those from an ordinary lamp filament. Since n is very large, this leads to an enormous increase in intensity.

Bohr's Theory of Hydrogen Atom

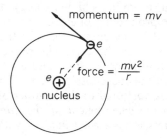

Figure 33.14 *Bohr's theory of hydrogen atom*

We conclude with a brief account of Bohr's theory of the hydrogen atom, which first gave important new ideas about atoms.

Bohr considered one electron of charge $-e$ and mass m, moving with speed v and acceleration v^2/r in an orbit round a central hydrogen nucleus of charge $+e$ (Figure 33.14). In classical physics, charges undergoing acceleration emit radiation and, therefore, they would lose energy. On this basis the electron would spiral towards the nucleus and the atom would collapse. Bohr, therefore, suggested that in those orbits where the angular momentum is a multiple of $h/2\pi$, the energy is constant. Twelve years later, de Broglie proposed that a particle such as the electron may be considered to behave as a *wave* of wavelength $\lambda = h/p$, where h is the Planck constant and p is the momentum of the moving particle.

If the electron can behave as a *wave*, it must be possible to fit a whole number of wavelengths around the orbit. In this case a *standing wave* pattern is set up and the energy in the wave is confined to the atom. A progressive wave would imply that the electron is moving from the atom and is not in a stationary orbit.

If there are n waves in the circular orbit and λ is the wavelength,

$$n\lambda = 2\pi r \qquad \qquad \qquad (1)$$

$$\therefore \lambda = \frac{2\pi r}{n} = \frac{h}{p} = \frac{h}{mv} \qquad \qquad (2)$$

So
$$mvr = \frac{nh}{2\pi} \qquad \qquad \qquad (3)$$

Now $mv \times r$ is the moment of momentum or *angular momentum* of the electron about the nucleus. So equation (3) states that the *angular momentum is a multiple of $h/2\pi$*. The quantisation of angular momentum is a key point in atomic theory.

In Figure 33.14, the electron moving round the nucleus has kinetic energy due to its motion and potential energy in the electrostatic field of the nuclear charge $+e$. Bohr calculated the total energy E of the electron in terms of its charge, mass, orbital radius and the number n which quantises the angular momentum. Thus for circular motion, (see p. 62)

$$mv^2/r = \text{force on electron} = e^2/4\pi\varepsilon_0 r^2 \qquad (4)$$

Also   potential energy of electron $= \dfrac{+e}{4\pi\varepsilon_0 r} \times -e = -\dfrac{e^2}{4\pi\varepsilon_0 r}$

and   kinetic energy of electron $= \tfrac{1}{2}mv^2 = \dfrac{e^2}{8\pi\varepsilon_0 r}$, from (4)

So   total energy of electron, $E = -\dfrac{e^2}{8\pi\varepsilon_0 r} \qquad \qquad (5)$

Using equations (3) and (4), v can be eliminated to find r. From (5) we then obtain

$$E_n \text{ (quantum number } n) = -\frac{me^4}{8\varepsilon_0^2 h^2} \cdot \frac{1}{n^2} \cdot \qquad (6)$$

The ground state, minimum energy E_0, corresponds to $n = 1$. If E is the energy value of a higher level corresponding to $n = n_1$, then

$$E - E_0 = \frac{me^4}{8\varepsilon_0^2 h^2}\left(1 - \frac{1}{n_1^2}\right) \qquad (7)$$

Bohr assumed that if the atom moves from a higher energy level to a lower level, the difference in energy is released in the form of radiation of energy hf, where h is the Planck constant and f is the frequency of the radiation. So

$$E - E_0 = hf = hc/\lambda$$

where λ is the wavelength of the radiation, since $c = f\lambda$ for the electromagnetic wave of speed c. From the values of the energy, Bohr was able to calculate the wavelength emitted and this agreed with the experimental value obtained.

Spectral Series of Hydrogen

Before Bohr's theory of the hydrogen atom it had been found that the wavelengths of the hydrogen spectrum could be arranged in a formula or series named after its discoverer. The visible spectrum was the Balmer series, the ultraviolet was the Lyman series and the infrared was the Paschen series.

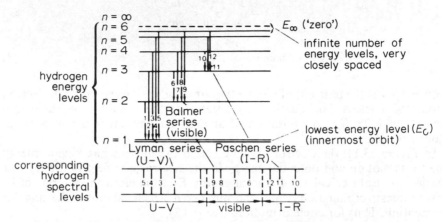

Figure 33.15 *Energy levels and spectra of hydrogen* (not to scale)

Bohr's theory of energy levels accounted for all the series. As shown in Figure 33.15, the hydrogen spectrum is obtained simply by using different numbers for n in calculating the energy levels. The ultraviolet series, for example, is obtained when the energy of the atom falls to the lowest energy level E_0 corresponding to $n = 1$; the visible spectrum is obtained for energy falls to a higher level corresponding to $n = 2$; and the infrared spectrum is obtained for energy falls to the higher level when $n = 3$. See also Figure 33.8.

From Figure 33.15, the energy change E for the ultraviolet series is greater than for the visible spectrum. Since $E = hf$, the frequency of ultraviolet radiation is greater than that of visible radiation. Wavelength, λ, is inversely-proportional to f, since $c = f\lambda$. Hence the ultraviolet wavelengths are *shorter* than those in the visible spectrum.

Bohr's theory of the hydrogen atom was unable to predict the energy levels in complex atoms, which had many electrons. Quantum or wave mechanics, beyond the scope of this book, is used to explain the spectral frequencies of these atoms. The fundamental ideas of Bohr's theory, however, are still retained; for example, the angular momentum of the electron has quantum values and the energy levels of the atom have only allowed separated values characteristic of the atom.

Franck–Hertz Experiment, Elastic and Inelastic Collisions

Bohr's ideas of energy levels in atoms were confirmed in a famous experiment due to Franck and Hertz. Figure 33.16 (i) shows the principle of the experiment.

Sodium vapour at a very low pressure of about a millimetre of mercury is inside a tube having a tungsten filament F, a grid plate G and a plate A. When F is heated electrons are emitted from it. The distance FG was arranged to be much greater than the mean (average) free path of the electrons in the sodium

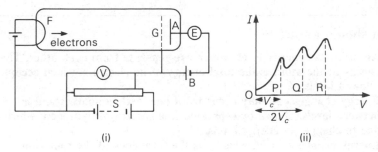

Figure 33.16 *Franck–Hertz experiment and results*

gas, so the electrons can make several collisions with the sodium atoms before reaching G. Plate A, which is near to G, can collect electrons.

The p.d. V between F and G can be varied by the potential divider arrangement S. V accelerates the electrons to kinetic energy values eV, which increases with V measured by a voltmeter as shown. The p.d. between A and G is made small, less than 1 volt. As shown by battery B, A is made negative in potential relative to G so that electrons with very low kinetic energy near A move in the direction AG and are not collected by A. Electrons reaching A produce a current I measured by the meter E.

When the accelerating p.d. V between F and G was increased from zero, the current I increased until V reached a value P, Figure 33.16 (ii). I then became *smaller* and reached a minimum value as V was increased more, but then I increased again to another peak or maximum as shown. As V was increased more, I decreased again to a minimum and then increased to a maximum at Q. With sodium atoms, the p.d. V_c between the maximum values was found to be about 2.1 V.

Elastic and inelastic collisions. The results of the experiment show that below the p.d. V_c starting from zero p.d., the electrons make *elastic collisions* with the atoms between F and G. The kinetic energy of the electrons are then unchanged and electrons can then reach A and produce a current I.

At the p.d. V_c, however, the electrons make *inelastic collisions* with the atoms so they lose some kinetic energy to the atoms. As we discussed earlier, the atoms are then excited to an energy level above the ground state. The electrons between A and G now have such low energy that they move away from A to G under the low retarding p.d. between A and G. So the current I decreases. As V is increased more, the retarding p.d. between A and G is overcome and I increases again. At a p.d. about $2V_c$ the electrons make two inelastic collisions with the atoms and so I decreases again as shown.

The results show that the energy of a normal atom is constant until the atom is excited to a particularly higher energy level than its ground state. With sodium atoms the excitation potential from zero was measured as 2·1 V. Using Bohr's theory, the frequency f and wavelength λ of the radiation emitted when the atom returns from its excited state to ground state is given, with the usual notation and values by

$$eV_c = hf = hc/\lambda$$

So $$\lambda = \frac{hc}{eV_c} = \frac{6 \cdot 6 \times 10^{-34} \times 3 \times 10^8}{1 \cdot 6 \times 10^{-19} \times 2 \cdot 1} = 5 \cdot 89 \times 10^{-7} \text{ m}$$

This is the wavelength of the yellow sodium wavelength in the atom's emission spectrum. It shows agreement with Bohr's idea of energy levels in atoms.

You should know:

1 An atom has energy levels which are separate from each other. The energy of the atom is the total energy of its electrons which occupy the allowed levels.

2 Energy changes can only occur from one level to another and not between levels. Shown by separated line spectra of hydrogen, which are due to changes in energy levels.

3 Energy change $E = hf$, where f is the frequency of the radiation emitted by the energy change from a high to a lower value. This change produces *emission* spectra. At high temperature, hydrogen produces ultraviolet, visible and infrared emission spectra.

4 The ground state of an atom is its lowest energy level. Electrons have least energy in the ground state so this is the case for a normal atom. The atom becomes *ionised* when an electron receives energy to just exceed its highest energy level and leave the atom. The highest energy level is given the value zero and lower energy values are therefore *negative*. Energy levels of the hydrogen atoms are given by $-13\cdot6/n^2$ eV, where $n = 1, 2, 3$.

5 From $E = hf$, high energy changes produces high frequency radiation or *short* wavelengths.

6 Radiation from the sun is seen crossed by dark lines. These are *absorption* spectra. Atoms in the cooler part of the sun absorb those photons in the sun's radiation which can excite the atoms to higher energy levels.

EXERCISES 33B Energy Levels in Atoms

Multiple Choice

1 An atom is excited to an energy level E_1 from its ground state energy level E_0. The wavelength of the radiation emitted is

A $(E_0 - E_1)/hc$ B $(E_1 - E_0)/h$ C $\dfrac{E_1}{hc} - \dfrac{E_0}{hc}$

D $\dfrac{hc}{(E_1 - E_0)}$ E $\dfrac{h}{(E_1 - E_0)}$

2 An atom X is excited to an energy level E_2 from the ground state E_0 by collision with an atom. Which of the following statements is correct?
1 The electrons in X lose energy just after the collision.
2 The excited atom X emits a radiation given by $hf = E_0 - E_2$.
3 The colliding atom must have lost energy to excite X.
A 1, 2, 3 are all correct   B 1, 2 only are correct
C 2, 3 only   D 1 only   E 3 only

3 Figure 33E (i) shows the energy levels of an atom. Which of the line spectra in A to E in Figure 33F (ii) correspond to the three energy level changes assuming energy levels are roughly to scale?

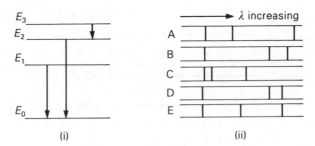

Figure 33F

Longer Questions

4 The ground state of the electron in the hydrogen atom may be represented by the energy − 13·6 eV and the first two excited states by − 3·4 eV and − 1·5 eV respectively, on a scale in which an electron completely free of the atom is at zero energy. Use this data to calculate the ionisation potential of the hydrogen atom and the wavelengths of three lines in the emission spectrum of hydrogen.

(Charge of the electron = − 1·6 × 10⁻¹⁹ C. Speed of light in a vacuum = 3·0 × 10⁸ m s⁻¹. The Planck constant = 6·6 × 10⁻³⁴ J s.) (*L.*)

5 Some of the energy levels of the hydrogen atom are shown (not to scale) in Figure 33G.
(a) Why are the energy levels labelled with negative energies?
(b) Mark on the diagram a transition which will result in the emission of radiation of wavelength 487 nm. Justify your answer by suitable calculation.
(c) What is likely to happen to a beam of photons of energy: (i) 12·07 eV; (ii) 5·25 eV when passed through a vapour of atomic hydrogen? (*O. & C.*)

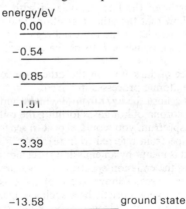

Figure 33G

6 Some of the energy levels of a mercury atom are shown in Figure 33H (page 850).
(i) Calculate the ionisation energy of a mercury atom in electronvolts (eV) and in joules (J).
(ii) Calculate the wavelength of radiation emitted when an electron moves from level 4 to level 2. In what part of the electromagnetic spectrum does this wavelength lie?
(iii) State what is likely to happen if a mercury atom in the unexcited state (level 1) is bombarded with electrons with energies 4·0 eV, 6·7 eV, and 11·0 eV respectively. (*O.*)

7 What are the chief characteristics of a line spectrum? Explain how lines spectra are used in analysis for the identification of elements.
Figure 33I, which represents the lowest energy levels of the electron in the hydrogen

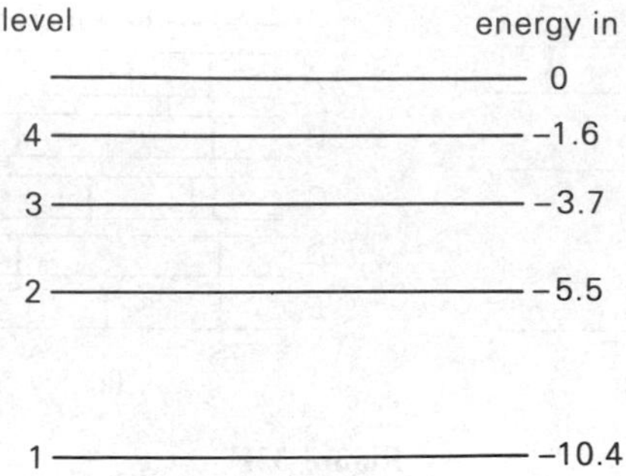

level   energy in eV

——————————————————— 0

4 ——————————————————— −1.6

3 ——————————————————— −3.7

2 ——————————————————— −5.5

1 ——————————————————— −10.4

Figure 33H

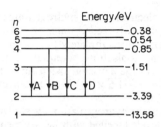

Figure 33I

atom, specifies the value of the principal quantum number n associated with each state and the corresponding value of the energy of the level, measured in electronvolts. Work out the wavelengths of the lines associated with the transitions A, B, C, D marked in the figure. Show that the other transitions that can occur give rise to lines that are either in the ultra-violet or the infra-red regions of the spectrum. (Take 1 eV to be 1.6×10^{-19} J; Planck constant h to be 6.5×10^{-34} J s; and c, the velocity of light *in vacuo*, to be 3×10^8 m s^{-1}.) (*O.*)

8 (a) Explain the presence of dark lines in the otherwise continuous spectrum of the sun in terms of the atomic processes involved.

   (b) In a complete solar eclipse the light from the main body of the sun is cut off and only light from the corona of hot gases forming the outer layer reaches the earth. State what kind of spectrum you would expect to see for this light. How would it compare with the spectrum referred to in (a)? (*N.*)

9 (a) Explain briefly what is meant by an emission spectrum. Describe a suitable source for use in observing the emission spectrum of hydrogen.

   (b) Figure 33J, which is to scale, shows some of the possible energy levels of the hydrogen atoms. (i) Explain briefly how such a diagram can be used to account for the emission spectrum of hydrogen. (ii) When the hydrogen atom is in its ground state, 13·6 eV of energy are needed to ionise it. Calculate the highest possible frequency in the line spectrum of hydrogen. In what region of the

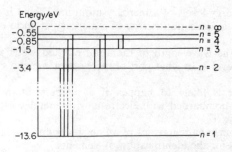

Figure 33J

electromagnetic spectrum does it lie? (The frequency range of the visible spectrum is from 4×10^{14} Hz to $7\cdot5 \times 10^{14}$ Hz.)   (iii) A number of transitions are marked on the energy level diagram. Identify which of these transitions corresponds to the lowest frequency that *would be visible*.

(c) The wavelengths of the lines in the emission spectrum of hydrogen can be measured using a spectrometer and a diffraction grating. If the grating has 5000 lines per cm, through what angle would light of frequency $4\cdot6 \times 10^{14}$ Hz be diffracted in the first order spectrum?

(1 eV $= 1\cdot60 \times 10^{-19}$ J. The Planck constant $= 6\cdot6 \times 10^{-34}$ J s, speed of light $= 3\cdot99 \times 10^{8}$ m s^{-1}.) (*L.*)

X-Rays

In 1895, Roentgen found that some photographic plates, kept carefully wrapped in his laboratory, had become fogged. Instead of merely throwing them aside he set out to find the cause of the fogging. He traced it to a gas-discharge tube, which he was using with a high voltage. This tube appeared to emit a radiation that could penetrate paper, wood, glass, rubber, and even thick aluminium. Roentgen could not find out whether the radiation was a stream of particles or a train of waves—Newton had the same difficulty with light—and he decided to call it *X-rays*.

Nature and Production of X-Rays

We now regard X-rays as waves, similar to light waves, but of much shorter wavelength, about 10^{-10} m or $0\cdot1$ nm. They are produced when fast electrons, or cathode rays, strike a target, such as the walls or anode or a low-pressure discharge tube.

In a modern X-ray tube there is no gas, or as little as high-vacuum technique can achieve; the pressure is about 10^{-5} mm Hg. The electrons are provided by thermionic emission from a white-hot tungsten filament. In Figure 33.17, F is the filament and T is the metal target embedded in a copper anode A. A cylinder C round F, at a negative potential relative to F, focuses the electron beam on

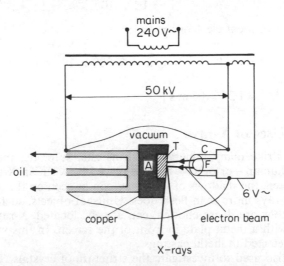

Figure 33.17 *An X-ray tube (diagrammatic)*

T. Because there is so little gas, the electrons on their way to the anode do not lose any appreciable amount of their energy in ionising atoms. From the a.c. mains, transformers provide about 6 volts for heating the filament, and about 50 000 volts for accelerating the electrons. On the half-cycles when the target is positive, the electrons bombard it, and generate X-rays. On the half-cycles when the target is negative, nothing happens at all—there is too little gas in the tube for it to break down. Thus the tube acts, in effect, as its own rectifier (p. 770), providing pulses of direct current between target and filament. The heat generated at the target by the electron bombardment is so great that the metal must be cooled. In large tubes this is done by circulating oil behind the anode, as shown. The target in an X-ray tube may be tungsten, for example, which has a high melting-point.

So X-rays (waves) are produced by bombarding matter with electrons (particles). The production of X-rays is therefore the inverse or opposite process to the photoelectric effect, where electrons (particles) are liberated from metals by incident light waves.

Example on Energy in X-ray Tube

An X-ray tube, operated at a d.c. potential difference of 40 kV, produces heat at the target at the rate of 720 W. Assuming 0·5% of the energy of the incident electrons is converted into X-radiation, calculate: (i) the number of electrons per second striking the target; (ii) the velocity of the incident electrons. (Assume charge–mass ratio of electron, $e/m_e = 1·8 \times 10^{11} \, \text{C kg}^{-1}$.)

(i) Heat per second at target = $99·5\% \times IV$, where I is the current flowing and V is the p.d. applied. So

$$0·995 \times I \times 40\,000 = 720$$

Then
$$I = \frac{720}{40\,000 \times 0·995} = 0·018 \text{ A (approx.)}$$

So number of electrons per second $= \dfrac{I}{e} = \dfrac{0·018}{1·6 \times 10^{-19}}$

$$= 1·1 \times 10^{17}$$

(ii) Energy of incident electrons = $\frac{1}{2}m_e v^2 = eV$

So
$$v = \sqrt{2V \times \frac{e}{m_e}} = \sqrt{2 \times 40\,000 \times 1·8 \times 10^{11}}$$

$$= 1·2 \times 10^8 \text{ m s}^{-1}$$

Effects and Uses of X-rays

When X-rays strike many minerals, such as zinc sulphide, they make them fluoresce. If a human—or other—body is placed between an X-ray tube and a fluorescent screen, the shadows of the bones can be seen on the screen, because they absorb X-rays more than flesh does. Unusual objects, such as swallowed safety-pins, if they are dense enough, can also be located. X-ray photographs can be taken with a metal plate in front of the screen. In this way cracks and flaws can be detected in metal castings.

X-rays are also used to investigate the structure of crystals. This important use of X-rays is discussed shortly.

Crystal Diffraction

The first proof of the wave-nature of X-rays was due to Laue in 1913, many years after X-rays were discovered. He suggested that the regular small spacing of atoms in crystals might provide a *natural diffraction grating* if the wavelengths of the rays were too short to be used with an optical line grating. Experiments by Friedrich and Knipping showed that X-rays were indeed diffracted by a thin crystal, and produced a pattern of intense spots round a central image on a photographic plate placed to received them, Figure 33.18. The rays had thus been scattered by interaction with electrons in the atoms of the crystal. The diffraction pattern obtained gave information on the geometrical spacing of the atoms. An X-ray diffraction pattern produced by a crystal is shown on p. 859.

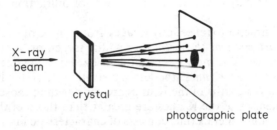

Figure 33.18 *Laue crystal diffraction*

Bragg Law

The study of the atomic structure of crystals by X-ray analysis was started in 1914 by Sir William Bragg and his son Sir Lawrence Bragg, with notable achievements. They soon found that a monochromatic beam of X-rays was reflected from a plane in the crystal rich in atoms, a so-called atomic plane, as if this acted like a mirror.

This important effect can be explained by Huygens' wave theory in the same way as the reflection of light by a plane surface. Suppose a monochromatic parallel X-ray beam is incident on a crystal and interacts with atoms such as A, B, C, D in an atomic plane P, Figure 33.19 (i). Each atom scatters the X-rays. Using Huygens's construction, wavelets can be drawn with the atoms as centres which all lie on a plane wavefronts reflected at an equal angle to the atomic plane P. When the X-ray beam penetrates the crystal to other atomic planes such as Q, R parallel to P, reflection occurs in a similar way, Figure 33.19 (ii). Usually, the beam or ray reflection from one plane is weak in intensity. If, however, the reflected beams or rays from all planes are *in phase* with each other, a strong reflected beam is produced by the crystal.

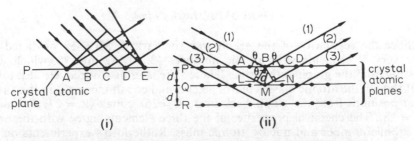

Figure 33.19 *Reflection (diffraction) at crystal atomic planes*

Suppose, then, that the glancing angle on an atomic plane ABCD in the crystal is θ, and d is the distance apart of consecutive parallel atomic planes, Figure 33.19 (ii). The path difference between the rays marked (1) and (2) $= LM + MN = 2LM = 2d \sin \theta$. So a strong X-ray beam is reflected when

$$2d \sin \theta = n\lambda$$

where λ is the wavelength and n has integral values. This is known as *Bragg's law*. Hence, as the crystal is rotated so that the glancing angle is increased from zero, and the beam reflected at an equal angle is observed each time, an intense beam is suddenly produced for a glancing angle θ_1 such as $2d \sin \theta_1 = \lambda$. When the crystal is rotated further, an intense reflected beam is next obtained for an angle θ_2 when $2d \sin \theta_2 = 2\lambda$. So several orders of diffraction images may be observed.

The intense diffraction (reflection) images from an X-ray tube are due to X-ray lines *characteristic of the metal used* as the target, or 'anti-cathode' as it was originally known. The more penetrating or harder X-ray lines emitted from the metal are called the *K lines*; the less penetrating or softer lines are called the *L lines*. The wavelengths of the K lines are shorter than those of the L lines, so that the frequencies of the K lines are greater than those of the L lines. The X-radiation from some metals thus consists of characteristic lines in the K series or L series or both.

Moseley's Law

In 1914 Moseley measured the frequency f of the characteristic X-rays from many metals, and found that, for a particular type of emitted X-ray such as K_α, whose origin is explained later, the frequency f varies in a regular way with the *atomic number* Z of the metal. When a graph of Z v. $f^{1/2}$ was plotted, an almost perfect straight line was obtained, Figure 33.20. Moseley therefore gave an empirical relation, known as *Moseley's law*, between f and Z as

$$f = a(Z - b)^2$$

where a, b are constants.

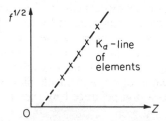

Figure 33.20 *Moseley's law*

Since the regularity of the graph was so marked. Moseley predicted the discovery of elements with atomic numbers 43, 61, 72 and 75, which were missing from the graph at that time. These were later discovered. He also found that though the atomic masses of iron, nickel and cobalt increased in this order, their positions from the graph were: iron ($Z = 26$), cobalt ($Z = 27$) and nickel ($Z = 28$). The chemical properties of the three elements agree with the order by atomic number and not by atomic mass. Rutherford's experiments on the scattering of α-particles (p. 882) had shown that the atom contained a central nucleus of charge $+Ze$ where Z is the atomic number, and Moseley's experiments confirmed the importance of Z in atomic theory.

Explanation of Emission Spectra

Since the frequency of the X-ray spectra of elements is related to Ze, the charge on the nucleus, Moseley's results showed that the radiation was due to energy changes of the atom resulting from the movement of *electrons close to the nucleus.*

We now know, in fact, that the electrons in atoms are in groups which have various energy states. These groups are called 'shells'. Electrons nearest the nucleus are in the so-called *K-shell*. The next group, which are further away from the nucleus, are in the so-called *L* shell. The *M* shell is farther from the nucleus than the L shell. Electrons farther from the nucleus have greater energy than those nearer, since more work is needed to move them farther away. So electrons in the L shell have greater energy than those in the K shell; electrons in the M shell have greater energy than those in the L shell.

In the X-ray tube, energetic electrons bombard the metal target and may eject an electron from the innermost shell, the K shell. The atom is then raised to an excited state since its energy has increased, and it is unstable. An electron from the L shell may now move into the vacancy in the K shell, so decreasing the energy of the atom. Simultaneously, radiation is emitted of frequency f, where $E =$ energy change of atom $= hf$. So $f = E/h$, and as E is very high for metals, the frequency f is very high and the wavelength is correspondingly short. It is commonly of the order of 10^{-9} m (1 nm) or less. The K series of X-ray lines is due to movement of electrons from the L (K_α line) or M (K_β line) shells to the K shell.

X-ray spectra are thus due to energy changes in electrons *close to the nucleus* of metals. In contrast, the optical spectra of the metals is due to energy changes in the outermost shell of the atom. Here the energy changes are about 1000 times smaller. So the frequencies of the optical lines are about 1000 times smaller, that is, their wavelengths are about 1000 times longer than those of X-rays.

Continuous X-ray Background Radiation, Minimum Wavelength

The characteristic X-ray spectrum from a metal is usually superimposed on a background of continuous, or so-called 'white', radiation of small intensity. Figure 33.21 illustrates the characteristic lines, K_α, K_β, of a metal and the continuous background of radiation for two values of p.d., 40 000 and 32 000

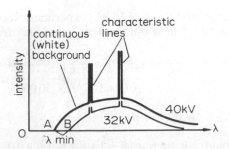

Figure 33.21 *X-ray characteristic lines and background*

volts, across an X-ray tube. It should be noted that: (i) the wavelengths of the characteristic lines are independent of the p.d.—they are characteristic of the metal, and (ii) the background of continuous radiation has increasing wavelengths which slowly diminish in intensity, but as the wavelength diminish they are cut off *sharply*, as at A and B.

When the bombarding electrons collide with the metal atoms in the target, most of their energy is lost as heat. A little energy is also lost in the form of electromagnetic radiation. Here the frequencies are given $E = hf$ with the usual notation, and the numerous energy changes produce the background radiation in Figure 33.21.

The existence of a *sharp* minimum wavelength at A of B can be explained only by the quantum theory. The energy of an electron before striking the metal atoms of the target is eV, where V is the p.d. across the tube. If a direct collision is made with an atom and *all* the energy is absorbed, then, on quantum theory, the X-ray quantum produced has maximum energy.

$$\therefore hf_{max} = eV$$

$$\therefore f_{max} = \frac{eV}{h} \qquad \qquad \qquad (1)$$

$$\therefore \lambda_{min} = \frac{c}{f_{max}} = \frac{ch}{eV} \qquad \qquad (2)$$

Verification of Quantum Theory

These conclusions are borne out by experiment. Thus for a particular metal target, experiment shows that the minimum wavelength is obtained for p.d.s of 40 kV and 32 kV at glancing angles of about $3 \cdot 0°$ and $3 \cdot 8°$ respectively. The ratio of the minimum wavelength is, from Bragg's law,

$$\frac{\lambda_1}{\lambda_2} = \frac{\sin 3 \cdot 0°}{\sin 3 \cdot 8°} = 0 \cdot 8 \text{ (approx.)}$$

From (2), $\lambda_{min} \propto 1/V$ using quantum theory.

$$\therefore \frac{\lambda_1}{\lambda_2} = \frac{32}{40} = 0 \cdot 8$$

With a tungsten target and a p.d. of 30 kV, experiment shows that a minimum wavelength of $0 \cdot 42 \times 10^{-10}$ m is obtained, as calculated from values of d and θ. From (2),

$$\therefore \lambda_{min} = \frac{ch}{eV} = \frac{3 \cdot 0 \times 10^8 \times 6 \cdot 6 \times 10^{-34}}{1 \cdot 6 \times 10^{-19} \times 30\,000} \text{ m}$$

$$= 0 \cdot 41 \times 10^{-10} \text{ m}$$

using $c = 3 \cdot 0 \times 10^8$ m s^{-1}, $h = 6 \cdot 6 \times 10^{-34}$ J s, $e = 1 \cdot 6 \times 10^{-19}$ C, $V = 30000$ V. This is in good agreement with the experimental result.

You should know:

1 X-rays are emitted from a metal when hit by high-energy electrons. The electrons penetrate close to the nucleus and displace electrons round the nucleus. The energy change $= hf$, where f is very high for the radiation emitted.

2 X-rays are electromagnetic waves of very short wavelength—about 100 times shorter than the wavelengths of visible light. The short wavelength is shown by X-ray diffraction at crystals, whose atomic planes have very small spacing.

3 X-ray tubes have (a) filament or cathode C emitting electrons when heated, (b) anode at high voltage V relative to C, (c) a target (metal) for X-ray emission, (d) a cooling arrangement behind the hot anode.

4 The energy of the incident electrons $= eV = \frac{1}{2}mv^2$, where v is the speed of the incident electrons of charge e and mass m. X-ray energy produced is less than 1% of electron energy—more than 99% of energy heats the anode, so cooling is necessary.

5 The graph of intensity against wavelength for X-rays emitted show (a) a high intensity line characteristic of the metal used and (b) a sharp minimum wavelength in the background radiation given by $eV = hf_{max} = hc/\lambda_{min}$.

EXERCISES 33C   X-Rays

Multiple Choice

1 The spacing of atomic layers in a crystal can be measured using X-ray wavelengths in metres of the order of

A 10^{-6} B 10^{-7} C 10^{-8} D 10^{-10} E 10^{-15}

2 X-rays are produced by energy changes in
A the nucleus   B electrons close to the nucleus
C electrons far from the nucleus   D electrons and protons
E electrons and neutrons.

3 In a X-ray tube, electrons of high energy E are incident on a target of tungsten. Which of the following three statements are correct?
1 All the energy is converted to X-rays.
2 The maximum X-ray frequency obtained is E/h.
3 The X-rays are diffracted by the tungsten metal.
A 1, 2, 3 are all correct   B only 1, 2   C only 2, 3
D 2 only   E 3 only

Longer Questions

4 (a) Explain briefly why a modern X-ray tube can be operated directly from the output of a step-up transformer.
(b) An X-ray tube works at a d.c. potential difference of 50 kV. Only 0·4% of the energy of the cathode rays is converted into X-radiation and heat is generated in the target at a rate of 600 W. Estimate: (i) the current passed into the tube; (ii) the velocity of the electrons striking the target. (Electron mass $= 9·00 \times 10^{-31}$ kg, electron charge $= -1·60 \times 10^{-19}$ C.) (N.)

5 In a modern X-ray tube electrons are accelerated through a large potential difference and the X-rays are produced when the electrons strike a tungsten target embedded in a large piece of copper.
(a) What is the purpose of the large piece of copper?
(b) Describe the energy changes taking place in the tube.
(c) If the accelerating voltage is 40 kV, calculate the kinetic energy of the electrons arriving at the target, and determine the minimum wavelength of the emitted X-rays.
(Electronic charge = $-1{\cdot}6 \times 10^{-19}$ C. Planck constant = $6{\cdot}6 \times 10^{-34}$ J s. Velocity of electromagnetic waves = $3{\cdot}0 \times 10^8$ m s^{-1}.) (AEB.)

6 Molybdenum K$_\alpha$ X-rays have wavelength 7×10^{-11} m. Find: (i) the minimum X-ray tube potential difference that can produce these X-rays, and (ii) their photon energy in electronvolts.
(The electronic charge, e = $-1{\cdot}6 \times 10^{-19}$ C; the Planck constant, h = $6{\cdot}6 \times 10^{-34}$ J s; speed of light, c = 3×10^8 m s^{-1}.) (W.)

7 Draw a labelled diagram of a modern X-ray tube. How may:
(a) the intensity;
(b) the penetrating power, of the X-rays be controlled?
An X-ray tube is operated with the anode potential of 10 kV and an anode current of 15·0 mA. (i) Estimate the number of electrons hitting the anode per second. (ii) Calculate the rate of production of heat at the anode, stating any assumptions made. (iii) Describe the characteristics of the emitted X-ray spectrum and account for any special features. (Electron charge e = $1{\cdot}60 \times 10^{-19}$ C; Planck constant, h = $6{\cdot}63 \times 10^{-34}$ J s; speed of light, c = $3{\cdot}00 \times 10^8$ m s^{-1}.) (C.)

8 Describe the properties of X-rays and compare them with those of ultraviolet radiation. Outline the evidence for the wave nature of X-rays.
The energy of an X-ray photon is hf joule where h = $6{\cdot}63 \times 10^{-34}$ J s and f is the frequency in hertz (cycles per second). X-rays are emitted from a target bombarded by electrons which have been accelerated from rest through 10^5 V. Calculate the minimum possible wavelength of the X-rays assuming that the corresponding energy is equal to the whole of the kinetic energy of one electron. (Charge of an electron = $1{\cdot}60 \times 10^{-19}$ C; velocity of electromagnetic waves *in vacuo* = $3{\cdot}00 \times 10^8$ m s^{-1}.) (O. & C.)

9 Describe the atomic process in the target of an X-ray tube whereby X-ray line spectra are produced. Determine the ratio of the energy of a photon of X-radiation of wavelength 0·1 nm to that of a photon of visible radiation of wavelength 500 nm. Why is the potential difference applied across an X-ray tube very much greater than that applied across a sodium lamp producing visible radiation? (JMB.)

10 In an X-ray tube the current through the tube is 1·0 mA and the accelerating potential is 15 kV. Calculate: (i) the number of electrons striking the anode per second; (ii) the speed of the electrons on striking the anode assuming they leave the cathode with zero speed; (iii) the rate at which cooling fluid, entering at 10°C, must circulate through the anode if the anode temperature is to be maintained at 35°C. Neglect any of the kinetic energy of the electrons which is converted to X-radiation. (Electronic charge = $1{\cdot}6 \times 10^{-19}$ C; mass of electron = $9{\cdot}1 \times 10^{-31}$ kg; specific heat capacity of liquid = $2{\cdot}0 \times 10^3$ J kg^{-1} K^{-1}.) (AEB.)

Wave–Particle Duality

Electron Diffraction

We have seen how the wave nature of X-rays has been established by X-ray diffraction experiments, as illustrated in Figure 33.22 (i). Similar experiments, first performed by Davisson and Germer, show that streams of *electrons* produce diffraction patterns and so these particles also show wave properties,

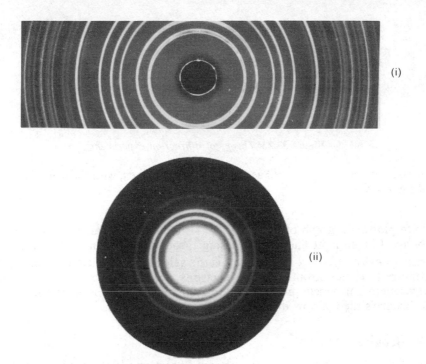

(i)

(ii)

Plate 33C (i) *X-ray diffraction rings produced by a crystal* (ii) *Electron diffraction rings produced by a thin gold film. Similarity of* (i) *wave and* (ii) *particle*

see Plate 33C (ii). Electron diffraction is now as useful a research tool as X-ray diffraction.

A teltron tube used for demonstrating electron diffraction is shown diagrammatically in Figure 33.22. A beam of electrons strikes a layer of carbon which is extremely thin, and a diffraction pattern, consisting of *rings*, is seen on the tube face. Sir George Thomson first obtained such a diffraction pattern using a very thin gold film, Figure 33.21 (ii). If the voltage V on the anode is

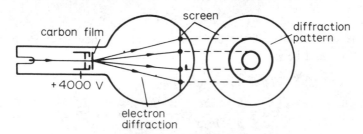

Figure 33.22 *Electron diffraction tube*

increased, the velocity v of the electrons is increased. The rings are then seen to become narrow, showing that the wavelength λ of the electron waves decreases with increasing v or increasing voltage V.

If a particular ring of radius R is chosen, the angle of deviation ϕ of the incident beam is given by $\phi = 2\theta$, where θ is the angle between the incident beam and the crystal planes, Figure 33.23. Now $\tan \phi = R/D$, and if ϕ is small, $\phi = R/D$ is a good approximation. Hence $\theta = R/2D$. If the Bragg law is true

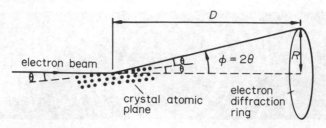

Figure 33.23 *Theory of diffraction experiment*

for electron diffraction as well as X-ray diffraction, then, with the usual notation, $2d \sin \theta = n\lambda$.

$$\therefore \lambda \propto \sin \theta \propto \theta \propto R . \qquad . \qquad . \qquad . \qquad (1)$$

On plotting a graph of R against $1/\sqrt{V}$ for different values of accelerating voltage V, a straight line graph passing through the origin is obtained. Now $\frac{1}{2}m_e v^2 = eV$, or $1/\sqrt{V} \propto 1/v$, where v is the velocity of the electrons accelerated from rest. Hence, from (1), the electrons appear to act as *waves* whose wavelength is inversely-proportional to their velocity. This is in agreement with de Broglie's theory, now discussed.

De Broglie's Theory

In 1925, before the discovery of electron diffraction, de Broglie proposed that

$$\lambda = \frac{h}{p} \qquad . \qquad . \qquad . \qquad . \qquad (2)$$

where λ is the wavelength of waves associated with particles of momentum p, and h is the *Planck constant*, $6\cdot63 \times 10^{-34}$ J s. The quantity h was first used by Planck in his theory of heat radiation and it is a constant which enters into all branches of atomic physics. It is easy to see that de Broglie's relation is consistent with the experimental result obtained with the electron diffraction tube. Here the gain in kinetic energy of the electron is eV so that

$$\frac{1}{2}m_e v^2 = eV$$

where v is the velocity of the electrons, assuming they start from rest. So $v = \sqrt{2eV/m_e}$ and hence

$$p = m_e v = \sqrt{2eVm_e}$$

$$\therefore \lambda = \frac{h}{p} = \frac{h}{m_e v} = \frac{h}{\sqrt{2eVm_e}} \propto V^{-1/2}$$

We can now estimate the wavelength of an electron beam. Suppose $V = 3600$ V. For an electron, $m = 9\cdot1 \times 10^{-31}$ kg, $e = 1\cdot6 \times 10^{-19}$ C, and $h = 6\cdot6 \times 10^{-34}$ J s.

$$\therefore \lambda = \frac{h}{\sqrt{2eVm_e}} = \frac{6\cdot6 \times 10^{-34}}{\sqrt{2 \times 1\cdot6 \times 10^{-19} \times 3600 \times 9\cdot1 \times 10^{-31}}}$$

$$= 2 \times 10^{-11} \text{ m}$$

This is about 30 000 times smaller than the wavelength of visible light. On this account electron beams are used in *electron microscopes*. These instruments, which use 'electron lenses', can produce resolving powers far greater than that of an optical microscope.

Wave Nature of Matter

Electrons are not the only particles which behave as waves. The effects are less noticeable with more massive particles because their momenta are generally much higher, and so the wavelength is correspondingly shorter. Since appreciable diffraction is observed only when the wavelength is of the same order as the grating spacing, the heavier particles, such as protons, are diffracted much less. Slow neutrons, however, are used in diffraction experiments, since the low velocity and high mass combine to give a momentum similar to that of electrons used in electron diffraction. The wave nature of α-particles is important in explaining α-decay.

Wave–Particle Duality

From what has been said, it is clear that particles can have wave properties, and that waves can sometimes behave as particles. As we saw in the photoelectric effect, electromagnetic waves appear to have a particle nature. Further, γ-rays behave as electromagnetic waves of very short, wavelength but on detection by Geiger–Müller tubes, which count *individual* pulses, they behave as particles (see p. 869). It would appear, therefore, that a paradox exists since wave and particle structure appear different from each other.

Scientists gradually realised, however, that the dual aspect of wave–particle properties are completely general in nature. All physical quantities can be described either as waves or particles; the description to choose is entirely a matter of convenience. The two aspects, wave and particle, are linked through the two relations

---

$$E = hf; \qquad p = h/\lambda$$

On the left of each of these relations, E and p refer to a particle description. On the right, f and λ refer to a wave description. Note that the Planck constant is the constant of proportionality in *both* these equations, a fact which can be predicted by *Einstein's special theory of relativity*. Further, from these equations, the frequency of the wave is proportional to the *energy* and the wavelength is inversely proportional to the *momentum*.

In the case of a photon of frequency f, we have $\lambda = c/f$. So $E = hf$ and $p = h/\lambda = hf/\lambda = E/c$. Hence

$$E = pc = mc^2$$

So, as a particle, the mass of a photon is E/c^2 or hf/c^2. About 1923, Compton investigated the scattering of X-rays (high-frequency photons) by matter. He showed that the experimental results agreed with the assumption that a *particle* of mass hf/c^2 collided elastically with an electron in the atom of the material.

Heisenberg Uncertainty Principle

Newton's laws o mechanics can be applied to large and heavy objects with great accuracy. They do not apply, however, to the microscopic world of the so-called 'elementary particles'.

Here laws of quantum mechanical theory apply, such as the important *Heisenberg uncertainty principle.* This states that, for example, there is a limit to the certainty with which we can locate simultaneously both the position of an elementary particle and its momentum. If Δx is the uncertainty in its position x and Δp is the uncertainty in its momentum p, then $\Delta x . \Delta p \geqslant h$, where h is the Planck constant. So $\Delta p \geqslant h/\Delta x$. This means that the smaller Δx, the uncertainty in x, the greater is Δp, the uncertainty in p. Energy–time and space–time measurements of elementary particles are also subject to the Heisenberg uncertainty principle.

EXERCISES 33D   Wave–Particle Duality

Multiple-Choice

1 Which of the statements **A** to **E** is correct?
   Electromagnetic radiation and electrons can each show both wave and particle properties because
   A Light can be refracted and diffracted.
   B X-rays can be reflected and diffracted by crystals.
   C Radio waves can be polarised and refracted.
   D Electrons have light mass and can be diffracted by thin metal films.
   E Electrons have light mass and move in a circular path in magnetic fields.

2 A beam of electrons X is accelerated from rest by a p.d. of 500 V and another beam of electrons is accelerated from rest by a p.d. of 2000 V. From the de Broglie relation, the ratio of their respective wavelengths $\lambda_X > \lambda_Y$ is

   A 1:2   **B** 1:4   **C** 2:1   **D** 4:1   **E** 8:1

Longer Questions

3 Calculate the wavelength associated with the following objects: (i) electron moving with velocity of 10^6 m s^{-1};   (ii) bullet of mass 0·01 kg with velocity 400 m s^{-1}; (iii) sprinter of mass 60 kg with velocity 10 m s^{-1}.

4 The atomic spacing in a thin metal film is of the order 10^{-10} m. For diffraction of an electron beam, what order or velocity is required?
   Calculate the accelerating voltage to produce this velocity.

5 A parallel beam of electrons moving with velocity v is incident normally on a thin graphite film of atomic spacing $d = 1·2 \times 10^{-10}$ m. The beam is diffracted through an angle θ of 11°, where $2d \sin \theta = \lambda$, the wavelength value.
   Calculate: (i) the wavelength;   (ii) the velocity v;   (iii) the accelerating voltage needed to produce this velocity.

6 Electrons are accelerated by a p.d. of: (i) 100 V;   (ii) 400 V. Calculate the wavelength associated with the electrons in each case.

7 When an electron beam accelerated on a p.d. V is incident normally on a thin metal film, several diffraction rings are seen on a luminous screen. If V increases, explain why the radii of the rings decrease and the brightness of the rings decreases.

8 An X-ray photon has a wavelength of $3·3 \times 10^{-10}$ m. Calculate the momentum, mass and energy of the particle associated with the photon, which moves with a velocity c.

9 Explain what is meant by:
   (a) excitation by collision;
   (b) ionisation by collision.
   Light or X-radiation may be emitted when fast-moving electrons collide with atoms. Using atomic theory, explain in each case how the radiation is emitted.
   Write down *two* differences and *one* similarity between optical emission spectra and X-ray emission spectra produced in this way.

Electrons are accelerated from rest through a p.d. of 5000 V. Calculate the wavelength of the associated electron waves. (Use $h = 6·6 \times 10^{-34}$ J s, $e = 1·6 \times 10^{-19}$ C, $m = 9·1 \times 10^{-31}$ kg.)

10 (a) When atoms absorb energy by colliding with moving electrons, light or X-radiation may subsequently be emitted. For each type of radiation, state typical values of the energy per atom which must be absorbed and explain in atomic terms how each type of radiation is emitted.

   (b) State *one* similarity and *two* differences between optical emission spectra and X-ray emission spectra produced in this way.

   (c) Electrons are accelerated from rest through a potential difference of 10 000 V in an X-ray tube. Calculate: (i) the resultant energy of the electrons in eV; (ii) the wavelength of the associated electron waves; (iii) the maximum energy and the minimum wavelength of the X-radiation generated. (Charge of electron $= 1·6 \times 10^{-19}$ C, mass of electron $= 9·11 \times 10^{-31}$ kg, Planck constant $= 6·62 \times 10^{-34}$ J s, speed of electromagnetic radiation *in vacuo* $= 3·00 \times 10$ m s^{-1}.) (*N.*)

34 Radioactivity, Nuclear Energy

In this chapter, we first discuss radioactive detectors and counters and the properties of α- and β-particles and γ-rays emitted by disintegration of the nucleus. We follow with half-life and its application in radioactive atoms. The discovery of the nucleus, proton and nucleon number, isotopes and the mass spectrometer are then discussed. The final part of the chapter describes nuclear energy and the Einstein mass–energy application, and the principles of nuclear fission and fusion and the nuclear reactor. The chapter ends with a brief account of elementary particles and a unified theory.

Plate 34A *Radiotherapy: the radiographer (standing) used a red laser beam to align the patient's head before treatment starts. For accurate delivery of the radiation beam, the patient's head is immobilised in a plastic face mask. Modern radiotherapy aims to apply a radiation dose that is sufficient to destroy a tumour, yet cause minimum damage to surrounding healthy tissue. This is achieved by training a beam of limited radiation dosage from different directions, over a number of sessions, to deliver a cumulative, effective dose.* (Photograph reproduced courtesy Tim Beddoe/Science Photo Library)

Radioactivity

In 1896 Becquerel found that a uranium compound affected a photographic plate wrapped in light-proof paper. He called the phenomenon *radioactivity*. We shall see later that natural radioactivity is due to one or more of three types of radiation emitted from heavy elements such as uranium whose nuclei are unstable. These were originally called α-, β- and γ-rays but α- and β-'rays' were soon shown to be particles.

α- and β-particles and γ-rays all produce ionisation as they move through a gas. On average, α-particles produce about 1000 times as many ions per unit length of their path as β-particles, which in turn produce about 100 times as many ions as γ-rays. There are numerous detectors of ionising radiations such as α- and β-particles and γ-rays. We begin by describing two detectors used in laboratories.

Geiger–Müller Tube

Geiger–Müller (GM) tubes are widely used for detecting ionising particles or radiation. In one form the GM tube contains a central thin wire A, the anode, insulated from a surrounding cylinder C, the cathode, which is metal or graphite-coated. The tube may have a very thin mica end window, Figure

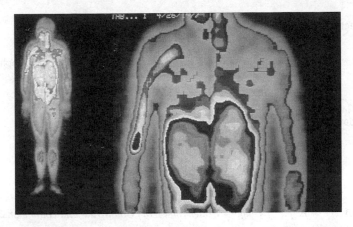

Plate 34B *Gamma camera scan showing the concentration of a radioactive tracer in the stomach. The image was recorded with the person resting, when the tracer uptake is stronger than in an active body. The radioactive tracer used, thallium-201, is often used to examine muscle tissue, in particular the cardiac (heart) muscle.* (Photograph reproduced courtesy CNRI/Science Photo Library)

Plate 34C *Using a Geiger counter to measure the natural radioactivity from uranium in granite rock at Edinburgh Castle, Scotland.* (Photograph reproduced courtesy UKAEA)

34.1 (i). A is kept at a positive potential V such as $+400$ V relative to C, which may be earthed.

When a single ionising particle enters the tube, a few electrons and ions are produced in the gas. If V is above the so-called *breakdown potential* of the gas the number of electrons and ions are multiplied enormously, as explained shortly. The electrons are attracted by and move towards A, and the positive ions move towards C. So a 'discharge' is suddenly obtained between A and C. The current flowing in the high resistance R produces a p.d. which is amplified and passed to a counter. In this way the counter registers the passage of an ionising particle or radiation through the tube.

Figure 34.1 (ii) shows how the count rate of a GM tube varies with the anode voltage V. Along BX, after ionisation by collision starts, not all particles entering the tube produce sufficient ionisation for a pulse voltage to record them. Along XY, the flat part (plateau) of the graph, however, *all* the particles produce a sufficiently high pulse voltage for counting, although their initial ionisation may be different. Beyond Y and YZ a continuous discharge may occur as an

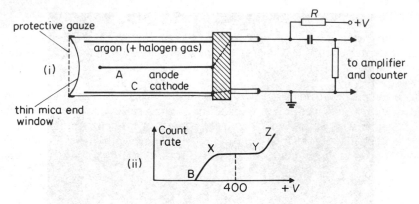

Figure 34.1 (*i*) *Principle of Geiger–Müller tube* (*ii*) *Count rate and voltage characteristic*

uncontrollable avalanche of ions and electrons are obtained. So the middle of XY (about 400 V) is the most suitable anode voltage *V*.

Quenching, Ionisation

The discharge persists for a short time, as secondary electrons are emitted from the cathode by the positive ions which arrive there. This would upset the recording of other ionising particles following fast on the first one recorded. The air in the tube is therefore replaced by argon mixed with a halogen vapour, whose molecules absorb the energy of the positive ions on collision. So the discharge is *quenched* quickly. Electrical methods are also used for quenching.

The anode wire A in the GM tube must be *thin*, so that the charge on it produces an intense electric field *E* close to its surface (*E* is inversely-proportional to the radius of the wire). An electron–ion pair, produced by an ionising particle or radiation, is then accelerated to high energies near the wire, thus producing, by collision with gas molecules, an 'avalanche' of more electron–ion pairs.

Solid State Detector

A *solid state detector*, Figure 34.2, is made from semiconductors. Basically, it has a p-n junction which is given a small bias in the non-conduction direction (p. 770).

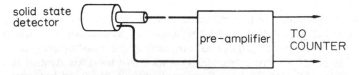

Figure 34.2 *Solid state detector*

When an energetic ionising particle such as an α-particle falls on the detector, more electron–hole pairs are created near the junction. These charge carriers move under the influence of the biasing potential and so a pulse of current is produced. The pulse is fed to an amplifier and the output passed to a counter.

The solid state detector is particularly useful for α-particle detection. If the amplifier is specially designed, β-particles and γ-rays of high energy may also be detected. This type of detector can then be used for all three types of radiation.

Dekatron Counter, Ratemeter

As we have seen, each ionising particle or radiation produces a pulse voltage in the external circuit of a Geiger–Müller or solid state detector. In order to measure the number of pulses from the detectors, some form of counter must be used.

A *dekatron counter* consists of two or more dekatron tubes, each containing a glow or discharge which can move round a circular scale graduated in numbers 0–9, together with a mechanical counter, Figure 34.3. Each impulse causes the glow in the first tube, which counts units, to move one digit. The circuit is designed so that on the tenth pulse, which returns the first counter to zero, a

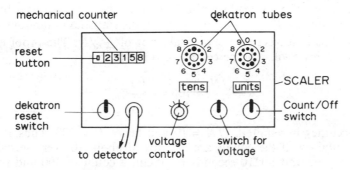

Figure 34.3 *Dekatron counter*

pulse is sent to the second tube. The glow here then moves on one place. The second tube thus counts the number of tens of pulses. After ten pulses are sent to the second tube, corresponding to a count of 100, the output pulse from the second tube is fed to the mechanical counter. This, therefore, registers the hundreds, thousands and so on. Dekatron tubes are used in radioactive experiments because they can respond to a rate of about 1000 counts per second. This is far above the count rate possible with a mechanical counter.

In contrast to a scaler, which counts the actual number of pulses, a *ratemeter* gives directly the average number of pulses per second or *count rate*. The principle is shown in Figure 34.4. The pulses received are passed to a capacitor C, which then stores the charge. C discharges slowly through a high resistor R and the average discharge current is recorded on a microammeter A. The greater the rate at which the pulses arrive, the greater will be the meter reading. The meter thus records a current which is proportional to the count rate.

A switch marked 'time constant' on most ratemeters allows the size of C to be chosen. If a large value of C is used, the capacitor will take a relatively long time to charge and correspondingly it will be a long time before the average

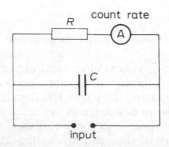

Figure 34.4 *Principle of a ratemeter*

count rate can be taken. The reading obtained, however, will be more accurate since the count rate is then averaged over a longer time (see below). For high accuracy, a small value of C may be used only if the count rate is very high.

Errors in Counting Experiments

Radioactive decay is random in nature (pp. 872–3). If the count rate is high, it is not necessary to wait so long before readings are obtained which vary relatively slightly from each other. If the count rate is low, successive counts, however, will have larger percentage differences from each other, unless a much longer counting time is used.

The accuracy of a count does not depend on the time involved but on *the total count obtained*. If N counts are received, the statistics of random processes show that this is subject to a statistical error of $\pm\sqrt{N}$. The proof is beyond the scope of this book. The percentage error is thus

$$\frac{\sqrt{N}}{N} \times 100 = \frac{100}{\sqrt{N}}\%$$

If 10% accuracy is required, $\sqrt{N} = 10$ and hence $N = 100$. Thus 100 counts must be obtained. If the counts are arriving at about 10 every second, it will be necessary to wait for 10 seconds to obtain a count of 100 and so achieve 10% accuracy. So a ratemeter circuit must be arranged with a time constant (CR) of 10 seconds, so that an average is obtained over this time. If, however, the counts are arriving at a rate of 1000 per second on average, it will be necessary to wait only 1/10 second to achieve 10% accuracy. So the 1-second time constant scale on the ratemeter will be more than adequate.

Detection of α-, β-particles and γ-rays

The existence of three different types of emission from radioactive substances can be shown by experiments such as those now given.

1. (a) When a radium-224 source S_1 is placed above a *spark counter*, which consists of a wire W close to an earthed metal base B, sparks are obtained between W and B, Figure 34.5 (i). If a sheet of thin paper is now introduced between S_1 and W, the sparks stop.

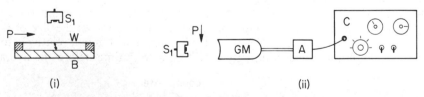

(i)   (ii)

Figure 34.5 *Detection by*   (i) *spark counter*   (ii) *GM tube*

(b) The radium source S_1 is now placed in front of a GM tube connected to a counter C (or ratemeter) through a low-noise pre-amplifier A, Figure 34.5 (ii). The count rate is measured and the paper P is then introduced between S_1 and the GM tube. Practically no difference in the count rate is observed.

We thus conclude that there is a type of radiation from S_1 which is detected by a spark counter but is absorbed by a thin sheet of paper—these are α-particles.

2. There are other types of radiation which do not affect a spark counter but affect a GM tube. We can show this by placing a strontium-90 source S_2 so that its radiation is shielded from the GM tube by a lead plate about 1 cm thick, Figure 34.6. The count rate is then extremely low. A strong magnetic field B, in a direction perpendicular to the radiation, is now introduced beyond the lead as shown. When the field B is directed into the paper, the count rate increases. So the magnetic field has now deflected the radiation towards the GM tube, which then detects it.

We conclude that there is a type of radiation which can be deflected by a strong magnetic field but is absorbed by lead 1 cm thick—these are β-particles.

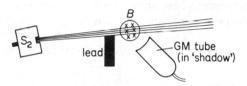

Figure 34.6 *Deflection of β-particles by magnetic field*

3. (a) We can repeat the experiment in Figure 34.6 with the radium source S_1 in place of S_2. This time the lead does not cut out all the radiation; some passes through it to the GM tube. But as before, there is an increase in the count rate when the magnetic field B is used as shown. So the radium source contains β-particles together with a third type of radiation which goes through 1 cm thickness of lead.

(b) Using a cobalt-60 source S directly in front of a GM tube as in Figure 34.5 (ii), the count rate is high. We can cut out any α-particles emitted by placing a thin sheet of paper between S and the GM tube. But using a fairly thick lead plate between S and the GM tube, the count rate continues to be high. Further, a magnetic field B between the lead and the GM tube makes no difference to the count rate. This type of radiation is called γ-rays.

Conclusion. There are at least three different types of emission from radioactive sources:

(1) A type recorded by a spark counter and cut off by thin paper—α-particles.

(2) A type recorded by a GM tube, cut out by lead, and deflected by a magnetic field—β-particles.

(3) A type recorded by a GM tube, not cut out except by very thick lead, and not affected by a magnetic field—γ-rays.

Further experiments show that the particles or radiation from most other radioactive sources fall into one of these three classes.

Alpha-particles

It is found that α-particles have a limited range in air at atmospheric pressure. This can be shown by slowly increasing the distance between a pure α-source and a detector. The count rate is observed to fall rapidly to zero at a separation greater than a particular value, which is called the 'range' of the α-particles. The range depends on the source and on the air pressure.

Using the apparatus of Figure 34.7, it can be shown that α-particles have *positive* charges. When there is no magnetic field, the solid state detector is placed so that the tube A is horizontal in order to get the greatest count. When the magnetic field is applied, the detector has to be moved *downwards* in order

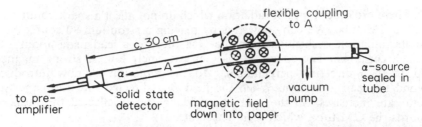

Figure 34.7 *Charge on an α-particle*

to get the greatest count. This shows that the α-particles are deflected by a small amount downwards. By applying Fleming's left-hand rule, we find that particles are *positively* charged. The vacuum pump is needed in the experiment, as the range of α-particles in air at normal pressures is too small.

Nature of α-particle

Lord Rutherford and his collaborators found by deflection experiments that an α-particle had a mass about four times that of a hydrogen atom, and carried a charge $+2e$, where e was the numerical value of the charge on an electron. The relative atomic mass of helium is about four. It was thus fairly certain that an *α-particle was a helium nucleus*, that is, a helium atom which has lost two electrons.

In 1909 Rutherford and Royds showed conclusively that α-particles were helium nuclei. Radon, a gas given off by radium which emits α-particles, was collected above mercury in a thin-walled tube P, Figure 34.8. After several days some of the α-particles passed through P into a surrounding vacuum Q. After

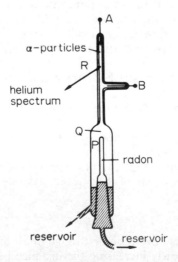

Figure 34.8 *Rutherford and Royd's experiment on α-particles*

about a week, the space in Q was reduced in volume by raising mercury reservoirs, and a gas was collected in a capillary tube R at the top of Q. A high voltage was then connected to electrodes at A and B. The spectrum of the discharge was observed to be exactly the same as the characteristic spectrum of helium.

Beta-particles and Gamma-rays

By deflecting β-particles with perpendicular magnetic and electric fields, their charge–mass ratio could be estimated. This is similar to Thomson's experiment, p. 751. These experiments showed that *β-particles are electrons moving at high speeds*. Generally, β-particles have a greater penetrating power of materials than α-particles. They also have a greater range in air than α-particles, since their ionisation of air is relatively smaller, but their path is not so well defined.

Using a strong bar magnet, it can be shown that β-particles are strongly deflected by a magnetic field. The direction of the deflection coresponds to a stream of *negatively*-charged particles, that is, opposite to the deflection of α-particles in the same field. This is consistent with the idea that β-particles are usually fast-moving electrons.

The nature of γ-rays was shown by experiments with crystals. Diffraction phenomena are obtained in this case, which suggest that *γ-rays are electromagnetic waves* (compare X-rays, p. 851). Measurement of their wavelengths, by special techniques with crystals, show they are shorter than the wavelengths of X-rays and of the order 10^{-11} m or less. γ-rays can penetrate large thicknesses of metals, but they have far less ionising power in gases than β-particles.

If a beam of γ-rays is allowed to pass through a very strong magnetic field no deflection is observed. This is consistent with the fact that γ-rays are electromagnetic waves and carry no charge.

The mass of an α-particle, a helium particle, is about 8000 times the mass of a β-particle, an electron. The α-particle is therefore massive compared with the β-particle—the ratio of their masses is similar to that of the mass of a small elephant compared to the mass of a new-born baby. The mass of the γ-ray is negligibly small compared with the α- and β-particles.

Inverse-square Law for γ-rays

If γ-rays are a form of electromagnetic radiation and undergo negligible absorption in air, their intensity I should vary inversely as the square of the

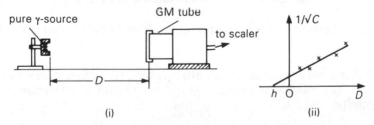

Figure 34.9 *Inverse square law for γ-rays*

distance between the source and the detector. The apparatus shown in Figure 34.9 (i) can be used to investigate if this is the case. A suitable γ-source is placed at a suitable distance from a GM tube connected to a scalar, and the intensity inside the tube will then be proportional to the count rate C.

Suppose D is the measured distance from a fixed point on the γ-source support to the front of the GM tube. To obtain the true distance from the source to the region of gas inside the tube where ionisation occurs, we need to add an unknown but constant distance h to D. Then, assuming an inverse-square law, $I \propto 1/(D + h)^2$. Thus

$$D + h \propto \frac{1}{\sqrt{I}} \propto \frac{1}{\sqrt{C}}$$

A graph of $1/\sqrt{C}$ is therefore plotted against D for varying values of D, Figure 34.9 (ii). If the inverse-square law is true, a straight line graph is obtained which has an intercept on the D-axis of $-h$. Note that if I is plotted against $1/D^2$ and h is not zero, a straight line graph is *not* obtained from the relation $I \propto 1/(D + h)^2$. Consequently we need to plot $1/\sqrt{I}$ against D.

If a pure β-source is substituted for the γ-source and the experiment is repeated, a straight-line graph is not obtained. The absorption of β-particles in air is thus appreciable compared with γ-rays.

Absorption of Radiation by Metals

A GM tube and counter or ratemeter can be used to investigate the absorption of γ-rays or β-particles by metals, using equal thicknesses of thin sheets of lead for γ-rays and of aluminium sheets for β-particles.

Figure 34.10 (i) shows roughly the results obtained in either case; the count rate C decreases with n, the number of sheets, along a curve. In the case of the β-particles, however, the count rate C reaches an almost constant low value (not

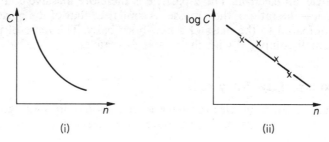

(i)   (ii)

Figure 34.10 *Absorption of radiation by metal plates*

shown) as n becomes bigger. The log C result in Figure 34.10 (ii) shows that

$$I = I_0\, e^{-\mu t}$$

where I is the intensity of the beam after passing through a thickness t of the metal, I_0 is the incident intensity, and μ is a constant which depends on the metal density.

Half-life and Decay Constant

Rate of disintegration

We now consider the laws governing the rate of disintegration of radioactive atoms. These laws lead to the identification of atoms and to useful applications.

Radioactivity, or the emission of α- or β-particles and γ-rays, is due to disintegrating nuclei of atoms (p. 885). The disintegration obey the *statistical law of chance*. Thus although we cannot tell which particular atom is likely to disintegrate next, the number of atoms disintegrating per second, dN/dt, is directly proportional to the number of atoms, N, present at that instant.

A statistical analogy is to say that with many cars on the roads on a national holiday, more accidents per hour will occur in proportion to the number of cars then on the road. So

$$\frac{dN}{dt} = -\lambda N \tag{1}$$

where λ is a constant characteristic of the atom concerned called the *radioactivity decay constant*. The negative sign indicates that N becomes *smaller* when t increases. So if N_0 is the number of radioactive atoms present at a time $t = 0$, and N is the number at the end of a time y, we have, by integration,

$$\int_{N_0}^{N} \frac{dN}{N} = -\lambda \int_{0}^{N} dt$$

$$\therefore \left[\ln N \right]_{N_0}^{N} = -\lambda t$$

$$\therefore N = N_0\, e^{-\lambda t} \qquad . \qquad . \qquad . \qquad . \qquad . \tag{2}$$

So the number N of undecayed atoms left decreases exponentially with the time t, and this is illustrated in Figure 34.11.

From (1), $\lambda = -\dfrac{1}{N}\dfrac{dN}{dt} = \dfrac{\text{no. of disintegrating atoms per second}}{\text{total no. of atoms}}$

So λ is the fraction per second of the decaying atoms or the *probability* of an atom decaying in the next second, though we can not know which atom in the group will disintegrate.

Since the rate of disintegration, or *activity A*, is proportional to the number N of disintegrating atoms, and $N = N_0\, e^{-\lambda t}$, then $A = A_0\, e^{-\lambda t}$, where A_0 is the initial rate of disintegration, or activity, of the atoms. So if the rate of disintegration is 8000 per second in a small sample of a radioactive element and the number of disintegrating atoms is 200 per second in a time t later, then $200 = 8000\, e^{-\lambda t}$, from $A = A_0\, e^{-\lambda t}$.

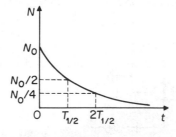

Figure 34.11 *Radioactive decay with time*

Half-life

The *half-life* $T_{1/2}$ of a radioactive element is defined as the time taken for the atoms to disintegrate to *half their initial number* (see Figure 34.11), that is, in a time $T_{1/2}$ the radioactivity of the element diminishes to half its value. So if

N_0 is the initial number of atoms, from (2),

$$\frac{N_0}{2} = N_0\,e^{-\lambda T_{1/2}}$$

$$\tfrac{1}{2} = 2^{-1} = e^{-\lambda T_{1/2}}$$

Taking logs to the base e on both sides of the equation,

$$-1\ln 2 = -\lambda T_{1/2}$$

$$\therefore\ T_{1/2} = \frac{1}{\lambda}\ln 2 = \frac{0\cdot693}{\lambda} \quad . \qquad . \qquad . \qquad . \qquad (3)$$

The half-life varies considerably for different radioactive atoms. For example, uranium has a half-life of the order of 4500 million years, radium has one about 1600 years, polonium about 138 days, radioactive lead about 27 minutes and radon about 1 minute.

The half-life is the same even if we start with different numbers of disintegrating atoms. So 8×10^6 atoms decay to 4×10^6 atoms in the half-life time $T_{1/2}$ and these decay to 2×10^6 atoms, half the number, in the same time $T_{1/2}$. A decay from 8×10^6 to 1×10^6 atoms takes place in a time $3T_{1/2}$ (note that $1/8 = 1/2^3$). A decay from N_0 atoms to $N_0/32$ takes place in a time $5T_{1/2}$, since $1/32 = 1/2^5$.

The rate of nuclear disintegration or decay of an element, also known as its *activity*, is now measured in the SI unit of the *becquerel*, symbol Bq. A rate of nuclear disintegration of 5000 per second is 5 kBq.

Measurement of Short and Long Half-life

The half-life of radon-220, a radioactive gas with a short half-life, can be measured by the apparatus shown in Figure 34.12 (i). C is a metal can or *ionisation chamber* containing a metal rod P insulated from C. B is a suitable d.c. supply such as 100 V connected between P and C, with a d.c. amplifier M joined in series as shown. This type of meter can measure very small currents.

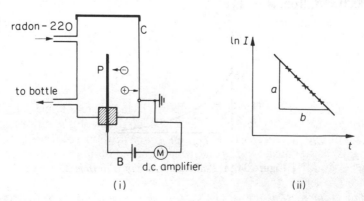

Figure 34.12 *Measurement of half-life*

A plastic bottle (not shown) containing the gas radon-220 is connected to the chamber C. The radioactive gas is passed into C by squeezing the bottle and when the atoms decay the α-particles produced ionise the air in C. The negative and positive ions produced move between P and C due to the supply voltage B and so a small current or ionisation current is registered on the meter M.

Starting with an appreciable deflection in M, the falling current I is noted at equal intervals of time t such as 15 seconds. A graph of $\ln I$ against t is then plotted from the results and a straight line AB is drawn through the points, Figure 34.12 (ii). Now $I = I_0\,e^{-\lambda t}$, since the number of disintegrations per second is proportional to the number of ions produced per second and hence to the current I. Taking logs to the base e,

$$\ln I = \ln I_0 - \lambda t$$

So the gradient a/b of the straight-line AB is λ numerically and this can now be found. The half-life $T_{1/2}$ is given by $T_{1/2} = 0.693/\lambda$ and so it can be calculated.

For a substance with a *long half-life* such as days, weeks or years, we can proceed as follows. Firstly, weigh a small mass of the specimen S, m say. If the relative molecular mass of S is M and S is monatomic, the number of atoms N in a mass is given by

$$N = \frac{m}{M} \times 6.03 \times 10^{23}$$

using the Avogadro constant.

Secondly, determine the rate of emission, dN/dt, all round the specimen S by placing S at a distance r from the *end face*, area A, of a GM tube connected to a scaler. Suppose the measured count rate through the area A is dN'/dt. Then the count rate all round S, in a sphere of area $4\pi r^2$, is given by

$$\frac{dN}{dt} = \frac{4\pi r^2}{A} \times \frac{dN'}{dt}$$

Since $dN/dt = -\lambda N$, the decay constant λ can be calculated as N and dN/dt are now known. The half-life is then obtained from $T_{1/2} = 0.693/\lambda$.

Carbon Dating

Carbon has a radioactive isotope ^{14}C. It is formed when neutrons react with nitrogen in the air, as below. The neutrons are produced by cosmic rays which enter the upper atmosphere and interact with air molecules.

$$^{14}_{7}N + ^{1}_{0}N \rightarrow ^{14}_{6}C + ^{1}_{1}H$$

The radioactive isotope ^{14}C is absorbed by living material such as plants or vegetation in the form of carbon dioxide. The amount of the isotope absorbed is very small and it reaches a maximum concentration so long as the plants are living. Experiment shows that the activity of the radioactive isotope in living materials is about 19 counts per minute per gram. The half-life of the isotope is about 5600 years.

When a plant dies, no more of the isotope is absorbed. Wood formed from dead or decaying plants or vegetation, which were alive thousands of years ago, containing the radioactive isotope but its activity is much less owing to decay of the atoms. Measuring the activity of the isotope in ancient wood or similar carbon materials can provide information about its age, a method known as *carbon dating*. For example, suppose the measured activity I in a piece of ancient wood is 14 counts per minute per gram. Then, from $I = I_0 e^{-\lambda t}$,

$$\frac{I}{I_0} = \frac{14}{19} = e^{-\lambda t}$$

Taking logs to the base e on both sides we obtain

$$-0.305 = -\lambda t$$

So
$$t = \frac{0.305}{\lambda}$$

But
$$\lambda = \frac{0.693}{T_{1/2}} = \frac{0.693}{5600 \text{ y}}$$

Hence
$$t = \frac{0.305 \times 5600 \text{ y}}{0.693} = 2465 \text{ y}$$

So the ancient wood is about 2500 y. By carbon dating the ancient material, Stonehenge in England was found to have a date of about 4000 years.

The following examples will show how to apply half-life values to calculate the number of disintegrations per second or rate of decay (activity) after a given time.

You should know:

1 **Rate of decay $(dN/dt) = -\lambda N$ (N = number of disintegrating atoms)**

2 $$N = N_0 e^{-\lambda t} \quad \text{or} \quad I = I_0 e^{-\lambda t}$$

3 $$T_{1/2} = \frac{0.693}{\lambda}$$

Examples on Half-life

1   A point source of γ radiation has a half-life of 30 minutes. The initial count rate, recorded by a Geiger counter placed 2·0 m from the source, is 360 s^{-1}. The distance between the counter and the source is altered. After 1·5 hours the count rate recorded is 5 s^{-1}. What is the new distance between the counter and the source? (*L.*)

$1.5 \text{ h} = 3 \times 30 \text{ min} = 3 \times$ half-life of source

So at *beginning* of 1·5 h, count rate $= 2 \times 2 \times 2 \times 5 \text{ s}^{-1} = 40 \text{ s}^{-1}$

$$= \frac{1}{9} \times 360 \text{ s}^{-1} = \frac{1}{9} \times \text{ initial rate}$$

$$\therefore \text{ intensity of radiation} = \frac{1}{9} \times \text{ initial intensity at 2·0 m}$$

But
$$\text{intensity} \propto \frac{1}{r^2}, \text{ where } r \text{ is the distance}$$

$$\therefore \text{ new distance} = 3 \times \text{ initial distance} = 6.0 \text{ m}$$

2 Lanthanum has a stable isotope ^{139}La and a radioactive isotope ^{138}La of half-life $1 \cdot 1 \times 10^{10}$ years whose atoms are $0 \cdot 1\%$ of those of the stable isotope. Estimate the rate of decay or activity of ^{138}La with 1 kg of ^{139}La. (Assume the Avogadro constant $= 6 \times 10^{23}$ mol^{-1}.)

$$\text{Decay rate } \frac{dN}{dt} = -\lambda N \qquad . \qquad . \qquad . \qquad . \qquad (1)$$

where λ is the decay constant and N is the number of atoms in ^{138}La.

Now number of atoms in 1 kg (1000 g) of ^{139}La $= \dfrac{6 \times 10^{23} \times 1000}{139}$

Since $0 \cdot 1\% = 10^{-3}$, then

$$\text{number of atoms in } {}^{138}\text{La, } N = \frac{10^{-3} \times 6 \times 10^{23} \times 1000}{139} = \frac{6 \times 10^{23}}{139}$$

Also
$$\lambda = \frac{0 \cdot 693}{T_{1/2}} = \frac{0 \cdot 693}{1 \cdot 1 \times 10^{10} \times 365 \times 24 \times 3600}$$

From (1),
$$\frac{dN}{dt} = \frac{0 \cdot 693 \times 6 \times 10^{23}}{1 \cdot 1 \times 10^{10} \times 365 \times 24 \times 3600 \times 139}$$
$$= 8600 \text{ s}^{-1}$$

3 At a certain instant, a piece of radioactive material contains 10^{12} atoms. The half-life of the material is 30 days.

(i) Calculate the number of disintegrations in the first second. (ii) What time will elapse before 10^4 atoms remain? (iii) What is the count rate at this time?

(i) We have $N = N_0 \, e^{-\lambda t}$

$$\therefore \frac{dN}{dt} = -N_0 \lambda \, e^{-\lambda t} = -\lambda N$$

Hence, when $N = 10^{12}$, $\dfrac{dN}{dt} = -\lambda 10^{12}$

Now
$$\lambda = \frac{0 \cdot 693}{T_{1/2}} = \frac{0 \cdot 693}{30 \times 24 \times 60 \times 60 \times} \text{ s}^{-1}$$

$\therefore$ number of disintegrations per second

$$= \frac{10^{12} \times 0 \cdot 693}{30 \times 24 \times 60 \times 60} = 2 \cdot 7 \times 10^5$$

(ii) When $N = 10^4$, we have

$$10^4 = 10^{12} \, e^{-\lambda t}, \quad \text{so } 10^{-8} = e^{-\lambda t}$$

Taking logs to base 10,

$$\therefore -8 = -\lambda t \log e$$

$$\therefore t = \frac{8}{\lambda} \frac{1}{\log e} = \frac{8T}{0 \cdot 693 \log e}$$

$$= 797 \text{ days (approx.)}$$

(iii) Since
$$\frac{dN}{dt} = -\lambda N$$

$\therefore$ number of disintegrations per hour $= \dfrac{0 \cdot 693}{30 \times 24} \times 10^4 = 9 \cdot 6$

Cloud Chambers

We conclude with a brief account of the *cloud chamber*, which can be used for detecting or photographing ionising particles and radiation. C. T. R. Wilson's cloud chamber, invented in 1911, was one of the most useful early inventions for studying radioactivity. It enabled photographs to be taken of α- or β-particles or γ-rays.

Basically, Wilson's cloud chamber consists of a chamber containing saturated water or alcohol vapour. Figure 34.13 (i) illustrates the basic principle of a cloud chamber C which uses alcohol vapour. An excess amount of liquid alcohol is placed on a dark pad D on a piston P. When the piston is moved down quickly, the air in C undergoes an adiabatic expansion and cools. The dust nuclei are all carried away after a few expansions by drops forming on them, and then the dust-free air in C is subjected to a controlled adiabatic expansion of about 1·31 to 1·38 times its original volume. The air is now supersaturated, that is, the vapour pressure is greater than the saturation vapour pressure at the reduced temperature reached but no vapour condenses. Simultaneously, the air is exposed to α-particles from a radioactive source S, for example. Water droplets immediately collect round the ions produced, which act as centres of formation. The drops are illuminated, and photographed by light scattered from them.

α-particles produce *short continuous straight trails*, as shown at the top of Figure 34.13. β-particles, which have much less mass, produce longer but *straggly paths* owing to collisions with gas molecules. Wilson's cloud chamber has proved of great value in the study of radioactivity and nuclear structure, see Figure 34.18.

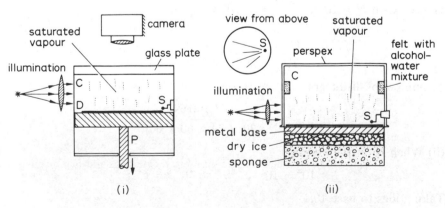

Figure 34.13 *Principle of* (i) *Wilson cloud chamber* (ii) *diffusion cloud chamber*

Diffusion cloud chamber. Figure 34.13 (ii) shows the principle of another type of cloud chamber. It has a Perspex chamber C, with a strip of felt at the top containing excess of a mixture of water and alcohol. The dark metal base is kept at a low temperature of about $-50°C$ by dry ice (solid carbon dioxide) packed below it. Vapour thus diffuses continuously from the top to the bottom of the chamber.

Above the cold metal base there is supersaturated vapour. A radioactive source S near the base, emitting α-particles, for example, produces vapour trails which can be seen on looking down through the top of C. These trails show the straight line paths of the emitted α-particles, see Figure 34.13. Unlike the Wilson cloud chamber, this type does not require adiabatic expansion of the gas inside. In both cases, the ions formed can be cleared by applying a suitable p.d. between the top and bottom of the chamber.

The random nature of radioactive decay can be seen by using this cloud chamber. Particles emitted by a radioactive substance do not appear at equal intervals of time but are entirely random. The length of the track of an emitted particle is a measure of its initial energy. The tracks of α-particles are nearly all the same, showing that the α-particles were all emitted with the same energy. Sometimes two different lengths of tracks are obtained, showing that the α-particles may have one of two energies on emission.

Safety in Experiments

Care must be taken with radioactive sources because radiation can damage cells in the body by ionisation produced. So (a) sources are always kept in lead boxes when they are not used and stored in a safe lead-lined box, (b) the hands and body must be kept as far as possible from radioactive sources. Tongs and rubber gloves are used in experiments with weak sources used in demonstrations, (c) screens of thick Perspex or glass help to shield β-particles.

EXERCISES 34A   Half-life, Activity

Multiple Choice

Students should note that knowledge of topics later in the chapter is also required in order to answer these questions.

1 A piece of a radioactive element has initially 8.0×10^{22} atoms. The half-life is 2 days. After 16 days, the number of atoms in the specimen is

  A 4.0×10^{22} **B** 2.5×10^{22} **C** 7.5×10^{21} **D** 5.0×10^{21} **E** 2.5×10^{20}

2 Which is a correct statement in **A** to **E**?
  The half-life of radon gas is about 60 s. This can be changed by
  A mixing with another radioactive gas
  B heating the radon
  C increasing the gas pressure
  D freezing the radon
  E none of these methods.

3 A nuclide $^{220}_{86}R$ decays to a new nuclide S by two α-emissions and two β-emissions. The nuclide S is

  A $^{218}_{84}S$ **B** $^{216}_{84}S$ **C** $^{212}_{82}S$ **D** $^{216}_{82}S$ **E** $^{212}_{84}S$

4 The table shows samples P, Q, R of different nuclides found in nuclear waste.

	no. of atoms	half-life
P	1×10^{20}	1 min
Q	3×10^{20}	24 days
R	8×10^{20}	1 year

Which of the three statements are correct?
1 P has the greatest decay constant.
2 Q has the greatest activity.
3 R eventually would be the most dangerous.

A 1, 2, 3 all correct **B** 1, 2 only **C** 1, 3 only
D 1 only   **E** 3 only

5 An α-particle incident on a nuclide X produces a nuclide Y and two β-particles. The reaction is represented by

$$^{88}_{37}X + ^{4}_{2}He \rightarrow ^{x}_{y}Y + 2\beta_{-}$$

Which is correct in **A** to **E**?
A $x = 90$, $y = 37$ **B** $x = 92$, $y = 35$ **C** $x = 90$, $y = 38$
D $x = 92$, $y = 40$ **E** $x = 92$, $y = 41$

Longer Questions

6 What is meant by the *half-life* of a radioactive element? Draw a labelled sketch of the relation $N = N_0 e^{-\lambda t}$ to illustrate your answer.
   The initial number of atoms in a radioactive element is 6.0×10^{20} and its half-life is 10 h. Calculate:
   (a) the number of atoms which have decayed in 30 h;
   (b) the amount of energy liberated if the energy liberated per atom decay is 4.0×10^{-13} J.
7 The radioactive isotope of iodine ^{131}I has a half life of 8·0 days and is used as a tracer in medicine. Calculate: (a) the number of atoms of ^{131}I which must be present in the patient when she is tested to give a disintegration rate of $6.0 \times 10^5 \text{ s}^{-1}$; (b) the number of atoms of ^{131}I which must be present in a dose prepared 24 hours before. (*N.*)
8 A point source of alpha particles, a tiny mass of the nuclide $^{241}_{95}Am$, is mounted 7·0 cm in front of a GM tube whose mica window has a receiving area of 3.0 cm^2,

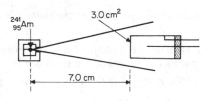

Figure 34A

Figure 34A. The counter linked to the GM tube records 5.4×10^4 counts per minute. Calculate:
(a) the number of disintegrations per second within the source;
(b) the number of $^{241}_{95}Am$ atoms in the source.
(The decay constant, λ, for $^{241}_{95}Am = 4.80 \times 10^{-11} \text{ s}^{-1}$.) (*L.*)
9 Explain what is meant by the *half-life* of a radioactive nuclide.
   Write down the relationship between the half-life and the decay constant.
   The graph in Figure 34B shows how $\ln A_d$ (where A_d is the activity per minute) changes with time, t, during the decay of a radioactive nuclide. Use the graph to find:
   (a) the initial activity;
   (b) the decay constant for the nuclide. (*L.*)
10 $^{32}_{15}P$ is a beta-emitter with a decay constant of $5.6 \times 10^{-7} \text{ s}^{-1}$. For a particular application the initial rate of disintegration must yield 4.0×10^7 beta particles every second. What mass of pure $^{32}_{15}P$ will give this decay rate? (*C.*)
11 A mixture of I^{131} (half-life 8 days) and I^{132} (half-life 2·3 hr) has an activity

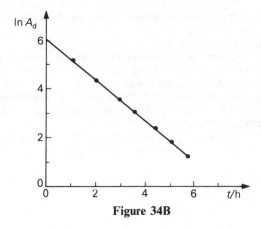

Figure 34B

of 5 kBq. After 12 days the activity is 1 kBq. What proportion of the 5 kBq was due to I^{132}?

(The activity of the products of disintegration may be neglected.) (*W.*)

12 What do you understand by *radioactivity* and *half-life*? Plot an accurate graph to show how the number of radioactive atoms of a given element (expressed as a percentage of those initially present) varies with time. Use a time scale extending over five half-lives.

The isotope $^{40}_{19}K$, with a half-life of 1.37×10^9 years, decays to $^{40}_{18}Ar$, which is stable. Moon rocks from the Sea of Tranquillity show that the ratio of these potassium atoms to argon atoms is 1/7. Estimate the age of these rocks, stating clearly any assumptions you make. Certain other rocks give a value of 1/4 for this ratio. By means of your graph, or otherwise, estimate their age. (*C.*)

13 Discuss the assumption on which the law of radioactive decay is based.

What is meant by the *half-life* of a radioactive substance?

A small volume of a solution which contained a radioactive isotope of sodium had an activity of 12 000 disintegrations per minute when it was injected into the bloodstream of a patient. After 39 hours the activity of 1.0 cm^3 of the blood was found to be 0.50 disintegrations per minute. If the half-life of the sodium isotope is taken as 15 hours, estimate the volume of blood in the patient. (*N.*)

14 List the chief properties of α-radiation, β radiation and γ-radiation. Describe a type of Geiger–Müller tube that can be used to detect all three types of radiation. If you were supplied with a radioactive preparation that was emitting all three types of radiation, describe and explain how you would use the tube to confirm that each of them was present.

A radon ($^{222}_{86}Rn$) nucleus, of mass 3.6×10^{-25} kg, decays by the emission of an α-particle of mass 6.7×10^{-27} kg and energy 8.8×10^{-13} J.

(a) Write down the values of the mass number A and the atomic number Z for the resulting nucleus.

(b) Calculate the momentum of the emitted α-particle.

(c) Find the velocity of recoil of the resulting nucleus. (*O.*)

15 What is meant by the *half-life period (half-life)* of a radioactive material?

Describe how the nature of α-particles has been established experimentally.

The half-life period of the body polonium-210 is about 140 days. During this period the average number of α-emissions per day from a mass of polonium initially equal to 1 microgram is about 12×10^{12}. Assuming that one emission takes place per atom and that the approximate density of polonium is 10 g cm^{-3}, estimate the number of atoms in 1 cm^3 of polonium. (*N.*)

16 (a) Describe how you would perform an experiment to investigate the inverse square law for γ rays. Assume that you are supplied with a suitable γ ray source of long half-life, a Geiger–Müller tube and the normal apparatus of a physics laboratory.

State how you would plot your results to obtain a straight line graph.

(b) A sample initially contains N_0 radioactive atoms of a single isotope of *radioactive decay constant* λ.

   (i) Explain the term in italics.

   (ii) Give an equation for N, the number of these radioactive atoms present after time t. Sketch a graph of N against t.

   (iii) State an equation relating λ and $T_{1/2}$.

(c) Living wood contains radioactive ^{14}C which is in equilibrium with ^{14}C in the atmosphere. $1 \cdot 00$ kg of carbon from living wood gives an average number of disintegrations of 255 per second.

   A sample of $2 \cdot 00 \times 10^{-2}$ kg of carbon from the wood of an ancient boat gives an average number of disintegrations of 270 per minute. The half-life of ^{14}C is 5570 years.

   Calculate the age of the boat in years. State any assumptions made in your calculation. (*N*.)

The Nucleus, Nuclear Energy

Discovery of Nucleus, Geiger–Marsden Experiment

In 1909 Geiger and Marsden, at Lord Rutherford's suggestion, investigated the scattering of α-particles by thin films of metal of high atomic mass, such as gold foil. The used a radon tube S in a metal block as a source of α-particles, and limited the particles to a narrow pencil, Figure 34.14. The thin metal foil A was placed in the centre of an evacuated vessel, and the scattering of the particles after passing through A was observed on a fluorescent screen B, placed at the focal plane of a microscope M. Scintillations were seen on B whenever it was struck by α-particles.

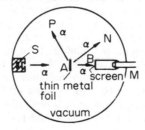

Figure 34.14 *Discovery of nucleus—Geiger and Marsden*

Geiger and Marsden found that α-particles struck B not only in the direction SA, but also when the microscope M was moved round to B and even to P. So though the majority of α-particles were scattered through small angles, some particles were scattered through very large angles. Rutherford found this very exciting news. It meant that some α-particles had come into the repulsive field of a highly concentrated positive charge at the heart or centre of the atom.

Paths of Scattered Particles

Rutherford assumed that an atom has a *nucleus*, in which all the positive charge and most of the mass is concentrated. The beam of α-particles incident on a thin metal foil are then scattered through various angles, as shown roughly in Figure 34.15. Those particles very close to the nucleus are deflected through a large angle, since the repulsive force is then very big.

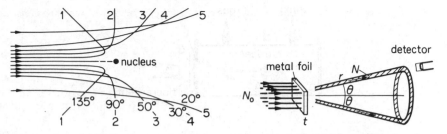

Figure 34.15 *Thin foil scattering of α-particles*   **Figure 34.16** *Rutherford scattering law*

Rutherford obtained a formula for the number N of α-particles scattered through an angle θ by thin metal foil of thickness t and detected at a distance r. In a series of experiments using a detector as illustrated diagrammatically in Figure 34.16, Geiger and Marsden verified the Rutherford formula and confirmed the existence of the nucleus.

Atomic Nucleus and Atomic (Proton) Number

In 1911 Rutherford proposed the basic structure of the atom which is accepted today, and which subsequent experiments by Moseley and others have confirmed. A neutral atom consists of a very tiny nucleus of diameter about 10^{-15} m which contains practically the whole mass of the atom. The atom is largely empty. If a drop of water was magnified until it reached the size of the earth, the atoms inside would then be only a few metres in diameter and the atomic nucleus would have a diameter of only about 10^{-2} millimetre.

The nucleus of hydrogen is called a *proton*, and it carries a charge of $+e$, where e is the numerical value of the charge on an electron. The helium nucleus has a charge of $+2e$. The nucleus of copper has a charge of $+29e$, and the uranium nucleus carries a charge of $+92e$. Generally, the positive charge on a nucleus is $+Ze$, where Z is the *atomic number* of the element and is defined as the number of protons in the nucleus (see also pp. 854–5). Under the attractive influence of the positively-charged nucleus, a number of electrons equal to the atomic number move round the nucleus and surround it like a negatively-charged cloud. These are called 'extra-nuclear' electrons, or electrons outside the nucleus.

Discovery of Protons in Nucleus, Mass (Nucleon) Number

In 1919 Rutherford found that energetic α-particles could penetrate nitrogen atoms and that *protons* were thrown out after the collision. The apparatus used is shown in Figure 34.17. A source of α-particles, A, was placed in a container D from which all the air had been pumped out and replaced by nitrogen. Silver foil, B, sufficiently thick to stop α-particles, was then placed between A and a fluorescent screen C, and scintillations were observed by a microscope M. The particles which have passed through B were shown to have a similar range, and the same charge, as protons. Plate 34D shows a photograph of a nuclear collision in a classic cloud chamber experiment.

Protons were also obtained with the gas fluorine, and with other elements such as the metals sodium and aluminium. It thus becomes clear that *the nucleus of all elements contain protons.*

The number of protons must equal the number of electrons surrounding the nucleus, so that each number is equal to the atomic number, Z, of the element.

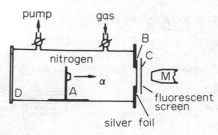

Figure 34.17 *Discovery of protons in the nucleus—Rutherford*

Plate 34D *The tracks of α-particle travelling through nitrogen. One α-particle has collided with a nitrogen nucleus—the tracks of the α-particle and the nitrogen nucleus after the collision are clearly visible*

A proton is represented by the symbol, $_1^1H$; the top number denotes the *mass number* or *nucleon number A* (a nucleon is a particle in the nucleus) and the bottom number is the *atomic* or proton number Z. The helium nucleus such as an α-particle is represented by $_2^4He$; its mass or nucleon number A is 4 and its proton number Z is 2.

One of the heaviest nuclei, uranium, can be represented by $_{92}^{238}U$; it has a nucleon number A of 238 and a proton number Z of 92.

Discovery of Neutron in Nucleus

In 1932, Chadwick found a new particle inside a nucleus, in addition to the proton. It had about the same mass as the proton but carried no charge and so Chadwick called the new particle a *neutron*. It is now considered that *all nuclei contain protons and neutrons*. The neutron is represented by the symbol $_0^1n$ as it has a mass number of 1 and zero charge.

We can now see that a helium nucleus, $_2^4He$, has 2 protons and 2 neutrons, a total mass or nucleon number of 4. The sodium nucleus, $_{11}^{23}Na$, has 11 protons and 12 neutrons. The uranium nucleus, $_{92}^{238}U$, has 92 protons and 146 neutrons. Generally, a nucleus represented by $_Z^AX$ has Z protons and $(A - Z)$ neutrons.

The nucleus has protons and neutrons. The *proton number Z* shows the chemical nature of the atom as this is equal to the number of electrons round the nucleus which take part in chemical reactions. The *nucleon number A* (protons + neutrons) shows the mass of the atom. A proton (hydrogen nucleus) has practically the same mass as a neutron.

Radioactive Disintegration, Uranium Series

Naturally occurring radioactive elements such as uranium, actinium and thorium disintegrate to form new elements, and these in turn are unstable and form other elements. Between 1902 and 1909 Rutherford and Soddy made a study of the elements formed from a particular 'parent' element. The *uranium series* is listed in the table on page 886.

Nuclear Reactions, Conservation of Mass and Charge

The new element formed after disintegration can be identified by considering the particles emitted from the nucleus of the parent atom. An α-particle, a helium nucleus, has a charge of $+2e$ and a mass number 4. Uranium I, of atomic number 92 and mass number 238, emits an α-particle from its nucleus of charge $+92e$, and so the new nucleus formed has an atomic number 90 and a mass number 234. This was called uranium X_1, and since the element thorium (Th) has an atomic number 90, uranium X_1 is actually thorium.

We write this *nuclear reaction* as:

$$^{238}_{92}U \rightarrow\ ^4_2He +\ ^{234}_{90}Th$$

The top numbers of the nuclei are their mass or nucleon numbers. The bottom numbers are their atomic or proton numbers, which also represent the charges on the nuclei in terms of the numerical value e of the charge on an electron. In any nuclear reaction, we always apply the conservation of mass and of charge. So

---

1 **Total mass (nucleon) number is constant before and after the reaction**
2 **Total charge (proton) number is constant before and after the reaction**

We can apply these rules to β-emission from a radioactive nucleus. A β-particle is usually an electron of negligible mass and charge $-e$ or -1 in terms of e. On rare occasions, however, a β-particle is emitted which has a positive charge $+e$ or $+1$ in terms of e. This particle is called a *positron* and denoted by β^+.

Let us assume that a β-particle and a γ-ray are emitted by thorium Th which has a nucleon number 234 and a proton number 90. Then we write the reaction as:

$$^{234}_{90}Th \rightarrow\ ^0_{-1}e +\ ^{234}_{91}Pa$$

Note that the proton number has *increased* from 90(Th) to 91(Pa). As explained later, if a nucleus has too many neutrons for stability, a neutron changes to a proton inside the nucleus and a β-particle is then emitted.

The uranium series contains *isotopes* of uranium (U), lead (Pb), thorium (Th) and bismuth (Bi), that is, elements which have the same proton or atomic numbers but different nucleon or mass numbers. We now discuss how isotopes were found.

Thomson Mass Spectrometer, Isotopes

When a chemist measures the density of a gas such as hydrogen or chlorine, he or she finds the volume and then the masses of *all* the atoms present in the gas sample. Individual atoms can not be identified in this measurement.

Element	Symbol	Atomic Number	Mass Number	Half-life Period (T)	Particle emitted
Uranium	U	92	238	4500 million years	α
Uranium X_1	Th	90	234	24 days	β, γ
Uranium X_2	Pa	91	234	1·2 days	β, γ
Uranium II	U	92	234	250 000 years	α
Ionium	Th	90	230	80 000 years	α, γ
Radium	Ra	88	226	1600 years	α, γ
Radon	Rn	86	222	3·8 days	α
Radium A	Po	84	218	3 minutes	α
Radium B	Pb	82	214	27 minutes	β, γ
Radium C*	Bi	83	214	20 minutes	β or α
Radium C′	Po	84	214	$1·6 \times 10^{-4}$ second	α
Radium C″	Tl	81	210	1·3 minutes	β
Radium D	Pb	82	210	19 years	β, γ
Radium E	Bi	83	210	5 days	β
Radium F	Po	84	210	138 days	α, γ
Lead	Pb	82	206	(stable)	

* Radium C exhibits branching; it produces Radium C′ on β-emission or Radium C″ on α-emission. Radium D is then produced from Radium C′ by α-emission or from Radium C″ by β-emission.

In 1911, Sir J. J. Thomson measured the masses of individual atoms. The atoms were first ionised in a hot vapour or gas, that is, they were stripped of one or more electrons to form an *ion* with a positive charge and which had a mass practically equal to that of the atom since electrons are very light.

Thomson used parallel electric and magnetic fields to deflect the fast-moving ions in a parabolic path. The equation of the parabola is $y = kx^2$, where k is a constant which depends on the *charge*–mass ratio (Q/M) for that ion. For a given value of x and charge Q, the deflection y is greatest for a hydrogen ion, since this has the smallest mass, and least for the ion of largest mass.

Ions of different gases have different parabolic paths. By comparing the deflections obtained in the y-direction, the masses of individual ions or atoms can be compared.

With chlorine gas, two parabolas were obtained which gave atomic masses of 35 and 37 respectively. So the atoms of chlorine have different masses but the same chemical properties, and these atoms are said to be *isotopes* of chlorine. In chlorine, there are three times as many atoms of mass 35 as there are of mass 37, so that the average atomic mass is $(3 \times 35 + 1 \times 37)/4$, or 35·5. The element xenon has as many as nine isotopes. One part in 5000 of hydrogen consists of an isotope of mass 2 called deuterium, or heavy hydrogen. An unstable isotope of hydrogen of mass 3 is called tritium. Hydrogen isotopes are use in nuclear energy experiments as we see later.

An *isotope* has the same chemical nature (proton number) as other atoms but a different mass (different number of neutrons) to most other atoms. Hydrogen has an isotope of mass number 2 (deuterium) and an isotope of mass number 3 (tritium).

Bainbridge Mass Spectrometer

Thomson's earliest form of mass spectrometer was followed by more sensitive forms. Bainbridge devised a mass spectrometer in which the ions were incident on a photographic plate after being deflected by a magnetic field.

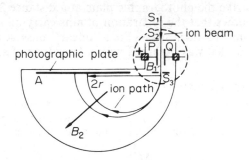

Figure 34.18 *Principle of Bainbridge mass spectrometer*

The principle of the spectrometer is shown in Figure 34.18. Positive ions were produced in a discharge tube (not shown) and admitted as a fine beam through slits S_1, S_2. The beam then passed between insulated plates P, Q, connected to a battery, which created an electric field of strength E. A uniform magnetic field B_1, perpendicular to E, was also applied over the region of the plates.

Velocity selector Suppose an ion of charge Q and mass M enters the region with a velocity v. If the magnetic force $B_1 Q v =$ the electric force EQ, the two forces cancel each other and so the ion travels *undeflected* through PQ. The velocity v is given by $v = E/B_1$ from this equation. Since this value of v does not depend on the charge Q or the mass M of an ion, it follows that *all* ions, even though their masses are different, pass through the plates PQ undeflected and through a slit S_3. This arrangement of perpendicular electric and magnetic fields is therefore called a 'velocity selector'—it only allows ions through S, which have the same velocity v equal to E/B_1.

Magnetic Analyser The selected ions were then deflected in a circular path of radius r by a uniform perpendicular magnetic field B_2, and an image was produced on a photographic plate A, as shown. In this case, if the mass of the ion is M,

$$\frac{Mv^2}{r} = B_2 Q v$$

$$\therefore \frac{M}{Q} = \frac{rB_2}{v}$$

But for the selected ions, $v = E/B_1$ from above

$$\therefore \frac{M}{Q} = \frac{rB_2 B_1}{E}$$

$$\therefore \frac{M}{Q} \propto r$$

for given magnetic and electric fields.

Since the ions strike the photographic plate at a distance $2r$ from the middle of the slit S_3, it follows that the separation of ions carrying the same charge is directly proportional to their mass. So a 'linear' mass scale is achieved. A resolution of 1 in 30 000 was obtained with a later type of spectrometer.

Example on Mass Spectrometer

In a mass spectrograph, an ion X of mass number 24 and charge $+e$ and an ion Y of mass 22 and charge $+2e$ both enter the magnetic field with the same velocity. The radius of the circular path of X is 0·25 m. Calculate the radius of the circular path of Y.

From above,
$$r \propto \frac{M}{Q}$$

For ion Y,
$$\frac{M}{Q} = \frac{22}{2e}$$

For ion X,
$$\frac{M}{Q} = \frac{24}{e}$$

So
$$\frac{r_Y}{0·25} = \frac{22/2e}{24/e} = \frac{11}{24}$$

$$\therefore r_Y = \frac{11}{24} \times 0·25 = 0·11 \text{ m (approx.)}$$

Einstein's Mass—Energy Relation

In 1905 Einstein showed from his Theory of Relativity that mass (m) and energy (E) can be changed from one form to the other. The energy ΔE produced by a change of mass Δm is given by the relation:

$$\Delta E = \Delta mc^2$$

where c is the numerical value of the velocity of light ΔE is in joules when Δm is in kg and c the numerical value 3×10^8, the speed of light in metres per second. So a change of mass Δm of 10^{-3} kg (1 g) would theoretically produce energy $\Delta E = 10^{-3} \times (3 \times 10^8)^2 = 9 \times 10^{13}$ J. Now 1 kilowatt-hour of energy is 1000×3600 or $3·6 \times 10^6$ joules, and so 9×10^{14} joules is $2·5 \times 10^7$ or 25 million kilowatt-hours. Consequently a change in mass of 1 g could be sufficient to keep the electrical lamps in a million houses burning for about a week in winter, on the basis of about seven hours' use per day.

In electronics and in nuclear energy, the unit of energy called and *electron-volt* (eV) is often used. This is defined as the energy gained by a charge equal to that on an electron moving through a p.d. of one volt. So

$$1 \text{ eV} = 1·6 \times 10^{-19} \text{ J}$$

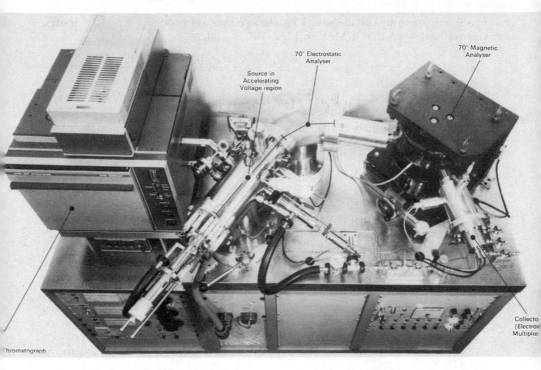

Plate 34E Industrial mass spectrometer. *Double-focusing by an electrostatic and a magnetic field is used, which produces mass and resolution superior to single focusing. The gas chromatograph (left) separates mixtures into their components. The ion sources are then passed into the front tube (left) to be accelerated by a high voltage. The ions are now deflected electrostatically by the analyser at the top and then deflected by the magnetic field of the second analyser. The separated ions are collected in the tube at the front. Here they are recorded electronically and their masses and relative abundance in the compound analysed are deduced from the readings.* (Courtesy of VG Micromass Limited)

The *megelectron-volt* (MeV) is a larger energy unit, and is defined as 1 million eV. So

$$1 \text{ MeV} = 1 \cdot 6 \times 10^{-13} \text{ J}$$

Unified Atomic Mass Unit

If another unit of energy is needed, then one may use a unit of mass, since mass and energy are interchangeable. The *unified atomic mass unit* (u) is defined as 1/12 of the mass of the carbon atom $^{12}_{6}C$. Now the number of molecules in 1 mole of carbon is $6 \cdot 02 \times 10^{23}$, the Avogadro constant, and since carbon is monatomic, there are $6 \cdot 02 \times 10^{23}$ atoms of carbon. These have a mass 12 g.

∴ mass of 1 atom of carbon

$$= \frac{12}{6 \cdot 02 \times 10^{23}} \text{ g} = \frac{12}{6 \cdot 02 \times 10^{26}} \text{ kg} = 12 \text{ u}$$

$$\therefore 1 \text{ u} = \frac{12}{12 \times 6 \cdot 02 \times 10^{26}} \text{ kg} = 1 \cdot 66 \times 10^{-27} \text{ kg (approx.)}$$

From our previous calculation, 1 kg change in mass produces 9×10^{16} joules; and we have seen that 1 MeV = $1 \cdot 6 \times 10^{-13}$ joules.

$$\therefore 1\,u = \frac{1 \cdot 66 \times 10^{-27} \times 9 \times 10^{16}}{1 \cdot 6 \times 10^{-13}}\text{ MeV}$$

$$\therefore \mathbf{1\,u = 931\ MeV\ (approx.)} \qquad . \qquad . \qquad . \qquad . \qquad . \qquad (1)$$

This relation is used to change mass units to energy units MeV, or the other way round, as we see shortly. An electron mass, $9 \cdot 1 \times 10^{-31}$ kg, corresponds to about 0·5 MeV.

Binding Energy

The protons and neutrons in the nucleus of an atom are called *nucleons*. The work or energy needed *just* to take all the nucleons apart so that they are *completely* separated is called the *binding energy* of the nucleus. Hence, from Einstein's mass–energy relation, it follows that the total mass of all the separated nucleons is greater than that of the nucleus, in which they are together. This is because the nucleons in the nucleus have gained some nuclear potential energy which has come from a mass change in all the separated individual nucleons. *So the difference in mass is a measure of the binding energy.*

As an example, consider a helium nucleus $^4_2$He. This has 4 nucleons, 2 protons and 2 neutrons. The mass of a proton is 1·0073 and the mass of a neutron is 1·0087 u.

$$\therefore \text{total mass of 2 protons plus 2 neutrons} = 2 \times 1 \cdot 0073 + 2 \times 1 \cdot 0087$$

$$= 4 \cdot 0320\ u$$

But the helium nucleus has a mass of 4·0015 u.

$$\therefore \text{binding energy} = \text{mass difference of nucleons and nucleus}$$

$$= 4 \cdot 0320 - 4 \cdot 0015 = 0 \cdot 0305\ u$$

$$= 0 \cdot 0305 \times 931\ \text{MeV} = 28 \cdot 4\ \text{MeV}$$

The *binding energy per nucleon* of a nucleus is binding energy divided by the total number of nucleons. In the case of the helium nucleus, since there are four nucleons (2 protons and 2 neutrons), the binding energy per nucleon is then 28·4/4 or 7·1 MeV.

Figure 34.19 shows roughly the variation of the binding energy per nucleon

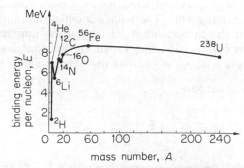

Figure 34.19 *Variation of binding energy per nucleon with mass number*

among the elements. Excluding the nuclei lighter than ^{12}C, we can see from Figure 34.19 that the average binding energy per nucleon, E/A, is fairly constant for the great majority of nuclei. The average value is about 8 MeV per nucleon. The peak occurs at approximately the iron nucleus ^{56}Fe, which is therefore one of the most stable nuclei.

Later we shall use the curve in Figure 34.19 to show that energy is produced when heavy elements such as uranium undergo *fission* to form two lighter masses or when very light elements such as hydrogen undergo *fusion* to form a heavier element.

---

1   **Energy and mass can be interchanged. Einstein's mass–energy relation is:**

$$\Delta E = \Delta m . c^2$$

2   **1 atomic mass unit (1 u) = 931 MeV of energy (about). This can be used to change mass to energy or energy to mass in nuclear reactions.**
3   **The nucleus of an atom has *binding energy*. When the nuclear particles come together and form the nucleus, the mass (*m*) of the nucleus is *less* than the total mass (*M*) of the separated individual particles. The binding energy = *M* − *m*.**
4   **The binding energy per nucleon of an atom varies with the atomic mass number A. The value is low for hydrogen and increases to a maximum, then decreases to higher values of A such as uranium.**

Nuclear Forces and Binding Energy

Inside the nucleus, the protons repel each other due to electrostatic repulsion of like charges. So for the nucleus to be stable there must exist other forces between the nucleons. These are called *nuclear forces*. They have short range, shorter than the interatomic distances, and are much stronger than electromagnetic interactions. They must provide a net attractive force greater than any repulsive electric forces.

A nucleus with a nucleon (mass) number $(N + Z)$ and proton (atomic) number Z has Z protons and N neutrons. When these particles come together in the nucleus there is an increase in potential energy due to the electrostatic forces of the protons but a *greater* decrease in potential energy due to the nuclear forces of the nucleons. There is therefore a net decrease in the potential energy of all the nucleons. This decrease per nucleon is the *binding energy per nucleon*.

So when the nucleons come together in the nucleus, there is a loss of energy equal to the binding energy. This results in a decrease in mass, from Einstein's mass–energy relation. As we have seen, the decrease in mass is the difference in the mass of the individual nucleons when they are completely separated and the mass of the nucleus when they are together; the so-called 'mass defect'. Expressed in symbols, suppose a nucleus has Z protons and N neutrons, so the mass number is $(N + Z)$. If we call the atom X, then the binding energy of its nucleus $^A_Z X$ or $^{N+Z}_Z X$ is given by:

Binding energy

$$= \text{mass of } N \text{ neutrons} + \text{mass of } Z \text{ protons} - \text{mass of nucleus } ^{N+Z}_Z X$$

Energy of Disintegration

It is instructive to consider, from an energy point of view, whether a particular nucleus is likely to disintegrate with the emission of an α-particle. As an illustration, consider radium F or polonium, $^{210}_{84}$Po. If an α-particle could be emitted from the nucleus, the reaction products would be the α-particle or helium nucleus, $^{4}_{2}$He, and a lead nucleus, $^{206}_{82}$Pb, a reaction which could be represented by:

$$^{210}_{84}\text{Po} \rightarrow {}^{206}_{82}\text{Pb} + {}^{4}_{2}\text{He} \qquad \cdot \qquad \cdot \qquad \cdot \qquad \cdot \qquad (1)$$

Here the sum of the mass numbers, 210, and the sum of the nuclear charges, $+84e$, of the lead and helium nuclei are respectively equal to the mass number and nuclear charge of the polonium nucleus, from the law of conservation of mass and of charge.

If we require to find whether energy has been released or absorbed in the reaction, we should calculate the total mass of the lead and helium nuclei and compare this with the mass of the polonium nucleus. It is more convenient to use atomic masses rather than nuclear masses, and since the total number of electrons required on each side of equation (1) to convert the nuclei into atoms is the same, we may use atomic masses in the reaction. These are as follows:

$$\text{lead, } {}^{206}_{82}\text{Pb} = 205{\cdot}969 \text{ u}$$

$$\text{α-particle, } {}^{4}_{2}\text{He, } = \quad 4{\cdot}004 \text{ u}$$

$$\therefore \text{ total mass} = 209{\cdot}973 \text{ u}$$

Now $\qquad\qquad$ polonium, $^{210}_{84}$Po, $= 209{\cdot}982$ u

So the atomic masses of the products of the reaction are together *less* than the original polonium nucleus, that is,

$$^{210}_{84}\text{Po} \rightarrow {}^{206}_{82}\text{Pb} + {}^{4}_{2}\text{He} + Q$$

where Q is the energy released. It therefore follows that polonium *can* disintegrate with the emission of an α-particle and a release of energy (see *uranium series*, p. 886), that is, the polonium is unstable.

Suppose we now consider the possibility of a lead nucleus, $^{206}_{82}$Pb, disintegrating with the emission of an α-particle, $^{4}_{2}$He. If this were possible, a mercury nucleus, $^{202}_{80}$Hg, would be formed. The atomic masses are as follows:

$$\text{mercury, } {}^{202}_{80}\text{Hg, } = 201{\cdot}971 \text{ u}$$

$$\text{α-particle, } {}^{4}_{2}\text{He, } = \quad 4{\cdot}004 \text{ u}$$

$$\therefore \text{ total mass} = 205{\cdot}975 \text{ u}$$

Now $\qquad\qquad$ lead, $^{206}_{82}$Pb, $= 205{\cdot}969$ u

Thus, unlike the case previously considered, the atomic masses of the mercury nucleus and α-particle are together *greater* than the lead nucleus, that is,

$$^{206}_{82}\text{Pb} + Q \rightarrow {}^{202}_{80}\text{Hg} + {}^{4}_{2}\text{He}$$

where Q is the energy which must be *given* to the lead nucleus to obtain the reaction products. It follows that the lead nucleus by itself is stable.

Generally, then, a nucleus would tend to be unstable and emit an α-particle if the sum of the atomic masses of the products are together *less* than that of the nucleus; and it would be stable if the sum of the atomic masses of the possible reaction products are together *greater* than the atomic mass of the nucleus.

Stable and Unstable Nuclei

Many factors contribute to the binding energy, E, of nuclei and therefore to their stability. The α-particle, ${}_2^4\text{He}$, appears to be particularly stable. Figure 34.20 (i) shows the binding energy per nucleon, E/A, for some light nuclei. Peaks

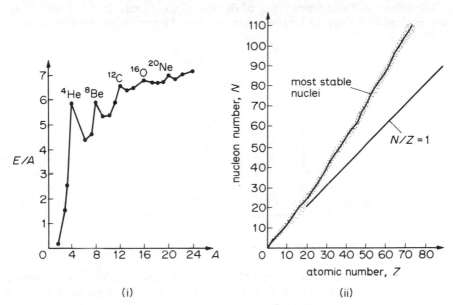

Figure 34.20 (*i*) *Variation of nuclear energy with atomic mass* (*ii*) *Most stable nuclei*

occur for nuclei such as ${}_4^8\text{Be}$, ${}_6^{12}\text{C}$, ${}_8^{16}\text{O}$, ${}_{10}^{20}\text{Ne}$. Each of these nuclei can be formed by adding an α-particle to the preceding nucleus.

Another factor affecting the stability of nuclei is the *neutron–proton ratio*. Figure 34.20 (ii) shows the number of neutrons, N, plotted against the number of protons (atomic number), Z, for all stable nuclei. It can be seen from the line of most stable nuclei that for light stable nuclei, such as ${}_6^{12}\text{C}$ and ${}_8^{16}\text{O}$, this ratio is 1. For nuclei heavier than ${}_{20}^{40}\text{Ca}$, the ratio N/Z increases slowly towards about 1·6. There are no stable nuclei above about $Z = 92$ (uranium).

Nuclear Emissions and Nuclear Stability

Unstable nuclei are radioactive. Their decay may occur in three main ways:

(1) *α-particle emission*. If the nucleus has excess protons, an α-particle emission would reduce the protons by 2 and the neutrons by 2. So, generally, if A is the nucleon (mass) number and Z is the proton (atomic) number of the atom X concerned, then

$$\,_Z^A\text{X} \rightarrow \,_{Z-2}^{A-4}\text{Y} + \,_2^4\text{He}$$

(2) *β^- particle (electron) emission*. If the nucleus has too many neutrons for stability, the neutron–proton ratio is reduced by β-particle (electron) emission. Here a neutron changes to a proton, so

$$\,_0^1\text{n} \rightarrow \,_1^1\text{H} + \,_{-1}^0\text{e (electron)}$$

Hence A remains unchanged but Z increases by 1.

(3) β^+ *particle* (*positron*) *emission.* If the nucleus is deficient in neutrons, a decay by β^+ (positron) emission may occur. A proton changes to a neutron:

$$^1_1\text{H} \rightarrow {}^1_0\text{n} + {}^0_1\text{e} \text{ (positron)}$$

So A is unchanged but Z decreases by 1.

The effects of these three types of decays, (1), (2) and (3), on A and Z are summarised in Figure 34.23; the boxes indicate diagrammatically unit changes in A and Z.

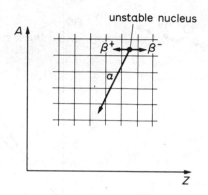

Figure 34.21 *Effect of particle emission on A and Z*

In nuclei lighter than the isotope ^{40}Co, the numbers of neutrons and protons are roughly equal. Heavier elements such as thorium or uranium have a greater number of protons which tend to repel each other and so make the nucleus less tightly bound. So most stable heavy nuclei have more neutrons than protons.

The nucleus formed by a decay may also be unstable. This gives rise to 'decay series or chains', which end when a stable nucleus, for example, ^{206}Pb (lead), is reached. A series consisting entirely of α-decays (α-particle emission) would increase the ratio of neutrons to protons as Z decreased. As an example, suppose $^{234}_{94}$Pu decays by α-emission to $^{240}_{92}$U. Now $^{240}_{92}$U has $N = 148$ and $Z = 92$, so the neutron–proton ratio, N/Z, is 1·609. *If* this nucleus were to decay by another α-emission to $^{236}_{90}$Th, the ratio would be increased to 146/90 or 1·622, which is too far from the line of stability shown in Figure 34.20 (ii). In fact, $^{240}_{92}$U decays to $^{240}_{93}$Np by β^- emission. The ratio N/Z is now 147/93 or 1·581, close to the line of the stability. In a natural decay series, then, there must always be some β^- decays in addition to α-decays.

β^+ (positron) decays are much rarer since they would result in an increase in N and a decrease in Z, so increasing the ratio N/Z.

In α-emission ($^4_2$He), the nucleon number A decreases by 4 and its proton number Z decreases by 2.
In β^- emission ($_{-1}^{0}$e), the nucleon number A is unchanged but Z increases by 1.
In β^+ emission ($_{+1}^{0}$e), the nucleon number A is unchanged but Z decreases by 1.
In γ emission, A and Z are both unchanged.

Artificial Disintegration, Nuclear Reactions

Uranium, thorium and actinium are elements which disintegrate naturally. The artificial disintegration of elements began in 1919, when Rutherford used α-particles to bombard nitrogen and found that protons were produced (p. 884). Some nuclei of nitrogen had changed into nuclei of oxygen, that is, transmutation of atoms had occurred.

From the laws of conservation of mass (nucleon number) and charge (proton number) the reaction can be represented by:

$$^{14}_{7}N + ^{4}_{2}He \rightarrow ^{17}_{8}O + ^{1}_{1}H$$

This reaction is often expressed briefly as

$$^{14}_{7}Na(\alpha, p) \quad \text{or} \quad ^{14}_{7}N(\alpha, p)^{17}_{8}O$$

which means that the *incident* particle on the $^{14}_{7}N$ nucleus is an α-particle ($^{4}_{2}He$), the *emitted* particle is a proton $p(^{1}_{1}H)$ and the nucleus formed after the reaction is $^{17}_{8}O$. Similarly, the reaction $^{9}_{4}Be(p, \alpha)$ or $^{9}_{4}Be(p, \alpha)^{6}_{3}Li$ is

$$^{9}_{4}Be + ^{1}_{1}H \rightarrow ^{4}_{2}He + ^{6}_{3}Li$$

In 1932 Cockcroft and Walton produced nuclear disintegrations by accelerating protons with a high voltage machine producing about half a million volts, and then bombarding elements with the high-speed protons. When the light element lithium was used, photographs of the reaction taken in the cloud chamber showed that two α-particles were produced. The particles shot in opposite direction from the point of impact of the protons, and as their range in air was equal, the α-particles had initially equal energy. This is a consequence of the principle of the conservation of momentum. The nuclear reaction was:

$$^{7}_{3}Li + ^{1}_{1}H \rightarrow ^{4}_{2}He + ^{4}_{2}He + Q \ . \qquad . \qquad . \qquad (1)$$

where Q is the energy released in the reaction.

To calculate Q, we should calculate the total mass of the lithium and hydrogen nuclei and subtract the total mass of the two helium nuclei. As already explained, however, the total number of electrons required to convert the nuclei to neutral atoms is the same on both sides of equation (1). So atomic masses can be used in the calculation in place of nuclear masses. The atomic masses of lithium and hydrogen are 7·016 and 1·008 u respectively, a total of 8·024 u. The atomic mass of the two α-particles is $2 \times 4·004$ u or 8·008 u. So

$$\text{energy released, } Q, = 8·024 - 8·008 = 0·016 \text{ u}$$

$$= 0·016 \times 931 \text{ MeV} = 14·9 \text{ MeV}$$

Each α-particle has therefore an initial energy of 7·4 MeV, and this theoretical value agreed closely with the energy of each α-particle measured from its range in air. The experiment was the earliest verification of Einstein's mass–energy relation $\Delta E = \Delta m \cdot c^2$.

Cockcroft and Walton were the first scientists to use protons for disrupting atomic nuclei after accelerating them by high voltage. Today, giant high-voltage machines are built at atomic energy centres for accelerating protons to enormously high speeds. The products of the nuclear explosion with light atoms such as hydrogen yield valuable information on the structure of the nucleus. For example, the proton itself contains particles called *quarks*.

Energy from Radioactive Isotopes

Energy from the decay of unstable radioactive isotopes is sometimes used where a continuous, powerful and compact energy source is required. Such isotopes have been used to provide power for the batteries of heart 'pacemakers' and for scientific apparatus used in space vehicles.

As an example, consider the isotope Po-210. This has a half-life of about 140 days and emits α-particles each of energy 5·3 MeV.

Since the molar mass is 210 g and the Avogadro constant is about 6×10^{23} mol^{-1}, each gram of the isotope contains $6 \times 10^{23}/210 = 2\cdot9 \times 10^{21}$ atoms. So in one-half-life or 140 days, about $1\cdot4 \times 10^{21}$ atoms decay.

These atoms release a total energy (using 1 MeV = $1\cdot6 \times 10^{-13}$ J)

$$= 1\cdot4 \times 10^{21} \times 5\cdot3 \times 1\cdot6 \times 10^{-13}\,\text{J}$$

$$= 1\cdot2 \times 10^9\,\text{J (approx.)}$$

So the mean output power per gram

$$= \frac{1\cdot2 \times 10^9}{140 \times 24 \times 3600} = 100\,\text{W (approx.)}$$

During the next half-life period, 140 days, the mean power output will only be 50 W, since half the remaining atoms decay in this time.

Energy Released in Fission, Chain Reaction

In 1934 Fermi began using neutrons to produce nuclear disintegration. These particles are generally more effective than α-particles or protons for this purpose, because they have no charge and are therefore able to penetrate more deeply into the positively-charged nucleus. Usually the atomic nucleus changes only slightly after disintegration, but in 1939 Frisch and Meitner showed that a uranium nucleus had disintegrated into two relatively-heavy nuclei. This is called *nuclear fission*, and as we shall now show, a large amount of energy is released in this case.

Natural uranium consists of about 1 part by mass of uranium atoms $^{235}_{92}\text{U}$ and 140 parts by mass of uranium atoms $^{238}_{92}\text{U}$. In a nuclear reaction with natural uranium and slow neutrons, it is usually the nucleus $^{235}_{92}\text{U}$ which is fissioned. If the resulting nuclei are lanthanum $^{148}_{57}\text{La}$ and bromine $^{85}_{35}\text{Br}$, together with several neutrons, then:

$$^{235}_{92}\text{U} + ^{1}_{0}\text{n} \rightarrow ^{148}_{57}\text{La} + ^{85}_{35}\text{Br} + 3^{1}_{0}\text{n} \qquad . \qquad . \qquad . \qquad (1)$$

Now $^{235}_{92}\text{U}$ and $^{1}_{0}\text{n}$ together have a mass of $(235\cdot1 + 1\cdot009)$ or 236·1 u. The lanthanum, bromine and neutrons produced together have a mass

$$= 148\cdot0 + 84\cdot9 + 3 \times 1\cdot009 = 235\cdot9\,\text{u}$$

$\therefore$ energy released = mass difference

$$= 0\cdot2\,\text{u} = 0\cdot2 \times 931\,\text{MeV} = 186\,\text{MeV}$$

$$= 298 \times 10^{-13}\,\text{J (approx.)}$$

This is the energy released per atom of uranium fissioned. In 1 kg of uranium there are about

$$\frac{1000}{235} \times 6 \times 10^{23} \quad \text{or} \quad 26 \times 10^{23}\,\text{atoms}$$

since the Avogadro constant, the number of atoms in a mole of any element, is $6\cdot02 \times 10^{23}$. So if all the atoms in 1 kg of uranium were fissioned, total energy

released

$$= 26 \times 10^{23} \times 298 \times 10^{-13} \, J = 8 \times 10^{13} \, J \text{ (approx.)}$$
$$= 2 \times 10^{7} \text{ kilowatts-hours}$$

which is the amount of energy given out by burning about 3 million tonnes of coal. The energy released per gram of uranium fissioned $= 8 \times 10^{10}$ J (approx.).

Fast neutrons are absorbed or captured by nuclei of U238 without producing fission. Slow or *thermal* neutrons in uranium, however, which have the same temperature as that of the uranium, produce fission when incident on nuclei of U235. About 2·5 neutrons per fission are released. When slowed, *each of these neutrons produce fission in another U235 nucleus*, and so on. So a rapid multiplying *chain reaction* can be obtained in the uranium mass, liberating swiftly an enormous amount of energy. This is the basic principle of the nuclear

Plate 34F *Nuclear research reactor, ZEUS. This view shows the heart of the reactor, containing a highly enriched uranium central core surrounded by a natural uranium blanket for breeding studies*

reactor or pile, Plate 34F. As we now describe, graphite can be used to slow down the speed of the fast neutrons released in fission.

1 Neutrons can penetrate into the uranium nucleus because they have no charge and are not repelled by the nuclear positive charge.
2 Nuclear fission occurs when a thermal neutron penetrates the uranium nucleus. The nucleus splits into two other lighter nuclei X and Y together with several neutrons. Energy released = mass of uranium nucleus − masses of X and Y and neutrons.
3 The neutrons cause more fission in other uranium nuclei and the new neutrons released start a chain reaction.

Nuclear Reactors

Figure 34.22 shows, in diagrammatic form, the principle of one type of nuclear reactor, used for commercial power generation.

Uranium fuel in the form of thick rods is encased in long aluminium tubes, which are air-tight to contain any gases released and prevent oxidation of the surrounding fuel. The tubes are lowered into hundreds of channels inside blocks.

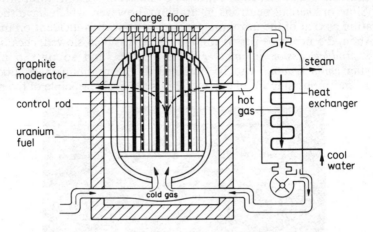

Figure 34.22 *Nuclear reactor principle*

As explained previously, the graphite is needed to moderate or reduce the speed of the neutrons released on fission until they become slow or thermal neutrons; they can then produce fission on collision with other U235 nuclei. The graphite *moderator* is in the form of blocks of pure carbon arranged in a stack.

Heavy water (deuterium oxide) has also been used as a moderator. It reduces the speed of a colliding neutron to about one-half, whereas graphite reduces the speed to about one-seventh. With a graphite moderator the separation of the uranium rods is about 20 cm. In this distance the fast neutron makes about 200 collisions with carbon nuclei and then becomes slow enough to produce fission on collision with a U235 nucleus. The separation of the uranium rods in a water moderator is less than a graphite moderator as the speed of the colliding neutron is reduced much more than a deuterium nucleus.

The power level reached by a reactor is proportional to the *neutron flux density*; this is the number of neutrons per second crossing unit cross-sectional area of the reactor. If the neutrons are produced too fast, the reactor may disintegrate. Ideally, the neutron reproduction factor should be just greater than 1 to make the the chain reaction self-sustaining. Boron-coated steel rods are used to control the rate of neutron production, Figure 34.22. The *control rods* are lowered or raised in channels inside the graphite block by electric motors operated from a control room until the neutron reproduction rate is just greater than 1, and the reactor is then said to go *critical*. In the event of an electrical failure or other danger, the rods fall and automatically shut off the reactor. Boron atoms have a high absorption cross-section for neutrons and so capture slow neutrons.

The energy produced by the nuclear reaction would make the reactor too hot. A *coolant* is therefore required. Water and gas such as carbon dioxide have been used as coolants. Molten sodium, which has a high value of (specific heat capacity × density) and a high thermal conductivity, has also been used as a coolant. In a *gas-cooled reactor* which produces power for the UK National Grid system, the hot gas is led from the reactor into a heat exchanger, Figure 34.22. Here the heat is transferred to water circulating through pipes so that steam is produced and this is used to drive turbines for electrical power generation.

Finally, it should be noted that if the mass of the uranium is too small, the neutrons will escape and a chain reaction is not produced. The *critical mass* is the least mass to make a self-sustaining chain reactions. The critical mass for a reactor in the shape of a sphere is less than that in the shape of a cube since the surface (neutron escape) area is a minimum for a given volume or mass of material. Fermi, a distinguished pioneer in nuclear research, made one of the first pilot reactors in 1942 in roughly a spherical shape.

Further details of reactors can be obtained from the United Kingdom Atomic Energy Authority or from specialist books.

Summarising, a nuclear reactor may contain:

1 Uranium rods as *fuel*
2 Graphite as a *moderator* of the neutron speeds for the fission process
3 Boron-coated steel rods to *control* the neutron reproduction rate
4 A *coolant* to reduce the excessive heat produced in the reaction
5 A *heat exchange* to remove the heat energy produced.

Radiation Hazards

Nuclear reactors may have waste matter of long half-lives. Because they decay slowly the waste matter is dangerous and so it is buried in lead-lined boxes very deeply below the surface of the earth. Waste matter with short half-lives is less dangerous because the decay is quicker but checks on the radiation intensity must be made at frequent intervals.

A nuclear diaster took place at the nuclear station in Chernobyl, Russia. Here radiation was spread into the air and blown by winds into central Europe thousands of kilometres away. It settled on land in countries such as Holland Animals feeding on the infected grass had to be destroyed to stop food they usually supplied from reaching people in these countries.

Health checks are made on workers in nuclear stations, discussed later.

Energy Released in Fusion

In fission, energy is released when a heavy nucleus is spit into two lighter nuclei. Energy is also released if light nuclei are *fused* together to form heavier nuclei, and fusion reaction, as we shall see, is also a possible source of considerable nuclear energy. As an illustration, consider the fusion of the nuclei of deuterium, ^2_1H. Deuterium is an isotope of hydrogen known as 'heavy hydrogen', and its nucleus is called a 'deuteron'. The fusion of two deuterons can result in a helium nucleus, ^3_2He, as follows:

$$^2_1H + {}^2_1H \rightarrow {}^3_2He + {}^1_0n$$

Now   mass of two deuterons

$$= 2 \times 2 \cdot 015 = 4 \cdot 030 \text{ u}$$

and   mass of helium plus neutron

$$= 3 \cdot 017 + 1 \cdot 009 = 4 \cdot 026 \text{ u}$$

∴ mass converted to energy by fusion

$$= 4\cdot030 - 4\cdot026 = 0\cdot004 \text{ u}$$

$$= 0\cdot004 \times 931 \text{ MeV} = 3\cdot7 \text{ MeV}$$

$$= 3\cdot7 \times 1\cdot6 \times 10^{-14} \text{ J} = 6\cdot0 \times 10^{-13} \text{ J}$$

∴ energy released per deuteron $= 3\cdot0 \times 10^{-13}$ J

6×10^{26} is the number of atoms in a kilomole of deuterium, which has a mass of about 2 kg. So if all the atoms could undergo fusion,

energy released per kg

$$= 3\cdot0 \times 10^{-13} \times 3 \times 10^{26} \text{ J}$$

$$= 9 \times 10^{13} \text{ J (approx.)}$$

Other fusion reactions can release much more energy, for example, the fusion of the nuclei of deuterium, $_1^2\text{H}$, and tritium, $_1^3\text{H}$, isotopes of hydrogen, released about 30×10^{13} joules of energy per kg according to the reaction:

$$_1^2\text{H} + {}_1^3\text{H} \rightarrow {}_2^4\text{He} + {}_0^1\text{n}$$

In addition, the temperature required for this fusion reaction is less than that needed for the fusion reaction between two deuterons given above, which is an advantage. Hydrogen contains about 1/5000 by mass of deuterium, or heavy hydrogen, needed in fusion reactions, and this can be obtained by electrolysis of sea-water, which is cheap and in plentiful supply.

Fusion and Binding Energy Curve, Thermonuclear Reaction

In Figure 34.19 (p. 890), the binding energy per nucleon is plotted against the mass number of the nucleons or nucleon number. Since the curve rises from hydrogen, the binding energy per nucleon of the helium nucleus, $_2^4\text{He}$, is *greater* than that for deuterium, $_1^2\text{H}$. So the binding energy of the helium nucleus, which has four nucleons, is *greater* than that of two deuterium nuclei, which also have four nucleons.

Now the binding energy of a nucleus is proportional to the difference between the mass of the individual nucleons and the mass of the nucleus (the so-called 'mass defect'), see p. 890. So the mass of the helium nucleus is *less* than that of the two deuterium nuclei. Hence if two deuterium nuclei can be used together to form a helium nucleus, the mass lost will be released as energy.

The rising part of the binding energy curve in Figure 34.19 shows that elements with low mass number can produce energy by fusion. In contrast, the falling part of the curve shows that very heavy elements such as uranium can produce energy by *fission* of their nuclei to nuclei of *smaller* mass number (see p. 896).

For fusion to take place, the nuclei must at least overcome electrostatic repulsion when approaching each other. Consequently, for practical purposes, fusion reactions can best be achieved with the lightest elements such as hydrogen, whose nuclei carry the smallest charges and hence repel each other least.

In attempts to obtain fusion, isotopes of hydrogen such as deuterium, $_1^2\text{H}$ and tritium, $_1^3\text{H}$, are heated to tens of millions of degrees Celsius. The thermal energy of the nuclei at these high temperatures is sufficient for fusion to occur. One technique of promoting this *thermonuclear reaction* is to pass enormously high currents through the gas, which heat it. A very high percentage of the atoms are then ionised and the name *plasma* is given to the gas. The gas discharge consists of parallel currents, carried by ions, and the powerful magnetic field round one current due to a neighbouring current draws the discharge together (see p. 349). The plasma, however, wriggles and touches the sides of the containing vessel, so losing heat. The main difficulty in thermonuclear experiments in the laboratory is to keep the heat in the gas for a sufficiently long time for a fusion reaction to occur. The stability of plasma is now the subject of considerable research.

It is believed that the energy of the sun is produced by thermonuclear reactions in the heart of the sun, where the temperature is many millions of degrees Celsius. Bethe has proposed a cycle of nuclear reactions in which, basically, protons are converted to helium by fusion in the sun, with the release of a considerable amount of energy.

Nuclear fission **occurs with heavy elements such as uranium which break up into lighter nuclei.**
Nuclear fusion **occurs with light elements such as hydrogen whose nuclei fuse together and form heavier nuclei such as helium. Nuclear fusion occurs in the sun and stars where temperatures are very high.**

Smoke Detectors

It may surprise you to know that radioactive materials can now be found in many homes around the world, because they are inside most common smoke detectors. The radioactive isotope (usually americium 241, an α-emitter with a half-life of 458 years) ionises the air in a small ionisation chamber (like the Geiger–Müller tube), which has a vent to the outside. The ionization current is measured using an electronic circuit, which also controls the alarm. If smoke enters the chamber, then the ionisation current is reduced, and the alarm is sounded.

This type of alarm is much more sensitive than the older heat detectors, which used to be fitted in offices and public buildings. Ionisation smoke detectors are perfectly safe, since the amount of radioactive isotope used is very small. They are also very cheap and simple to produce, which explains why they are becoming commonplace in many domestic households throughout the United Kingdom, so greatly increasing fire safety.

Radiation Explosure and Doses

People all over the world are always exposed to radiation. If the level of radiation is above a small value, the radiation may cause ionisation in parts of the body and seriously affect health.

Doses of radiation are measured today in units of sieverts, named after Sievert, a famous Swedish pioneer of radiation protection. A dose of about 1000 microsieverts, 1000 μSv, is an average annual dose from the radioactive gas radon received by people in the UK, which is not harmful. We have already met the unit becquerel, symbol Bq, used for the activity of radiation or the rate of atomic disintegration per second. In air, a concentration of 20 Bq m^{-3} in a volume gives a dose of radiation of 1000 μSv or 1 mSv (millisievert).

Sources of Radiation

There are two main sources of radiation which affect people: (a) natural sources; (b) artificial (man-made) sources.

Natural sources. Radon gas is a natural source which is about 50% of the total of natural sources. It comes from uranium buried in rocks and in the soil. Uranium has a half-life of millions of years. It is therefore long-lived and its radiation was first detected in 1896 by Becquerel. Radon from the ground spreads through the air. Its concentration level in the air is low but in a closed space such as a house it is much higher and radiation protection is needed where the level is more than 200 Bq m^{-3} or a dose of 10 mSv.

Another natural source of radiation is due to cosmic rays from space when they interact with atoms in the upper atmosphere. This gives an exposure of about 250 μSv annually to people on the ground. Air travellers, at high altitude, are exposed to much higher radiation doses from cosmic rays than people on the ground. Gamma radiation is also obtained from radioactive nuclides in the soil, rocks and building materials. Food, such as fish, also contains radioactive nuclides taken into the body; the average annual dose here is about 300 μSv. Coal miners are exposed to radiation from rocks underground.

Summarising, the main natural sources of radiation are radon, cosmic rays, gamma radiation from rocks and soil, radioactive nuclides in food and drink.

Artificial sources. Patients are exposed to radiation in X-ray examinations when radiographs are taken of injuries to limbs or stomach disorders. These medical examinations produce radiation doses per year of about 350 μSv. Similar doses are obtained in medical treatment such as radiotherapy where patients may have cancer.

Fallout in radiation from the atmosphere takes place when nuclear weapons are tested. Fallout occurred in Russia at the Chernobyl nuclear station when this went out of control. Milk and animal food from a large area of farms had high doses of radiation at first but tests in the UK now show that is small, such as 0·1 μSv. Workers at nuclear power stations are exposed to radiation from waste products and the doses are monitored for health safety. At a particular station in the UK the annual dose is less than 5 μSv. Liquid nuclear waste is discharged into the sea round the UK under regulations which protect fish—an average annual dose is about 0·1 μSv. In a bedroom a smoke detector has an annual dose to a sleeper of less than 0·1 μSv.

Summarising, the main artificial sources of radiation are medical, fallout and discharges from nuclear stations.

For further details of radiation exposure, the reader is referred to specialist information from the National Radiological Protection Board publications in the UK.

Particles, Forces and the Unified Theory. Superstrings[1]

The results of experiments at very high energies in accelerators, such as the Large Electron Positron collider (LEP) at CERN (the European laboratory for nuclear research) in Geneva, have lead to the discovery of a wealth of so-called 'elementary' particles. The proton and neutron are now known to be composed of 'quarks', which come in several varieties. In addition to several kinds of

---

[1] The author of this section is Professor Michael Green FRS, Cambridge University, an authority on the theory of Superstrings.

quarks and the well-known electron and the photon there are a number of other particles (the μ, the τ, several kinds of neutrinos and the W and Z bosons). This variety of particles interact with each other via the weak, electromagnetic and strong forces (the weak and electromagnetic forces having been unified into a single 'electro-weak force' in 1967). This apparently haphazard collection of particles and forces fit together beautifully into a theoretical structure called the *Standard Model*, that accounts for almost all the experimental data.

However, since these forces arise quite separately the standard model cannot be viewed as a truly unified theory. Furthermore, the most familiar force of all, the force of gravity, is not included at all in the model. The modern theory of gravity, Einstein's general theory of relativity, is well-tested on cosmological scales. For some purposes it is satisfactory to ignore it when describing microscopic particles since it is by far the weakest of the forces. However, there is a fundamental difficulty in combining general relativity with quantum theory in order to describe the force of gravity at small scales. A simple application of Heisenberg's uncertainty principle—the embodiment of quantum mechanics —at incredibly small distances leads to the unpalatable conclusion that at distance scales of around 10^{-35} metres (the 'Planck' length) there are huge fluctuations of energy that are sufficiently big to create tiny black holes. Space is not empty but full of these tiny black hole fluctuations that come and go on time scales of around 10^{-43} seconds (the Planck time). The vacuum is full of *space–time foam* and classical theories are in trouble.

This problem has prompted radical proposals. Currently, the most promising is 'superstring theory', in which the so-called 'elementary' particles are described as vibrations of tiny (Planck-length) closed loops of string. This not only has the potential for unifying all the observed particles, which now arise as different harmonics (where the frequency of the string's vibration is related to its energy, and hence the mass of the particle) but it also avoids the previous problem of uniting quantum theory and general relativity. In this theory the classical laws of physics, such as electromagnetism and general relativity, are modified at tiny distances comparable to the length of the string.

The theory is promising but not yet well understood. Current research is aimed at a deeper formulation which not only describes the particles and their interactions in a quantum mechanical manner but also the geometry of the space–time through which they are moving. This notion of 'quantum space– time' is the goal of any unified theory of the physical forces.

EXERCISES 34B   Nuclear Reactions and Energy

Multiple Choice

1 Isotopes of uranium have the same
   A nucleon number   **B** number of neutrons   **C** decay rate
   D half-life   **E** proton number.
2 In the scattering of α-particles by thin metal films, the path of the particle can be

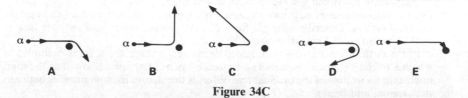

Figure 34C

3 Which statement from **A** to **E** is correct?
A A neutron can produce nuclear fusion.
B An alpha-particle can produce nuclear fission.
C Heavy elements such as radium can combine and produce nuclear fusion.
D A neutron can produce nuclear fission.
E At low temperatures light elements can produce nuclear fission.

4 In a nuclear reactor, the ^{235}U nucleus breaks into two nuclei. Which statements are also correct?
1 Control rods vary the neutron rate of absorption.
2 Low energy neutrons are more likely to produce nuclear fission than high energy neutrons.
3 The two nuclei produced have respectively very low mass and very high mass.
A 1, 2, 3 are all correct   **B** 1, 2 only   **C** 2, 3 only
D 1 only   **E** 3 only

Longer Questions

5 Figure 34D shows a gold foil mounted across the path of a narrow, parallel beam of alpha particles. The fraction of incident alpha particles reflected back through more than 90° is very small.

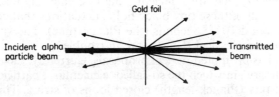

Figure 34D

How does this result lead to the idea that an atom has a nucleus
(a) whose diameter is small compared with the atomic diameter, and
(b) which contains more of the atom's mass? (*L.*)

6 When a nucleus of deuterium (hydrogen-2) fuses with a nucleus of tritium (hydrogen-3) to give a helium nucleus and a neutron, $2 \cdot 88 \times 10^{-12}$ J of energy are released.
The equation of the reaction is:

$$^2_1\text{H} + {}^3_1\text{H} \rightarrow {}^4_2\text{He} + {}^1_0\text{n}$$

Calculate the mass of the helium nucleus produced.
(Mass of deuterium nucleus $= 3 \cdot 345 \times 10^{-27}$ kg, mass of tritium nucleus $= 5 \cdot 008 \times 10^{-27}$ kg, mass of a neutron $= 1 \cdot 675 \times 10^{-27}$ kg, speed of light in a vacuum $= 3 \cdot 00 \times 10^8$ m s^{-1}.) (*L.*)

7 (a) State the results of experiments on the scattering of alpha particles by a thin foil. Explain the importance of these results.
(b) In such experiments explain: (i) why the foil should be thin, and point out any differences in the results for foils of different materials; (ii) why the alpha particles incident on the foil should be in a narrow parallel beam; (iii) how the scattered alpha particles might be detected.
(c) An alpha particle travels directly towards a nucleus which may be assumed to remain stationary. Describe in qualitative terms how, during the motion, the kinetic energy and the potential energy of the alpha particle vary with the separation between the alpha particle and the nucleus. (*N.*)

8 Define *nucleon number* (*mass number*) and *proton number* (*atomic number*) and explain the term *isotope*. Describe a simple form of mass spectrometer and indicate how it could be used to distinguish between isotopes.
    In the naturally occurring radioactive decay series there are several examples in which a nucleus emits an α-particle followed by two β-particles. Show that the final nucleus in an isotope of the original one. What is the change in mass number between the original and final nuclei? (*L.*)

9 Explain the term *nuclear binding energy*. Sketch a graph showing the variation of binding energy per nucleon with nucleon number (mass number) and show how both nuclear fission and nuclear fusion can be explained from the shape of this curve.

Calculate in MeV the energy liberated when a helium nucleus ($^4_2$He) is produced:
(a) by fusing two neutrons and two protons, and
(b) by fusing two deuterium nuclei ($^2_1$H). Why is the quantity of energy different in the two cases?

(The neutron mass is 1·008 98 u, the proton mass is 1·007 59 u, the nuclear masses of deuterium and helium are 2·014 19 u and 4·002 77 u respectively. 1 u is equivalent to 931 MeV.) (*L.*)

10 Using the information on atomic masses given below, show that a nucleus of uranium 238 can disintegrate with the emission of an alpha according to the reaction:

$$^{238}_{92}U \rightarrow {}^{234}_{90}Th + {}^4_2He$$

Calculate:
(a) the total energy released in the disintegration;
(b) the kinetic energy of the alpha particle, the nucleus being at rest before disintegration.

Mass of ^{238}U = 238·124 92 u, mass of ^{234}Th = 234·116 50 u, mass of ^4He = 4·003 87 u, 1 u is equivalent to 930 MeV. (*N.*)

11 Explain briefly what is meant by *nuclear fusion*.

Write down the equation for a fusion reaction, making clear what the various symbols mean. What is the great difficulty in producing fusion in the laboratory? Where in nature does fusion occur continuously? (*W.*)

12 In the fusion reaction $^2_1$H + $^3_1$H = $^4_2$He + $^1_0$n, how much energy, in joules, is released?

Mass of $^2_1$H = 3·345 × 10^{-27} kg, $^3_1$H = 5·008 × 10^{-27} kg, $^4_2$He = 6·647 × 10^{-27} kg, $^1_0$n = 1·675 × 10^{-27} kg, speed of light = 3·0 × 10^8 m s^{-1}.) (*L.*)

13 In the Rutherford α-particle scattering experiment, α-particles of mass 7 × 10^{-27} kg and speed 2 × 10^7 m s^{-1} were fired at a gold foil. What was the closest distance of approach between an α-particle and a gold nucleus? How could those α-particles which made such a closest approach be identified experimentally?

(The atomic number for gold is 79; the electronic charge is −1·6 × 10^{-19} C; $1/4\pi\varepsilon_0$ = 9 × 10^9 F^{-1} m.) (*W.*)

14 (a) Explain the meaning of the term *mass difference* and state the relationship between the mass difference and the *binding energy* of a nucleus.
   (b) Sketch a graph of nuclear binding energy per nucleon versus mass number for the naturally occurring isotopes and show how it may be used to account for the possibility of energy release by nuclear fission and nuclear fusion.
   (c) The sun obtains its radiant energy from a thermonuclear fusion process. The mass of the sun is 2 × 10^{30} kg and it radiates 4 × 10^{23} kW at a constant rate. Estimate the life time of the sun, in years, if 0·7% of its mass is converted into radiation during the fusion process and it loses energy only by radiation. (1 year may be taken as 3 × 10^7 s.) The speed of light, c = 3 × 10^8 m s^{-1}. (*N.*)

15 (a) Describe in terms of nuclear structure the three different stable forms of the element neon, which have nucleon numbers 20, 21 and 22. Why is it impossible to distinguish between these forms chemically?

   A beam of singly ionised atoms is passed through a region in which there are an electric field of strength E and a magnetic field of flux density B at right angles to each other and to the path of the atoms. Ions moving at a certain speed v are found to be undeflected in traversing the field region. Show (i) that $v = E/B$ and (ii) that these ions can have any mass.

   If the emerging beam contained ions of neon, suggest how it might be possible to show that all three forms of the element were present.
   (b) Explain what is meant by the *binding energy* of the nucleus. Calculate the binding energy per nucleon for $^4_2$He and $^3_2$He. Comment on the difference in these binding energies and explain its significance in relation to the radioactive decay of heavy nuclei. (Mass of $^1_1$H = 1·007 83 u, mass of $^1_0$n = 1·008 67 u, mass of $^3_2$He = 3·016 64 u, mass of $^4_2$He = 4·003 87 u, 1 u ≡ 931 MeV.) (*L.*)

Answers to Exercises

Mechanics

Exercises 1A (p. 13)
1 D
2 B
3 C
4 C
5 (i) 15 s (ii) 62·5 m (iii) 17 m s^{-1}
6 (i) 4 s (ii) 20 m
7 3 s
8 (i) 5 m s^{-2} (ii) 1·4 s (iii)3 s (iv) 22·5 m
9 (i) 3 s (ii) 30 m (iii) 72°

Exercises 1B (p. 17)
1 B
2 C
3 E
4 D

Exercises 1C (p. 20)
1 17·3, 10 m s^{-1}
2 (a) 500 N (b) 200 N (c) 866 N
3 C
4 E
5 5 m s^{-1}, 53° to Y-direction
6 316 m s^{-1},71° to X
7 36 m s^{-1}
8 500 N
9 (a) 3·9 m s^{-1} (b) 300 km s^{-1}

Exercises 1D (p. 28)
1 C
2 C
4 2000 N, 3000 N
5 550 N, 500 N
6 (i) 600 N (ii) 400 N
7 (a)(i) 0·38 m (ii) 3 m s^{-2}
  (b)(i) 5750 N (ii) 5000 N
8 (a) 4 m s^{-2} (b) 7·1 m s^{-1}
9 (a)(i) $t = 0$
10 (a) 0·0192 kg (b) 0·022 m s^{-2}
11 (b)(ii) 11·3° (iii) 17°

Exercises 1E (p. 39)
1 D
2 (i) 10 N s (ii) 100 N
3 2·0 N
4 2·5 m s^{-2}
5 200 N
6 (i) 4·9 N (ii) 4·9 N (iii) 3·5 m s^{-1}
7 24 N

Exercises 1F (p. 43)
1 C
2 D
3 E
4 C
5 B
6 D
7 8 N
8 10 m s^{-1}
9 2 m s^{-1}
10 2 m s^{-1}, 4 m s^{-1}
11 10 m s^{-1}, 2·0 s
12 (a)(i) 5·1 × 10^{-4} N s (ii) 0·128 Pa
   (c) 2 times
13 (a)(i) 250 m (ii) 4000 N
   (iii) 20 000 N, 2000 N (b) 8 m s^{-1}
14 0·9v, 30°

Exercises 1G (p. 56)
1 C
2 E
3 D
4 E
5 D
6 E
7 180 J, 60 N s; 3 s, 9 m
8 (i) 4 s (ii) 20 m (iii) 10 J, 10 J
9 (i) 5·2 m s^{-1}, 58 J (ii) 1·2 m s^{-1},
   314 J
10 20 J, 15 J
11 1000 s
12 4 kW, 14 kW
13 (i) B, C, E, G (ii) A, D, F, H
14 (i) 3 m s^{-2}(ii) 45 000 J (iii) 30 kW
15 2/3
16 6 × 10^5 m s^{-1}, 2·1 × 10^{-15} J
17 (b)(i) 400 kW (ii) 656 kW
18 (a) 10 m s^{-1} (b) 24 J, 12·65 m s^{-1}
19 (a) 10/3 N (b) 5/9 W (c) 5/18 W

Exercises 2A (p. 67)
1 D
2 B
3 C
4 E
5 E
6 (i) 2 rad s^{-1} (ii) 96 N
7 (i) 118 N (ii) 32°
8 42°, 13 450 N
9 224 N, 64 N

10 15·8 N
11 $7·3 \times 10^{-5}$ rad s^{-1}, 0·068 N
12 (i) 2·3 N (ii) 1·7 m s^{-1}
13 1·6 rev s^{-1}
14 (i) 4·2 m s^{-1} (ii) 11 N (iii) 7 N
15 90 N
16 0·675 rev s^{-1}
17 (b) 2·625 N (c) 1·125 N
18 (a) 0·09 J (b)(i) 2·7 m s^{-1}
(ii) 1·8 m s^{-1} (iii) 0·55 N
19 (a) 250 N (b) 4·7 m s^{-1}
20 (ii) 1·6 m s^{-1} (iii) 0·63 m
21 7·7 rad s^{-1}, 122 cm away
22 (a) 8·0 N (b) 8·0 N

Exercises 2B (p. 87)
1 (a) A (b) E (c) E
2 D
3 E
4 A
5 (i) 0·05 s (ii) $3200\pi^2$ cm s^{-2}
(iii) 80π cm s^{-1}
6 (a) 0·08 m s^{-1} (b) 1·57 s
7 (a)(i) 20 mm (ii) 0·2 s
(iii) 628 mm s^{-1} (b) $7·9 \times 10^{-2}$ J
8 (i) 0·4 s (ii) 0·5 N (iii) 0·005 J
9 A, D, E, F
10 (i) 1·26 s (ii) 0·1 m s^{-2}
(iii) 4×10^{-5} J (iv) 4×10^{-5} J
11 10·0 m s^{-2}, 4·5 m; 2·45:1
12 (a) 0·25 m s^{-1} (b) 15·8 mJ
13 (a)(i) 0·05 m (ii) $\pi/3$ s, 1·8 m s^{-2}
14 6·3 cm
15 (a)(i) π rad s^{-1} (ii) 0·30 m s^{-2}
16 (a) 2·43 s (b) 0·15 J, 0·39 m s^{-1}
(c) 0·56 (d) 0·067 m
17 (i) 0·27 s (ii) $1·25 \times 10^{-3}$ J
(iii) $1·25 \times 10^{-3}$ J

Exercises 3A (p. 101)
1 E
2 A
3 B
4 E
5 (i) 0·8 N m (ii) 0·4 N m; 10·1 J
6 690 N, 1390 N
7 500 N m; 167 N
8 22·6 cm
10 41·7 N, 41·7 N
11 180 N, 159 N
12 0·206 nm from N atom
13 1710 N

Exercises 3B (p. 109)
1 D
2 C
3 A
4 (i) 3:2 (ii) 40:9
5 1·62 N
6 5·17 cm
7 38 cm
8 773·8 mmHg
9 $W\rho/\sigma$
10 (a) 750 kg (b) 270 kg (c) 0·42 m s^{-2}

Exercises 4A (p. 123)
1 E
2 D
3 E
4 B
5 (i) 2000 J (ii) 200 kg m^2 s^{-1}
(iii) 3·2 rev s^{-1}
6 (i) 8 rad s^{-1} (ii) 25 133 J
7 6·4 rev, 8 s
8 50 rad s^{-1}, 25 000 J
9 4 rev s^{-1}
10 (i) 2 rad s^{-1} (ii) 15 J
11 (i) 6 rad s^{-1} (ii) 12 N m (iii) 14·3 rev
12 (i) 2·2 rad s^{-1} (iii) 3·2 rad s^{-1}
13 $1·3 \times 10^{-4}$ kg m^2
14 18·3 rad s^{-1}
15 (a) 20 rad s^{-2} (b) 0·32 N m
16 $7·3 \times 10^{-4}$ kg m^2
17 (i) 2×10^7 J (ii) 10 km
18 (i) 2304 J, 38·4 kg m^2 s^{-1}
(ii) 0·42 N m (iii) 92·2 s

Exercises 4B (p. 103)
1 D
2 C
3 1·5 kg s^{-1}
4 9 m s^{-1}
6 1 m s^{-1}, $9·25 \times 10^4$ Pa
7 4 m s^{-1}, 12 kg s^{-1}
8 (i) 2·8 m s^{-1} (ii)
$5·7 \times 10^{-3}$ m^3 s^{-1}

Exercises 5 (p. 139)
1 2828 V
3 (a)(i) 5645 kWh (ii) 4788 kWh
(iii) £133·06 (iv) £207·90
(b) $6·66 \times 10^9$ J

Elasticity, Molecules, Solid Materials

Exercises 6A (p. 151)
1 D
2 B
3 C
4 (i) 1/2000 (ii) 10^{11} Pa (iii) 0·025 J
5 (i) Y
6 6·4 × 10^6 Pa, × 10^{-5}, 1·1 × 10^{11} Pa
7 67 N
8 0·08 mm
9 0·04 J, 0·08 J
10 8·3 × 10^{-3} J
11 (a)(i) 100 N (ii) 10 J (b) 57·7 N
   (c) 57·7 N
12 (a) 1·5 N (b) 6 mm, 4 mm (c) 780 N
13 (a) 2·3 × 10^{-6} m (b) 5·7 × 10^{-5} J
14 20 m s^{-1}
15 (i) 50 N (ii) 0·0018 m (iii) 0·0008 m
   (iv) 0·12 m

Exercises 6B (p. 160)
1 D
2 A
6 2·8 × 10^{-10} m (a) 2 × 10^8 Pa
   (b) 38 J (approx.) (c) 1·4 × 10^{11} Pa,
   3 mm
7 (i) 1000 N (ii) 8·7 × 10^{-5} m^2

Exercises 6C (p. 182)
1 D
2 C
3 D
4 B
14 (i) A (ii) B (iii) B
15 (a)(i) 2 × 10^{11} Pa

Exercises 7 (p. 203)
1 E
2 D
3 D
4 C
5 E
7 250 N
8 0·25 g
10 $M^{-1} L^3 T^{-2}$
11 Y
12 9910 s
13 6 × 10^{24} kg
14 9·9 m s^{-2}
15 4·5 × 10^{-9} rad s^{-1}
16 (a) 7852 m s^{-1} (b) 5250 s
17 889 N
18 14·4 N, 24·5 h

19 (c) 0·012
20 1·3 × 10^{11} − (8 × 10^{17})/R_0
21 2·2 × 10^{11}m, 6·3 × 10^7 s,
   1·6 × 10^{30} kg
22 (a)(ii) 5·5 × 10^{24} kg (b)(ii) 1740 km

Electricity

Exercise 8 (p. 234)
1 E
2 D
3 E
4 B
5 2 × 10^4 V m^{-1}, 1·6 × 10^{-14} N
7 1·25 × 10^{-18} C
9 (i) 0·075 m
11 1·6 × 10^{-19} C
12 100 V
13 (a) 1·45 × 10^{-10} N, 1·45 × 10^{-11} J
14 1·33 × 10^{-5} C, 6 × 10^5 V, yes

Exercises 9 (p. 261)
1 C
2 D
3 B
4 D
5 E
6 0·2 μF, 2·5 × 10^{-4} J
7 (a) 9 × 10^{-4} C, 18 × 10^{-4} C;
0·135 J, 0·27 J (b) 6 × 10^{-4} C,
6 × 10^{-4} C; 0·06 J, 0·03 J
8 (a) 6 × 10^{-8} C (b) 3 × 10^{-6} J
(c) 6 × 10^{-6} J; 1 × 10^{-6} J
10 (a) 60 μF (b) 150 μF
11 33·9 m
12 133·3 μC, 133·3 μC
13 (a) 5·2 × 10^{-9} J (b) 10·4 × 10^{-9} J;
1 × 10^{-6} N
14 4
15 (i) 1·5 μF (ii) 1·08 × 10^{-4} J (iii) 9 V
16 (a) 0·1 C (b) 10^{-2} F (c) 5 kΩ
17 171 kΩ (i) 9 mJ (ii) 500 μF
18 (ii) 200 μF, 43·2 s

Exercises 10A (p. 289)
1 D
2 C
3 D
4 D
5 E
6 C
7 3·2 V, 10 Ω
8 (i) 0·4 A (ii) 10 Ω (iii) 12 V

9 5 V, 0; 5·8 V, 4·8 V
10 14·4 V, 9·6 V; 8 V; 5·3 V
11 $5·2 \times 10^{-5}$ m s^{-1}
13 (a) 2 Ω (b) 1·44 W
14 8 V, 38 Ω
15 (a) 5 A, 100 V (b) 2·1 Ω
16 (i) 0·6 A, 0·2 A, 0·4 A (ii) 6 V
    (iii) 5·8 W (iv) 7·2 W

Exercises 10B (p. 303)
1 C
2 D
3 A
4 B
5 0·01 A
6 2:1
7 14 m, 1 m
8 0·16 Ω
9 $R_1 = 16·7$ Ω, $R_2 = 12·5$ Ω
10 (a) 1 Ω (b) 20 Ω or 0·05 Ω
11 18 Ω
12 $\frac{1}{2}$, 1, 2 kW

Exercises 10C (p. 309)
1 C
2 D
3 C
4 B
5 0·004 K^{-1}, 50°C
6 14·9 Ω
7 0·367 m
8 1·4:1, 1500 s
11 $8·8 \times 10^{-5}$ K^{-1}

Exercises 10D (p. 316)
1 D
2 C
3 1·53 V, 60·0 cm
4 0·12 A
5 1197 Ω, 3 mV
6 15·8 mV
7 0·825 m, 1·29 V

Exercises 11A (p. 336)
1 C
2 C
3 B
4 A
5 E
6 (i) 0·2 N (ii) 60°
7 5 A
8 5×10^{-3} N m, $2·5 \times 10^{-3}$ N m
11 21·8°
12 (a) shunt 0·0125 Ω (b) series 9995 Ω

Exercises 11B (p. 343)
1 $1·6 \times 10^{-14}$ N
2 $2·6 \times 10^{22}$ m^{-3}
3 $3·75 \times 10^{-24}$ N
4 $2·1 \times 10^{22}$

Exercises 12 (p. 360)
1 C
2 A
3 E
4 (b) 1×10^{-5} T (c) 3×10^{-5} T,
   greater
5 0·8 A
6 (b) 0·04 m
7 6×10^{-6} N away from X
8 9 A, $1·44 \times 10^{-4}$ N
10 $3·7 \times 10^{-3}$ V
11 6 cm from 12 A wire
12 (i) $1·6 \times 10^{-4}$ T (ii) 0

Exercises 13A (p. 389)
1 D
2 A
3 D
4 B
5 B
8 (a) 0 (b) 0·6 V. End P
9 (a) 0 (b) 0·1 V; 10^{-3} C
10 3·1 V (i) 3·1 V (ii) 1·6 V (iii) 0
12 $1·2 \times 10^{-4}$ Wb per turn
15 44·2 rev s^{-1}
16 0·53 V
17 1·05 V

Exercises 13B (p. 399)
1 B
2 D
3 C
4 4 mH
5 25 μF
6 (a)(i) 4·0 A (ii) 1·0 A s^{-1}
   (iii) 0·5 A s^{-1} (b) 26·8 div
8 (a) 25 000 V (b) 2×10^{-2} A
9 (a) 4 A (b) 2 A s^{-1}

Exercises 14 (p. 424)
1 C
2 E
3 E
4 C
5 (i) 2·1 A (ii) 2·1 A (iii) 3 A
6 0·35 A
7 (a)(i) 25 Hz (ii) 1·41 A
   (iv) $1·8 \times 10^{-4}$ W (b) $4·4 \times 10^{-5}$ W
8 (c)(i) 0 (ii) π(180°)

9 0·5 A

10 3·46 A

11 (i) 0·4 mA, 3·2 µF (ii) 40 mA

12 (i) 2 A (rms) (ii) 0·016 A (rms)
(iii) 0·0031 A (rms)

13 5 V, 53·1°

Optics

Exercises 15A (p. 439)

1 35·3°

2 41·8°

3 (i) 26·3° (ii) 56·4°

4 (i) 41·8° (ii) 48·8° (iii) 62·5°

6 1·60

7 1·50

8 (i) 60°04′ (ii) 48°27′

Exercises 15B (p. 445)

1 D

2 B

6 (a) 61° (b) 51°

7 (i) 0·4 µs (iii) 2·5 MHz

Exercises 16A (p. 456)

1 (i) 12 cm, $m = 1$ (ii) 12 cm, $m = 3$

2 13·3, 40 cm

3 5 cm

4 1·95 cm

5 13·3 cm

6 (b)(i) $-2D$ (ii) 0·28 m

Exercises 16B (p. 465)

1 C

2 C

3 (i) infinity (ii) 20

5 11·25 cm

8 60 mm

9 5°

Exercises 16C (p. 471)

1 (i) 4·2 cm (ii) 6·5

3 110 mm

Waves. Wave Optics

Exercises 17 (p. 503)

1 D

2 E

3 D

4 C

5 D

6 (i) 1·33 m (ii) 400 Hz

7 0·42 m s^{-1}

9 (i) $5\pi/3$ (ii) 6 cm

12 3400 Hz

13 (b) 10^{10} Hz

14 (i) 100 Hz (ii) 1·7 m (iii) 170 m s^{-1}

Exercises 18 (p. 514)

1 D

2 C

3 B

4 C

7 18·6°

8 47°10′, 41°48′ to vertical at oil surface

9 (a)(i) 7·5°, 45·6°

11 34·8°, 34%

13 4·0 × 10^{-7} m

Exercises 19 (p. 533)

1 C

2 D

3 C

6 1·8 × 10^{-3} m; 1·5 × 10^{-3} rad

8 0·34 mm

9 (a)(i) 9·375, 15 (b) 2·74 mm,
1536 nm, 5·14 mm

Exercises 20 (p. 555)

1 D

2 B

3 C

4 D

5 (a) 2·5 × 10^{-3} mm (b) 13·9° (c) 4

8 2 × 10^{-6} m

10 97 000 m^{-1}

13 600 nm, 285 000 m^{-1}, 43·3°

14 7; 0°, ±17·1°, ±36·1°, ±62·1°

Exercises 21 (p. 567)

1 D

2 C

3 E

5 67·5°

Exercises 22A (p. 576)

1 D

2 C

3 E

5 25 W

6 1·83 s

10 (a) 4:1, 2500 Hz (b) 9:1 (c) 1:1
(e) 12·4 cm

Exercises 22B (p. 587)

1 1091 Hz, 909 Hz

2 (a) away (b) 4 × 10^5 m s^{-1}

3 425 Hz

4 514 Hz, 545 Hz

5 (b)(i) 66 cm (ii)(1) 550 (2) 545
(3) 600 Hz

6 2.9×10^{-6} rad s^{-1}
7 45 132 Hz; 45 265 Hz; 265 Hz;
0·02 m s^{-1}

Exercises 23A (p. 598)
1 D
2 A
3 B
4 D
5 (i) 0·24 m (ii) 12 m s^{-1} (iii) 1·15 N
6 1.6×10^{-3} kg m^{-1}
8 (a) 2 m (b) 100 Hz
9 -3.2%, $+6.8\%$
10 300 Hz
12 (b)(i) 60 Hz (ii) 90 Hz

Exercises 23B (p. 613)
1 C
2 C
3 C
4 (i) $\lambda/2$ (ii) $\lambda/4$ (iii) $\lambda/2$, 567 Hz
5 (ii) 341 m s^{-1}, 12 mm (iii) 213 Hz
6 (a) 0·322 m (b) 0·645 m
7 (i) 320 m s^{-1} (ii) 10 cm (iii) 40 cm
(iv) 1200 Hz

Thermal Properties of Matter
Exercises 24 (p. 626)
1 D
2 B
3 D
4 16·6°C, 17·0°C
6 1·43°C
7 12·5°C, 7·1°C
8 881°C
9 385°C

Exercises 25 (p. 641)
1 B
2 C
3 E
4 (i) 100 s (ii) 0·2 kg
5 (a)(i) 36 000 J (ii) 0·016 kg
(b)(ii) 29°C
6 0·0935 kg
7 6.76×10^{-20} J molecule^{-1}
8 3.78×10^5 J kg^{-1}
9 360 J K^{-1}, 0·035 kg
10 2.23×10^6 J kg^{-1}
11 1.6×10^5 J kg^{-1}

Exercises 26A (p. 654)
1 E
2 D
3 C

4 E
5 (i) 2.4×10^5 Pa (ii) 0·24 m^3
(iii) 0·13 m^3
7 1·1 kg
8 100°C
9 0·014 m^3
10 (c) 2.45×10^5 Pa (d) 2.5×10^5 Pa,
2.08×10^5 Pa; 31

Exercises 26B (p. 663)
1 C
2 C
3 E
4 C
5 (b) 285 s (c) 160 J, 7%
6 1250 cm^3, 137·5 kPa
7 (i) 2000 J (ii) 450 K (iii) 60 J

Exercises 27 (p. 679)
1 B
2 D
3 C
4 C
6 (i) 1732 m s^{-1} (ii) 433 m s^{-1}
7 (i) 1039 m s^{-1} (ii) 900 m s^{-1}
9 (b) 4·62 mol (c) 517 m s^{-1}
10 (d)(i) 50·1 J (ii) 35·5 J
11 4.35×10^{10} m^{-3}
12 (a) 1·07 (b) 4:1 (c) 0·87
13 461 m s^{-1}, 64 rad s^{-1}
14 0·21 mmHg
15 Kr
16 (i) 508 m s^{-1} (ii) 400–475 m s^{-1}

Exercises 28A (p. 694)
1 D
2 B
3 E
4 C
5 (i) 4.5×10^7 C (ii) 1.8×10^8 J
7 42°C, 274 W
8 51°C
10 0·5 W m^{-2} K^{-1}, 0·01 W m^{-1} K^{-1}
11 (i) 5645 kWh (ii) 4788 kWh
(iii) £133 (iv) £208
12 (a) 90 W m^{-2} (b) 3 m, 19·5°C,
5·5°C, 1/32

Exercises 28B (p. 702)
1 D
2 D
3 C
4 5·5°C

Exercises 29A (p. 706)
1 (ii) 8·3 J mol^{-1} K^{-1}
2 (i) 576 J (ii) 410 J (iii) 166 J
3 (b) 1·06 MJ
5 8·3 J mol^{-1} K^{-1}, 25·9 J
6 164 J

Exercises 29B (p. 711)
3 6904 J, 6904 J
4 9 × 10^{-8} m
5 5 × 10^5 m^2 s^{-2}, 4 times
6 (i) 1·2 × 10^{19} m^{-3}
  (ii) 2·7 × 10^{25} m^{-3} (iii) 0·043 Pa

Exercises 29C (p. 716)
6 3·35 × 10^{-29} m^3

Exercises 29D (p. 720)
1 3450 J, 3450 J
2 (a) 11 350 J (b) 2220 J (c) 20% (d) 0
4 160 000 J
5 9·6 kg

Exercises 29E (p. 734)
2 1070 K
3 6023 K, 4·4 × 10^{-9} kg s^{-1}, 280 K
4 0·19 nm, 1·07 mm
5 1105 K
6 5290 K
7 6·0 × 10^7 J

Electrons, Electronics, Atomic Physics

Exercises 30A (p. 755)
1 D
2 C
3 D
4 E
5 B
6 6 × 10^6 m s^{-1}, 1·7 × 10^{-2} T
7 33 mm
8 4·2 × 10^4 V m^{-1}
9 3·2 mm
10 0·08 m
11 (a)(i) 3·6 × 10^7 m s^{-1}
  (ii) 2 × 10^{-3} T (b) 0
12 0·23 m
13 0·5 × 10^{-4} T

Exercises 30B (p. 763)
1 C
2 2·2 × 10^8 m s^{-1}
3 14·1 V, 200 Hz
4 60 km
5 (d) 30°

Exercises 31A (p. 783)
1 D
2 E
3 C
5 40 Ω, 40 Ω

Exercises 31B (p. 791)
1 D
2 E
3 E
4 C
5 (b)(i) 6800 Hz (ii) 199·6–199·7 kHz
  (iii) 200·3–200·4 kHz (c)(i) 240 ms
  (ii) 28 ms

Exercises 32A (p. 808)
1 (i) S/R (ii) 180°
2 0·2 V
3 −0·32 V
4 10
5 101
6 (a) + 5·4 V (b) >5·4 V (c) 1·03 kΩ
7 50
8 (i) −46·8 (ii) 10 300
9 3·9:1
10 (i) 5 (ii) 250
11 (a) 10
12 0·34 V

Exercises 32B (p. 820)
1 B
2 C
3 A
4 A
9 AND gate
10 X = 1 Y = 0

Exercises 33A (p. 834)
1 D
2 B
3 C
4 E
6 (i) 3·6 × 10^{-14} J (ii) 2·8 × 10^{17} s^{-1}
  (iii) 9%
7 5·1 × 10^{14} Hz
8 6·7 × 10^{-34} J s
10 2·0 × 10^{-7} m
11 (ii) 6·9 × 10^{-20} J (iii) 6·9 × 10^{-34} J s
12 7·8 × 10^{-19} J, 3·6 × 10^{-19} J

Exercises 33B (p. 848)
1 D
2 E
3 A
4 13·6 eV

6 (i) 10·4 eV, 21·8 × 10^{-19} J
(ii) 3·2 × 10^{-7} m, UV
7 A, B, C, D = 6·5, 4·8, 4·3,
4·0 × 10^{-7} m
9 (b) 3·3 × 10^{15} Hz (c) 19°

Exercises 33C (p. 857)
1 D
2 B
3 C
4 (i) 0·012 A (ii) 1·33 × 10^8 m s^{-1}
5 6·4 × 10^{-5} J, 3·1 × 10^{-11} m
6 (i) 17·7 kV, 17·7 keV
7 (i) 9·375 × 10^{16} (ii) 150 W
8 2·0 × 10^{-7} m
9 5000:1
10 (i) 6·25 × 10^{15} (ii) 7·3 × 10^7 m s^{-1}
(iii) 3 × 10^4 kg s^{-1}

Exercises 33D (p. 862)
1 D
2 A
3 (i) 6·6 × 10^{-10} m (ii) 1·7 × 10^{-34} m
(iii) 1·1 × 10^{-36} m
4 10^7 m s^{-1}, 136 V
5 (i) 4·6 × 10^{-11} m
(ii) 1·4 × 10^7 m s^{-1} (iii) 450 V
6 (i) 1·2 × 10^{-10} m (ii) 5·8 × 10^{-11} m
8 2·2 × 10^{-24} N s, 6·2 × 10^{-33} kg,
6 × 10^{-16} J
9 1·7 × 10^{-11} m
10 (i) 10^4 eV (ii) 1·23 × 10^{-11} m
(iii) 1·6 × 10^{-15} J, 1·24 × 10^{-10} m

Exercises 34A (p. 879)
1 D
2 E
3 C
4 C
5 E
6 (a)5·25 × 10^{20} (b) 2·1 × 10^8 J
7 (a) 6·0 × 10^{11} (b) 6·5 × 10^{11}
8 (a) 18·5 × 10^4 Bq (b) 3·8 × 10^{15}
9 (a) 403·4 (b) 6·88 × 10^{-5} s^{-1}
10 3·8 × 10^{-9} g
11 0·43:1
12 4·1 × 10^9 y, 3·2 × 10^9 y
13 6000 cm^3
14 (a) 218, 84 (b) 1·08 × 10^{-19} N s
(c) 3·1 × 10^5 m s^{-1}
15 3·36 × 10^{22}
16 (c) 1006 y

Exercises 34B (p. 903)
1 E
2 C
3 D
4 B
6 6·65 × 10^{-27} kg
9 (a) 28·3 MeV (b) 23·8 MeV
10 (a) 4·23 MeV (b) 4·16 MeV
12 2·8 × 10^{-12} J
13 2·6 × 10^{-14} m
14 1 × 10^{11} y
15 6·78, 2·39 MeV/nucleon

ANGLO-EUROPEAN COLLEGE OF CHIROPRACTIC

Index

ANGLO-EUROPEAN COLLEGE OF CHIROPRACTIC

ANGLO-EUROPEAN COLLEGE OF CHIROPRACTIC